W9-AGN-171

356-7965

MODERN EARTH SCIENCE

Robert J. Sager
William L. Ramsey
Clifford R. Phillips
Frank M. Watenpaugh

HOLT, RINEHART AND WINSTON
Harcourt Brace & Company
Austin • New York • Orlando • Atlanta • San Francisco • Boston • Dallas • Toronto • London

Portions of this work were published in previous editions.

Printed in the United States of America

ISBN 0-03-050609-3

3 4 5 6 7 041 01 00 99 98 97

Acknowledgments

Academic Reviewers

Richard Busch, Ph.D.
Associate Professor of Geology and Science Education
Boucher Science Center
West Chester University
West Chester, Pennsylvania

Larry Davis, Ph.D.
Associate Professor of Geology
Department of Geology
Washington State University
Pullman, Washington

Albert B. Dickas, Ph.D.
Vice Chancellor for Research
Professor of Geology
University of Wisconsin
Superior, Wisconsin

Lydia K. Fox, Ph.D.
Associate Professor of Geology
University of the Pacific
Stockton, California

George Greenstein, Ph.D.
Professor of Astronomy
Amherst College
Amherst, Massachusetts

Frederick Heck, Ph.D.
Associate Professor of Geology
Ferris State University
Big Rapids, Michigan

Mary K. Hemenway, Ph.D.
Astronomy Department
University of Texas
Austin, Texas

Jan Hopmans, Ph.D.
Associate Professor, Hydrologic Science
University of California
Davis, California

Richard Jackson, Ph.D.
Professor of Geography
Brigham Young University
Provo, Utah

James Kaler, Ph.D.
Professor of Astronomy
University of Illinois
Urbana, Illinois

Timothy Lincoln, Ph.D.
Professor of Geology
Albion College
Albion, Michigan

Thomas Myers, Ph.D.
Professor of Chemistry Emeritus
Kent State University
Kent, Ohio

James O'Connor
Associate Professor of Geoscience
University of the District of Columbia
Washington, D.C.

Anne Pasch
Professor of Geology
University of Alaska Anchorage
Anchorage, Alaska

Bobby E. Price, Ph.D., P.E.
Professor, Department of Civil Engineering
Louisiana Tech University
Ruston, Louisiana

Kenneth Rubin, Ph.D.
Assistant Professor, Department of Geology and Geophysics
University of Hawaii at Manoa
Honolulu, Hawaii

Donald de Sylva, Ph.D.
Professor of Marine Biology and Fisheries
University of Miami
Miami, Florida

Baxter E. Vieux, Ph.D., P.E.
Associate Professor
School of Civil Engineering and Environmental Science
University of Oklahoma
Norman, Oklahoma

Teacher Reviewers

Robert W. Avakian
Earth Science Teacher
Alamo Junior High School
Midland, Texas

Dennis W. Cheek, Ph.D.
Coordinator of Mathematics, Science and Technology
Rhode Island Department of Elementary and Secondary Education
Providence, Rhode Island

Richard Kadish
Earth Science Teacher
Chesterton High School
Chesterton, Indiana

Kathleen Kaska
Life and Earth Science Instructor
Lake Travis Middle School
Austin, Texas

Janis Lariviere
Science Teacher
Westlake High School
Austin, Texas

James B. Pulley
Science Teacher
Metropolitan College
Kansas City, Missouri

Kevin Reel
Head of Science Department
The Thacher School
Ojai, California

Bill Schrandt
Math and Science Teacher
Albuquerque Public Schools
Albuquerque, New Mexico

David C. Tucker
Science Teacher
Mt. Baker High School
Deming, Washington

Government and Industry Reviewers

Steven A. Bryson
Director of Geology
WARD'S Natural Science Establishment, Inc.
Rochester, New York

Gretchen M. Gillis
Geologist
Oryx Energy Company
Dallas, Texas

Cliff Morrison
Chief Meteorologist
K•EYE TV
Austin, Texas

Joel L. Morrison, Ph.D.
Chief, Geography Division
U.S. Bureau of the Census
Washington, D.C.

CONTENTS

UNIT 3 Composition of the Earth 136

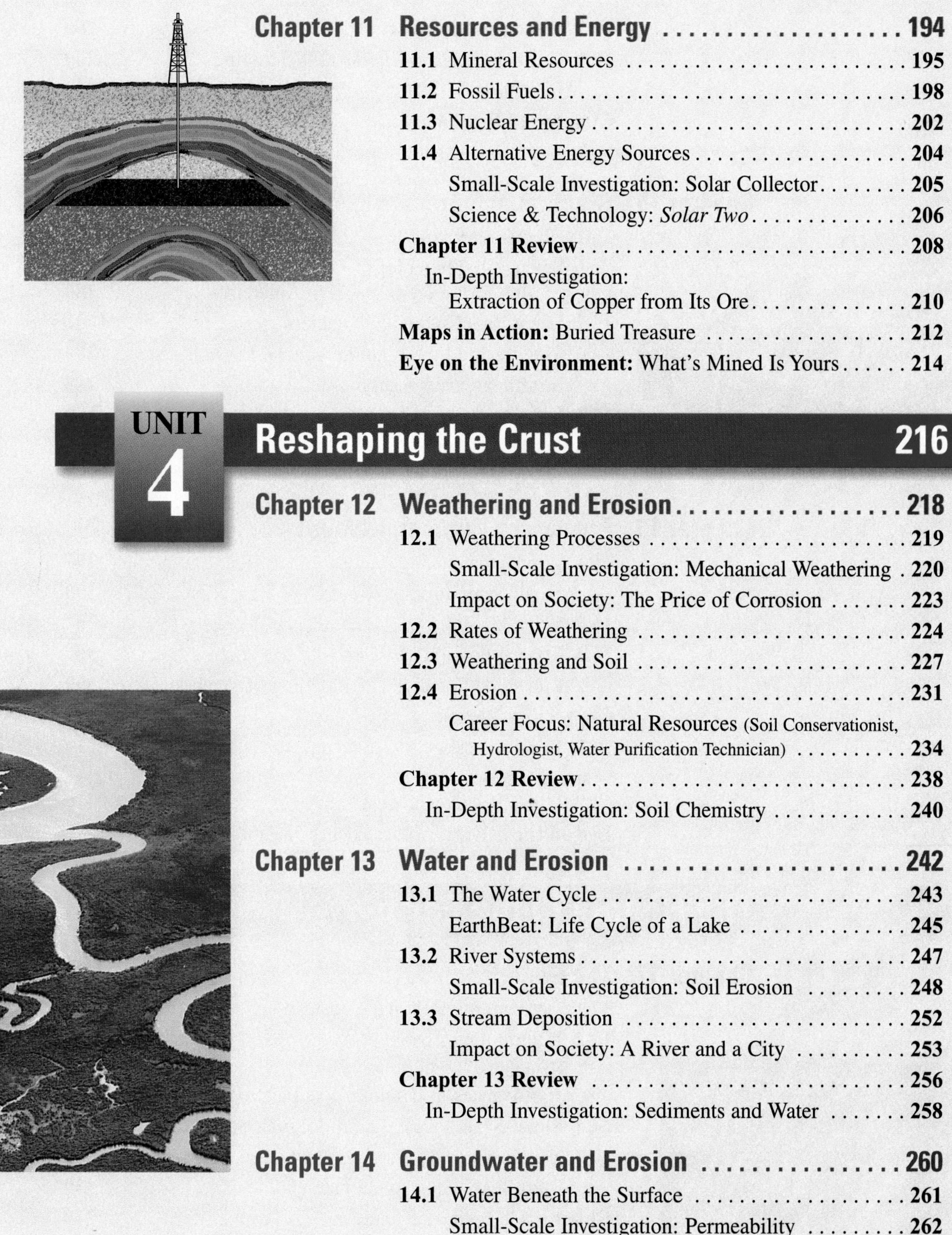

UNIT 4 Reshaping the Crust 216

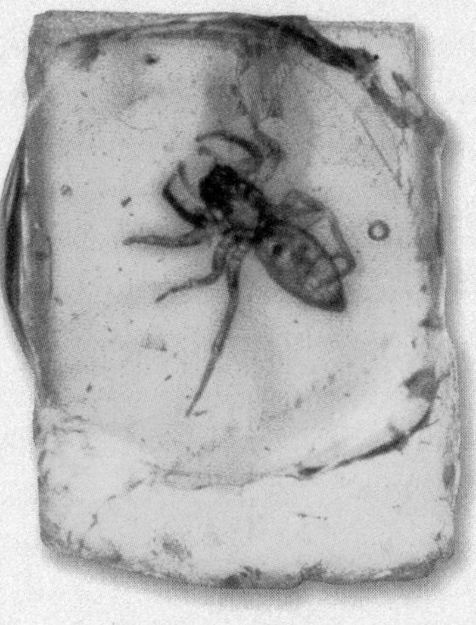

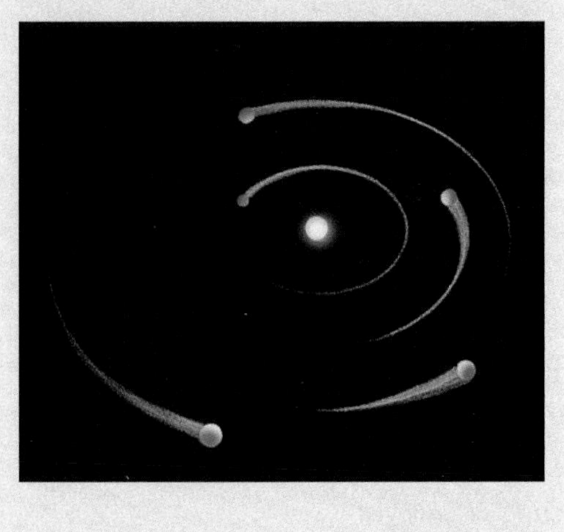

Small-Scale Investigations

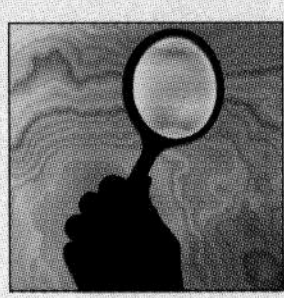

In-Depth Investigations

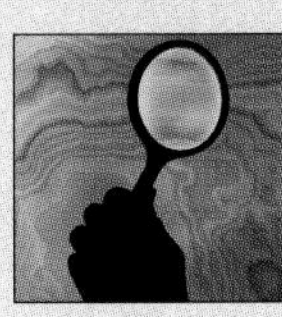

Long-Range Investigations

Maps in Action

Eye on the Environment

EarthBeat

Science & Technology

Impact on Society

Career Focus

Introducing Earth Science

Studying the earth is like putting together a complex jigsaw puzzle that is made up of pieces that are constantly changing shape. As time passes, many of the pieces change so much they don't even resemble their original shape. Or worse yet, some of the pieces are even destroyed or replaced with different pieces. The only constant in this puzzle is change.

Does this description surprise you? Perhaps you think of the earth as being constant and unchanging. After all, the earth is made up of some very durable materials, such as rock. Compared to most things, rock does not seem to change much. Appearances, however, can be deceiving. The earth, in fact, undergoes incredible changes. It's just that many of these changes happen very, very slowly.

Of course, not everything on earth occurs slowly. Earthquakes, tornadoes, tidal waves, and lightning strikes, for example, occur so quickly, they pose their own sets of problems for earth scientists. Imagine the task of studying a phenomenon that occurs randomly, at an unknown location, and only lasts for 30 seconds. That's exactly what earth scientists face when they study earthquakes. Seem impossible? It is difficult, but not impossible. Such study requires creativity and an innovation.

This site was a seabed about 200 million years ago. Much has happened here since then. Today it is known as El Capitan of the Guadalupe Mountains.

Earth scientists use elaborate systems to pinpoint the location of earthquakes. Yet they still do not know enough to predict when and where they will occur.

Earth scientists also beat the odds when they study remote locations. Consider the difficulty of exploring the bottom of the ocean at depths of more than 10 kilometers. Or consider the exploration of other planets and outer space. Earth Scientists use technology in new and incredible ways to achieve such difficult observations and studies.

Oceanographers often go to great depths and take great risks to add to their understanding of the oceans.

Meteorologist, geologist, oceanographer, and astronomer are all specialties within the area of earth science. Together, these scientists attempt to understand our physical home within the universe. Such understanding will help us make wise decisions, both now and in the future.

Hot situations such as this one are not unusual for a volcanologist—an earth scientist who studies volcanoes.

Begin your own journey of understanding by turning the page and entering the realm of earth science. I am proud to be your guide on what I hope will be a fascinating and enlightening experience.

Robert J. Sager
Chair, Department of Earth Sciences
Pierce College
Lakewood, Washington

Lab and Field Safety

The study of science is challenging and fun, but it can also be dangerous. Don't take any chances! Follow the guidelines listed here, as well as safety information provided in the particular Investigation you are doing. Also, follow your teacher's instructions and don't take shortcuts—even when you think there is little or no danger.

Accidents can be avoided. The major causes of laboratory accidents are carelessness, lack of attention, and inappropriate behavior. These things reflect a person's attitude. By adopting a positive attitude and by following all safety guidelines, you can greatly reduce your chances of having an accident. Even a minor accident in a science laboratory can cause major injuries, so be very careful.

Safety Guidelines

General

Always get your teacher's permission before attempting any laboratory or field investigation. Read the procedures carefully, paying particular attention to safety information and caution statements. If you are unsure about what a safety symbol means, look it up here or ask your teacher. You cannot be too careful when it comes to safety. If an accident does occur, inform your teacher immediately, regardless of how minor you think the accident is.

Safety Equipment

Know the location of and how to use the nearest fire alarms and any other safety equipment, such as fire blankets and eyewash fountains, as identified by your teacher.

Neatness

Keep your work area free of all unnecessary books and papers. Tie back long hair, and secure loose sleeves or other loose articles of clothing, such as ties and bows. Remove dangling jewelry. Don't wear open-toed shoes or sandals in laboratory situations. Never eat, drink, or apply cosmetics in a laboratory setting; food, drink, and cosmetics can easily become contaminated with dangerous materials.

Cleanup

Before leaving, clean up your work area. Put away all equipment and supplies. Dispose of all chemicals and other materials as directed by your teacher. Make sure water, gas, burners, and electric hot plates are turned off. Hot plates and other electrical equipment should also be unplugged. Wash your hands with soap and water after working in a laboratory situation.

Electrical Safety

- Never handle electrical equipment with wet hands. Work areas, including floors and tables, should be dry.
- Never overload an electrical circuit.
- Make sure all electrical equipment is properly grounded.
- Keep electrical cords away from areas where someone may trip on them, or where the cords can tip over laboratory equipment.

Fire Safety

- Make sure that fire extinguishers and fire blankets are available in the laboratory.
- Tie back long hair and confine loose clothing.
- Wear safety goggles when working with flames.
- Never reach across an open flame.

Gas Precaution

- Do not inhale fumes directly. When instructed to smell a substance, wave fumes toward your nose and inhale gently.
- Use flammable liquids only in small amounts and in a well-ventilated room or under a fume hood.
- Always use a fume hood when working with toxic or flammable fumes.
- Do not breathe pure gases such as hydrogen, argon, helium, nitrogen, or high concentrations of carbon dioxide.

Glassware Safety

- Check the condition of glassware before and after using it. Inform your teacher about any broken, chipped, or cracked glassware; it should not be used.
- Air-dry glassware; do not dry by toweling. Do not use glassware that is not completely dry.
- Do not pick up broken glass with your bare hands.
- Never force glass tubing into rubber stoppers.
- Never place glassware near edges of your work surface.

Proper Waste Disposal

- Clean up the laboratory after you are finished; dispose of paper towels, etc.
- Follow your teacher's directions regarding proper procedures for waste disposal, especially for chemical disposal.

Heating Safety

- Use proper procedures when lighting Bunsen burners.
- Turn off hot plates, Bunsen burners, and other open flames when not in use.
- Heat flasks or beakers on a ring stand with a wire gauze between the glass and the flame.
- Store hot liquids in heat-resistant glassware. Heat materials only in heat-resistant glassware.
- Turn off gas valves when not in use.

Chemical Safety (Poison)

- Never taste any substance in the laboratory. Do not eat or drink from laboratory glassware.
- Do not eat or drink in the laboratory.
- Properly label all bottles and test tubes containing chemicals.
- Never transfer substances with a mouth pipet; use a suction bulb.

Chemical Safety (Caustic Substances)

- Alert your teacher to any chemical spills.
- Do not let acids and bases touch your skin or clothing. If a substance gets on your skin, rinse it immediately with cool water, and alert your teacher.
- Wear your laboratory apron to protect your clothing.
- Never add water to acids; always add acids to water.
- When shaking or heating a test tube containing chemicals, always point the test tube away from yourself and other students.

Explosion Danger

- Use safety shields or screens if there is a potential danger of an explosion or implosion of apparatus.
- Never use an open flame when working with flammable liquids such as ether or alcohol.
- Follow a water-bath procedure to heat solids. Never risk an explosion by heating rocks or minerals directly.

Hand Safety (Avoiding Injuries)

- Always wear gloves when cutting, fire polishing, or bending glass tubing.
- Use tongs when heating test tubes. Never hold them in your hand.
- Always allow heated materials, including glassware, to cool before handling them.

Hand Safety (Hygienic Care)

- Always wash your hands after completing an investigation.
- Keep your hands away from your face and mouth.

Clothing Protection

- Wear laboratory aprons in the laboratory.
- Confine loose clothing.

Eye Safety

- Wear approved safety goggles in the laboratory.
- Make sure an emergency eyewash station is available in the laboratory.
- Never look directly at the sun, even for short periods of time. Laboratory goggles will not protect your eyes from the sun.

Water Safety

- When working near water, always work with a partner or adult.
- Always wear a life jacket.
- Do not work near water during stormy weather.

Concept Mapping *A Way to Bring Ideas Together*

What Is a Concept Map?

Have you ever tried to tell someone about a book or a chapter you've just read, and you find that you can remember only a few isolated words and ideas? Or maybe you've memorized facts for a test, and then weeks later you're not even sure what topic those facts are related to.

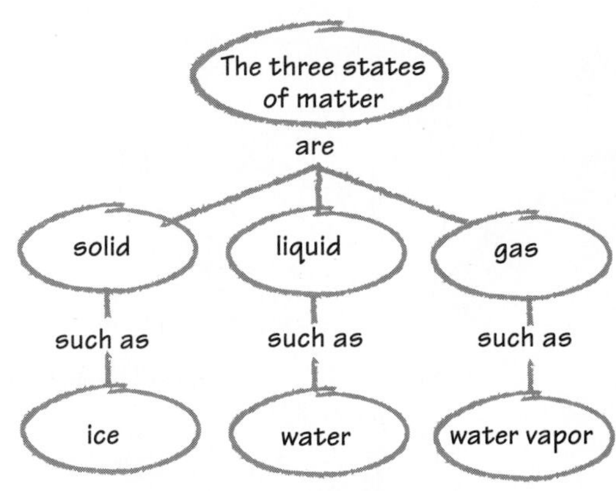

In both cases, you may have understood the ideas or concepts by themselves, but not in relation to one another. If you could somehow link the ideas together, you would probably understand them better and remember them longer. This is something a concept map can help you do. A concept map is a visual way of choosing how ideas or concepts fit together. It can help you see the "big picture."

How to Make a Concept Map

1. Make a list of the main ideas or concepts. It might help to write each concept on its own slip of paper. This will make it easier to rearrange the concepts as many times as you need to before you've made sense of how the concepts are connected. After you've made a few concept maps this way, you can go directly from writing your list to actually making the map.

2. Spread out the slips on a sheet of paper, and arrange the concepts in order from the most general to the most specific. Put the most general concept at the top and circle it. Ask yourself, "How does this concept relate to the remaining concepts?" As you see the relationships, arrange the concepts in order from general to specific.

3. Connect the related concepts with lines.

4. On each line, write an action word or short phrase that shows how the concepts are related.

Look at the concept map on this page and then see if you can make one for the following terms: The Solar System, The Sun. Planets, Star, Earth, Jupiter.

An answer is provided below, but don't look at it until you try the concept map yourself.

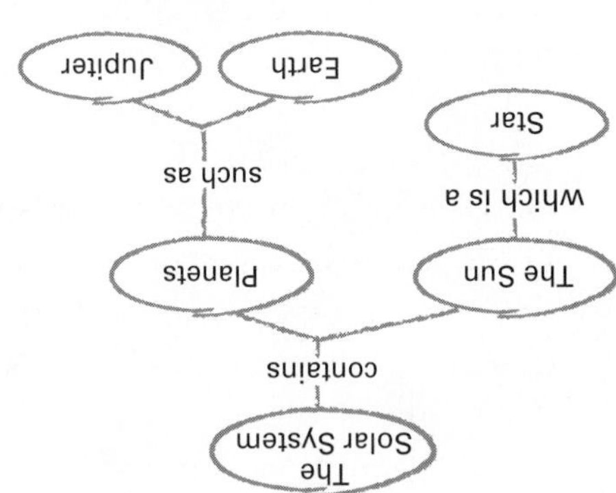

UNIT 1

Studying the Earth

Astronaut working outside a space shuttle (background); Oceanographer studying a coral reef in the Red Sea (inset)

Introducing Unit One

Communication is the key to progress in science. Without a way to build on past knowledge and pass new knowledge on to others, each new generation would have to rediscover facts about the planet on which we live.

When I was a student majoring in urban and economic geography, I discovered one of our most valuable aids to scientific communications—maps. Maps are a delightful blend of art and science. They give us an efficient, graphic way to present volumes of knowledge in a single picture. By looking at a map, you can view the world or part of it at a glance. By making a map, you can preserve your knowledge for future use.

In this unit, you will discover how other earth scientists use both words and pictures to communicate important information about our amazing planet.

Elizabeth Ann Olson

Elizabeth Ann Olson
Cartographer
U.S. Geological Survey

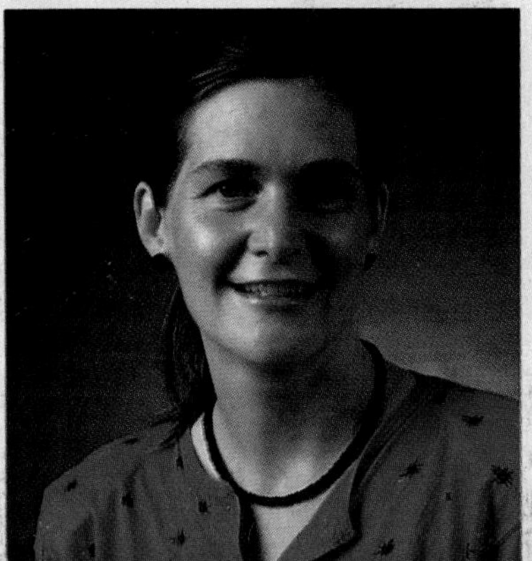

In this UNIT

Chapter 1

Introduction to Earth Science

The natural world challenges our understanding with questions about why seasons change, rivers flood, and planets spin through space. Unlocking the secrets of the natural world takes earth scientists to the rainforests of equatorial Africa and to ice caves such as this one in Canada.

This chapter describes the origins of earth science and some of the methods earth scientists use to learn about the world around us.

Chapter Outline

◀ These climbers are poised at the mouth of an ice cave in Canada.

1.1 What Is Earth Science?

Section Objectives

- **Name the four main branches of earth science.**
- **Discuss the relationship between earth science and ecology.**

Since the beginning of human history, people have observed the world around them and wondered about the forces that shaped that world. As early humans watched a volcano erupt, felt the earth tremble beneath them, or saw the moon darken during an eclipse, they asked why. To explain these natural phenomena, ancient people forged myths and legends, attributing such events to powerful supernatural forces. Angry goddesses hurled fire from volcanoes; giants wrestled underground, causing the earth to shake.

Not until people began to make careful observations and to search for natural causes to natural phenomena did the scientific study of the earth begin. The ancient Chinese, for example, began keeping written records of earthquakes as early as 780 B.C. The ancient Greeks compiled a catalogue of rocks and minerals in the third century B.C. Other ancient people, such as the Mayas in Central America, kept track of the movements of the sun, the moon, and the planets. They used these observations to create accurate calendars.

At first, scientific discoveries were limited to observations made with the unaided eye. Then, in the seventeenth century, the invention of instruments such as the microscope and the telescope extended human observation to previously hidden worlds.

Eventually, people accumulated an organized body of knowledge about the earth, and the field of **earth science** was born. Earth science is the study of the earth and of the universe around it. Earth science, like other modern sciences, is based on the assumption that the causes of natural phenomena can be discovered through careful observation and experimentation.

Branches of Earth Science

As the technology for studying the earth improved, the range of human observation increased dramatically. With the help of special equipment, scientists began to explore the dark ocean depths, the

Figure 1-1. El Caracol, an observatory built by the ancient Mayas of Mexico, is the oldest known observatory in the Americas.

EARTHBEAT

The Earth's Circumference

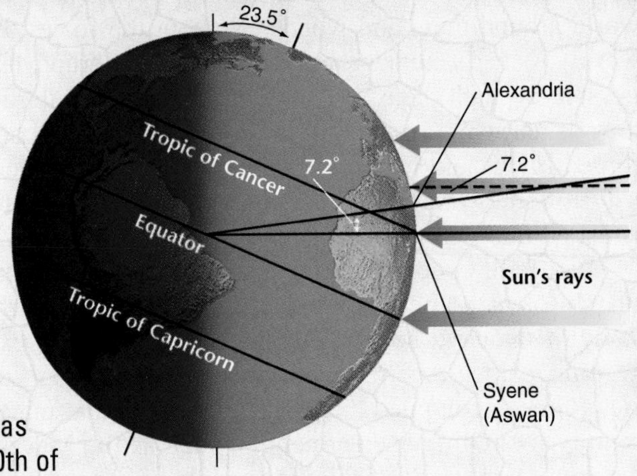

More than 22 centuries ago, a Greek mathematician named Eratosthenes used careful observations and simple geometry to determine the circumference of, or distance around, the earth.

Eratosthenes read of a well in Syene, Egypt—modern-day Aswan, Egypt—where the sun's rays reached the bottom once each year, at noon on June 21. Furthermore, the sun cast no shadows in Syene at that time. Eratosthenes thought these facts meant that the sun was directly overhead at Syene on June 21. Eratosthenes knew that the sun *did* cast shadows in Alexandria, a city to the north of Syene, on June 21, indicating that the sun's rays were striking that city at an angle.

Eratosthenes determined that the angle of the sun's rays in Alexandria at noon on June 21 was 7.2°. He thought that meant that the distance between the two cities was also 7.2°, or 1/50th of the 360° circumference of the earth. Eratosthenes knew that the distance between Syene and Alexandria was about 5,000 stadia (925 km). By multiplying the known distance between Alexandria and Syene by 50, Eratosthenes calculated the polar circumference of the earth to be 250,000 stadia (46,250 km).

With the aid of modern technology and sophisticated instruments, scientists now calculate the earth's circumference to be about 40,000 km. This value, which is considered accurate, differs from Eratosthenes' calculation of 46,250 km by only 6,250 km.

Suppose that city A is 3,335 km north of city B and that on June 21 the angle of the sun's rays in city A is 0° while the angle of the rays in city B is 30°. Use Eratosthenes' method to find the earth's circumference.

earth's unknown interior, and the vastness of space. Their discoveries have created an immense body of knowledge about the earth.

Because one person cannot keep up with the developments in all areas of earth science, most earth scientists today specialize. Currently, earth scientists specialize in one of the following four major areas of study: the solid earth, the oceans, the atmosphere, and the universe beyond the earth. *Career Focus,* a special feature that appears in each unit of this book, offers detailed information about careers in earth science.

Geology

The study of the origin, history, and structure of the solid earth and the processes that shape it is called **geology.** Geology is a broad field that includes many areas of specialization. Some geologists explore the earth's crust in search of new deposits of coal, oil, gas, and other valuable resources; some geologists study the forces within the earth in order to better understand and forecast earthquakes and volcanic eruptions; and some geologists study fossils to learn more about the earth's past. Units 1, 2, 3, 4, and 5 of this book deal with topics of primary concern to geologists.

Oceanography

Vast oceans cover nearly three-fourths of the earth's surface. The study of the earth's oceans is called **oceanography.** Some oceanographers work on research ships equipped with special instruments for studying the sea. Other oceanographers study waves, tides, and ocean currents. Some oceanographers explore the ocean floor for clues to the earth's history and to locate mineral deposits. Other oceanographers study marine plant and animal life. A discussion of the earth's oceans is presented in Unit 6.

Meteorology

The study of the earth's atmosphere is called **meteorology.** Using satellites, radar, and other modern technology, meteorologists study the variations in atmospheric conditions that produce weather. Many meteorologists work as weather observers, measuring such factors

Figure 1–2. A geologist uses special equipment to study erupting volcanoes (left). Oceanographers prepare to enter a submersible to study the ocean floor (above).

Figure 1–3. Two astronomers use a telescope to observe distant stars (left). A meteorologist uses a computerized system to track storms (right).

as wind speed, temperature, and rainfall. This weather information is then used to prepare detailed weather maps. Other meteorologists use weather maps, satellite images, and computer data to make weather forecasts. You will learn more about meteorology in Unit 7 of this book.

Astronomy

The study of the universe beyond the earth is called **astronomy.** It is one of the oldest branches of earth science. In fact, the ancient Babylonians charted the positions of planets and stars nearly 4,000 years ago. Modern astronomers use earth-based and space-based telescopes and other instruments to study the universe. Space probes, such as *Pioneer, Voyager, Galileo,* and *Ulysses* have also provided much useful data. Unit 8 of this book presents information about the moon, the planets, the sun, the stars, and the universe.

The Importance of Earth Science

Powerful forces are at work on the earth. Volcanoes erupt and earthquakes shake the ground. These events not only shape the earth, but also affect life on the earth. A volcanic eruption may bury a town in ash. An earthquake may produce huge waves that destroy shorelines. By understanding how natural forces shape our environment, earth scientists can better forecast potential disasters and help save lives and property.

Observations made by earth scientists have contributed greatly to our knowledge of the world around us. For example, information gathered by astronomers studying distant galaxies has led to theories about the origins of this solar system. Geologists studying rock layers have found clues to the earth's past environments and to the evolution of life on this planet.

The earth also provides many valuable resources that enrich the quality of people's lives. For example, the fuel that powers a jet, the metal used to make surgical instruments, and the paper and ink in this book all come from the earth. The study of earth science can help people gain access to the earth's resources and teaches them to use those resources wisely.

Ecology

Earth scientists primarily study the **geosphere,** the solid earth; the **hydrosphere,** its water; and the **atmosphere,** the gases surrounding the earth. Other scientists, called biologists, study the living world. An area of science in which biology and earth science are closely linked is called **ecology.** Ecology is the study of the complex relationships between living things and their environment. Most ecologists have backgrounds in either earth science or biology.

Organisms on the earth inhabit many different environments. A community of organisms and the environment they inhabit is called an **ecosystem.** The terms *ecology* and *ecosystem* come from the Greek word *oikos,* meaning "house." Each ecosystem is a physically distinct, self-supporting system. An ecosystem may be as large as an ocean or desert or as small as a tide pool or rotting log.

The largest ecosystem is called the **biosphere.** The biosphere encompasses all life on earth and the physical environment that supports it. The biosphere extends from the ocean depths to the atmosphere a few kilometers above the earth's surface.

A tropical rain forest is one example of a large ecosystem within the biosphere. Plants in the rain forest use sunlight to produce food through a process known as *photosynthesis*. The plants are then eaten by animals, which are in turn eaten by other animals. When

Figure 1–4. This small tidal pool on the coast of Maine, in Acadia National Park, and the vast ocean are both ecosystems.

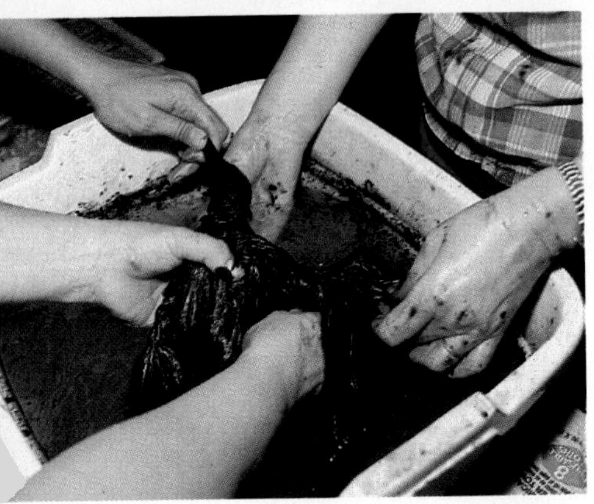

Figure 1–5. People try to help the environment by cleaning up the shore after an oil spill (top). Pollution also harms wildlife. Volunteers remove oil from a bird's feathers (bottom).

rain-forest plants and animals die, their bodies are decomposed by microorganisms. The resulting chemicals enter the soil to nourish other plants and animals. Thus, the system is practically self-supporting. What other examples of ecosystems can you name?

Environmental Pollution

Each ecosystem is delicately balanced. When that fragile ecological balance is upset, the survival of the ecosystem, and in some cases the entire biosphere, is threatened. One serious threat to ecosystems today is **pollution,** the contamination of the environment with waste products or impurities.

Some waste products are **biodegradable.** As such, they can be broken down by microorganisms into harmless substances that can then be used by other organisms. Biodegradable waste products pose little threat to the environment, and in many cases they contribute to the well-being of the environment. For example, the chemicals found in such biodegradable wastes as banana peels and eggshells make excellent plant fertilizer.

Many modern waste products, such as most plastics, are not biodegradable. Some ecosystems are threatened by the large quantities of nonbiodegradable wastes. For example, plastic wastes dumped in oceans or lakes can harm the animals there. When particles of plastic are ingested, they clog the digestive tracts of fish, birds, and turtles. Ducks and other birds can starve to death when they become tangled in plastic litter.

Protecting the Environment

Pollution poses serious problems for all living organisms. To help protect the environment from pollution, ecologists often work together with earth scientists in other fields such as meteorology.

For example, in the early 1970's, meteorologists found that the level of ozone, a form of oxygen, in the upper atmosphere was decreasing. This discovery was alarming to ecologists and earth scientists. They knew that ozone helps protect the earth's plant and animal life from the harmful ultraviolet rays of the sun. Further research revealed that the ozone layer was being destroyed by chlorofluorocarbons (CFCs), chemical compounds commonly used as propellants in aerosol sprays. To reduce this threat to the environment, the United States and other countries have agreed to abide by international treaties limiting the production and use of such ozone-depleting chemicals.

Section 1.1 Review

1. What are the four major branches of earth science?
2. Describe the work of meteorologists.
3. What is ecology?
4. Give an example of an ecosystem and explain how it is self-supporting.
5. How might the study of earth science contribute to the survival of the biosphere?

1.2 Paths to Discovery: Scientific Methods

Section Objectives

- **Identify the steps that make up scientific methods.**
- **Explain how the meteorite-impact hypothesis developed.**

Through research, scientists seek to explain natural phenomena and solve mysteries of the earth. Over the years, the scientific community has developed organized, logical approaches to scientific research, called **scientific methods.** Scientific methods are not a set of sequential steps that scientists invariably follow. Rather, they are guides to scientific problem solving.

State the Problem

Scientific inquiry often begins as a result of **observation.** Simply put, observation is using the senses of sight, touch, taste, hearing, and smell to gather information about the world. When you notice thunderclouds forming in the summer sky, that is an observation. So, too, is feeling the cool, smooth surface of polished marble or hearing the roar of rapids around the bend of a river.

Observations often lead to questioning. What causes tornadoes to form? Why is oil found only in certain locations and not in others? What causes a river to change its course? Asking questions like these is one way of stating the problem to be investigated through scientific methods.

One problem that has long puzzled scientists is the extinction of the dinosaurs. For more than 135 million years, these huge reptiles dominated the earth. Then, about 65 million years ago, the dinosaurs and three-fourths of all the other species on the earth died out. Scientists wondered what could have caused such a mass extinction.

Gather Information

To investigate a problem, such as the extinction of the dinosaurs, scientists gather information. An important means of gathering information is **measurement.** Measurement involves the comparison of

Figure 1-6. One means of gathering information is through careful measurement. In this photo, geologists are measuring a crack in the earth's surface caused by an earthquake.

Figure 1–7. Geologists found a clay layer with high-iridium content in Montana.

some aspect of an object or phenomenon with a standard unit, such as a meter, a Celsius degree, or a kilogram. For example, when you measure a rock and find that it is 20 cm long, you are comparing the length of the rock with a unit of measure—one centimeter. What other units of measure can you name?

Accuracy is important in scientific measurements. Inaccurate measurements can lead to an incorrect conclusion. Scientists often use special tools such as micrometers and calipers to help them make precise measurements.

In the case of the dinosaurs, scientists examined the fossil record for clues to what happened 65 million years ago. They studied rock layers throughout the world that date from the time when the dinosaurs disappeared. The scientists discovered that in certain locations these layers contain iridium, a substance that is uncommon in earth rocks, but common in meteorites. Scientists then measured the amount of iridium in the rock layers. They found that the rock layers in these particular locations contained nearly 160 times the amount of iridium normally found in earth rocks. The scientists searched for an explanation that would relate the iridium measurements to the disappearance of the dinosaurs.

Science & Technology

The Internet

Do you want to go for a spin on the information highway? Do you want to venture into cyberspace and "surf the Net"? Then take a trek on the Internet, a powerful information system that brings the electronic frontier to your doorstep.

The Internet began as an experimental computer network created by the United States Department of Defense during the Cold War. Its purpose was to safeguard scientific and military research in the event of a nuclear attack. If one or two computing centers were lost, the remaining sites in the network would continue to process and communicate vital data.

Today the backbone of the Internet consists of more than 2 million host computers in about 60 countries worldwide. At least 20 million computer users access these Internet computers using ordinary telephone lines. Once connected, they are able to send and retrieve information around the world with just a few keystrokes or mouse clicks.

Access to so much information makes the Internet an exciting place for scientific research. From on-line photographs of earthquake damage to space shuttle experiments displayed in real time, the information on the Internet is

Form a Hypothesis

Once a problem has been stated and information gathered, a scientist may propose a **hypothesis,** (HIE-POTH-uh-sus, pl. hypotheses), a possible explanation or solution to the problem. A hypothesis is based on facts, which are often established through observation.

For example, the scientists who discovered the iridium-laden rock layers proposed the meteorite-impact hypothesis to explain the extinction of the dinosaurs. This hypothesis states that about 65 million years ago, a giant meteorite crashed into the earth. The impact of the collision raised enough dust to block the sun's rays for many years. The earth probably became colder, plant life began to die, and many animal species, including the dinosaurs, became extinct. As the dust settled over the earth, it formed a layer of iridium-laden rock.

Test the Hypothesis

Once a hypothesis has been proposed, it should be tested. A hypothesis will not be accepted by the scientific community unless there is evidence to support it.

This Web page is home to the National Geophysical Data Center (www.ngdc.noaa.gov).

extremely current. Much of this information is difficult or impossible to find elsewhere.

The Internet also makes scientific collaboration easier than ever before. For the cost of an Internet subscription, anyone can quickly share research with an international audience. Various discussion groups and electronic message (e-mail) services help scientists stay informed about new discoveries and make valuable contacts with other professionals in their field.

The best way to get to know the Internet is to explore it firsthand. One popular vehicle for doing this is the World Wide Web, which combines text with graphics such as video clips or photographs. The Web enables users to travel the Internet by simply clicking on highlighted text or icons.

For the new user, there is on-line help available as well as numerous books and magazines about the Internet. Throughout this textbook, you will find Web addresses directing you to Web sites known as home pages. These home pages relate to specific earth science topics and contain links to other useful Web sites.

How might the Internet enhance scientific methods?

On-line information about the Internet and the World Wide Web is available at many sites, including the University of Maryland at **www.cs.umd.edu** and Pacific Lutheran University at **www.plu.edu.** Please note that the standard **http://** prefix has been omitted from all Web addresses in this textbook.

A hypothesis is tested by **experimentation.** An experiment is a scientific procedure carried out according to certain guidelines. An experiment enables scientists to test each **variable** that might prove or disprove the hypothesis. A variable is a factor in an experiment that can be changed. An experiment set up to test a variable is called a *controlled experiment*.

To ensure that only one variable is tested in an experiment, scientists will also run a control. The control will have the same conditions as the experiment except for the variable being tested. For example, to test the effects of sunlight on a green plant, a scientist would grow two identical plants. To control the experiment, the scientist would vary the amount of sunlight reaching one plant, while keeping the amount of sunlight constant on the other plant. The scientist would then observe both plants and record the observations. What is the variable in this experiment?

In the study of earth science, setting up controlled experiments to test a hypothesis is often difficult, and sometimes impossible. The scientists studying the disappearance of the dinosaurs, for instance, cannot bombard the earth with a giant meteorite to see if it produces life-threatening conditions.

Recently, however, scientists have developed computer models that enable them to test hypotheses by simulating certain conditions. For example, scientists have entered information into a computer about the possible climatic conditions during the period when the dinosaurs became extinct. They found that a dust cloud resulting

Figure 1-8. According to the meteorite-impact hypothesis, a huge meteorite crashed into the earth 65 million years ago. Dust from the impact blocked out the sun and led to the extinction of the dinosaurs.

Figure 1–9. Meteor Crater, in the northern Arizona desert, is 1,300 m in diameter and nearly 200 m deep. The crater is visible proof of the explosive power of a meteorite hitting the earth.

from the collision of a meteorite 10 km in diameter would have been sufficient to lower the earth's temperature considerably.

When experimentation is impossible, scientists often make more observations to gather evidence that will either support or discredit the hypothesis. The hypothesis is then tested by examining how well it fits or explains all the known observations.

To test the meteorite-impact hypothesis further, scientists had to find additional evidence that the iridium in the rock layers on earth had once come from meteorites. Scientists again examined the rock layers and this time they found strangely deformed particles of the mineral quartz. Previously this type of quartz had been found only near meteorite craters, at nuclear-testing sites, and in moon rocks. Scientists concluded that such quartz particles could only have been produced by an extremely powerful explosion. The collision of the earth with a huge meteorite, they reasoned, could have produced such an explosion.

INVESTIGATE!

To learn more about scientific methods, try the In-Depth Investigation on pages 20–21.

State a Conclusion

After many experiments and observations, scientists generally reach conclusions regarding the correctness of the hypothesis being considered. Depending on how well the hypothesis fits the known facts, it may be accepted as stated, altered slightly, or discarded altogether.

The fossil evidence for the meteorite-impact hypothesis does not prove that a meteorite was responsible for the extinction of the dinosaurs. The evidence does show, however, that an abnormally high amount of meteorite dust reached the earth at that time. Thus, until new evidence is found or a better hypothesis is proposed, the meteorite-impact hypothesis serves as one possible explanation of why the dinosaurs disappeared.

Section 1.2 Review

1. What are scientific methods?
2. Define *hypothesis*.
3. How do scientists test hypotheses?
4. Summarize the evidence scientists found to support the meteorite-impact hypothesis.
5. How have scientific methods contributed to the development of modern science?

PANDEMAGGIO

Section Objectives

- **Distinguish between a hypothesis, a theory, and a scientific law.**
- **Describe the Doppler effect.**
- **Summarize the big bang theory of the origin of the universe.**
- **List evidence for the big bang theory.**

1.3 Birth of a Theory: The Big Bang

Scientific methods are useful tools for the study of earth science. However, the development and testing of a hypothesis is just one step along the way to scientific understanding. Once a hypothesis has been tested and generally accepted, it may lead to the development of a **theory.** A theory is a hypothesis or a set of hypotheses that is supported by the results of experimentation and observation. A theory provides a general explanation for scientific observations that is consistent with known facts.

Once a theory is well established through research and experimentation, it may become a **scientific law.** A scientific law is a rule that correctly describes a natural phenomenon. To become a law, a theory must be proven correct every time it is tested. For example, the law of conservation of mass and energy, which states that the total amount of matter and energy in the universe does not change, has been tested again and again. It has never been found to fail.

Light and the Doppler Effect

One of the most exciting theories of modern science—how the universe began—has its roots in observations made more than 300 years ago. In 1665, the British scientist Isaac Newton observed that sunlight passing through a glass prism produced a rainbow of colors: red, orange, yellow, green, blue, and violet. Newton named this display of colors the **spectrum** (pl. spectra).

Light travels in waves. The distance from the crest of one wave to the crest of the next is a **wavelength.** Each color in the spectrum has a different wavelength. Red light has the longest wavelength, and violet light has the shortest. As light passes through a prism, each wavelength is bent at a different angle, and the band of colors results.

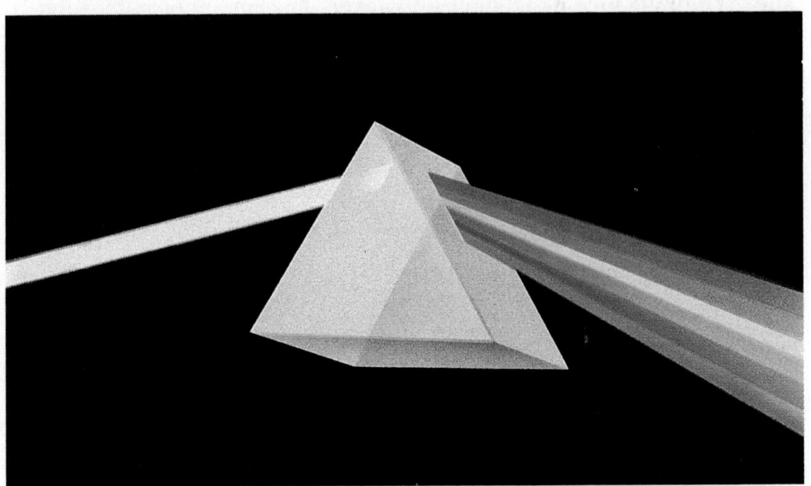

Figure 1–10. A prism bends the different wavelengths of light, separating white light into a rainbow of colors.

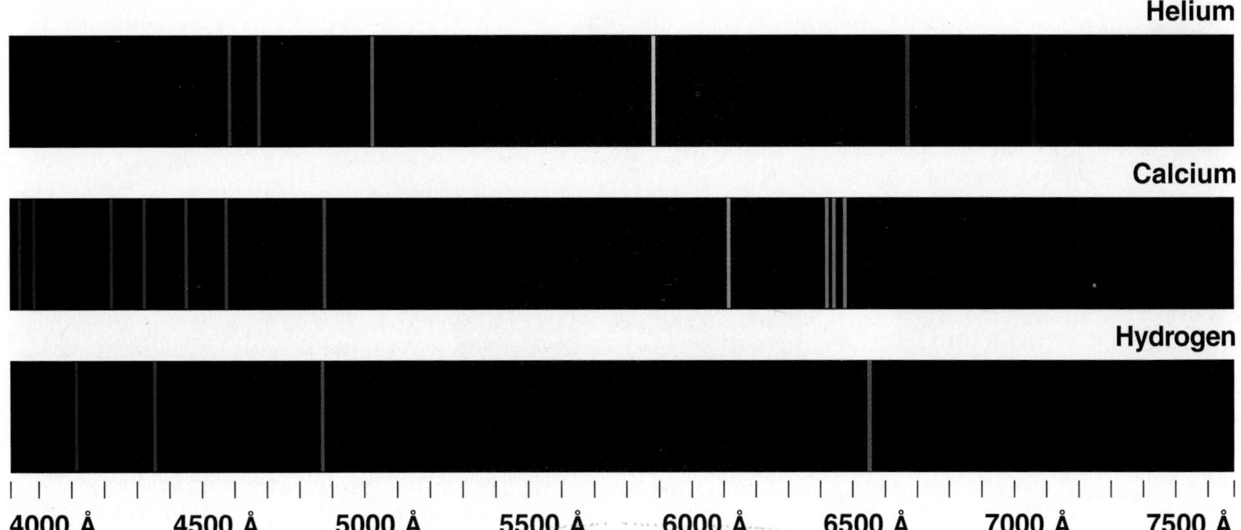

Figure 1–11. Each element produces a bright-line spectrum when heated.

In the late nineteenth century, research revealed that when chemical **elements** are heated, they too produce spectra. An element is a substance, such as hydrogen or iron, that cannot be broken down into a simpler form by ordinary chemical means. Instead of a full spectrum of continuous colors, like that produced by sunlight, a heated element produces only a series of thin colored lines spaced at uneven intervals. This series of colored lines, called a *bright-line spectrum,* indicates that the light source is sending out only certain wavelengths of light. An example of a bright-line spectrum is shown in Figure 1–11. Each element produces its own bright-line spectrum, as unique as a set of fingerprints.

Scientists also discovered that when a light source is moving toward an observer, the wavelengths of the light produced appear shorter to the observer. As a result, the spectral lines of the light source appear to shift slightly toward the shorter wavelengths, or the blue end of the spectrum. When a light source is moving away from an observer, the light waves appear longer to the observer. The spectral lines of the light source appear to shift toward the longer wavelengths, or the red end of the spectrum. The faster a light source is moving, the greater the shift of its spectrum. The apparent shift in the wavelengths of energy emitted by an energy source moving away from or toward an observer is called the **Doppler effect.**

Evidence: Red Shift

Using an instrument called a **spectroscope,** scientists studied starlight to determine what elements were present in the stars. A spectroscope contains a prism, which splits starlight into a spectrum of different colors and wavelengths. By comparing the spectrum produced by each star with the spectra of known elements, scientists were able to determine the chemical makeup of various stars. The sun, for example, was found to be about 92 percent hydrogen and almost 8 percent helium, with traces of nearly 100 other elements.

The study of starlight spectra revealed surprising information about our universe. Scientists found that the spectra of most *galaxies,*

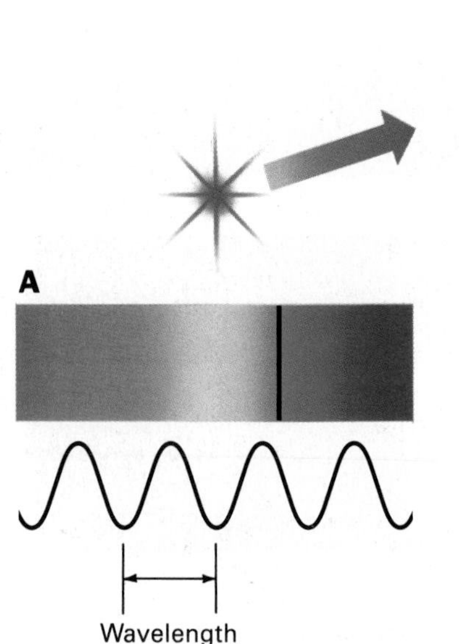
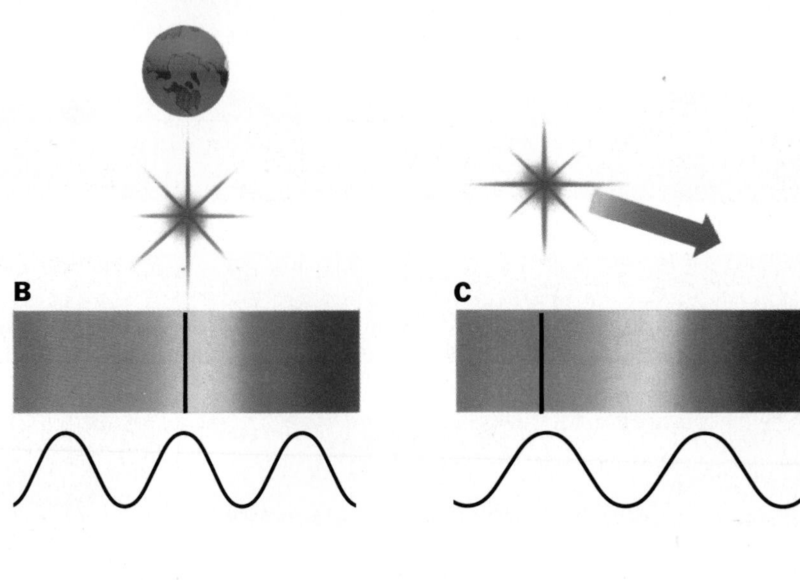

Figure 1-12. The wavelengths of light produced by star A, moving toward the earth, appear shorter. Therefore, the spectral lines of star A are shifted toward the blue end of the spectrum. The wavelengths of light produced by star C, moving away from the earth, appear longer. Therefore, the spectral lines of star C are shifted toward the red end of the spectrum. Star B is stationary.

or large systems of stars, tested were shifted toward the red end of the spectrum. Only a few close galaxies showed a shift toward the blue end of the spectrum. The red shift indicates that almost all of the galaxies in the universe are moving away from the earth.

By examining the degree of red shift, scientists were also able to determine the speed at which the galaxies were traveling. Scientists found that the most distant galaxies showed the greatest red shift and thus were moving away the fastest. From these observations of spectra, most scientists have concluded that the universe is expanding.

A Theory Emerges

Observations of red shift led scientists to propose a hypothesis to explain the expanding universe. The hypothesis states that billions of years ago, all the matter and energy in the universe was compressed into an extremely small volume. About 17 billion years ago, a sudden event called the *big bang* sent all the matter and energy hurtling outward in a giant cloud. As the cloud expanded, some of the matter gathered into clumps that evolved into galaxies. Today the universe is still expanding, and the galaxies continue to move apart from one another. This movement causes the apparent red shift in the spectra of galaxies.

Despite the evidence of red shift in the spectra of the galaxies, a number of scientists did not accept the big bang hypothesis. They argued that if the big bang had taken place as the hypothesis proposed, the energy left from the explosion would be found evenly distributed throughout the expanding universe. If this energy could not be found, they insisted, then there was little reason to accept the big bang hypothesis.

An important discovery in the 1960's finally convinced most scientists who had doubted the evidence of red shift. Using radio telescopes, researchers detected low levels of energy, called **background radiation,** evenly distributed throughout the universe. The presence of this energy convinced most scientists that the big bang hypothesis was correct. Because of the abundant evidence and

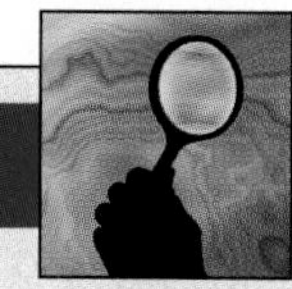

Small-Scale Investigation

The Big Bang Theory

According to the big bang theory, almost all galaxies are moving outward from all other galaxies. You can demonstrate the principles of this expansion with a simple model.

Materials

large (6–7 cm), uninflated round balloon; water-based felt-tip pen; string, 30 cm long; ruler

Procedure

1. Mark a pair of dots 0.5 cm apart across the middle of the uninflated balloon. Label them **A** and **B**. Mark a third dot 5.0 cm away from **B**. Label this dot **C**.
2. Blow into the balloon for 2–3 seconds. Record your elapsed time. Pinch the end of the balloon between your fingers to keep it inflated, but do not tie the neck.
3. Use the string and ruler to measure the distance between **A** and **B** and between **C** and **B**.
4. Calculate the rate of change in the distances between **A** and **B** and between **C** and **B**. To calculate the rate, subtract the original starting distance between the dots from the distance measured after inflation. Divide this number by the number of seconds you blew into the balloon.

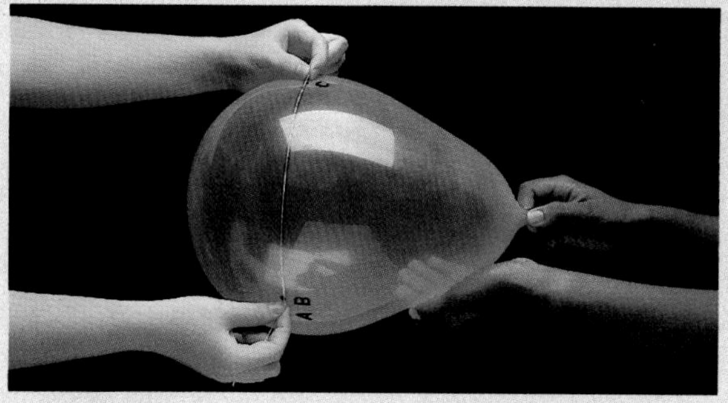

5. With the balloon still inflated from Step 2, blow into the balloon for an additional 2–3 seconds.
6. Measure and calculate the rate of change in the distances between **A** and **B** and between **C** and **B**. To calculate the rate, use the distance measured in Step 3 as the "original" distance.

Analysis and Conclusions

1. Did the distance between **A** and **B** or between **C** and **B** show the greatest rate of change?
2. Did the rate of change for either set of dots differ in Steps 4 and 6?
3. Suppose dots **C** and **A** represent galaxies and dot **B** represents the earth. How does the distance between the galaxies and the earth relate to the rate at which they are moving apart?

widespread acceptance of the big bang hypothesis, this explanation of the origin of the universe became known as the **big bang theory.**

Like any theory, the big bang theory must continue to be tested against each new discovery about the universe. As new information about our universe emerges, the big bang theory may be revised, or a new theory may take its place.

Section 1.3 Review

1. What is a scientific theory?
2. Describe the Doppler effect for light.
3. Describe the universe before the big bang.
4. What evidence supports the big bang theory?
5. If scientists discovered blue shift in the spectra of some distant galaxies, how might this information affect the big bang theory?

Chapter 1 Review

Key Terms

- astronomy (6)
- atmosphere (7)
- background radiation (16)
- big bang theory (17)
- biodegradable (8)
- biosphere (7)
- Doppler effect (15)
- earth science (3)
- ecology (7)
- ecosystem (7)
- elements (15)
- experimentation (12)
- geology (4)
- geosphere (7)
- hydrosphere (7)
- hypothesis (11)
- measurement (9)
- meteorology (5)
- observation (9)
- oceanography (5)
- pollution (8)
- scientific law (14)
- scientific methods (9)
- spectroscope (15)
- spectrum (14)
- theory (14)
- variable (12)
- wavelength (14)

Key Concepts

The four main branches of earth science are geology, oceanography, meteorology, and astronomy. **See page 3.**

Earth scientists and ecologists often work together to protect the environment. **See page 8.**

Scientific research attempts to solve problems logically through scientific methods. **See page 9.**

The meteorite-impact hypothesis provides a possible explanation for the mass extinction of the dinosaurs. **See page 11.**

Through observation and experimentation, scientists develop hypotheses, theories, and laws to describe natural phenomena. **See page 14.**

The Doppler effect describes the shift in the spectrum of a moving energy source, including light. **See page 14.**

The big bang theory provides a possible explanation of the origin of the universe. **See page 16.**

Evidence for the big bang theory comes from the study of starlight spectra and background radiation. **See page 16.**

Review

On your own paper, write the letter of the term that best completes each of the following statements.

1. The study of the solid earth is called
a. geology. b. oceanography.
c. meteorology. d. astronomy.

2. The earth scientist most likely to study storms is
a. a geologist. b. an oceanographer.
c. a meteorologist. d. an astronomer.

3. The study of the complex relationships between living things and their environment is called
a. geology. b. meteorology.
c. ecology. d. astronomy.

4. An example of a nonbiodegradable waste product is
a. an apple core. b. a plastic milk jug.
c. a pile of rotting leaves. d. an eggshell.

5. Usually the first step in scientific problem solving is to
a. form a hypothesis. b. state the problem.
c. gather information. d. state a conclusion.

6. A possible explanation for a scientific problem is called
a. an experiment. b. an observation.
c. a theory. d. a hypothesis.

7. The development of the meteorite-impact hypothesis began with the observation of
   a. blue shift in the spectra of stars.
   b. red shift in the spectra of stars.
   c. background radiation.
   d. iridium in earth rocks.
8. A statement that consistently and correctly describes some natural phenomenon is a scientific
   a. hypothesis. b. observation.
   c. law. d. control.
9. The apparent change in the wavelengths of a moving energy source is called
   a. the big bang. b. the Doppler effect.
   c. the spectrum. d. background radiation.
10. Scientists have found that as a light source moves toward a stationary observer, the wavelengths of the light source appear
    a. longer. b. shorter.
    c. higher. d. lower.
11. The big bang theory states that the galaxies in the universe are
    a. moving away from one another.
    b. moving towards one another.
    c. remaining stationary.
    d. being bombarded by meteorites.
12. Evidence for the big bang theory includes
    a. iridium in earth rocks.
    b. deformed quartz particles in earth rocks.
    c. blue shift in the spectra of galaxies.
    d. red shift in the spectra of galaxies.

Critical Thinking

On your own paper, write answers to the following questions.

1. A meteorite lands in your back yard. Which earth scientist would you call to study the meteorite? Why?
2. A stream that feeds a small pond gradually dries up. How might this change affect the ecosystem of the pond?
3. Some scientists have hypothesized that meteorites have periodically bombarded the earth, causing mass extinctions every 26 million years. How might this hypothesis be tested?
4. Imagine that you are on another planet in a galaxy far from the earth. If you used a spectroscope to examine the spectrum of the sun, would you expect to find red shift, blue shift, or no shift at all? Why?
5. According to the big bang theory, the original big bang took place about 17 billion years ago. How might scientists have been able to determine this?

Application

1. You find a yellow rock and wonder if it is gold. How could you apply scientific methods to this problem?
2. A scientist observes that each eruption of a volcano is preceded by a series of small earthquakes. The scientist then makes the following statement: Earthquakes cause volcanic eruptions. Is the scientist's statement a hypothesis or a theory? Why?
3. Construct a **concept map** using 10 of the new terms listed on page 18 by making connections that illustrate the relationship among the terms. See page xxi for instructions on making concept maps.

Extension

1. Assume that you must lose all the benefits of one of the earth sciences. Which would you give up first? Why? Defend your selection in an essay.
2. Choose an ecosystem and draw a diagram showing some of the interrelationships between the living things and their environment. Identify the living things according to their roles as producers, consumers, or decomposers.
3. Find out about the organizations in your community that are working to protect the environment. Write a short profile describing the work of each organization. Put the information together in a directory.

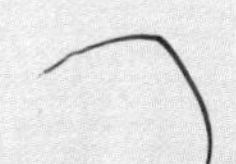

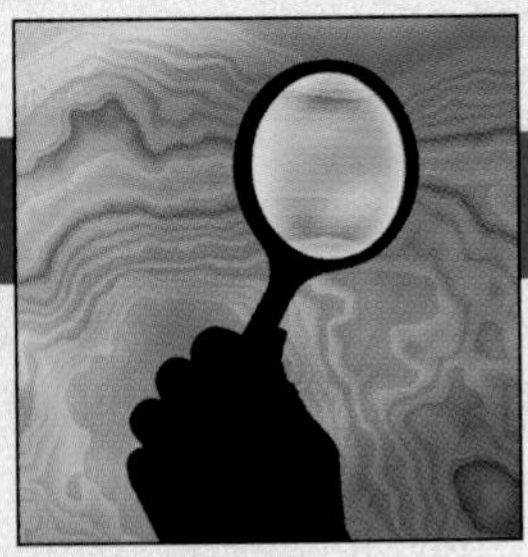

In-Depth Investigation

Scientific Method

Materials

- meter stick
- hand lens

Introduction

Not all scientists think alike; nor do they always agree about various theories. However, all scientists use *scientific method,* part of which includes the skills of observing, inferring, and predicting. In this investigation, you will apply scientific method as you examine a place where puddles often form after rainstorms. You can study the puddle area even when the ground is dry, but it would be best to observe the area again when it is wet. Since water is the most effective agent of change in our environment, you will be able to make many observations.

Prelab Preparation

1. Review Section 1.2, pages 9–13.
2. Find out the difference between a quantitative observation and a qualitative observation.

Procedure

1. Examine the area of the puddle and the surrounding area carefully. On a sheet of paper, write the heading "Observations." Make a numbered list of what can be seen, heard, smelled, or felt. Sample observations: "The ground where the puddle forms is lower than the surrounding area," "There are cracks in the soil," and so forth. *Note: Avoid any suggestions of causes.*
2. On another sheet of paper, write the heading "Inferences." Review your observations and write possible causes for those observations. Sample: "Cracks in the soil (Observation 2) may have been caused by lack of rain (Observation 5)."

3. Review your observations and possible causes, and place them into similar groups if possible. Can one cause or set of causes explain several observations? Is each cause reasonable when compared with the others? Does any cause contradict any of the other observations?
4. Start a new page labeled "Hypotheses." You have learned that a hypothesis is a possible explanation to a problem or an occurrence. Create a hypothesis for each group of causes and observations above. Look for the underlying and general mechanisms that produce and link these events. Example: "The clays in the soil shrink as they lose water, thereby causing cracks to form."
5. Based only on your hypotheses, make some predictions about what will happen at the puddle as conditions change. Describe the changes you expect and your reasoning. Sample prediction: "The cracks will grow wider as the puddle dries. Any added water will shrink the cracks."
6. Revisit the puddle several times to see if the changes you observe match your predictions.

Analysis and Conclusions

1. Which of your senses did you use most to make your observations? How could you improve observations using this sense?
2. What could you have used to measure, or put into numbers, many of your observations? Is quantitative observation better than qualitative observation? Explain.
3. Can inferences generally be relied on as true? Explain.
4. If your predictions are found to be incorrect, was the act of forming your inferences a waste of time? Explain.
5. When knowledge is derived from observation and prediction, this process is called "scientific method." After reporting the results of a prediction, how might a scientist continue his or her research?

Extensions

1. Decide whether each of the following statements is an observation (O) or an inference (I). *Note: It does not matter whether the statement happens to be true or false.*
   a. Grass is present inside the puddle.
   b. The grass surrounding the puddle is greener and taller than that inside the puddle.
   c. During a rainstorm, some soil is washed into the puddle.
   d. Water always runs downhill.
   e. Gravity causes the water to run downhill.
   f. The soil that washes out of the puddle will eventually become part of a stream.
   g. Brownish water contains suspended soil particles.
   h. The soil particles are suspended because water is flowing fast.
   i. When the rain stops, the puddle water looks clear.
   j. Mud cracks result from drying the soil.
2. Choose another small area to examine, but look for changes caused by a different factor, such as wind. Follow the steps outlined in the investigation to predict changes that will occur in the area. Use scientific method to design an experiment you can test. Briefly describe your experiment and how you tested it.

During one of the Apollo missions to the moon that began in 1969, an astronaut snapped this picture of the earth above the surface of the moon. For the first time in history, people could view the beautiful blue earth suspended in the blackness of space. Since that time, space exploration has added much to the knowledge of the earth. This chapter describes the earth, its movements through space, and the use of spacecraft to explore the earth.

Chapter Outline

The blue earth, viewed from the Apollo II spacecraft orbiting the moon

2.1 Earth: A Unique Planet

Photos of the earth from space reveal a blue sphere covered with white clouds. Surrounded by the blackness of space, the earth appears beautiful, yet fragile. In fact, the earth is unique in the solar system. It is the only known planet with liquid water on its surface and an atmosphere that contains a large amount of oxygen. Most important, the earth is the only planet known to support life.

Section Objectives

- **List the characteristics of the earth's three major zones.**
- **Explain how studies of seismic waves have provided information about the earth's interior.**
- **Define *magnetosphere* and identify the possible source of the earth's magnetism.**
- **Summarize Newton's law of gravitation.**

Statistics

Viewed from space, the earth appears to be a perfect sphere. On a perfect sphere, the circumference, or distance around, is the same no matter where it is measured. However, careful measurements reveal that the earth's circumference varies slightly, depending on where it is measured. The circumference measured around the poles is 40,007 km. The circumference measured around the *equator* is 40,074 km. The equator is an imaginary line that divides the earth into the Northern and Southern Hemispheres.

Because of these differences in the earth's circumference, the earth is more accurately described as an *oblate spheroid,* a slightly flattened sphere. The spinning of the earth on its **axis** causes the polar regions to flatten and the equatorial zone to bulge. The earth's axis is an imaginary straight line running through the earth from the North Pole to the South Pole.

Seen from space, the earth's surface also appears to be smooth. Given the size of the planet, its surface is relatively smooth. The difference between the height of the tallest mountains and the depth of the deepest ocean trenches is only about 20 km. This distance is very small compared to the earth's average diameter of 12,735 km.

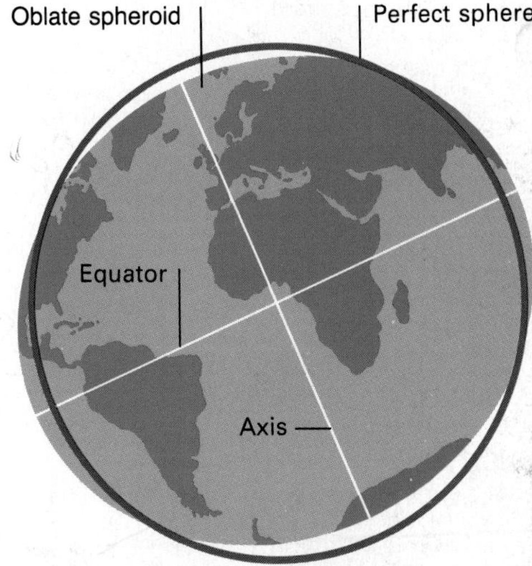

Figure 2–1. The earth's shape is really an oblate spheroid, not a perfect sphere. The red line shows the shape of a perfect sphere.

The Hydrosphere and the Atmosphere

The earth is 71 percent covered by water, and about 97 percent of that water is in the salty oceans. The remaining 3 percent of earth's water is fresh water that is found in lakes, rivers, and streams and that is frozen in glaciers and the polar ice sheets. All the earth's water makes up the *hydrosphere*.

The earth is surrounded by a blanket of gases called the *atmosphere*. The atmosphere provides the air you breathe and shields the earth from the sun's harmful radiation. The atmosphere is made up of 78 percent nitrogen and 21 percent oxygen. The remaining 1 percent includes other gases, such as argon, carbon dioxide, and helium.

The Earth's Interior

Direct observation of the earth's interior is impossible, so scientists must rely on indirect methods to study it. For example, scientists have made important discoveries about the earth's interior through studies of **seismic** (SIZE-mik) **waves.** Seismic waves are vibrations that travel through the earth. Earthquakes and explosions on or near the earth's surface produce seismic waves. By studying seismic waves as they travel through the earth, scientists have determined that the earth is made up of three major zones.

Zones of the Earth

The thin, solid outermost zone of the earth is the outer covering or the **crust.** The crust makes up only 1 percent of the earth's *mass.* Mass is the amount of matter in an object. Beneath the oceans, the crust is called *oceanic crust.* Oceanic crust is only 5 km to 10 km thick. The part of the crust on which the continents rest is called *continental crust.* The continental crust varies in thickness from 32 km to 70 km. Continental crust is thickest beneath mountains.

Below the crust lies the **mantle,** a zone of rock nearly 2,870 km thick. The mantle makes up almost two thirds of the earth's mass and is divided into two different regions.

The uppermost part of the mantle is solid. This solid portion of the mantle and the crust above it make up the *lithosphere,* a rigid layer 65 km to 100 km thick. Just below the lithosphere is a region of the mantle called the *asthenosphere.* The asthenosphere is approximately 200 km thick. Because of enormous heat and pressure, the solid rock of the asthenosphere has the ability to flow. The ability of a solid to flow is called *plasticity.*

Below the mantle is the **core.** It forms the center of the earth and is made mostly of iron. From seismic studies, scientists think the outer core is a dense liquid layer about 2,190 km thick. The inner core is a dense solid about 2,680 km in diameter. The inner and outer core together make up nearly one third of the earth's mass.

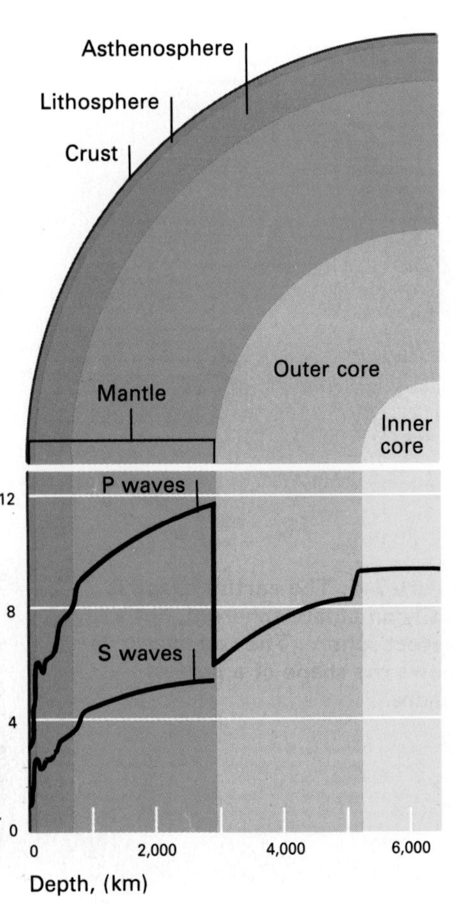

Figure 2–2. Changes in the speed of seismic waves were used to determine the location of the earth's different zones.

Seismic Wave Studies

There are two types of seismic waves—primary waves, or *P waves,* and secondary waves, or *S waves.* They are useful to scientists exploring the earth's interior. P waves and S waves behave differently. P waves travel through liquids, solids, and gases. S waves travel only through solids. P waves also travel faster than S waves. The speed and direction of both types of waves are affected by the composition of the material through which they travel. Both P waves and S waves travel faster through more-rigid materials. The speed and direction of seismic waves reveal much about the earth's interior.

The Moho In 1909 Andrija Mohorovičić (MOE-huh-ROE-vuh-CHICH), a Croatian scientist, discovered that the speed of seismic waves increases abruptly 32 km to 70 km beneath the earth's surface. As Figure 2–2 shows, this change in the speed of the waves marks the boundary between the crust and the mantle. This boundary is called the *Mohorovičić discontinuity,* or the **Moho.** The increase in speed at the Moho indicates that the earth's mantle is denser than its crust.

Below the Moho, at a depth of about 100 km, a decrease in seismic-wave speed marks the boundary between the lithosphere and

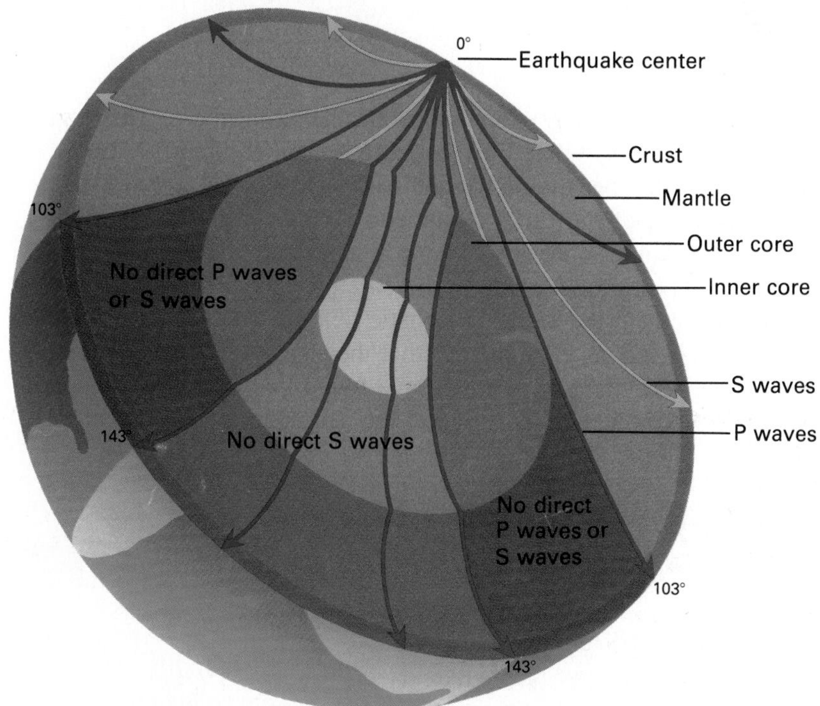

Figure 2–3. No direct S waves can be detected in locations more than 103° from the earthquake epicenter. No direct P waves can be detected in locations between 103° and 143° from the earthquake center.

the less rigid asthenosphere. Seismic waves then increase in speed until, at a depth of about 2,900 km, P waves slow down again, while S waves disappear entirely. These changes in the seismic waves mark the boundary between the mantle and the outer core. Since S waves cannot travel through liquids and P waves slow down in less-rigid materials, scientists think the outer core may be a dense liquid. At a depth of 5,150 km, P waves speed up again, marking the boundary between the outer core and the inner core. This increase in speed suggests that the inner core is a dense, rigid solid.

Shadow Zones Recordings of seismic waves around the world reveal **shadow zones** on the earth's surface. Shadow zones are locations on the earth's surface where neither S waves nor P waves are detected or where only P waves are detected.

Shadow zones occur because the materials that make up the earth's interior are not uniform in rigidity. When seismic waves travel through materials of differing rigidities, their speed changes, causing the waves to bend and change direction.

As Figure 2–3 shows, a large S-wave shadow zone covers the side of the earth that is opposite an earthquake. S waves do not reach the S-wave shadow zone because they are blocked by the liquid outer core. Although P waves can travel through all the layers, the speed and direction of the waves change as the waves pass through each layer. The waves bend in such a way that a P-wave shadow zone forms.

The Earth as a Magnet

If you have ever used a magnetic compass to find direction, you know that the earth acts as a giant magnet. Like a bar magnet, the earth has two magnetic poles. As Figure 2–4 shows, the lines of force of the earth's magnetic field extend between the North Magnetic Pole and the South Magnetic Pole. The earth's magnetic field

Figure 2–4. The flow of charged particles from the sun compresses and shapes the magnetosphere so that it flares out on the side of the earth away from the sun.

also affects an area that extends beyond the atmosphere. This region of space, which is affected by the earth's magnetic field, is called the **magnetosphere** (mag-NEET-uh-SFIR).

The source of the earth's magnetic field may be the liquid iron in the earth's outer core. Because iron is a good conductor, scientists hypothesize that motions within the core produce electrical currents that in turn create the earth's magnetic field.

Research, however, indicates that there may be another source for the magnetic field. Scientists have found that both the sun and the moon have magnetic fields. Yet the sun contains no iron, and

CAREER FOCUS: *Earth Structure*

"As a geologist, you know that no matter what problem you're trying to solve, the answer is out there just waiting to be discovered."

Ken Pierce

Field Geologist

Talking to Denver-based field geologist Ken Pierce is one thing. Finding him is quite another.

"During the field season, I can be out in the field area for as long as four months at a time," he says. "Though I mostly drive a jeep and hike, I have made some month-long pack trips on horseback as well."

Like other field geologists, Pierce, pictured at left, studies the composition, structure, and history of the earth's crust. Pierce specializes in the study of the most recent geologic time.

After collecting samples, including sand, gravel, and glacial deposits, Pierce determines which samples merit further study. He brings them to the lab, where they are X-rayed, studied under high-powered microscopes, and subjected to chemical analysis. Pierce also performs experiments to test his geologic theories.

"Although I have always liked being outdoors, it wasn't until I was in college that I really became interested in learning why the surface of the earth looks the way it does and knew I wanted a career in geology."

Pierce holds undergraduate and graduate degrees in geology with an emphasis on geomorphology, the study of the origin of the earth's surface features. He says he enjoys the never-ending challenge of his work.

"You know if you do enough research, make the right observations, and ask the right questions, you'll find your answer. It is mostly a matter of perseverance."

Instrumentation Technician

Instrumentation technicians, such as the one pictured below, help

Instrument technician ▶

the moon does not have a liquid core. Discovering the sources of the magnetic fields of the sun and moon may help scientists identify the source of the earth's magnetic field.

The Earth's Gravity

The seventeenth-century British scientist Isaac Newton made many contributions to the fields of mathematics, physics, and astronomy. Among the most important were his studies of **gravity,** the force of attraction that exists between all matter in the universe.

develop the instruments that earth scientists use to study the earth. Many of these devices aid in the search for fossil fuels and geothermal energy. Others are used in the fields of soil conservation, pollution control, and oceanography.

In addition to helping design new machinery, instrumentation technicians may also test, install, operate, and service earth-science equipment.

Instrumentation technicians generally have completed a technical degree program or received on-the-job training.

▲ This cliff presents a challenge to surveyor field parties.

Surveyor

Surveyors measure the height and shape of the land and mark waterways, boundary lines, and distances. Surveyors usually work in small groups called field parties. One surveyor may operate a *transit,* which is a small telescope or laser that measures angles and distances. Meanwhile, another member of the field party holds the rod that marks the height of land being surveyed.

The results of the survey are recorded, and the data are verified. Surveyors then prepare the sketches, maps, and reports needed to establish official land and water boundaries.

Although some states require surveyors to have a four-year college degree, most people enter the field through technical training programs.

For Further Information

For more information on careers that deal with the earth's structure, write the American Geological Institute, 4220 King St., Alexandria, VA 22303. Or check out their Web site at www.agiweb.org.

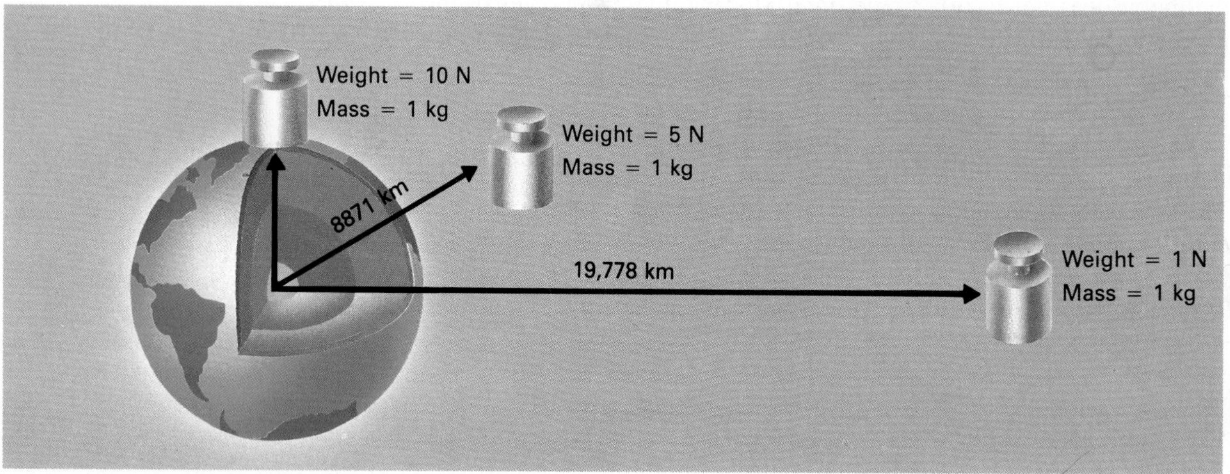

Figure 2–5. As the distance from the earth's center increases, the weight of an object decreases. The mass of an object stays the same.

Newton described the effects of gravity in his **law of gravitation.** This law states that the force of attraction between any two objects depends upon their masses and the distance between them. The larger the masses of two objects and the closer together they are, the greater will be the force of gravity between them.

The mass of the earth exerts a force of gravity that pulls objects toward the center of the earth. *Weight* is a measure of the strength of the pull of gravity on an object. Weight is measured in *newtons* (N). On the earth's surface, a kilogram of mass weighs about 10 N. What is the weight of a 2-kg mass on the earth's surface? the weight of a 10-kg mass?

Weight and mass are not the same. Mass is the amount of matter in an object. Weight is the force of gravity on that matter. The mass of an object does not change with location, but its weight does. The weight of an object depends on its mass and its distance from the earth's center. As the law of gravitation states, the force of gravity decreases as the distance from the earth's center increases. For example, at 8,871 km from the center of the earth (2,500 km above the surface), a 1-kg mass weighs about 5 N, half of its weight on the earth's surface. At 19,778 km from the center of the earth (13,400 km above the surface), a 1-kg mass weighs only about 1 N.

Weight also varies according to location on the earth's surface. As you have read, the earth spins on its axis, and this motion causes the earth to bulge out slightly near the equator. Therefore, the distance between the earth's surface and its center is greater at the equator than it is at the poles. This difference in distance means that your weight at the equator would be about 0.3 percent less than your weight at the North Pole.

Section 2.1 Review

1. How thick is the mantle?
2. Describe the earth's core.
3. Compare the behavior of P waves and S waves.
4. Define magnetosphere.
5. Explain why you weigh less on a mountain than at sea level.

2.2 Movements of the Earth

Section Objectives

- **Describe the earth's revolution and rotation.**
- **Tell why the seasons change.**
- **Explain how the sun is used as a basis for measuring time.**

The earth is constantly in motion. You are traveling with the earth around the sun at an average speed of 106,000 km/hr. The movement of the earth around the sun is called **revolution.** Each complete revolution takes 365.24 days, or one year.

As it revolves around the sun, the earth also spins on its axis. This spinning motion is called **rotation.** Each complete rotation takes about 24 hours, or one day.

The Rotating Earth

For everyone on earth, the most observable effects of the earth's rotation on its axis are day and night. As the earth rotates from west to east, the sun appears to rise in the east in the morning. It travels across the sky and sets in the west. At any given moment, it is daytime on the hemisphere of earth facing the sun. It is nighttime on the hemisphere of the earth facing away from the sun.

The Revolving Earth

The earth's orbit, or path around the sun, is slightly elliptical, or oval-shaped. Therefore, the earth is not always the same distance from the sun. At its closest point to the sun, the earth is said to be at **perihelion** (PER-uh-HEEL-yuhn). At its farthest point, the earth is at **aphelion** (a-FEEL-yuhn). As shown in Figure 2–6, the earth reaches perihelion on January 3 and aphelion on July 4.

The earth's aphelion distance is 152 million kilometers. Its perihelion distance is 147 million kilometers. The average distance between the earth and the sun is 150 million kilometers. When the orbit of the earth is drawn as in Figure 2–6, it appears to be perfectly circular. However, it is actually slightly elliptical.

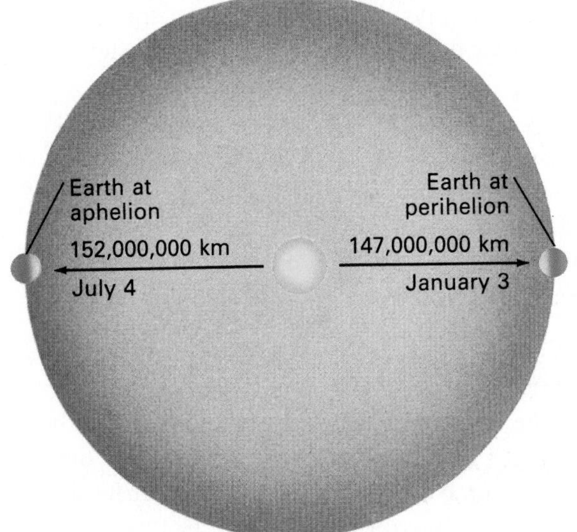

Figure 2–6. Each year, the earth is farthest from the sun in July and closest to the sun in January.

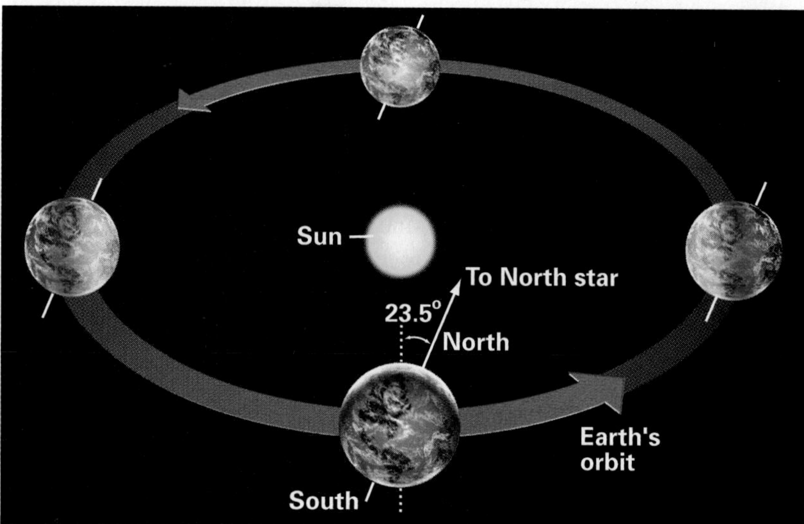

Figure 2–7. The earth's orbit forms an elliptical plane. The axis is tilted 23.5° from the perpendicular to the orbital plane. The direction of tilt of the earth's axis remains the same throughout the earth's orbit, always pointing toward the North Star.

As shown in Figure 2–7, the earth's orbit lies in a plane. The earth's axis is tilted 23.5° from the perpendicular (90°) to the plane of its orbit. As the earth revolves around the sun, the direction its axis points does not change; it always points toward the North Star. Consequently, during each revolution, the North Pole tilts at times toward the sun, and at other times away from it.

When the North Pole is tilted toward the sun, the Northern Hemisphere has longer periods of daylight. When the North Pole is tilted away from the sun, the Southern Hemisphere has longer periods of daylight.

The sun's rays are nearly parallel to one another as they reach the earth. Because the earth's surface is curved, however, the sun's rays strike different parts of the earth at different angles. The amount of solar heat an area receives depends on the angle at which the sun's rays strike that area of the earth's surface.

When the sun is directly overhead, its rays strike the earth at a 90° angle. The closer to 90° the rays are, the more concentrated they are and the greater is the heat they produce on the earth's surface. As the angle of the rays decreases, the rays spread out. When this happens, the solar heat reaching the surface of the earth becomes less intense.

The angle at which the sun's rays strike each part of the earth's surface changes as the earth moves through its orbit. When the North Pole is tilted toward the sun, the sun's rays strike the Northern Hemisphere at higher angles. When the North Pole is tilted away from the sun, the sun's rays strike the Southern Hemisphere at higher angles.

The Seasons

Changes in the angle at which the sun's rays strike the earth's surface and changes in the amount of daylight cause the seasons. For example, when the North Pole is tilted away from the sun, the angle of the sun's rays falling on the Northern Hemisphere is lower. As a result, there are fewer hours of daylight. The weak rays of the sun and the short hours of daylight produce the cool winter season in the Northern Hemisphere. At the same time, the angle of the sun's rays striking the Southern Hemisphere is higher, and there are more hours of daylight. The sun's strong rays and the longer hours of daylight produce the warm summer season there.

Summer and Winter Solstices

Because of the earth's orbit, each year on June 21 or 22, the North Pole tilts toward the sun. On this day, the sun's rays strike the earth at a 90° angle along the Tropic of Cancer. This day is called the **summer solstice.** It marks the beginning of summer in the Northern Hemisphere. *Solstice* means "sun stop" and refers to the fact that in the Northern Hemisphere the sun follows its highest path across the sky on that day. Figure 2–8 shows this path as it appears in the Northern Hemisphere at the summer solstice.

The Northern Hemisphere has the most hours of daylight at the summer solstice. The farther north of the equator you are, the longer the period of daylight you have. North of the Arctic Circle, there are 24 hours of weak daylight at the summer solstice. At the other extreme, south of the Antarctic Circle, there are 24 hours of darkness.

By December the earth is halfway through its orbit, and the North Pole tilts away from the sun. On December 21 or 22, the sun's rays strike the earth at a 90° angle along the Tropic of Capricorn. This day is called the **winter solstice.** It marks the beginning of winter in the Northern Hemisphere.

At winter solstice, the Northern Hemisphere has the fewest daylight hours. The sun follows its lowest path across the sky, as Figure 2–8 shows. Places north of the Arctic Circle have 24 hours of darkness; those south of the Antarctic Circle have 24 hours of daylight.

Autumnal and Vernal Equinoxes

On September 22 or 23 of each year, the sun's rays strike the earth at a 90° angle along the equator. This day is called the **autumnal equinox** and marks the beginning of the fall season in the Northern Hemisphere. *Equinox* means "equal night" and refers to the fact

Figure 2–8. In the Northern Hemisphere, the sun appears to follow its highest path across the sky on the summer solstice.

DAN DEMAGGIO

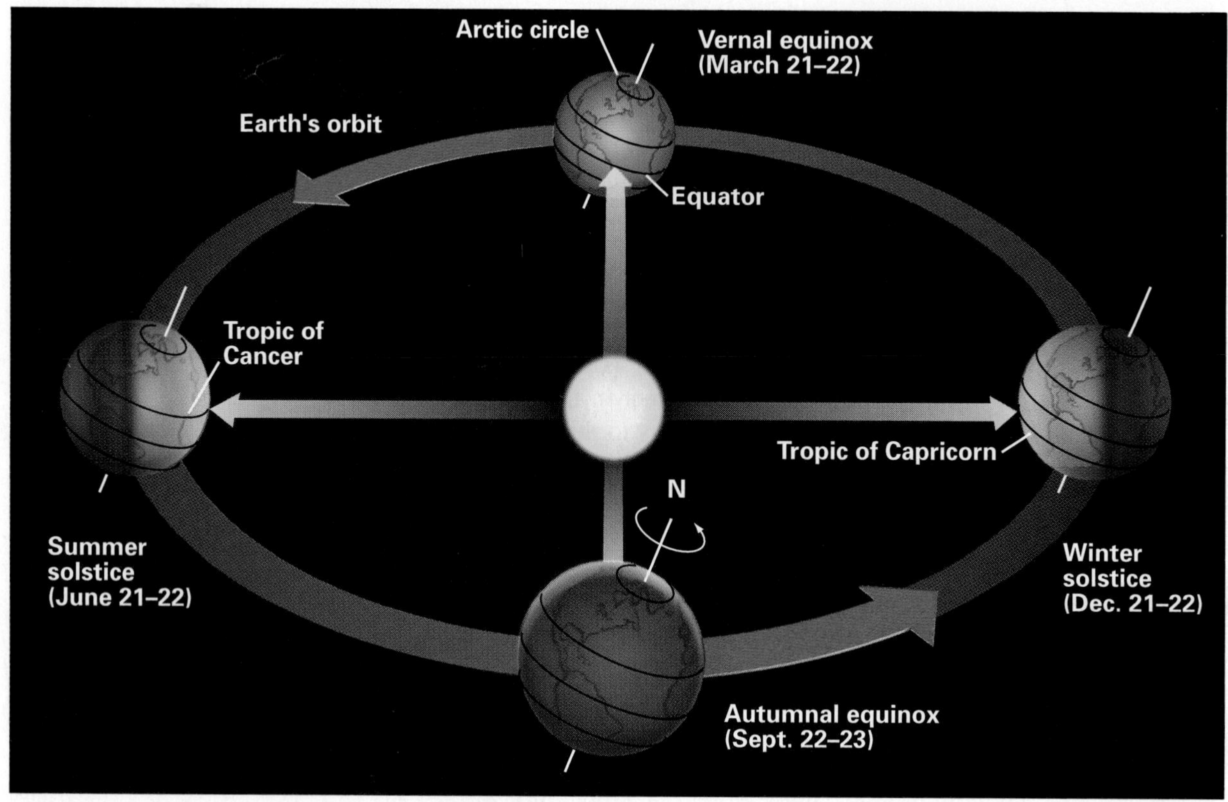

Figure 2–9. At the solstices, the sun's rays hit the Tropic of Cancer or the Tropic of Capricorn at a 90° angle. At the equinoxes, they hit the equator at a 90° angle.

that the hours of daylight and darkness are equal everywhere on the earth on that day. The hours of daylight and darkness are equal because at the equinox the North Pole tilts neither toward nor away from the sun. This position of the North Pole in relation to the sun is illustrated in Figure 2–9.

On March 21 or 22, the sun's rays again strike the earth at a 90° angle along the equator. This day is called the **vernal equinox** and marks the beginning of spring in the Northern Hemisphere. As during the autumnal equinox, the hours of daylight and darkness are equal everywhere on the earth. On this day, the North Pole tilts neither toward nor away from the sun.

Precession

The illustration in Figure 2–9 also shows that as the earth completes its revolution of the sun, its axis of rotation continuously points toward Polaris, the North Star. This has not always been the case, however. As the earth rotates about its axis, the direction in which the axis points slowly changes in relation to the distant stars. This change results from a circular motion of the earth's axis, called **precession.** Precession is caused by forces acting on a spinning body, in this case by the gravitational pull exerted on the rotating earth by the moon, the sun, and the other planets. Precession causes the earth's axis to move slowly in a circle, much like a top does as it spins on a table. The earth's axis, however, completes only one full circle every 26,000 years, so Polaris will continue to be our North Star for many years to come.

INVESTIGATE!

To learn more about the motion of the sun and earth, try the In-Depth Investigation on pages 40–41.

Time Zones

Using the sun as the basis for measuring time, 12:00 noon is defined as the time when the sun is highest in the sky. Because of the sun's apparent movement from east to west, the sun appears highest over different locations at different times. For example, assume the sun is highest over New York City at 12:00 noon. In Philadelphia, a short distance west of New York City, the sun would appear highest a few minutes later. In Baltimore, just west of Philadelphia, the sun would appear highest a few minutes after that. If these communities were to set their clocks precisely by the sun, the clocks in each community would mark slightly different times. To avoid problems created by different local times, the earth's surface is divided into 24 **standard time zones.** In each zone, noon is set as the time when the sun is highest over the center of that zone.

Biological Clocks

*O*ur earliest ancestors lived in harmony with the cycles of the sun. They were awakened by the sun's first rays, and they ended their workday at sunset. With the development of artificial light sources, however, humans have become less dependent on the availability of sunlight to set their daily routines. Yet research reveals surprising evidence that human behavior is still closely tied to solar rhythms.

Scientists have discovered that many body processes occur in 24-hour cycles called *circadian rhythms. Circa* is Latin for "about"; *dies* is Latin for "day." No one understands exactly what controls circadian rhythms, but the human body seems to have a number of internal clocks. They regulate patterns of sleeping and waking, daily changes in body temperature, hormone secretions, heart rate, and blood pressure. Even moods, coordination, and memory have their own circadian rhythms.

Studies indicate that the cycle of darkness and light caused by the earth's rotation sets and resets the clocks of the body. It more or less keeps vital processes on a 24-hour schedule.

When the internal clocks get out of sync with the sun's cycle, problems arise. One graphic example is jet lag, the combination of exhaustion, irritability, and insomnia that travelers often suffer after a long flight across several time zones.

Based on your knowledge of time zones and circadian rhythms, explain the cause of jet lag.

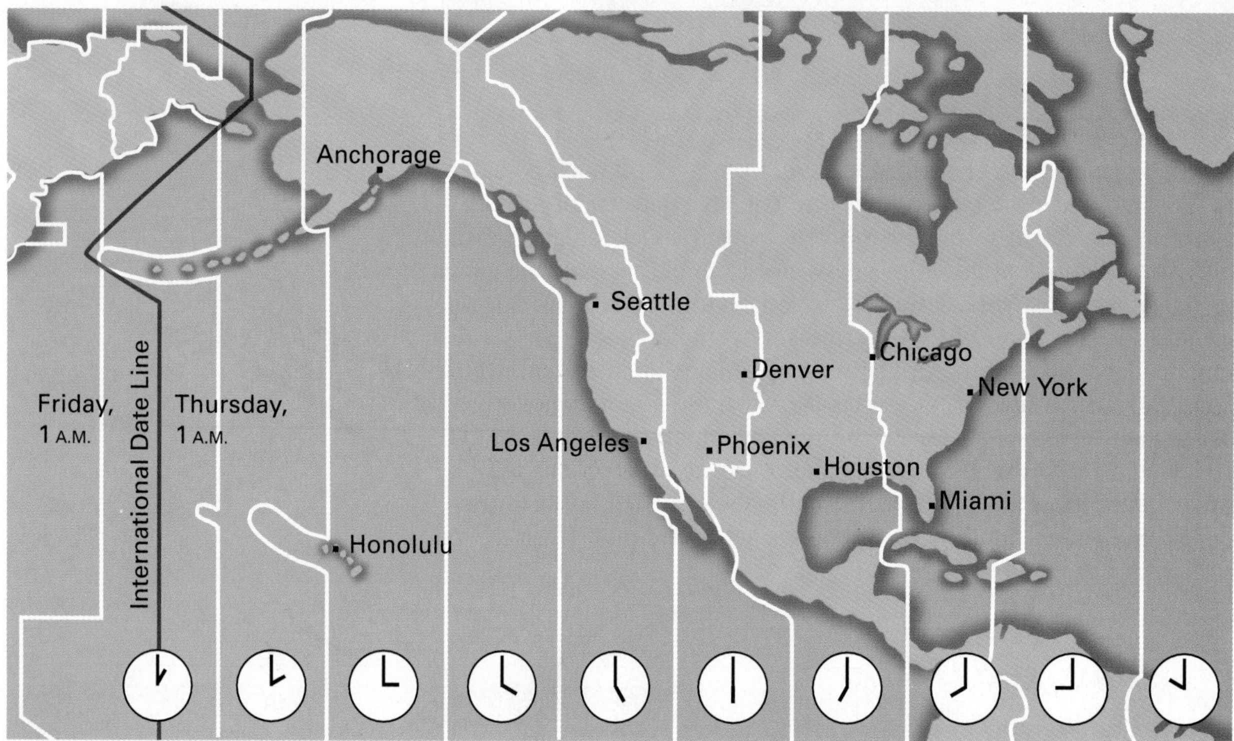

Figure 2–10. The earth has been divided into 24 standard time zones. Going east, travelers must set their clocks ahead one hour for each time zone crossed. Going west, travelers must set their clocks back one hour.

Because the earth is nearly spherical, its circumference equals 360°. Dividing 360° by the 24 hours needed for one rotation, you find that the earth rotates at a rate of 15° each hour. Therefore, each of the earth's 24 standard time zones covers about 15°. The time in each zone is one hour earlier than the time in the zone to its east. Figure 2–10 shows the standard time zones in the United States. At 6 P.M. in New York City, what time is it in Los Angeles? in Denver?

There are 24 standard time zones and 24 hours in a day. But there must be some point on the earth's surface where the date changes from one day to the next. To prevent confusion, the **International Date Line** has been established. The International Date Line is a line running from north to south through the Pacific Ocean. When it is 8:00 A.M. Friday west of the International Date Line, it is 8:00 A.M. Thursday east of the line. The line is drawn so that it does not cut through islands and continents in the Pacific Ocean. Thus, the people living within one country have the same date. Note how this is done between Alaska and Siberia in Figure 2–10.

During the summer, most of the United States uses **daylight saving time.** Under this system, clocks are set one hour ahead of standard time in April, which provides an additional hour of daylight during the evening. For example, if the sun sets at 7 P.M. standard time, it would set at 8 P.M. daylight saving time. In October, clocks are set back one hour, returning to standard time.

Section 2.2 Review

1. Describe the position of the earth during the summer solstice. Where do the sun's rays strike the earth at a 90° angle?
2. What causes winter in the Northern Hemisphere?
3. What are the advantages in using daylight saving time?

2.3 Artificial Satellites

Section Objectives

- **Compare two types of satellite orbits.**
- **Discuss ways in which satellites are used to study the earth.**

If you drop a ball, it will fall straight down because the force of gravity pulls objects toward the center of the earth. If you throw a ball horizontally, it will follow a curving path. The force of gravity still pulls the ball toward the earth, but the ball also moves horizontally as it falls. The greater the speed at which the ball is thrown, the farther it will travel before gravity pulls it to the earth's surface. For example, a ball thrown at about 8 km/s and not slowed down by air resistance would follow the curve of the earth. As shown in Figure 2–11, the ball would fall toward the earth but never reach the earth's curved surface. Instead the ball would become an artificial earth **satellite.** A satellite is any object in orbit around another body with a larger mass.

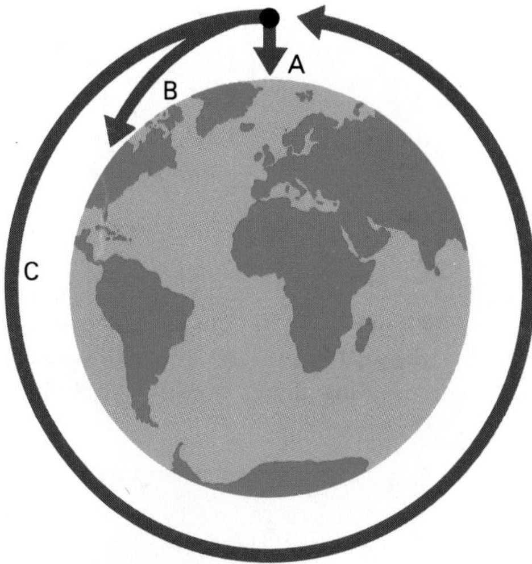

Figure 2–11. *A* is the path of a ball that is dropped. *B* is the path of a ball thrown horizontally. *C* is the path of a ball moving fast enough to go into orbit.

Satellites and Orbits

Today hundreds of artificial satellites orbit the earth. Meteorological satellites gather and transmit weather information. Communications satellites relay radio, telephone, and television signals to and from various locations on earth. Navigation satellites send out radio signals that help pilots of ships and aircraft determine their locations. Scientific satellites and orbiting telescopes are outfitted with special instruments that allow scientists to study the distant reaches of space.

Satellites are put into orbit by powerful computer-guided rockets. A rocket carries the satellite to an appropriate altitude above the earth. The rocket then aims the satellite at the angle necessary for the desired orbit. Once the satellite is in place, the rocket automatically detaches from the satellite. A smaller rocket on the satellite may provide the extra speed necessary to send the satellite orbiting around the earth. The earth's gravity holds the satellite in orbit.

The altitude of a satellite determines the speed necessary to keep it in orbit. As the distance from the earth increases, the force of earth's gravity decreases. Air resistance is also reduced in the earth's thin upper atmosphere. Therefore, the higher the orbit of a satellite, the lower will be the speed needed for it to stay in orbit.

At an altitude of 36,100 km, a satellite completes one revolution in 24 hours. At this altitude, a satellite in orbit directly above the earth's equator and moving in the direction of the earth's rotation is in **geosynchronous orbit.** A satellite in geosynchronous orbit always remains at the same point above the equator and appears to be stationary in the sky. Satellites used for communications are usually put into geosynchronous orbits. High above the earth, these satellites can act as antennas, relaying electronic signals over great distances. Some weather satellites use this high stationary orbit to continuously track hurricanes and other storms.

A satellite can also be placed in a **polar orbit.** A polar orbit carries the satellite over the earth's North and South poles. As the earth rotates beneath it, a satellite in polar orbit passes over a different portion of the earth's surface during each revolution. After a certain number of revolutions, the satellite will have surveyed the entire surface of the earth. A polar orbit is useful for mapping the earth's surface and for tracking weather. Why might a polar orbit be better than a geosynchronous orbit for some weather satellites?

Most satellites follow slightly elliptical orbits around the earth. The point closest to the earth in the orbit is called *perigee*. The point farthest from the earth in the orbit is called *apogee*. A satellite travels fastest at perigee and slowest at apogee.

To stay in orbit, a satellite must maintain a speed adequate for its altitude. If the orbit comes too close to the earth at perigee, friction with the earth's upper atmosphere will slow the satellite down. This decrease in speed means that each successive orbit will bring the satellite closer to the earth at perigee. Eventually, the increasing friction of the earth's atmosphere will generate too much heat, and the satellite will burn up.

Exploring the Earth by Satellite

Satellites have given scientists new ways of studying the earth. Some of the most fascinating information has come from a series of American scientific satellites called *Landsat*. Each Landsat orbits the earth in a polar orbit and collects information, using both television cameras and electronic sensors. The resulting images are used to identify features on the earth's surface, such as cities, vegetation regions, and even rock types.

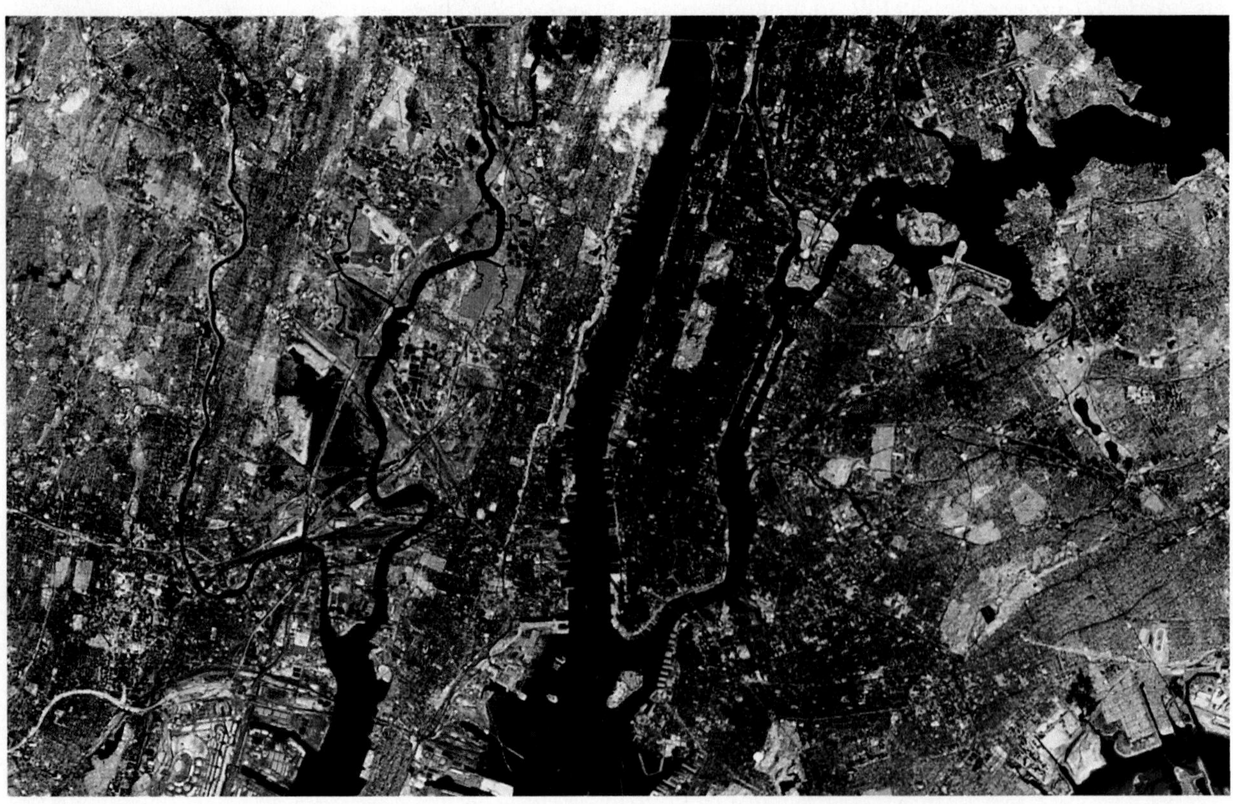

Figure 2–12. This Landsat image shows Manhattan and the surrounding areas of New York and New Jersey.

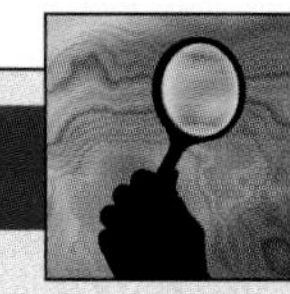

SMALL-SCALE INVESTIGATION

Gravity and Orbits

A moving body tends to move in a straight line at constant speed unless some outside force acts on it. Gravity is the outside force that acts on satellites and keeps them in orbit around the earth. You can investigate the effect of gravity on a moving body with a simple model.

Materials

strip of flexible cardboard, 3 × 30 cm; transparent tape; sheet of white paper; a marble

Procedure

1. Form a hoop with the cardboard strip and fasten the two ends together with tape.
2. Place the hoop in the center of your paper and trace a circle along the outside edge of the hoop as shown in the photo. Mark four points at equal distances around the circle. Number the points 1, 2, 3, and 4.
3. Place a marble inside the cardboard hoop. Slowly swirl the hoop clockwise until the marble rolls smoothly around the inside edge of the hoop. Stop swirling the hoop as the marble approaches point l; then quickly lift the hoop, allowing the marble to escape. You may have to practice this step several times to get the marble released at the right time.

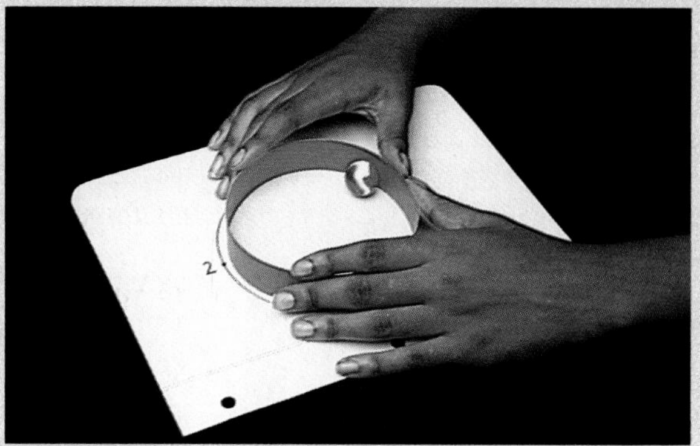

4. Observe and record the path of the marble as it exits the hoop.
5. Repeat Steps 3 and 4, stopping the hoop and releasing the marble at points 2, 3, and 4.

Analysis and Conclusions

1. What path does the marble take when the hoop is removed? Is the pattern of the path the same for all four exit points?
2. In what direction does the hoop push the marble? What force does the hoop represent?
3. Compare the motion of the marble with that of a satellite around the earth. How are they alike? How are they different?

The Global Positioning System, or GPS, is a network of satellites used for accurate navigation of ships and aircraft. Even with a small hand-held receiver, the system can locate virtually any point on the earth's surface with an accuracy within several meters.

Most satellites are designed simply to receive information and to send electronic information back to earth. The space shuttle, however, is an exception. This temporary satellite is designed to carry cargo, orbit the earth, and then return to the earth's surface. While in orbit, the shuttle can release or pick up other satellites.

Section 2.3 Review

1. Compare a geosynchronous orbit with a polar orbit.
2. How are satellites used to study the earth?
3. Why is a polar orbit useful for surveying purposes?

Chapter 2 Review

Key Terms

- aphelion (29)
- autumnal equinox (31)
- axis (23)
- core (24)
- crust (24)
- daylight saving time (34)
- geosynchronous orbit (35)
- gravity (27)
- International Date Line (34)
- law of gravitation (28)
- magnetosphere (26)
- mantle (24)
- Moho (24)
- perihelion (29)
- polar orbit (35)
- precession (32)
- revolution (29)
- rotation (29)
- satellite (35)
- seismic wave (23)
- shadow zone (25)
- standard time zone (33)
- summer solstice (31)
- vernal equinox (32)
- winter solstice (31)

Key Concepts

The solid earth consists of three major zones: the crust, the mantle, and the core. **See page 24.**

Studies of seismic waves provide information about the thickness and composition of the zones of the earth's interior. **See page 24.**

The earth has magnetic properties that probably originate in the earth's core. **See page 25.**

Newton's Law of Gravitation states that the strength of the gravitational force between two objects depends on their mass and the distance between them. **See page 28.**

The earth revolves around the sun and rotates on its axis. **See page 29.**

The seasons of the year are caused by two factors. **See page 30.**

The apparent motion of the sun across the sky is the basis for measuring time. **See page 33.**

Many satellites orbit the earth in geosynchronous orbits or in polar orbits. **See page 35.**

Satellites collect information on features of the earth's surface, using equipment such as television cameras and other electronic sensors. **See page 36.**

Review

On your own paper, write the letter of the term that best completes each of the following statements.

1. The zone that makes up nearly two-thirds of the earth's mass is the
a. crust. b. mantle. c. core. d. lithosphere.

2. The boundary between the earth's crust and mantle is called the
a. shadow zone. b. asthenosphere.
c. Moho. d. magnetosphere.

3. Both P waves and S waves can travel through
a. liquids and solids. b. solids.
c. liquids. d. gases.

4. The possible source of the earth's magnetism is the earth's
a. crust. b. mantle. c. core. d. lithosphere.

5. The amount of matter in an object is the object's
a. mass. b. weight.
c. gravity. d. plasticity.

6. As the distance from the center of the earth increases, the force of gravity
a. decreases. b. increases.
c. stays the same. d. doubles.

7. The measured weight of an object is slightly less at the equator than it is at the poles because of the earth's
   a. orbit. b. axis. c. shape. d. tilt.
8. The point closest to the sun in the earth's orbit is called
   a. apogee. b. perigee. c. aphelion. d. perihelion.
9. At noon on the winter solstice, the sun's vertical rays strike the earth along the
   a. Tropic of Cancer. b. Tropic of Capricorn.
   c. equator. d. North Pole.
10. At noon on the vernal equinox, the sun's vertical rays strike the earth along the
   a. Tropic of Cancer. b. Tropic of Capricorn.
   c. equator. d. North Pole.
11. When the sun's rays reach their highest angle in the Northern Hemisphere, the season there is
   a. spring. b. summer. c. fall. d. winter.
12. The wobbling motion made by the earth's axis as it turns in space is called
   a. precession. b. gravity.
   c. plasticity. d. equinox.
13. A person crossing the International Date Line gains or loses
   a. 2 hours. b. 8 hours.
   c. 12 hours. d. 24 hours.
14. A satellite in geosynchronous orbit is always directly above the
   a. equator. b. North Pole.
   c. South Pole. d. International Date Line.
15. Landsat provides information about the
   a. earth's surface. b. moon's surface.
   c. sun's surface. d. stars.

Critical Thinking

On your own paper, write answers to the following questions.

1. Is a hard-boiled egg a good model of the earth's different zones? Why or why not?
2. Explain why the weight of an object might increase when the object moves from point *A* to point *B* on the earth's surface.
3. Although the earth's orbit brings it closest to the sun in January, the Northern Hemisphere is having winter at that time of the year. Explain.
4. If the earth ceased rotating as it revolved around the sun, how would periods of daylight and surface temperatures on the earth be affected?
5. Suppose the earth's rotation slowed to one rotation every 48 hours. How might that change affect timekeeping systems?

Application

1. A scientist proposes the hypothesis that the moon has a liquid core. How might this hypothesis be tested?
2. If scientists discovered that the earth's magnetic field had weakened substantially, what might they suspect to be the cause of this change?
3. A group of scientists is planning to launch a satellite into orbit around the earth. The satellite will be used to survey the entire earth's surface in search of oil deposits. Which type of orbit would you recommend for such a satellite? Why?

Extension

1. Scientists made recent discoveries about the earth's core using seismic tomography. Research seismic tomography and write a report on its uses.
2. Find out how the Lapps and the Eskimos have adapted to periods of continuous daylight and continuous darkness.
3. Draw a simple map of the world. Label Washington, D.C., with the time and date 1:00 p.m., January 1. Label the correct time and date for five other cities.

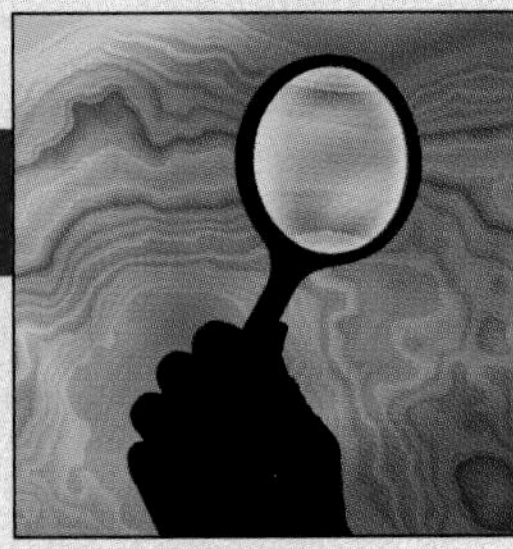

IN-DEPTH INVESTIGATION

Earth-Sun Motion

Materials

- magnetic compass
- clock or watch
- metric ruler
- wooden board (9″ × 12″)
- pencil
- wooden dowel (about 12″ long and 1/4″ diameter)
- masking tape

Introduction

During the course of a day, the sun seems to move across the sky. This apparent motion is due to the earth's rotation. In ancient times, one of the earliest devices used by people to study the sun's motion was the shadow stick. The shadow stick is a primitive form of a sundial. Before clocks were invented, sundials were the only means of telling time.

In this investigation, you will construct a shadow stick in order to identify how changes in a shadow are related to the earth's rotation. You will also determine how a shadow stick can be used to measure time.

Prelab Preparation

1. Review Section 2.2, pages 29–32.
2. Constructing a shadow stick: Using a hand or power drill, make a hole in the board at the location shown in the illustration below, just large enough to hold the wooden dowel.

CAUTION: *If you are using hand or power tools, it is best to allow your parents or another adult skilled in the tool's operation to help you.* Place the dowel in the hole. Place a sheet of three-hole notebook paper on the base of the shadow stick. Slip the middle hole of the paper over the stick, and slide the paper down so that it rests on the base. Fasten the paper securely to the base with tape. The shadow stick should be just long enough to cast a shadow nearly across the piece of notebook paper.

Procedure

1. Set up the shadow stick in a sunny spot outdoors. **CAUTION:** *You should never look directly at the sun.* Align the stick so that its shadow is parallel to the lines of the notebook paper. See the figure to the left.

2. Place the compass on the paper above the shadow. In one corner of the paper indicate north with an arrow. Label the arrow with a capital N. Why is it necessary to indicate the direction north on the paper?
3. Make a pencil dot at the end of the shadow cast by the wooden dowel. Write the time above the dot. Do this two more times at five-minute intervals. *Note: Do not move the base after you begin to make measurements.*
4. After your last five-minute measurement, wait 10 minutes. Make another dot to show the position of the end of the shadow. Do this two more times at five-minute intervals. Be sure to record the time above each dot. Then return the shadow stick to your classroom.
5. Remove the notebook paper from the base of the shadow stick. Connect the dots with a thin pencil line. Is the line connecting the dots on the paper a straight line?
6. Draw an arrowhead on the end of the line to show the direction in which the shadow moved. In what direction did the shadow move?
7. Make two measurements of the shadow's length in centimeters. The first measurement should be from the center of the paper hole to the first dot. The second measurement should be from the paper hole to the last dot. Record the two lengths.
8. Measure and record the length, in centimeters, of the line connecting the dots. This is the distance the shadow moved in 30 minutes.

Analysis and Conclusions

1. In what direction did the sun appear to move in the 30-minute period?
2. In what direction does the earth rotate?
3. If you made your shadow stick half as long, would its shadow move the same distance in 30 minutes? Explain.
4. How might a shadow stick be used for telling time?

Extension

Repeat this investigation at different hours of the day. Do it early in the morning, early in the afternoon, and early in the evening. Record the results and any differences that you observe. Explain how shadow sticks can be used to tell direction.

Aerial photographs, such as this one, reveal much about the earth's surface. For example, an aerial photograph can show the course of a river, the curve of a shoreline, or the general shape of other landforms. Sometimes, however, photographs do not provide the specialized information that earth scientists need. As a result, earth scientists often rely on maps as models of the earth's surface. This chapter explains how various types of maps are made and used.

Chapter Outline

◀ This aerial photograph was taken over Maurelle Island Wilderness, Tongass National Park, Alaska.

3.1 Finding Locations on the Earth

Section Objectives

- **Distinguish between latitude and longitude.**
- **Explain how latitude and longitude can be used to locate places on the earth.**
- **Explain how a magnetic compass can be used to find directions on the earth.**

The earth is very nearly a sphere. A sphere has no top, bottom, or sides to use as reference points for finding locations on its surface. The earth does rotate, however. The points on the surface of the earth that are intersected by the earth's axis of rotation are used as reference points for establishing direction. These reference points are called the *North* and *South Geographic Poles*. Halfway between the poles, a circle called the *equator* divides the earth into the Northern and Southern Hemispheres. Based on these reference points, an entire system of intersecting circles has been established to locate places on the earth's surface.

Latitude

One set of circles describes positions north and south of the equator. These are called **parallels** because they are circles that run east and west around the world parallel to the equator.

The angular distance north or south of the equator is called **latitude.** Latitude is measured in degrees, beginning at the equator with 0°. A full circle has 360°. Since the distance from the equator to either of the poles is one-quarter of a circle, the latitude of both the North and South Poles is ¼ of 360°, or 90°. See Figure 3–1. In actual distance, one degree of latitude equals 1⁄360 the earth's circumference (over 40,000 km), or about 111 km.

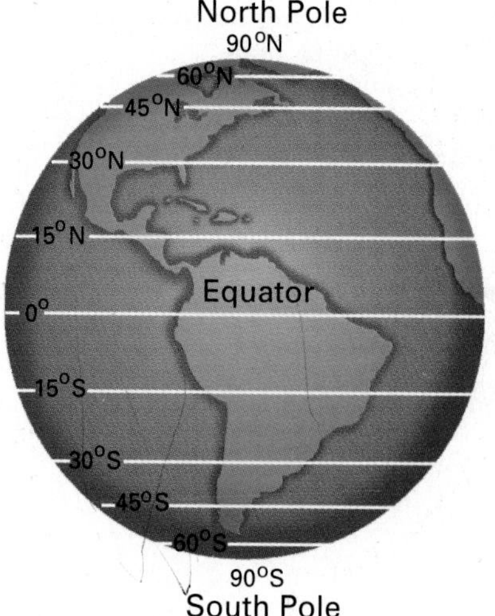

Figure 3–1. Parallels are circles describing positions north and south of the equator. Each parallel forms a complete circle around the globe.

Parallels of latitude north of the equator are labeled *N;* those south of the equator are labeled *S*. For example, in the Northern Hemisphere, Washington, D.C., is located near a parallel of latitude that is 39° north of the equator. The latitude of Washington, D.C., is 39° N. In the Southern Hemisphere, Melbourne, Australia, has a latitude of 39° S.

Each degree of latitude consists of 60 equal parts, called *minutes*. One minute (symbol: ′) of latitude equals 1.85 km. A more precise latitude for Washington, D.C., is 38°53′ N. For even greater precision, each minute is divided into 60 equal parts, called *seconds* (symbol: ″). Using degrees, minutes, and seconds, the latitude of Washington, D.C., is 38°53′51″ N. How much distance on the earth's surface does one second of latitude equal?

Longitude

The latitude of a particular place indicates only its position north or south of the equator. To determine the specific location of a place, you also need to know how far east or west the location is along its circle of latitude. **Meridians** are used to establish east-west locations. As shown in Figure 3–2, each meridian is a semicircle (half of a circle) running from pole to pole.

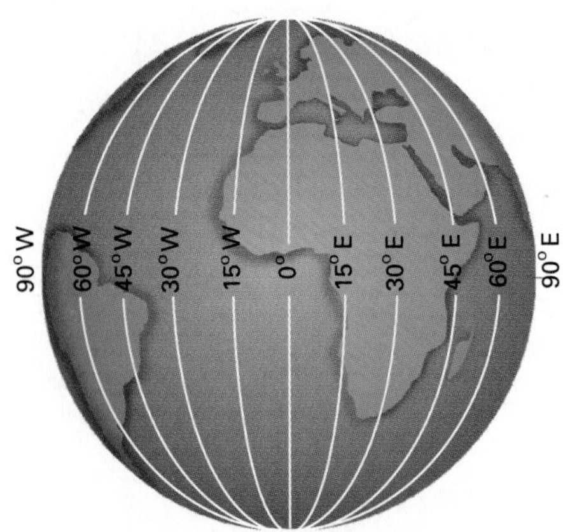

Figure 3–2. Meridians are semicircles reaching around the earth from pole to pole.

By international agreement, one meridian was selected to be 0°. This meridian, called the **prime meridian,** passes through Greenwich, England. **Longitude** is the angular distance, measured in degrees, east or west of the prime meridian. Since a full circle is 360°, the meridian opposite the prime meridian, halfway around the world, is labeled *180°*.

All locations east of the prime meridian have longitudes between 0° and 180° E. All locations west of the prime meridian have longitudes between 0° and 180° W. Washington, D.C., which lies west of the prime meridian, has a longitude of 77° W. As with latitude, longitude can be expressed in degrees, minutes, and seconds. Therefore, a more precise location for Washington, D.C., is 38°53′51″ N, 77°0′33″ W.

The distance covered by a degree of longitude depends on where the degree is measured. At the equator, 0° latitude, a degree of longitude equals approximately 111 km. However, all meridians meet at the poles. Thus, the distance measured by a degree of longitude decreases as you move from the equator to the poles. At a latitude of 60° N, for example, one degree of longitude equals about 55 km. At 80° N, one degree of longitude equals only about 20 km.

Great Circles

A **great circle** is often used in navigation, especially by long-distance aircraft. A great circle is any circle that divides the globe into halves. Any circle formed by two meridians of longitude directly across from each other on opposite sides of the globe is a great circle. The equator, however, is the only parallel of latitude that is a great circle. Great circles can run in any direction around the globe. Just as a straight line is the shortest distance between two points on a plane, a great-circle route is the shortest distance between two points on a sphere. As a result, air and sea routes often follow along great circles.

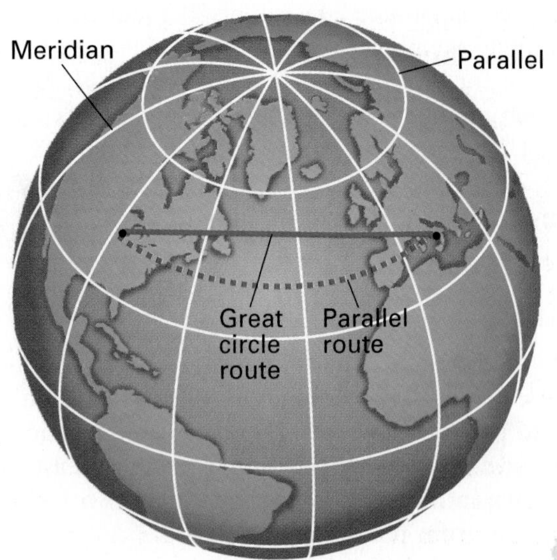

Figure 3–3. As the illustration shows, a great-circle route from Chicago to Rome is much shorter than a route following a parallel.

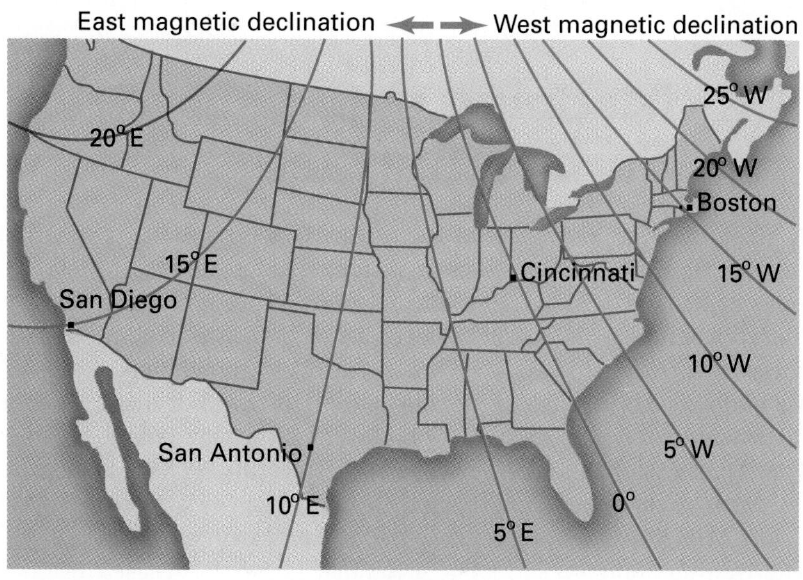

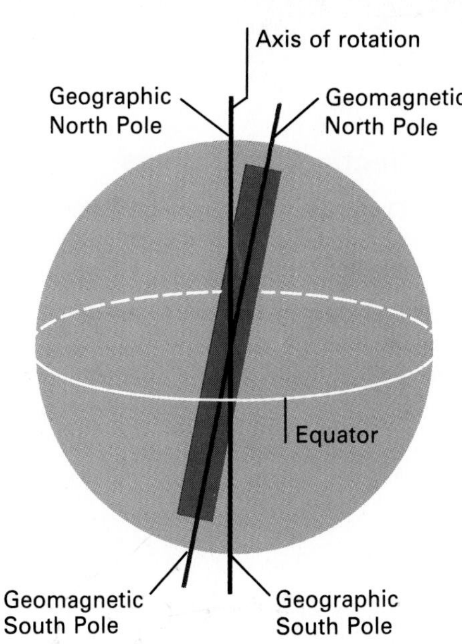

Figure 3–4. The red lines on the map (left) connect points with the same magnetic declination. Note that the earth's magnetic poles are at an angle to the earth's axis of rotation (right).

Finding Direction

One way to find direction on the earth is to use a magnetic compass. A magnetic compass can indicate direction because the earth has magnetic properties, as if a powerful bar-shaped magnet were buried inside. As you can see in Figure 3–4, the earth's imaginary magnet is at an angle to the earth's axis of rotation.

The points on the earth's surface just above the poles of the imaginary magnet are called the **geomagnetic poles.** Because of the tilt of the imaginary magnet inside the earth, the geomagnetic poles and the geographic poles are located in different places. A compass needle points to the geomagnetic North Pole.

The angle between the direction of the geographic pole and the direction in which the compass needle points is called **magnetic declination.** In the Northern Hemisphere, magnetic declination is measured in degrees east or west of the geographic North Pole. As Figure 3–4 shows, a compass needle at Boston, Massachusetts, points 15° west of **true north,** the direction of the geographic North Pole. At Cincinnati, Ohio, the compass needle lines up with the geographic and geomagnetic North Poles, so the declination is zero. At San Antonio, Texas, the declination is 10° east of true north, while at San Diego, California, it is 15° east of true north.

Magnetic declination has been determined for points all over the earth. By adjusting a measurement of magnetic north, a person can determine geographic north for any place on earth. Locating geographic north is important in navigation and map making. The magnetic declination for most of the United States is shown in Figure 3–4. What is the magnetic declination for western Illinois?

Section 3.1 Review

1. Define parallels. What do they measure?
2. Define meridians. What do they measure?
3. Name one way of determining direction on the earth.
4. Why is a great-circle route often used in navigation?

counter Right
cm back ps

Section Objectives

- **Describe the characteristics and uses of three types of map projections.**
- **Define *scale*, and explain how scale can be used to find distance on a map.**

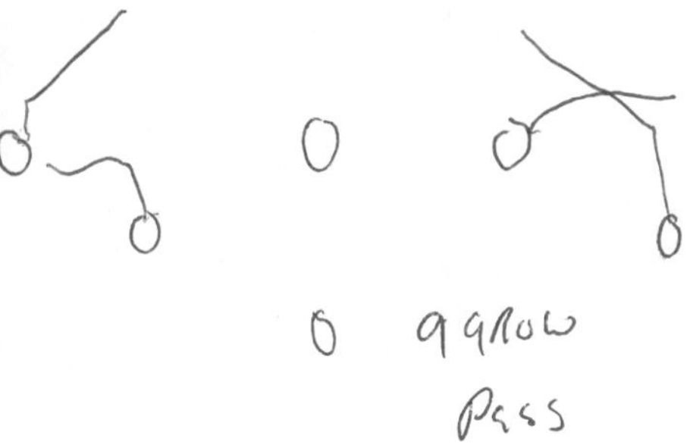

3.2 Mapping the Earth's Surface

A globe is a familiar model of the earth. Because a globe is spherical like the earth, the locations of surface features and their relative areas and shapes can be represented accurately. A globe is especially useful in studying the larger surface features, such as continents and oceans. However, most globes are too small to show many details of the earth's surface, such as streams and highways. For that reason, a fantastic variety of maps have been developed for studying the earth. The science of map making is called **cartography.** It is a subfield of the earth sciences and geography.

A map is a flat representation of the earth's curved surface. Transferring a curved surface to a flat map, however, causes distortion. For example, if you remove the peel from an orange and attempt to flatten the peel, it will stretch and tear. The larger the piece of peel, the more its shape is distorted as it is flattened. Also distorted are distances between points on the orange peel. Similarly, an area shown on a map may be distorted in size, shape, distance, and direction. The larger the surface area being shown, the greater the distortion. A map of the entire earth, like the peel of an entire orange, would show the greatest distortion. A map of a small area, such as a city, would show only slight distortion.

Map Projections

Over the years, map makers, or *cartographers,* have developed several ways of transferring the curved surface of the earth onto flat maps. A flat map that represents the three-dimensional curved surface of a globe is called a **map projection.** To understand how map projections are made, imagine the earth as a transparent globe with a light inside. If you hold a piece of paper against the globe, shadows appear on the paper that reflect markings on the globe, such as continents, oceans, parallels, and meridians. The way the paper is held against the globe determines the kind of projection made.

Figure 3–5. A light at the center of a transparent globe would project lines on a cylinder of paper (left), producing a Mercator projection (right).

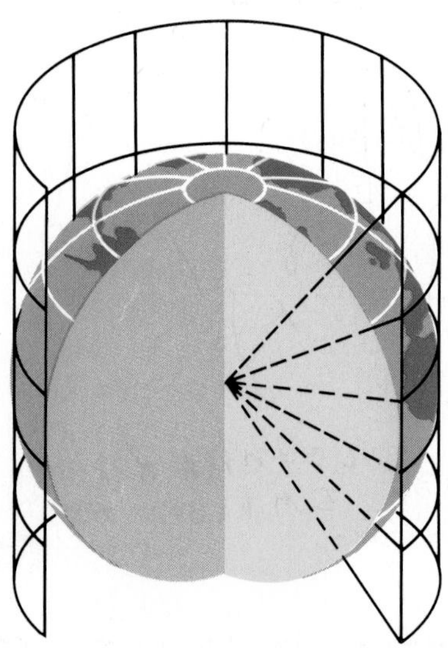

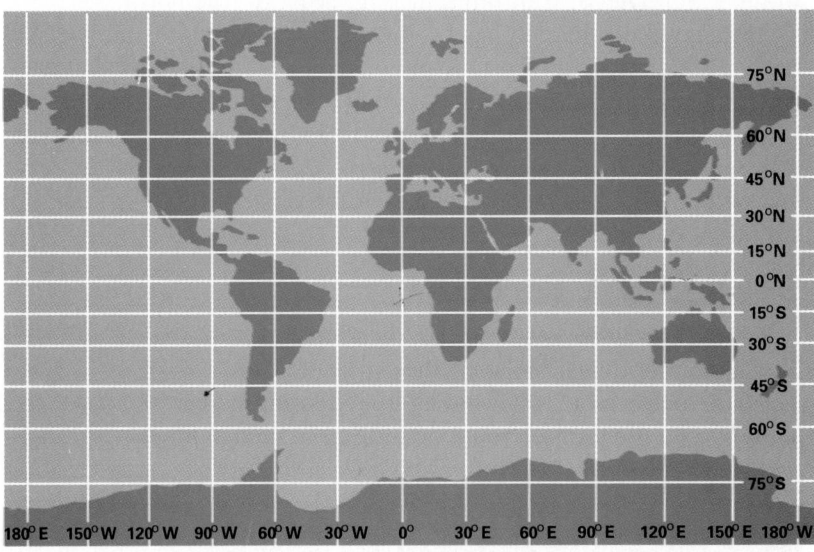

Three common types of map projections are **Mercator, gnomonic** (noe-MON-ick), and **conic projections.** None of these map projections is an entirely accurate representation of the earth's surface. However, each kind of projection has certain advantages and disadvantages that must be considered when choosing a map.

Mercator Projection

If you were to wrap a cylinder of paper around a lighted globe, a Mercator projection like the one shown in Figure 3–5 would result. The meridians on a Mercator projection appear as straight parallel lines with an equal amount of space between lines. On a globe, however, the meridians come together at the poles. A Mercator projection, though accurate near the equator, distorts distances between regions of land and distorts the sizes of areas near the poles.

Though distorted, a Mercator projection has several advantages. All compass directions appear as straight lines with the four main points of the compass located along the four sides of the map. All parallels and meridians are shown clearly, and latitude and longitude are easily measured with a ruler. Also, the shapes of the land and water bodies are shown correctly. These qualities make the Mercator projection a valuable tool for navigation. The major disadvantage is that areas far from the equator, such as Alaska and Norway, have an exaggerated size and appear much larger than they do on a globe.

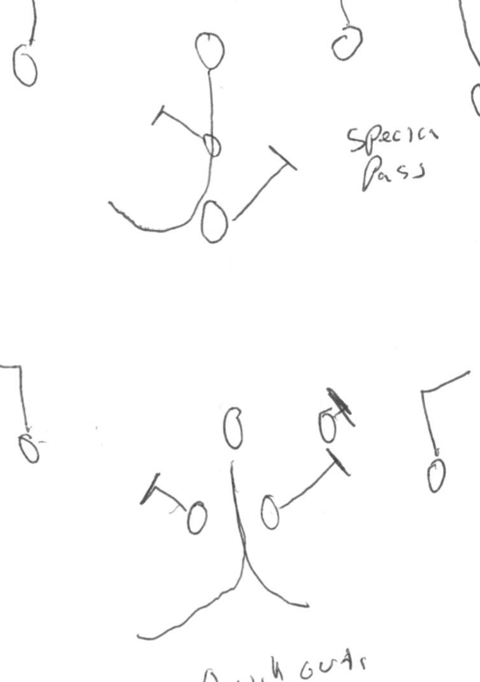

Gnomonic Projection

A sheet of paper touching a lighted globe at only one point produces a gnomonic projection, as shown in Figure 3–6. On a gnomonic projection, little distortion occurs at the point of contact, which is usually one of the poles. However, the unequal spacing between parallels causes a distortion in both direction and distance that increases as the distance from the point of contact increases.

Despite distortion, a gnomonic projection is a great help to navigators in plotting routes used in air travel. As you know, a great circle is the shortest distance between two points on the globe. When projected onto a gnomonic projection, it appears as a straight line. Therefore, by drawing a straight line between any two points on a gnomonic projection, navigators can readily find the great-circle route.

Figure 3–6. This gnomonic projection (left) is produced as points on a globe are projected onto a sheet of paper in contact with the North Pole of the globe.

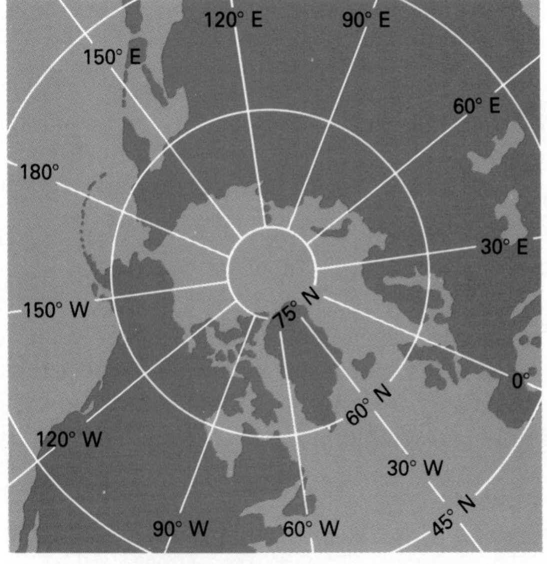

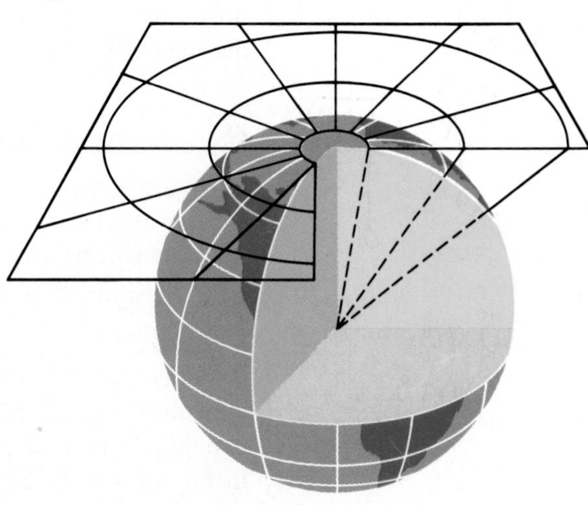

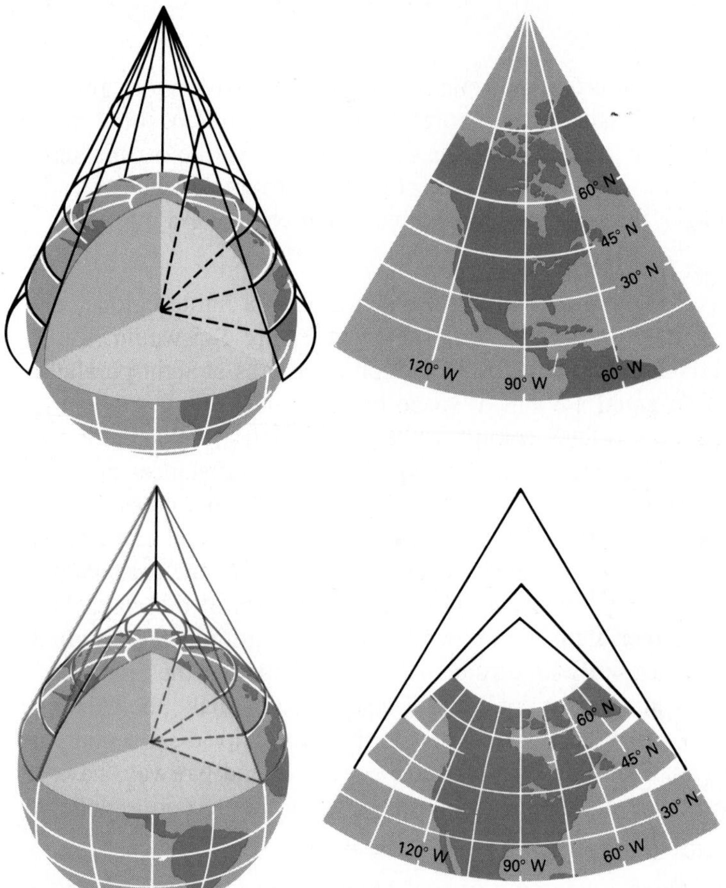

Figure 3–7. In the top left illustration, the cone comes in contact with the globe along the parallel of latitude at 30° N, producing the conic projection shown top right. In the bottom left illustration, a conic projection is made using a series of cones, each touching the globe at a different parallel. The conic projections are then assembled to form the polyconic projection shown bottom right.

Conic Projection

A paper cone placed over a lighted globe so that the axis of the cone aligns with the axis of the globe produces a conic projection. The cone touches the globe along one parallel of latitude. As shown in Figure 3–7, areas near the parallel of latitude where the cone and globe are in contact are distorted the least.

A series of conic projections may be used to map a number of neighboring areas. Each cone touches the globe at a slightly different latitude, as Figure 3–7 shows. Fitting the adjoining areas together then produces a continuous map. Maps made in this way are called **polyconic projections.** The relative size and shape of small areas on the map are nearly the same as those on the globe.

Reading a Map

Maps are models of the earth's surface. They provide information through the use of symbols. To read a map, you must understand the symbols and be able to find directions and calculate distances.

Symbols

Maps often have symbols for features such as cities and rivers. The symbols are explained in the map **legend,** a list of the symbols and their meanings. Some symbols resemble the features they represent. Others are more abstract, such as those for towns and urban areas. In Figure 3–8, which symbols resemble the features they represent?

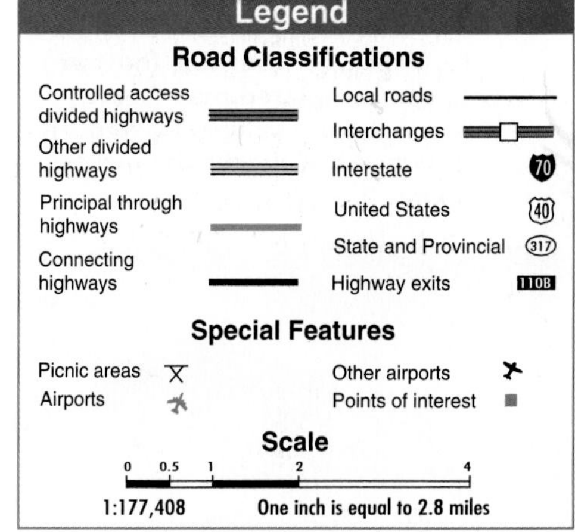

Figure 3–8. When using the road map shown above, a map scale is needed. Three types of map scales are shown on this legend. The map legend also explains the other symbols used on this map.

Direction on a Map

Once you understand the legend, you must determine the compass directions. Maps are usually drawn with north at the top, east at the right, west at the left, and south at the bottom. Parallels run from side to side, and meridians run from top to bottom. Directions should always be determined in relation to the parallels and meridians.

Map Scale

To be accurate, a map must be drawn to **scale.** The scale of a map indicates the relationship between distance as shown on the map and actual distance. As Figure 3–8 shows, a map scale can be expressed as a graphic scale, a fractional scale, or a verbal scale.

A graphic scale is a printed line divided into equal parts and labeled. The line represents a unit of measure, such as kilometers or miles. Each part of the scale represents a specific distance on the earth. To find the actual distance between two points on the earth, you first measure the distance between the points as shown on the map. Then you compare that measurement with the map scale.

A second way of expressing scale is a ratio, or a fractional scale. For example, a fractional scale such as 1:25,000 indicates that 1 unit of distance on the map represents 25,000 of the same unit on the earth. A fractional scale remains the same with any system of measurement. In other words, the scale 1:100 could be read as 1 in. equals 100 in. or as 1 cm equals 100 cm.

A verbal scale may be the statement, "One centimeter equals one kilometer." This means that 1 cm on the map represents 1 km on the earth. What would the fractional scale be on such a map?

Section 3.2 Review

1. Name three common map projections.
2. What is a map legend?
3. Explain why all maps are inaccurate representations in some way.

Section Objectives

- **Explain how elevation and topography can be shown on a map.**
- **Interpret a topographic map.**

3.3 Topographic Maps

A type of map that is especially useful in earth science is called a **topographic map.** Topographic maps show the surface features, or **topography,** of the earth. Most topographic maps show both natural features, such as rivers and hills, and constructed features, such as buildings and roads.

The top illustration in Figure 3–9 shows a side view of an island in the ocean. The drawing shows a hill on the island, but it does not indicate the size of the island or the height of the hill. The middle illustration shows the same island on a political map. The map shows the shape of the island at sea level and the relative length and width of the island. However, the political map gives no information about the height of the island, the steepness of its slopes, or the shape of the land above sea level. A topographic map provides more detailed information about the surface of the island than either the drawing or the political map does.

Making a Topographic Map

A topographic map of the island shows the **elevation,** or height above sea level, of various island locations. Elevation is measured from **mean sea level,** the point midway between the highest and lowest tide levels of the ocean. The elevation at mean sea level is zero. Other elevations are measured as distances above or below mean sea level.

Contour Lines

On topographic maps, **contour lines** are used to show elevation. Each contour line connects all points on the map that have the same elevation. For example, one contour line would connect all points on the map that have an elevation of 100 m. Another line would connect all points with an elevation of 200 m. Because all points at any given elevation are connected, the shape of the contour lines reflects the shape of the land.

Figure 3–9. A drawing gives little information about the surface of the island (top). A political map shows only the shape of the island (middle). To start making a topographic map of the island, a topographer connects points at an elevation of 20 m above sea level to form a contour line (bottom). In the completed map (right), additional contour lines have been drawn. An *X* marks the highest point.

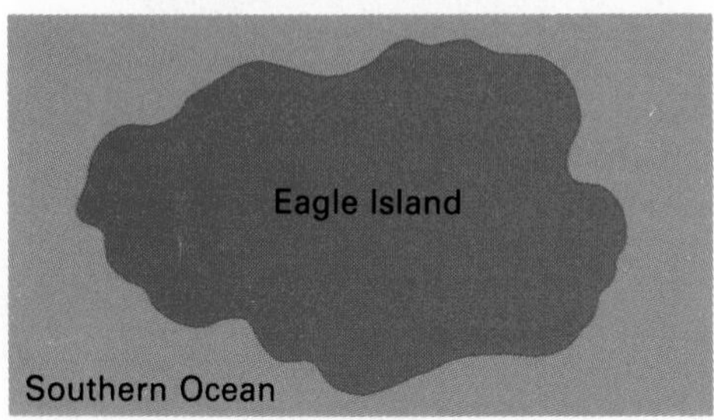

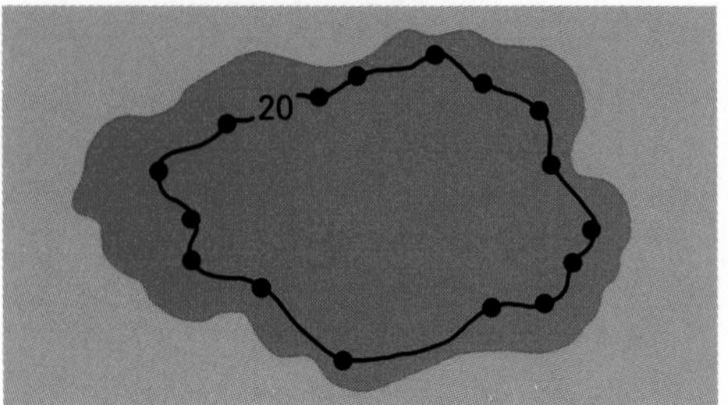

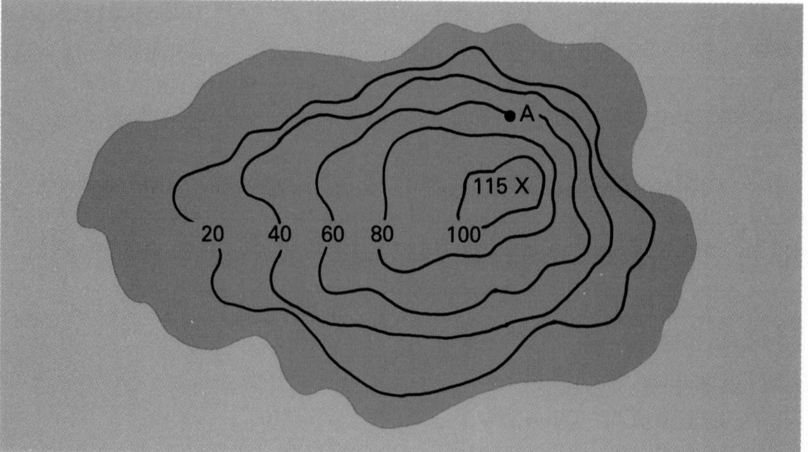

Contour Intervals

The difference in elevation between one contour line and the next is called **contour interval.** A cartographer chooses a contour interval suited to the size of the map and the **relief** of the land. Relief is the difference in elevation between the highest and lowest points of the area being mapped. On maps of mountainous areas where the relief is high, the contour interval may be as large as 50 m or 100 m. Where the relief is low, the interval may be only 1 m or 2 m.

In mapping the island shown in Figure 3–9, the map maker chooses a contour interval of 20 m. The map maker then marks a series of points surveyed at 20 m above sea level and connects the points with a contour line. Next the map maker marks points and contour lines for elevations of 40, 60, 80, and 100 m. The shoreline serves as the contour line for points at sea level. The completed topographic map in Figure 3–9 shows the elevation of the island and the general shape of the land above sea level. Find point *A* on the map. What is its elevation?

Interpreting a Topographic Map

Just as printed words on a page transmit ideas, contour lines and other symbols on topographic maps give a picture of the earth's surface. Because of the specialized nature of topographic maps, some training and practice are needed in order to read and interpret these maps accurately.

The United States Geological Survey (USGS), a branch of the federal government, has made topographic maps for virtually all of the United States. These detailed maps are called *topographic sheets,* or *quadrangles*. The earlier series of USGS maps represent

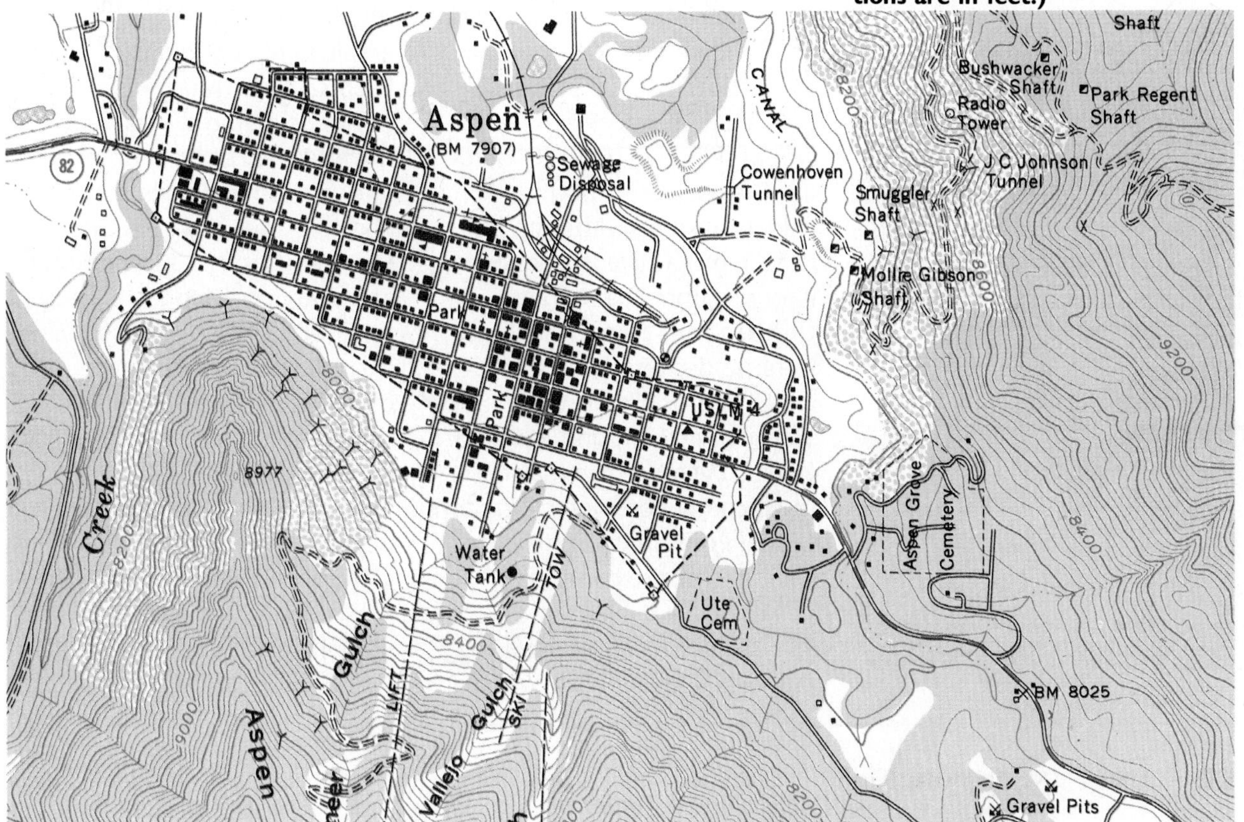

Figure 3–10. This portion of a topographic sheet produced by the USGS shows the area around Aspen, Colorado. (Note: elevations are in feet.)

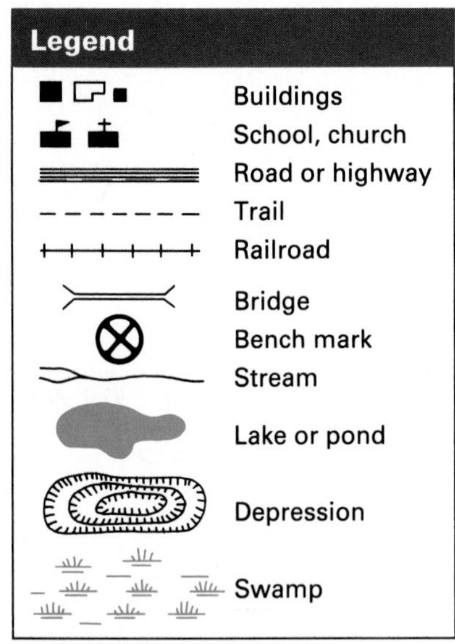

Figure 3–11. Symbols used on topographic maps represent both natural and constructed features.

quadrangles that cover 15′ of latitude and 15′ of longitude. The newer series of maps represent quadrangles that cover 7.5′ of latitude and 7.5′ of longitude. The 7.5′ topographic sheets show a smaller area than the 15′ sheets, but with greater detail.

As shown in Figure 3–11, symbols are used to show certain features on topographic maps. Different colors are used for different types of symbols. In general, constructed features, such as buildings, roads, and railroads, are shown in black. Major highways are shown in red. Bodies of water are shown in blue, and forested areas are printed in green. Contour lines are shown in brown.

On most topographic maps, direction is found by following lines of latitude and longitude. On maps published by the USGS, north is located at the top of the map and marked by a parallel of latitude. The southern boundary, at the bottom of a map, is also marked by a parallel. At least two additional parallels are usually drawn in or indicated by cross hairs (+) at 2.5′ intervals.

Meridians of longitude indicate the eastern and western boundaries of USGS maps. Additional meridians may also be shown. All parallels and meridians shown on these maps are labeled in degrees and minutes.

Science & Technology

Navstar: A Global Positioning System

Have you ever been lost in the woods? Even with a compass and a topographic map, most people would still have difficulty guiding themselves through a wilderness area without a trail. Realizing that this is true for military personnel as well, the United States Department of Defense designed *Navstar,* which stands for *Nav*igation *S*atellite *T*racking *a*nd *R*anging. Over 20 years in the making, the recently completed *Navstar* Global Positioning System, or GPS, may make compasses obsolete.

Navstar provides a way to pinpoint a location anywhere in the world with a small, relatively inexpensive radio receiver. The *Navstar* system consists of 24 satellites that orbit the earth at an altitude of about 20,000 km. Each satellite transmits high-frequency radio waves, or microwaves, containing precise information about its position and the time of day. A GPS receiver uses this information to calculate its distance, or range, to several satellites. The receiver then determines its three-dimensional location by finding the common intersection point of these ranges.

Since *Navstar*'s introduction, many new applications have been explored. For example, meteorologists are able to use GPS signals to measure the temperature and water content

▲ GPS receivers equipped with minicomputers have revolutionized surveying techniques.

Distance on Topographic Maps

As on other maps, distance on topographic maps is determined by referring to the map scale. A common scale used on USGS maps is 1:24,000. Based on this scale, 1 in. on the map is equal to 2,000 ft. on the earth's surface. You can use a ruler to measure distances on the map and then convert the inches to feet or miles. If a graphic scale is used, you can mark off distance on a piece of paper and then compare it with the scale.

Elevation on Topographic Maps

On a topographic map, the contour interval determines the elevation at which each contour line is drawn. If the contour interval is 10 m, contour lines will be shown for elevations of 10, 20, 30, 40, 50, 60 m, and so on. To make reading the map easier, every fifth contour line is printed bolder than the others. These lines are called **index contours** and are labeled by elevation. A point located between two contour lines has an elevation somewhere between the elevations of the two lines. For example, if a point is located halfway between the 50-m and 100-m contour lines, its elevation is approximately 75 m. Exact elevations are marked by an × and labeled.

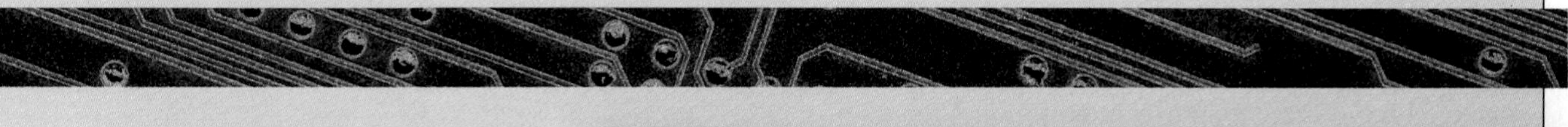

of the atmosphere, surveyors are able to monitor deformation of the earth's crust caused by tectonic motion, and geologists can cheaply and precisely map changes in the Greenland ice sheet to understand global climatic changes.

More common applications of *Navstar* include an in-car navigation system that combines a GPS receiver with a computerized map display. The system not only determines the best route to follow but also tells the driver when to turn. Similar systems may one day serve as navigation aids for visually impaired people.

How does the information provided by a compass differ from that provided by a GPS receiver?

▲ Finding the intersection point of three spheres is the geometric basis for a GPS.

Landforms on Topographic Maps

By studying the spacing and the direction of contour lines, you can also determine the shapes of landforms shown on a topographic map. Contour lines spaced widely apart indicate that the change in elevation is gradual and that the land is relatively level. Closely spaced contour lines indicate that the change in elevation is rapid and that the slope is steep. Contour lines that are almost touching indicate a very steep slope or cliff. Evenly spaced contour lines indicate that the slope increases at about the same angle over a great distance. Find point *B* on the map in Figure 3–12. How would you describe the change in elevation in the area around point *B?*

Contour lines that bend to form a V-shape indicate a valley. The V points toward the higher end of the valley. If a stream or river flows through the valley, the V in the contour lines will point upstream, or in the direction from which the water flows. A river al-

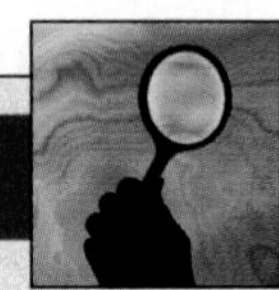

SMALL-SCALE INVESTIGATION

Topographic Maps

Contour lines show elevation and landforms on topographic maps. You can use contour lines to make a topographic map of a model mountain.

Materials

modeling clay (1–2 lb.); paper clip; large waterproof container, at least 8 cm deep; water; ruler; pencil; adhesive or masking tape

Procedure

1. Make a mountain 6–8 cm high out of modeling clay. Work on a flat surface and smooth out the mountain's shape, making the mountain slightly steeper on one side.
2. Run a paper clip down one side of the model to form a valley several millimeters wide.
3. Place the model in the center of the container. Tape the ruler upright in the container with one end resting on the bottom of the container. Make sure the container is resting on a level surface.
4. Add water to the container to a depth of 1 cm, using the ruler as a guide. With a sharp pencil, trace around the model and inscribe the clay along the waterline as shown above.
5. Raise the water level 1 cm at a time until you reach the top of the model. Each time you add water to the container, inscribe another contour line in the clay where the waterline meets the model.
6. When finished, carefully drain the water and remove the model from the container.

Analysis and Conclusions

1. What is the contour interval on your model?
2. Observe your model from directly above. Try to duplicate the size and spacing of the contour lines on a sheet of paper.
3. Compare the contour lines on a steep slope with those on a gentle slope. How do they differ?
4. How is a valley represented on your topographic map?

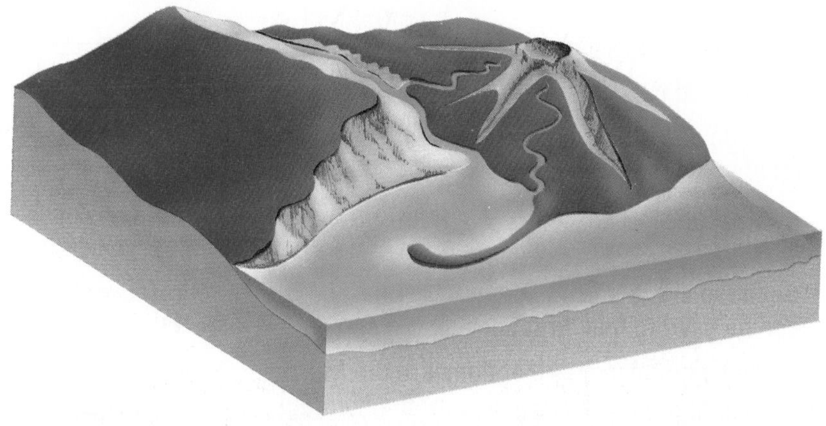

INVESTIGATE!

To learn more about topographic maps, try the In-Depth Investigation on pages 58–59.

Figure 3–12. The features of the coastal valley shown in the diagram above are represented by contour lines on the topographic map below. What are the elevations at *A* and *X*?

ways flows from higher to lower elevation. The steeper the course of the river or stream, the closer together are the contour lines that cross it. The width of the valley is shown by the width of the V formed by the contour lines.

Contour lines that form closed loops indicate a hilltop or a depression. To avoid confusion, a depression is usually shown by **depression contours,** which are marked with short, straight lines. These lines are drawn along the inside of the loop and point toward its center, indicating the direction of depression. Find point *C* on the map in Figure 3–12. Is it located on a hilltop or in a depression?

Section 3.3 Review

1. How is elevation shown on a topographic map?
2. Define *contour interval*.
3. How would you determine the elevation of a point that falls between two contour lines?
4. How is a depression shown on a contour map?
5. Why are topographic maps useful to someone who wishes to study earth science?

Chapter 3 Review

Key Terms

- cartography (46)
- conic projection (47)
- contour interval (51)
- contour line (50)
- depression contour (55)
- elevation (50)
- geomagnetic pole (45)
- gnomonic projection (47)
- great circle (44)
- index contour (53)
- latitude (43)
- legend (48)
- longitude (44)
- magnetic declination (45)
- map projection (46)
- mean sea level (50)
- Mercator projection (47)
- meridian (43)
- parallel (43)
- polyconic projection (48)
- prime meridian (44)
- relief (51)
- scale (49)
- topographic map (50)
- topography (50)
- true north (45)

Key Concepts

Latitude and longitude are a system of intersecting circles used to locate places on the earth's surface. **See page 43.**

Parallels of latitude run east-west around the earth. Meridians of longitude run north-south from pole to pole. **See page 43.**

Because of the earth's magnetic properties, a magnetic compass can be used to find directions on the earth. **See page 45.**

Three common map projections are the Mercator, gnomonic, and conic projections. Each has certain advantages and disadvantages. **See page 46.**

Map scale is used to find distances on a map. **See page 49.**

Contour lines can be used to show topography on a map. **See page 50.**

The spacing and direction of contour lines on a topographic map indicate the shapes of landforms. **See page 50.**

Review

On your own paper, write the letter of the term that best completes each of the following statements.

1. A point whose latitude is 0° is located on the
   a. North Pole. b. South Pole. c. equator. d. prime meridian.
2. One degree of latitude equals
   a. $\frac{1}{90}$ the earth's circumference.
   b. $\frac{1}{100}$ the earth's circumference.
   c. $\frac{1}{360}$ the earth's circumference.
   d. $\frac{1}{720}$ the earth's circumference.
3. A point whose longitude is 0° is located on the
   a. North Pole. b. South Pole. c. equator. d. prime meridian.
4. A point halfway between the equator and the South Pole has a latitude of
   a. 45° N. b. 45° S. c. 45° E. d. 45° W.
5. The distance in degrees east or west of the prime meridian is
   a. latitude. b. longitude. c. declination. d. projection.
6. The distance covered by a degree of longitude
   a. is $\frac{1}{180}$ the earth's circumference.
   b. is $\frac{1}{360}$ the earth's circumference.
   c. increases as you approach the poles.
   d. decreases as you approach the poles.
7. The needle of a magnetic compass points toward the
   a. geomagnetic pole. b. geographic pole.
   c. parallels. d. meridians.

8. In the Northern Hemisphere, a declination of 10° E indicates that the compass needle points 10° east of the
   a. geomagnetic North Pole. b. geographic North Pole.
   c. equator. d. prime meridian.
9. On a Mercator projection, distortion is greatest near the
   a. poles. b. great circles.
   c. meridians. d. parallels.
10. Compass directions are shown as straight lines on a
   a. gnomonic projection. b. conic projection.
   c. Mercator projection. d. polyconic projection.
11. The shortest distance between any two points on the globe is along
   a. the equator. b. a line of latitude.
   c. the prime meridian. d. a great circle.
12. A navigator can find the shortest distance between two points by drawing a straight line between any two points on a
   a. Mercator projection. b. gnomonic projection.
   c. conic projection. d. polyconic projection.
13. The relationship between distance on a map and actual distance on the earth is called the
   a. legend. b. scale. c. elevation. d. relief.
14. If 1 m on a map equals 1 km on the earth, the fractional scale would be written
   a. 1:1. b. 1:10. c. 1:100. d. 1:1,000.
15. On a topographic map, elevation is shown by means of
   a. great circles. b. contour lines.
   c. verbal scale. d. fractional scale.
16. Closely spaced contour lines indicate a
   a. gradual slope. b. flat area.
   c. steep slope. d. valley.

Critical Thinking

On your own paper, write answers to the following questions.

1. What is wrong with the following locations: 135° N, 185° E?
2. As you move from point A to point B in the Northern Hemisphere, the length of a degree of longitude progressively decreases. In which direction are you moving?
3. Imagine you are at a location where the magnetic declination is 0°. Describe your position in relation to magnetic north and true north.
4. You examine a topographic map on which the contour interval is 100 m. In general, what type of terrain is shown on the map?
5. Concept map ? ? ? Selecting from the list of new terms on the previous page, which one term would most likely be found at the top of a concept map designed for this chapter? Explain.

Application

1. One expedition is preparing to explore the South Pole; another is preparing to explore the equator. To which expedition would you recommend the Mercator projection? Explain why.
2. A cartographer has to draw one map for use in three different countries that do not share a common unit of measure. Which type of scale should this map maker use? Why?
3. You are using a topographic map to plan a hike. Along path A, the contour lines are widely spaced. Along path B, the contour lines are almost touching. Which path would probably be easier and safer? Why?
4. How could you use contour lines on a topographic map to help you locate the source of a river?

Extension

1. Do some research on the navigation instrument known as the sextant. Make a diagram explaining how the sextant can be used to determine latitude.
2. On a piece of unlined paper, draw an estimated topographic map of an area near your school or home. If possible, choose an area that includes various surface features, such as hills, cliffs, or a river. Be sure to use the correct colors for the features on your map. Refer to the chart of symbols on page 52.

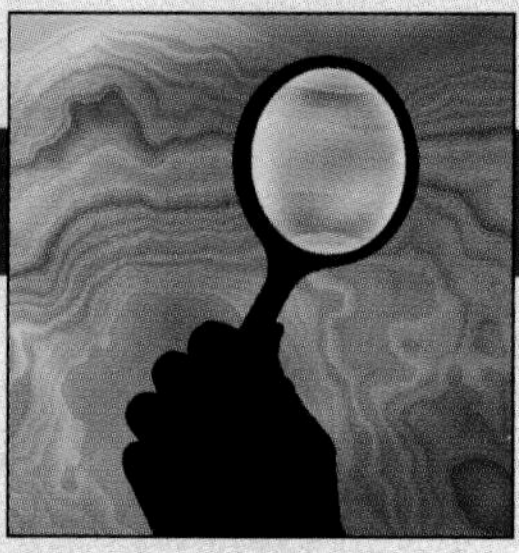

IN-DEPTH INVESTIGATION

Contour Maps: Island Construction

Materials
- modeling clay (4 lb.)
- flat basin or large pan (8 cm deep)
- scissors
- water
- plastic knife
- thick wooden dowel or rolling pin
- white paper
- pencil
- metric ruler

Introduction
A map is a drawing that shows a simplified version of some detail of the earth's surface. Many different types of maps are available. Each type has its own special features and purpose. One of the most basic and useful maps is the topographic map, or contour map. This type of map shows the elevation of the landscape as well as other important features. A contour map is made after a careful survey and photographic study of the area it represents.

Prelab Preparation
1. Review Section 3.3, pages 50–55.
2. Study the features of the island contour map shown on page 674. The elevation is measured in meters.

Procedure
1. Record the contour interval used on the island contour map. Then count the number of contour lines that appear on the map.
2. Press out as many flat squares of clay as the number of contour lines counted in Step 1. Make each layer about 1 cm thick. Also make the clay layers the same size or larger than the island of your contour map.

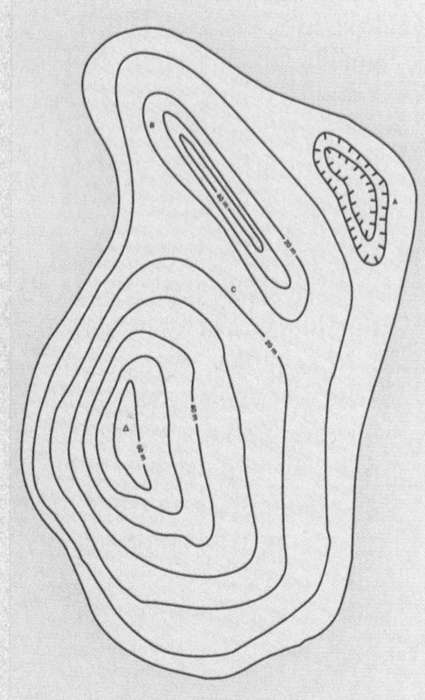

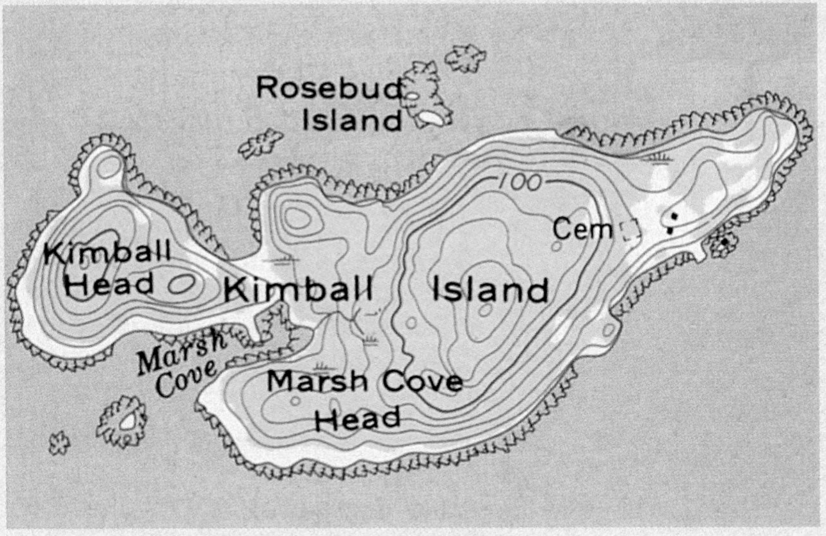

3. On a blank sheet of paper, trace the island contour map. Using the scissors, cut out the island from your copy of the contour map at the outermost contour line.
4. Place this cutout on top of one of the squares of clay. Trace the edge of the cutout in the clay and remove all clay outside the tracing.
5. Cut out again along the next contour line. Place the paper ring of the contour line that you just cut out on top of the first layer of clay so that the edges line up.
6. Repeat Step 4 with the island map and a new layer of clay.
7. Stack the second layer of clay on the first layer so it fits inside the contour ring you placed on the first layer of clay. This will give you the same contour spacing as shown on the original map. Remove the paper ring.
8. Continue Steps 4–7 for each of the contour layers.
9. Use leftover clay to smooth the terraced edges into a more natural profile.
10. Make a mark inside the pan approximately 1 cm down from the rim. Put the clay model of the island into the pan and add water to a depth of 1 cm. Compare the shoreline with the lines on the contour map. Continue to add water at 1 cm intervals until the water reaches the mark on the pan.

Analysis and Conclusions

1. What is the contour interval of your map?
2. How could you tell the steepest slope from the gentlest slope by observing the spacing of the contour lines?
3. What is the elevation above sea level for the highest point of your model?
4. How do you know if there are any points on your model that are below sea level? If there is such an area, where is this location and what is its elevation?
5. What landscape feature is located at *C* on your model?
6. What is the elevation of point *B* on your model?

Extension

Based on observations of your model, what conclusions can you make about where people might live on this island? Explain.

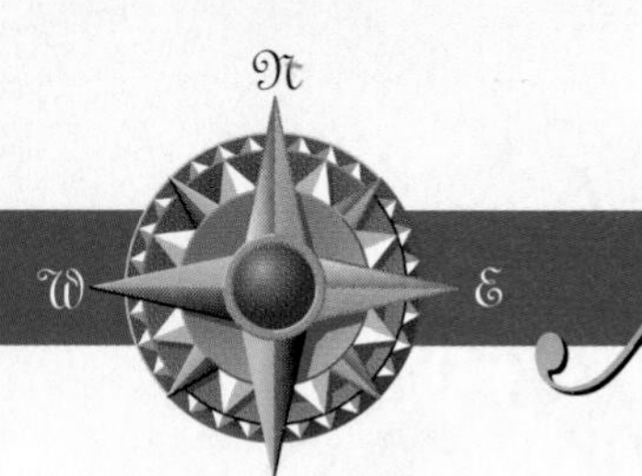

Maps in Action

Journey to Red River

Materials

- metric ruler
- magnetic compass with degree markings (optional)

How do you get from one place to another when you don't know the route? Whether you are planning a trip on foot or by car or boat, a map can be very handy. The ability to read a map can help you arrive quickly and safely at your destination instead of becoming disoriented and lost.

The topographic map on the facing page shows the area around Red River, New Mexico. Imagine that you are traveling to Red River to do some camping, hiking, and sightseeing. Study the map for a few moments and note the locations of roads, creeks, hills, and other features. Then answer the questions below.

1. Red River lies in the eastern half of New Mexico near the Colorado border. Use the map on page 45 to estimate the magnetic declination in this area. How does this compare with the magnetic declination in your hometown? Draw a diagram that shows how you would adjust a compass to determine true north in Red River. Why is it important to distinguish true north from geomagnetic north?
2. You set up a tent at Mallette Campground. If you walk in a straight line from your campsite to the cemetery at the base of Graveyard Canyon, how far will that be? Show your work.
3. You decide to hike from St. Edwin Chapel to location **A**. How much higher are you when you reach your destination? (Keep in mind that the elevations on the map are given in feet.)
4. Notice that the road in the lower-right corner of the map winds back and forth, making a series of hairpin turns. Why do you think the road designers planned it this way?
5. Most of the USGS maps like this one were originally created in the early 1960's. Like most towns, Red River has undergone some changes in the last few decades. Which features of your map might not reflect the way Red River looks today? Which features are probably still accurate?

Going Further

Cartography is the science of making maps. You can find out about new developments in the field of cartography by visiting the USGS Web site at www.usgs.gov.

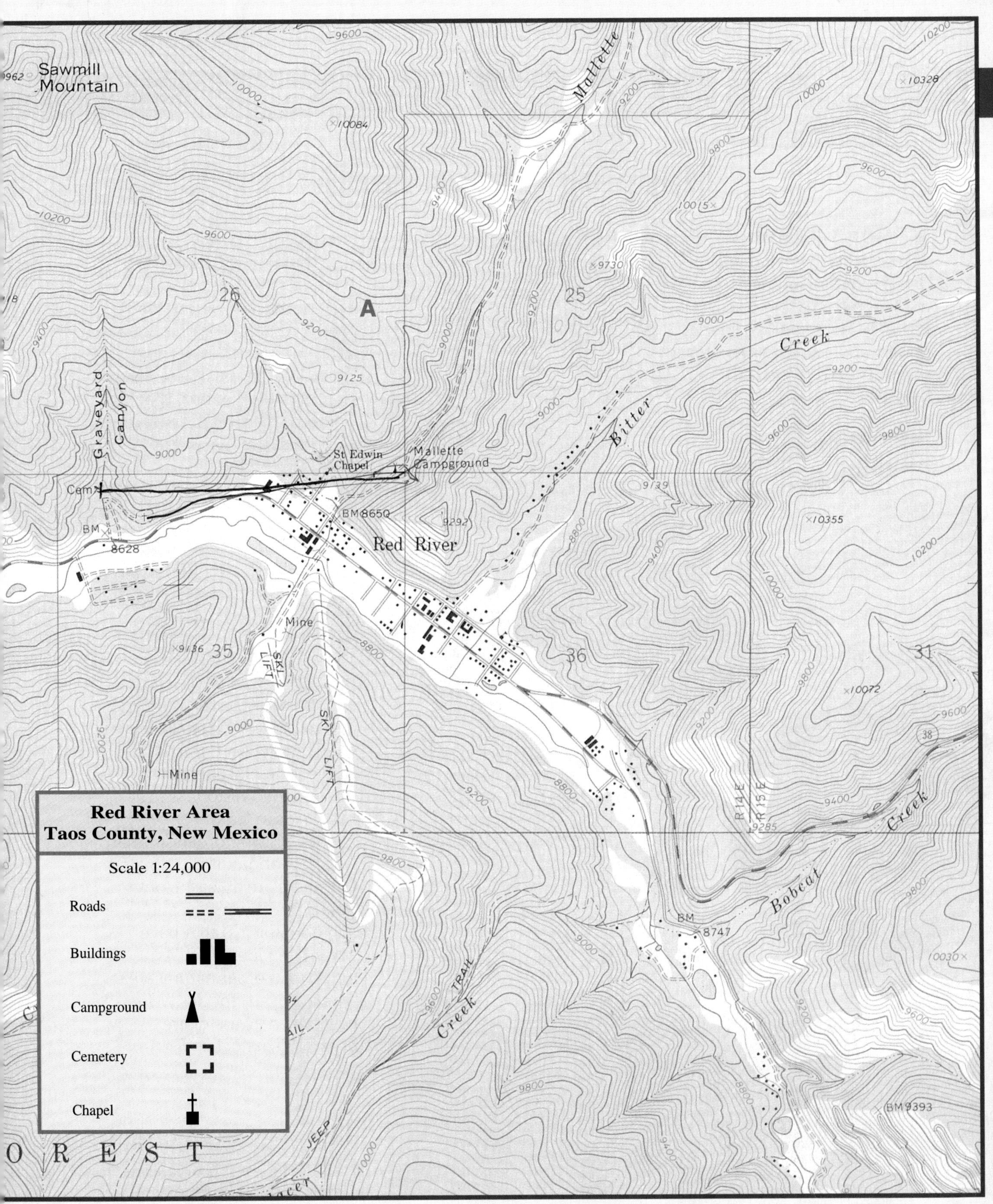

Red River Area
Taos County, New Mexico
Scale 1:24,000
Roads
Buildings
Campground
Cemetery
Chapel
Sawmill Mountain
Mallette
Graveyard Canyon
St Edwin Chapel
Mallette Campground
Red River
Bitter Creek
Bobcat Creek
Mine
SKI LIFT
TRAIL
JEEP
O R E S T
A

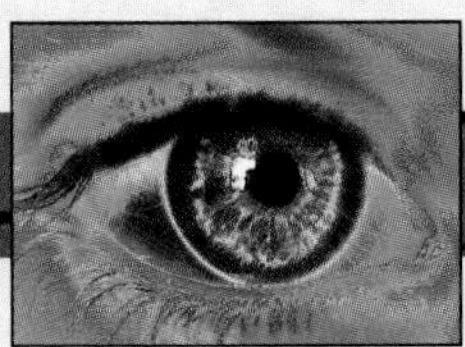

Eye on the Environment

Mapping Life on Earth

In 1992, a group of researchers working in a dense forest in Southeast Asia made some amazing discoveries. They identified two species of deer and a species of ox that scientists had never seen before!

A Call to Action

The group reported its findings and quickly made plans for further study. Unfortunately, there was a problem. The forest was being cut down at an incredibly fast rate. If the forest was destroyed, the world would lose these newly discovered animals as well as any other new resources that the forest might contain. Fortunately, citizens, other scientists, and politicians got involved. Now, thanks to their efforts, Vietnam and Laos have begun measures to protect the remaining forests in the region.

▲ **Scientists had never seen the Vu Quang Ox before 1992.**

Why All the Fuss?

Although discoveries of new mammals are rare, stories like the one above are not uncommon. New species are continually being discovered. The *biosphere,* which includes only a thin slice of the earth and its atmosphere, is the area on our planet where all life exists. Although humans have studied the biosphere throughout recorded history, we still have a lot to learn. For every species of living organism that we know about, it is estimated that there are at least eight others (mostly insects and microorganisms) that have never been identified. As forests and other natural areas are destroyed, species we've never been aware of are becoming extinct.

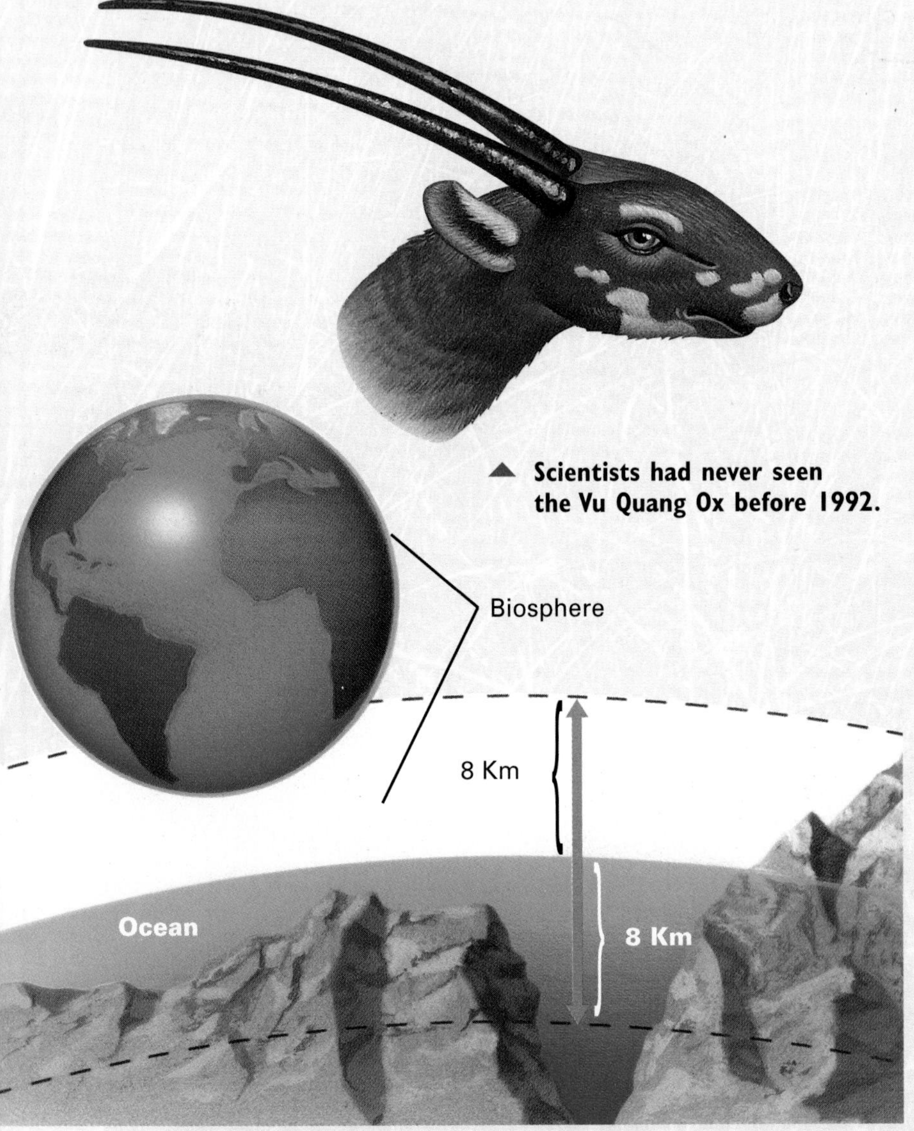

◀ **All of the earth's organisms exist within the biosphere.**

Help From Above

Given the huge number of natural areas that have never been thoroughly studied, how do scientists identify ecosystems that are endangered? Some scientists are approaching this problem with a little help from above. Using detailed photographs from satellites, these scientists can identify the makeup and size of the earth's remaining natural areas. At the same time, ground surveys are conducted to observe living organisms, and scientists interview local residents to find out about the changes that have occurred in the past and that occur at different times of the year. By pooling these data, the researchers then create a rough biodiversity map.

Biodiversity is the term used to describe the range of different organisms that live in an area. Rain forests and coral reefs have a very high biodiversity, which means that a large number of different organisms live in a small area. Other areas with high biodiversity include rivers, coastal regions, and natural grasslands. Biodiversity maps show which parts of a region have the highest biodiversity and are thus in the most need of protection.

Short-Term Success Story

Combining satellite images with brief ground surveys has already been very successful. In Jamaica, for instance, biodiversity maps helped officials identify areas that might benefit from being turned into national parks. The maps also highlighted some offshore areas of high biodiversity, and officials were able to close some of these areas to fishermen. Along the coast of Virginia, similar techniques allowed scientists to monitor and eventually control the invasion of an exotic plant species from Australia.

Although biodiversity maps are an exciting new development in environmental science, they are no substitute for the traditional methods of studying our biosphere. In Jamaica, Virginia, and other areas, long-term studies will be necessary to fully understand what species are present and how they interact.

▲ **This map of Jamaica was used in determining where national parks should be located.**

Think About It

How is creating a biodiversity map different than creating a topographic map? How is it similar? Could you create a biodiversity map that is also topographic? Why or why not?

Extension: Using string or tape, mark off a 1 m × 1 m plot near the edge of a natural area near your home or school. Divide the area into squares of 20 cm × 20 cm. Being careful not to disturb the organisms in your plot, count the number of plant species that you find within each square. Using these data, create a biodiversity map of your plot. What observations can you make about your map?

UNIT 2

The Dynamic Earth

Black Rock Mountain State Park, Georgia (background);
Medicine Bow National Forest, Wyoming (inset)

Introducing Unit Two

Earthquakes are one of the many mysteries of the planet Earth that we are beginning to solve. During one year alone, 13,400 earthquakes were located throughout the world. In each case, the only thing we knew in advance was that the potential for damage to lives, property, and our natural surroundings was catastrophic. As seismologists, we don't start earthquakes and we can't stop them. No one can. What we can do is save lives and minimize suffering by providing early alerting services. When the alarms go off, we're ready—24 hours a day, every day—to respond to one of the earth's most violent forces.

The unit you are about to study discusses earthquakes and other forces that shape the earth. Understanding these forces is one key to unlocking the mysteries of our planet.

Waverly J. Person
Seismologist
National Earthquake Information Service

In this UNIT

Chapter 4 Plate Tectonics

The earth is in constant motion. The planet spins on its axis and orbits the sun. Earth scientists also believe that the surface of the earth itself is in motion, broken up into plates that drift slowly around the planet, a process called plate tectonics. Plate tectonics has shaped the San Francisco Bay area, which includes crust from the ocean floor as well as sediments possibly from past continents. In this chapter, you will learn how plate tectonics causes changes in the earth's crust.

Chapter Outline

◀ San Francisco is built on crust from the ancient ocean floor and possibly past continents.

4.1 Continental Drift

Section Objectives

- **Explain Wegener's hypothesis of continental drift.**
- **List evidence for Wegener's hypothesis of continental drift.**
- **Describe seafloor spreading.**

One of the most exciting recent theories in earth science began with observations made more than 400 years ago. As explorers such as Christopher Columbus and Ferdinand Magellan sailed the oceans of the world, they brought back information about new continents and their coastlines. Mapmakers used the information to chart the new discoveries and to make the first reliable world maps.

As people studied the maps, they were impressed by the similarity of the continental shorelines on either side of the Atlantic Ocean. The continents looked as though they would fit together, like parts of a giant jigsaw puzzle. The east coast of South America seemed to fit perfectly into the west coast of Africa. Greenland seemed to fit between North America and northwestern Europe.

These observations soon led to questions. Were the continents once part of the same huge landmass? If so, what caused this landmass to break apart? What caused the continents to move to their present locations? These questions eventually led to the formulation of hypotheses.

In 1912, a German scientist, Alfred Wegener, proposed a hypothesis called **continental drift,** which stated that the continents had moved. Wegener hypothesized that the continents once formed part of a single landmass, which he named **Pangaea** (pan-JEE-uh), meaning "all lands." Surrounding Pangaea was a huge ocean, **Panthalassa,** meaning "all seas." According to Wegener, about 200 million years ago, Pangaea began breaking up into smaller continents, which drifted to their present locations. Wegener speculated that this motion may have crumpled the crust in places, producing mountain ranges such as the Andes on the western coast of South America.

Evidence of Continental Drift

In addition to the similarities in the coastlines of the continents, Wegener soon found other evidence to support his hypothesis. If the continents had once been joined, he reasoned, research should

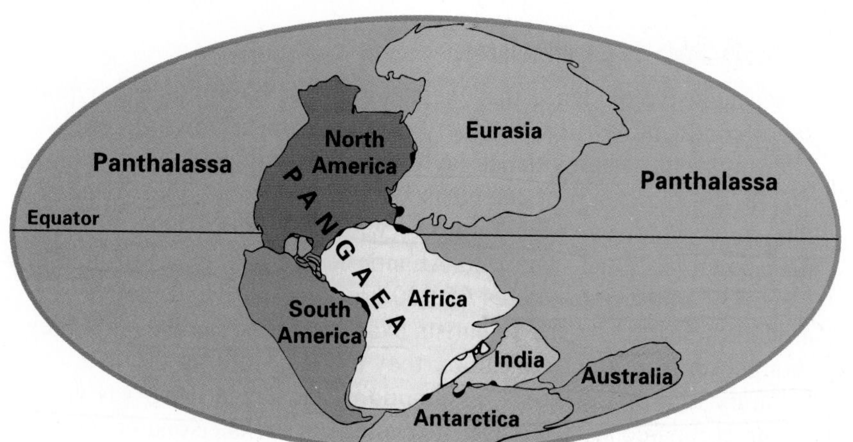

Figure 4–1. This map shows Pangaea as Alfred Wegener envisioned it.

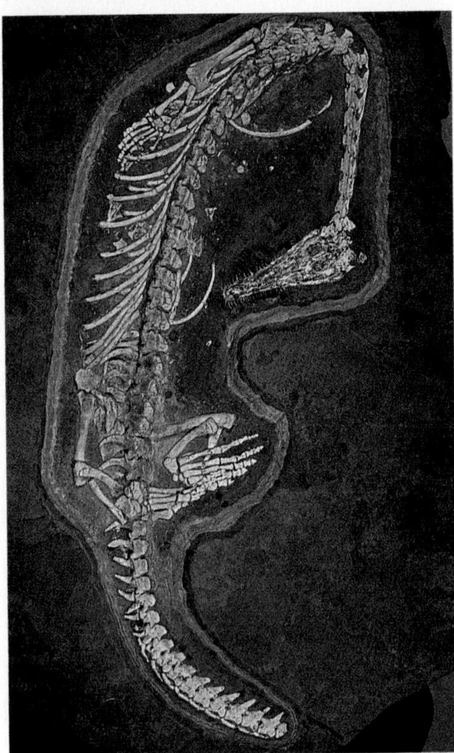

Figure 4–2. These *Mesosaurus* bones were discovered in Sao Paulo, Brazil. Identical fossil bones of *Mesosaurus* were found in western Africa, giving scientists strong evidence of continental drift.

uncover fossils of the same plants and animals in areas that had been adjoining parts of Pangaea. Wegener knew that identical fossil remains of *Mesosaurus,* a small, extinct land reptile that lived 270 million years ago, had already been found in both eastern South America and western Africa. Wegener knew that it was impossible for these reptiles to have swum across the Atlantic. And there was no evidence of any land bridges that might have connected the continents at some earlier time. Wegener thus concluded that South America and Africa must have been joined at one time.

Geologic evidence also supported Wegener's hypothesis of continental drift. The age and type of rocks in the coastal regions of widely separated areas, such as western Africa and eastern Brazil, matched closely. Mountain chains that ended at the coastline of one continent seemed to continue on landmasses across the ocean. The Appalachians, for example, extend northward along the eastern United States, while mountains of similar age and structure are found in Greenland and northern Europe. If these three landmasses are assembled in a model of Pangaea, the mountains fit together in one continuous chain.

Evidence of changes in climatic patterns added strength to Wegener's hypothesis. Geological research revealed layers of debris from glaciers in southern Africa and South America, areas that today have much warmer climates. Other fossil evidence—such as the coal deposits in the eastern United States, Europe, and Siberia—indicated that tropical or subtropical swamps covered much of the land area in the Northern Hemisphere. If the continents were once joined and positioned over the South Pole, Wegener suggested, these climatic differences would be easy to explain.

Despite the evidence supporting the hypothesis of continental drift, Wegener's ideas met with strong opposition. Many scientists rejected the hypothesis because it did not satisfactorily explain the force causing continental drift. In an effort to convince the scientific community that his hypothesis was valid, Wegener spent the rest of his life searching for evidence of that force. Unfortunately, Wegener died in 1930, while on an expedition to Greenland. He never found an explanation of what caused continents to move.

Seafloor Spreading

The conclusive evidence that Wegener sought to support his hypothesis of continental drift was finally discovered nearly two decades after his death. The evidence lay on the ocean floor.

In 1947, a group of scientists set out to map the **Mid-Atlantic Ridge,** an undersea mountain range with a steep, narrow valley running down its center. The Mid-Atlantic Ridge is one part of an entire system of **mid-ocean ridges** 65,000 km long that wind their way around the earth. As the scientists examined rock samples that they brought up from the ocean floor, they made a startling discovery. Contrary to most scientists' assumptions, the ocean floor was very young compared with the age of continental rocks. None of the oceanic rocks found were more than 150 million years old. The oldest continental rocks are about 4 billion years old. Why do you suppose scientists found this information surprising?

The Renewal of the Ocean Floor

After analyzing the data gathered from the Mid-Atlantic Ridge, Harry Hess, a geologist at Princeton University, suggested the following hypothesis. Suppose, he said, that the valley at the center of the ridge was actually a break, or rift, in the earth's crust and that molten rock, or magma, from deep inside the earth was welling up through the rift. This upwelling would be possible, Hess reasoned, if the ocean floor was moving away from both sides of the ridge. As the ocean floor moved away from the ridge, it was replaced by rising magma that cooled and solidified into new rock. Another

EARTHBEAT

Evidence of Seafloor Spreading

Deep under the Atlantic Ocean lies a mountain range so vast that it dwarfs the Himalayas. This mountain range is called the Mid-Atlantic Ridge, the mid-ocean ridge at the diverging boundary between the North American and Eurasian plates and the South American and African plates.

Explorations of the Mid-Atlantic Ridge have enabled scientists to obtain firsthand evidence of seafloor spreading. For 15 years, scientists aboard the *Glomar Challenger,* shown at left, traveled more than 575,000 km and collected nearly 100 km of core samples from 635 drill holes. Scientists examining the fossils embedded in the samples discovered that cores drilled closest to the ridge had the youngest fossils. The fossils proved to be progressively older as samples were taken farther from the ridge. This find supported the idea that new crust forms along the ridge as older crust moves aside.

With the help of deep-diving submarines, scientists were able to explore the rift valley that bisects the mid-ocean ridge. The crew on board the submarine *Archimede* were among the first to witness magma bubbling up from the mantle. A year later, cameras on board the *Alvin* photographed lava formations that had emerged from the mantle and hardened in the rift valley. One of these photographs is shown above. The lava formations provided further evidence of seafloor spreading.

If you had six core samples collected from the ridge outward, how many samples would you have to test to determine whether the samples on the right or the left were closer to the mid-ocean ridge? Why?

geologist, Robert Dietz, named this movement **seafloor spreading.** Hess suggested that if the ocean floor was moving, the continents might also be moving. Perhaps seafloor spreading was the force that Wegener had failed to find to support his hypothesis of continental drift.

Still, Hess's ideas were just hypotheses. The proof would come years later, in the mid-1960's, and would be discovered through paleomagnetism, the study of the past magnetic properties of rocks.

Paleomagnetism of the Ocean Floor

If you have ever used a compass to determine direction, you know that the earth acts as a giant magnet, with both a north and a south pole. The compass needle aligns with the field of magnetic force that extends from one pole to the other.

Living on the Mid-Atlantic Ridge

Although most of the earth's mid-ocean ridges lie completely underwater, part of the Mid-Atlantic Ridge rises above sea level just south of the Arctic Circle. This exposed section of the Mid-Atlantic Ridge forms the island country of Iceland. Since its founding by Vikings over 1,000 years ago, the inhabitants of Iceland have had to contend with the constant geological activity associated with seafloor spreading.

Separation of the earth's crust along the Mid-Atlantic Ridge affects Iceland's landscape in several ways. Tectonic motion, for example, causes frequent earthquakes. Iceland is also one of the most volcanically active areas in the world. It contains about 200 volcanoes and averages one eruption every five years. Magma flowing up from the mantle creates numerous hot springs, geysers, and sulfurous gas vents. Scientists estimate that one-third of the total lava flow from the earth in the last 500 years has occurred on Iceland.

Despite its numerous volcanoes, much of Iceland's lava comes not from isolated eruptions but rather from cracks, or fissures, in the crust. In a recent rifting episode that lasted nearly 10 years, a series of fissures spit out 35 km^2 of molten basalt. Over the course of this event, individual fissures grew as much as 8 m in width. At present, seafloor spreading adds an average of 2.5 cm of new material each year to Iceland. At this rate, Iceland will grow 25 km in width during the next million years.

If geologists want to locate the youngest rocks on Iceland, where should they look? Where should they look to find the oldest rocks? Why?

A similar phenomenon occurs when magma cools and solidifies. Certain iron-bearing minerals within the rock become magnetized. When the rock hardens, the magnetic orientation of the minerals becomes permanent and points to the north.

Scientists discovered, however, that this was not always the case. From the beginning of the nineteenth century, they had been finding rocks with magnetic orientations that pointed south. Some scientists concluded that the earth's magnetic field must have reversed itself at times during the earth's history. This conclusion was verified by dating rocks with different magnetic orientations. All the rocks with magnetic fields pointing north fell into the same time periods—periods of normal polarity. All the rocks with magnetic fields pointing south also fell into similar time periods—periods of reverse polarity. The scientists discovered that throughout the earth's history the magnetic field has reversed itself many times.

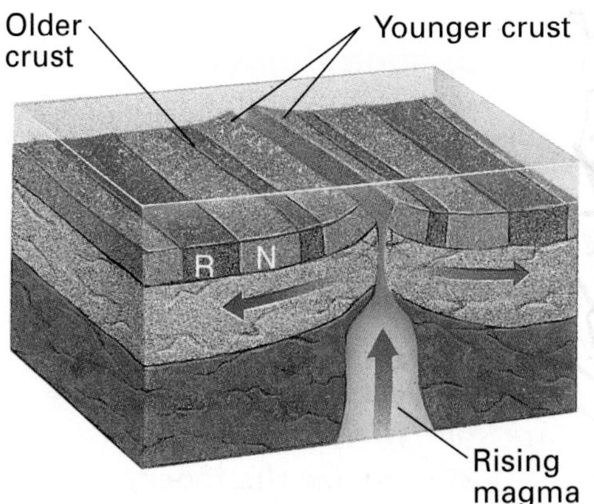

Figure 4–3. The stripes in the crust are shown here to illustrate the earth's alternating magnetic field. Dark stripes represent the ocean floor with reversed polarity (R), while the lighter stripes show normal polarity (N).

At the same time these discoveries about the earth's magnetic field were being made, scientists were also finding puzzling magnetic patterns on the ocean floor. These patterns, when drawn in on maps of the ocean floor, showed alternating bands of normal and reversed magnetism. As molten rock rises from the rift in a mid-ocean ridge, it quickly cools and hardens and its magnetic orientation becomes fixed. Its magnetic orientation will reflect the polarity of the earth's magnetic field at that time, either normal or reversed.

Scientists' confidence in the validity of this idea grew when they discovered that the striped patterns of magnetism on one side of a ridge are mirror images of the striped patterns on the other side of the ridge. This discovery supported Hess's idea that molten rock from a rift cools, hardens, and then moves away in opposite directions on both sides of the ridge. The ocean floor, it seemed, was indeed spreading.

Finally, in 1965, two groups of scientists working independently of each other discovered a previously unknown reversal in the earth's magnetic field. One group discovered the reversal in rocks on land, and the other group discovered the reversal in rocks on the ocean floor. The dates of both reversals were exactly the same. This was clear evidence that the earth's magnetic polarity does reverse itself and that the ocean floor does spread. Scientists reasoned that seafloor spreading provides a way for the continents to be moved over the surface of the earth. Here, at last, were the discoveries that Wegener had sought, the scientific evidence he needed to verify his hypothesis of continental drift.

Section 4.1 Review

1. What observation first led to Wegener's hypothesis of continental drift?
2. What types of evidence support Wegener's hypothesis?
3. Describe the process of seafloor spreading.
4. Explain how scientists know that the earth's magnetic poles have reversed themselves many times during earth's history.

Section Objectives

- **Summarize the theory of plate tectonics.**
- **Compare the characteristic geologic activities that occur along the three types of plate boundaries.**
- **Explain the possible role of convection currents in plate movement.**
- **Summarize the theory of suspect terranes.**

4.2 The Theory of Plate Tectonics

By the 1960's, accumulated evidence supporting the hypothesis of continental drift and seafloor spreading led to the formulation of a more far-reaching theory. This theory is called **plate tectonics.** The theory of plate tectonics not only describes continental movement but also proposes a possible explanation of why and how continents move. The term *tectonics* comes from the Greek word *tektonikos*, meaning "construction." Tectonics is the study of the formation of features in the earth's crust.

The earth's crust consists of two types—**oceanic crust** and **continental crust.** Material on the ocean floor forms oceanic crust. Continental crust makes up the continental landmasses.

The oceanic and continental crust and the rigid upper mantle make up the **lithosphere.** It forms the thin outer shell of the earth. Beneath the lithosphere lies the **asthenosphere,** a layer of plastic rock, that is, solid rock that slowly flows (like putty) when under pressure. According to the theory of plate tectonics, the lithosphere is broken into separate plates that ride on the denser asthenosphere much like blocks of wood float on water. The continents and oceans are carried along as "passengers" on the moving lithospheric plates. Most lithospheric plates are composed of both continental and oceanic crust.

To date, about 30 lithospheric plates have been identified. Some plates are moving toward each other, some are moving apart, and some are sliding past each other. This constant movement has created the earth's major surface features, such as mountain ranges and deep-ocean trenches.

Figure 4–4. This map shows the location and movement of various lithospheric plates.

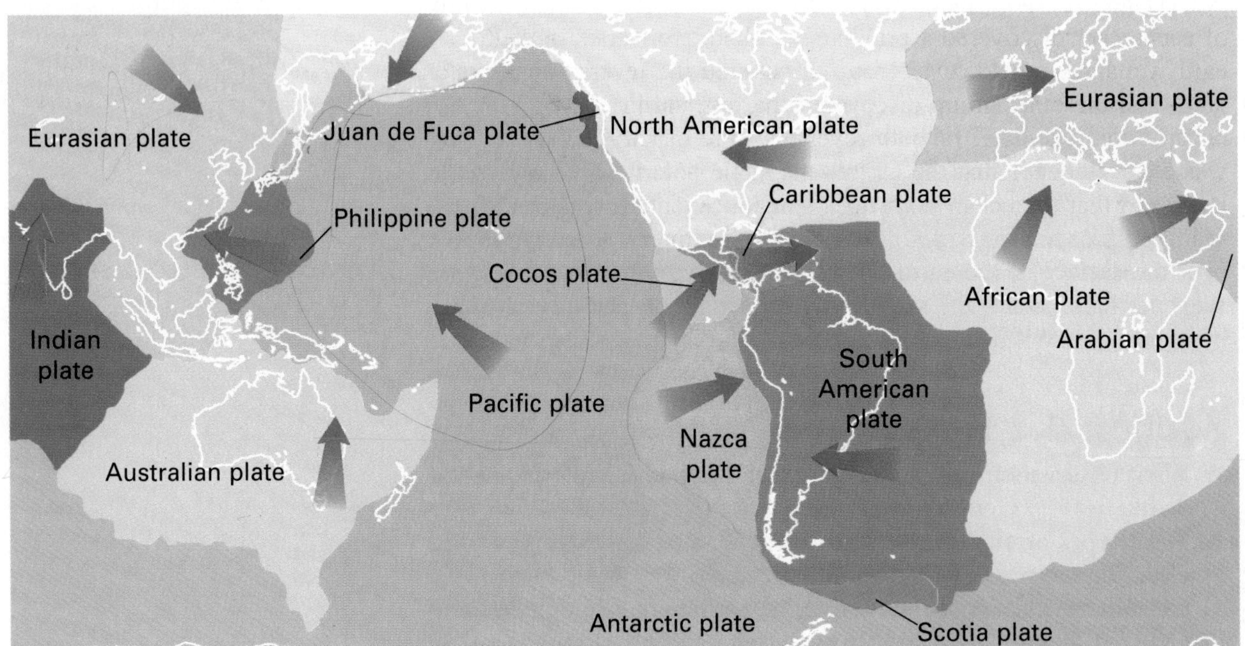

Lithospheric Plate Boundaries

Many changes in the earth's crust originate along lithospheric plate boundaries. The boundaries of the plates are not always easy to identify. As you can see in Figure 4–4, the familiar outlines of the continents and oceans depicted on maps do not always resemble the outlines made by the plate boundaries. Plate boundaries may be in the middle of the ocean floor, around the edges of continents, or within continents. There are three types of plate boundaries, each of which is associated with a characteristic type of geologic activity.

Divergent Boundaries

The geologic activity that occurs along plate boundaries differs according to the way plates move in relation to each other. For example, two plates moving away from each other form a **divergent** (die-VUR-junt) **boundary.** As the plates move apart, molten rock from the asthenosphere rises and fills the space between the plates. As the molten rock cools, it hardens onto the edges of the separating plates and creates new oceanic crust. Most divergent boundaries are found on the ocean floor. The locations of these spreading boundaries follow the mid-ocean ridges.

In the center of a mid-ocean ridge is a narrow valley formed as the plates separate. This formation is called a **rift valley.** Other rift valleys may form where continents are separated by plate movement. For example, the Red Sea occupies a huge rift valley formed by the separation of the African plate and the Arabian plate.

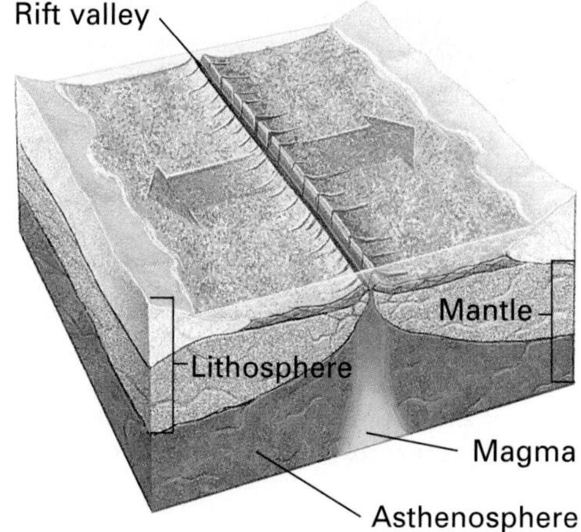

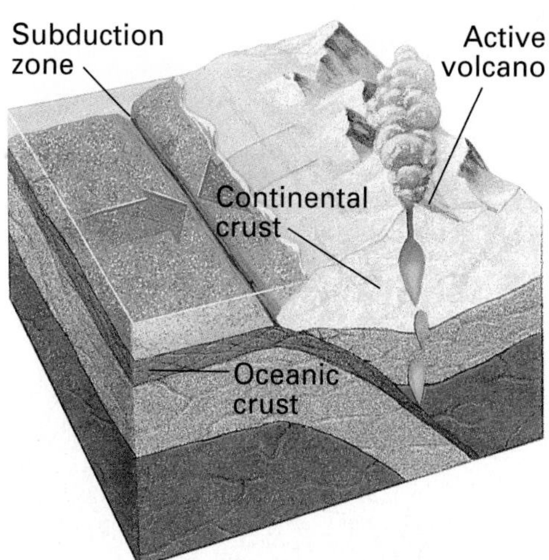

Figure 4–5. At divergent boundaries (top), plates separate. Plates collide at convergent boundaries (bottom).

Convergent Boundaries

As seafloor spreading pulls plates apart at one boundary, those plates push into neighboring plates at other boundaries. The direct collision of one plate with another makes another type of plate boundary—a **convergent** (kun-VUR-junt) **boundary.**

Three types of collisions can occur at convergent boundaries. One type occurs when a plate with oceanic crust at its leading edge collides with a plate with continental crust at its edge. Because oceanic crust is denser, it is *subducted,* or forced under the less dense continental crust, as shown in Figure 4–5. Scientists refer to the region along a plate boundary where one plate moves under another plate as a **subduction** (sub-DUK-shun) **zone.** A deep **ocean trench** generally forms along a subduction zone. As the oceanic plate moves down into a subduction zone, it melts and becomes part of the mantle material. Some of the magma formed rises to the surface through the continental crust and produces volcanic mountains.

A second type of collision occurs when two plates with continental crust at their leading edges come together. During this type of collision, neither plate is subducted because the two plates have the same density. Instead, the colliding edges are crumpled and uplifted, producing large mountain ranges. Scientists are convinced that the Himalayas were formed by this type of collision.

The third type of collision along convergent boundaries occurs between oceanic crust and oceanic crust. A deep ocean trench also

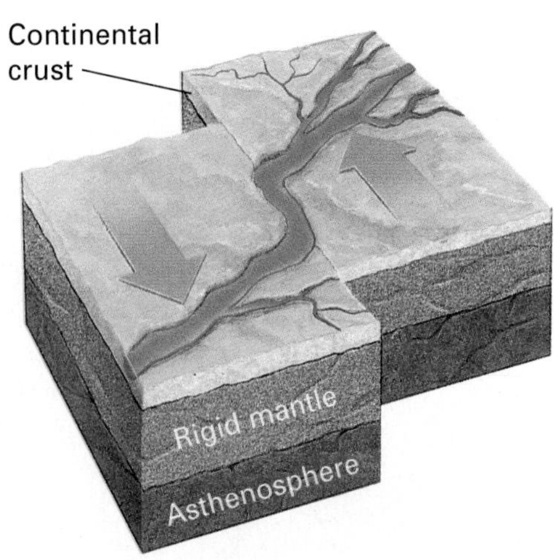

Figure 4–6. Plates scrape past each other at transform fault boundaries. Note the change in the course of the river as the plates move past each other.

forms when one of these plates is subducted. Part of the subducted plate melts, and the resulting molten rock rises to the surface along the trench to form a chain of volcanic islands, called an **island arc.**

Transform Fault Boundaries

A plate boundary called a **transform fault boundary** forms where two plates are grinding past each other. The plate edges usually do not slide along smoothly. Instead, they scrape together and move in a series of sudden spurts of activity separated by periods of little or no motion. A major transform fault boundary is the San Andreas fault in California.

Causes of Plate Motion

Many earth scientists think that the movement of lithospheric plates is due to **convection,** the transfer of heat through the movement of heated fluid material. This same process occurs when you place a pot of water on the stove to boil. As the water at the bottom of the pot heats up, it expands and becomes less dense than the cool water above it. The cool water, which is now denser than the warm water, sinks and forces the warm water to the surface. This cycle of warm water rising and cool water sinking to replace it is called a **convection current.**

Scientists think a similar process results in convection currents within the asthenosphere. Heat from the earth's core and mantle causes some material in the lower asthenosphere to become hotter and therefore less dense than the material above it. The hot material rises. When this hot material reaches the base of the lithosphere, it cools. When the molten material cools, it becomes more dense and starts to sink. The cooling material is pushed to the side by new hot

Figure 4–7. Scientists think that convection currents are the mechanism that moves lithospheric plates

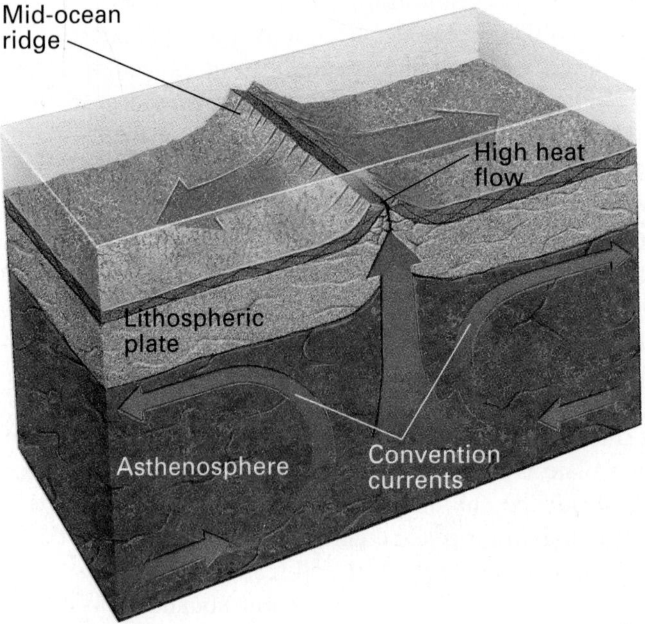

material that rises. As the process continues, the lithospheric plate is carried along with the moving material, as shown in Figure 4–7.

Evidence for the existence of convection currents in the asthenosphere comes from recent studies of the ocean floor. Scientists have measured the amount of heat leaving rocks at various points in the lithosphere. They have found this heat flow to be higher along plate boundaries where two plates are moving apart than it is elsewhere on the ocean floor. If hot convection currents are rising along these plate boundaries as the theory suggests, these temperature differences can be explained.

Though convection currents can explain some aspects of plate movement, questions remain. Scientists asked whether convection currents alone are strong enough to move the plates at the rates suggested by geological evidence. If not, they speculated, another mechanism may be responsible.

INVESTIGATE!

To learn more about convection currents, try the In-Depth Investigation on pages 80–81.

SMALL-SCALE INVESTIGATION

Lithospheric Plate Boundaries

The movement of lithospheric plates has created many of the earth's topographical features. You can demonstrate the results of plate movement by using clay models of lithospheric plates.

Materials

ruler, paper, scissors, rolling pin or rod, modeling clay (2–3 lb.), plastic knife, lab apron

Procedure

1. Draw two 10 × 20 cm rectangles on your paper, and cut them out.
2. Use a rolling pin to flatten out two pieces of clay until they are about 1 cm thick. Cut each piece into a 10 × 20 cm rectangle. Place a paper rectangle on each piece of clay.
3. Place the two clay models side by side on a flat surface, paper side down. Place your hands directly on top of each piece, as shown, and slowly push the models together until the edges begin to buckle and rise off the surface of the table.
4. Turn the clay models around so that the unbuckled edges are touching. If these edges have been slightly deformed during Step 3, smooth them out before proceeding.

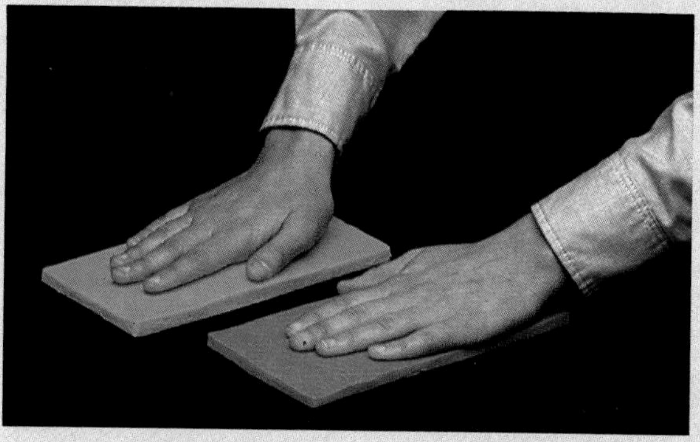

5. Place one hand on each clay model. Apply only slight pressure toward the seam. Slide one clay model forward and the other model backward about 7 cm.
6. Repeat Step 5 three more times, alternating the direction in which you push each model.

Analysis and Conclusions

1. What type of plate boundary are you demonstrating with the model in Step 3?
2. What type of plate boundary are you demonstrating in Steps 5 and 6?
3. How does the appearance of the facing edges of the models in the two processes compare? How do you think these processes might affect the appearance of the earth's surface?

Suspect Terranes

Alfred Wegener's hypothesis of continental drift was an attempt to explain how the continents arrived at their present locations. The theory of plate tectonics refined Wegener's hypothesis, suggesting the actual mechanisms by which the continents might move. Neither continental drift nor plate tectonics, however, can explain how the continents were formed.

New discoveries are providing some possible explanations of how continents formed. These new discoveries provide the basis for the **theory of suspect terranes.** Simply put, this theory suggests that the continents are actually a patchwork of **terranes**—pieces of lithosphere, each with its own distinct geologic history. Each terrane has three identifying characteristics. First, a terrane contains rock and fossils that differ from the rock and fossils of neighboring terranes. Second, there are major faults at the boundaries of a terrane. Finally, the magnetic properties of a terrane do not match those of neighboring terranes.

Geologists have found evidence to support the suspect terrane theory. Northern California is a good place to observe this evidence. Geologists have found 10 different terranes in the San Francisco Bay area alone. For example, in the hills of Palo Alto, there is fossil evidence of coral atolls—tropical ocean islands made up of coral, the skeletons of sea organisms. Farther south lies a terrane called Permanette. The limestone of Permanette contains fossils that prob-

Figure 4–8. The rocks of the Slide Mountain terrane are fragments of the ancient ocean floor that are now part of North America.

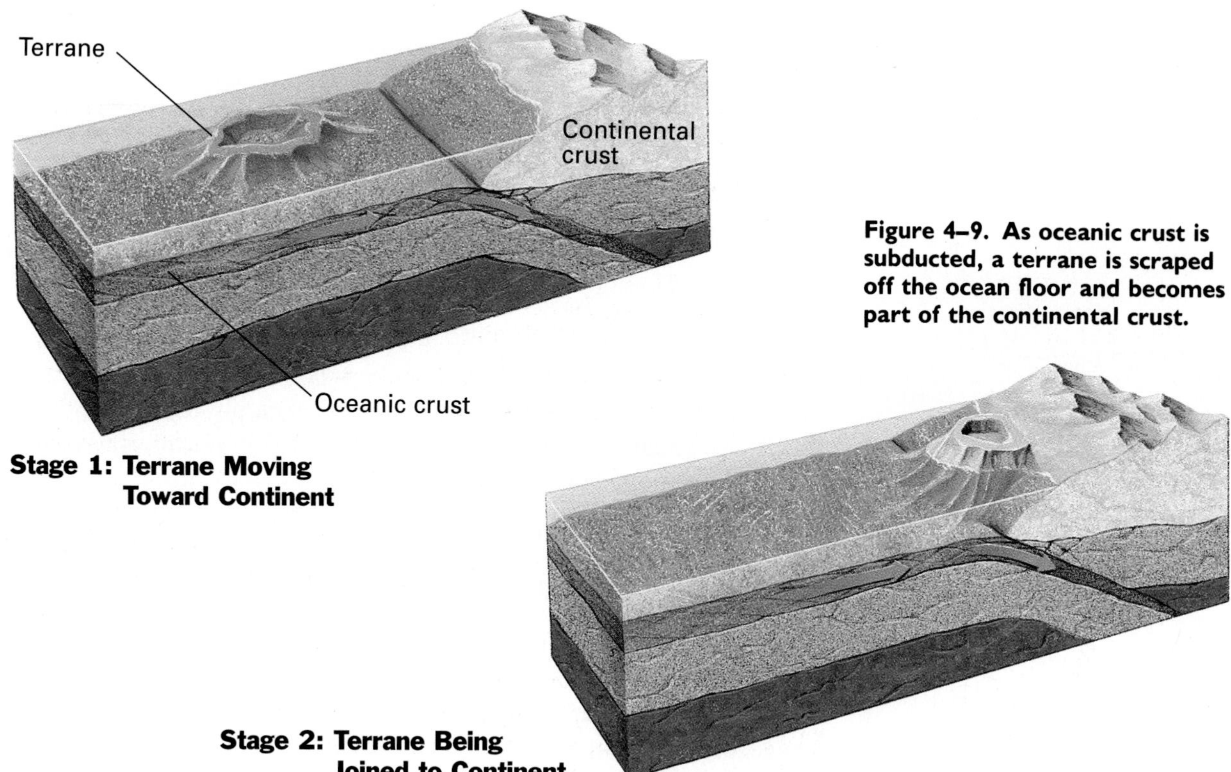

Figure 4–9. As oceanic crust is subducted, a terrane is scraped off the ocean floor and becomes part of the continental crust.

ably originated from the ocean depths near the equator. The theory of suspect terranes explains how tropical coral atolls and equatorial ocean fossils became part of the geology of northern California.

According to the suspect terrane theory, blocks of terranes are carried along on the ocean floor by the action of seafloor spreading to a lithospheric plate boundary where subduction is occurring. As the plate with oceanic crust moves under the plate with continental crust, the terranes are scraped off the descending ocean floor, as shown in Figure 4–9. Some terranes may form mountains, while others simply add to the surface area of a continent.

When Alfred Wegener first proposed his hypothesis of continental drift, he could not have imagined the explosion of scientific inquiry it would inspire. Like many hypotheses, continental drift raised more questions than it answered. The theories of plate tectonics and suspect terranes are attempts to answer some of those questions.

Section 4.2 Review

1. Summarize the theory of plate tectonics.
2. Name and describe the three types of plate boundaries.
3. Describe the three types of plate collisions that occur along convergent boundaries.
4. How might convection currents cause plate movement?
5. Explain how mountains on land can be composed of rocks that contain fossils of animals that lived in the ocean.

Chapter 4 Review

Key Terms

- asthenosphere (72)
- continental crust (72)
- continental drift (67)
- convection (74)
- convection current (74)
- convergent boundary (73)
- divergent boundary (73)
- island arc (74)
- lithosphere (72)
- Mid-Atlantic Ridge (68)
- mid-ocean ridges (68)
- ocean trench (73)
- oceanic crust (72)
- Pangaea (67)
- Panthalassa (67)
- plate tectonics (72)
- rift valley (73)
- seafloor spreading (70)
- subduction zone (73)
- terrane (76)
- theory of suspect terranes (76)
- transform fault boundary (74)

Key Concepts

Wegener's hypothesis of continental drift states that the continents were once a single landmass. **See page 67.**

Scientists have found fossil, geological, paleomagnetic, and climatic evidence to support the hypothesis of continental drift. **See page 67.**

New ocean floor is constantly being produced through seafloor spreading. **See page 68.**

The theory of plate tectonics proposes that changes in the earth's crust are caused by the very slow movement of large lithospheric plates. **See page 72.**

The geological activity that occurs along the three types of plate boundaries differs according to the way plates move in relation to each other. **See page 73.**

Convection currents may be responsible for plate movements. **See page 74.**

The theory of suspect terranes proposes that portions of the continents are a patchwork of terranes scraped off subducting lithospheric plates. **See page 76.**

Review

On your own paper, write the letter of the term that best completes each of the following statements.

1. The German scientist Alfred Wegener proposed the existence of a huge landmass called
a. Panthalassa. b. rift valley.
c. Mesosaurus. d. Pangaea.

2. Support for Wegener's hypothesis of continental drift includes evidence of changes in
a. climatic patterns. b. Panthalassa.
c. terranes. d. subduction.

3. New ocean floor is constantly being produced through the process known as
a. subduction. b. continental drift.
c. seafloor spreading. d. terranes.

4. An underwater mountain chain formed with molten rock from seafloor spreading is called a
a. divergent boundary. b. subduction zone.
c. mid-ocean ridge. d. convergent boundary.

5. The term *tectonics* comes from a Greek word meaning
a. "movement." b. "plate."
c. "continent." d. "construction."

6. The layer of mantle with plastic rock that underlies and moves the plates is called the
   a. lithosphere. b. asthenosphere.
   c. oceanic crust. d. terrane.
7. To date, scientists have identified approximately
   a. 5 plates. b. 30 plates.
   c. 15 plates. d. 50 plates.
8. Two plates moving away from each other form a
   a. transform fault boundary. b. convergent boundary.
   c. fracture. d. divergent boundary.
9. The collision of one lithospheric plate with another forms a
   a. convergent boundary. b. transform fault boundary.
   c. rift valley. d. divergent boundary.
10. The region along lithospheric plate boundaries where one plate is moved beneath another is called a
   a. rift valley. b. transform fault boundary.
   c. subduction zone. d. convergent boundary.
11. Two plates grind past each other at a
   a. transform fault boundary. b. convergent boundary.
   c. subduction zone. d. divergent boundary.
12. Convection occurs because heated material becomes
   a. less dense and rises. b. more dense and rises.
   c. more dense and sinks. d. less dense and sinks.
13. Scientists think that the convection currents that are responsible for the movement of lithospheric plates are found in the
   a. lithosphere. b. asthenosphere.
   c. terranes. d. rift valleys.
14. Geologists think that portions of the continents are made up of formerly separate pieces of lithosphere called
   a. terranes. b. plates.
   c. continental crust. d. oceanic crust.

Critical Thinking

On your own paper, write answers to the following questions.

1. In what ways might the concept of continental drift be compared to a jigsaw puzzle?
2. If Alfred Wegener had found identical fossil remains of plants and animals that had lived no more than 10 million years ago in both eastern Brazil and western Africa, what might he have concluded about the breakup of Pangaea?
3. Assume that the total surface area of the earth is not changing. If new material is being added to the earth's crust at one boundary, what would you expect to find happening at another boundary?
4. Explain the following statement: Due to the action of convection currents, the ocean floor is constantly renewing itself.

Application

1. Explain the role of technology in the progression from the hypothesis of continental drift to the theory of plate tectonics.
2. If you wanted to prove that a suspect terrane had been scraped onto the North American plate, what kind of evidence would you search for?
3. Construct a **concept map** using 10 of the new terms listed on the previous page by making connections that illustrate the relationship among the terms.

Extension

1. Copy a world map showing the outlines of the continents. Cut out the continents and assemble a model of Pangaea. Compare your model to Wegener's model of Pangaea that is shown on page 67.
2. In the late 1800's, Eduard Suess, an Austrian scientist, developed the concept of a supercontinent called Gondwanaland. Look up information about Suess's ideas and compare them with the hypothesis of continental drift proposed by Wegener.
3. Plate tectonics is a relatively new theory. Conduct library research to find out how early scientists accounted for changes in the continents long before the theory of plate tectonics. Report your findings to the class.

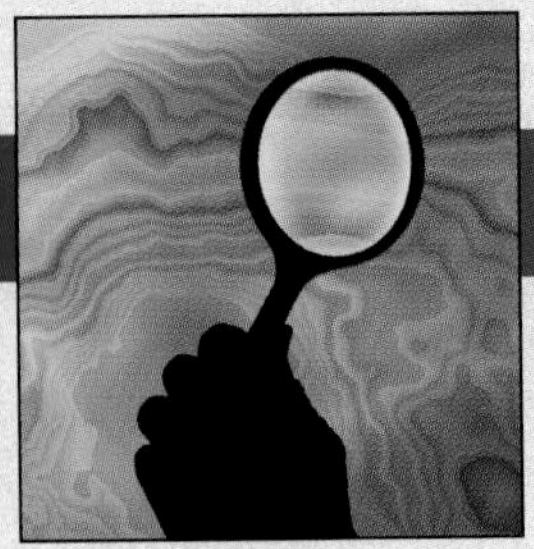

In-Depth Investigation

A Model of Convection Currents

Materials

- rectangular aluminum pan (at least 23 × 33 × 5 cm)
- 2 Bunsen burners with flame spreader attachments
- 2 craft sticks
- cold water
- metric ruler
- beaker (1 L or larger)
- food coloring with dropper
- 4 ring stands and clamps
- pencil
- lab apron
- 3 thermometers

Introduction

Mid-ocean ridges are places where heat from the mantle reaches the earth's surface. Many geologists think that mid-ocean ridges mark the location of the rising portions of convection currents in the semi-molten asthenosphere. The drag on the lithosphere from such currents might provide the energy for plate motion.

In this investigation, you will demonstrate the action of convection currents and attempt to model how convection currents may be the cause of plate motion.

Prelab Preparation

1. Review Section 4.2, pages 72–77.
2. Review the safety guidelines for heating safety and glassware safety.
3. Draw and label a cross section of a mid-ocean ridge system that lies above a convection current in the asthenosphere.

Procedure

1. Set up the equipment as shown in the illustration below. The bottom of the pan should be horizontal and about 6 cm above the top of the flame spreaders of the Bunsen burners.

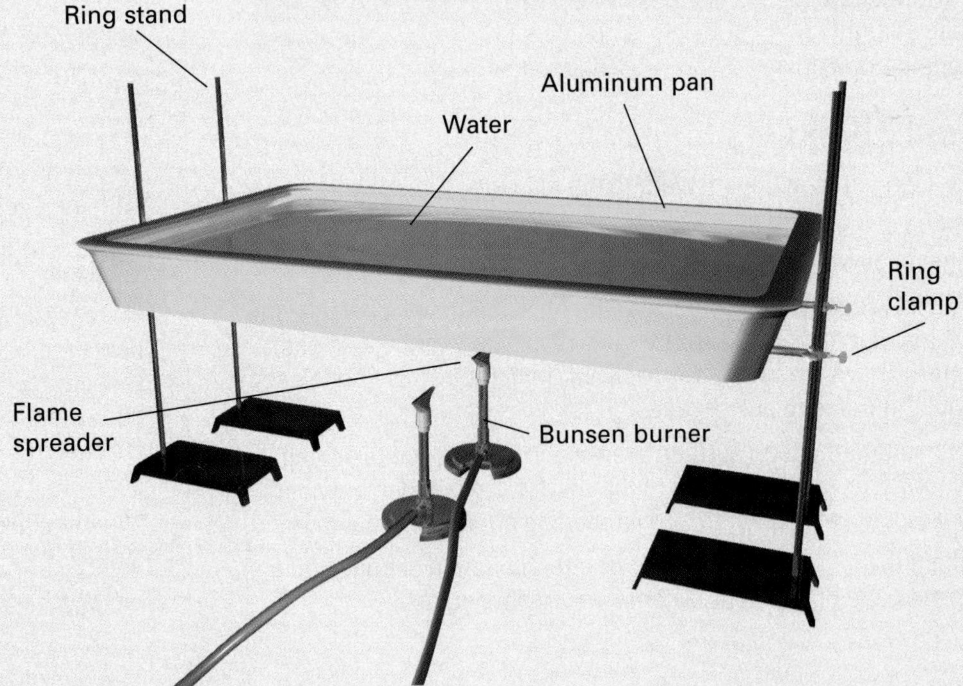

2. Use the beaker or another container to fill the pan with cold water to a depth of at least 4 cm.
3. Imagine a line down the length of the pan that would divide it in half. Place one drop of food coloring into the water about one-third of the way along the imaginary line, and place another drop two-thirds of the way. What is the motion of the food coloring?
4. Light and adjust the flame of each Bunsen burner. **CAUTION:** *The flames of the burners should be low.* Check to make sure the burners are still in position and centered along the bottom width of the pan. Refer back to the illustration on the previous page.
5. After a minute or two, tiny bubbles will appear in the water above the Bunsen burners. When this occurs, place another drop of food coloring into each of the positions chosen in Step 3. Describe any differences in the movement of the food coloring compared with earlier movements.
6. After another minute or so, gently place one of the craft sticks onto the water's surface about 3 cm to the left of the center of the pan. Now place the second stick to the right of the center of the pan. Make the strips parallel to each other and parallel to the ends of the pan. Use the pencil to line up the sticks.
7. As soon as you observe the sticks moving, place two more drops of food coloring in the same positions as in Step 3. Describe the motion of the food coloring when the sticks began to move. What is the relationship between the motion of the sticks and the motion of the water?
8. With the help of a partner, hold one thermometer bulb just under the water's surface at the center of the pan. Hold the other two thermometers in similar positions near the ends of the pan. Record the temperatures. How does the temperature of the water relate to the motion of the food coloring in the water?
9. Turn off the Bunsen burners. After the pan has completely cooled, carefully empty the water into a sink.

Analysis and Conclusions

1. In relation to plate tectonics, what do the Bunsen burners, water, and craft sticks represent?
2. Explain how the motion of the water affected the motion of the craft sticks.
3. Does the temperature of the water at different places relate to the flow pattern that you observed with the food coloring? Explain.

Extensions

1. Suggest some reasonable modifications of the investigation that would make it a more realistic illustration of the process of seafloor spreading.
2. Find a diagram that shows measurements of heat flow from ocean crust. Do the readings show a pattern of high heat flow at the mid-ocean ridges?

Chapter 5

Deformation of the Crust

If you were to look carefully among the rocks on the peaks of some mountains, you might be surprised to find fossils of animals that lived in the sea. A closer inspection might also reveal waves of folded, twisted, or fractured rock. The forces that deform the earth's crust and make mountains out of the flat ocean bottom are mainly the result of plate tectonics, the movement and collision of lithospheric plates. In this chapter, you will learn how plate movement builds and alters the topographic features of the earth.

Chapter Outline

◀ **The action of moving plates raised the Colorado Rockies.**

5.1 How the Crust Is Deformed

Section Objectives

- **Predict isostatic adjustments that will result from changes in the thickness of the earth's crust.**
- **Identify sources of stress in crustal rock.**

The Himalayas, the Rockies, the Andes—these are the names of some of the earth's most majestic mountain ranges. These imposing landforms are visible reminders that the shape of the earth's surface is always changing. Such changes result from **deformation,** the bending, tilting, and breaking of the earth's crust. Plate tectonics, the movement of the earth's lithospheric plates, is the major cause of crustal deformation. However, plate movement is not the only force that shapes the earth's crust.

Isostatic Adjustment

Some changes in the earth's crust occur because of changes in the weight of some part of the crust. The crust rides on top of the mantle. When parts of the crust become thicker and heavier, they will sink more deeply into the mantle. If the crust becomes thinner and lighter, it will rise higher on the mantle.

These movements can be compared to the behavior of a block of wood floating on water, as illustrated in Figure 5–1. When a metal weight is placed on the block, the weight forces the block to float more deeply in the water. What will happen to the block of wood when the weight is lifted?

The up-and-down movements of the crust occur because of two opposing forces. The crust presses down on the mantle. The mantle presses up on the crust. When these two forces are balanced, the crust moves neither up nor down. However, when weight is added to the crust, it sinks until a balance of the forces is again reached. The balancing of these two forces is called **isostasy.** The up-and-down movements of the crust to reach isostasy are called **isostatic adjustments.** As these isostatic adjustments occur, areas of the crust are bent up and down. Pressure created by this bending causes the rocks in that area of the crust to deform.

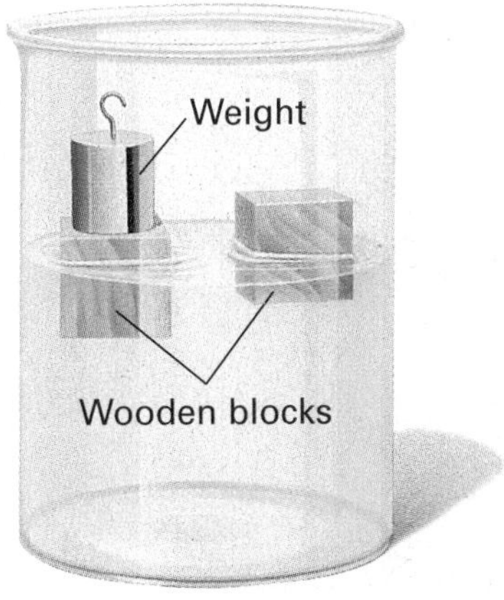

Figure 5–1. The thicker, heavier parts of the crust sink more deeply into the asthenosphere than the thinner, lighter parts (left). The illustration to the right shows isostatic adjustment. Adding weight to the wooden block on the left forces it to float more deeply in the water.

Isostatic adjustments are constantly occurring in areas of the crust with mountain ranges. Over millions of years, the wearing away of rocks can significantly reduce the overall height and weight of a mountain range, such as the Appalachians of the eastern United States. As the crust becomes lighter, the region may rise.

Another example of isostatic adjustment can be found in areas where rivers flow into large bodies of water carrying large amounts of mud, sand, and gravel. When a river flows into the ocean, this material is deposited on the nearby ocean floor. The added weight of the material causes the floor to sink, allowing more material to be deposited. A very thick accumulation of these deposits has formed in the Gulf of Mexico at the mouth of the Mississippi River.

Isostatic adjustments can also be found in areas where glaciers once covered the land. The weight of the ice caused the crust underneath it to sink. Much of the glacial ice has now retreated. The land that earlier was covered with ice is slowly rising again in response to its reduced weight. Thus, recently glaciated regions in Canada and northern Europe are now undergoing another isostatic adjustment.

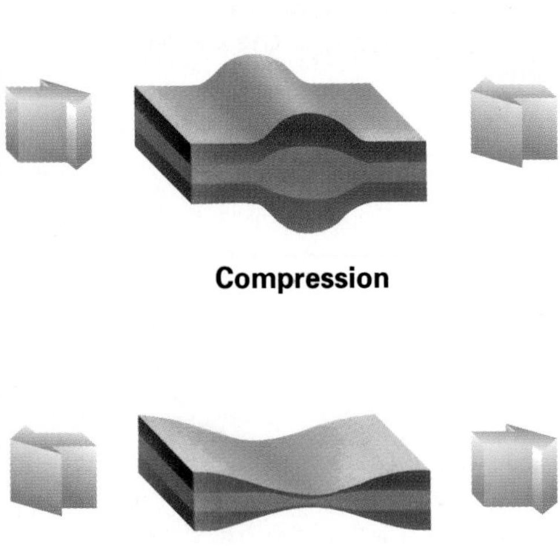

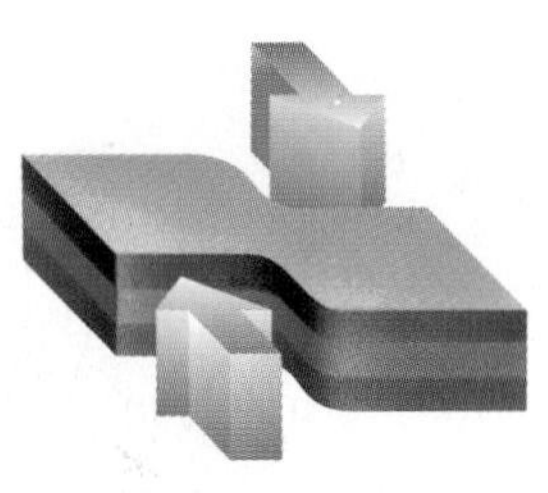

Figure 5–2. The forces that deform rock are compression, tension, and shearing.

Stress

Isostatic adjustment and plate movement cause **stress** in the rocks that make up the earth's crust. Stress is a force that causes pressure in the rocks of the crust. For example, when the weight of an ice sheet causes part of the crust to sink, stress increases on the rocks of the sinking crust. As the ice sheet disappears, a resulting rise in the crust also causes stress. Similarly, crustal stress occurs when lithospheric plates collide, separate, or rub together. This stress causes **strain** in crustal rocks. Strain is a change in the shape or volume of rocks that results from the stress of being squeezed, twisted, or pulled apart.

There are three main types of stress, as illustrated in Figure 5–2. **Compression** occurs when crustal rocks are squeezed together. Compression reduces the volume of rocks. It also tends to push the rocks higher up or deeper down into the crust.

Another type of stress is **tension.** Tension is the force that pulls rocks apart. When rocks are pulled apart by tension, they tend to become thinner.

A third type of stress is called **shearing.** Shearing pushes rocks in opposite horizontal directions. Sheared rocks bend, twist, or break apart as they slide past each other.

Section 5.1 Review

1. Explain the principle of isostatic adjustment.
2. Define *stress* and *strain*.
3. Name and describe the three main types of stress.
4. Contrast the isostatic adjustment that might result from the melting of glacial ice with the isostatic adjustment that a large river emptying into the ocean might cause.

5.2 The Results of Stress

Section Objectives

- **Compare folding and faulting as responses to stress.**
- **Describe four types of faults.**

The high pressures and temperatures caused by stress in the crust generally deform rocks. When stress is applied slowly, the deformed rock may return to its original shape as the force is removed. There is a limit, however, to the amount of force each type of rock can withstand and still retain its shape. If the force exceeds that limit, the shape of the rock changes permanently. In places of extreme stress, rock becomes so deformed that it may break.

Folding

When rock responds to stress by becoming permanently deformed without breaking, the result is **folding.** Folding is most easily observed where flat layers of rock are under severe compression and are squeezed inward from the sides. The layers move into new, folded positions without breaking. Cracks may appear, but the rock layers remain intact.

Folds, which appear as wavelike structures in rock layers, vary greatly in size. Some folds are small enough to be contained in a hand-held rock specimen. Others cover thousands of square kilometers and can be viewed only from the air.

The three general types of folds—**anticline** (ANT-ih-kline), **syncline** (SIN-kline), and **monocline** (MON-uh-kline)—are shown in Figure 5–3. Anticlines are upcurved folds in the layers, and synclines are downcurved folds. Monoclines are gently dipping bends in horizontal rock layers.

Generally, wherever large folds occur, an anticline will form a ridge, and a syncline will form a valley. Landforms of this type are commonly found among the ridges and valleys of the Appalachian Mountains.

Figure 5–3. Study the rock fold shown in the photo below on the left. Compare it with the three types of rock folds illustrated below right. Which type of rock fold is shown in the photo?

Anticline

Syncline

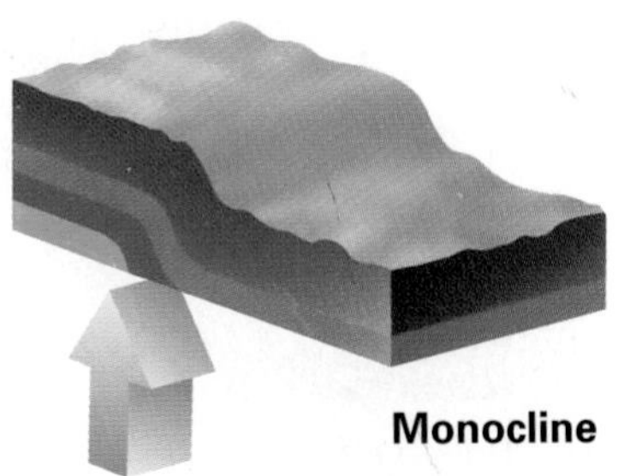

Monocline

Faulting

Rock does not always respond to stress by folding. Cooler temperatures and lower pressure near the earth's surface often cause rock to respond to stress by breaking. The difference in how rock near the earth's surface and rock deep inside the earth's crust respond to stress can be compared to the behavior of a heated glass rod. If the rod is heated until it is red hot, it can be bent easily. What would happen if you tried to bend the rod without heating it?

Breaks in rocks are divided into two categories. When there is no movement in the rocks along either side of a break, it is called a **fracture.** When the rocks do move, it is called a **fault.** A **normal fault** is one where the **fault plane**—the surface of a fault—is at a steep angle or almost vertical. The rocks above the normal fault plane, called the **hanging wall,** move down relative to the rocks below the fault plane, called the **footwall.** Normal faults occur along divergent boundaries, where the crust is being pulled apart due to tensional stress. Normal faults usually occur in a series of parallel fault lines, forming steep steplike landforms. The Great Rift Valley of East Africa is an area of large-scale normal faulting.

Another type of steep or nearly vertical fault is called a **reverse fault.** A reverse fault occurs when compression causes the hanging wall to move up relative to the footwall, as shown in Figure 5–4.

A **thrust fault** is a special type of reverse fault. The fault plane of a thrust fault is at a low angle or nearly horizontal. Because of

Figure 5–4. The illustration below shows four basic types of faults.

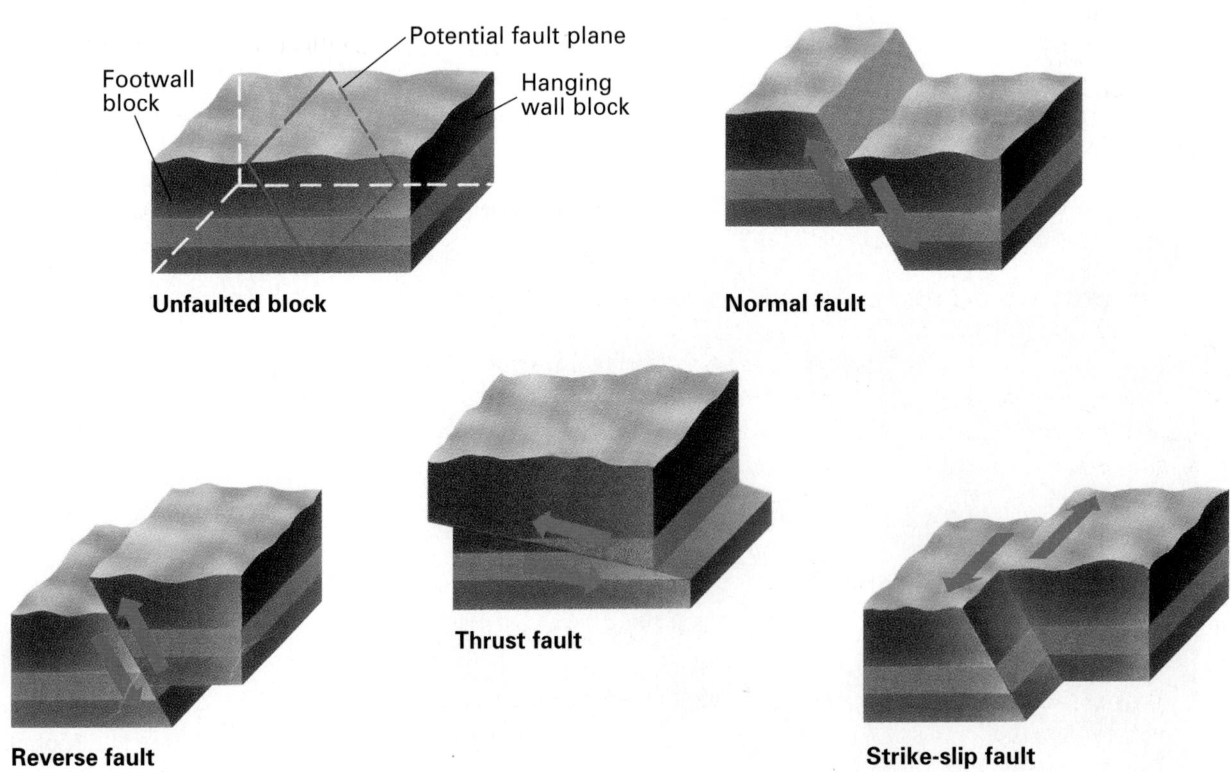

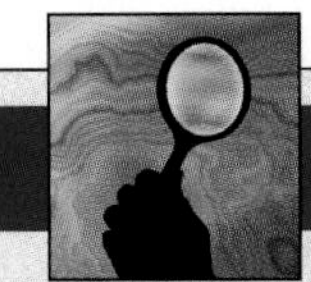

Small-Scale Investigation

Folds and Fractures

You can use some common objects to demonstrate the factors that govern the ways rock responds to stress from plate movements.

Materials

safety goggles; soft wood dowel, 2 mm × 15 cm; 2 books; plastic play putty

Procedure

1. **Put on the safety goggles.** Lay the dowel on a table. Place a book at each end of the dowel.
2. Place one hand on each book and gently and slowly slide the books against the ends of the dowel until the dowel bends slightly.
3. Move the books back to their original position. Observe and record what happens to the dowel as the books are moved.
4. Again move the books toward each other with the dowel between them. This time move the books quickly and forcefully. Record what happens to the dowel. Remove the safety goggles.
5. Roll a piece of play putty into a cylinder about 15 cm long and about the same diameter as the dowel.

6. Repeat Steps 1 through 3, using the putty in place of the dowel.
7. Re-form the putty into a cylinder. Grasp one end of the cylinder in each hand and pull quickly and sharply on both ends of the putty.

Analysis and Conclusions

1. Compare the responses of the dowel and the putty in Step 3. What two responses of rock to stress are represented?
2. Compare the response of the dowel in Step 4 with the response of the putty in Step 7.
3. What two factors influence the way the items respond to stress in this investigation? How do these factors influence the way rock responds to stress? Explain your answer.

the low angle of the fault plane, the rocks in the hanging wall are pushed up and over the rocks in the footwall. Reverse and thrust faults are common in steep mountains such as the Rockies and Alps.

Along a **strike-slip fault,** the rock on either side of the fault plane slides horizontally. Strike-slip faults often occur at transform fault boundaries. The San Andreas Fault, which runs through California, is the best known example of a strike-slip fault.

Section 5.2 Review

1. What results when rock responds to stress by permanently deforming without breaking?
2. Why is faulting more likely to occur near the surface than deep within the earth?
3. Describe four types of faults.

Section Objectives

- **Identify the types of plate collisions that build mountains.**
- **Identify four types of mountains and discuss the forces that shaped them.**

5.3 Mountain Formation

Mount Everest has the highest elevation of any mountain on earth, rising 8 km above sea level. Mount St. Helens became the most newsworthy mountain in the United States after it erupted in 1980.

Despite the fame of these mountains, neither stands alone. Each is part of a **mountain range,** a group of adjacent mountains with the same general shape and structure. Mount Everest is in the Great Himalaya Range, and Mount St. Helens is part of the Cascade Range.

Just as a group of individual mountains make up a range, a group of adjacent mountain ranges make up a **mountain system.** The Great Smoky, the Blue Ridge, the Cumberland, and the Green mountain ranges make up the Appalachian mountain system in the eastern United States.

The largest mountain systems are part of two still larger systems called **mountain belts.** The two major mountain belts on earth, the circum-Pacific belt and the Eurasian-Melanesian belt, are shown in Figure 5–5. The circum-Pacific belt forms a ring around the Pacific Ocean. The Eurasian-Melanesian belt runs from the Pacific islands through Asia and southern Europe and into northwestern Africa.

Plate Tectonics and Mountains

Both the circum-Pacific mountain belt and the Eurasian-Melanesian mountain belt are located along convergent plate boundaries. Scientists think that the location of these two mountain belts is evidence that most mountains were formed when lithospheric plates collided. Some mountains, such as the Appalachians, do not lie along active convergent plate boundaries. However, evidence indicates that these ranges formed where plates collided in the past.

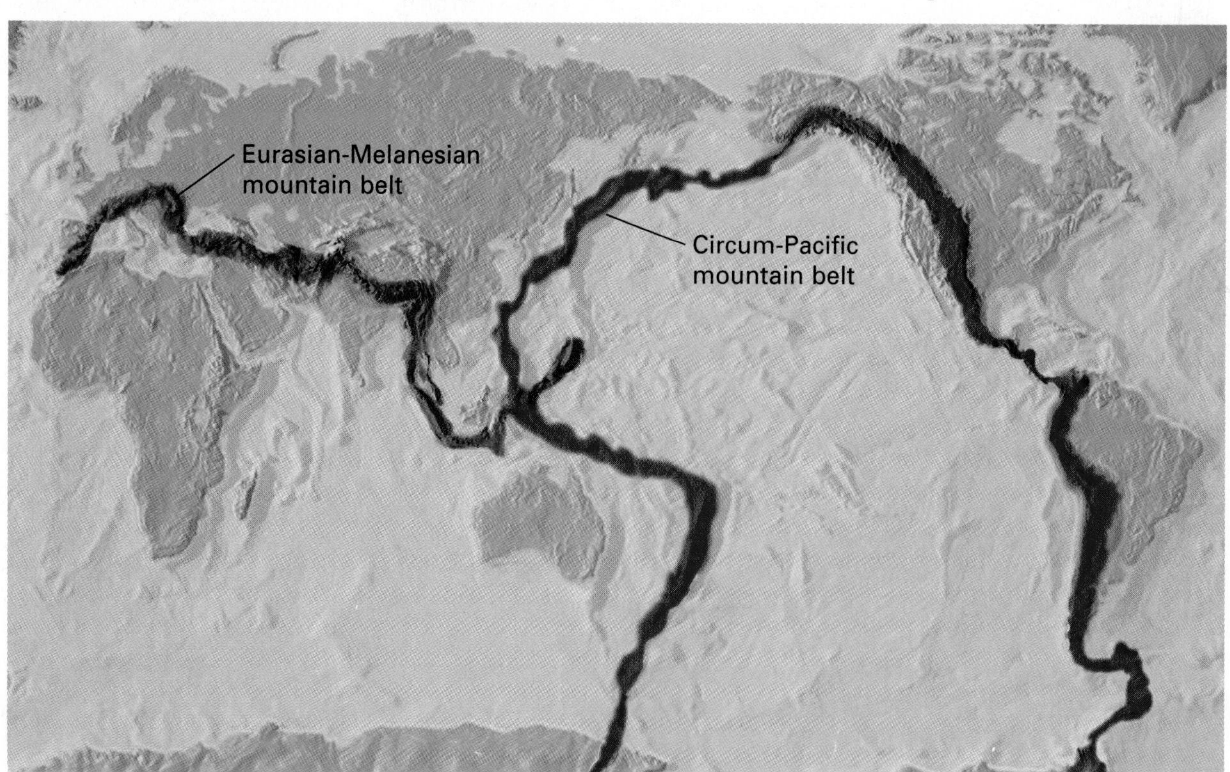

Figure 5–5. Most of the great mountain ranges of the world lie along two mountain belts.

Collisions Between Continental and Oceanic Crust

Some mountains form when oceanic crust and continental crust collide at convergent plate boundaries. When the moving plates collide, the oceanic crust is subducted beneath the continental crust. This type of collision produces such large-scale deformation of rock that high mountains are pushed up. In addition, the subducted oceanic crust partially melts, producing the magma that might eventually erupt and form volcanic mountains on the earth's surface. The mountains of the Cascade Range in the Pacific Northwest were formed in this way.

Some mountains at the boundary between oceanic crust and continental crust might have formed by a different process. As the oceanic crust was subducted, pieces of crust called *terranes* were scraped off. According to the suspect terrane theory, they became part of the continental crust. Some of these terranes formed mountains. To learn more about terranes, see Chapter 4.

> **INVESTIGATE!**
>
> *To learn more about continental collisions, try the In-Depth Investigation on pages 96–97.*

Collisions Between Oceanic Crust and Oceanic Crust

Volcanic mountains sometimes form where two plates with oceanic crust at their edges collide. The oceanic crust of one plate subducts beneath the oceanic crust of the other plate. As the oceanic crust plunges deeper into the mantle, the intense heat melts the crustal material, forming magma. The magma rises and breaks through the oceanic crust. These eruptions form an arc of volcanic mountains on the ocean floor. The Mariana Islands, in the North Pacific Ocean, are the peaks of volcanic mountains that rose above sea level.

Collisions Between Continents

Mountains can also form when two continents collide. The Himalaya Mountains were formed by just such a collision. According to the theory of plate tectonics, India was at one time a separate

Figure 5–6. The Himalayas, shown below, were formed when India collided with Eurasia.

continent riding on the Indian plate, which was moving north toward Eurasia. The oceanic crust of the Indian plate was subducted beneath the Eurasian plate. It continued to move until the continental crust of India collided with the continental crust of Eurasia. This collision produced intense deformation of both continents. As a result, the Himalayas formed and India was connected to the Eurasian continent. Severe earthquakes still occur in the Himalayas. What can you infer about the movement of the Indian plate?

Types of Mountains

Mountains are more than just elevated parts of the earth's crust. They are complicated structures with rock formations that yield evidence of the forces that created them. Scientists classify mountains according to the way in which the crust was deformed and shaped by mountain-building forces.

Folded Mountains and Plateaus

The highest mountain ranges in the world are made up of **folded mountains** and are commonly found where continents have collided.

SCIENCE & TECHNOLOGY

Tracking Plate Movements

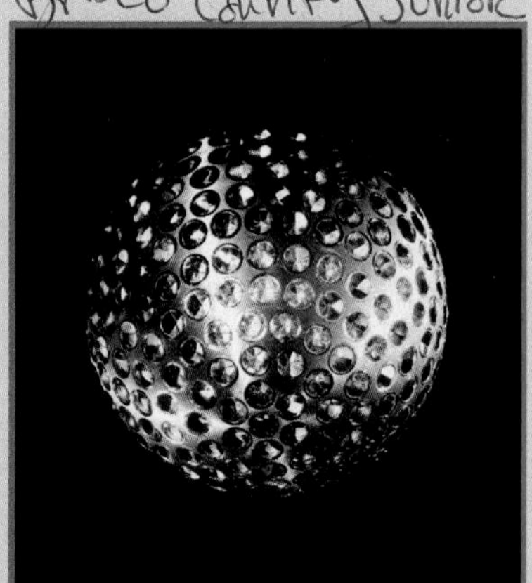

Each *Lageos* sphere has a diameter of 60 cm and a mass of 405 kg. The spheres' high density will allow them to maintain stable orbits for centuries.

In what appears to be a scene right out of a science fiction movie, scientists are firing laser beams at satellites orbiting 6,000 km above the earth. These scientists, however, do not intend to disable the satellites. Instead, they are gathering data on lithospheric plate motion. The scientists will use this information to understand better how plate tectonics shapes the earth's surface.

The target satellites, called *Lageos* (*La*ser *Geo*dynamic *S*atellites), resemble huge golf balls. Each *Lageos* sphere is made of aluminum with a brass core and is covered with 426 light retroreflectors called *corner cubes.* A corner cube is a specially designed mirror that reflects light back along exactly the same path by which it came.

When scientists at a laser station, such as the one shown below, fire a pulse of light at a *Lageos* sphere, the corner cubes reflect the light directly back to the station. The scientists record the time it takes for the light to reach the satellite and return. Then they determine the exact

International cooperation has played a crucial role in maintaining more than 35 laser stations worldwide.

In folded mountains, tectonic movements have squeezed rock layers together like an accordion. Parts of the Alps, the Himalayas, the Appalachians, and Russia's Ural Mountains consist of very large and complex folds. These long mountain chains also show evidence of faulting.

The same forces that form folded mountains also uplift **plateaus.** Plateaus are large areas of flat-topped rocks high above sea level. Most plateaus are formed when thick, horizontal layers of rock are slowly uplifted. The area is pushed up gently enough so that the layers remain flat rather than faulting and folding as do mountains. Most plateaus are found next to mountain ranges. For example, the Tibetan Plateau is next to the Himalayan Mountains, and the Colorado Plateau is next to the Rockies. A number of plateaus also were formed when layers of molten rock hardened and piled up on the earth's surface.

Figure 5–7. The Canadian Rockies are a dramatic example of folded mountains.

Fault-Block Mountains and Grabens

Some mountains have been formed by faults where parts of the earth's crust have been broken into large blocks. These blocks were then lifted above the surrounding crust. Faulting tilted the blocks,

location of the sphere at the time they fired their laser. The scientists use this information to calculate the speed at which the laser station is drifting, as well as its direction. Because the station rests on a lithospheric plate, the scientists are essentially calculating the speed and direction of movement of the plate.

Scientists are interested in the *Lageos* data for several reasons. Knowing which way the earth's plates are moving enables scientists to model the future geography of the earth. The data could also help scientists predict geological activity caused by plate interactions. Scientists may find ways of using the *Lageos* data to study minute changes in the earth's gravity field, in the rotation of the earth, and in isostatic adjustments.

Message in a Bottle

In 1992, astronauts aboard the space shuttle *Columbia* deployed a *Lageos* sphere that contains a message intended for explorers from other planets. Inside *Lageos II* are two stainless steel plates etched with identical messages. Each plate contains a picture of the *Lageos* sphere, a binary code representing the numbers one through ten, and sketches of the earth. The sketches depict the earth's continents as they might have looked 268 million years ago, as they look now, and as they might look 8.5 million years in the future.

What are some specific predictions scientists might make when they know the direction and speed of plate movements?

▲ Each laser station is equipped with a receiving telescope capable of detecting individual photons

creating **fault-block mountains,** as shown in Figure 5–8. The Sierra Nevada range of California is an example of fault-block mountains. The desert regions of Nevada, Arizona, western Utah, southern Oregon, northern New Mexico, and southeastern California are broken up by fault-block mountains. These mountains form nearly parallel ranges that average 80 km in length.

The same type of faulting that forms fault-block mountains also forms long, narrow valleys called **grabens.** Grabens develop when steep faults break the crust into blocks and a block slips downward relative to the surrounding blocks. Death Valley in California is a graben.

Volcanic Mountains

Mountains that form when molten rock erupts onto the earth's surface are called **volcanic mountains.** They may develop on land or on the ocean floor. The Cascade Range of Washington, Oregon, and northern California is composed of volcanic mountains. Some of the largest volcanic mountains are found along divergent plate boundaries, which form the mid-ocean ridges. These mid-ocean ridges are actually huge volcanic mountain chains that run through the centers of the Atlantic, eastern Pacific, and Indian oceans. The peaks of the highest mountains sometimes rise above sea level to form volcanic islands, such as Iceland and the Azores.

Other large volcanic mountains are formed on the ocean floor over *hot spots*. Hot spots are pockets of magma beneath the earth's crust that erupt onto the surface. The Hawaiian Islands are the tips of high volcanic mountains that were formed over a hot spot on the seafloor. The main island of Hawaii is actually a mountain that reaches almost 9 km above the ocean floor, with a base that is more than 160 km wide. Almost 4 km of this mountain is above sea level.

Dome Mountains

An unusual type of mountain is formed when molten rock rises through the crust and pushes up the rock layers above it. The result is a circular dome on the earth's surface. The molten rock that

Figure 5–8. In addition to folded mountains, there are fault-block mountains such as the Sierra Nevadas (top), volcanic mountains such as Mount Hood and Mount Adams (below, left), and dome mountains such as the Solitario Uplift in southwest Texas (below, right).

EarthBeat

The Disappearing Mediterranean

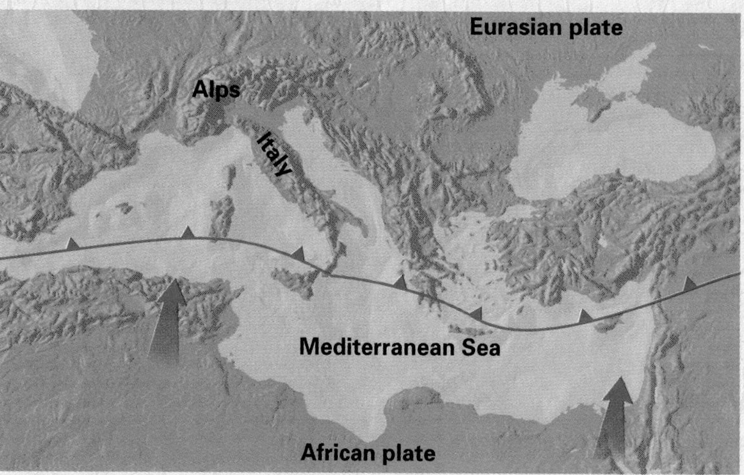

Two interesting regions in the world are the Alps and the Mediterranean Sea. The Alps, considered to be among the earth's most beautiful mountains, have become a vast natural playground for skiers, hikers, and climbers. The Mediterranean plays host to travelers from around the world who wish to sample its diverse cultures, balmy climate, and famous beach resorts. More important, the people who live nearby depend on the sea for their economic well-being.

The same natural forces that produced the Alps are slowly swallowing up the Mediterranean. The Alps were formed, and are still being shaped, by the collision of two lithospheric plates. Italy, part of which rides on the African plate, collided sometime in the past with Eurasia. The collision formed the Alps, but it did not stop the movement of the African plate. The northern oceanic crust of the African plate, which is actually the sea floor of the Mediterranean, is still subducting beneath the continental crust of Eurasia. As more oceanic crust subducts, the Mediterranean Sea will become smaller. Italy, which continues to be pushed into Eurasia, will eventually cease to exist as we know it. When the northern coast of the African continent finally collides with Eurasia, the Mediterranean Sea will become just a memory, a photograph, perhaps, in some history book.

What do you think will happen to the Alps as the African plate continues to push northward?

pushed up the rock layers eventually cools and forms hardened rock. When the pushed-up rock layers are worn away over time, the hardened rock is exposed. This rock wears away in places, leaving separate high peaks, or **dome mountains.** The Black Hills of South Dakota, and the Adirondack Mountains of New York State are dome mountains.

Section 5.3 Review

1. Describe the types of lithospheric plate collisions that build mountains.
2. Name the four types of mountains and explain how each is formed.
3. Explain how plateaus are formed.
4. How do volcanic mountains grow?

Chapter 5 Review

Key Terms

- anticline (85)
- compression (84)
- deformation (83)
- dome mountain (93)
- fault (86)
- fault plane (86)
- fault-block mountain (92)
- folded mountain (90)
- folding (85)
- footwall (86)
- fracture (86)
- graben (92)
- hanging wall (86)
- isostasy (83)
- isostatic adjustment (83)
- monocline (85)
- mountain belt (88)
- mountain range (88)
- mountain system (88)
- normal fault (86)
- plateau (91)
- reverse fault (86)
- shearing (84)
- strain (84)
- stress (84)
- strike-slip fault (87)
- syncline (85)
- tension (84)
- thrust fault (86)
- volcanic mountain (92)

Key Concepts

The movements of lithospheric plates and isostatic adjustments are sources of stress that cause deformation in crustal rock. **See page 83.**

Stress can squeeze rocks together, pull them apart, and bend and twist them. **See page 84.**

Rock responds to stress by bending into folds and by fracturing or faulting. **See page 85.**

Four major types of faults occur in rock: normal faults, reverse faults, thrust faults, and strike-slip faults. **See page 86.**

Mountain building is often the result of the collision of lithospheric plates. **See page 88.**

Mountains are classified according to the way in which the crust was deformed and shaped by mountain-building forces. **See page 90.**

Review

On your own paper, write the letter of the term that best completes each of the following statements.

1. The state of balance between the thickness of the crust and the depth at which it rides on the asthenosphere is called
a. stress. b. isostasy.
c. strain. d. shearing.

2. The increasing weight of mountains causes the crust to
a. sink. b. fold.
c. rise. d. fracture.

3. The force that changes the shape and volume of rocks is
a. footwall. b. isostasy.
c. rising. d. stress.

4. The type of stress that squeezes rock together is
a. compression. b. tension.
c. shearing. d. faulting.

5. The type of stress that pulls rocks apart, making them thinner, is
a. folding. b. compression.
c. tension. d. isostasy.

6. Shearing
a. bends, twists, or breaks rocks.
b. squeezes rock together.
c. causes rock to melt.
d. pulls rock apart.

7. High pressure and high temperature will cause rocks to
a. fracture. b. adjust. c. plateau. d. deform.

8. Upcurved folds in rock are called
   a. anticlines. b. monoclines.
   c. geosynclines. d. synclines.
9. Downcurved folds in rock are called
   a. geosynclines. b. monoclines.
   c. anticlines. d. synclines.
10. Gently dipping bends in rock formations with horizontal layers are called
   a. monoclines. b. geosynclines.
   c. synclines. d. anticlines.
11. When no movement occurs along the sides of a break in a rock structure, the break is called a
   a. normal fault. b. fracture.
   c. fold. d. hanging wall.
12. The rock above the fault plane makes up the
   a. tension. b. footwall.
   c. hanging wall. d. compression.
13. A nearly vertical fault in which the rock on either side of the fault plane moves horizontally is called a
   a. normal fault. b. reverse fault.
   c. strike-slip fault. d. thrust fault.
14. The largest mountain systems are part of still larger systems called
   a. continental margins. b. ranges.
   c. belts. d. synclines.
15. Mount St. Helens in Washington State is an example of a
   a. folded mountain. b. volcanic mountain.
   c. fault-block mountain. d. dome mountain.

Critical Thinking

On your own paper, write answers to the following questions.

1. Suppose glaciers, which are vast fields of slow-moving ice, were to cover much of the earth's surface once again. What would you expect to happen to those parts of the continents that were covered by ice? Explain.
2. When the Indian plate collided with the Eurasian plate, producing the Himalaya Mountains, which type of stress most likely occurred? Which type of stress is most likely occurring along the Mid-Atlantic Ridge? Which type of stress would you expect to find along the San Andreas fault? Use your knowledge of stress and plate tectonics to explain your answers.
3. If the force that is causing a rock to be slightly deformed begins to ease, what might happen to the rock? What would happen if the force causing the deformation became greater?
4. Why do you suppose dome mountains do not become volcanic mountains?

Application

1. Suppose that a new highway is being planned. This proposed road would intersect a transform fault boundary. What would happen to the highway if a strike-slip fault occurred along the boundary? Why?
2. A geologist discovers that part of a mountain range along the west coast of the United States contains the fossil remains of animals that do not match any other fossils from North America. What is the most likely explanation for this phenomenon?
3. Construct a **concept map** that illustrates the relationship between crustal deformation and types of mountains.

Extension

1. Using modeling clay to represent crustal rock, create examples of rock that have been subjected to the forces of compression, tension, and shearing.
2. Look for examples of folding and faulting in the area where you live. Take photographs of the examples you find. Use your photos, or drawings if you prefer, to create a poster explaining folding and faulting.
3. Do some research about different types of mountains in the United States. On an outline map of the United States, locate the folded, dome, volcanic, and fault-block mountains in this country.

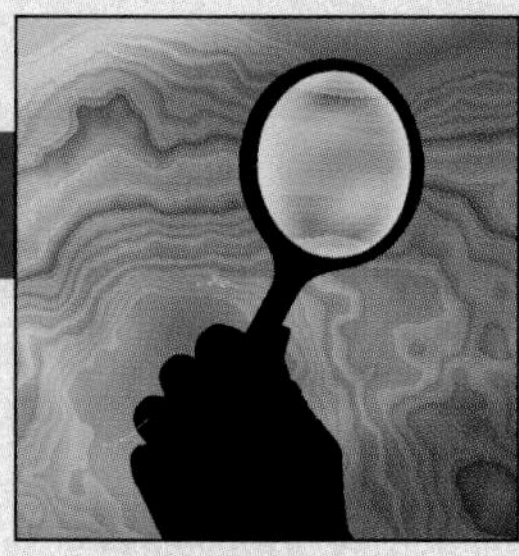

In-Depth Investigation

Continental Collisions

Materials

- thick cardboard (at least 15 × 15 cm)
- wood blocks (2.5 × 2.5 × 6 cm)
- adding-machine paper (6 × 35 cm)
- 5 long bobby pins
- paper napkins (light and dark)
- masking tape
- metric ruler
- scissors

Introduction

When the subcontinent of India broke away from Africa and began to move northward toward Eurasia, the oceanic crust on the northern side of India began to subduct beneath the Eurasian plate. The deformation of the crust resulted in mountain formation, and the Himalayas grew higher and higher. Earthquakes in the Himalayan region suggest that India is still pushing against Eurasia.

In this investigation, you will create a model to help explain how the Himalaya Mountains formed as a result of the collision of these two continental plates.

Prelab Preparation

1. Review Section 4.2, pages 72–77, and Section 5.3, pages 88–93.
2. Make a cross-sectional diagram to show what happens when a plate carrying oceanic and continental crust collides with a plate carrying continental crust at its edge.

Procedure

1. To assemble the continental-collision model, cut a 7-cm slit in the cardboard about 6 cm from the end of the cardboard and parallel to it. Cut the slit open just wide enough so that the adding-machine paper will feed through it without being loose.
2. Securely tape one wood block lengthwise along the slit between the slit and the near end of the cardboard. Tape the other block to the paper strip lengthwise about 6 cm from one end of the paper. Both blocks should be parallel to one another as shown in the illustration on the next page.

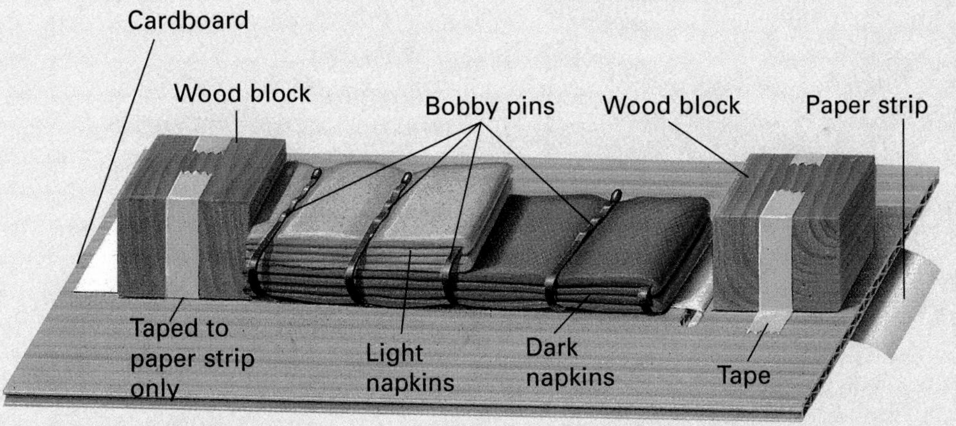

3. Cut two strips of the light-colored paper napkin about 6 cm wide and 8 cm long. Cut two strips of the dark-colored paper napkin about 6 cm wide and 16 cm long. Fold all four strips in half along their width.
4. Stack the napkin strips on top of each other with all the folds along the same side. Place the two dark-colored napkins on the bottom.
5. Place the napkin strips lengthwise on the paper strip with the ends butted up against the wood block.
6. Attach the napkins to the paper strip using the bobby pins as shown in the illustration above.
7. Push the long end of the paper strip through the slit in the cardboard until the first fold of the napkin rests against the other wood block. Hold the cardboard at about eye level and pull gently downward on the paper strip. You may need a partner's help. What happens as the dark-colored paper napkin comes in contact with the fixed block of wood? What happens as you continue to pull downward on the strip of paper? Stop pulling when you feel resistance from the paper strip.

Analysis and Conclusions

1. Explain what is represented by the dark napkins, the light napkins, and the wood blocks.
2. What plate-tectonics process is represented by the motion of the paper strip in the model? Explain.
3. What type of mountains would result from the kind of collision shown by the model?
4. Explain the differences between the model and the real Himalaya Mountains.

Extensions

1. Obtain a world map of earthquake epicenters. Study the map. Describe the pattern of epicenters in the Himalayan region. Does the pattern suggest that the Himalayas are still growing?
2. Read about the breakup of Gondwanaland and the movement of India toward the northern hemisphere. Write about stages in India's movement. List the time frame at which each important event occurred.

Chapter 6 Earthquakes

Stress causes rocks along a fault to fracture and shift. The result is an earthquake. Some earthquakes are mild, but others are exceedingly violent and destructive. In 1556, for example, in Shaanxi Province in north-central China, more than 800,000 people perished during a single earthquake.

In this chapter, you will learn about the causes and effects of earthquakes, the way they are measured, and scientists' attempts to predict earthquakes.

Chapter Outline

An earthquake can cause the ground to split apart.

6.1 Earthquakes and Plate Tectonics

Section Objectives

- **Discuss the elastic rebound theory.**
- **Explain why earthquakes generally occur at plate boundaries.**

Vibrations of the earth's crust are called **earthquakes.** Earthquakes usually occur when rocks under stress suddenly shift along a fault. As you read in Chapter 5, stress is a force that can change the size and shape of rocks.

Normally the rocks along both sides of a fault are pressed together tightly. For a time, friction prevents the rocks from moving past each other. In this immobile state, a fault is said to be "locked." Parts of a fault remain locked until the stress becomes so great that the rocks suddenly grind past each other. This slippage causes the trembling and vibrations of an earthquake.

Elastic Rebound Theory

Geologists explain many earthquakes by the **elastic rebound theory.** According to this theory, the rocks on each side of a fault are moving slowly. If the fault is locked, stress in the rocks increases. When they are stressed past a certain point, however, the rocks fracture, separate at their weakest point, and spring back to their original shape, or rebound.

As they fracture and slip into new positions, rocks along a fault release energy in the form of vibrations called *seismic waves*. This release of energy often increases the stress in other rocks along the fault, causing them to fracture and spring back. This reaction is the reason that major earthquakes are usually followed by a series of smaller tremors called **aftershocks.**

As you can see in Figure 6–1, the area along a fault where slippage first occurs is called the **focus** of an earthquake. The point on the earth's surface directly above the focus is called the **epicenter** (EP-ih-SENT-ur). When an earthquake occurs, seismic waves radiate

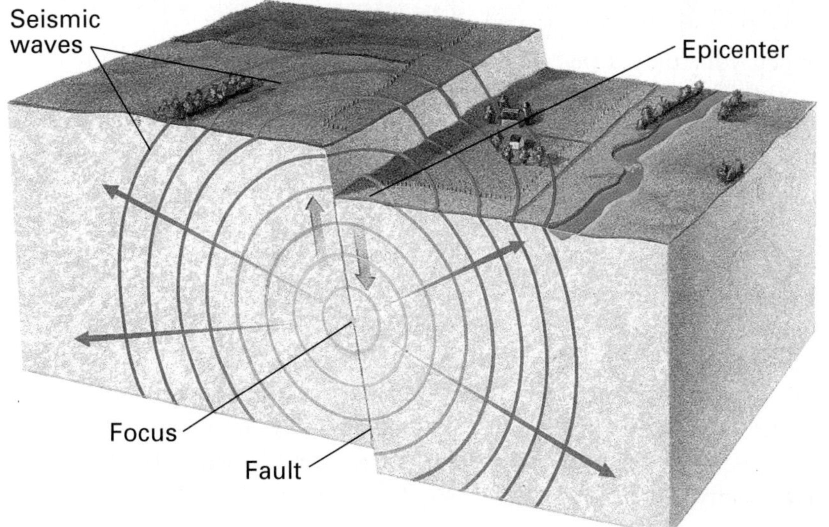

Figure 6–1. The epicenter of an earthquake is the point on the surface directly above the focus.

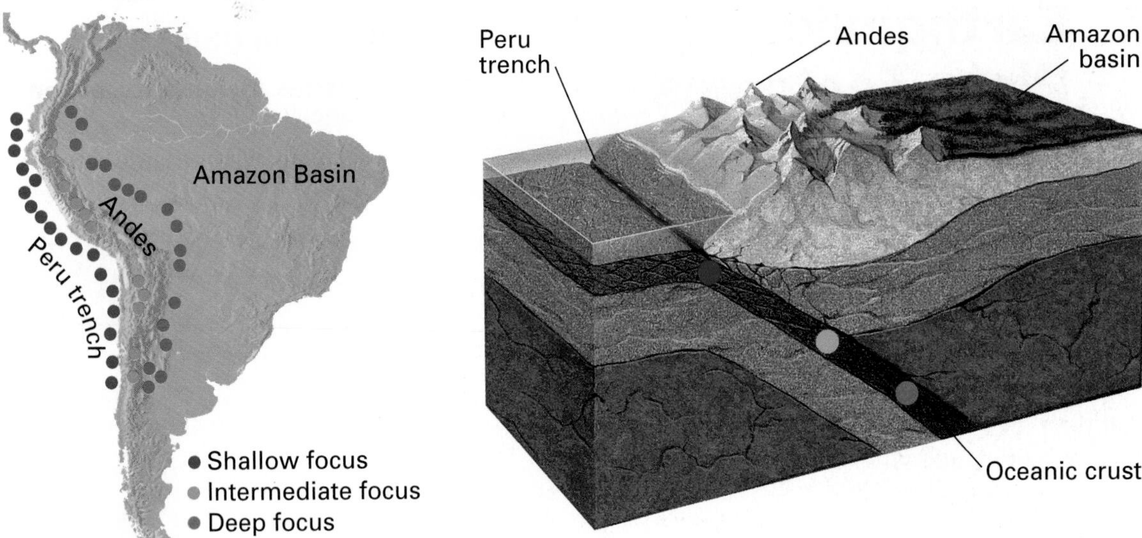

Figure 6–2. Earthquakes at various depths have occurred along the west coast of South America in nearly parallel lines (left). The illustration (right) shows a subduction zone along the west coast of South America. The pattern of the earthquakes is characteristic of an area where one plate has been subducted.

outward in all directions from the focus. The action is similar to what occurs when you drop a stone into a pool of still water and circular waves ripple outward from the center.

Although the focus depths of earthquakes vary, about 90 percent of continental earthquakes have a shallow focus. Shallow-focus earthquakes occur within 70 km of the earth's surface. Intermediate-focus earthquakes occur at a depth of between 70 km and 300 km. Deep-focus earthquakes occur in subduction zones at a depth of between 300 km and 650 km. As you can see in Figure 6–2, deep-focus earthquakes usually occur farther inland from the subduction zone than do shallow-focus or intermediate-focus earthquakes. Why do earthquakes not occur deeper than 650 km within the earth?

By the time the vibrations from an intermediate-focus or deep-focus earthquake reach the surface, much of their energy has been used up. For this reason, the earthquakes that usually cause the most damage have a shallow focus.

Major Earthquake Zones

When you study a map that shows the locations of earthquakes worldwide, you can see the link between earthquakes and plate tectonics. Most earthquakes occur along or near the edges of the earth's lithospheric plates. It is at these moving plate boundaries that stress is greatest and rocks experience the greatest amount of strain.

The earth has three major earthquake zones, as shown in Figure 6–3. The first is a large area known as the **Pacific Ring of Fire.** The Pacific Ring of Fire includes the west coasts of North and South America, the east coast of Asia, and the western Pacific islands of the Philippines, Indonesia, New Guinea, and New Zealand. Along this ring, most plates are being subducted, while some plates scrape past each other. Many earthquakes occur along the Pacific Ring of Fire because the plate movements cause stress to build up in rocks. Eventually the rocks fracture and shift, and an earthquake occurs.

The second major earthquake zone is along the mid-ocean ridges. Earthquakes occur along mid-ocean ridges because oceanic crust is pulling away from both sides of each ridge. This spreading motion creates stress in the rocks along the major ocean ridges.

The third major earthquake zone is the Eurasian-Melanesian mountain belt. You read in Chapter 5 that the mountains along this belt were formed by the collision of the Eurasian plate with the African and Indian plates. These plates are still colliding, and the same forces that are pushing up the mountains also produce numerous earthquakes.

At some plate boundaries there are groups of interconnected faults called **fault zones.** Fault zones form at plate boundaries because of the intense stress that results when the plates separate, collide, subduct, or slide past each other. One such fault zone includes the San Andreas Fault, which extends almost the length of California. The San Andreas Fault zone has formed where the edge of the Pacific plate slips northwest along the North American plate. Movement could occur along one or more of the individual faults in the San Andreas Fault zone. Any such movement could cause a major earthquake within the zone.

Not all earthquakes, however, result from movement along plate boundaries. For example, the most widely felt series of earthquakes in the history of the United States did not occur near any active plate boundary. Rather, these earthquakes occurred in the middle of the continent, near New Madrid, Missouri, in 1812. The vibrations from the earthquakes that rocked New Madid were so strong that they rang church bells as far away as Boston, Massachusetts. After the discovery of plate tectonics, scientists were puzzled by earthquake activity in the center of the United States. After all, they reasoned, New Madrid is far away from any active plate boundaries.

Figure 6–3. The map below illustrates the earth's three major earthquake zones: the Pacific Ring of Fire, the mid-ocean ridges, and the Eurasian-Melanesian belt.

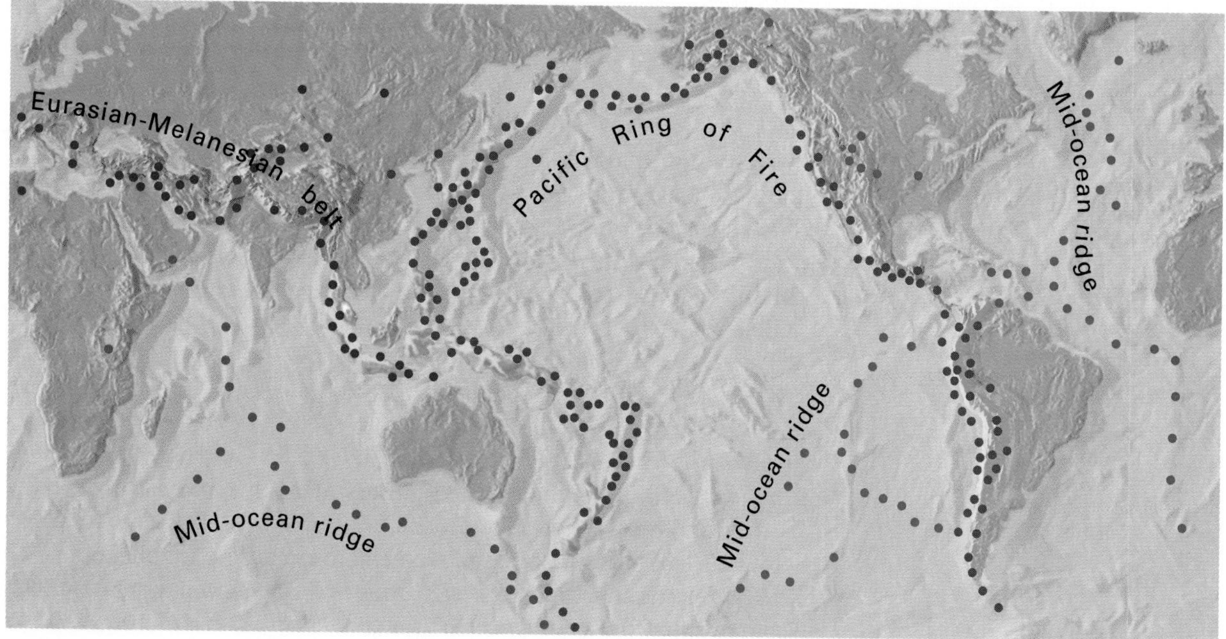

EarthBeat

The Big Squeeze

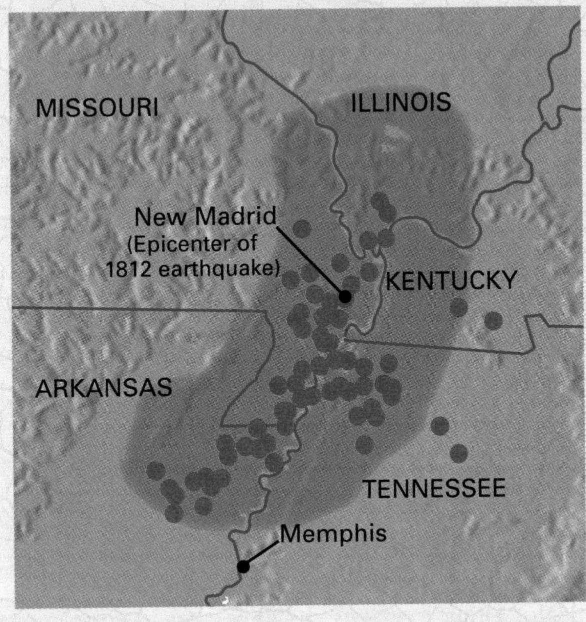

Scientists think that about 600 million years ago the North American continent began to break apart into two or three pieces. For reasons yet unknown, this tectonic activity stopped and the continent remained intact.

One clue to this past activity is a fault zone buried deep underground in the central United States. This fault zone, discovered in the late 1970's, consists of three interlocking faults that stretch about 240 km from northern Arkansas to southern Illinois. Scientists think that within this fault zone was the epicenter of the 1812 earthquake that shook New Madrid, Missouri.

When these faults were first discovered, one question puzzled scientists. How, they wondered, could a fault zone far from any plate boundary accumulate enough stress to become active—to fracture and rebound, causing an earthquake. The answer seems to lie in plate tectonics.

Scientists have found that the North American plate is being compressed—squeezed together—by the force of the spreading Atlantic ocean floor. This stress may eventually be powerful enough to activate the centrally located fault zone.

China has also had earthquakes that do not originate at plate boundaries, and a similar hypothesis has been suggested to explain those quakes. China is being squeezed together on three sides, by the North American plate, the Indian plate, and the Pacific plate.

If stress continues to increase along the fault zone in the central United States, what might result?

Not until the late 1970's did studies of the Mississippi River region reveal an ancient fault zone deep within the earth's crust. This zone appears to be part of a major fault in the North American plate. Scientists have calculated that the fault formed at least 600 million years ago and that it was later buried under many layers of sediment and rock.

Section 6.1 Review

1. Explain the elastic rebound theory.
2. In earthquakes that cause the greatest damage, at what depth would slippage most likely occur?
3. What four types of plate movements can cause earthquakes?
4. If an earthquake occurs in the center of Brazil, what can you infer about the geology of that area?

6.2 Recording Earthquakes

Seismic waves can be detected and recorded by using an instrument called a **seismograph** (SIZE-muh-GRAF). A seismograph consists of three separate sensing devices. One device records the vertical motion of the ground. The other two record horizontal motion in the east-west and north-south directions. A seismograph records motion by tracing wave-shaped lines on paper or by translating the motion into electronic signals. The electronic signals can be recorded on magnetic tape or directly loaded into a computer that analyzes the seismic waves.

Section Objectives

- **Compare the three types of seismic waves.**
- **Discuss the method scientists use to pinpoint an earthquake.**
- **Discuss the method most commonly used to measure the magnitude of earthquakes.**

Types of Seismic Waves

Scientists have determined that every earthquake produces three major types of seismic waves. Each type of wave travels at a different speed and causes different movements in the earth's crust.

Primary waves, or **P waves,** move the fastest and are therefore the first to be recorded by a seismograph. P waves moving through the earth can travel through solids and liquids. The more rigid the material, the faster the P waves travel through it. P waves cause rock particles to move together and apart along the direction of the waves.

Secondary waves, or **S waves,** are the second waves to be recorded on a seismograph. Unlike P waves, S waves can travel only through solid material. You may recall from Unit 1 that S waves cannot be detected on the side of the earth that is opposite the earthquake's epicenter. The S waves cannot be detected because they do not penetrate the liquid part of the earth's outer core. S waves cause rock particles to move at right angles to the direction in which the waves are traveling.

When P waves and S waves reach the earth's surface, their energy is converted into a third type of seismic wave. *Surface waves,* also called *long waves* or **L waves,** are the slowest-moving waves

INVESTIGATE!

To learn more about earthquake waves, try the In-Depth Investigation on pages 114–115.

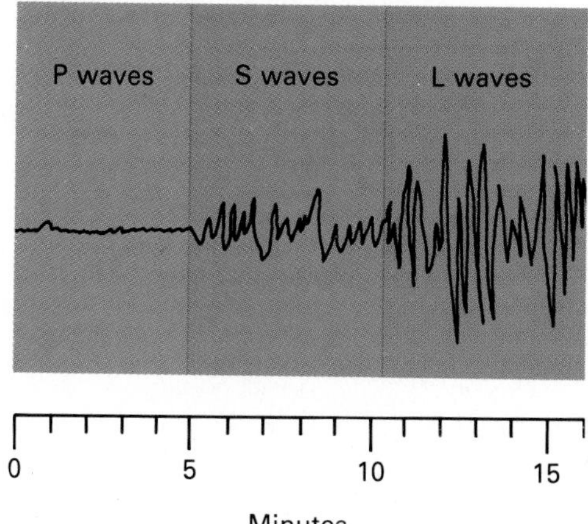

Figure 6–4. Most seismograph stations, like the one shown (left), have banks of seismographs to record earth tremors. Each type of seismic wave (right) leaves a unique "signature" on a seismograph.

and therefore the last to be recorded on a seismograph. L waves travel slowly over the earth's surface in a movement similar to that of ocean waves. L waves, which cause the surface to rise and fall, are particularly destructive when traveling through loose earth. Why do L waves cause the greatest damage during an earthquake?

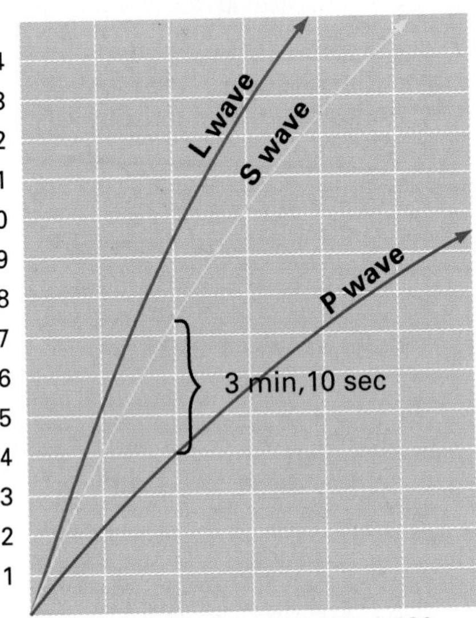

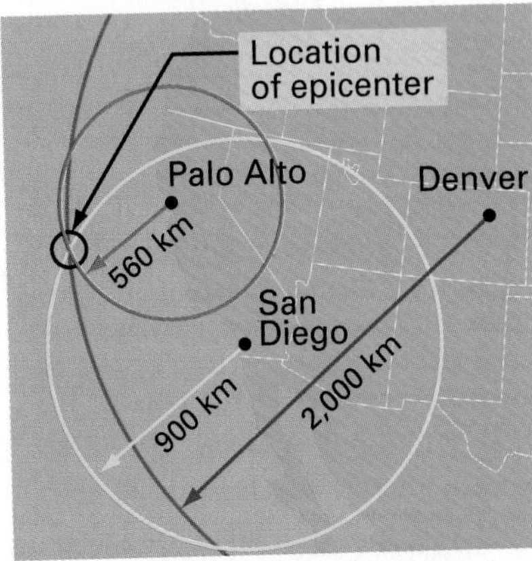

Figure 6–5. The time-distance graph (top) is a standard graph used to find the distances of an earthquake from seismograph stations. The distances can be plotted on a map (bottom) to find the location of the earthquake.

Locating an Earthquake

To find the epicenter of an earthquake, scientists analyze the difference between the arrival times of the P waves and the S waves. P waves travel about 1.7 times faster than S waves. Thus, if the S waves arrive soon after the P waves, the earthquake must have originated fairly close to the seismograph station. If the S waves arrive a long time after the P waves, the earthquake must have occurred farther away. To determine how far an earthquake is from a given seismograph station, scientists plot the difference between the arrival times of the two waves. Then they consult a standard graph that translates the difference in arrival times into distance from the epicenter.

For scientists to locate the epicenter of the earthquake, they need information from at least three seismograph stations at different locations. Using the example shown in Figure 6–5, suppose that the epicenter of an earthquake is 560 km away from one station, 2,000 km away from another, and 900 km away from a third. Three circles are drawn on a map that shows all three stations. Each station is at the center of a circle. The distance from a particular station to the epicenter is the radius of that circle. The epicenter of the earthquake is located very near the point at which the three circles intersect.

Earthquake Measurement

Scientists use the **Richter scale** to express the magnitude of an earthquake. Magnitude is a measure of the energy released by an earthquake. Each increase of one whole number of magnitude represents the release of 31.7 times more energy than that of an earthquake measuring one whole number lower. Thus, an earthquake with a magnitude of 8 releases 31.7 × 31.7—or about 1,000 times—as much energy as an earthquake with a magnitude of 6.

The largest earthquake so far recorded registered a magnitude of 9.6. A major earthquake, one that causes widespread damage, has a magnitude of 7 or above. A moderate earthquake has a magnitude between 6 and 7, and a minor earthquake, between 2.5 and 6. Earthquakes with magnitudes of less than 2.5 are called **microquakes** and usually are not felt by people.

The **Mercalli scale** expresses the **intensity** of an earthquake, or the amount of damage it causes, by Roman numerals from I to XII and a description. The Mercalli scale describes an earthquake with a rating of II, which has a low intensity, as follows: "Felt only by a few persons at rest, especially on upper floors of buildings. Delicately suspended objects may swing." An earthquake with a higher

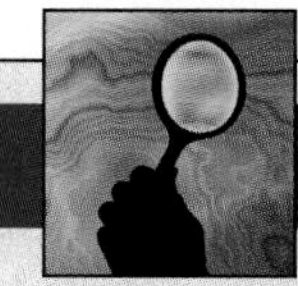

Small-Scale Investigation

Seismographic Record

A seismograph records the energy release of an earthquake. You can observe energy release by making a model seismograph.

Materials

shoe box, large plastic bag, sand (15–20 lb.), felt-tip pen, rubber band or masking tape, pad of paper or clipboard, four objects with various masses (max. 1–2 kg), balance, metric ruler, newspaper

Procedure

1. Line the shoe box with the plastic bag, and fill the box with sand. Put on the lid.
2. Mark an *X* near the center of the lid.
3. Fasten the felt-tip pen to the lid of the box with a tight rubber band or masking tape so that it just extends beyond the edge of the box.
4. Measure and record each test object's mass.
5. Have a partner hold the pad of paper so that it just touches the pen.
6. Hold your first object directly over the *X* at a height of about 30 cm. As your partner slowly moves the paper horizontally past the pen, drop the object on the *X*.
7. Label the resulting line with the mass of the object dropped and the material in the box.

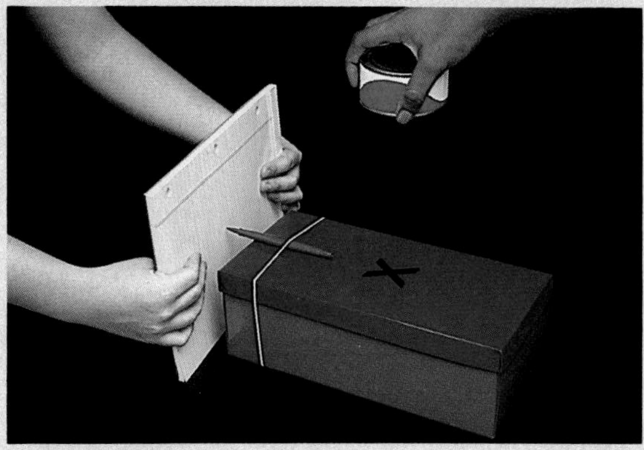

8. Repeat Steps 5–7 with three more objects of different masses. Move the starting position of the paper up about 2 cm each time so you will end up with four lines at 2 cm intervals.
9. Replace about 2/3 of the sand with crumpled newspaper. Reassemble the box and repeat Steps 5–8.

Analysis and Conclusions

1. What do the lines on the paper represent?
2. What do the sand and newspaper represent?
3. Compare the lines made in Steps 6–8 with those made in Step 9. Explain any differences.
4. What can you infer about the relationship between seismographic records and the energy released by an earthquake?

intensity, for example one designated by Roman numeral X, is described as follows: "Some well-built wooden structures destroyed; most masonry and frame structures destroyed with foundations; ground badly cracked. Rails bent. Landslides considerable from riverbanks and steep slopes. Shifted sand and mud. Water splashed over banks." The highest intensity earthquake is designated by Roman numeral XII and is described as "total destruction."

Section 6.2 Review

1. What instrument is used to record seismic waves?
2. Explain the three types of seismic waves.
3. How is the epicenter of an earthquake located?
4. How do scientists measure the magnitude of an earthquake?
5. Why do P waves travel faster through the lithosphere than through the asthenosphere?

Section Objectives

- **Describe possible effects of a major earthquake on buildings.**
- **Discuss the relationship of tsunamis to earthquakes.**
- **List safety rules to follow when an earthquake strikes.**
- **Identify changes in the earth's crust that may signal earthquakes.**

6.3 Earthquake Damage

During a severe earthquake, you would be much safer in an open, level field than in a city of skyscrapers. Movement of the ground itself seldom causes many deaths or injuries. Instead, most injuries result from the collapse of buildings and other structures or from falling objects and flying glass. Other dangers include landslides, fires, explosions caused by broken electric and gas lines, and floodwaters released from collapsing dams. Duration can affect the damage caused by an earthquake. A moderate earthquake that continues for a long time often causes more damage than an earthquake of higher magnitude that lasts only a short time.

Destruction to Buildings and Property

Most buildings are not designed to withstand the swaying motion caused by earthquakes. Buildings with weak walls may completely collapse. Very tall buildings may sway so violently that they tip over and fall onto lower neighboring structures.

The type of ground beneath a building can affect the way the building responds to seismic waves. For example, a building constructed on loose soil and rock is much more likely to be damaged during an earthquake than one built on more solid ground. During an earthquake, the loose soil and rock can vibrate like jelly. Buildings constructed on top of this kind of ground experience exaggerated motion and sway violently. On what kind of ground should a tall building in an earthquake-prone region be constructed?

Tsunamis

A major earthquake with an epicenter on the ocean floor sometimes causes a giant ocean wave called a **tsunami** (su-NAH-mee). This name comes from the Japanese word for "harbor wave." Scientists think that most tsunamis are caused by two events related to undersea earthquakes: faulting and underwater landslides. Faulting may cause a sudden drop or rise in the ocean floor. A large mass of sea water also drops or rises with the ocean floor. This mass of water churns up and down as it adjusts to the change in sea level. The violent water movement sets into motion a series of long, low waves that develop into tsunamis.

An earthquake may also trigger severe underwater landslides. The water above a landslide is thrown into an up-and-down motion, thereby creating a series of tsunamis.

The tsunami that accompanied the 1964 Alaskan earthquake caused heavy damage to towns near the Gulf of Alaska. Most of the fishing fleet off Kodiak, Alaska, was destroyed. Many fishing vessels were swept into the business district of the town by the tsunami. The tsunami caused 107 deaths, whereas only 9 people died as a direct result of the earthquake.

Disastrous earthquakes and tsunamis have encouraged the expansion and improvement of the Seismic Sea Wave Warning System (SSWWS). This network of seismograph stations around and in the Pacific Ocean alerts scientists to the location and magnitude of earthquakes. If a tsunami seems possible, scientists estimate its arrival times at different locations. They can then issue warnings immediately to these areas. However, there will not be enough time to issue tsunami warnings to areas very near the epicenter of the earthquake.

Earthquake Safety

A destructive earthquake may occur in any region of the United States. However, destructive earthquakes are more likely to occur in certain geographic areas, such as California or Alaska. People living or visiting near active faults should be ready to follow a few

The Great Hanshin Earthquake

At 5:46 A.M. on January 17, 1995, residents of Kobe, Japan, suffered a terrifying awakening when tremors from the Great Hanshin Earthquake shot through this historic port city. The main tremor lasted less than a minute and measured 7.2 on the Richter scale. It toppled highway overpasses, tore apart railway lines, and mangled 190,000 buildings. Fires fed by ruptured gas lines raged on for days, and property losses exceeded 100 billion dollars.

As the city slowly recovered from the disaster, seismologists and engineers began to survey the damage. They found that older structures fared poorly because they were never refitted to meet current building codes. According to today's codes, buildings must have a combination of strength and elasticity, which helps them withstand quakes by flexing and then returning to their original shape. Although they may become permanently deformed by severe quakes, such buildings save lives by not collapsing on their occupants.

Besides failing today's building standards, many of Kobe's buildings were constructed on soft, unstable land. This type of land not only shifts and settles during an earthquake but also amplifies the shock waves from the quake.

The Hanshin earthquake is a grim reminder of the damage that even a moderate earthquake can cause if it hits an urban area. The earthquake killed 5,500 people, injured tens of thousands more, and left 310,000 homeless in near-freezing temperatures. If the quake had occurred later in the day, many more people could have died.

What steps might Japanese officials take to limit future earthquake damage?

simple earthquake safety rules. These safety rules may help prevent death, injury, and property damage.

Before an earthquake occurs, be prepared. Keep on hand a supply of canned food, bottled water, flashlights, batteries, and a portable radio. Plan what you will do if an earthquake strikes while you are at home, in school, or in a car. Discuss these plans with your family and friends. Learn how to turn off the gas, water, and electricity in your home.

During an earthquake, stay calm. There are usually a few seconds between tremors, during which you can move to a safer posi-

CAREER FOCUS: *Earth Forces*

"I get . . . satisfaction when I can find out something that no one has discovered before . . . and come up with logical arguments to support my findings."

Wayne Thatcher

Tectonophysicist

Back in California after one month of field work in Russia, China, and Japan, tectonophysicist Wayne Thatcher describes the rewards of his work.

"When I can close the door, cut out the distractions, and think up new projects to understand how the earth works, I'm a very happy man," he says. "Sometimes I actually feel lucky to get paid for doing something I enjoy so much."

Thatcher, pictured at left, studies the structure of the earth's crust and the forces that shape the crust. He is particularly interested in understanding the causes of earthquakes.

"I had a high school physical geography teacher who really got me started in the field. He was interested in theories about how the continents were formed and whether their drifting was important to show how the earth's surface formed and changed."

Thatcher's work as project chief for earthquake hazards with the U.S. Geological Survey involves both administrative duties and research. He uses data that is gathered by observation and by seismic networks worldwide.

"Although we know earthquakes are almost exclusively the result of plate tectonics, more research is needed to define the actual processes involved."

Volcanologist

Volcanologists use modern equipment and techniques to monitor volcanoes and predict

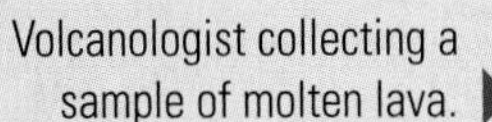

Volcanologist collecting a sample of molten lava.

tion. If you are indoors, protect yourself from falling debris by standing in a doorway or crouching under a desk or table. Stay away from windows, heavy furniture, and other objects that might topple over. Do not run outside. If you are in school, follow the instructions given by your teacher or principal. If you are in an automobile, stop in a place away from tall buildings, tunnels, power lines, and bridges, and remain in the car until the tremors cease.

After an earthquake, be cautious. Check for fire and fire hazards. Always wear shoes when walking among broken glass, and avoid downed powerlines and objects touched by downed wires.

their activity. Using a seismograph, for example, a volcanologist, pictured lower left, can detect earthquakes, which often signal eruptions. Currently, most monitoring studies are conducted on volcanoes in Hawaii, Italy, Japan, New Zealand, and Russia.

Volcanologists also study rock formations and past lava flows to determine which way lava might flow in future eruptions. Their findings may help save hundreds of lives and homes from destruction.

Most volcanologists begin their career preparation with a bachelor's degree in geology or geophysics.

▲ White Island, New Zealand, is an active volcano.

Exploration Geophysicist

Exploration geophysicists study the earth's subsurface. They may use their skills to locate sources of fresh water or solve environmental problems. Most exploration geophysicists, however, search for sources of petroleum.

To help them in their work, these exploration geophysicists often rely on seismic surveys. In seismic surveys, sound waves are used to detect rock layers where petroleum might be found.

After collecting data about a specific area, exploration geophysicists analyze the data, prepare maps and reports, and determine the best drilling locations.

Entry-level positions in exploration geophysics require a minimum of a bachelor's degree in geophysics or geology.

For Further Information

For more information on careers in geophysics, write the American Geophysical Union, 2000 Florida Ave., NW, Washington, DC 20009 (see also www.agu.org) or the Society of Exploration Geophysicists, 8801 South Yale, Tulsa, OK 74137.

Earthquake Warnings and Predictions

Humans have long dreamed of being able to accurately predict earthquakes. One of the earliest means of predicting earthquakes was to observe the behavior of animals. People knew that just before an earthquake, some animals appeared nervous and restless, almost as if they could sense the coming catastrophe. Chinese research into strange animal behavior as an effective earthquake warning continues. Some areas experience major earthquakes at regular intervals. Using records of past earthquakes, scientists can make approximate predictions of future quakes. However, these predictions may be off by years.

To make more accurate predictions, scientists are trying to detect changes in the earth's crust that can signal an approaching earthquake. Faults near many population centers have been located and mapped. Instruments placed along these faults measure small changes in rock movement around the fault and can detect an increase in strain. Along some faults, scientists have identified zones of immobile rock called **seismic gaps.** A seismic gap is a place where the fault is locked and unable to move. The strain in the surrounding rock has increased, and no major earthquake has occurred in this location for at least 30 years. Scientists think seismic gaps are likely locations of future earthquakes. Several gaps that exist along the San Andreas Fault will probably be sites of major earthquakes in the future.

Figure 6–6. Earthquakes have periodically occurred along the San Andreas Fault throughout its life of about 15–20 million years. Seismologists predict that future great earthquakes along the fault will not strike suddenly, but will probably be preceded by an increase in seismic activity.

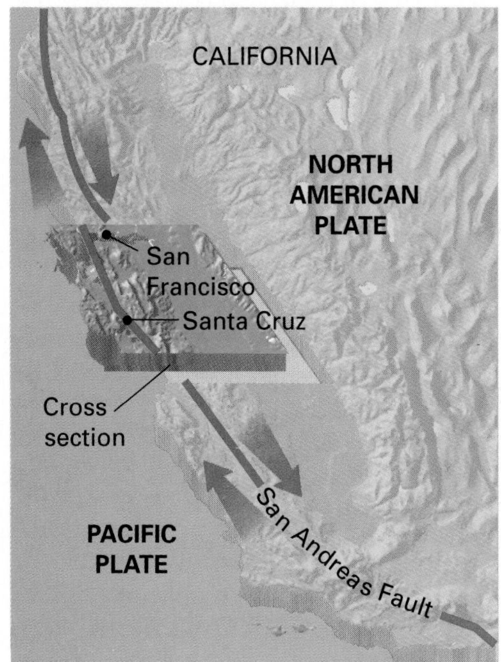

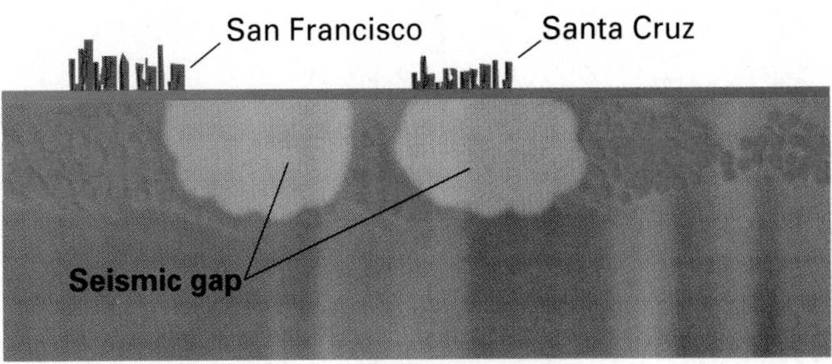

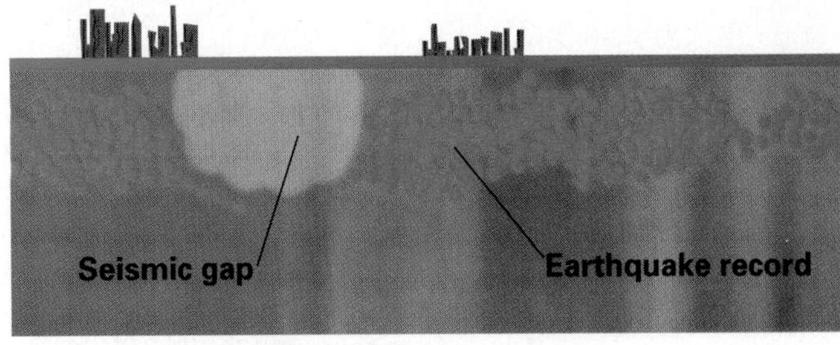

Figure 6–7. Recent earthquakes have occurred at seismic gaps along the San Andreas Fault in California. Other seismic gaps may indicate the location of future earthquakes.

Sometimes scientists detect a slight tilting of the ground shortly before an earthquake. They are also able to detect the strain and cracks in rocks caused by the stress that builds up just before an earthquake. The magnetic and electrical properties of rock change when these cracks fill with water. As a result, scientists might be able to detect small changes in the earth's magnetic field and in the way electricity is conducted by rocks prior to an earthquake. Scientists can also detect increased natural gas seepage from strained or fractured rocks.

Some earthquakes are preceded by a decrease in the speed of local P waves. The P waves being measured have traveled from distant earthquakes. The decrease in the speed of local P waves may last for several days or several years preceding an earthquake. Just before the earthquake occurs, the speed of the local P waves suddenly returns to normal. Evidence even suggests that the longer the decrease in speed lasts, the stronger the earthquake will be.

Scientists would like to be able to control the force of earthquakes. A better understanding of earthquakes and their causes may enable them to develop a system of earthquake control. In fact, tests at Rangely, Colorado, showed that when water was injected along a fault, friction was reduced and earthquakes were less severe.

Section 6.3 Review

1. How do tall buildings usually respond during a major earthquake?
2. What causes tsunamis?
3. What should you do if an earthquake strikes while you are at home? In a car?
4. What are some early warning signs of earthquake activity?
5. What type of building construction and location regulations should be included in the building code of a city located near an active fault?

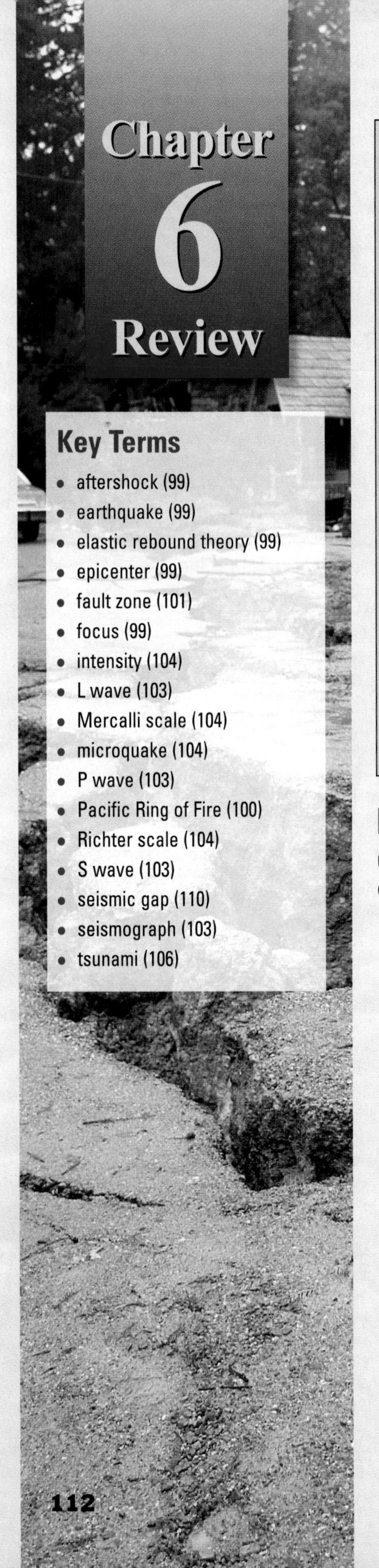

Chapter 6 Review

Key Terms

- aftershock (99)
- earthquake (99)
- elastic rebound theory (99)
- epicenter (99)
- fault zone (101)
- focus (99)
- intensity (104)
- L wave (103)
- Mercalli scale (104)
- microquake (104)
- P wave (103)
- Pacific Ring of Fire (100)
- Richter scale (104)
- S wave (103)
- seismic gap (110)
- seismograph (103)
- tsunami (106)

Key Concepts

According to the elastic rebound theory, stress builds in rocks along a fault until they break and snap into new positions. **See page 99.**

Most earthquakes occur near plate boundaries. **See page 100.**

There are three types of seismic waves: P waves, S waves, and L waves. **See page 103.**

The difference in the time that it takes P waves and S waves to arrive at a seismograph station helps scientists locate the epicenter of an earthquake. **See page 104.**

The Richter scale expresses the energy released by an earthquake. **See page 104.**

Most earthquake damage is caused by the collapse of buildings and other structures. **See page 106.**

Tsunamis often accompany ocean-floor earthquakes. **See page 106.**

People in areas around active faults should know and follow safety rules in case of possible earthquakes. **See page 107.**

Gaps in fault movement, tilting ground, and variations in seismic waves are among the changes in the earth's crust that scientists use in predicting earthquakes. **See page 110.**

Review

On your own paper, write the letter of the term that best completes each of the following statements.

1. Vibrations in the earth caused by the sudden movement of rock are called
   a. epicenters. b. earthquakes. c. faults. d. tsunamis.
2. The elastic rebound theory states that as a rock becomes stressed, it first
   a. deforms. b. melts. c. breaks. d. shifts position.
3. The point along a fault where an earthquake begins is called the
   a. fracture. b. epicenter. c. gap. d. focus.
4. The point on the earth's surface directly above the point where an earthquake occurs is called the
   a. focus. b. epicenter. c. fracture. d. fault.
5. A characteristic of earthquakes that causes the most severe damage is
   a. a deep focus. b. an intermediate focus.
   c. a shallow focus. d. a deep epicenter.
6. Most severe earthquakes occur
   a. in mountains. b. along major rivers.
   c. at plate boundaries. d. in the middle of plates.

7. The boundary of the Pacific plate scrapes against that of the North American plate and forms
   a. a single fault.
   b. a subduction zone.
   c. a volcano.
   d. a fault zone.
8. P waves travel through
   a. solids only.
   b. liquids and gases only.
   c. both solids and liquids.
   d. liquids only.
9. S waves cannot pass through
   a. solids.
   b. the mantle.
   c. the earth's outer core.
   d. the asthenosphere.
10. By analyzing the difference in the time it takes for P waves and S waves to arrive at a seismograph station, scientists can determine an earthquake's
   a. epicenter. b. L waves. c. fault zone. d. intensity.
11. The Richter scale expresses an earthquake's
   a. magnitude. b. location. c. duration. d. depth.
12. Most injuries during earthquakes are caused by
   a. the collapse of buildings.
   b. cracks in the earth's surface.
   c. the vibration of S waves.
   d. the vibration of P waves.
13. If an earthquake strikes while you are in a car, you should
   a. continue driving.
   b. get out of the car.
   c. park the car under a bridge.
   d. stop the car in a clear space and remain in the car.
14. An earthquake is frequently preceded by
   a. a temporary change in the speed of local P waves.
   b. a temporary change in the speed of the surface waves.
   c. landslides.
   d. tsunamis.

Critical Thinking

On your own paper, write answers to the following questions.

1. Earthquakes with a very deep focus cannot be explained by the elastic rebound theory. Explain why.
2. If a seismograph station measures P waves but no S waves from an earthquake, what can you conclude about the earthquake's location ?
3. Two cities are struck by earthquakes. The cities are the same size, are built on the same type of ground, and have the same types of buildings. The city in which the quake measured 4 on the Richter scale suffered $1 million in damage. The city in which the quake measured 6 on the scale suffered $50 million in damage. What might account for this great difference in damage costs?
4. Would an earthquake in the Colorado Rockies be likely to form a tsunami? Explain why.

Application

1. You are going to choose a building site for a home. You would like a high place with a view, but you are concerned about earthquakes. What information do you need to make an informed decision about the site?
2. Why is it wise to stand in a doorway when an earthquake strikes?
3. Imagine that you are monitoring a seismograph station along the San Andreas Fault. An earthquake occurs in Mexico, and you notice that the P waves you are recording from that quake have a velocity that is less than normal. What does this tell you about the area around your seismograph station?

Extension

1. Research major earthquake activity along the San Andreas Fault over the past five years. Write a brief report on your findings.
2. Find out how and why the worldwide network of seismograph stations was formed. Also find out how all the stations in the network work together. Report on your findings.
3. Find out if there is a building code designed to minimize earthquake damage in your city. Summarize the regulations within the code.

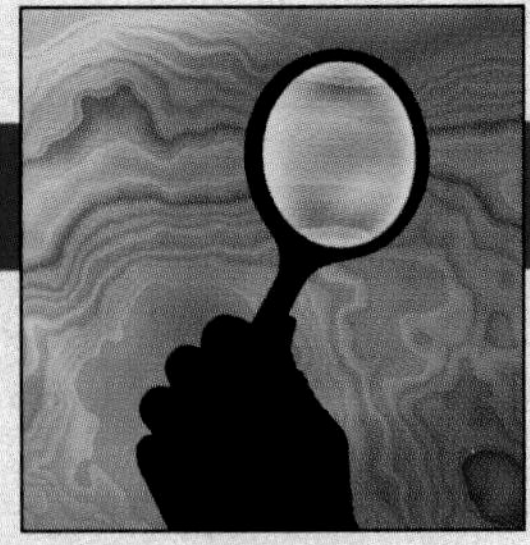

In-Depth Investigation

Earthquake Waves

Materials
- drawing compass
- ruler
- calculator

Introduction
An earthquake releases energy that travels through the earth in all directions. This energy is in the form of waves. Two kinds of earthquake waves are primary waves (P waves) and secondary waves (S waves). Primary waves travel faster than secondary waves and are the first to reach and be recorded at a seismograph station. The secondary waves arrive sometime after the P waves. The time difference between the arrival of the P waves and the S waves increases as the waves travel farther from their origin. This difference in arrival time, called *lag time,* can be used to find the distance to the epicenter of the earthquake. Once the distance from three different locations is determined, scientists can find the exact location of the epicenter.

Prelab Preparation
1. Review Section 6.2, pages 103–105.
2. Calculate the following problems, which apply to information in the procedure. The average speed of P waves is 6.1 km/s. The average speed of S waves is 4.1 km/s. To calculate the time it takes seismic waves to travel a given distance, divide that distance by the average speed of each wave.
   a. How long would it take P waves to travel 100 km? How long would it take them to travel 200 km?
   b. How long would it take S waves to travel 100 km? How long would it take them to travel 200 km?
   c. What is the lag time between the arrival of P waves and S waves over a distance of 100 km? What is the lag time for a distance of 200 km?

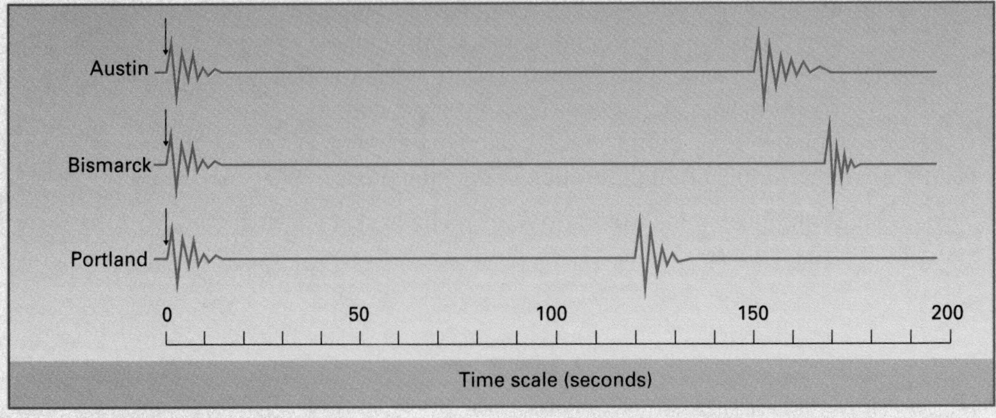

Procedure

1. The illustration on page 114 shows seismograph records made in three cities following an earthquake. These traces begin at the left, and arrows indicate the arrival of the P waves. Use the time scale provided to find the lag time between the P waves and the S waves for each city. Be sure to measure the time from the arrival of the P wave to the arrival of the S wave. Record this information in a table of your own with columns for city, lag time, and distance from epicenter.

2. Find the distance from each city to the epicenter of the earthquake. To calculate these distances, use the lag times you found in Step 1, information from the prelab preparation, and the following formula.

$$\text{distance} = \frac{\text{measured lag time (s)} \times 100 \text{ km}}{\text{lag time for 100 km (s)}}$$

Record this information in your table.

City	Lag time (seconds)	Distance from city to epicenter
Austin		
Bismarck		
Portland		

3. Copy the map at right, which shows the location of the three cities with a scale in kilometers. Using the map scale on your copy, adjust the compass so that the radius of the circle with Austin at the center is equal to the distance from the epicenter of the earthquake to Austin as calculated in Step 2. Put the point of the compass on Austin. Draw a circle on your copy of the map.

4. Repeat Step 3 for Bismarck and then for Portland. The epicenter of the earthquake is located near the point at which the three circles intersect.

Analysis and Conclusions

1. The location of the earthquake epicenter is closest to what city?

2. Why must there be measurements from three different locations to find the epicenter of an earthquake?

Extensions

1. What is the probability of a major earthquake occurring in the area where you live?

2. If an earthquake did occur in your area, what would be its probable cause?

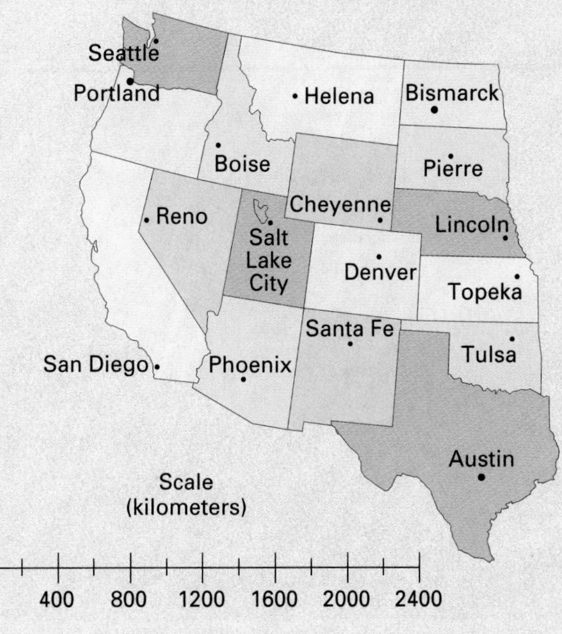

Chapter 7

Volcanoes

Mount Vesuvius, Krakatau, Mount St. Helens, Olympus Mons—these are some of the best-known volcanoes in the solar system. Volcanoes are powerful and sometimes destructive reminders that the earth and other planetary bodies have been, and in some cases still are, geologically active.

In this chapter, you will learn about the mechanics, the power, and the effects of volcanic activity.

Chapter Outline

7.1 Volcanoes and Plate Tectonics
- Volcanism
- Major Volcanic Zones

7.2 Volcanic Eruptions
- Kinds of Eruptions
- Volcanic Rock Fragments
- Volcanic Features
- Predicting Volcanic Eruptions

7.3 Extraterrestrial Volcanism
- The Moon
- Mars
- Io

Mount St. Helens, in Washington State, exploded in a volcanic eruption.

7.1 Volcanoes and Plate Tectonics

Section Objectives

- **Describe the formation and movement of magma.**
- **Define *volcanism*.**
- **List three locations where volcanism occurs.**

Scientists have no direct way to measure temperatures deep within the earth. However, analysis of seismic waves and computer modeling allow scientists to estimate those temperatures. Figure 7–1 shows estimates of the earth's inner temperatures and pressures. As this graph shows, the combined temperature and pressure in the lower part of the mantle keep the rocks located there below their melting point.

Despite the high temperature in the asthenosphere, most of this zone remains solid because of the great pressure of the surrounding rock. Sometimes, however, areas of the solid rock will melt, forming **magma,** or liquid rock. Geologists think that magma forms in areas where the surrounding rock exerts less pressure than normal. The lower pressure allows the rock particles to move more quickly; thus, the rock becomes liquid.

Volcanism

Any activity that includes the movement of magma toward or onto the surface of the earth is called **volcanism.** Pockets of magma grow due to melting of some of the surrounding rock. As more rock melts, the magma pockets expand. The magma slowly pushes upward into the crust because it is less dense than solid crustal rock. The magma slowly rises, forcing its way into cracks in the overlying rock. This process causes large blocks of overlying rock to break off and melt, adding still more material to the magma pockets.

Most magma forms at plate boundaries, where one lithospheric plate, usually of oceanic crust, is subducted beneath another plate, often of continental crust. The subducting plate moves deep into the hot asthenosphere, where parts of it melt to become magma.

Sometimes magma breaks through to the surface of the earth. Magma that erupts onto the earth's surface is called **lava.** The opening through which the molten rock flows onto the surface is called a **vent.** The vent and the volcanic material that builds up on the earth's surface around the vent is called a **volcano.**

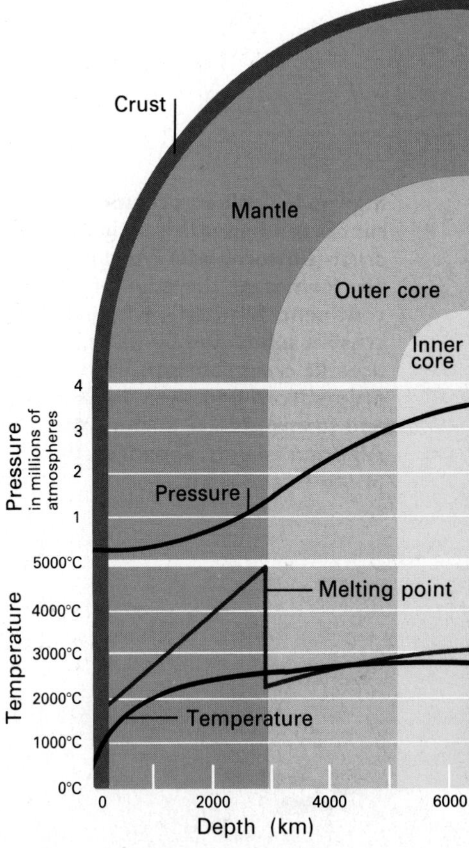

Figure 7–1. Temperature and pressure increase as the depth beneath the earth's surface increases.

Major Volcanic Zones

If you were to plot the location of the 600 or so volcanoes that have erupted within the past 50 years, you would see that they form a pattern across the earth. Most of these active volcanoes are found in zones near both convergent and divergent boundaries of the lithospheric plates.

Subduction Zones

Many volcanoes are located along subduction zones, where one plate moves under another plate. When a plate with oceanic crust meets a plate with continental crust, the oceanic crust, which is more dense, moves beneath the continental crust. A deep trench forms on the ocean floor along the edge of the continent where the plate is being subducted. The plate with the continental crust buckles and folds, forming a line of mountains along the edge of the continent.

The subducted plate dives deep into the hot asthenosphere. There, heat from the mantle and from the friction between the moving plates begins to melt the subducted plate into magma. Some of this magma erupts through the earth's surface, forming volcanic mountains near the edge of the continent.

A major zone of active volcanoes is caused by subducting plates encircling the Pacific Ocean. This zone, called the Pacific Ring of Fire, results from plates subducting along the Pacific coasts of North America, South America, Asia, and the islands of the western Pacific Ocean. As you read in Chapter 6, the Pacific Ring of Fire is also one of earth's three major earthquake zones.

If two plates with oceanic crust at their boundaries collide, one of the plates is subducted, forming a deep trench. As the plate descends into the mantle, it melts. Some of the resulting magma breaks through to the surface along the trench. In time, a string of volcanic islands, called an *island arc,* forms along the trench. The early stages of this type of subduction produce an arc of small volcanic islands. One example is the Aleutian Islands, which stretch across the North Pacific Ocean between Alaska and Siberia. As more magma surfaces, more islands appear, become larger, and join, such as the group of volcanic islands that make up Japan.

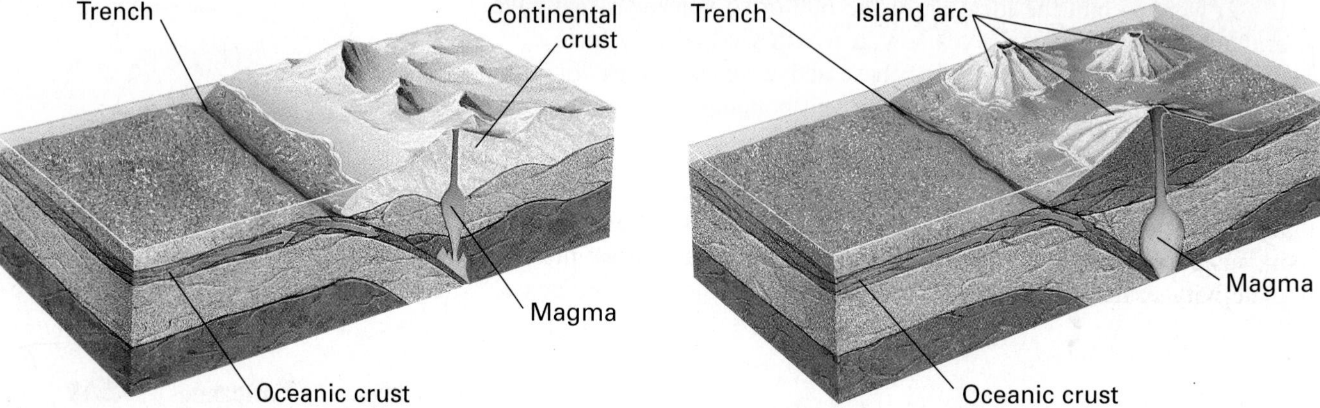

Figure 7–2. When oceanic crust is subducted beneath continental crust (bottom, left), volcanoes often form near the edge of the continent. Similarly, when oceanic crust is subducted beneath oceanic crust (bottom, right), lava erupts from undersea volcanoes and forms islands, such as the Aleutian Islands, shown in the photo to the right.

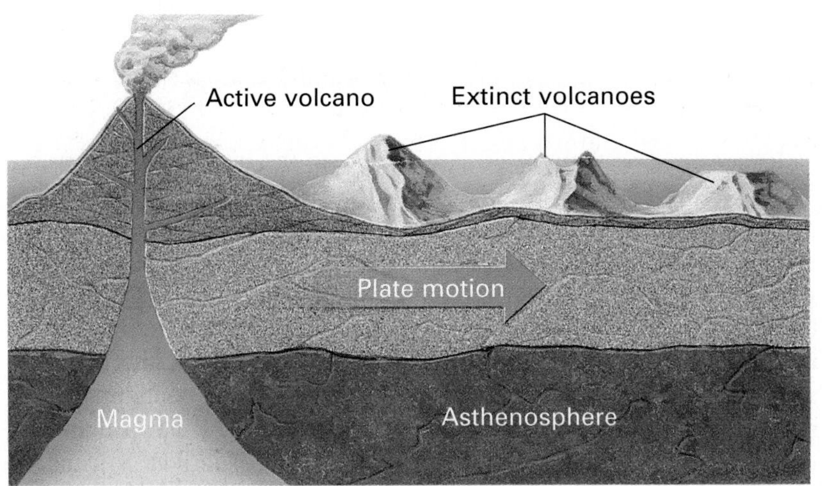

Figure 7–3. Lithospheric plate movement over a hot spot produces a chain of volcanic islands. The islands farthest away from the hot spot are the oldest. Eventually, plate movement will carry these older islands beneath the surface of the ocean.

Mid-Ocean Ridges

The greatest amount of magma comes to the surface where plates are moving apart along mid-ocean ridges. Thus, a major zone of volcanoes is the interconnected mid-ocean ridges that circle the earth. Fractures between the spreading plates along the ridges reach down to the asthenosphere. Magma rises through the fractures and comes to the surface along long, narrow cracks called *rifts*. The emerging lava forms new ocean floor and adds material to the ridges.

Most volcanic eruptions along the mid-ocean ridges go unnoticed because they take place deep beneath the surface of the ocean. An exception is found on Iceland. Iceland is a part of the Mid-Atlantic Ridge that is above sea level. One-half of Iceland is on the North American plate and is moving westward. The other half is on the Eurasian plate and is moving eastward. The middle of Iceland is cut by large **fissures,** cracks through which lava flows. What do you think causes the fissures in Iceland?

Hot Spots

Not all volcanoes develop along plate boundaries. Sometimes magma works its way to the earth's surface within the interiors of lithospheric plates. These areas of volcanism within plates are called **hot spots.** Hot spots appear to remain stationary. However, the lithospheric plate above a hot spot continues drifting slowly. As a result, the volcano on the surface is eventually carried away from the hot spot. The activity of the volcano ceases because there is no longer any hot-spot magma underneath to feed it. A new volcano then forms where new crust has moved over the hot spot, as shown in Figure 7–3. The Hawaiian Islands are an example of a chain of volcanic islands formed over a hot spot.

INVESTIGATE!

To learn more about hot spots and volcanoes, try the In-Depth Investigation on pages 130–131.

Section 7.1 Review

1. What must happen for volcanism to occur?
2. Explain how magma reaches the surface.
3. How does subduction produce magma?
4. What would be a likely explanation for the onset of volcanic activity in the central United States?

Section Objectives

- **Summarize the relationship between lava types and the force of volcanic eruptions.**
- **Describe the major types of tephra.**
- **Identify the three main types of volcanic cones.**
- **Summarize the events that may signal a volcanic eruption.**

7.2 Volcanic Eruptions

Volcanoes can be thought of as windows to the interior of the earth. The lava that erupts from them provides an opportunity for scientists to study firsthand the materials that form deep within the earth's mantle. By analyzing the types of minerals found in hardened lava, geologists have concluded that there are two general types of lava. One type of lava is dark colored when hardened and is rich in magnesium and iron. This type of lava is called **mafic lava** and is usually of oceanic crust origin. The second type of lava contains much silica, with lesser amounts of iron and magnesium. It has a lighter color when hardened and is called **felsic lava.** Felsic lava is usually from melted continental crust. Other lavas have a range of compositions that fall between the mafic and the felsic varieties.

Thin, mafic lava usually hardens with a wrinkled surface. This type of solidified lava is known as **pahoehoe** (puh-HOEEE-HOEEE), which means "ropey" in Hawaiian. Rapid cooling on the surface of the lava forms a crust that breaks into jagged chunks as the liquid below continues to flow. The lava deposit that remains is described by its Hawaiian name **aa** (AH-ah), which refers to the sharp, blocky shapes into which the hardened lava breaks. Sometimes the outer part of a mafic lava flow cools so rapidly that it forms a hardened shell around a liquid interior that flows out, leaving tunnels in the hardened lava shell. When lava flows out of fissures on the ocean floor, it cools rapidly, often in rounded shapes. This is called **pillow lava.** Pillow lava is commonly found along mid-ocean ridges.

Kinds of Eruptions

The composition of the lava that reaches the surface largely determines the force with which a particular volcano will erupt. Why would lava that contains large amounts of trapped dissolved gases usually produce a more explosive eruption than lava that contains few dissolved gases?

Figure 7–4. Pahoehoe lava has a wrinkled, "ropey" surface.

Figure 7–5. Lava flows from a quiet eruption (left) like a red-hot river. During an explosive eruption, lava, steam, ash, and other volcanic material are ejected violently from the volcano (right).

Oceanic volcanoes, both those that erupt on the ocean floor and those that erupt on oceanic islands, usually are produced by mafic lava. Mafic lava is very hot and thin, and it flows almost as easily as water. Because gases can easily escape from mafic lava, eruptions from oceanic volcanoes like those in Hawaii are usually quiet. That is, the lava flows out from the volcanic opening like a red-hot river.

In contrast to the fluid lavas produced by oceanic volcanoes, the felsic lavas of continental volcanoes, such as Mount St. Helens, tend to be cooler and thicker. They also contain large amounts of trapped gases, mostly water vapor and carbon dioxide. When a vent or fissure opens up, the dissolved gases within the lava boil out explosively, sending molten and solid particles shooting into the air.

Volcanic Rock Fragments

Unlike mafic lava, which tends to flow quietly, felsic lava explodes, throwing **tephra** into the air. Tephra, sometimes called **pyroclastic material,** is rock fragments ejected from a volcano. Some tephra forms when cooling magma breaks into fragments because of the rapidly expanding gases within it. Other tephra forms when a spray of lava cools and solidifies as it flies through the air.

Tephra particles less than 2 mm in diameter make up **volcanic ash.** Particles less than 0.25 mm in diameter are called **volcanic dust.** After a volcanic eruption, wind may carry dust and ash far from the volcano. Most of the volcanic dust and ash settles on the land immediately surrounding the volcano. Some of the smallest dust particles, however, may travel completely around the earth in the upper atmosphere.

Larger tephra particles, less than 64 mm in diameter, are called **lapilli** (luh-PIL-ie), from a Latin word that means ''little stones.'' Lapilli generally fall near the vent.

Large clots of lava are thrown out of an erupting volcano while they are red-hot. As the clots spin through the air, they cool and develop a round or spindle shape. These tephra particles are called **volcanic bombs.** The largest tephra, formed from solid rock blasted from the fissure, is known as **volcanic blocks.** Some volcanic blocks are as big as houses.

Volcanic Features

Volcanic activity produces a variety of characteristic features on the earth's surface. These features are formed during both quiet and explosive eruptions. For example, the lava and tephra ejected during volcanic eruptions can build up around the vent. These piles of volcanic materials, known as *volcanic cones,* are classified into three main types.

Volcanic Cones

Volcanic cones that are broad at the base and have gently sloping sides are called **shield cones.** A shield cone covers a wide area and generally results from quiet lava eruptions. Layers of hot mafic lava flow out around the vent, harden, and slowly build up to form the cone. The Hawaiian Islands are a chain of shield cones built up from the ocean floor by a hot spot.

Explosive eruptions form different types of volcanic cones. Some explosive eruptions form **cinder cones.** A cinder cone is made

Figure 7–6. Shield cones like Mauna Kea (top) have broad bases and gentle slopes. Cinder cones like Sunset Crater (middle) have narrow bases and very steep slopes. Composite cones like Mount Fuji (bottom) are both broad and steep.

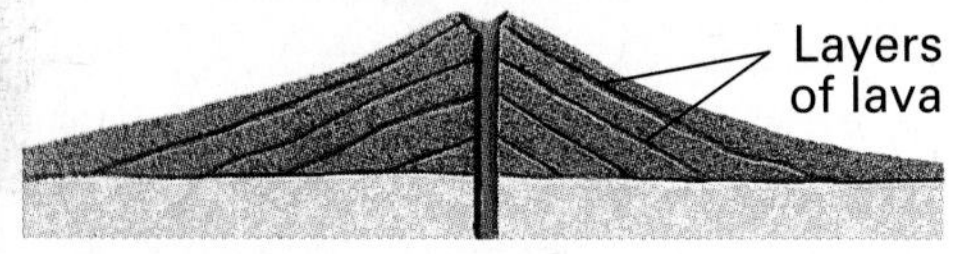

Shield cone

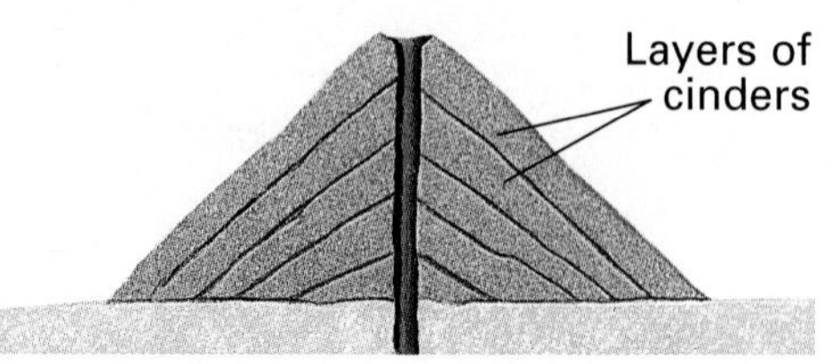

Cinder cone

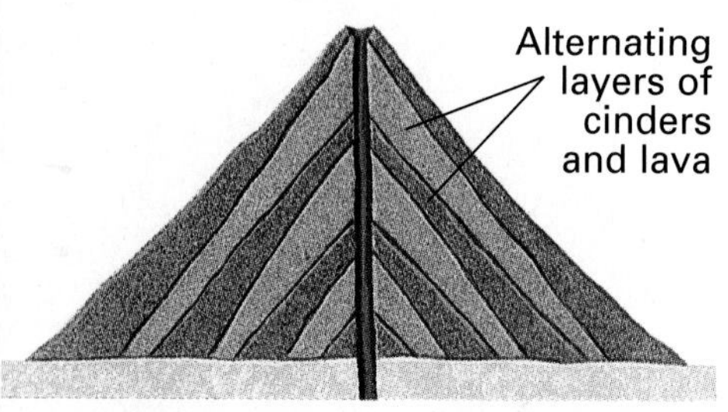

Composite cone

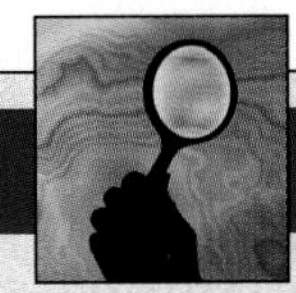

SMALL-SCALE INVESTIGATION

Volcanic Cones

A cinder cone is formed by material thrown out during an explosive volcanic eruption. A shield cone is made up of layers of lava that have flowed out during a quieter eruption. You can demonstrate that the very different shapes of these cones are the result of the different materials from which they are made.

Materials

plaster of Paris; measuring cup; water; mixing spoon; metric ruler; 2 paper plates; protractor; dry cereal or potato flakes, 8 oz.; graduated cylinder, 100 mL

Procedure

1. Pour 1/2 cup (about 4 oz.) of plaster of Paris into the measuring cup. Gently tap the cup so that the plaster settles to the 1/2 cup level.
2. Use the graduated cylinder to measure out 60 mL of water, and add it to the dry plaster in the measuring cup. Use a mixing spoon to blend the mixture until it is smooth and uniform.
3. Hold the measuring cup about 2 cm over a paper plate. Pour the contents slowly and steadily onto the center of the plate.
4. When the cone has hardened (15–20 min.), remove it from the plate. Measure the average slope angle with a protractor, as shown here.

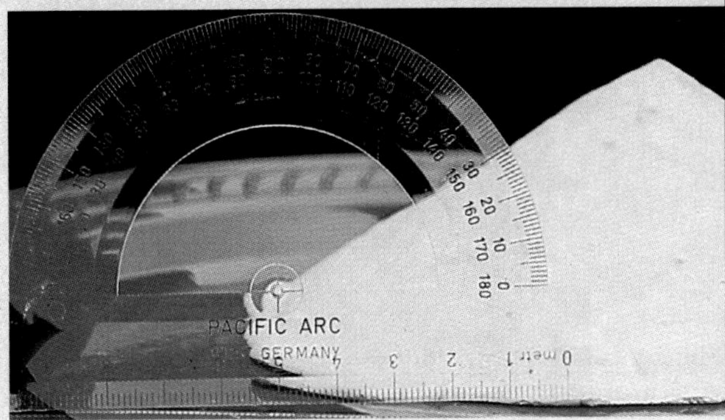

5. Pour dry cereal or potato flakes slowly onto the center of a clean paper plate until the mound is approximately 5 cm high.
6. Without disturbing the mound, measure its slope with a protractor.

Analysis and Conclusions

1. Which cone that you formed represents a cinder cone? a shield cone? How do the angles formed by these cones compare?
2. How would the slope be affected if the cereal were rounder? thicker?
3. Suppose you formed a cone by pouring alternating layers of wet plaster of Paris and dry cereal. How would the shape and overall size of this cone differ from the two cones you have already formed? Which type of volcano would such a cone look like?

of solid fragments ejected from the volcano. Most cinder cones have very steep slopes, often close to 40°, and are rarely more than a few hundred meters high.

Many volcanoes have both quiet eruptions and explosive eruptions. During a quiet eruption, the cone is formed mainly by lava flows. Then an explosive eruption occurs, depositing large amounts of tephra around the vent. The explosive eruption is followed again by quiet lava flows. Thus, the resulting cone is formed of alternating layers of hardened lava flows and tephra. These **composite cones,** also known as **stratovolcanoes,** often develop into high volcanic mountains. Some of the best-known volcanic mountains in the world are composite volcanoes. Among these are Mount Fuji in Japan and Mount Rainier, Mount Hood, Mount Shasta, and Mount St. Helens in the United States.

Figure 7–7. When calderas fill with water, they form scenic lakes such as the one at Crater Lake National Park, Oregon, shown above.

Craters and Calderas

The funnel-shaped pit at the top of a volcanic vent is known as the **crater.** The crater is formed when material is blown out of the volcano by explosions. A crater usually becomes wider as magma melts and breaks down the walls of the crater, allowing loose materials to collapse back into the vent. Sometimes a smaller cone forms within a crater from materials erupting from the vent.

When the magma chamber below a volcano is emptied, the volcanic cone may collapse. Explosions may also completely destroy the upper part of the cone, leaving a large, basin-shaped depression called a **caldera** (cal-DER-uh). Krakatau, a volcanic island in Indonesia, is an example of a caldera. When the volcanic cone exploded in 1883, a caldera with a diameter of 6 km formed. This huge caldera changed the shape of the entire island.

Predicting Volcanic Eruptions

A volcanic eruption can be one of the earth's most destructive natural phenomena. When Mount St. Helens exploded in 1980, the surrounding area, including forests and lakes, was completely devastated. The eruption of Krakatau exploded away more than half the island and produced huge tsunamis that killed more than 30,000 people. The entire city of Pompeii was stilled by the power of a volcano. If volcanic activity could be predicted far enough in advance, many lives might be saved.

Scientists have made progress in their efforts to predict the eruptions of active volcanoes. They use sensitive instruments to detect geological events that may signal the beginning of an eruption. For example, seismographs are used to monitor one of the most important warning signals—small earthquakes. These small earthquakes result from the growing pressure on the surrounding rocks as magma works its way upward. Temperature changes within the rock and the actual fracturing of the rock surrounding a volcano also contribute to the small earthquakes. The number of earthquakes often increases until they occur almost continuously just before an eruption. An increase in the strength of such earthquakes may also be a signal that an eruption is about to occur.

Another signal of volcanic activity that scientists watch for is a slight bulging of the surface of a volcano. Before an eruption, the upward movement of magma beneath the surface may push out the surface of the volcano. These bulges slightly alter the distance between two marked points on the slope, just as spots on a balloon would move apart as the balloon is inflated. Special instruments can measure these small changes in distance as well as changes in the tilt of the ground surface. Another possible signal of an increase in volcanic activity that is being investigated is changes in the composition of gases given off by volcanoes.

Predicting the eruption of a particular volcano also requires some knowledge of its previous eruptions. To provide the best forecast, scientists compare the volcano's past behavior with current

Krakatau

For centuries, Krakatau, a volcanic island between Java and Sumatra, Indonesia, was geologically inactive. In late May of 1883, however, that calm was shattered by the first of a series of volcanic eruptions. The eruptions continued throughout the summer, growing increasingly violent. The climax came on August 27, when a tremendous explosion destroyed the island, propelling ash and rock 80 km into the sky. The blast was heard 3,500 km away in Australia.

The eruption had a catastrophic impact. Twenty cubic kilometers of rock and great quantities of ash fell over an area of 800,000 km^2. A thick cloud of ash kept the area in total darkness for two and a half days. Volcanic ash buried all life on what was left of the island.

The Krakatau eruption triggered a series of tsunamis. One wave, estimated at 36 m high, took 36,000 lives in Java and Sumatra.

The effects of the blast continued long after the initial eruption. Fine dust and ash, which reflected sunlight, circled the earth, causing dramatic sunrises and sunsets for two years. The dust and ash also blocked some of the sun's rays for several years, causing global temperatures to drop by as much as 0.27°C. The climate worldwide remained below normal well into the 1890's, possibly as a result of the spectacular destruction of Krakatau.

If you examined the remains of Krakatau that lie beneath the surface of the ocean, what would you expect to find?

daily measurements of earthquakes, surface bulges, and changes in gases. Unfortunately, only a few of the active volcanoes in the world have been scientifically studied long enough to establish any patterns in their activity. Also, volcanoes that have been dormant for long periods of time may, with little warning, suddenly become active.

Section 7.2 Review

1. Explain the difference between mafic and felsic lava.
2. Define *tephra* and give some examples.
3. Name and compare the three types of volcanic cones.
4. What event forms a caldera?
5. Name three events that may precede a volcanic eruption.
6. Would quiet eruptions or explosive eruptions be more likely to increase the height of a volcano? Why?

Section Objectives

- **Summarize the evidence for extraterrestrial volcanism.**
- **Explain the differences between volcanism on earth and on Io.**

7.3 Extraterrestrial Volcanism

Many people think of the other planets and moons in the solar system as geologically dead worlds. However, evidence gathered by human exploration of the moon and data on the planets sent back by various spacecraft show otherwise. Indications are that many of the planets and moons in the solar system, including the earth's moon, were volcanically active in the past. Some of these planets and moons remain active. At least one of them, Io, a moon of Jupiter, has more volcanic activity than does the earth.

The Moon

The earth's moon is covered with basaltic lava flows. This is evidence that sometime in the past, active volcanoes dotted the surface of the moon. Most of the craters on the lunar surface result from meteorite bombardment. However, some of them have smooth, lava-flow interiors and gently sloping, channelled exteriors. These features suggest that the craters were once active volcanoes.

Questions yet to be resolved by scientists are how magma formed in the lunar interior and how it reached the surface. There is no evidence of plate tectonics or convection currents on the moon, so the magma must have been produced in other ways. Great heat would have been needed to produce both the magma in the upper layers of the moon and the cracks through which it flowed onto the lunar surface. Some scientists think this heat may have come from a long period of intense meteorite bombardment. How might meteorite bombardment have produced such great heat?

Mars

Spacecraft have orbited Mars and have sent back photographs that show numerous volcanoes and volcanic features on the Martian surface. The largest of these is the shield volcano Olympus Mons.

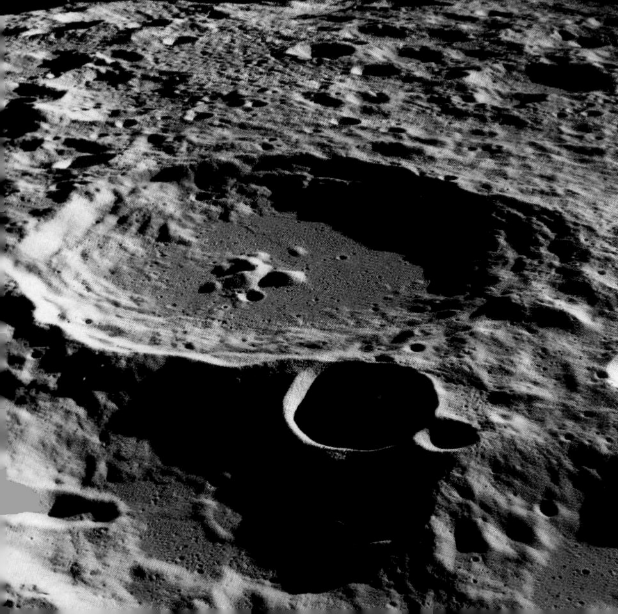

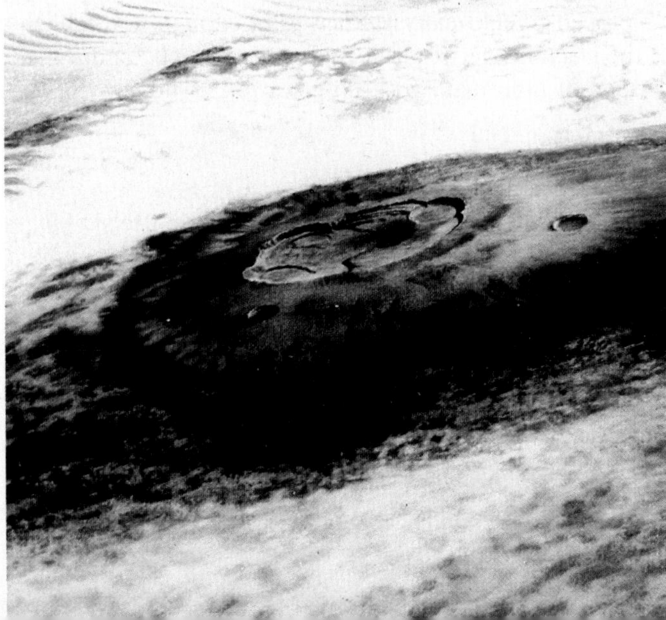

Figure 7–8. The structure of lunar craters, like the one shown on the left, suggests past volcanism on the moon. The immense Martian volcano Olympus Mons (right) is the largest known volcano in the solar system.

Olympus Mons rises nearly 28 km above the Martian surface. Its base measures 600 km across, and its caldera, 70 km across. Scientists think the volcano has grown to such a tremendous size because, unlike the earth's crust, the Martian crust does not drift. Consequently, Olympus Mons stayed over the lava source for perhaps millions of years with the lava continuously building the size of the volcano.

Whether Martian volcanoes are still active is a question scientists have yet to answer. However, Mars does seem to be seismically active. A Viking landing craft searching Mars for signs of life detected two geological events that produced waves similar to those of an earthquake. These "marsquakes" might also mean that the "red planet" is still volcanically active.

Figure 7–9. This computer-enhanced photo provided some of the first visible evidence of active volcanism on Io, one of the moons of Jupiter. Note the erupting volcano in the upper right of the photo.

Io

Io, one of the many moons of Jupiter, is the first planetary body (other than earth) on which active volcanoes have been sighted. In 1979, the spacecrafts *Voyager 1* and *Voyager 2* flew by the moons of Jupiter. The photographs and other data they transmitted back to the earth showed nine volcanoes erupting on Io.

Io, about the size of the earth's moon, is probably the most volcanically active body in the solar system. Scientists calculate that volcanoes on Io eject several thousand metric tons of material each second. This means that every month the surface of Io is covered with volcanic material equal to that ejected by Mount St. Helens in May 1980.

Unlike volcanoes on earth, the material that erupts from the volcanoes on Io is neither mafic nor felsic. Because Io is colored a brilliant yellow-red, scientists think the volcanic material is primarily sulfur and sulfur dioxide.

Volcanoes on Io appear to be much more powerful than those on earth. Plumes of volcanic material reach heights of hundreds of kilometers. In general, the plumes are shaped like giant umbrellas.

Io moves inward and outward in its orbit around Jupiter because of the gravitational pull of the other moons of Jupiter. As Io is pulled back and forth, its surface also moves in and out. Heat from the friction caused by this surface movement probably results in the melting of the interior of Io and leads to volcanism.

Section 7.3 Review

1. What might be responsible for most of the craters on earth's moon?
2. What is the most likely explanation for the growth of Olympus Mons to its present size?
3. Why do scientists think Io, one of the moons of Jupiter, is made of sulfur and sulfur dioxide?
4. What evidence is there for volcanism on other planets and moons in the solar system?

Chapter 7 Review

Key Terms

- aa (120)
- caldera (124)
- cinder cone (122)
- composite cone (123)
- crater (124)
- felsic lava (120)
- fissure (119)
- hot spot (119)
- lapilli (121)
- lava (117)
- mafic lava (120)
- magma (117)
- pahoehoe (120)
- pillow lava (120)
- pyroclastic material (121)
- shield cone (122)
- stratovolcano (123)
- tephra (121)
- vent (117)
- volcanic ash (121)
- volcanic block (122)
- volcanic bomb (122)
- volcanic dust (121)
- volcanism (117)
- volcano (117)

Key Concepts

Magma develops in the asthenosphere. **See page 117.**

Volcanism is the movement of magma toward the earth's surface. **See page 117.**

Volcanism is common at the boundaries of lithospheric plates. **See page 117.**

The collision of lithospheric plates might produce continental volcanoes or oceanic island arcs. **See page 118.**

There are two main types of lava, each with a very different chemical composition. **See page 120.**

Hot, thin lava is associated with quiet eruptions; cooler, thicker, gas-containing lava is associated with explosive eruptions. **See page 121.**

A volcano ejects a variety of tephra during an explosive eruption. **See page 121.**

Volcanic cones are classified into three categories according to composition and form. **See page 122.**

Events that might signal a volcanic eruption include changes in earthquake activity, surface changes in the volcano, and changes in the composition of gases given off by the volcano. **See page 124.**

Evidence of volcanism is common throughout our solar system. **See page 126.**

Volcanism on Io differs significantly from volcanism on earth. **See page 127.**

Review

On your own paper, write the letter of the term that best completes each of the following statements.

1. Factors that allow magma to push its way upward include temperature and
a. color. b. density. c. crust. d. thickness.

2. Activity caused by the movement of magma is called
a. extraterrestrial. b. tephra.
c. volcanism. d. subduction.

3. The belt of volcanoes that encircles the Pacific Ocean is called
a. the subduction zone. b. an island arc.
c. a hot spot. d. the Pacific Ring of Fire.

4. Island arcs are formed by the collision of
a. two plates with continental crust at their edges.
b. two calderas.
c. two volcanic bombs.
d. two plates with oceanic crust at their edges.

5. Areas of volcanism within plates are called
a. hot spots. b. calderas. c. cones. d. fissures.

6. Lava that breaks into jagged chunks when it is subjected to rapid cooling is called
a. aa. b. pahoehoe. c. pillow lava. d. felsic lava.

7. Explosive volcanic eruptions result from
a. mafic lava. b. tephra lava. c. felsic lava. d. pahoehoe lava.

8. Tephra that forms into rounded or spindle shapes as it flies through the air is called
a. ash. b. lapilli.
c. volcanic bombs. d. volcanic blocks.

9. The Hawaiian Islands are formed from
a. shield cones. b. cinder cones.
c. composite cones. d. calderas.

10. A cone formed only by solid fragments built up around a volcanic opening is a
a. shield cone. b. cinder cone.
c. composite cone. d. stratovolcano.

11. The depression that results when a cone collapses into an empty magma chamber is a
a. crater. b. vent. c. caldera. d. fissure.

12. Shortly before a volcano erupts, magma may cause its surface to
a. bulge out. b. cave in. c. get darker. d. melt.

13. Scientists have discovered that before an eruption, earthquakes
a. completely stop. b. increase in number.
c. bear no relation to volcanism. d. decrease in number.

14. Olympus Mons, the largest known volcano in the solar system, is found on
a. the moon. b. Io. c. Mars. d. Venus.

15. The material ejected from volcanoes on Io is probably
a. sulfur and sulfur dioxide. b. basalt.
c. granite. d. lapilli.

Critical Thinking

On your own paper, write answers to the following questions.

1. Most lava that forms on the earth's surface goes unnoticed and unobserved. Why does this happen?
2. Can you assume that every mountain located along the edge of a continent is a volcano? Why or why not?
3. Why would felsic lava not produce a natural tunnel within itself as it cools?
4. If craters on a planetary body have channels in their sloping sides, are the craters more likely the result of meteorite bombardment or volcanism? Explain.

Application

1. While exploring a volcano that has been dormant, you observe volcanic ash first and lapilli later. Are you more likely to be moving toward or away from the volcanic opening? Explain your answer.
2. If you see a steep volcanic cone that is only 300 m high, what can you assume about the type of cone and its composition?
3. To predict a volcanic eruption, what kinds of information would you seek?
4. Concept map ? ? ? Construct a **concept map** using as many new terms as possible from the list on the previous page to illustrate the concepts in this chapter.

Extension

1. Draw a series of diagrams that depicts the usual order of events in the formation of a subduction-zone volcano. Start with lithospheric plates colliding, and end with a cone that has been worn away.
2. Use modeling clay of various colors to build models of the three major types of volcanic cones. Label each part.
3. Using Internet resources, prepare a report on the latest evidence for extraterrestrial volcanism.

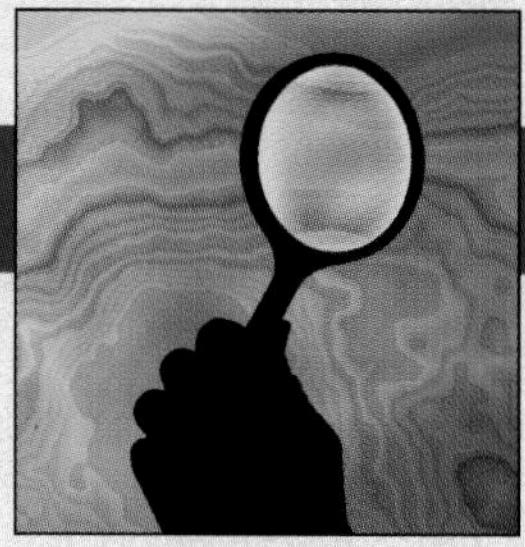

In-Depth Investigation

Hot Spots and Volcanoes

Materials

- cardboard (about 20 × 30 cm)
- red-colored gelatin (about 16 mL)
- white poster board (about 15 × 25 cm)
- eyedropper or pipet
- plastic squeeze bottle (16 mL capacity)
- metric ruler
- sharp pencil
- lab apron
- scissors

Introduction

Geologists predict that there are about 100 hot spots on the earth. A hot spot is a place where an extraordinary amount of heat rises from the asthenosphere on a plume of magma. Volcanoes grow where the magma reaches the surface.

The cause of the hot spots is not fully understood. However, it is thought that hot spots maintain their position in the asthenosphere or move very slowly in comparison with plate movement. Therefore, lines of volcanic islands show how the lithospheric plates have moved over the hot spots. By dating the volcanic rocks of such islands, the rate and direction of plate motion are revealed.

In this investigation you will construct a model to demonstrate how the movement of the Pacific plate is revealed by the orientation of the Hawaiian Islands and others. You will also demonstrate the relationship between hot spots and volcanoes.

Prelab Preparation

1. Review Section 7.1, pages 117–119; Section 4.2, pages 72–77; and Section 5.3, pages 88–93.
2. Obtain a map showing the Pacific plate and its relation to surrounding plates. Study the North Pacific Ocean on the map. Locate the Emperor Seamount chain and the Hawaiian Islands. Count and record the number of individual mountains shown on the map.
3. Mix the gelatin the day prior to the lab.

Procedure

1. Draw a circle about 3 mm in diameter at the center of the cardboard. Using a sharp pencil, make a hole in the cardboard from the underside. *Note: Make the hole slowly and carefully. Do not try to punch or jab at the cardboard with the point.*
2. Again using the sharpened pencil, carefully make six holes in the poster board in the pattern shown in the illustration at left. Each hole should be about 3 mm in diameter. Be sure to make the holes from the underside of the poster board. Write the name "Pacific Plate" on the bottom of the paper.

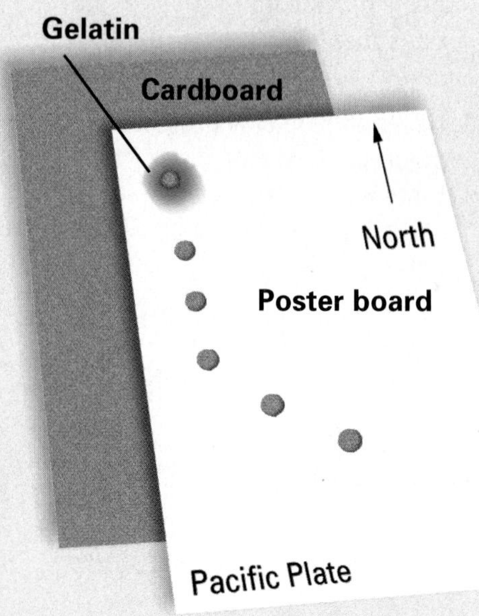

Draw an arrow toward the top of the paper, and label it "North," as shown in the illustration.

3. Cut the tip of the squeeze bottle so that the opening is about 3 mm in diameter. Fill the bottle with gelatin using an eyedropper.
4. While your partner holds the cardboard, insert the tip of the squeeze bottle through the hole in the cardboard from underneath.
5. Place the upper left hole in the poster board over the top of the squeeze bottle. Squeeze the bottle so that about 2 mL of gelatin are deposited on the paper around the tip of the bottle. Make certain the bottle tip remains tightly against the hole while squeezing and that the tip is inserted from the underside of the cardboard.
6. Move the poster board up and left to the next hole, and repeat the procedure. Continue the same procedure for the third and fourth holes.
7. Continue to the fifth and sixth holes, but deposit about 4 mL of gelatin at each hole.
8. Remove the bottle, and place the poster board on your lab table.
9. Write Midway Island, Kauai, and Hawaii next to the appropriate pile of gelatin. Then label the Emperor Seamounts and the Hawaiian Islands.
10. Write the age "70 million years" next to the first volcano that you made. Label the third volcano "40 million years." Label the last volcano "Presently active."
11. Clean up your work space, and store the equipment as directed by your teacher.

Analysis and Conclusions

1. Describe the direction of motion of the Pacific plate over the last 70 million years as demonstrated by the model.
2. The hot-spot model represents only a small number of volcanoes that exist in the Emperor Seamount chain and the Hawaiian Islands. Based on the count of volcanoes you completed in the prelab preparation, does it seem that the hot spot now under the Hawaiian Islands has been active for at least 70 million years? Explain.
3. The distance from the southernmost tip of the island of Hawaii to Midway Island is about 2,400 km.
   a. What is the rate of motion of the Pacific plate (in centimeters per year), assuming that the plate traveled that distance in 40 million years?
   b. Geologists estimate that the current rate of motion of the Pacific plate is 2 cm/yr. If this rate is different from the rate that you determined, what is the explanation for the difference?
   c. Explain the possible relation between the rate of plate motion and the size of the volcanic islands in the Hawaiian Islands.

Extension

Obtain a map of Atlantic Ocean seafloor topography to answer the following. Both Iceland and the Azores Islands are located over the Mid-Atlantic Ridge. What are the differences and similarities between these islands and the Hawaiian Island chain?

A Case of the Tennessee Shakes

Materials

road map of Tennessee

Did you know that almost 10,000 earthquakes occur every day? In fact, there is a good chance that an earthquake is occurring right now somewhere in the world. Fortunately, less than 2 percent of the earthquakes that seismographs record are strong enough to do serious damage.

Earthquake Frequency (Based on Observations Since 1900)

Descriptor	Magnitude	Average occurring annually
Great	8.0 and higher	1
Major	7.0 – 7.9	18
Strong	6.0 – 6.9	120
Moderate	5.0 – 5.9	800
Light	4.0 – 4.9	about 6,200
Minor	3.0 – 3.9	about 49,000
Very minor	2.0 – 2.9	about 1,000 per day
	1.0 – 1.9	about 80,000 per day

▲ **A strong earthquake struck Kobe, Japan, on January 17, 1995. This earthquake had a magnitude of about 7.2 on the Richter scale.**

You might think that scientists are most interested in the strongest earthquakes. In fact, weak earthquakes can tell a *seismologist* (someone who studies earthquakes) even more than strong ones can.

1. Examine the maps on pages 72 and 101. Which lithospheric plates are involved with the earthquakes that occur in North America? Why do earthquakes occur mostly along the boundaries of these plates?

2. The map on page 101 indicates that some earthquakes occur in the interior of the northwestern United States. Propose a hypothesis that explains these earthquakes in terms of plate tectonics.

3. At the boundaries of a lithospheric plate, the epicenters of earthquakes tend to be packed together closely, or densely distributed. Describe the patterns of distribution that you observe on the map shown here.

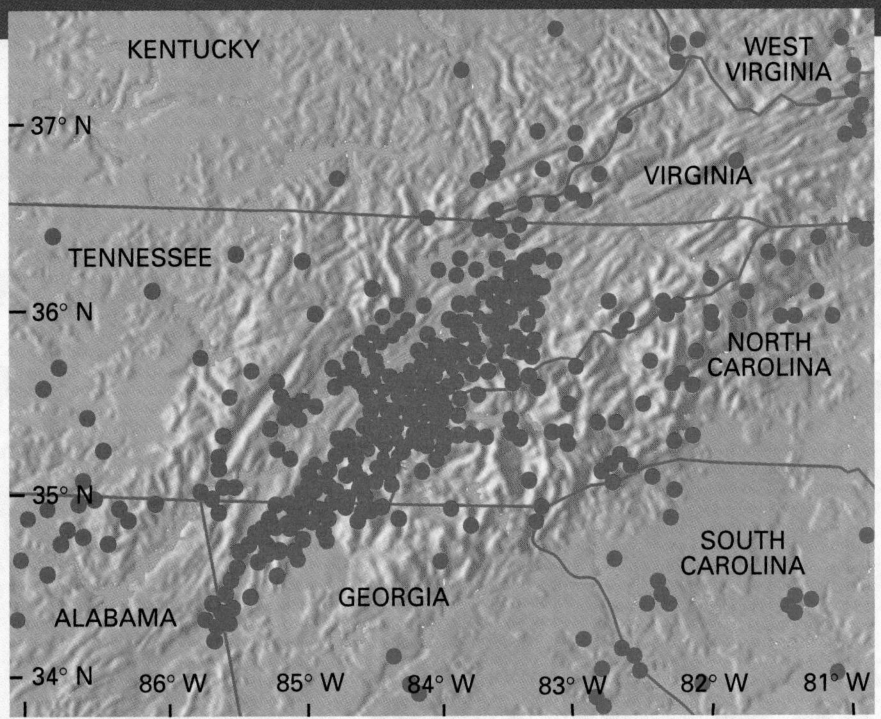

Each dot on this map represents the epicenter of an earthquake. Most of these earthquakes, which occurred over a 20-year period, were too weak to be felt by people.

Review the EarthBeat feature on page 102. It describes the New Madrid earthquake zone, which passes through southeastern Missouri and western Tennessee. Scientists think that the earthquakes in the New Madrid zone are the result of an ancient fault that has been reactivated. Now look again at the map on this page, which shows the epicenters of earthquakes in eastern Tennessee. Some scientists think that these earthquakes are also the result of an ancient fault. Other scientists disagree. Christine Powell, who has analyzed several decades of earthquake data, has proposed that a *new* fault zone may be forming in eastern Tennessee. If this is true, there is a possibility that eastern Tennessee may experience a major earthquake in this zone.

4. Describe the location of the eastern Tennessee seismic zone (ETSZ) in terms of longitude and latitude.

5. Using a road map of Tennessee, name at least two major cities that are located in the ETSZ. How could a major earthquake affect these cities?

6. Two nuclear power plants are located in the ETSZ. Imagine that a company has plans to build a plant near McMinnville, Tennessee. The USGS has hired you to advise this company about the risk of a major earthquake. Briefly describe what you would say in a letter to the company, explaining your reasoning as clearly as possible.

Going Further

One of the organizations that monitors earthquakes in Tennessee is the Center for Earthquake Research and Information at the University of Memphis. You can find out about recent earthquakes in the area by visiting CERI online at www.ceri.memphis.edu.

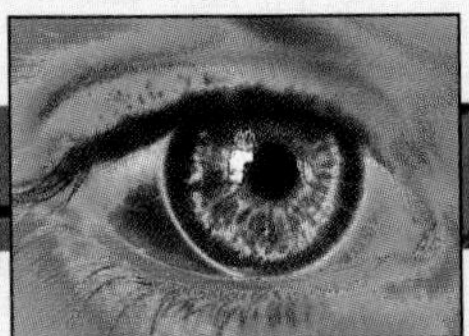

Eye on the Environment

Volcanoes Put a Chill in the Air

Imagine world temperatures rising to the point that ecosystems across the globe are drastically changed. Or consider the possibility of coastal beaches and cities permanently covered by a rising sea. Some scientists predict that such scenarios could become realities during your lifetime.

Hot Topics

When humans burn fossil fuels, such as gasoline and coal, the carbon dioxide that is released into the air disrupts the delicate balance of gases in the atmosphere. This added carbon dioxide traps extra heat at the earth's surface, increasing the natural "greenhouse effect." Because of this, some scientists think that average temperatures around the globe will rise at least 2°C by the year 2050. This could alter the earth's weather, cause melting at the polar icecaps, and cause sea levels to rise. In order to understand the possible effects of human activity on the earth's environment, scientists are studying another source of atmospheric disturbance—volcanoes.

▲ **During and after the eruption of Mount Pinatubo in the Philippines**

The lava, gases, dust, and ash that spew into the atmosphere during a large volcanic eruption can darken the skies for hundreds of kilometers in all directions. In the months that follow an eruption, some of this material settles to the earth, but much of it remains in the atmosphere, where wind currents disperse it around the world. At this point it is no longer visible to the human eye, yet the dust and gases from a single volcano may nonetheless affect the climate of the entire planet.

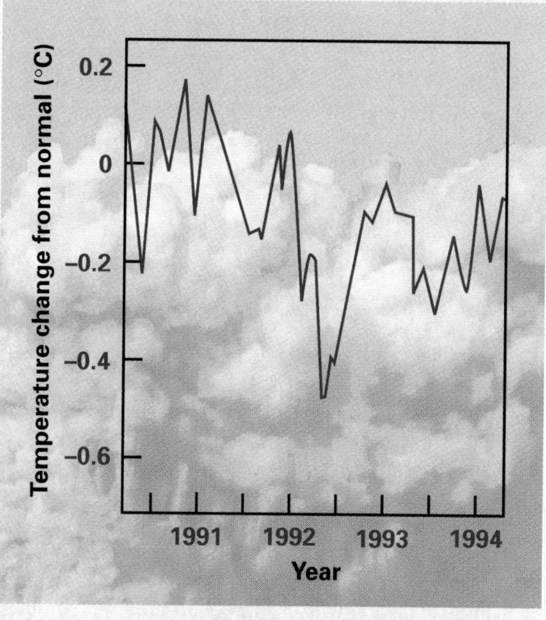

▲ **The low temperatures in 1992 are most likely the result of Mount Pinatubo's eruption.**

Cool Ideas About Volcanoes

In June of 1991, Mount Pinatubo began a long series of eruptions, filling the skies over the Philippines. The resulting lava, ash, and mud flows devastated 20,000 km^2 of land, destroying homes, crops, livestock, roads, bridges, and forests. In addition, about 20 million tons of sulfur dioxide was released into the atmosphere. Once in the atmosphere, sulfur dioxide combines with water to become sulfuric acid. Whereas carbon dioxide traps the sun's energy within the atmosphere, sulfuric acid seems to reflect the sun's energy back out into space. Because the sun provides warmth as well as light, scientists predicted that adding large amounts of sulfur dioxide to the atmosphere would have a cooling effect on the earth's surface.

Scientists carefully observed the effects of Mount Pinatubo to find out if their ideas about volcanic gases in the atmosphere were correct. As predicted, average global surface temperatures dropped about 0.6°C by late 1992 and began to recover slowly after that. (That may not seem like much, but remember that temperatures during the last ice age were only about 5°C cooler than they are today!) These dramatic temperature decreases served to offset global warming, if only temporarily.

Analyzing the Past and the Future

Using data collected from the eruptions of Mount Pinatubo and other volcanoes, scientists are developing computer models that may help them better understand and predict global climate changes. Learning how the atmosphere is affected by such natural disasters will also help scientists understand how pollution from human sources affects the atmosphere. This knowledge may be instrumental in ensuring the future health of our atmosphere and all of the life that depends on it.

Think About It

How would you determine if the 1991 Mount Pinatubo eruption affected the average monthly temperatures in your community?

Extension: The 1980 eruption of Mount St. Helens had local and widespread effects. Look into how this eruption changed nearby ecosystems, and find out which effects are still noticeable today.

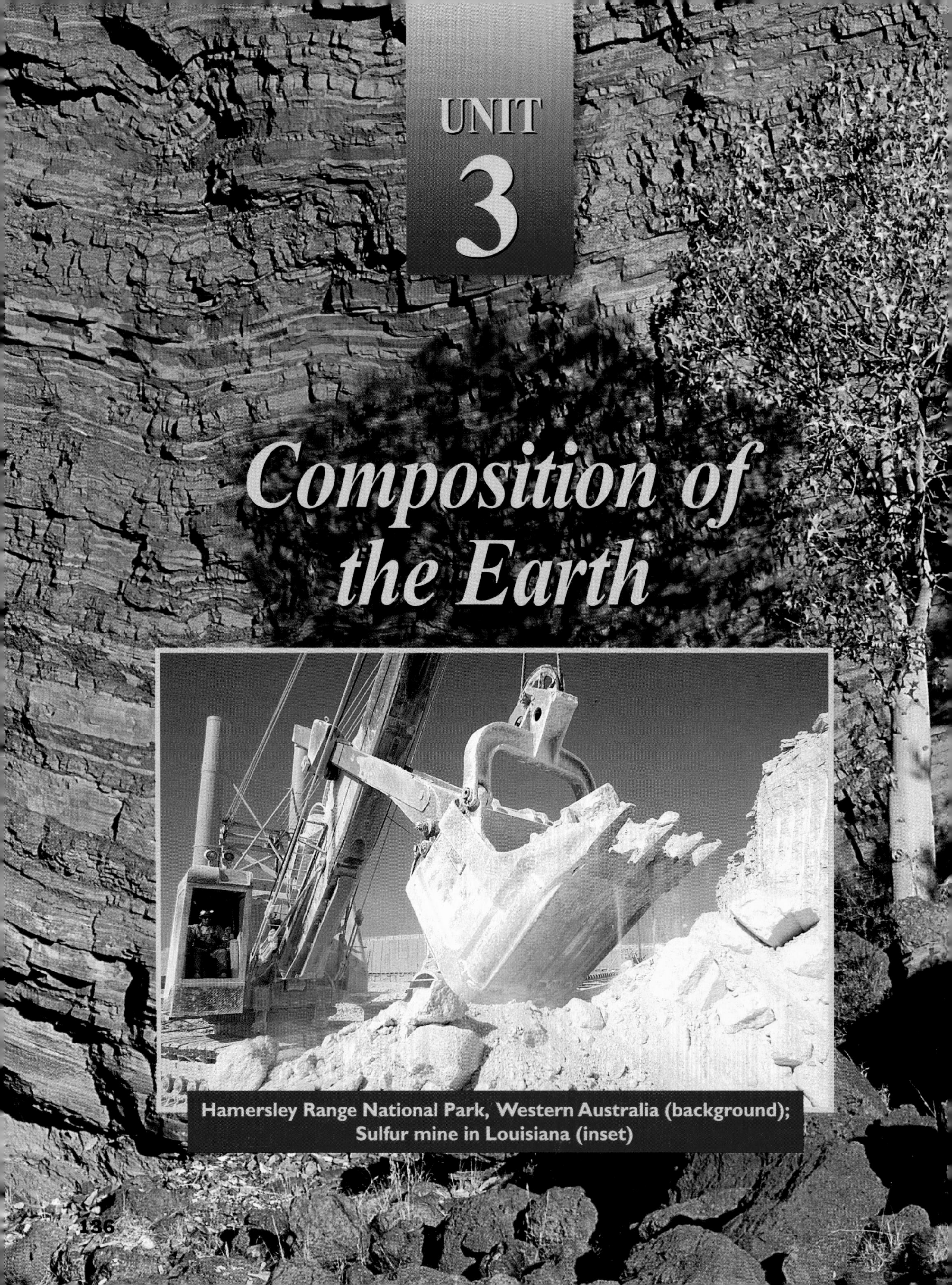

Hamersley Range National Park, Western Australia (background); Sulfur mine in Louisiana (inset)

Introducing Unit Three

Scientists at the U.S. Geological Survey were among the first to describe the minerals found in lunar samples collected during the *Apollo II* Mission. I will never forget the excitement at the first Lunar Science Conference. Hundreds of dignitaries and scientists from around the world had come to hear what our findings would be.

When you learn something brand new about the earth's chemistry and composition, there's that same kind of excitement. Although scientists have been studying our planet for a long time, the possibilities for new discoveries remain endless. The earth still poses so many questions. The answer to one question often leads to a dozen more questions to take its place.

In this unit, you will explore some of the answers that have been discovered about the composition of the earth and will explore some of the questions that remain.

Malcolm Ross

Malcolm Ross
Research Mineralogist
U.S. Geological Survey

In this UNIT

Chapter 8

Earth Chemistry

Scientists have long sought to understand the universe by examining increasingly smaller parts of it. Scientists have studied parts as large as planet earth. They have classified the material that makes up the earth into solids, liquids, and gases. In their quest to better understand the earth, scientists search for smaller and smaller parts and have turned their attention to studying atoms and their interactions.

Chapter Outline

◀ This scene is from Geyser Basin in Yellowstone National Park, Wyoming.

8.1 Matter

Section Objectives

- **State the distinguishing characteristic of an element.**
- **Describe the basic structure of an atom.**
- **Define atomic number and mass number.**
- **Explain what an isotope is.**
- **Compare solids, liquids, and gases.**

Every object in the universe is made of particles of some kind of substance. Scientists use the word **matter** to describe the substance of which an object is made. Matter is anything that takes up space and has *mass*. The amount of matter in any object determines the mass of that object. Scientists are able to identify the kind of matter that makes up a substance by observing the properties of that substance.

All matter has two major distinguishing properties. **Physical properties** are those characteristics that can be observed without changing the composition of the substance. For example, physical properties include density, color, hardness, freezing point, boiling point, and the ability to conduct electricity. **Chemical properties** are those characteristics that describe how a substance interacts with other substances to produce different kinds of matter. For example, a chemical property of iron is that it interacts with oxygen to form rust. A chemical property of helium is that it usually does not interact with other substances. Understanding the chemical properties of a substance requires some basic information about the particles that make up all substances.

Atoms and Elements

All matter is made up of *elements*. An element is a substance that cannot be broken down into a simpler form by ordinary chemical means. Figure 8–1 shows the most common elements in the earth's crust. Notice that there is a universally understood symbol of one or two letters that represents each element. Over 90 elements occur

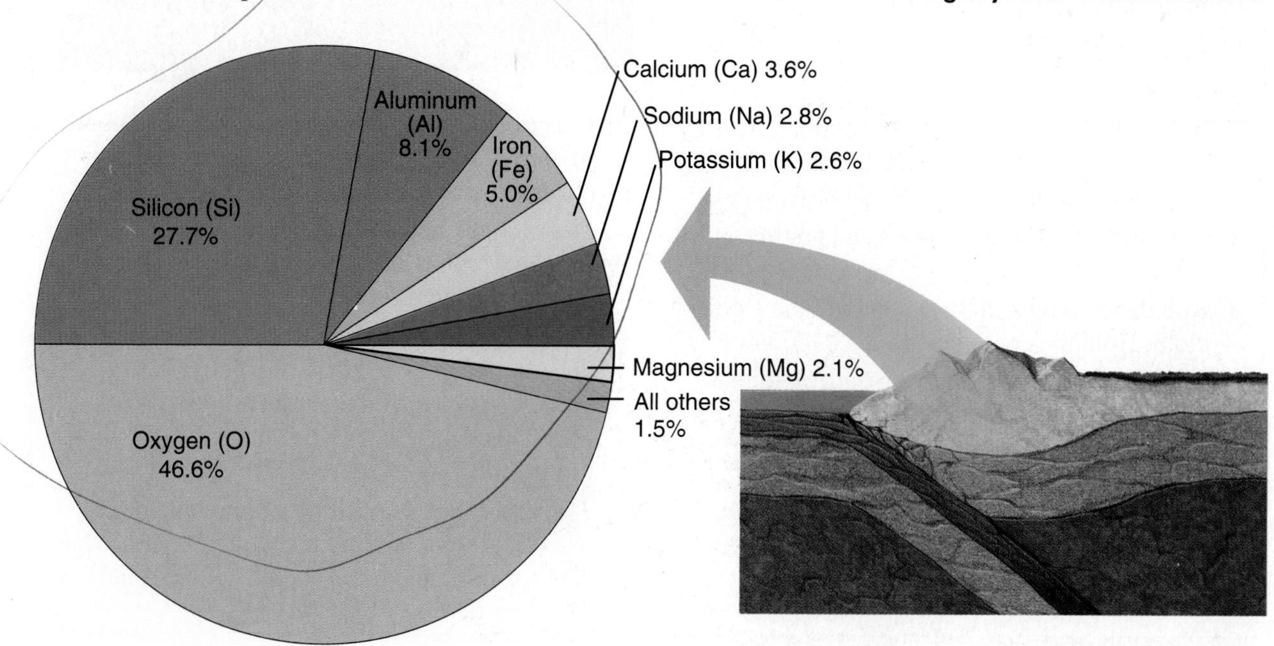

Figure 8–1. The earth's crust is made up primarily of these elements. The graph shows the percentage by mass of each element.

Figure 8–2 Quartz is made of silicon and oxygen, the two most abundant elements in the earth's crust.

naturally in the earth. Another dozen or so have been created in the laboratory. Of the natural elements, only eight make up over 98 percent of the earth's crust. In fact, two elements—silicon and oxygen—make up almost 75 percent of the earth's crust.

Elements consist of **atoms**. An atom is the smallest unit of an element and has all the properties of that element. The word *atom* comes from a Greek word for "indivisible." You can identify an element by its atoms. The atoms of any one element are significantly different from the atoms of any other element. Atoms cannot be broken down into smaller particles by ordinary physical or chemical processes.

An atom is so small that the size of a single atom is difficult to imagine. To get an idea of how small an atom is, look at the thickness of this page. More than a million atoms lined up side by side would about equal that thickness.

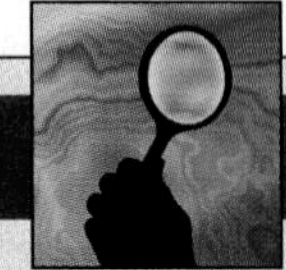

SMALL-SCALE INVESTIGATION

Electrically Charged Objects

Electrically charged objects behave in predictable ways. Objects with unlike charges attract each other; objects with like charges repel each other. Objects with a neutral charge neither attract nor repel each other. You can demonstrate this behavior with several common objects.

Materials

balloon, thread, ruler, rubber or plastic comb, drinking glass, wool cloth, plastic bag, books

Procedure

1. Inflate the balloon and hang it from the ruler by a piece of thread, as shown in the photo. Hold the comb next to the balloon and observe.
2. Repeat Step 1 using the glass and not the comb.
3. Rub the balloon and the comb with the wool cloth. Hold the comb next to the balloon.
4. Rub the glass with the plastic bag. Hold the glass next to the balloon.

Analysis and Conclusions

1. Describe the behavior of the balloon, comb, and glass in Steps 1 and 2. What can you conclude about their electrical charges?
2. Describe the behavior of the balloon and comb in Step 3. What can you conclude about their electrical charges?
3. Describe the behavior of the balloon and glass in Step 4. What can you conclude about their electrical charges?
4. Rubbing the glass with the plastic bag produces a positive electrical charge on the glass. What is the charge of the balloon in Steps 3 and 4? What is the charge of the comb in Step 3?
5. Two ions, such as positive sodium and negative chloride, may form a chemical bond. What can you infer about the charges of the ions?

Atomic Structure

As tiny as atoms are, they are made up of even smaller parts called *subatomic particles*. The three major kinds of subatomic particles are **electrons, protons,** and **neutrons.** Electrons carry a negative electrical charge, protons carry a positive charge, and neutrons are neutral, having no net electrical charge.

Subatomic particles are arranged in a similar way within all atoms. The protons and neutrons of an atom are packed close to one another. Together they form the **nucleus,** a small region in the center of an atom. Because neutrons do not carry an electrical charge, the protons give a nucleus a positive charge. Compared with the size of the whole atom, the nucleus takes up about the same space as a gumdrop in a football stadium!

The electrons of an atom move in a certain region of space around the nucleus known as an **electron cloud.** Because unlike electrical charges attract, the negatively charged electrons are attracted to the positively charged nucleus. Consequently, the electrons tend to remain relatively close to the nucleus of an atom. However, the constant motion of the electrons keeps them from falling into the nucleus.

An atom of a specific element is distinguished from the atoms of all other kinds of elements by the number of its protons. For example, an atom of the element lithium contains three protons. No other element has three protons.

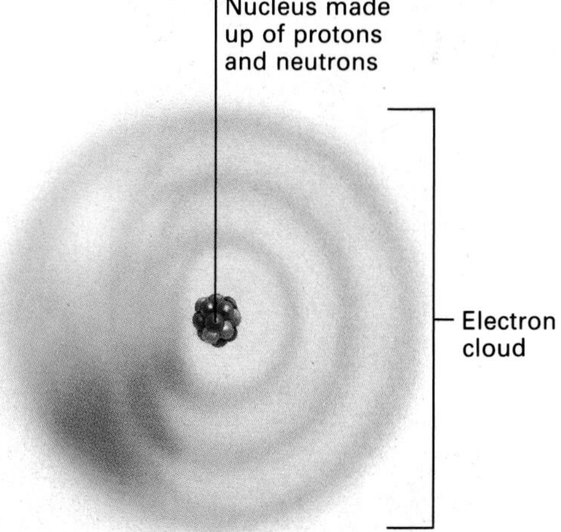

Figure 8–3 The nucleus of the atom contains the protons and neutrons. The protons give the nucleus a positive charge. The negatively charged electrons are in the electron cloud around the nucleus.

Atomic Number

To help in the identification of elements, scientists have assigned an **atomic number** to each kind of atom. All atoms of any one element have the same atomic number. The atomic number is equal to the number of protons in the atom. An uncharged atom has an equal number of protons and electrons and so has neither a positive nor negative charge. Therefore, the atom is electrically neutral. Thus, the atomic number also equals the number of electrons in an uncharged atom of that element. For example, the atomic number of oxygen, which is 8, shows that an oxygen atom has eight protons and eight electrons. The atomic number of 1 for hydrogen shows that each atom has only one proton and one electron.

Atomic number is one of the components listed for each element in the **periodic table.** The periodic table is a system for classifying elements.

Mass Number

Each atom has a **mass number.** The mass number of any atom represents the sum of the number of protons and neutrons in that atom. The actual mass of a subatomic particle is so small that, rather than grams, a special unit called the *atomic mass unit* (u) is used. Protons and neutrons each have an atomic mass close to 1 u. In contrast, electrons are much less massive than protons and neutrons. In fact, it takes the combined mass of about 1,840 electrons to equal the mass

The Periodic Table

Metals

Atomic mass	Element symbol	Element name	Atomic number	Electrons
12.01	C	Carbon	6	2, 4

Period	1	2	3	4	5	6	7	8	9
1	1.01 **H** Hydrogen 1 1								
2	6.94 **Li** Lithium 3 2, 1	9.01 **Be** Beryllium 4 2, 2							
3	22.99 **Na** Sodium 11 2, 8, 1	24.30 **Mg** Magnesium 12 2, 8, 2							
4	39.10 **K** Potassium 19 2, 8, 8, 1	40.08 **Ca** Calcium 20 2, 8, 8, 2	44.96 **Sc** Scandium 21 2, 8, 9, 2	47.88 **Ti** Titanium 22 2, 8, 10, 2	50.94 **V** Vanadium 23 2, 8, 11, 2	52.00 **Cr** Chromium 24 2, 8, 13, 1	54.94 **Mn** Manganese 25 2, 8, 13, 2	55.85 **Fe** Iron 26 2, 8, 14, 2	58.93 **Co** Cobalt 27 2, 8, 15, 2
5	85.47 **Rb** Rubidium 37 2, 8, 18, 8, 1	87.62 **Sr** Strontium 38 2, 8, 18, 8, 2	88.91 **Y** Yttrium 39 2, 8, 18, 9, 2	91.22 **Zr** Zirconium 40 2, 8, 18, 10, 2	92.91 **Nb** Niobium 41 2, 8, 18, 12, 1	95.94 **Mo** Molybdenum 42 2, 8, 18, 13, 1	(98) **Tc** Technetium 43 2, 8, 18, 13, 2	101.1 **Ru** Ruthenium 44 2, 8, 18, 15, 1	102.9 **Rh** Rhodium 45 2, 8, 18, 16, 1
6	132.9 **Cs** Cesium 55 2, 8, 18, 18, 8, 1	137.33 **Ba** Barium 56 2, 8, 18, 18, 8, 2	Lanthanide Series 57–71 See below	178.5 **Hf** Hafnium 72 2, 8, 18, 32, 10, 2	180.9 **Ta** Tantalum 73 2, 8, 18, 32, 11, 2	183.8 **W** Tungsten 74 2, 8, 18, 32, 12, 2	186.2 **Re** Rhenium 75 2, 8, 18, 32, 13, 2	190.2 **Os** Osmium 76 2, 8, 18, 32, 14, 2	192.2 **Ir** Iridium 77 2, 8, 18, 32, 15, 2
7	(223) **Fr** Francium 87 2, 8, 18, 32, 18, 8, 1	(226) **Ra** Radium 88 2, 8, 18, 32, 18, 8, 2	Actinide Series 89–103 See below	(261) ***** 104 2, 8, 18, 32, 32, 10, 2	(262) ***** 105 2, 8, 18, 32, 32, 11, 2	(263) ***** 106 2, 8, 18, 32, 32, 12, 2	(262) ***** 107 2, 8, 18, 32, 32, 13, 2	(265) ***** 108 2, 8, 18, 32, 32, 14, 2	(266) ***** 109 2, 8, 18, 32, 32, 15, 2

Lanthanide Series

138.9 **La** Lanthanum 57 2, 8, 18, 18, 9, 2	140.1 **Ce** Cerium 58 2, 8, 18, 20, 8, 2	140.9 **Pr** Praseodymium 59 2, 8, 18, 21, 8, 2	144.2 **Nd** Neodymium 60 2, 8, 18, 22, 8, 2	(145) **Pm** Promethium 61 2, 8, 18, 23, 8, 2	150.4 **Sm** Samarium 62 2, 8, 18, 24, 8, 2	152.0 **Eu** Europium 63 2, 8, 18, 25, 8, 2
(227) **Ac** Actinium 89 2, 8, 18, 32, 18, 9, 2	232.0 **Th** Thorium 90 2, 8, 18, 32, 18, 10, 2	231.0 **Pa** Protactinium 91 2, 8, 18, 32, 20, 9, 2	238.0 **U** Uranium 92 2, 8, 18, 32, 21, 9, 2	237.0 **Np** Neptunium 93 2, 8, 18, 32, 22, 9, 2	(244) **Pu** Plutonium 94 2, 8, 18, 32, 24, 8, 2	(243) **Am** Americium 95 2, 8, 18, 32, 25, 8, 2

Actinide Series

Nonmetals

10	11	12	13	14	15	16	17	18
								4.00 **He** Helium 2 (2)
			10.81 **B** Boron 5 (2, 3)	12.01 **C** Carbon 6 (2, 4)	14.01 **N** Nitrogen 7 (2, 5)	16.00 **O** Oxygen 8 (2, 6)	19.00 **F** Fluorine 9 (2, 7)	20.18 **Ne** Neon 10 (2, 8)
			26.98 **Al** Aluminum 13 (2, 8, 3)	28.09 **Si** Silicon 14 (2, 8, 4)	30.97 **P** Phosphorus 15 (2, 8, 5)	32.07 **S** Sulfur 16 (2, 8, 6)	35.45 **Cl** Chlorine 17 (2, 8, 7)	39.95 **Ar** Argon 18 (2, 8, 8)
58.69 **Ni** Nickel 28 (2, 8, 16, 2)	63.55 **Cu** Copper 29 (2, 8, 18, 1)	65.39 **Zn** Zinc 30 (2, 8, 18, 2)	69.72 **Ga** Gallium 31 (2, 8, 18, 3)	72.61 **Ge** Germanium 32 (2, 8, 18, 4)	74.92 **As** Arsenic 33 (2, 8, 18, 5)	78.96 **Se** Selenium 34 (2, 8, 18, 6)	79.90 **Br** Bromine 35 (2, 8, 18, 7)	83.80 **Kr** Krypton 36 (2, 8, 18, 8)
106.4 **Pd** Palladium 46 (2, 8, 18, 18, 0)	107.9 **Ag** Silver 47 (2, 8, 18, 18, 1)	112.4 **Cd** Cadmium 48 (2, 8, 18, 18, 2)	114.8 **In** Indium 49 (2, 8, 18, 18, 3)	118.7 **Sn** Tin 50 (2, 8, 18, 18, 4)	121.8 **Sb** Antimony 51 (2, 8, 18, 18, 5)	127.6 **Te** Tellurium 52 (2, 8, 18, 18, 6)	126.9 **I** Iodine 53 (2, 8, 18, 18, 7)	131.3 **Xe** Xenon 54 (2, 8, 18, 18, 8)
195.1 **Pt** Platinum 78 (2, 8, 18, 32, 17, 1)	197.0 **Au** Gold 79 (2, 8, 18, 32, 18, 1)	200.6 **Hg** Mercury 80 (2, 8, 18, 32, 18, 2)	204.4 **Tl** Thallium 81 (2, 8, 18, 32, 18, 3)	207.2 **Pb** Lead 82 (2, 8, 18, 32, 18, 4)	209.0 **Bi** Bismuth 83 (2, 8, 18, 32, 18, 5)	(209) **Po** Polonium 84 (2, 8, 18, 32, 18, 6)	(210) **At** Astatine 85 (2, 8, 18, 32, 18, 7)	(222) **Rn** Radon 86 (2, 8, 18, 32, 18, 8)
(269) * 110 (2, 8, 18, 32, 32, 17, 1)	(272) * 111 (2, 8, 18, 32, 32, 18, 1)	(277) * 112 (2, 8, 18, 32, 32, 18, 2)						

* Elements synthesized, but not officially named.

A value given in parentheses denotes the mass number of the isotope of longest known half-life.

157.2 **Gd** Gadolinium 64 (2, 8, 18, 25, 9, 2)	158.9 **Tb** Terbium 65 (2, 8, 18, 27, 8, 2)	162.5 **Dy** Dysprosium 66 (2, 8, 18, 28, 8, 2)	164.9 **Ho** Holmium 67 (2, 8, 18, 29, 8, 2)	167.3 **Er** Erbium 68 (2, 8, 18, 30, 8, 2)	168.9 **Tm** Thulium 69 (2, 8, 18, 31, 8, 2)	173.0 **Yb** Ytterbium 70 (2, 8, 18, 32, 8, 2)	175.0 **Lu** Lutetium 71 (2, 8, 18, 32, 9, 2)
(247) **Cm** Curium 96 (2, 8, 18, 32, 25, 9, 2)	(247) **Bk** Berkelium 97 (2, 8, 18, 32, 27, 8, 2)	(251) **Cf** Californium 98 (2, 8, 18, 32, 28, 8, 2)	(252) **Es** Einsteinium 99 (2, 8, 18, 32, 29, 8, 2)	(257) **Fm** Fermium 100 (2, 8, 18, 32, 30, 8, 2)	(258) **Md** Mendelevium 101 (2, 8, 18, 32, 31, 8, 2)	(259) **No** Nobelium 102 (2, 8, 18, 32, 32, 8, 2)	(262) **Lr** Lawrencium 103 (2, 8, 18, 32, 32, 9, 2)

of 1 proton. Because electrons add so little to the total mass of an atom, their mass may be ignored in calculating the mass of an atom.

You can use an element's atomic number and mass number to determine the number of protons, neutrons, and electrons in an atom of the element. For instance, if you find sodium on the periodic table, you will see that its atomic number is 11. Since the atomic number equals the number of protons or electrons, sodium has 11 protons and 11 electrons. Notice that sodium has an atomic mass of 22.99, which rounds to 23—the mass number. Since the mass number equals the sum of the protons and neutrons, the mass number minus the atomic number equals the number of neutrons. Therefore, subtracting 11 from 23, you find that sodium has 12 neutrons. What is the number of protons, electrons, and neutrons for nitrogen?

Isotopes

Although all atoms of a given element contain the same number of protons, they do not always contain the same number of neutrons. Because the mass number is equal to the sum of the protons and

SCIENCE & TECHNOLOGY

The Smallest Particles

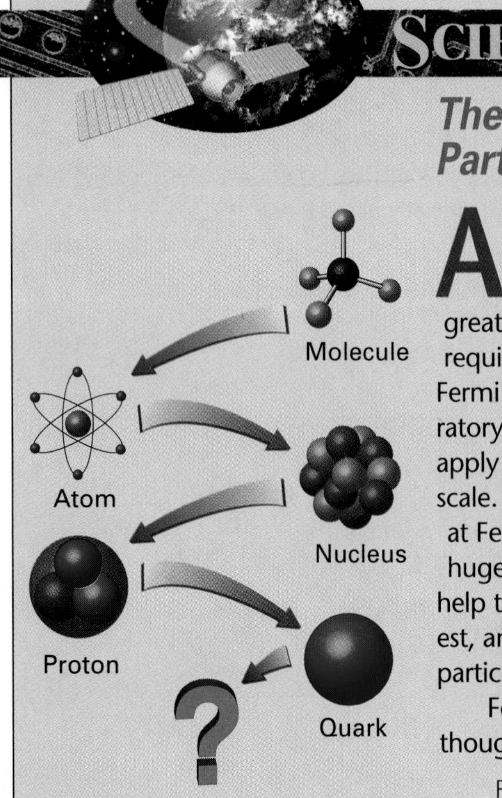

A fundamental principle of physics holds that the smaller an object is, the greater the amount of energy required to see it. Scientists at Fermi National Accelerator Laboratory near Chicago, Illinois, apply this principle on a grand scale. More than 850 physicists at Fermilab rely on Tevatron, a huge particle accelerator, to help them search for the smallest, and therefore most basic, particles of matter.

For a long time, scientists thought that the most basic particles of matter were atoms. Then when the protons, neutrons, and electrons found in atoms were discovered, these particles were thought to be the most basic. Now the honor goes to particles called *quarks* and *leptons.* The illustration to the left shows these increasingly smaller particles. As you can see,

Particle accelerators are so expensive to build and operate that they require international cooperation.

neutrons in a nucleus, each additional neutron increases the mass number. Atoms of the same element that differ from each other in mass number are called **isotopes** (IE-suh-topes). Isotopes of any given atom have the same number of protons but a different number of neutrons.

Hydrogen is an example of an element with several isotopes. All hydrogen atoms have the atomic number 1. That means a hydrogen atom has one proton in its nucleus and one electron moving around the nucleus. However, some hydrogen atoms also have one neutron in the nucleus. Because one neutron adds one unit to the mass, hydrogen atoms with one neutron in the nucleus have a mass number of 2. A very rare form of hydrogen has two neutrons in its nucleus. It has a mass number of 3. Figure 8–4 shows the three isotopes of hydrogen.

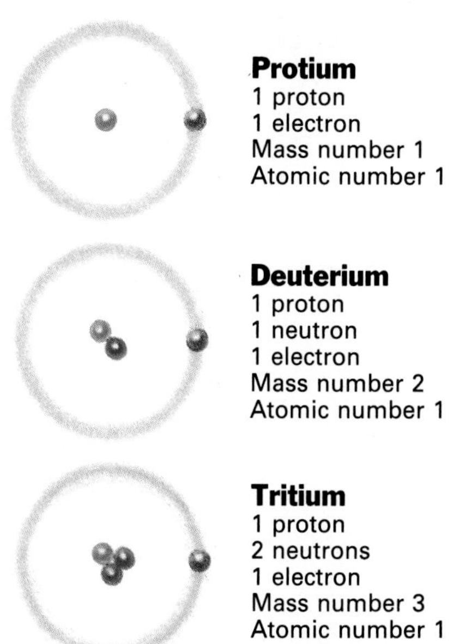

Figure 8–4 Why does tritium, an isotope of hydrogen, have a mass number of 3?

Solids, Liquids, and Gases

There are several ways to classify matter. One way is to classify it into three physical forms—**solid, liquid,** and **gas.** The particles that make up a solid are packed tightly together in fixed positions and are

however, recent evidence suggests that quarks may also be composed of something even smaller.

Because the amount of energy needed to see subatomic particles does not exist naturally, scientists have to build up enough energy to break apart these particles so individual parts can be isolated for study.

Tevatron, shown above, is an underground, ring-shaped tunnel, 6.4 km in circumference. It is built of 1,000 superconducting magnets that move beams of tiny particles at increasingly higher speeds. As they gain speed, the particles build up energy.

When the energized particles are moving near the speed of light (299,792,458 m/s), they are directed to hit either a fixed target or particles moving in an opposite direction. At impact, the particles split. The byproducts of the particles are separated and scattered. The smaller particles, quarks and leptons among them, are isolated by a collider detector, shown in the photo at far left.

For now, the nature of subatomic particles may seem far removed from earth science. However, research into matter's most basic particles may one day influence the work of earth scientists by changing basic theories ranging from the ultimate structure of matter to the origin of the universe.

You have learned that the word atom *comes from the Greek word meaning "indivisible." Is an atom really indivisible? Explain your answer.*

Solid

Liquid

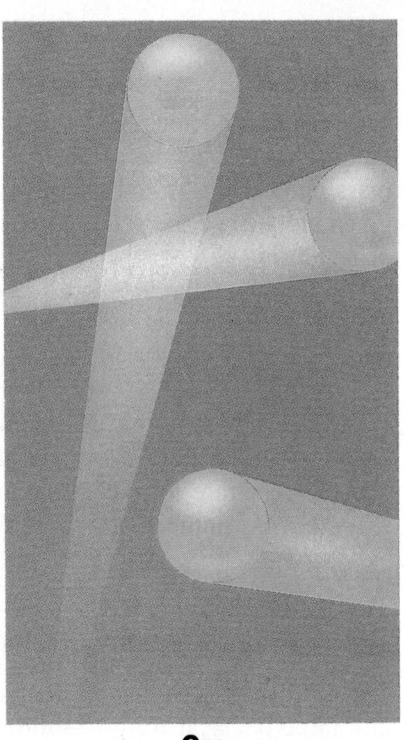

Gas

Figure 8–5 The relative motion of particles in solids, liquids, and gases is shown in these diagrams.

not free to move very much. Therefore, a solid has a definite shape and volume. A liquid has a definite volume but does not have a definite shape. Instead, a liquid takes the shape of the container that holds it. The particles that make up a liquid are also tightly packed, but they are more free to move in relation to each other than those in a solid. A gas does not have a definite volume or shape. The particles of a gas are farther apart and move faster and more freely than those of a liquid. A gas is thus a formless collection of particles that tends to expand in all directions at the same time. If a gas is not confined, the space between its particles will continue to increase. Look at Figure 8–5, which shows the relative motions of atoms in solids, liquids, and gases.

To melt a solid material, heat must be added. Heat causes the particles in the solid to move faster. When enough energy has been added, the individual particles move so rapidly that they break away from their fixed positions in relation to each other. This results in the material becoming a liquid. Additional heat causes the particles in a liquid to move even faster. The result is the formation of a gas.

Section 8.1 Review

1. How do chemical properties differ from physical properties?
2. What is an element?
3. Name the three basic subatomic particles.
4. Compare atomic number with mass number.
5. Explain how isotopes of an element differ.
6. In what physical form—solid, liquid, or gas—are particles farthest apart?
7. Which element has more electrons, iron (Fe) or carbon (C)? Explain your answer.

8.2 Combinations of Atoms

Section Objectives

- **Explain how atoms join together and form compounds.**
- **Describe two ways that electrons form chemical bonds between atoms.**
- **Read and interpret chemical formulas.**
- **Explain the difference between compounds and mixtures.**

Elements rarely occur in pure form in the earth's crust. Instead, they generally are found in combination with other kinds of elements. When the atoms of two or more elements are chemically united, the resulting substance is called a **compound.** A compound is a new substance with properties different from those of the elements that compose it. For example, water is a compound formed when atoms of hydrogen (H) and oxygen (O) combine. Sodium chloride, table salt, is a compound of sodium (Na) and chlorine (Cl) atoms.

The smallest complete unit of a compound is called a **molecule.** In the compound water, for example, every individual water molecule is made up of two atoms of hydrogen and one atom of oxygen.

Some elements exist naturally as **diatomic** molecules, molecules made up of two atoms. For example, hydrogen occurs naturally only as a diatomic molecule. Free oxygen, oxygen that is not part of a compound, also occurs as a diatomic molecule. The oxygen in the air you breathe is the diatomic molecule O_2. The *O* in this notation is the symbol for oxygen. The subscript *2* indicates the number of atoms of oxygen.

Electron Energy Levels

Different kinds of atoms join together and form compounds based upon the way their electrons are arranged. Remember that the electrons in an atom move in an electron cloud outside the nucleus. Within the electron cloud of an atom, electrons are arranged in **energy levels.** The electrons in each energy level have a specific amount of energy, and each energy level can hold only a certain maximum number of electrons. For example, the first energy level can hold a maxium of two electrons, and the second can hold up to eight electrons.

An electron occupies only one energy level at a time, although it may go to higher or lower levels. Also, the electrons of an atom may fill several different levels. For example, the electrons of a large atom such as radium may occupy seven levels. However many energy levels are occupied in an atom, the outermost level can never hold more than eight electrons. What is the number of electrons that an element with one energy level can have?

Not all atoms have the maximum number of electrons in their outermost energy level. However, the most stable atoms are those in which the outermost energy level is filled. In other words, the outermost level in these atoms holds the maximum number of electrons, or eight electrons. Atoms with filled outer energy levels do not easily lose or gain electrons, and so they do not easily form compounds with other elements.

Atoms of certain elements give up electrons more easily than atoms of certain other elements. Atoms with only one, two, or three electrons in the outermost level give up electrons easily. The

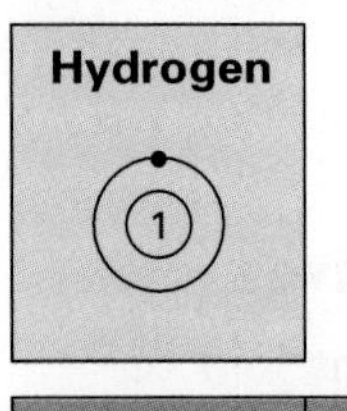

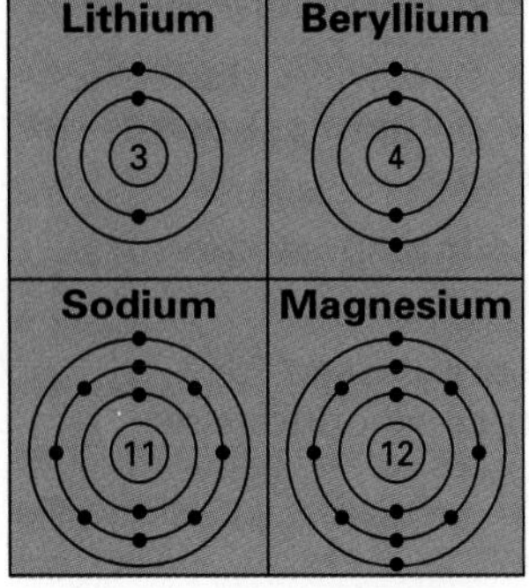

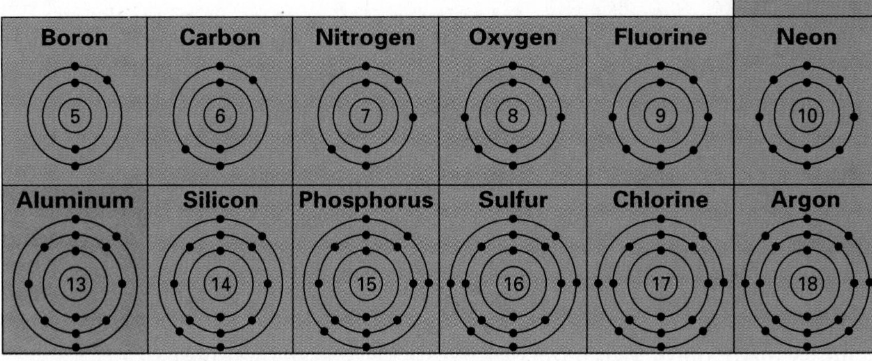

Figure 8–6 The diagrams above show the electron energy levels of the first 18 elements in the periodic table. The number at the center of each diagram indicates the number of protons in the nucleus.

elements that have these limited numbers of electrons in the outermost level are the same elements that have metallic properties and are classified as metals. Examples of metals are aluminum (Al), cobalt (Co), copper (Cu), gold (Au), iron (Fe), lead (Pb), and tin (Sn). Metals have a number of useful characteristics. Most metals are good conductors of heat and electricity. Many metals are malleable, which means they can be hammered into thin sheets. These metals are also ductile—they can be drawn out into wire.

Atoms with four, five, or six electrons in the outermost level do not lose electrons easily. These are elements without metallic properties and are classified as nonmetals. Carbon (C), nitrogen (N), and oxygen (O) are examples.

Chemical Bonds

The atoms that combine to make up a compound are held together by forces called **chemical bonds.** A chemical bond is produced by the interaction of electrons from the outermost energy levels of two or more atoms. Scientists can predict which kinds of atoms will form chemical bonds with each other. They compare the number of electrons in the outermost energy level of the atoms with the maximum number possible for that energy level. For example, a hydrogen atom has only one electron in its single energy level. Because this one energy level can hold two electrons, you can predict that hydrogen will accept another electron if it is available.

Atoms can form chemical bonds by sharing electrons or by transferring electrons from one atom to another. In both cases, the attraction that joins the atoms is the pull of an electrical charge. When electrons are transferred from one atom to another, the bond is an **ionic bond.**

Ionic Bonds

A compound formed through the transfer of electrons is called an **ionic compound.** Most ionic compounds are formed by the transfer of electrons from a metal to a nonmetal. A familiar example is sodium chloride, or common table salt. A sodium atom has 11 electrons, with 2 electrons in its first energy level, 8 electrons in its second energy level, and 1 electron in its outermost energy level. A chlorine atom has 17 electrons, with 2 electrons in its first energy level, 8 electrons in its second energy level, and 7 in its outer level. With only 7 electrons in its outer energy level, chlorine can accept 1 more electron. Because of the number of electrons in their respective outermost energy levels, sodium atoms have a strong tendency to lose 1 electron and chlorine atoms tend to gain 1 electron. Therefore, a sodium atom will give up an electron to a chlorine atom, as shown in Figure 8–7.

When an electron is transferred from one atom to another, both atoms become electrically charged. An **ion** is an atom or group of atoms that carries an electrical charge. A sodium atom becomes a positive ion (Na^+) when it gives up an electron. The loss of an electron leaves a sodium atom with 11 protons in its nucleus but only 10 electrons. In a similar way, a chlorine atom becomes an ion when it gains an electron. The additional electron gives a chlorine atom 18 electrons, one more than its 17 protons. The extra electron changes the neutral chlorine atom into a negatively charged chloride ion (Cl^-).

As you have read, sodium and cloride ions combine and form the compound sodium chloride. The positively charged sodium ions and negatively charged chloride ions are held together because oppositely charged atoms attract one another. The attraction between the sodium and chloride ions creates a crystal arrangement with a definite cubic shape. This arrangement can easily be seen by observing some table salt in the palm of your hand.

Figure 8–7. The ionic compound sodium chloride is formed by the transfer of an electron.

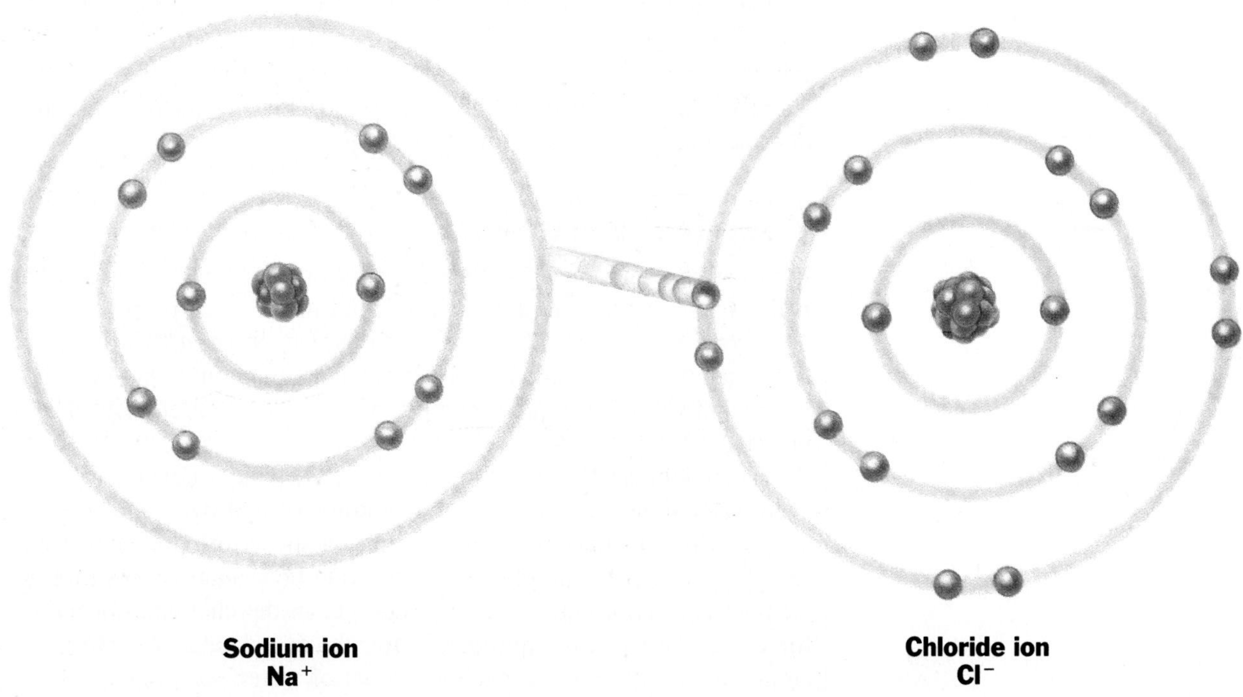

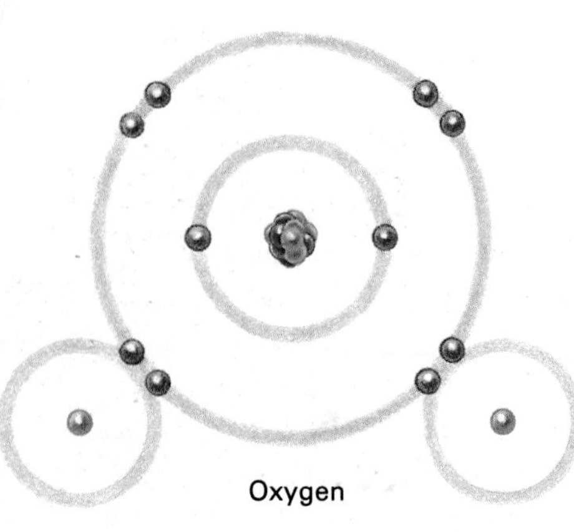

Figure 8–8 In a water molecule, hydrogen and oxygen atoms are joined by covalent bonds.

Covalent Bonds

A bond based on the attraction between atoms that share electrons is a **covalent bond.** Water is an example of a **covalent compound**–that is, a compound formed by the sharing of electrons. Two hydrogen atoms can share their single electrons with an oxygen atom, giving oxygen a stable number of eight electrons in the outermost energy level. At the same time, the oxygen atom shares two of its electrons, one with each hydrogen atom, giving each hydrogen atom a stable number of electrons in its outermost level of electrons. Thus, a water molecule consists of two atoms of hydrogen combined with one atom of oxygen. Each covalent compound consists of only one kind of molecule, just as each element consists of only one kind of atom.

When atoms share electrons, the positive nucleus of each atom is attracted to the negative electrons being shared. The pull between the positive and negative charges is the force that keeps the atoms joined. In diatomic gases such as nitrogen (N_2) and oxygen (O_2), the most abundant gases in the earth's atomsphere, there are covalent bonds between the two atoms of each diatomic molecule. Because the nuclei are identical, they each pull with the same strength on the shared electrons. In a compound, the covalent bonds are between two different kinds of atoms. The nucleus of one atom pulls more strongly than the other. The electrons are shared but remain closer to the nucleus that pulls more strongly. In a water molecule, for example, the oxygen nucleus attracts the electrons more strongly than does either hydrogen nucleus. As a result, the molecule has a slightly negative charge at its oxygen end and a slightly positive charge at its hydrogen end.

The various molecules that make up a covalent compound such as water are usually held together by weak forces of attraction. These forces are far weaker than the powerful attraction between ions in an ionic compound such as sodium chloride. For this reason, ice, a covalent compound, melts at a much lower temperature than does sodium chloride, an ionic compound.

Chemical Formulas

In any given compound—either covalent or ionic—the elements that make up the compound are always found in the same proportion. Therefore, a compound is represented by a fixed **chemical formula.** A chemical formula indicates what elements the compound contains. It also indicates the relative number of atoms in each element that is present.

The chemical formula for water is H_2O, which indicates that each water molecule consists of two atoms of hydrogen and one atom of oxygen. In a chemical formula, a subscript number is used after the symbol for an element to indicate how many atoms of that element are in a compound. For example, in the chemical formula for water, the subscript number *2* after the *H* indicates that two atoms of hydrogen are in each molecule of water.

Figure 8–9. Dust and fumes in the air form the unhealthful mixture known as smog, shown at left. Sandstone, shown at right, is a mixture of sand grains, which are usually quartz, held together by silica or calcium carbonate.

Mixtures

On the earth, elements and compounds are generally found mixed together. A **mixture** is material that contains two or more substances that are not chemically combined. The substances in a mixture keep their individual properties. Therefore, unlike a compound, a mixture can be separated into its parts by physical means. For example, you can use a magnet to separate a mixture of powdered sulfur (S) and iron (Fe) filings. The magnet will attract the iron and leave behind the sulfur, which is not magnetic.

Many substances studied in earth science are mixtures. Rocks and soil, for example, are mixtures of many different compounds. The atmosphere is a mixture of gases, including oxygen (O_2), nitrogen (N_2), carbon dixoide (CO_2), and water vapor (H_2O). Air pollution forms when dust and chemicals are added to this mixture. Dust and fumes from industrial smokestacks and car exhausts form an unhealthy mixture called **smog.**

Sea water is an example of a **solution,** a mixture in which one substance is uniformly dispersed in another substance. Sodium chloride (NaCl) and many other ionic compounds are dissolved in sea water. The ability of water to dissolve salt and other substances is related to what you have already read about the distribution of electrical charge in a water molecule. The positive end of a water molecule attracts negative chloride ions, and the negative end of the water molecule attracts positive sodium ions.

You probably think of solutions as liquids, but gases and solids may also be solutions. For example, an **alloy** is a solution of two or more metals. Common alloys include brass, a mixture of copper (Cu) and zinc (Zn); and bronze, a mixture of copper and tin (Sn).

INVESTIGATE!

To learn more about chemical compounds, try the In-Depth Investigation on pages 154–155.

Section 8.2 Review

1. Define the term *compound*.
2. Which type of bonding involves the sharing of electrons?
3. Write out the words for this formula: $ZnCl_2$.
4. How does a mixture differ from a compound?

Chapter 8 Review

Key Terms

- alloy (151)
- atom (140)
- atomic number (141)
- chemical bond (148)
- chemical formula (150)
- chemical property (139)
- compound (147)
- covalent bond (150)
- covalent compound (150)
- diatomic (147)
- electron (141)
- electron cloud (141)
- energy level (147)
- gas (145)
- ion (149)
- ionic bond (148)
- ionic compound (148)
- isotope (145)
- liquid (145)
- mass number (141)
- matter (139)
- mixture (151)
- molecule (147)
- neutron (141)
- nucleus (141)
- periodic table (141)
- physical property (139)
- proton (141)
- smog (151)
- solid (145)
- solution (151)

Key Concepts

An element is a substance that cannot be broken down into a simpler form by ordinary chemical means. **See page 139.**

An atom consists of electrons surrounding a nucleus made up of protons and neutrons. **See page 141.**

The atomic number of an atom equals the number of protons in the atom. The mass number equals the sum of the protons and neutrons in the atom. **See page 141.**

Isotopes are atoms of the same element that have different numbers of neutrons. **See page 144.**

A solid holds its shape because it has particles in fixed positions. A liquid has no definite shape because the particles can move past each other. A gas is a formless collection of particles that tends to expand in all directions when not confined. **See page 145.**

Different kinds of atoms join together and form compounds based on the way their electrons are arranged in energy levels within the electron cloud. **See page 147.**

Chemical bonds between atoms are formed by the sharing or transfer of electrons between atoms. **See page 148.**

A chemical formula tells what elements a compound contains and the relative number of atoms of each element present. **See page 150.**

A compound consists of atoms from two or more elements that are chemically united. A mixture consists of two or more substances that are not chemically united. **See pages 147, 151.**

Review

On your own paper, write the letter of the term that best completes each of the following statements.

1. Color and hardness are examples of an element's
a. physical properties. b. chemical properties.
c. atomic structure. d. molecular properties.

2. A substance that cannot be broken down into a simpler form by ordinary chemical means is
a. a mixture. b. a gas. c. an element. d. a compound.

3. The smallest unit of an element is
a. a molecule. b. an atom. c. an ion. d. an electron.

4. Particles in atoms that do not carry an electrical charge are called
a. neutrons. b. nuclei. c. protons. d. ions.

5. The number of protons in the nucleus indicates the atom's
a. mass number. b. electrical charges.
c. isotope. d. atomic number.

6. The mass number of an atom is equal to its
   a. total number of protons.
   b. total number of electrons and protons.
   c. total number of neutrons and protons.
   d. total number of neutrons and electrons.
7. Atoms of the same element that differ in mass are
   a. ions. b. isotopes. c. neutrons. d. molecules.
8. A material with a definite shape and volume is a
   a. compound. b. liquid. c. gas. d. solid.
9. A liquid does not have a definite
   a. shape. b. volume.
   c. chemical formula. d. mass.
10. If a gas is not confined, the space between its particles will
    a. decrease slowly. b. decrease rapidly.
    c. increase. d. not change.
11. Atoms of two or more elements that are chemically united form
    a. a mixture. b. a nucleus. c. an ion. d. a compound.
12. An atom does not easily lose or gain electrons if it has
    a. many protons. b. a filled outer energy level.
    c. many energy levels. d. few neutrons.
13. A molecule of water, or H_2O, has one atom of
    a. hydrogen. b. helium. c. oxygen. d. osmium.
14. A material that contains two or more substances that are not chemically combined is
    a. a mixture. b. a compound. c. an ion. d. a molecule.

Critical Thinking

On your own paper, write answers to the following questions.

1. Oxygen combines with hydrogen to form water. Is this process a result of the physical or chemical properties of oxygen?
2. What distinguishes a particular atom from all other kinds of atoms?
3. Why do isotopes of an element have different mass numbers?
4. The mercury in a thermometer has a volume that varies with temperature. It takes the shape of the glass tube that holds it. Is the mercury in a thermometer a solid, a liquid, or a gas?
5. Calcium chloride is an ionic compound. Carbon dioxide is a covalent compound. Which of these compounds would you expect to have a lower melting point? Explain your answer.
6. Is a diatomic molecule more likely to be held together by a covalent bond or an ionic bond? Explain why you think this is so.
7. What happens to the properties of a substance when it becomes part of a mixture?

Application

1. How many neutrons does a potassium atom have if its atomic number is 19 and its atomic mass is 39?
2. The atomic number of calcium is 20, and the atomic number of copper is 29. Which has more electrons, a calcium atom or a copper atom? How do you know?
3. A helium atom has two electrons in its first and only energy level. Would you predict that helium easily forms compounds with other elements? Why or why not?
4. The chemical formula for quartz is SiO_2. How many atoms of silicon and oxygen does a molecule of quartz contain?

Extension

1. Choose one of the elements that is abundant in the earth's crust. Prepare a report on the atomic structure of the element, its physical and chemical properties, and its economic importance. Illustrate the report with photographs, if possible.
2. Use the library or the Internet to research some common forms of air and water pollution. Find out the source of the pollution and what the pollution is made up of. Find out what lawmakers are doing to control these forms of pollution. Compile your findings in a chart.

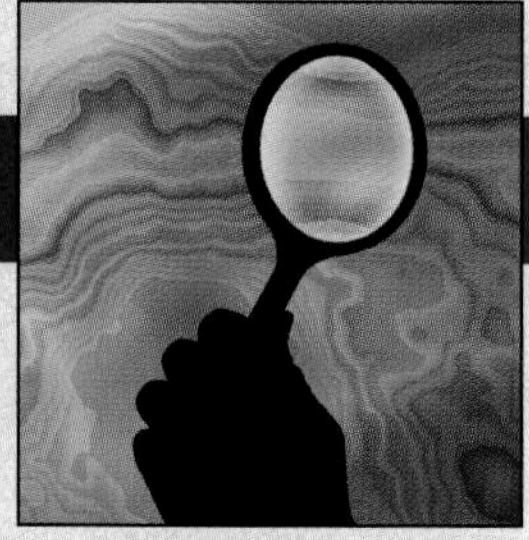

IN-DEPTH INVESTIGATION

Chemical Analysis

Materials

- lab apron
- electrodes (stainless steel)
- wood splints
- 2 connecting wires
- beaker (400-mL)
- water
- 6-V battery
- Epsom salts (300 g)
- matches
- 2 test tubes (13 × 100 mm)
- safety goggles
- latex gloves
- masking tape
- stirring rod
- balance

Introduction

Chemists use a process called *chemical analysis* to learn about the nature and composition of the materials around us. While mixtures can be separated into their component parts by physical means, compounds must be broken down by means of a chemical reaction. Elements cannot be divided by any ordinary chemical means. Therefore, to find out if a substance is a compound or an element, you must carry out a chemical reaction. Chemical reactions involve energy in the form of heat, light, or electricity.

In this investigation, you will use some of the same techniques to demonstrate whether water is an element or a compound.

Prelab Preparation

1. Review Section 8.1, pages 139–140, and Section 8.2, pages 147–150.
2. Review general lab safety procedures and proper use of eye and hand protection.
3. Water is a pure substance that is abundant on earth. Do you think water is a compound or an element? Make a hypothesis. How could you find out?

Procedure

1. Set up the apparatus as shown in the diagram below. Fill the beaker about three-fourths full with tap water. Connect the battery terminals to the electrodes and observe what happens. What evidence do you see that a reaction is taking place? Is the speed of the reaction the same at each electrode?
2. The rate of the reaction taking place in the beaker is slow. Remove the electrodes from the beaker. *Put on your safety goggles and gloves.* Next, measure out approximately 300 g of Epsom salts, and slowly add it to the water in the beaker. Stir the mixture until all or most of the salt has dissolved. The Epsom salts will speed up the reaction because it increases the electrical conductance of the water.

3. Fill two test tubes with some of the water–Epsom salt mixture. Disconnect one terminal of the battery, and put the electrodes back in the beaker.
4. Place your gloved index finger over the open end of the first test tube. Now carefully invert the tube and place it below the surface of the mixture in the beaker. Release you finger and note if any air has entered the tube. If so, repeat the inversion process. Follow the same procedure for the second test tube.
5. Place one test tube over each electrode, and reconnect the battery. See the figure above. To speed up the reaction rate, lift the test tubes so that the mouth of each tube just comes to the base of the exposed metal electrode. Use masking tape to hold each tube in this position along the beaker wall. Allow gas to collect until one of the test tubes is full of gas. Disconnect the battery. Which electrode, positive or negative, gives off more gas?
6. Test to identify the gases. Hydrogen burns with a colorless flame and makes a popping noise. Oxygen does not burn, but a glowing wood splint will burst into flames when thrust into oxygen. Test the gas that was collected from the negative electrode first. Slowly remove the test tube from the beaker, keeping it vertical so that any remaining solution can drain into the beaker when it breaks the surface. Once this occurs, quickly place your finger over the mouth of the test tube to keep the gas from escaping. Keep the tube upside down and your finger over the opening until you or your partner brings a lit match to the mouth of the test tube. Be sure to hold the test tube firmly. What happens? What gas is in the test tube?
7. Test the gas in the test tube that was over the positive electrode. Remove the tube and drain any remaining solution as in Step 6, but this time quickly turn the test tube right side up and cover the open end with your finger as before. Uncover the tube, and thrust a glowing splint into the open end. What happens? What gas is in the test tube?

Analysis and Conclusions

1. Is water a compound or an element? Explain.
2. Many tests have been performed to break down hydrogen and oxygen into simpler substances by chemical means. All these tests have failed. Are hydrogen and oxygen compounds or elements? Explain.

Extensions

1. Baking soda and some antacid tablets fizz when put in water. Is this a physical or a chemical change? How do you know? How might you be able to support your answer?
2. When substances such as sugar or salt dissolve in water, is the resulting substance a mixture or a compound? How do you know? How might you support your answer?

Chapter 9

Minerals of the Earth's Crust

You probably would not recognize the blue substance in the rock pictured on the left as containing copper. Yet scientists have a number of methods for identifying copper and other minerals. Copper, like every other mineral, has its own chemical composition. Minerals can also be identified by properties such as structure, color, and hardness. These differences among minerals result from the way the atoms of each mineral are held together. This chapter explains how atoms combine to form the minerals of the earth's crust.

Chapter Outline

This blue peacock copper ore is from Arizona.

9.1 What Is a Mineral?

Section Objectives

- **Define a mineral and distinguish between the two main mineral groups.**
- **Identify the elements found most abundantly in common minerals.**
- **Name six types of nonsilicate minerals.**
- **Distinguish among four main arrangements of silicon-oxygen tetrahedra found in silicate minerals.**

A ruby, a gold nugget, and a grain of salt look very different from one another, but they all have one thing in common. They are all **minerals,** the basic materials of the earth's crust. A mineral is a natural, **inorganic,** crystalline solid. An inorganic substance is one that is not made up of living things or the remains of living things. Every mineral has a characteristic chemical composition and can be either an element or a compound. Most rocks that make up the crust are mixtures of various minerals.

To determine whether a substance is a mineral or a nonmineral, scientists ask four basic questions about the substance. If the answer is yes to all four questions, the substance is a mineral. First, scientists ask whether the substance is inorganic. Coal, for example is *organic*—it is composed of the remains of ancient plants. Thus it is not a mineral. Magnetite, composed of the inorganic substances iron (Fe) and oxygen (O), is a mineral.

Second, scientists ask whether the substance occurs naturally. The minerals quartz (SiO_2), silver (Ag), and sulfur (S), for example, all occur naturally in the earth. Manufactured substances, such as steel or brass, are not minerals.

Third, scientists determine whether the substance is a solid in crystalline form. Petroleum and natural gas are naturally occurring substances. They are not solids, however, and thus are not minerals. In addition, both are made up of the remains of plants and animals.

The fourth question is whether the substance has a definite chemical composition. The mineral gold (Au) is an element with only gold atoms. The mineral fluorite is a compound, made up of only calcium (Ca) and fluoride (F) ions in a specific crystalline pattern. A chunk of concrete, however, is made up of several substances. When concrete is made, the amounts of these substances vary according to the intended purpose of the concrete. Is concrete a mineral? Explain how you know.

Kinds of Minerals

Earth scientists have identified more than 3,000 different minerals, but fewer than 20 of them are common. The common minerals are called **rock-forming minerals** because they form the rocks of the earth's crust. Of the 20 rock-forming minerals, 10 are so common

Figure 9–1. Plagioclase feldspar (left), muscovite mica (center), and orthoclase feldspar (right) are 3 of the 20 common rock-forming minerals.

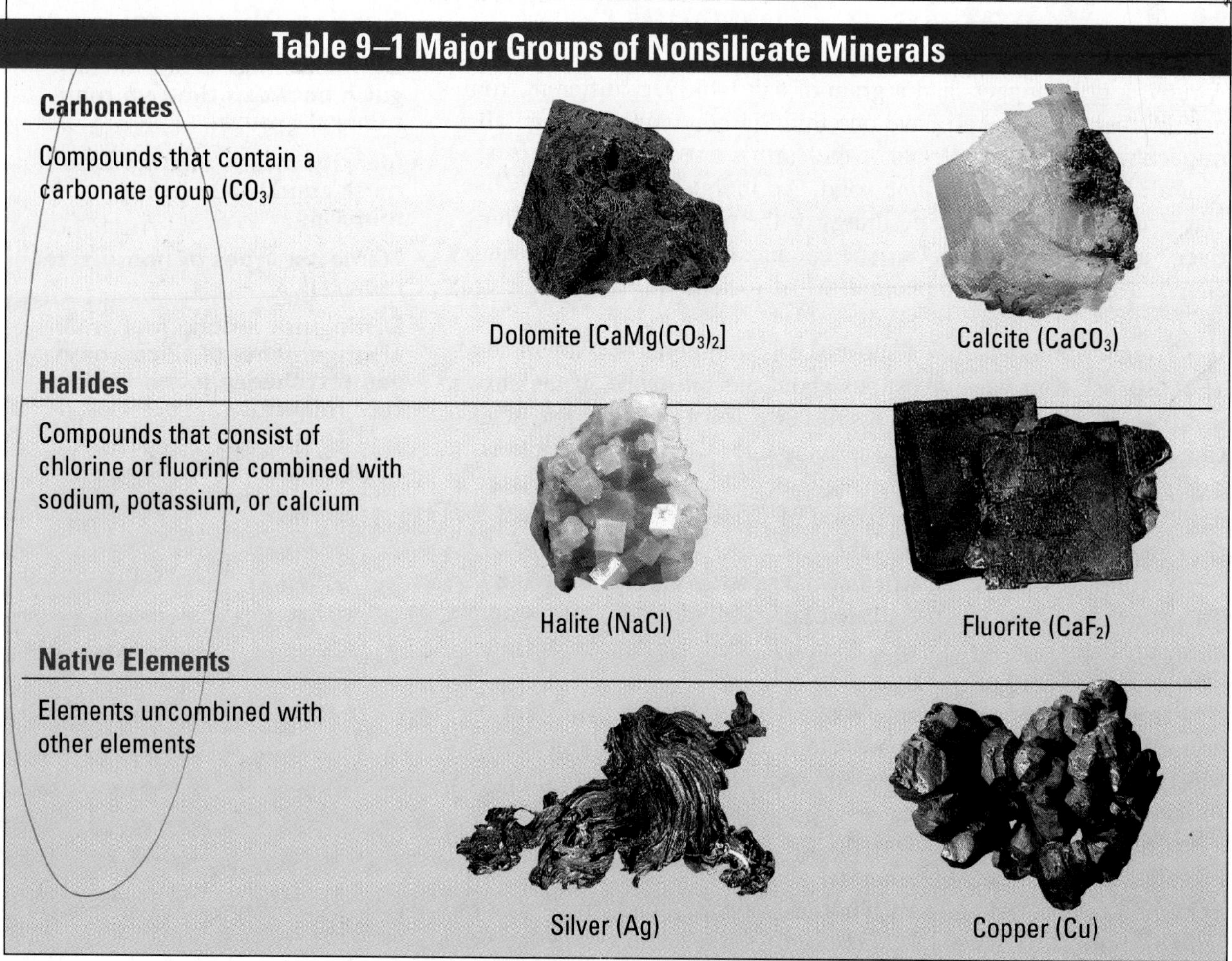

Table 9–1 Major Groups of Nonsilicate Minerals

Group	Description	Examples	
Carbonates	Compounds that contain a carbonate group (CO_3)	Dolomite [$CaMg(CO_3)_2$]	Calcite ($CaCO_3$)
Halides	Compounds that consist of chlorine or fluorine combined with sodium, potassium, or calcium	Halite ($NaCl$)	Fluorite (CaF_2)
Native Elements	Elements uncombined with other elements	Silver (Ag)	Copper (Cu)

that they make up 90 percent of the mass of the earth's crust. These minerals are quartz, orthoclase, plagioclase, muscovite, biotite, calcite, dolomite, halite, gypsum, and ferromagnesian minerals, which include olivines, pyroxenes, and amphiboles. All minerals, however, can be classified into two main groups based on their chemical composition—**silicate minerals** and **nonsilicate minerals.**

Silicate Minerals

All silicate minerals contain atoms of silicon (Si) and oxygen (O). The common mineral quartz consists of only silicon and oxygen atoms. However, most silicate minerals also contain one or more other kinds of atoms. Feldspars are the most common silicate minerals. The type of feldspar that forms depends on which metal combines with the silicon and oxygen atoms. Orthoclase results when the metal is potassium (K). Plagioclase forms when the metal is sodium (Na), calcium (Ca), or both. Besides quartz and the feldspars, there are ferromagnesian minerals rich in iron (Fe) and magnesium (Mg). They include hornblende, olivine, muscovite, and biotite. Silicate minerals make up 96 percent of the earth's crust. Feldspar and quartz alone make up more than 50 percent of the crust.

Group	Description	Examples	
Oxides	Compounds that contain oxygen and an element other than silicon	Corundum (Al_2O_3)	Hematite (Fe_2O_3)
Sulfates	Compounds that contain a sulfate group (SO_4)	Gypsum ($CaSO_4 \cdot 2H_2O$)	Anhydrite ($CaSO_4$)
Sulfides	Compounds that consist of one or more elements combined with sulfur	Galena (PbS)	Pyrite (FeS_2)

Nonsilicate Minerals

Four percent of the earth's crust consists of nonsilicate minerals—that is, minerals that do not contain silicon. Based on their chemical composition, nonsilicate minerals are classified into six major groups: carbonates, halides, native elements, oxides, sulfates, and sulfides. Table 9–1 describes the characteristics of each group and gives representative examples.

Crystalline Structure

All minerals in the earth's crust have a crystalline structure. Each type of crystalline material is characterized by a specific geometric arrangement of its atoms or ions. A **crystal** is a natural solid with a definite shape. A large mineral crystal displays the characteristic geometry of its internal structure. The conditions under which minerals are produced, however, do not usually allow large single crystals to grow. As a result, minerals more commonly consist of masses of crystals so small that you can see them only with a microscope. But if a crystalline mineral forms unrestricted by surrounding material, the mineral will develop into a single, large crystal in one of six crystal shapes. The crystal shapes are helpful in identifying minerals. You will read more about these crystal shapes later in this chapter.

Scientists use X rays to study the structure of crystals. X rays passing through a crystal and striking a photographic plate produce an image that shows the geometric arrangement of the atoms or ions that make up the crystal.

The Crystalline Structure of Silicate Minerals

Although there are many kinds of silicate minerals, their crystalline structure is made up of the same basic building blocks. Each of

CAREER FOCUS: *Mineral Resources*

"You can work for years to find one simple answer and only come up with more questions. I really enjoy that challenge."

Lani Boldt

Mining Engineer

Civil engineer Lani Boldt often talks to high school students about her work in mining research. But each time her message is the same.

"You don't have to be a super brain to make it through an engineering curriculum. What you do need is interest, perseverance, and plain old-fashioned hard work," she says.

Like other engineers in mining, who must have at least a bachelor of science degree in civil engineering, Boldt's main objective is to make all kinds of mining safer, easier, and more profitable. In the future, this may include seabed and space mining. Boldt, pictured at left, has been working to develop methods for using the wastes that are produced during the mining process.

The majority of Boldt's work involves both laboratory and field investigations using computers, soil and rock testing equipment, and environmental test chambers, devices used in the lab to simulate the natural environment.

"The most exciting thing about my work is its variability," she says. "I never know from one day to the next where I'll be or what I'll be doing. I just know I'll be learning something new and, I hope, helping someone in the meantime."

Gem Cutter

A gem cutter, shown below, evaluates and grades all colored gemstones in their rough state. The gem cutter will then determine how to cut and polish the stones to best reveal their color and luster.

The process begins when the gem cutter saws or slices jewelry-quality minerals into usable pieces. Those pieces are

these building blocks consists of four oxygen atoms arranged in a pyramid with one silicon atom in the center. Figure 9–2 shows this four-sided structure. The structure is known as a **silicon-oxygen tetrahedron.**

The silicon-oxygen tetrahedra combine in different arrangements to produce the many diverse silicate minerals. The various arrangements are the result of the kinds of bonds formed between the oxygen atoms of the tetrahedra and other atoms. The bonds may form between the oxygen atoms and silicon atoms of neighboring tetrahedra, or they may form between the oxygen atoms in the tetrahedra

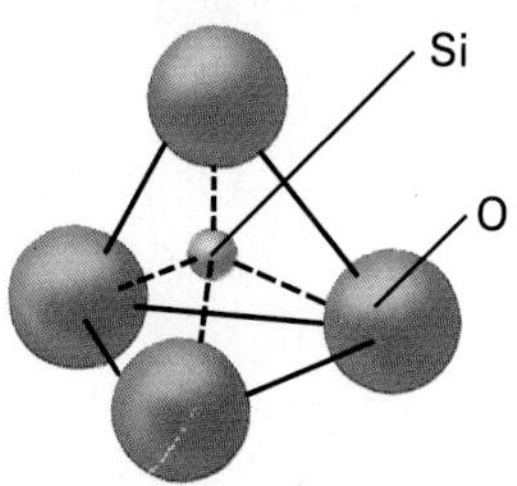

Figure 9–2. The silicon-oxygen tetrahedron is the basic structure of all silicate minerals.

then ground or cut into one of the two basic styles—cabochon or faceted. While cabochon cuts result in rounded surfaces, faceted cuts produce many small, flat surfaces organized in intricate patterns. In both cutting procedures, a knowledge of the form, structure, properties, and classification of crystals is necessary.

Gem cutters enter the field through a variety of on-the-job and technical training programs.

Mine Inspector

Mine inspectors evaluate various kinds of mines, including coal, limestone, uranium, and copper mines, to ensure the health and safety of mine workers. Some mines are deep underground. Other mines, such as the copper mine shown above, are open. Among the concerns of mine inspectors are explosive gases, poisonous fumes, unstable roof and ground structures, and harmful dust.

In addition to gathering information on health and safety conditions, a mine inspector enforces established health and safety regulations by notifying mine management of any violations and then seeing that those conditions are corrected. Mine inspectors also investigate and report on mine accidents and may occasionally become involved in directing rescue attempts.

Entry-level positions are available through apprenticeships, work experience, or four years of related education after high school.

▲ This open-pit copper mine is regularly checked by certified mine inspectors.

For Further Information

For more information on careers in mining, write the Office of Minerals Information, U.S. Geological Survey, 983 National Center, Reston, VA 22092, or peruse their web site at www.usgs.gov.

Single Chain (pyroxenes)

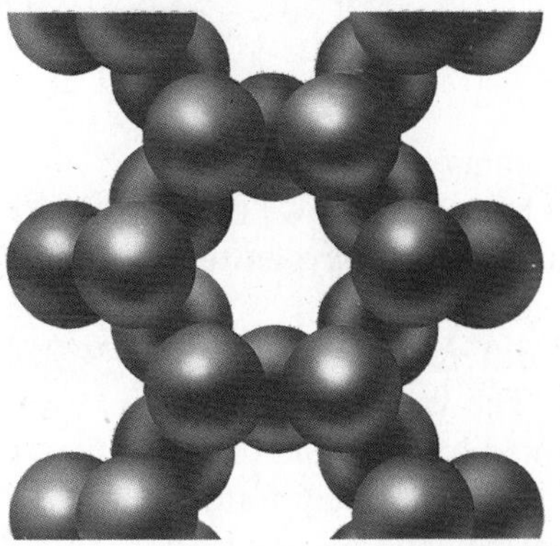

Double Chain (amphiboles)

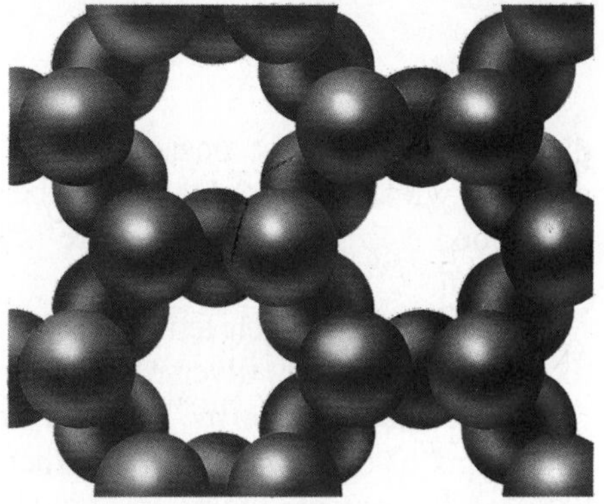

Sheets (micas)

Figure 9–3. The silicon-oxygen tetrahedra shown above combine to form single chains (left), double chains (center), and sheets (right).

and other elements outside the tetrahedra. Figure 9–3 illustrates three kinds of arrangements formed by tetrahedra.

Ionic, Chain, and Sheet Silicates

Silicon-oxygen tetrahedra linked only by atoms of elements other than silicon and oxygen make up ionic silicate materials. For example, olivine is a ferromagnesian mineral formed when the oxygen atoms of tetrahedra bond to magnesium (Mg) and iron (Fe) atoms.

In the bonding arrangement called a single-chain silicate, each tetrahedron is bonded to two others by shared oxygen atoms. In double-chain silicates, two single chains of tetrahedra bond to each other. Minerals made up of single chains are called pyroxenes, and those made up of double chains are amphiboles.

In tetrahedral sheets, each tetrahedron shares three oxygen atoms with other tetrahedra. The fourth oxygen atom bonds with an atom of potassium (K) or aluminum (Al), which joins one sheet to another. These sheets separate easily because the three bonds among the tetrahedra are much stronger than the bond with potassium or aluminum. The micas are examples of silicate sheets.

Network Silicates

In the more complex arrangement known as network silicate, each tetrahedron is bonded to four neighboring tetrahedra. Networks that contain only silicon-oxygen tetrahedra form the mineral quartz, which has the chemical formula SiO_2. Quartz is a very hard mineral because all of its atoms are tightly bonded together.

The feldspars are also network silicates. Unlike quartz, however, some of the tetrahedra in the feldspars have atoms of aluminum or other metals instead of silicon atoms. The bonds between these atoms are weaker than those between silicon and oxygen. Thus, a feldspar can be broken down more easily than quartz.

Section 9.1 Review

1. What are the two main groups of minerals?
2. Which two elements are most commonly found in minerals?
3. Are carbonates and halides silicate or nonsilicate minerals?
4. How do sheet silicates differ from network silicates?
5. Silver is a naturally occurring inorganic substance that is formed in the earth's crust. It is a solid and has a definite chemical composition. Is silver a mineral? Explain your answer.

9.2 Identifying Minerals

Section Objectives

- **Describe some characteristics that help distinguish one mineral from another.**
- **List four special properties that may help identify certain minerals.**

If you have ever found an unfamiliar mineral, you may have wondered how to go about identifying it. Earth scientists called **mineralogists** conduct tests with special equipment to properly identify minerals. Mineralogists work in laboratories and in the field to identify minerals—from the most common to the most rare and precious.

Characteristics of Minerals

Each mineral has specific properties that are a result of its chemical composition and crystal structure. These properties provide useful clues for identifying minerals. You can identify some of these properties by simply looking at a sample of the mineral. You can determine other properties through simple tests.

Color

A property you can easily observe is the color of a mineral. Some minerals have very distinct colors. For example, sulfur is bright yellow, and azurite is a deep blue. The mineral cinnabar is red, while serpentine is green. Color alone, however, is generally not a reliable clue in identifying a mineral sample. Many minerals are similar in color, and very small amounts of certain elements may greatly affect the color. For example, corundum is a colorless mineral composed of aluminum and oxygen atoms. However, corundum that contains traces of chromium (Cr) forms ruby, a rare red mineral. Sapphire, a rare blue mineral, consists of corundum with traces of cobalt (Co) and titanium (Ti). Quartz exists in many colors. Amethyst, for example, is purple because it contains tiny amounts of the elements manganese (Mn) and iron (Fe).

Color is also an unreliable identification clue because weathered surfaces may hide the color of minerals. For example, iron pyrite is the color of gold, but it appears dark yellow when it is weathered. When you examine a mineral for color, be sure to inspect a freshly exposed surface.

Figure 9–4. Pure quartz (left) is colorless. Amethyst (above) is a variety of quartz that is purple because of the presence of manganese and iron.

Luster

Light reflected from the surface of a mineral is called **luster.** Minerals that reflect light like polished metal are said to have a *metallic luster.* All other minerals have a *nonmetallic luster.* Mineralogists distinguish several types of nonmetallic luster. Transparent quartz and other minerals that look like glass have a glassy luster. Minerals with an appearance like the surface of candle wax have a waxy luster. Some minerals, such as the micas, have a pearly luster. Diamond is an example of a mineral with a brilliant luster. A mineral that lacks any kind of shine has a dull (earthy) luster.

Figure 9–5. All minerals have either a metallic luster like platinum (top) or a nonmetallic luster like muscovite mica (bottom).

Streak

A more reliable clue to the identity of a mineral is the color of that mineral in powdered form, which is called its **streak.** The easiest way to observe the streak of a mineral is to rub some of the mineral against a piece of unglazed ceramic tile called a streak plate. Because the streak is the powdered form of the mineral, it may not be the same color as the larger piece of the mineral. Metallic minerals generally have a dark streak. For example, the streak of gold-colored pyrite is black. For most nonmetallic minerals, however, the streak is either colorless or a very light shade of the normal color of the mineral.

Cleavage and Fracture

Some minerals tend to split easily along certain flat surfaces. This property, called **cleavage,** is related to the types of bonds in the internal structure of the mineral. The surface along which cleavage occurs runs parallel to a plane in the crystal where bonding is relatively weak. For example, the micas, which are made of tetrahedral sheets, tend to split into parallel sheets. Mineralogists use cleavage to identify and describe some minerals. The mineral galena breaks into small cubes because the three cleavage directions are at right angles to each other.

Many minerals, however, do not break along cleavage planes. Instead, they *fracture,* or break, unevenly into curved or irregular pieces. Mineralogists describe a fracture according to the appearance of the broken surfaces. For example, a rough surface is called *uneven* or *irregular.* A broken surface that looks like a piece of broken wood is called *splintery* or *fibrous.* Curved surfaces on a fractured mineral are called *conchoidal.*

Figure 9–6. The diagram on the left demonstrates cleavage in three directions. Galena is a mineral that cleaves in three directions. The diagram on the right demonstrates cleavage in one direction. The micas are minerals that have cleavage in only one direction.

Hardness

The measure of the ability of a mineral to resist scratching is called **hardness.** Hardness does not mean resistance to cleavage or fracture. A diamond, for example, is extremely hard, but it can be split along cleavage planes more easily than can calcite, a softer mineral.

The hardness of an unknown mineral can be determined by scratching it against the minerals on **Mohs hardness scale,** shown in Table 9–2. This scale lists 10 minerals in order of increasing hardness. The softest mineral is talc, with a hardness of 1. The hardest mineral is diamond, with a hardness of 10. The difference in hardness between two consecutive minerals is about the same throughout the scale except for the difference between the two hardest minerals. Diamond (10) is much harder than corundum (9), the mineral just before it on the scale. Care must be taken in testing hardness. For example, the mark usually left by talc on an unknown mineral may appear to be a scratch. Actually, it is the streak made by talc, and it is easily rubbed off. A true scratch will remain when a harder mineral rubs a softer mineral.

To test an unknown mineral for hardness, you must determine which is the hardest mineral on the scale that it can scratch. For example, galena can scratch gypsum but not calcite. Between which two numbers on Mohs scale does galena fall? If neither of two minerals scratches the other, they have the same hardness.

The hardness of a mineral is largely determined by the strength of the bonds between the atoms or ions that make up its internal structure. Both diamond and graphite consist exclusively of carbon atoms. Diamond has a hardness of 10, however, while the hardness of graphite is only between 1 and 2. A diamond's hardness results from a strong crystal structure in which each carbon atom is firmly bonded to four other carbon atoms. In contrast, the carbon atoms in graphite are arranged in layers that are held together by weak forces.

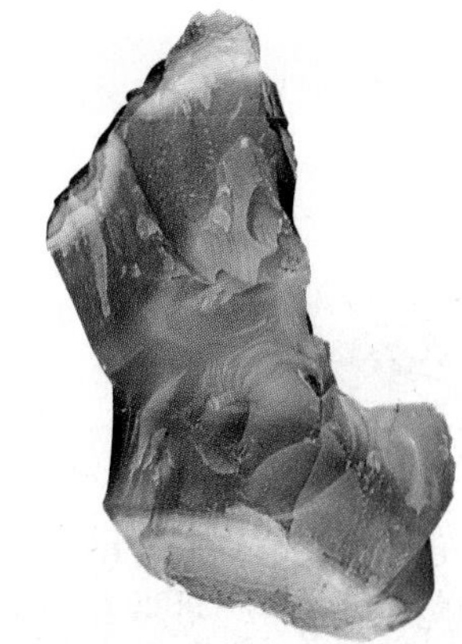

Figure 9–7. Conchoidal fracture is a type of fracture that looks like broken glass and that helps to identify some minerals.

Table 9–2 Mohs Hardness Scale

Minerals	Hardness	Common Test
Talc	1	Easily scratched by fingernail
Gypsum	2	Can be scratched by fingernail
Calcite	3	Barely can be scratched by copper penny
Fluorite	4	Easily scratched with steel file or glass
Apatite	5	Can be scratched by steel file or glass
Feldspar	6	Scratches glass with difficulty
Quartz	7	Easily scratches both glass and steel
Topaz	8	Scratches quartz
Corundum	9	No simple tests
Diamond	10	Scratches everything

Table 9–3 The Six Basic Crystal Systems

Isometric or Cubic System

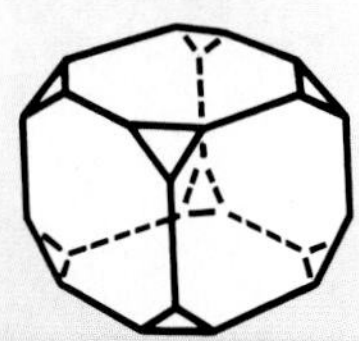
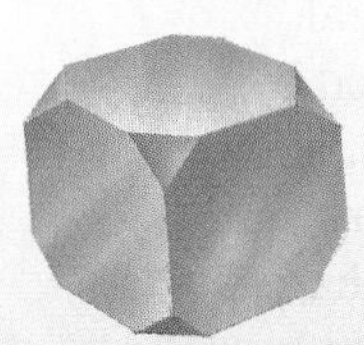

Three axes of equal length intersect at 90° angles. Examples: galena, halite, and pyrite

Orthorhombic System

Three axes of different lengths intersect at 90° angles. Examples: olivine, topaz, and staurolite

Triclinic System

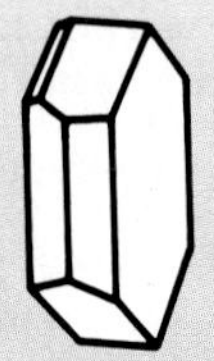

Three axes of unequal length are oblique to one another. Examples: plagioclase feldspars, turquoise, and axinite

Monoclinic System

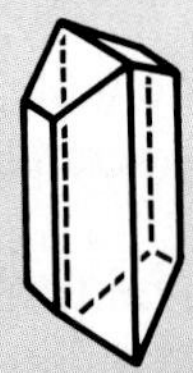
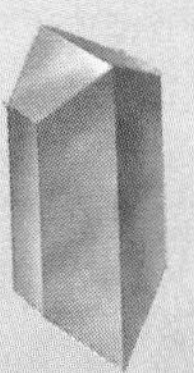

Three axes of different lengths, two intersect at 90° angles. The third axis is oblique to the others. Examples: micas, gypsum, and augite

Hexagonal System

Three horizontal axes the same length intersect at 60° angles. The vertical axis is longer or shorter than the horizontal axes. Examples: calcite, hematite, and quartz

Tetragonal System

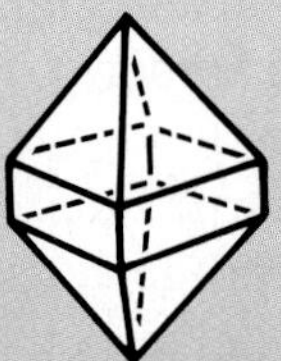
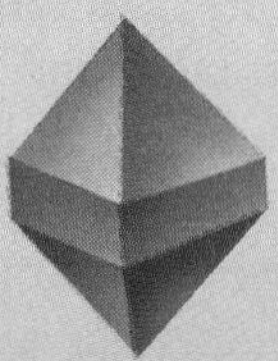

Three axes intersect at 90° angles. The two horizontal axes are of equal length. The vertical axis is longer or shorter than the horizontal axes. Examples: cassiterite, chalcopyrite, and zircon

Crystal Shape

As explained in Section 9.1, mineral crystals will form in one of six basic shapes. These shapes are illustrated in Table 9–3. Each kind of mineral is characterized by crystals of a specific shape. A certain mineral always has the same general shape because the atoms or ions that form its crystals always combine in the same geometric pattern.

Density

When handling equal-sized specimens of various minerals, you may notice that some feel heavier than others. For example, a piece of galena feels heavier than a piece of quartz of the same size. One way to compare these minerals is to lift, or heft, mineral samples of the same size. However, a more precise comparison can be made by measuring the **density** of a sample. Density is the ratio of the mass of a substance to its volume. The units of density are grams per cubic centimeter (g/cm^3). For example, if the mass (m) of a mineral sample is 85 g and its volume (V) is 34 cm^3, its density (D) is found by solving the following equation:

$$D = \frac{m}{V} = \frac{85\ g}{34\ cm^3} = 2.5\ g/cm^3$$

The density of a mineral depends on the kinds of atoms it contains and how closely they are packed. Most of the common minerals in the earth's crust have densities in the narrow range between 2 and 3 g/cm^3. However, the densities of minerals containing such heavy metals as lead, uranium, gold, and silver range from 7 to 20 g/cm^3. Therefore, density helps identify the heavier minerals more readily than it does the lighter ones.

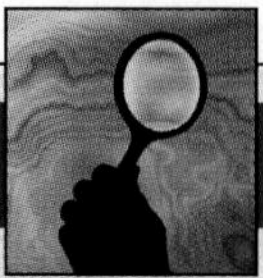

SMALL-SCALE INVESTIGATION

Mineral Identification

Most minerals can be identified by two or three physical properties. You can use some basic tests of mineral properties and a table of mineral characteristics to identify an unknown mineral.

Materials

mineral specimen, streak plate, Mohs hardness scale on page 165, copper penny, steel file, glass plate, Guide to Common Minerals on pages 666–667

Procedure

1. Make a data table on a separate piece of paper similar to the one shown here.

Color	
Luster	
Streak	
Hardness	

2. Find a mineral specimen that seems to be the same color and makeup throughout. Record the color of your specimen.
3. If your specimen shines like polished metal, consider its luster metallic; otherwise, nonmetallic. Record your observations.
4. Rub your specimen across the streak plate. Record the color of the streak.
5. Study Mohs hardness scale. Use your fingernail, a penny, a glass plate, and finally a steel file to scratch your specimen. Match the item that scratched your specimen with those listed beside the scale. Record your results.
6. Compare your results with the Guide to Common Minerals. List all those minerals that fit the description of your specimen.
7. Refer to the guide again and make a list of five minerals that clearly do not fit the description of your specimen.

Analysis and Conclusions

1. Which tests were the most useful in selecting the minerals that fit the description of your specimen? Tentatively identify the type of mineral you have.
2. Suggest other tests that would positively identify your specimen.

Figure 9–8. Notice the change in color of the fluorescent minerals calcite and willemite as they go from ordinary light (left) to ultraviolet light (right).

INVESTIGATE!

To learn more about identifying minerals, try the In-Depth Investigation on pages 172–173.

Special Properties of Minerals

All minerals exhibit the properties described earlier in this section. In addition to those properties, however, a few minerals have some special properties that can aid in their identification.

Magnetism

A magnet passed through some sand or loose soil may attract small particles of iron-containing minerals. Magnetite is the most common among this group of magnetic minerals. Lodestone is a form of magnetite that acts as a magnet. When lodestone happens to be in an elongated shape, it is polarized, with a north pole at one end and a south pole at the other, just like a bar magnet. The needles of the first magnetic compasses used in navigation were made of tiny slivers of lodestone.

Fluorescence and Phosphorescence

The mineral calcite is usually white in ordinary light, but under ultraviolet light it often appears red. This ability to glow under ultraviolet light is referred to as **fluorescence.** Fluorescent minerals absorb ultraviolet light and then produce visible light of various colors. For example, willemite is light brown in ordinary light, but under ultraviolet light it appears green. You can see the effect of fluorescence on minerals in Figure 9–8.

Some minerals subjected to ultraviolet light will continue to glow after the ultraviolet light is cut off. Minerals that continue to glow have the property called **phosphorescence.**

Figure 9–9. The mineral calcite exhibits double refraction.

Double Refraction

Light rays bend as they pass through transparent minerals. This bending of light rays as they pass from one substance, such as air, to another, such as a mineral, is called **refraction.** Crystals of calcite and some other transparent minerals bend light in such a way that they produce a double image of any object viewed through them. This property is called **double refraction.** Double refraction occurs because the light ray is split into two parts as it enters the crystal.

Radioactivity

Some minerals have a property known as *radioactivity.* You learned in Chapter 8 that certain atoms have unstable electron arrangements. Some atoms also have unstable arrangements of protons and neutrons in their nuclei. Radioactivity results as unstable nuclei decay

Mineral Resources

Minerals, from on and within our planet, serve as the raw materials for building an ever more sophisticated and technological world. The towering skyscrapers that rise above the urban skylines of the world's major cities, for example, would be impossible to build without the high-strength steel girders that support them. Steel is a manufactured substance made up mostly of iron mined from various iron-rich minerals. The stone facings on these buildings and even the glass that makes up their windows are also products of the earth's crust.

Automobiles, which have made our society the most mobile in history, are stamped out of corrosion-resistant steel. The aircraft that criss cross the planet depend on lightweight, high-strength metals such as alloys of magnesium and aluminum to form their skins. The high-performance jet engines that propel them require metals like titanium and molybdenum to operate at high temperatures for long periods of time. All of these metals come from mineral ores that are mined from the crust of the earth.

Gemstones, too, play a role in modern technology. Diamonds, the hardest of the minerals, are usually thought of as desirable additions to jewelry. But industrial-grade diamonds are less glamorously used to coat cutting and grinding tools. Diamond films are also used to coat high-resolution optical lenses on cameras and telescopes.

Steel, which is essential to the building of automobiles and skyscrapers, is not a mineral. How do you know it is not a mineral? What element makes up much of our steel?

over time into stable nuclei by releasing particles and energy. Uranium (U) and radium (Ra) are examples of radioactive elements that occur in mineral deposits. Pitchblende is the most common mineral containing uranium. Other uranium-bearing minerals include carnotite, uraninite, and autunite.

Section 9.2 Review

1. What are the two main types of luster?
2. How would you determine the hardness of an unidentified mineral sample?
3. Why is color not a reliable clue to the identity of a mineral?
4. What is fluorescence?
5. Gypsum is a 2 on Mohs hardness scale, while topaz is an 8. Can gypsum scratch topaz? Explain your answer.

Chapter 9 Review

Key Terms

- cleavage (164)
- crystal (159)
- density (167)
- double refraction (168)
- fluorescence (168)
- hardness (165)
- inorganic (157)
- luster (164)
- mineral (157)
- mineralogist (163)
- Mohs hardness scale (165)
- nonsilicate mineral (158)
- phosphorescence (168)
- refraction (168)
- rock-forming mineral (157)
- silicate mineral (158)
- silicon-oxygen tetrahedron (161)
- streak (164)

Key Concepts

A mineral is a natural inorganic crystalline solid with a characteristic chemical composition. **See page 157.**

Oxygen and silicon are the most abundant elements in common minerals. **See page 158.**

The six major groups of nonsilicate minerals are composed of carbonates, halides, native elements, oxides, sulfates, and sulfides. **See pages 158–159.**

Silicate minerals consist of pyramid-shaped units called silicon-oxygen tetrahedra, which may be arranged in a variety of ways. **See page 161.**

Clues to identifying minerals are revealed by simple tests for streak, hardness, density, and other properties. **See pages 163–167.**

Special properties, such as magnetism, fluorescence, double refraction, and radioactivity, can aid in the identification of certain minerals. **See page 168.**

Review

On your own paper, write the letter of the term that best completes each of the following statements.

1. A natural inorganic crystalline solid with a characteristic chemical composition is called
   a. an atom. b. a gemstone. c. a mineral. d. a tetrahedron.
2. Minerals that contain silicon and oxygen are
   a. sulfide minerals. b. sulfate minerals.
   c. ores. d. silicate minerals.
3. The most common silicate minerals are the
   a. feldspars. b. halides.
   c. carbonates. d. sulfates.
4. Ninety-six percent of the earth's crust is made up of
   a. sulfur and lead. b. silicate minerals.
   c. copper and aluminum. d. nonsilicate minerals.
5. The basic structural units of all silicate minerals consist of
   a. tetrahedral networks. b. silicon-oxygen tetrahedra.
   c. single chains. d. double chains.
6. An example of a mineral with a basic structure consisting of single tetrahedra linked by atoms of other elements is
   a. mica. b. olivine.
   c. quartz. d. feldspar.
7. When two single chains of tetrahedra bond to each other, the result is called a
   a. single-chain silicate. b. sheet silicate.
   c. network silicate. d. double-chain silicate.

8. The appearance of the light reflected from the surface of a mineral is called
   a. color.   b. streak.
   c. luster.   d. fluorescence.
9. The words *waxy, pearly,* and *dull* describe a mineral's
   a. luster.   b. hardness.
   c. streak.   d. fluorescence.
10. The words *uneven* and *splintery* describe a mineral's
   a. cleavage.   b. fracture.
   c. hardness.   d. luster.
11. Mohs Scale is used in measuring a mineral's
   a. hardness.   b. specific gravity.
   c. color.   d. luster.
12. The ratio of the mass of a mineral to its volume is the mineral's
   a. atomic weight.   b. density.
   c. mass.   d. weight.
13. The needles of the first magnetic compasses used in navigation were made of the magnetic mineral
   a. iron pyrite.   b. silver.
   c. cinnabar.   d. lodestone.
14. When calcite absorbs ultraviolet light and gives off red light, it is displaying the property of
   a. radioactivity.   b. double refraction.
   c. magnetism.   d. fluorescence.
15. A mineral that is radioactive probably contains the element
   a. uranium.   b. silicon.
   c. fluorine.   d. calcium.
16. Double refraction is a distinctive property of crystals of
   a. mica.   b. feldspar.
   c. calcite.   d. galena.

Critical Thinking

On your own paper, write answers to the following questions.

1. Natural gas is a substance that occurs naturally in the earth's crust. Is it a mineral? Explain how you know.
2. Which of the following are you most likely to find in the earth's crust: the silicates feldspar and quartz or the nonsilicates copper and iron? Explain your answer.
3. Which, if any, of the following mineral groups contain silicon: carbonates, halides, or sulfates? Explain how you know.
4. Describe the tetrahedral arrangement of olivine, an ionic silicate material.
5. Can you determine conclusively that an unknown substance contains magnetite using only a magnet? Explain.

Application

1. Why is it difficult to identify a mineral simply by its color?
2. Iron pyrite (FeS_2) is called *fool's gold* because it looks very much like gold. What simple test could you use to determine whether a mineral sample is gold or pyrite? Explain what it would show.
3. A mineral sample has a mass of 51 g and a volume of 15 cm^3. What is the density of the mineral sample?
4. Concept map ? ? ? Construct a **concept map** starting with the term *minerals* and using as many new terms as possible from the list on the previous page. Make connections that illustrate the classification of minerals.

Extension

1. Use an encyclopedia and other reference sources to research properties of 10 common rock-forming minerals. Make a chart showing the name of each mineral, its chemical formula, its hardness, its luster, its color, and its density. If possible, include its streak, how it breaks, and what special properties it has.
2. Use the atlas in your school library to find a mineral map of the United States. Find out what minerals are most common in the United States. List these minerals and the states in which they are found. Present your findings to the class.

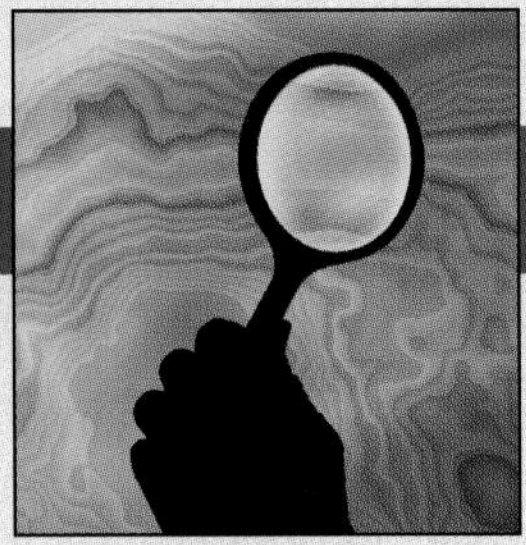

In-Depth Investigation

Mineral Identification

Materials

- copper penny
- quartz sand
- streak plate
- glass square
- steel file
- hand lens
- mineral samples (5)

Introduction

A mineral identification key can be used to compare the properties of minerals so that unknown mineral samples can be identified. Mineral properties that are often used in mineral identification keys are color, hardness, streak, luster, cleavage, and fracture. Color is probably the most obvious property of minerals. Hardness is determined by a scratch test. The Mohs hardness scale classifies minerals from 1 (soft) to 10 (hard). Streak is the color of a mineral in a finely powdered form. The streak shows less variation than the color of a sample, and it is more useful in identification. The luster of a mineral is either metallic (having an appearance of metals) or nonmetallic. Some minerals break along defined planes. The planes may be in several directions. This is called cleavage. Other minerals break into irregular fragments. This is called fracture.

In this investigation, you will classify several mineral samples using a mineral identification key.

Prelab Preparation

1. Review Section 9.2, pages 163–169.
2. Most sand grains are composed of the mineral quartz. Look at a sample of quartz sand grains with a hand lens. Are all the sand grains the same color? Is it possible to use only the property of color to identify minerals such as quartz sand?
3. Review Mohs hardness scale in Table 9–2, page 165. What is the hardness of a mineral sample that is scratched by a copper penny but not by a fingernail?

Procedure

1. Study the Mineral Identification Key on page 668. Use the key to help you identify the mineral samples. Remember that samples of the same mineral will vary somewhat. Because no specific mineral sample will exactly fit all the properties listed in an identification key, match each sample with the properties that best describe it.
2. Make a table with columns for sample number, color/luster, hardness, streak, cleavage/fracture, and mineral name. Then observe and record the color of each mineral in your table. Note whether the luster of each mineral is metallic or nonmetallic.

Sample number	Color/ Luster	Hardness	Streak	Cleavage/ Fracture	Mineral name
1					
2					
3					
4					
5					

3. Rub each mineral against the streak plate and determine the color of the mineral's streak. Record your observations.
4. Using a fingernail, copper penny, glass square, and steel file, test each mineral to determine its hardness, based on the hardness scale provided in Table 9–2, page 165. Arrange the minerals in order of hardness. Record your observations in your table.
5. Determine whether the surface of each mineral displays cleavage or fracture. Record your observations.

Analysis and Conclusions

1. Identify your mineral samples. Describe the properties that helped you identify each one.
2. Although color is the most obvious property of a mineral, it is difficult to identify a mineral by its color alone. Explain.
3. What is the difference between a scratch test and a streak test? Why do some very hard minerals leave no streak?
4. Compare the streak with the color of the mineral. Which minerals have the same color as their streak? Which do not?

Extensions

1. Diamonds and graphite are both made of the element carbon, but they are not considered the same mineral. Explain.
2. Corundum, rubies, and sapphires all have the chemical formula Al_2O_3, and they are considered the same types of mineral. Explain.

Have you ever collected rocks? Sooner or later, as the rock collection grows, even a casual collector finds a need to organize the collection. Rock samples may be grouped by color, shape, or size. They may also be organized into scientific categories used by geologists. This chapter describes these scientific categories and explains how each kind of rock forms and changes.

Chapter Outline

◀ **Mount Whitney is in the Sierra Nevada of California.**

10.1 Rocks and the Rock Cycle

Section Objectives

- **Identify the three major types of rock, and explain how each is formed.**
- **Summarize the steps in the rock cycle.**

Hot, molten rock, or *magma,* from the earth's interior is the parent material for all rocks. From the time magma cools and hardens at or near the surface of the earth, the resulting rock begins to change. In time, the rock formed from the original magma is altered many times. Geologists study not only the crust of the earth but also the forces and processes that act upon the rocks of the crust. Based on these studies, geologists have classified rocks into three major types based on the way the rocks are formed.

Three Major Types of Rock

Studies of volcanic activity provide information about the formation of one rock type—**igneous rock.** The word *igneous* is derived from a Latin term meaning "from fire." Igneous rock forms when magma cools and hardens. Magma is called *lava* if it cools at the earth's surface.

Forces such as wind and waves break down all types of rock into small fragments. Rock, minerals, and organic matter that have been broken into fragments are known as **sediment.** Sediment is carried away and deposited by water, ice, and wind. When sediment deposits harden after being compressed and cemented together, they form a second type of rock. This type of rock is called **sedimentary rock.**

Certain forces and processes, including tremendous pressure, extreme heat, and chemical processes, also can change the form of existing rock. The existing rock can thus be changed into a third

Figure 10-1. When hot molten rock, or magma, reaches the surface of the earth, it is called *lava*.

type of rock, called **metamorphic rock.** The word *metamorphic* means "changed form." Figure 10–2 shows an example of each major type of rock.

Any of the three major types of rock can be changed into another type. Various geological forces and processes cause rock to change from one major type to another and back again. This series of changes is called the **rock cycle.**

The Rock Cycle

As you have read, cooled and hardened magma forms igneous rock. Igneous rocks thus provide a good beginning for an examination of the rock cycle. Study Figure 10–3, which shows the steps in the rock cycle.

Once a body of igneous rock has formed, a number of processes on the earth's surface break down the igneous rocks into sediments. When the sediments from the igneous rocks are compacted and hardened, they form sedimentary rocks. If the resulting sedimentary rocks are subjected to extremely high temperature and great pressure within the earth, they become metamorphic rocks. If the heat and pressure become even more intense, the metamorphic rock will melt and form magma. This magma may then cool and form new rock. What kind of rock—igneous, sedimentary, or metamorphic—will this new rock be?

All the rocks in the earth's crust have probably passed through the rock cycle many times during the earth's history. However, as

Figure 10–2. Examples of the three major rock types. Rhyolite (left) is an igneous rock. Sandstone (center) is a sedimentary rock. Marble (right) is a metamorphic rock.

Rhyolite (igneous)

Sandstone (sedimentary)

Marble (metamorphic)

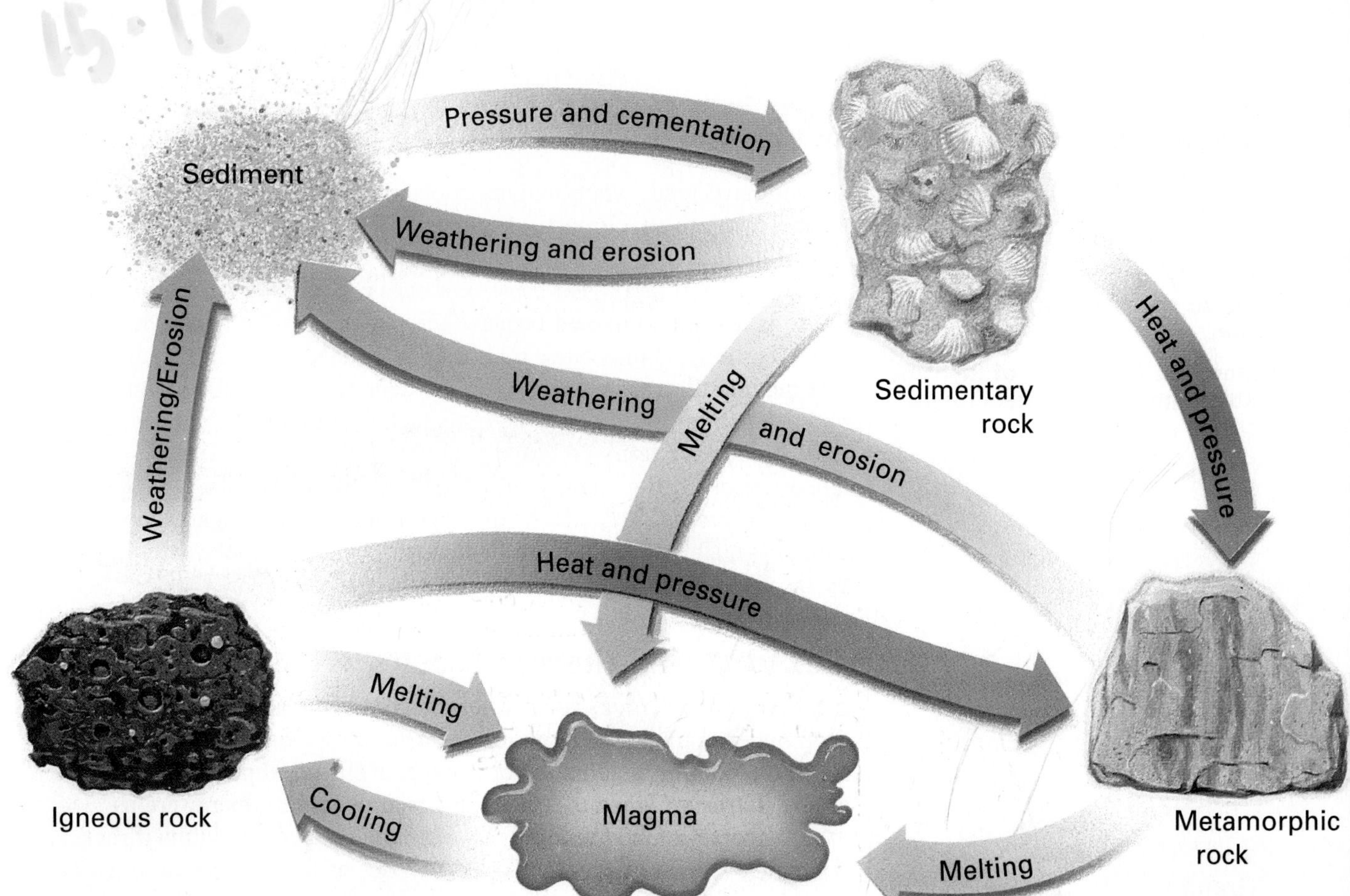

Figure 10–3 shows, a particular body of rock does not always pass through each stage of the complete rock cycle. For example, igneous rock may never be exposed at the earth's surface where it would be changed into sediments. Instead, the igneous rock may be changed directly into metamorphic rock while still beneath the earth's surface. Igneous and sedimentary rock may melt and become magma without first becoming metamorphic rock. All three of the major rock types may form sediments when they are exposed to processes at the surface of the earth.

Figure 10–3. The rock cycle illustrates the changes that sedimentary, igneous, and metamorphic rocks undergo.

Section 10.1 Review

1. Which major type of rock—igneous, sedimentary, or metamorphic—forms from magma that cools and hardens?
2. Which major type of rock is composed of cemented fragments of rock or minerals?
3. Which major type of rock forms from other rocks as a result of intense heat, pressure, or chemical processes?
4. What is the rock cycle?
5. Does every rock go through the complete rock cycle, from igneous rock to sedimentary rock to metamorphic rock and back to igneous rock, each time around? Explain your answer.
6. The sedimentary rock limestone is changed into marble, a metamorphic rock. Name the three processes that cause this change.

Section Objectives

- **Describe how the cooling rate of magma and lava affects the structure of igneous rocks.**
- **Classify igneous rocks according to their mineral composition.**
- **Describe a number of identifiable igneous rock structures.**

10.2 Igneous Rock

As you have read, when magma cools and hardens, it forms igneous rock. There are two groups of igneous rocks. They are classified according to where the molten rock cools and hardens. The cooling of magma deep below the crust produces **intrusive igneous rocks.** These rocks are so named because the magma that forms them intrudes, or enters, into other rock masses beneath the earth's surface. The magma then slowly cools and hardens to form intrusive igneous rocks. The rapid cooling of lava, or melted rock on the earth's surface, produces **extrusive igneous rocks.** Intrusive and extrusive igneous rocks differ mainly in the size of their crystalline mineral masses or grains. The size of the crystalline grains in igneous rock is called its *texture*. The texture of igneous rocks is largely determined by the cooling rate of the magma or lava that formed the rock.

Texture of Igneous Rocks

Intrusive igneous rocks form when magma cools and hardens slowly deep underground. The slow loss of heat allows time for the minerals in the cooling magma to form large, well-developed crystalline grains. Intrusive igneous rocks composed of large mineral grains have a coarse-grained texture. An example of a coarse-grained rock is granite, shown in Figure 10–4. The core of the continental crust is made of granite.

Extrusive igneous rocks form when lava cools rapidly on the earth's surface. The rapid loss of heat to air or sea water does not allow time for large crystalline grains to form, and thus it produces fine-grained rock. Most extrusive igneous rocks are composed of small mineral grains that cannot be seen by the unaided eye. An example of a common fine-grained igneous rock is basalt. Figure 10–5 shows Devils Tower in Wyoming, which is composed of basalt. The oceanic crust is composed of basalt.

Some intrusive igneous rocks form from magma that cools slowly at first and then more rapidly as it nears the surface. This type of cooling produces large crystals embedded within a mass of smaller ones. An igneous rock with a mixture of large and small crystals is called a **porphyry** (POR-fuh-ree).

Extremely rapid cooling produces a kind of igneous rock in which crystals were unable to form. The obsidian, or volcanic glass, shown in Figure 10–5 illustrates the glassy appearance of this type of rock. During extremely rapid cooling, gases escaping from the molten material may become trapped in the rock and form many small bubbles. These are evident in the pumice in Figure 10–5.

Granite

Figure 10–4. Notice the coarse-grained texture of this sample of granite, an intrusive igneous rock.

Composition of Igneous Rocks

The mineral composition of an igneous rock is determined by the chemical composition of the magma from which the rock develops. Different types of igneous rocks have similar mineral compositions.

Figure 10–5. Devils Tower (left) in Wyoming, is composed of basalt, a fine-grained igneous rock. Obsidian (top right) has no crystalline structure and thus has a glassy texture. Gases trapped inside magma as it cools give pumice (bottom right) its spongelike appearance. Some pumice samples may actually float in water.

Geologists divide igneous rocks into three families—granite, basalt, and diorite—based on mineral composition.

Rocks in the granite family form from magmas called *felsic*, which are high in silica. Rocks in this family have the light coloring of their main mineral components, orthoclase feldspar and quartz. These rocks may also contain plagioclase feldspar, hornblende, and muscovite mica. In addition to the coarse-grained intrusive rock known as granite, several other rocks are part of the granite family. The fine-grained extrusive rock rhyolite also has the mineral composition that makes it a member of the granite family. Obsidian, or volcanic glass, is of the granite family. It may be bluish, black, or red, depending on its mineral composition.

Rocks in the basalt family form from magmas called *mafic*, which are low in silica but rich in iron. The main mineral components of rocks in this family are plagioclase feldspar and augite. These rocks may also include dark-colored ferromagnesian minerals such as olivine, biotite mica, and hornblende. These three components, along with augite, give basaltic rocks a dark color. The dark gray or black extrusive rock known as basalt is a fine-grained member of this family. Coarse-grained intrusive gabbro is also a member of the basalt family.

The medium-colored rocks of the diorite family are made up of the minerals plagioclase feldspar, hornblende, augite, and biotite mica. Components of rocks in this family include little or no quartz. Rocks in the diorite family include the coarse-grained diorite and the fine-grained andesite. If you know only that an igneous rock has coarse grains, can you identify the family it is in? Why?

Igneous Rock Structures

A number of identifiable rock formations are formed of igneous rock. The underground rock masses made up of intrusive igneous rocks are called *intrusions*. The surface rock masses made up of extrusive igenous rocks are called *extrusions*.

Intrusions

The largest of all intrusions are called **batholiths.** Batholiths are very large masses of igneous rock that cover over 100 square kilometers. The word *batholith* means "deep rock." Batholiths were once thought to extend to great depths. However, studies have shown that many batholiths do have lower boundaries, and they are several thousand meters thick. Batholiths form the cores of many major mountain ranges, such as the Sierra Nevada in California. The largest batholith in North America forms the core of the Coast

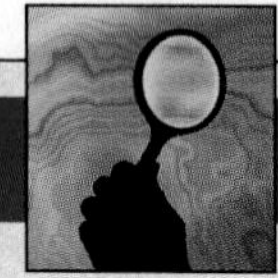

SMALL-SCALE INVESTIGATION

Crystal Formation

The rate of cooling affects the size of the crystals of minerals found in igneous rock. You can demonstrate this relationship by cooling crystals of Epsom salts at three different rates.

Materials

3 glasses or glass jars, water, ice cubes, 3 large test tubes, small pan, spoon or stirring rod, measuring cup, Epsom salts, stove or hot plate, tongs or test-tube clamp, cork, clock or watch

Procedure

1. In a small sauce pan, mix 120 mL of Epsom salts in 120 mL of water. Heat over low heat. Do not let the mixture boil. Stir until no more crystals will dissolve.
2. Add the following until each glass is 2/3 full: glass 1—water and ice cubes; glass 2—water at room temperature; glass 3—hot tap water
3. Carefully pour equal amounts of the Epsom salts mixture into the 3 test tubes. Use the tongs to steady the test tubes as you pour. Drop a few crystals of Epsom salt into each test tube and gently shake. Place one test tube into each glass.
4. Observe what happens to the solutions as they cool at different rates over the next 10–15 minutes. Let the glasses sit overnight, and examine the solutions again after 24 hours.

Analysis and Conclusions

1. In which test tube are the crystals the largest? the smallest?
2. How does the rate of cooling affect the size of crystals formed? Explain your answer.
3. How would you change the procedure to obtain even larger crystals of Epsom salts? Why?
4. Some igneous rocks that are thrown out of a volcano contain large crystals surrounded by very small ones. Based on your observations in this activity, explain why the crystals are different sizes.

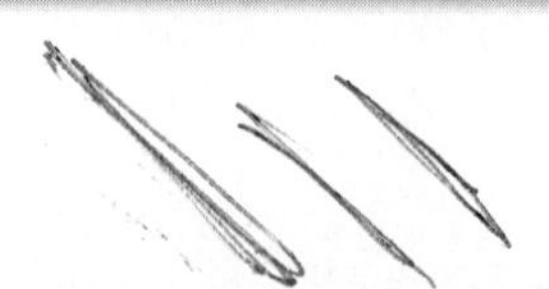

Range in British Columbia. A **stock,** illustrated in Figure 10–6, is an intrusion similar to a batholith that covers less than 100 square kilometers.

When magma flows between rock layers and spreads upward, it sometimes pushes the overlying rock layers into an arc. The floor of the intrusion is parallel to the rock layer beneath it. This type of intrusion is called a **laccolith.** *Laccolith* means ''lake of rock.'' Laccoliths are frequently found in groups. You can sometimes identify them by the small dome-shaped mountains they push up on the earth's surface. Many laccoliths are located beneath the Black Hills of South Dakota.

When a sheet of magma flows between the layers of rock and hardens, a **sill** is formed. A sill lies parallel to the rock layers surrounding it, even if the layers are tilted. Sills vary in thickness from a few centimeters to hundreds of meters and can extend laterally for several kilometers. Big Bend National Park in Texas has excellent examples of sills.

Magma sometimes forces its way through rock layers by following existing vertical fractures or by creating new ones. When the magma solidifies, a **dike** is formed. Dikes differ from sills in that they cut across rock layers rather than lying parallel to the rock layers. Dikes are common in areas of volcanic activity.

Extrusions

When lava erupts onto the earth's surface, it often forms a *volcano.* A volcano is a cone of extrusive rock surrounding a central vent through which the lava flows. When the eruption stops, the lava in the vent cools and solidifies. When a volcano stops erupting for a long period, its cone gradually wears away. Eventually the softer parts of the cone are carried away by wind and water, and only the hard, solidified rock in the vent remains. The solidified central vent is called a **volcanic neck.** Narrow dikes that sometimes radiate out from the neck may also be exposed. A dramatic example of a volcanic neck, called *Shiprock,* is located in New Mexico.

Extrusive rock may also take other forms. Many extrusions are simply flat masses of rock called *lava flows.* Some extrusions, however, take the form known as a **lava plateau.** A lava plateau develops from lava that flows out of long cracks in the earth's surface. The lava then spreads over a vast area, filling in valleys and covering hills. When the lava hardens, it forms a plateau.

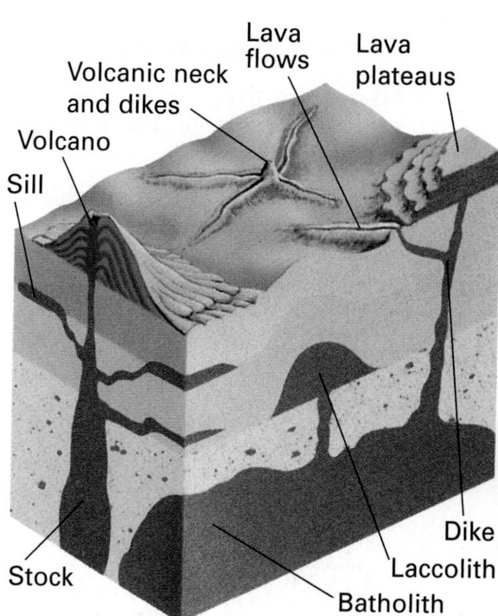

Figure 10–6. Intrusions and extrusions formed from igneous rocks create a number of characteristic landforms (top). Shiprock, in New Mexico, is an example of a volcanic neck and dikes exposed by erosion.

Section 10.2 Review

1. What determines whether an igneous rock will have large crystals or small crystals?
2. Name the three families of igneous rocks.
3. What is a batholith?
4. An unidentified light-colored igneous rock is made up of orthoclase feldspar and quartz. To what family of igneous rocks does it belong? Explain your answer.

Section Objectives

- **Name the three main types of sedimentary rock and give an example of each.**
- **Describe several identifiable sedimentary rock features.**

10.3 Sedimentary Rock

Sedimentary rock is made up of accumulations of various types of sediments. **Compaction** and **cementation** are the processes that form sedimentary rock. During compaction, the weight of overlying sediments causes pressure, pushing the fragments together and squeezing out air and water from the fragments. In cementation, water carries dissolved minerals through the sediments. These minerals are left between the fragments of sediment and provide a cement to hold the fragments together. Geologists classify sedimentary rocks according to the kind and size of sediments that form them.

Formation of Sedimentary Rocks

One class of sedimentary rock is made up of rock fragments carried away from their source by water, wind, or ice and left as deposits elsewhere. Over time, the separate fragments may become compacted and cemented into solid rock. The rock formed from these deposits is called **clastic sedimentary rock.** A second class of sedimentary rock, called **chemical sedimentary rock,** forms from minerals that have been dissolved in water. A third class, called **organic sedimentary rock,** forms from the remains of organisms.

Clastic Sedimentary Rocks

Clastic sedimentary rocks are classified by the size of the sediments they contain. One group consists of gravel-sized fragments that are cemented together by minerals. Rock composed of rounded gravel-sized fragments, or pebbles, is called a **conglomerate.** If the fragments are angular and have sharp corners, it is called a **breccia** (BRECH-ee-uh). In conglomerates, as shown in Figure 10–7, and breccias, the individual pieces of sediment can be easily seen.

Figure 10–7. Conglomerate rock is composed of rounded, pebble-sized fragments held together by a cement (left). The formation on the right is made up of conglomerate rock.

Conglomerate

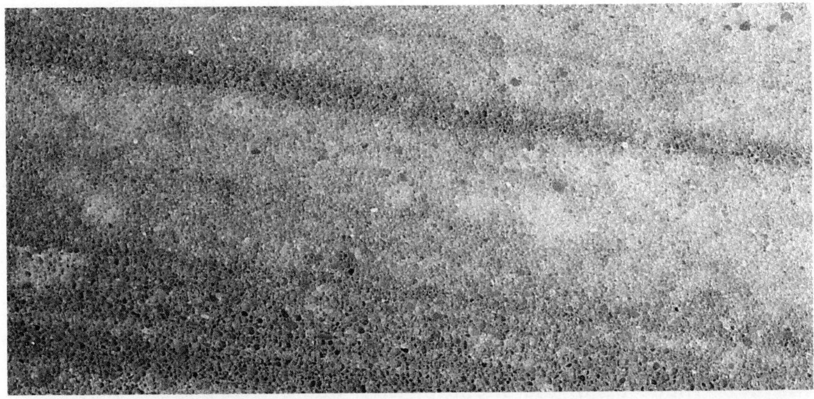

Figure 10–8. Magnified view of sandstone and shale. Sandstone (top) is made up of small quartz grains cemented together by the mineral calcite. The flaky clay particles in shale (bottom) compress into flat layers.

Another group of clastic sedimentary rocks is made up of sand-sized grains that have been cemented together. These rocks are the sandstones. Since most sediments of sand-grain size are made of quartz, quartz is the major component of sandstone. Many sandstones have pores between the sand grains through which liquids, such as groundwater and crude oil, can move.

A third group of clastic sedimentary rock is shale, which consists of clay-sized particles cemented and compacted under pressure. The flaky clay particles are usually pressed into flat layers that will easily split apart. Figure 10–8 shows these characteristic layers.

Chemical Sedimentary Rocks

Some sedimentary rocks are not made up of rock fragments but are chemical in origin. These rocks form from minerals that were once dissolved in water. Some form from dissolved minerals that precipitate, or settle, out of the water as a result of a change in temperature. For example, a certain type of chemical limestone forms when cool currents lower the temperature of warm ocean water, and calcite precipitates, settles, and eventually solidifies on the ocean floor.

Another type of sedimentary rock results when water evaporates and leaves behind the minerals dissolved in the water. The dissolved minerals left behind form rocks called **evaporites.** Gypsum and halite, or rock salt, are two examples of sedimentary rocks formed by rapid evaporation. An example of extensive evaporite deposits can be found on the Bonneville Salt Flats near Great Salt Lake in Utah.

Organic Sedimentary Rocks

The third class of sedimentary rocks is *organic,* which means "formed from the remains of living things." Coal and some limestones are both examples of organic sedimentary rocks. Coal forms from decayed plant remains that are buried and compacted into matter that is mostly carbon. You have learned that chemical limestones can be precipitated as a chemical sediment. However, the formation of organic limestones begins when the mineral calcite is removed from sea water by marine organisms, such as coral, clams, oysters, and plankton. These organisms use the calcite to make their shells. When they die, their shells become limestone. Chalk is a type of limestone made up of the shells of tiny, one-celled marine organisms. The chalk originally formed as mud at the floor of an ancient sea. The white cliffs of Dover, in England, are made up of chalk.

Sedimentary Rock Features

Sedimentary rocks have a number of easily identifiable features. These include stratification (layering), ripple marks, mud cracks, fossils, and concretions.

SCIENCE & TECHNOLOGY

Landsat Maps the World

Landsat 7 is the latest in the Landsat series of earth observation satellites.

Landsat satellites have been recording images of the earth for over two decades. In that time, these earth-scanning satellites have logged more than a million images. As Landsat satellites periodically rescan regions, they create a visual history of the earth's changing landscapes. It is the longest continuous record of the earth's landmasses ever created.

Landsat images resemble aerial photographs. Each image records about 30,000 km^2 of the earth's surface. Landsat images are not ordinary photographs, however. Each satellite uses a scanning sensor system called a *thematic mapper* (TM) to create multispectral images. The TM sensors detect not only visible light, the light recorded by an ordinary camera, but also other parts of the electromagnetic spectrum that the human eye cannot detect, such as infrared light. This capability gives Landsat images much more detail than conventional photographs would have.

The thematic mapper detects seven segments, or bands, of the electromagnetic spectrum besides visible light. It assigns a different color to each band to produce false-color images. The use of bright artificial colors makes it easier to interpret the images. Vegetation appears pink or red, land that is

Stratification

Layering, or **stratification,** of sedimentary rock occurs when there is a change in the kind of sediment being deposited. The type of deposit varies for a number of reasons. Changes in river currents or sea level, for example, may result in a different kind of sediment. The stratified layers, or beds, vary in thickness depending on how long each type of sediment was being laid down. Although most water-deposited strata are laid down horizontally, some sediments deposited by wind, such as sand dunes, are characterized by cross-bedding. Cross-bedding is shown in Figure 10–9.

When various sizes and kinds of materials are deposited within one layer, a type of stratification called graded bedding will occur. Graded bedding occurs as different sizes and shapes of sediment settle to different levels. Where in the layer do you think the largest and heaviest grains of sediment settle?

Figure 10–9. Notice the cross-bedding in the rock layers. This type of pattern generally indicates wind deposits, such as sand dunes, or a rapidly changing stream bed.

Ripple Marks and Mud Cracks

Some sedimentary rocks clearly display *ripple marks*. Ripple marks are formed by the action of wind or water on sand. When the sand becomes sandstone, the ripple marks may be preserved.

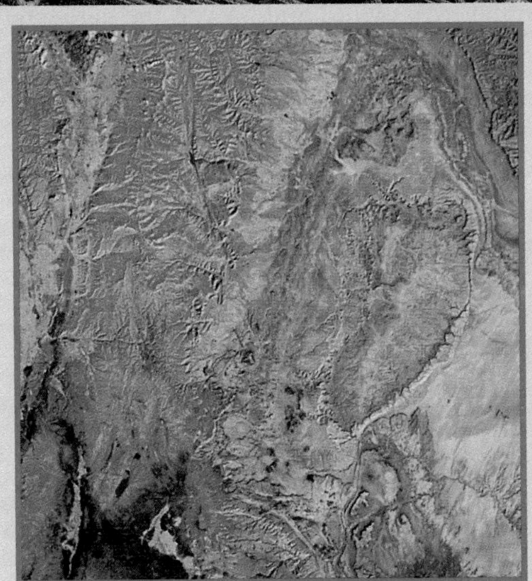

Enhanced-color Landstat image of the uranium-rich San Rafael Swell, 200 km southeast of Salt Lake City.

rich in iron oxides appears yellow or orange, and urban areas appear grayish blue.

Once a satellite radios an image back to earth, cartographers can use computers to create a thematic map. A thematic map is a map that illustrates a particular subject or feature. Selecting different combinations of bands allows cartographers to highlight features such as river deltas, geologic faults, and mineral deposits.

Landsat images appeal to earth scientists in a variety of fields. Landsat images have enabled cartographers to map remote areas of the world. They have allowed hydrologists to find unmapped lakes. Landsat images have also helped geologists discover oil in the Sudan, tin in Brazil, and copper in Mexico.

More recently, ecologists have begun to use Landsat images to see changes in the earth's environment. With a series of images of the same region, taken years apart, ecologists can monitor the effects of processes such as urbanization, deforestation, and soil erosion. With so many useful applications, Landsat satellites are likely to continue working well into the next century.

How might cartographers use Landsat images to check the accuracy of maps?

Figure 10-10. Features of sedimentary rock may include ripple marks (top, left), mud cracks (top, right), and geodes (bottom).

The ground in Figure 10–10 shows *mud cracks,* another feature of sedimentary rock. Mud cracks result when muddy deposits dry and shrink. The shrinking causes the dried mud to crack. A river flood plain or dry lake bed is a likely place to find mud cracks. Once the area is again flooded, new deposits fill in the cracks and preserve them as the mud hardens to solid rock.

Fossils

Fossils are the remains or traces of ancient plants and animals. They are usually preserved in sedimentary rock. As sedimentary deposits pile up, plant and animal remains are buried. Harder parts of these remains may be preserved in the rock. More often, even the harder parts dissolve, leaving only an impression in the rock. Fossils are valuable in helping earth scientists learn about the development of the earth's crust. Fossils are discussed in detail in Unit 5.

Concretions

Sedimentary rocks sometimes contain lumps, or nodules, of rock with a composition different from that of the main rock body. These nodules are known as **concretions.** Concretions form when minerals precipitated from solutions build up around an existing rock particle.

Groundwater sometimes deposits dissolved quartz or calcite inside cavities in sedimentary rock. The quartz or calcite crystallizes into beautiful forms inside the cavities, as shown in Figure 10–10. The crystal cavities are called *geodes.*

Section 10.3 Review

1. How does clastic sedimentary rock differ from chemical sedimentary rock?
2. What kind of sedimentary rock forms from the remains of decaying organisms?
3. What term describes the remains or impressions of plants and animals in sedimentary rock?
4. You suspect that a rock you have found is sedimentary rock. What features would you look for to confirm your identification?

10.4 Metamorphic Rock

The changing of one type of rock to another by heat, pressure, and chemical processes is called **metamorphism.** Most metamorphic rock forms deep beneath the surface of the earth. All metamorphic rock is formed from existing igneous, sedimentary, or metamorphic rock.

Section Objectives

- **Distinguish between regional and contact metamorphism.**
- **Distinguish between foliated and unfoliated metamorphic rocks and give an example of each.**

Formation of Metamorphic Rocks

During metamorphism, heat, pressure, and hot fluids can cause certain minerals to change into other chemicals. Minerals may also change in size or shape or separate into parallel bands that give the rock a layered appearance. Hot fluids from magma may circulate through the rock, changing the mineral composition by dissolving some materials and adding others. All of these changes are part of metamorphism. Two types of metamorphism occur in the crust of the earth. One type of metamorphism occurs when rocks come into direct contact with magma. The other type occurs due to the heat and pressure created by tectonic activity.

When hot magma pushes through existing rock, the heat from the magma can change the structure and mineral composition of the surrounding rock. This type of metamorphism is called **contact metamorphism** because only rocks near or actually touching the hot magma are metamorphosed by its heat. Hot chemical fluids working through fractures may also cause changes in the surrounding rock during contact metamorphism.

Metamorphism sometimes occurs over an area of thousands of square kilometers during periods of tectonic activity. This type of metamorphism is called **regional metamorphism.** The movement of one tectonic plate against another creates tremendous heat and pressure in the rocks at the plate edges. This heat and pressure causes chemical changes in the minerals of the rock. Most metamorphic rock is formed by regional metamorphism. However, volcanism and magma movement often accompany tectonic activity. Thus, rocks formed by contact metamorphism often are found where regional metamorphism has occurred.

Classification of Metamorphic Rocks

Metamorphic rocks are classified according to their structure. Metamorphic rocks have either a **foliated** structure or an **unfoliated** structure. Rocks with a foliated structure have visible parallel bands. Rocks without visible bands are unfoliated.

INVESTIGATE!

To learn more about classifying rocks, try the In-Depth Investigation on pages 192–193.

Foliated Rocks

Foliated rocks can form in one of two ways. Extreme pressure may flatten the mineral crystals in the original rock and push them into parallel bands. Foliation also occurs as minerals of different densities separate into bands, producing a series of alternating dark and light bands.

Common metamorphic rocks with foliated structure include slate, schist, and gneiss. Slate, shown in Figure 10–11, is formed by pressure acting on the sedimentary rock shale, which contains clay minerals of a flaky consistency. The fine-grained minerals in slate are compressed into thin layers, which split easily into flat sheets. Flat sheets of slate are often used for building materials, such as roof tiles or walkway stones.

A greater amount of heat and pressure change slate into the coarser-grained metamorphic rock known as *schist,* shown in Figure 10–11. Deep underground, intense heat and pressure cause the elements in schist to change to very coarse-grained minerals separated into bands of different densities. This greatly metamorphosed rock with bands of light and dark minerals is *gneiss,* which you can also see in Figure 10–11.

Figure 10–11. Increasing heat and pressure change slate (left) into the metamorphic rocks schist (center) and gneiss (right), which shows definite foliation.

Unfoliated Rocks

Unfoliated metamorphic rocks do not have bands of crystals. One common unfoliated rock is quartzite, the result of metamorphism of the sedimentary rock sandstone. During metamorphism, the sandstone is compacted so tightly that the spaces between the particles

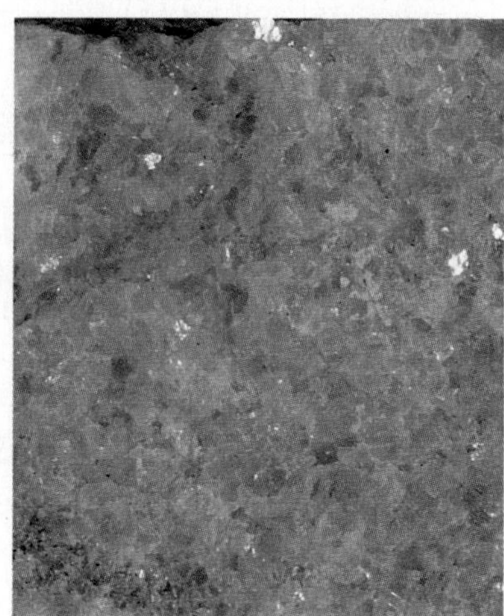

Figure 10–12. Unfoliated metamorphic rocks do not have defined bands of crystals, as shown by these samples of quartzite (left) and marble (right).

EarthBeat

Moon Rocks

NASA's Apollo moon missions have brought almost 400 kg of lunar rocks back to earth. Soon after the *Apollo 11* astronauts set foot on the moon in 1969, they began to fill two boxes with brown and gray moon rocks. Later expeditions included vehicles that allowed a total of about 2,000 specimens to be collected.

Geologists discovered that moon rocks are similar to earth rocks in composition and in the way they were formed. Geologists use their knowledge of earth rocks to analyze the moon rocks and learn about the geological history of the moon.

Much of the rock matter brought back from the moon was in a powdery form. This pulverized rock, called *regolith,* covers much of the surface of the moon. Rock-dating methods indicate that regolith is among the oldest of the moon rocks. From this information, geologists concluded that during the first billion years of the moon's existence, a shower of meteorites pulverized most of the then-existing moon rocks.

The solid rocks on the moon are of two types—highland rocks and mare rocks. The highland rocks are igneous rocks high in plagioclase feldspar content.

Mare (Latin for "sea") refers to dark areas on the moon. Mare rock formed after meteor showers made craters on the moon. Lava from within the moon poured onto the crater floors, covering wide areas of the lunar surface. The lava then cooled and hardened into basalt about 4 billion years ago.

Is mare rock igneous, metamorphic, or sedimentary? Explain your answer.

disappear. Because quartzite is very hard and durable, it remains to form a large part of many hills and mountains after weaker rocks have been worn away. Marble, the beautiful building stone used for monuments and statues, is a metamorphic rock formed from the compression of limestone.

Section 10.4 Review

1. Which kind of metamorphism affects only those rocks near or actually touching the hot magma?
2. What is a foliated structure? In what two ways do rocks get a foliated structure?
3. What is an unfoliated structure? Explain how quartzite gets its unfoliated structure.
4. The metamorphic rock phyllite breaks into flat sheets. Is phyllite foliated or unfoliated? Explain your answer.

Chapter 10 Review

Key Terms

- batholith (180)
- breccia (182)
- cementation (182)
- chemical sedimentary rock (182)
- clastic sedimentary rock (182)
- compaction (182)
- concretion (186)
- conglomerate (182)
- contact metamorphism (187)
- dike (181)
- evaporites (183)
- extrusive igneous rocks (178)
- foliated (187)
- fossil (186)
- igneous rock (175)
- intrusive igneous rocks (178)
- laccolith (181)
- lava plateau (181)
- metamorphic rock (176)
- metamorphism (187)
- organic sedimentary rock (182)
- porphyry (178)
- regional metamorphism (187)
- rock cycle (176)
- sediment (175)
- sedimentary rock (175)
- sill (181)
- stock (181)
- stratification (185)
- unfoliated (187)
- volcanic neck (181)

Key Concepts

Rocks are classified into three major types based on how they form. These types are igneous rock, sedimentary rock, and metamorphic rock. **See pages 175–176.**

In the rock cycle, rocks change from one type into another. **See page 176.**

The rate at which magma and lava cools determines the crystal size of igneous rock. **See page 178.**

Igneous rocks are classified into three families based on their mineral composition. These families are granite, basalt, and diorite. **See page 179.**

Igneous rock structures take two basic forms. They are intrusions and extrusions. **See pages 180–181.**

Sedimentary rock forms in one of three ways. It may form from rock fragments, from minerals once dissolved in water, or from the remains of organisms. **See page 182.**

Sedimentary rocks have a number of identifiable features, including stratification, ripple marks, mud cracks, fossils, and concretions. **See page 184.**

Metamorphic rock is formed by heat and pressure caused by hot magma or tectonic plate movement. See **page 187.**

Metamorphic rocks can have a foliated or unfoliated structure. **See page 187.**

Review

On your own paper, write the letter of the term that best completes each of the following statements.

1. Rock that is formed from magma is called
   a. igneous. b. metamorphic.
   c. sedimentary. d. clastic.

2. The process in which rock changes from one type to another and back again is called
   a. a rock family. b. the rock cycle.
   c. contact metamorphism. d. foliation.

3. Intrusive igneous rocks are characterized by a coarse-grained texture because they contain
   a. heavy elements.
   b. small crystals.
   c. large crystals.
   d. fragments of different sizes and shapes.

4. Light-colored igneous rocks are part of the family called
   a. basalt. b. hornblende.
   c. granite. d. diorite.

5. Magma that solidifies underground forms rock masses that are known as
   a. extrusions. b. volcanic cones.
   c. lava plateaus. d. intrusions.
6. One example of an extrusion is a
   a. stock. b. dike.
   c. batholith. d. lava plateau.
7. Sedimentary rock formed from rock fragments is called
   a. organic. b. chemical.
   c. clastic. d. granite.
8. One example of a chemical sedimentary rock is
   a. evaporites. b. coal.
   c. gneiss. d. breccia.
9. Contact metamorphism is a result of
   a. plate movement. b. hot magma.
   c. sedimentation. d. lava flows.
10. Regional metamorphism is a result of
   a. plate movement. b. hot magma.
   c. cementation. d. compaction.
11. The splitting of slate into flat layers illustrates its
   a. contact metamorphism. b. formation.
   c. sedimentation. d. foliation.

Critical Thinking

On your own paper, write answers to the following questions.

1. What type of rock will be formed from a sedimentary rock that comes under extreme pressure and heat but does not melt? Explain your answer.
2. Explain how metamorphic rock can change into either of the other two types of rocks through the rock cycle.
3. A certain rock is made up mostly of plagioclase feldspar and augite. It also includes olivine, biotite, and hornblende. Will the rock have a light or dark coloring? Explain your answer.
4. Some of the powdery rock found on the moon serves as the cementing agent for sedimentary moon rocks. What type of sedimentary rocks are these? How do you know?
5. Imagine that you have found a piece of limestone, a sedimentary rock, with strange-shaped lumps on it. Will the lumps have the same composition as the limestone? Explain your answer.
6. Which would be easier to break, the foliated rock slate or the unfoliated rock quartzite? Explain your answer.

Application

1. Suppose you found an igneous rock with a coarse texture. Would the magma that formed the rock have cooled slowly or quickly? Explain how you know.
2. There is a huge batholith in the northwestern part of Idaho. What can you say about the landscape in that area? Explain your answer.
3. If you know that a certain area in South Dakota has a number of laccoliths, what might you expect the landscape to look like?
4. The western part of California is located on a boundary between two tectonic plates. Would most of the metamorphic rock in that area occur in small patches or wide regions? How do you know?

Extension

1. Imagine that you have decided to start a rock collection. List the first 10 kinds of rocks you would like to acquire. Use the encyclopedia entry "Rocks" for help. Include on your list a method for classifying the rocks in your collection. Explain your system of classification to the class.
2. Find out what types of rock are most abundant in your state. Draw a chart of the rock cycle, and indicate where on the chart the rocks fall. Present your findings to the class.

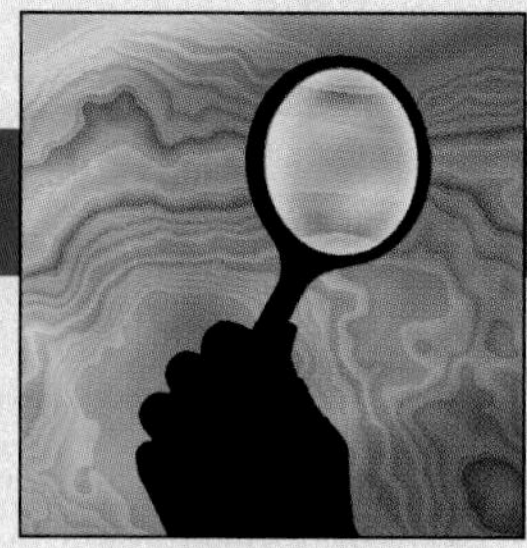

IN-DEPTH INVESTIGATION

Classification of Rocks

Materials

- hand lens
- rock samples
- safety goggles
- medicine dropper
- 10% dilute hydrochloric acid

Introduction

There are many different types of igneous, sedimentary, and metamorphic rocks. Therefore, it is necessary to know important distinguishing features of the rocks in order to classify them. The classification of rocks is generally based on their mode of origin, their mineral composition, and the size and arrangement (or texture) of their minerals.

The many types of igneous rocks differ in the minerals they contain and the sizes of their crystalline mineral grains. Igneous rocks composed of large mineral grains have a coarse-grained texture. Some igneous rocks have small mineral grains that cannot be seen with the unaided eye. These types of rocks have a fine-grained texture.

Many sedimentary rocks are made of fragments of other rocks compressed and cemented together. Some sedimentary rocks have a wide range of sediment sizes, while others may have only one size. Other common features of sedimentary rocks include parallel layers, ripple marks, cross-bedding, and the presence of fossils.

Metamorphic rocks often look similar to igneous rocks, but they may have bands of color. Metamorphic rocks with a *foliated* texture have minerals arranged in bands. Metamorphic rocks that do not have bands of minerals are *unfoliated.*

In this investigation, you will use a rock identification table to identify various rock samples.

Prelab Preparation

1. Study the Rock Identification Table on page 669.
2. Review the safety guidelines for eye safety and the handling of caustic substances.

Procedure

1. In your notebook, make a table with columns for sample number, description of properties, rock class, and rock name. List the numbers of the rock samples you were given by your teacher.

Specimen	Description of Properties	Rock Class	Rock Name

2. Using a hand lens, study the rock samples. Look for characteristics such as the shape, size, and arrangement of the mineral crystals. For each sample, list in your table the distinguishing features that you observe.
3. Refer to the Rock Identification Table on page 669. Compare the properties for each rock sample that you listed with the properties listed in the identification table. If you are unable to identify certain rocks, examine these rock samples again.
4. Certain rocks react with acid, indicating that they are composed of calcite. If a rock contains calcite, it will bubble, releasing carbon dioxide. Using a medicine dropper and 10% dilute hydrochloric acid, test various samples for their reactions. **CAUTION:** *Wear goggles when working with hydrochloric acid.*
5. Complete your table by identifying the class of rocks—igneous, sedimentary, or metamorphic—that each rock sample belongs to, and then name the rock.

Analysis and Conclusions

1. What properties were most useful in identifying each rock sample?
2. Were there any samples that you found difficult to identify? Explain.
3. Were there any characteristics common to all the rock samples?
4. How can you distinguish between a sedimentary rock and a foliated metamorphic rock when both have observable layering?

Extensions

1. Name properties that distinguish the following pairs of rocks from one another.
   a. granite and limestone
   b. obsidian and sandstone
   c. pumice and slate
   d. conglomerate and gneiss
2. Collect a variety of rocks in your area. Use the Rock Identification Table to see how many you can classify. How many rocks did you collect that were igneous? How many were sedimentary rocks? How many were metamorphic rocks? After you identify the class of each rock, try to name the rock.

Supplies of valuable natural resources are diminishing rapidly. The world supply of essential minerals, for example, is limited. Energy resources, such as coal, petroleum, and natural gas, are also limited. As these resources dwindle, scientists search for new, renewable, and alternative resources.

Chapter Outline

This hydroelectric dam is on the Colorado River in Arizona.

11.1 Mineral Resources

Section Objectives

- **Explain what ores are and how they form.**
- **Discuss the wide variety of uses for mineral resources.**

The earth's crust contains a wealth of useful mineral resources. Earth scientists have identified over 3,000 minerals. These minerals, however, are **nonrenewable resources**—their supply is limited and they cannot be replaced once they are used. The processes that form these resources may take millions of years. **Renewable resources,** such as air, water, and plants, can be replaced within a human lifetime or as they are used. Trees, for example, are a renewable resource because more trees can be planted to replace those cut down.

Mineral resources may be *metals,* such as gold (Au), silver (Ag), and aluminum (Al), or *nonmetals,* such as sulfur (S) and quartz (SiO_2). Metals can be identified by their shiny surfaces, ability to conduct heat and electricity, and tendency to bend easily when in thin sheets. Nonmetals have a dull surface and are poor conductors of heat and electricity. Which would be a better insulator for a hot-water pipe, a metal or a nonmetal? Why?

Formation of Ores

Gold, silver, copper (Cu), and a number of other metals are found in the earth's crust as native, or uncombined, elements. However, most metallic and nonmetallic elements are found in chemically combined forms as minerals in the crust. Deposits of minerals from which metals and nonmetals can be removed profitably are called **ores.** For example, iron (Fe) can be obtained from magnetite and hematite ores. Mercury (Hg) can be separated from cinnabar, and aluminum can be separated from bauxite.

Ores and Cooling Magma

Ores form within the earth's crust in a variety of ways. Chromium (Cr), nickel (Ni), and lead (Pb) ores may form within cooling magma. As magma cools, dense metallic minerals sink to the bottom of the body of magma. Layers of these minerals accumulate and form ore deposits within the hardened magma.

Ores and Contact Metamorphism

Some lead, copper, and zinc (Zn) ores form through contact metamorphism when hot magma comes into contact with existing rock. Heat and chemical fluids from the magma change the surrounding rock. The amount of contact metamorphism that takes place depends on such things as the temperature of the magma, pressure, and the composition and permeability of the existing rock.

A similar but far more extensive process forms a different type of ore deposit. Hot mineral solutions spread through many small cracks in a large mass of rock and may deposit valuable minerals in narrow fingerlike bands called **veins.** Figure 11–1 shows mineral veins. A large number of thick mineral veins form a deposit called a **lode.** Many gold miners were hoping to find the "Mother Lode" in their search for that valuable ore during California's gold rush.

Figure 11–1. Hot mineral solutions have left veins of quartz in this rock. Lead, silver, gold, and copper are also found in veins.

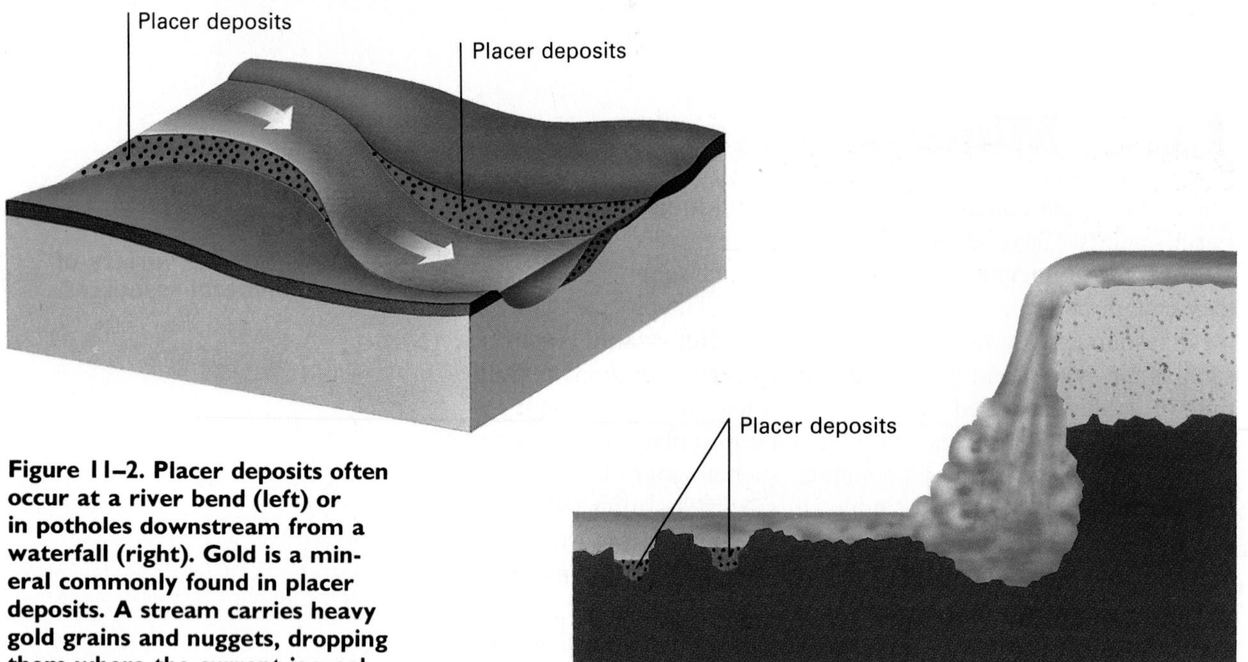

Figure 11–2. Placer deposits often occur at a river bend (left) or in potholes downstream from a waterfall (right). Gold is a mineral commonly found in placer deposits. A stream carries heavy gold grains and nuggets, dropping them where the current is weak.

Ores and Moving Water

The movement of water helps to form ore deposits in two ways. In one process, fragments of native metals such as gold are released as rock breaks down due to natural processes. Streams carry the fragments until, because of their great density, they are deposited where the currents are weak. The fragments become concentrated at the bottom of stream beds in layers called **placer deposits,** as shown in Figure 11–2. In the second process, water dissolves minerals as it flows through cracks in rocks on the earth's surface. Dissolved minerals accumulate in placers, forming ore deposits. Ores of heavy minerals, such as gold, tin, lead, copper, and platinum, often form in placer deposits.

Uses of Mineral Resources

Some metallic ores, such as gold, platinum, and silver, are prized for their beauty and rarity. Other metallic ores are sources of valuable elements. Some nonmetallic minerals are **gemstones,** rare mineral crystals that display extraordinary brilliance and color when specially cut for jewelry. Other nonmetallic minerals, such as calcite and gypsum, are used as building materials. Table 11–1 shows various uses for metallic and nonmetallic minerals.

Mineral Conservation

Most mineral resources are being consumed at an increasing rate each year. Every person in the growing population of the world represents a need for additional mineral resources. One source of untapped mineral resources lies beneath the ocean floor. However, these deposits are difficult to locate and even more difficult to remove. The time may come when these deep-ocean mineral deposits will have to be mined.

INVESTIGATE!

To learn more about metals and ores, try the In-Depth Investigation on pages 210–211.

Table 11–1 Minerals and Their Uses

Metallic Minerals	Uses
Hematite and magnetite (iron)	in making steel
Galena (lead)	in car batteries; in solder
Gold, silver, and platinum	in electronics and dental work; as objects such as coins, jewelry, eating utensils, and bowls
Chalcopyrite (copper)	as wiring, in coins and jewelry, as building ornaments
Sphalerite (zinc)	in making brass
Bauxite (aluminum)	as wiring, in cans, in aircraft and spacecraft
Nonmetallic Minerals	**Uses**
Diamond	in drill bits and saws (industrial grade), in jewelry (gemstone quality)
Calcite	in cement, as building stone
Halite (salt)	in food preparation
Kaolinite (clay)	in ceramics, cement, and bricks
Quartz (sand)	as glass
Sulfur	in gunpowder, medicines, and rubber
Gypsum	in plaster and wallboard

The only sure way to preserve mineral resources is through conservation. One way to conserve minerals is to use other, more abundant or renewable materials in place of minerals. One such substitute is *plastics*. Another way to conserve minerals is by recycling them. Recycling is using materials over again. Some metals, such as iron, copper, and aluminum, can be recycled. Glass, from quartz sand, is also successfully recycled. Many building materials, which contain minerals, can also be reused.

Section 11.1 Review

1. What is an ore?
2. Explain how some ores form through the process of contact metamorphism.
3. What are some uses of the metals gold, silver, platinum, and copper?
4. Explain what is meant by renewable and nonrenewable resources.

Section Objectives

- **Explain why coal is a fossil fuel.**
- **Describe how petroleum and natural gas are formed and how they are removed from the earth.**
- **Discuss the importance of fossil fuels as a source of energy and of petrochemical products.**
- **Explain that fossil fuels are nonrenewable resources that must be used wisely.**
- **Describe some of the effects that the use of fossil fuels has on the environment.**

11.2 Fossil Fuels

Buried within the earth's crust lie the sources of much of the energy you use every day. These natural resources—coal, petroleum, and natural gas—formed from the remains of living things. They have developed and accumulated over long periods of time. Because of their organic origin, coal, petroleum, and natural gas are called **fossil fuels.** Fossil fuels consist primarily of compounds of carbon and hydrogen called **hydrocarbons.** These compounds contain energy originally obtained from sunlight by plants and animals that lived millions of years ago. When hydrocarbons are burned, energy is released as heat and light that can be used.

Coal

Coal is a dark-colored, organic rock. Complex chemical and physical changes produced coal from the remains of plants such as giant ferns that flourished in prehistoric swamps millions of years ago. Usually, dead plants and other organisms are decomposed by microorganisms. However, if oxygen in a swamp is limited, microorganisms can only partially decompose the plant remains, and coal may form.

Formation of Coal

The vast coal deposits of today are the remains of plants that have undergone **carbonization.** Carbonization may occur when partially decomposed trees and other plants are buried in swamp mud. Bacteria consume some of the plant material. These bacteria then release marsh gas, which includes methane (CH_4) and carbon dioxide (CO_2). As the gas escapes, the original complex chemical compounds present in the plants gradually change, and only carbon remains.

Types of Coal

The partial decomposition of plant remains produces a brownish-black material called **peat.** Over time, peat deposits are covered by layers of sediments. The weight of these overlying sediments

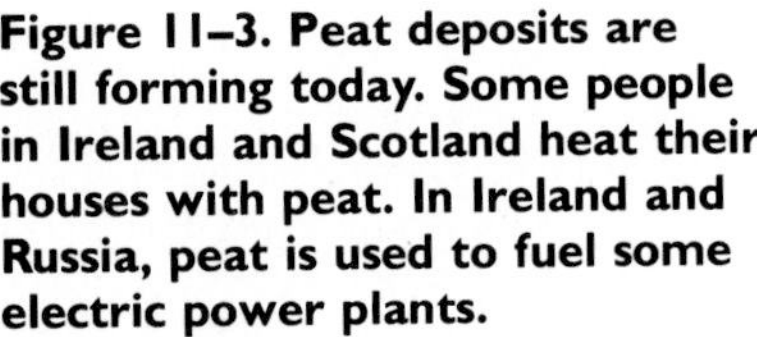

Figure 11–3. Peat deposits are still forming today. Some people in Ireland and Scotland heat their houses with peat. In Ireland and Russia, peat is used to fuel some electric power plants.

Figure 11–4. Coal that is deposited in horizontal beds, or layers, near the surface is strip mined, as shown on the left. Large earth-moving machines strip away soil and rock to expose the coal.

squeezes out water and gases from the peat. It then becomes a denser material called **lignite,** or brown coal. The pressure of more deposited sediments further compresses the lignite and forms **bituminous coal,** or soft coal. Bituminous coal is the most abundant type of coal. Where the folding of the earth's crust produces extremely high temperatures and pressure, bituminous coal is changed into **anthracite,** the hardest of all forms of coal. Bituminous coal and anthracite consist of 80 to 90 percent carbon and produce a great amount of heat when they burn. Is anthracite like an igneous, sedimentary, or metamorphic rock? Why?

Petroleum and Natural Gas

Petroleum and natural gas are mixtures of hydrocarbons. The hydrocarbons formed largely from microorganisms that lived in oceans or large lakes millions of years ago. Petroleum, also called oil, consists of liquid hydrocarbons. Natural gas is made up of hydrocarbons in gaseous form.

Formation of Petroleum and Natural Gas

When microorganisms died in shallow prehistoric oceans and lakes, their remains accumulated on the ocean floor and lake bottoms and were buried by marine sediments. As in coal formation, these sediments limit the available oxygen supply and prevent the remains from decomposing completely. As more and more sediments accumulated, the heat and pressure on the buried organisms increased. When the heat and pressure became great enough, chemical changes occurred that converted the remains into petroleum and natural gas.

Petroleum and Natural Gas Deposits

The sedimentary rocks from which petroleum is collected have many interconnected spaces between rock particles. Liquids can flow through these spaces. Rock that liquids can easily flow through is called *permeable rock.* As sedimentary rock becomes deeply buried under overlying sediments, pressure increases. It forces water and hydrocarbons out of the pores and up through the layers of permeable rock. The petroleum and water continue to move upward until they meet a layer of impermeable rock, or rock through which liquids cannot flow. This impermeable rock layer, most commonly shale, is called *cap rock*. Petroleum that accumulates beneath the cap rock forms an oil reservoir by saturating all the spaces between the rock particles. Because petroleum is less dense than water, it floats on top of any trapped water. If natural gas is present, it floats on top of the petroleum, since natural gas is less dense than both oil and water.

Geologists explore the earth's crust to discover the kinds of rock structures, like the anticline in Figure 11–5, that may contain oil or gas. When a well is drilled into an oil reservoir, the petroleum and natural gas often flow to the surface. The pressure that causes water and hydrocarbons to move up through the permeable rock forces the petroleum up and out through the well.

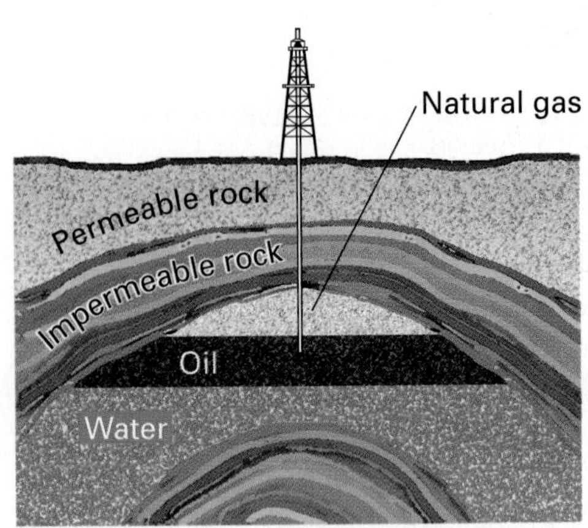

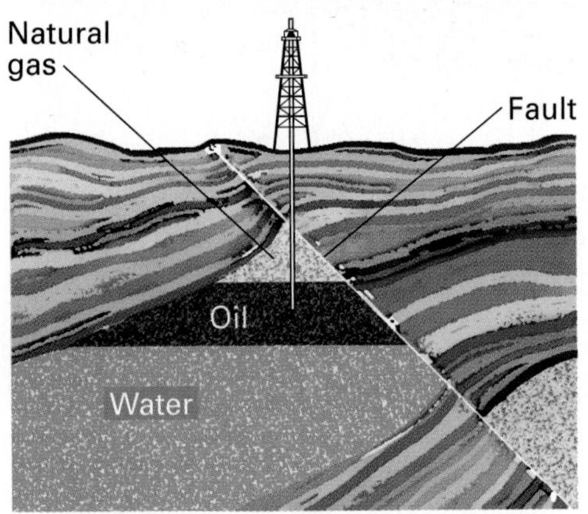

Figure 11–5. Many oil traps are anticlines, or upward folds, in rock layers (top). Another common type of oil trap is a fault, or fracture, in the earth's crust (bottom).

Uses of Fossil Fuels

Fossil fuels are the main sources of energy for transportation, farming, and industry. **Crude oil,** or unrefined petroleum, is also the source of many useful products, as shown in the graph in Figure 11–6. Notice the term **petrochemical** on the graph. Petrochemicals are chemicals derived from petroleum. Petrochemicals are the essential components of over 3,000 products. These products include plastics, synthetic fabrics, medicines, tars, waxes, synthetic rubber, insecticides, chemical fertilizers, detergents, and shampoos.

Fossil Fuel Supplies

Fossil fuels, like minerals, are nonrenewable resources. Coal is the most abundant fossil fuel in the world. Every continent has coal, but almost two-thirds of the known deposits are found in three countries—the United States, Russia, and China. At the present rate of use, scientists estimate that the worldwide coal reserves will last about 200 years.

The United States has been explored for petroleum more completely than any other part of the world. Scientists estimate that at least 75 percent of all the petroleum in the United States has already been discovered. Much of the undiscovered supply is thought to lie under the ocean floor along the edges of North America.

Petroleum production in the United States has almost certainly reached its peak and will decline in the future. In some areas of the world, however, more than 90 percent of the petroleum known to exist is still in the ground. Scientists also think that much natural gas remains to be discovered, but some of the gas may be deeper than 4,600 m below the earth's surface. Oil shale also contains petroleum. But the cost of recovering the oil from shale is far greater than the present cost of recovering oil from conventional wells.

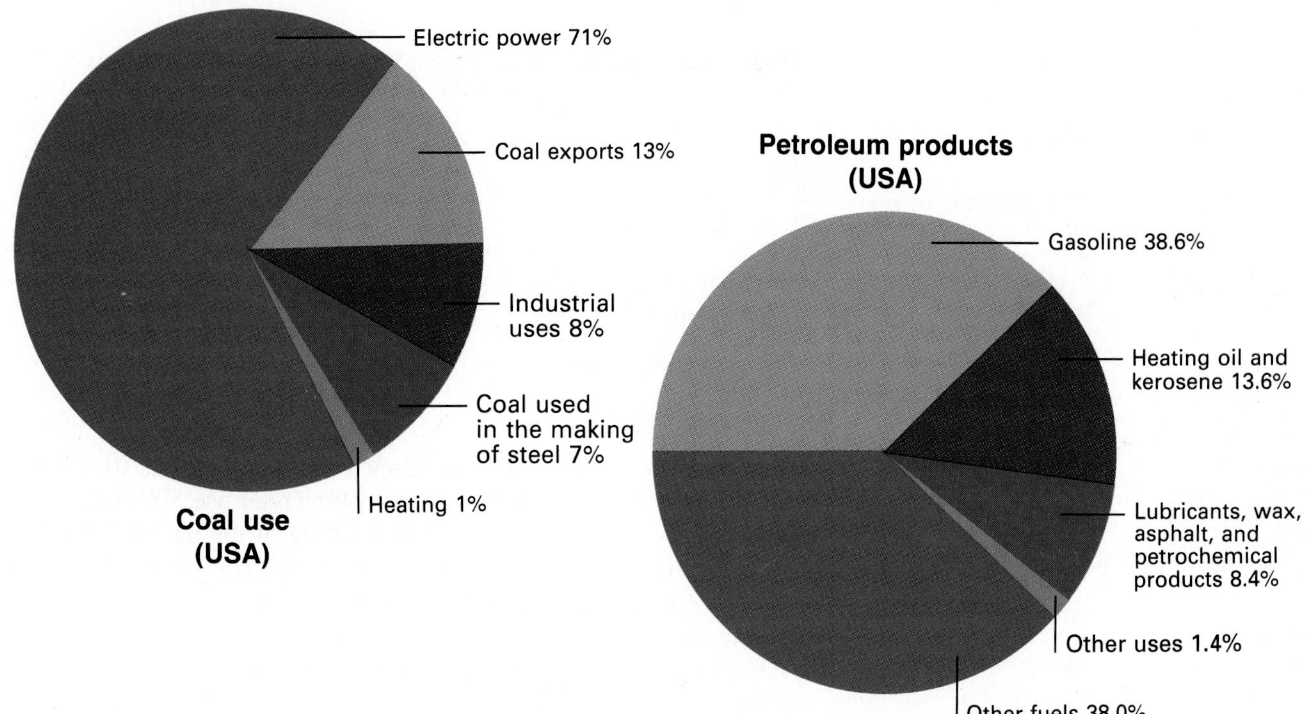

Figure 11–6. The pie graphs show the uses of coal in the United States (left) and the products derived from petroleum (right).

Fossil Fuels and the Environment

The use of any fossil fuel has an impact on the environment. Strip mining of coal, shown in Figure 11–4, leaves deep ditches where coal is removed. Rocks and topsoil that are displaced to expose the coal are left in steep slopes. Without plants and topsoil to protect it, the exposed land will wear away. When wet, rocks that are exposed during mining give off acids. Rain may carry the acids into nearby rivers and streams, causing harm to aquatic life. As a result, United States legislators have passed laws to limit the damage done by strip mining.

Fossil fuels can also cause air pollution. The burning of coal with a high sulfur content releases large amounts of sulfur dioxide into the atmosphere; this is the main source of acid precipitation. Although petroleum and natural gas are cleaner-burning fuels than coal, they too can damage the environment. The burning of gasoline in automobiles, for example, is a major contributor to air pollution. Other parts of the environment are also susceptible to pollution from fossil fuels. You may have heard of spills from oil wells, tankers, or pipelines that have polluted the ocean and harmed wildlife.

Section 11.2 Review

1. List the three types of coal.
2. Describe the kind of rock structures in which petroleum reservoirs are formed.
3. Are fossil fuels renewable or nonrenewable resources?
4. Name two environmental problems associated with the mining and use of coal.
5. Why would manufacturing less plastic help conserve petroleum?

Section Objectives

- Summarize the process of nuclear fission.
- Summarize the process of nuclear fusion.

11.3 Nuclear Energy

You read about the structure of atoms in Chapter 8. Remember that only the outermost electrons of atoms are involved in the formation of chemical bonds. The chemical changes that bond elements into molecules and compounds have no effect on the nucleus of the atom. The atomic nuclei, however, can undergo changes that release great amounts of energy.

Nuclear Fission

Scientists and engineers have developed a technology that uses nuclear reactions to produce energy for commercial use. This technology is based on **nuclear fission,** the splitting of the nucleus of a large atom into two or more smaller nuclei. Fission involves a release of a tremendous amount of heat energy, which can be used to generate electricity.

Only one kind of naturally occurring element can be used for nuclear fission. It is a rare isotope of the element uranium called *uranium-235* (U-235). Before U-235 can be used for nuclear fission, it must be separated from deposits of uranium ore. Next, U-235 is mixed with radioactive U-238 and formed into pellets. The uranium is then shaped into rods called *fuel rods*. Bundles of these fuel rods are bombarded by neutrons. When struck by a neutron, the U-235 nuclei in the fuel rods split. When one U-235 nucleus in these fuel rods splits, it releases neutrons and more energy. A chain reaction occurs as these neutrons strike neighboring U-235 atoms and cause their nuclei to undergo fission. In turn, the neutrons from these nuclei strike other U-235 atoms, and the chain reaction continues. The nuclear chain reaction causes the fuel rods to become very hot. What do you suppose accounts for this heat?

Figure 11–7. The chain reaction that produces energy by nuclear fission can be harnessed and controlled in reactors like this one.

Water is pumped around the fuel rods to absorb and carry away the heat. The resulting hot water or steam becomes a source of energy for turning the turbines that run electric generators. Once the water gives off its heat, it is again pumped around the fuel rods to absorb more heat.

The chain reaction that occurs during nuclear fission can be controlled. The neutron flow can be regulated so that it is slowed down, speeded up, or stopped. The equipment in which controlled nuclear fission is carried out is called a *nuclear reactor.* A reactor is shown in Figure 11–7.

The production of nuclear power has a number of drawbacks. The waste products of nuclear fission give off dangerous radiation. The radiation can destroy plant and animal cells and cause harmful changes in the genetic material within the cells of living things. In the past, nuclear waste was dumped into the ocean. However, these wastes may be harmful to marine life and may also be harmful to the human beings who eat this marine life. A method used currently to dispose of nuclear waste is to store it in salt mines located deep

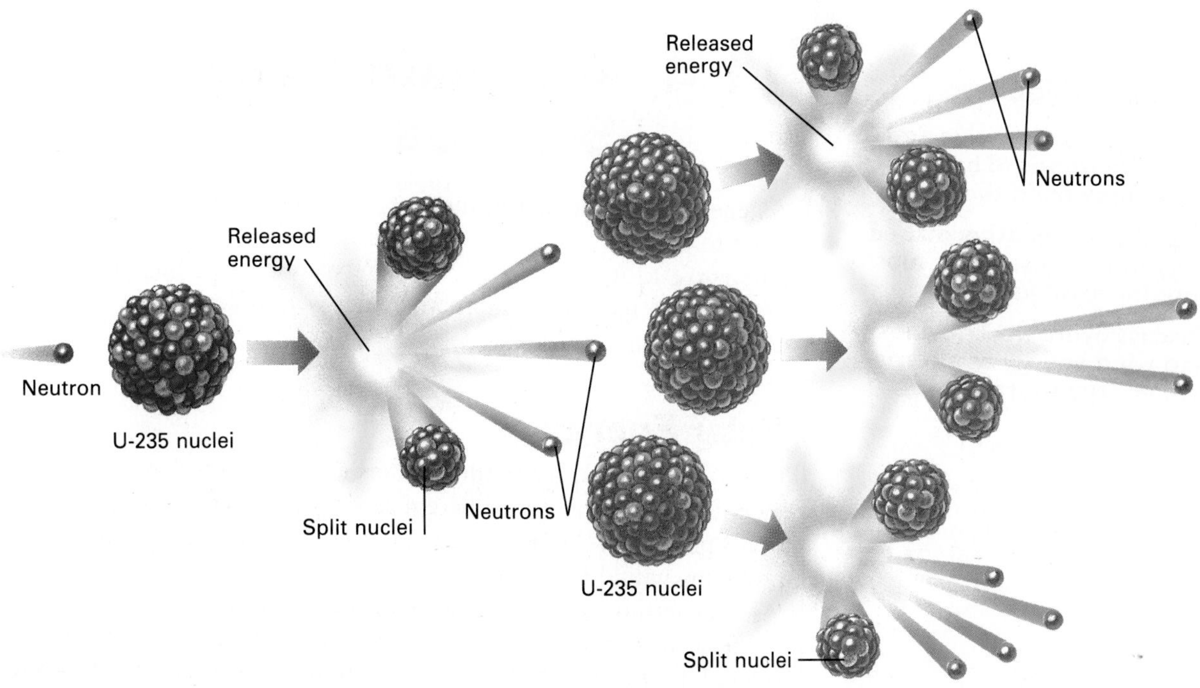

Figure 11–8. The splitting of nuclei increases rapidly in a chain reaction during nuclear fission.

underground. Some radioactive waste remains dangerous for thousands of years, and no sure solution to the nuclear waste problem has yet been found.

Nuclear Fusion

All of the energy that reaches the earth from the sun is produced by another kind of nuclear reaction, which is called **nuclear fusion.** In nuclear fusion, the nuclei of smaller atoms combine and form larger nuclei. For example, nuclear fusion in the sun releases energy when hydrogen nuclei combine and form helium nuclei. These reactions can take place only at temperatures of more than 15 million degrees Celsius.

Scientists are trying to harness nuclear fusion for the production of large amounts of electricity. Much more research is needed before a commercial fusion reactor can be built. If such a reactor is built in the future, hydrogen atoms from ocean water could be used as the fuel. Thus, the amount of energy available from nuclear fusion on earth would be almost limitless. The wastes from fusion would also be less dangerous than those from fission.

Section 11.3 Review

1. What is the naturally occurring element that must be used for nuclear fission? From where is this element obtained?
2. What occurs during a fission chain reaction?
3. Through which kind of nuclear reaction does the sun produce energy?
4. Can the waste products of nuclear fission be safely disposed of in rivers and streams? Explain why.

Section Objectives

- **Describe both passive and active methods of harnessing energy from the sun.**
- **Explain how geothermal energy may be used as a substitute for fossil fuels.**
- **Discuss hydroelectric energy and wind energy as alternative energy sources.**

11.4 Alternative Energy Sources

As energy needs increase, the world supply of fossil fuels will continue to be used at a much faster rate. Nuclear energy provides an alternative to fossil fuels but has safety concerns. Thus, nations are looking into the use of alternative energy in order to find safe and renewable energy resources.

Solar Energy

Every 15 minutes, the earth receives enough energy from the sun to meet the energy needs of the entire world for one year. If even a small part of this solar energy could be captured, the world's energy problems would be solved.

Converting sunshine into useful heat can be done in two ways. A house with windows facing toward the sun, for example, is considered a *passive system* because it requires no working parts. Sunlight enters the house and warms the building material, which may be stone or brick, and stores some heat for the evening. An *active system* involves the use of solar collectors. A **solar collector** is a box with a glass top, usually placed on the roof. Within the box are tubes that circulate water through the building. The sun heats the water as it moves through the tubes, providing heat and hot water. On cloudy days, however, there may not be enough sunlight to heat the water and the system must use the heat it has stored.

Photovoltaic cells convert sunshine directly into electricity. This works very well for small objects like calculators. Producing electricity for cities from these cells is now in the experimental stage.

Geothermal Energy

In many locations, water flows within the earth's crust far beneath the surface. This water may flow through rock that is heated by a nearby body of magma or by hot gases released by magma. The

Figure 11–9. Geothermal steam fields produce electricity in New Zealand (shown at right), Italy, Japan, Mexico, Russia, Iceland, and the United States.

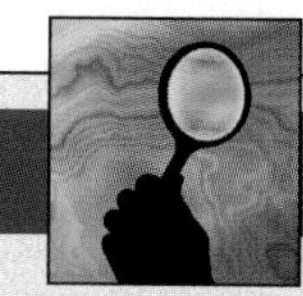

SMALL-SCALE INVESTIGATION

Solar Collector

You can demonstrate the principles of solar collection with a simple model.

Materials

small shallow pan, sheet of black plastic, thermometer, adhesive tape, water at room temperature, plastic wrap, rubber bands, clock or watch

Procedure

1. Line the inside of the pan with black plastic. Tape the thermometer to the inside of the pan as shown. Fill the pan with water at room temperature just enough to cover the end of the thermometer. Fasten plastic wrap over the pan with a rubber band. Be sure you can read the thermometer.
2. Place the pan in a sunny area. Record the temperature every 5 minutes until it stops rising. Discard the water.
3. Repeat the procedure in Steps 1 and 2, but do not cover the pan with plastic wrap.
4. Repeat the procedure in Steps 1 and 2, but this time do not line the pan with black plastic.
5. Repeat the process once again, but do not line the pan with black plastic and do not cover the pan with plastic wrap.

6. Calculate the rate of temperature change for each trial. To do so, subtract the beginning temperature from the ending temperature. Divide this number by the number of minutes it took until the temperature stopped rising.

Analysis and Conclusions

1. What are the variables in this investigation? Which model had the greatest rate of temperature change? the least?
2. Which variable that you tested has the most significant effect on temperature change?
3. What materials would you use to make an efficient solar collector of your own design?

water is heated as it flows through the rock. The hot water, or the steam that results, is the source of a huge supply of energy. This energy is called **geothermal energy,** which means ''energy from the heat of the earth's interior.''

Energy experts have harnessed some of this energy by drilling wells to reach the hot water. A steady stream of very hot water or steam is pumped up through the wells. Sometimes water must first be pumped down into the hot rocks if water does not already flow through them. The resulting steam and hot water can be used as a direct source of heat. The steam and hot water also serve as sources of power to drive turbines, which generate electricity. The city of San Francisco, for example, gets some of its electricity from a geothermal power plant located in nearby mountains. In Iceland, 80 percent of the homes are heated by geothermal energy, and Italy and Japan have developed power plants using geothermal energy.

Energy from Water and Wind

Two additional renewable resources can help satisfy the growing energy needs of the world. These resources are wind and water. Energy from water can come from the running water of rivers and streams or from ocean tides.

Energy from Running Water

Some of the world's electrical energy needs can be met by **hydroelectric energy,** the energy produced by running water. Today, 11 percent of the electricity in the United States comes from hydroelectric power plants. At a hydroelectric plant, massive dams hold back running water and channel it through the plant. Inside the plant, the force of the water spins turbines of electrical generators, which in turn produce electricity.

Energy from Tides

Twice each day, water in the oceans moves toward and then away from the shore. These movements are called *tides*. High tide occurs when the water reaches its highest point along the shore; low tide occurs when it reaches its lowest point. To make use of this tidal

SCIENCE & TECHNOLOGY

Solar Two

▲ Solar Two's power tower and surrounding heliostats

On June 5, 1996, an innovative power plant called Solar Two went on-line in California's Mojave Desert. Solar Two is the largest and most advanced example of what scientists refer to as solar thermal, or "power tower," technology. If successful, Solar Two will serve as a model for even larger solar thermal plants.

Solar Two uses 1,926 sun-tracking mirrors, or *heliostats,* to focus sunlight on a single solar collector. A computer program that tracks the sun's position throughout the day controls electric motors attached to each heliostat. The computer adjusts the angle of the mirrors every few seconds to maximize the amount of sunlight hitting the collector.

Solar Two's collector rests on top of a central receiving tower over 90 m high. The collector is similar to rooftop designs that convert solar energy into heat. It does not use photovoltaic cells, which convert sunlight directly into electricity.

Solar Two's collector contains a series of stainless steel pipes coated with energy-absorbing paint. As the heliostats concentrate sunlight on the collector, molten salt flowing through the pipes reaches a temperature of over 565°C. The hot molten salt is then used to change water into

flow, dams are built to trap the water at high tide and release it at low tide. Based on what you have read about hydroelectric energy, how does the channeled tide water generate electricity?

Energy from Wind

Wind energy is now being used to produce electricity in certain locations. Small wind-driven generators can meet some of the energy needs of individual homes. Wind farms, with hundreds of giant propeller wind turbines, can produce enough energy to meet the electricity needs of entire communities. There are only a few places, however, where the wind generators are practical. Even in the most favorable locations, such as in windy mountain passes, the wind does not always blow. Because of this, wind energy cannot be depended on as an alternative energy source for every situation.

Section 11.4 Review

1. How does a solar collector work?
2. How do most of the people in Iceland heat their homes?
3. Is tidal power renewable or nonrenewable? Why?

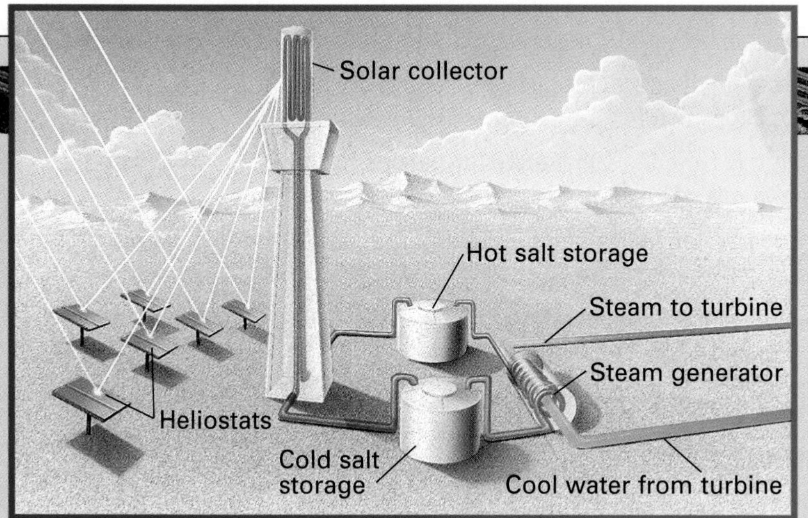

Energy from the sun is collected by heliostats and focused on a solar collector. The hot salts flowing through the collector are then used to create the steam needed to turn the turbines of an electrical generator.

steam that turns an electrical generator's turbines.

The unique feature of Solar Two is its use of molten salt to store solar energy. The salt is a mixture of sodium and potassium nitrates commonly known as saltpeter. At operating temperatures between 280°C and 565°C, the mixture remains a liquid.

The molten salt mixture absorbs and retains heat extremely well. Solar Two can store hot molten salt for up to a week in its well-insulated tank. This technology allows Solar Two to generate electricity during long periods of cloud cover and even at night.

At present, Solar Two produces enough electricity to service 10,000 homes. Solar thermal plants in the future will likely provide 20 times this amount of electricity. By using conventional steam generators and an inexpensive way to store solar energy, Solar Two shows that solar energy can provide electricity cheaply enough and on a scale large enough to compete with fuel-burning power plants.

What advantages do solar thermal plants have over fuel-burning power plants?

Chapter 11 Review

Key Terms

- anthracite (199)
- bituminous coal (199)
- carbonization (198)
- crude oil (200)
- fossil fuel (198)
- gemstone (196)
- geothermal energy (205)
- hydrocarbon (198)
- hydroelectric energy (206)
- lignite (199)
- lode (195)
- nonrenewable resource (195)
- nuclear fission (202)
- nuclear fusion (203)
- ore (195)
- peat (198)
- petrochemical (200)
- placer deposit (196)
- renewable resource (195)
- solar collector (204)
- vein (195)

Key Concepts

Ores are mineral deposits from which metals and nonmetallic minerals can be profitably removed. **See page 195.**

Minerals are important sources of many useful and valuable materials. **See page 196.**

Millions of years ago, chemical and physical changes turned the remains of plants into coal. **See page 198.**

Petroleum and natural gas formed from the remains of ancient microorganisms. Both are found beneath impermeable layers of rock. **See page 199.**

Coal, petroleum, and natural gas provide much of the energy for developed societies. Numerous additional products are made of chemicals derived from petroleum. **See page 200.**

Fossil fuels are nonrenewable resources, which must be used wisely. **See page 200.**

The extraction and use of fossil fuels may damage the environment. **See page 201.**

Nuclear fission can produce energy to generate electricity. Nuclear fission, however, may release dangerous radiation. **See page 202.**

If scientists could produce nuclear fusion, the energy available would be limitless. **See page 203.**

Energy from the sun can be harnessed by both passive and active methods. **See page 204.**

Experts can utilize geothermal energy—energy from the heat of the earth's interior. **See page 204.**

Alternative forms of energy include hydroelectric and wind energy. **See page 206.**

Review

On your own paper, write the letter of the term that best completes each of the following statements.

1. Metals are known to
   a. have a dull surface.
   b. provide fuel.
   c. conduct heat and electricity well.
   d. be found in shale.

2. Aluminum can be taken out of bauxite, which is
   a. an ore. b. an energy source.
   c. a renewable resource. d. a fossil fuel.

3. Hot mineral solutions that spread through cracks in rock form bands called
   a.strata. b. crystals. c. veins. d. silicates.

4. Energy resources that have formed from the remains of living things are called
a. minerals. b. metals. c. gemstones. d. fossil fuels.

5. At the top of an oil reservoir is a layer of
a. coal. b. cap rock. c. peat. d. water.

6. Plastics, synthetic fabrics, and synthetic rubber are composed of chemicals derived from
a. anthracite. b. peat. c. petroleum. d. shale.

7. The most abundant fossil fuel is
a. coal. b. petroleum. c. natural gas. d. shale.

8. One problem caused by the strip mining of coal is
a. increased rainfall.
b. loss of soil.
c. the release of sulfur dioxide into the air.
d. the drying up of rivers.

9. The splitting of the nucleus of an atom to produce energy is called
a. geothermal energy. b. nuclear fission.
c. nuclear fusion. d. hydroelectric power.

10. Hydrogen atoms may someday provide fuel for
a. nuclear fission. b. hydroelectric power.
c. geothermal energy. d. nuclear fusion.

11. Most solar collectors require
a. coal. b. water. c. fission. d. wind.

12. Energy experts have harnessed geothermal energy by
a. building dams. b. building wind generators.
c. drilling wells. d. burning coal.

13. In a hydroelectric plant, running water produces energy by spinning a
a. turbine. b. fan. c. windmill. d. reactor.

Critical Thinking

On your own paper, write answers to the following questions.

1. What might a geologist from a mining company look for in rock masses to identify possible copper deposits? Explain your answer.
2. Do you think it would be profitable for a mining company to mine hematite? Explain your answer.
3. A certain area has extensive deposits of shale. Why might a petroleum geologist be interested in the area?
4. If the United States continues using petroleum in vast amounts, it will become even more dependent on foreign sources for this resource. Why?
5. A certain company in your area produces U-235 pellets and fuel rods. With which energy source is the company involved? How do you know?

Application

1. Imagine that you are on a committee to reduce air pollution in a crowded city. Would you recommend the use of high-sulfur coal as a fuel? Why or why not?
2. Imagine that your senator is thinking of proposing a bill to cut off funding for nuclear fusion research. What might you say in a letter to change the senator's mind?
3. If your school were converting to solar energy, how would you decide the best location on the roof to place a solar collector?
4. If you were to construct a **concept map** using the new terms listed on the previous page, what other word would you use as the top entry?

Extension

1. Choose one important mineral and research its various uses. Present your findings to the class in a chart. Indicate the original ore, the mineral extracted from it, and a number of uses of the mineral.
2. Using Figure 11–5 as a reference, make a clay model of an oil trap. Use clay of various colors to show the rock layers in the oil trap. Present your model to the class.
3. Research the effects of automobile fuel on the quality of air. Report your findings to the class.

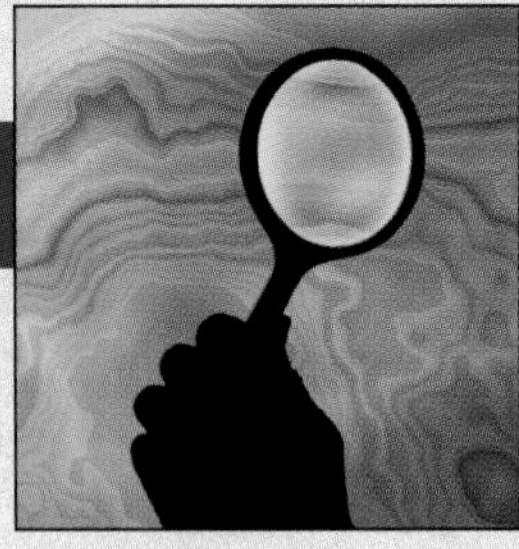

IN-DEPTH INVESTIGATION

Extraction of Copper from Its Ore

Materials
- 2 test-tubes (13 × 100 mm)
- copper (cupric) carbonate
- iron filings
- funnel
- Bunsen burner
- test-tube rack
- water
- dilute sulfuric acid
- test-tube clamp
- safety goggles
- lab apron
- heat-resistant gloves

Introduction
Most metals are found in the earth's crust combined with other elements. A material in the crust that is a profitable source of metal is called an *ore*. Malachite, a native copper ore, is the basic carbonate of copper, which can be represented by the formula $CuCO_3Cu(OH)_2$. The green corrosion that forms on copper due to weathering has the same composition as malachite. The reactions of malachite ore are similar to those of copper carbonate, $CuCO_3$.

In this investigation, you will extract copper from copper carbonate in much the same way that copper is extracted from malachite ore.

Prelab Preparation
1. Review Section 9.1, pages 157–162, and Section 11.1, pages 195–197.
2. Review the safety guidelines for fire, heating, and chemical safety, and for handling caustic substances.

Procedure
1. **CAUTION:** *Wear your laboratory apron and safety goggles throughout this investigation.* Fill one of the test tubes about one-fourth full of copper carbonate. Record the color of the copper carbonate in your notebook.
2. Light the Bunsen burner and adjust the flame.
3. Heat the copper carbonate by holding the tube over the flame with a test-tube holder as shown in the photograph at left.

   CAUTION: *When heating a test tube, always point it away from yourself and other students.* **CAUTION:** *Gently move the test tube over the flame to heat it slowly and prevent it from breaking.* As you heat the copper carbonate, observe any changes in color.
4. Continue heating the test tube over the Bunsen-burner flame for five minutes. The following chemical reaction is occurring in the test tube:

$$CuCO_3 \rightarrow CuO + CO_2$$

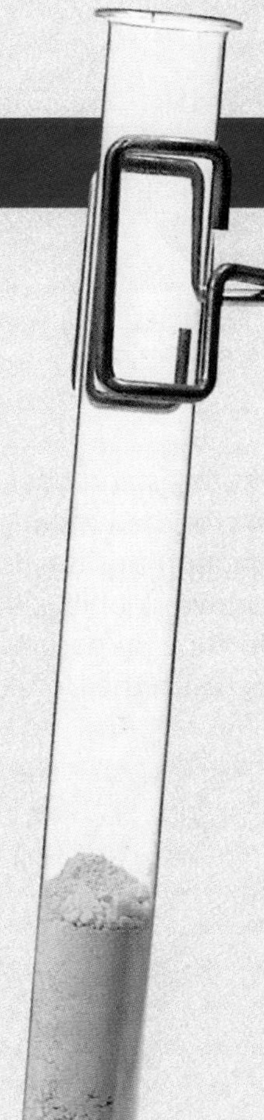

5. Allow the test tube to cool. Then place it in the test-tube rack. Insert a funnel in the test tube and add dilute sulfuric acid until the test tube is 3/4 full. **CAUTION:** *Avoid touching the sides of the test tube; it may get hot. If any of the acid gets on your skin or clothing, rinse immediately with cool water and alert your teacher.*

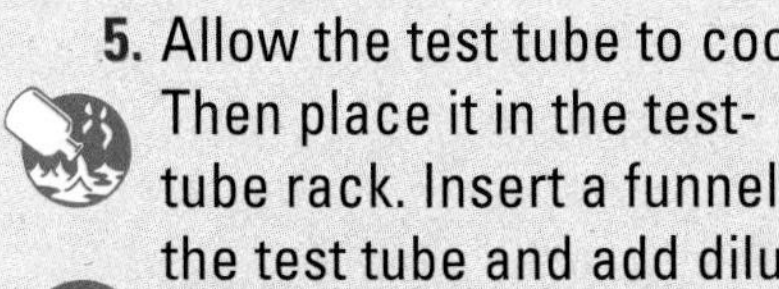

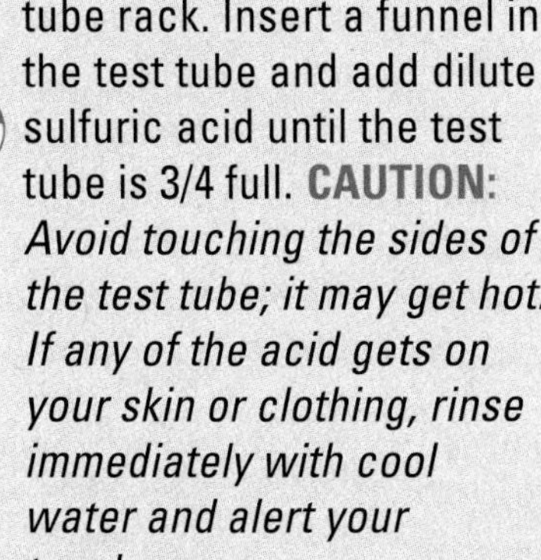

6. Allow the test tube to stand until some of the substance at the bottom of the test tube dissolves. After the sulfuric acid has dissolved some of the solid substance, note the color of the solution.
7. Use the funnel to add more sulfuric acid to the test tube until it is nearly full. Allow the test tube to stand until more of the substance at the bottom of the test tube dissolves. Pour this solution (copper sulfate) into the second test tube.
8. Add a small amount of iron filings to the second test tube. Observe what happens.
9. Clean all laboratory equipment and dispose of the sulfuric acid as directed by your teacher.

Analysis and Conclusions

1. a. Disregarding any condensed water on the test-tube walls, what is the new substance formed in the first test tube called?
   b. Does the new substance take up as much space in the test tube as did the copper carbonate? Explain.
2. a. When the iron filings were added to the second test tube, what indicated that a chemical reaction was taking place?
   b. Describe any change to the iron filings.
   c. Describe any change in the solution.
3. In the actual process of extracting copper from its ore, the copper sulfate solution is allowed to flow over cast iron scrap metal. The loose layer of copper that forms on the scrap metal is then separated and pressed into bars or redissolved for purification. What do the iron filings represent in the actual process of extracting copper from its ore?

Extension

Suppose that a certain deposit of copper ore contains a minimum of 1.0% copper by mass and that copper sells for 30 cents per kilogram. Approximately how much could you spend to mine and process the copper out of 100 kg of copper ore and remain profitable?

Buried Treasure

By many standards, the United States is one of the wealthiest countries in the world. Although this is largely due to the ingenuity and hard work of its citizens, it is also due to good fortune. In particular, the earth's crust that lies beneath the United States holds a huge supply of natural resources. These include ores and precious metals, such as copper, silver, and gold, as well as fossil fuels, such as petroleum, coal, and natural gas. The availability and distribution of these resources has been important in shaping the history of our country.

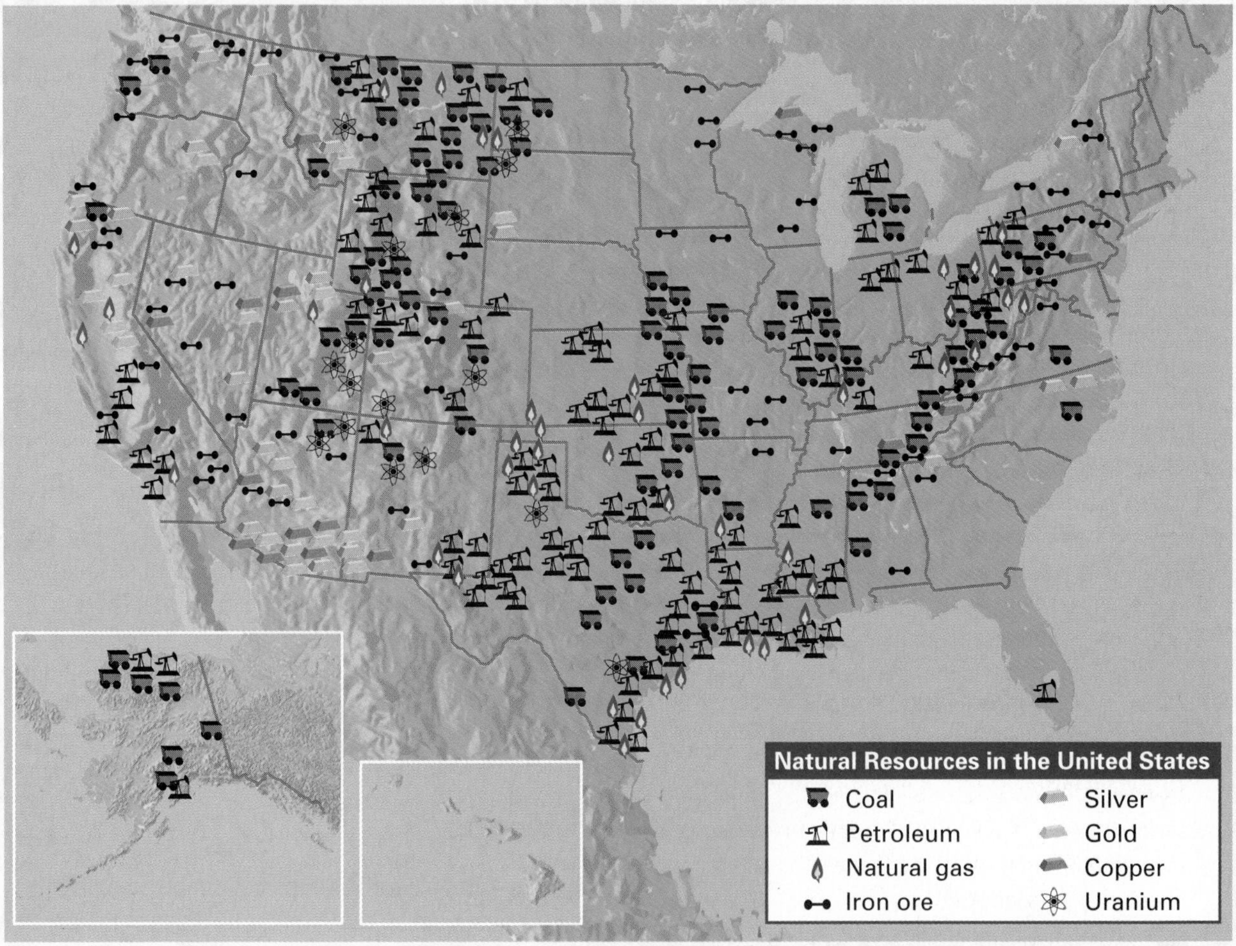

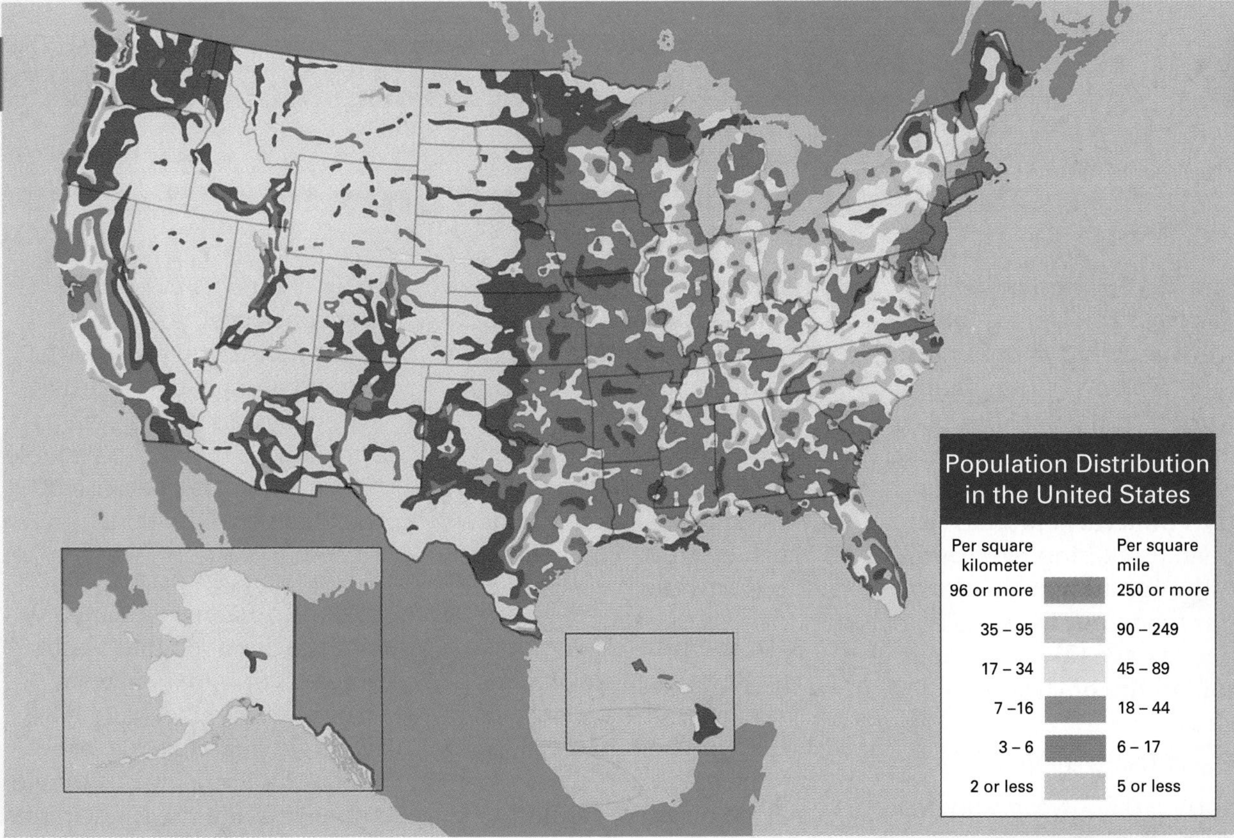

1. Find your state on the map of natural resources. What is produced in your state?

2. Look at the locations of the different resources. Which ones have similar distributions? Why do certain resources have similar distributions?

3. When tectonic plates collide, pockets of hot magma may shift around and come into contact with cooler, solid rock. Based on this information and what you learned on page 195, why does it make sense that Washington has more iron-ore deposits than Nebraska?

4. Make a sketch of what the United States might have looked like millions of years ago when the petroleum and natural-gas deposits were in the process of forming.

5. Compare the map on this page with the map on the previous page. What areas have both high concentrations of people and a large reserve of natural resources? What areas have many people but few resources? How do you think resources could be transported to areas that need them?

6. Based on these two maps, would you say that most cities have grown up around places where natural resources are located? Why or why not? What other factors could have been involved?

7. Imagine that you work for a company that builds electrical equipment made primarily of copper. Why might southern Arizona be a good place to locate a new plant? What might be a disadvantage of locating your plant there?

Going Further

The mining industry is a fast-moving and exciting field. You can find out about current events, new mining sites, and even careers in mining by visiting INFOMINE on the World Wide Web at www.info-mine.com.

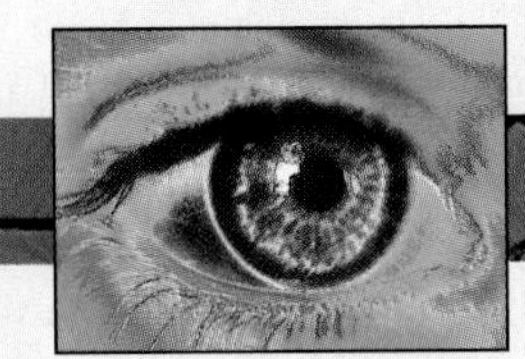

Eye on the Environment

What's Mined Is Yours

In 1918, when industries across the United States were growing at an amazing rate, a record 177 million tons of coal were produced from mines in Pennsylvania. The demand for this country's mineral resources has been increasing ever since. Although new technology has made our lives easier, our needs have taken a serious toll on the environment.

A Mining Nation

About 10,000 years ago, residents of what is now Michigan dug underground for copper. Today, in northern Michigan and several other states, tons of copper ore are extracted from large underground reserves. Fertilizer companies produce phosphate ore from large surface mines in Florida. Uranium is produced throughout the Rocky Mountain states. Wyoming recently produced 171 million tons of coal to become number one in coal production. Only recently have we begun to understand how much environmental damage these activities have caused.

In a typical mining operation, huge bulldozers clear away layers of soil, destroying vegetation and drastically altering the landscape. Streams may be diverted, causing droughts in some areas and floods in others. Nearby residences may become infested with snakes and other pests that are fleeing from the disturbance.

The long-term effects are even more serious. When the discarded soil from a mine is exposed to air and rainfall, chemical reactions occur. As a result, acidic moisture seeps into the ground and pollutes nearby waterways. Years later, plants and animals may still be unable to live in the area because of the changed conditions.

In Pennsylvania, 1,700 streams have been polluted by the acids and other waste products from mines. Many mining sites in that state have undergone *reclamation,* the process of returning the terrain and plants at a site to their initial condition. Even so, about 200,000 acres remain unreclaimed. In every state, centuries of mining have left the environment scarred. National parks alone contain more than 4,000 abandoned mining sites in 45 states.

A piece of copper ore

An open-pit copper mine in Utah

Taking Responsibility

Most mining companies take environmental issues very seriously. New techniques and a better understanding of natural processes have made returning land to its original condition more efficient than ever. For instance, recent studies have focused on the roles of grasses, native trees, and even earthworms in reclamation.

Unfortunately, reclamation is very expensive. Depending on the location, returning land to its original state may cost as much as $20,000 per acre. As environmental laws become tougher on mining companies, fewer of them are able to make a profit. Many go out of business. In the last 20 years, over 800 coal-mining companies have shut down in Pennsylvania alone. As a result, thousands of people have lost their jobs, and local economies have suffered greatly.

▲ **This recently harvested corn field is located on a reclaimed mining site.**

No Easy Answers

Ironically, some workers who lost their jobs in the mines have found work with reclamation companies. Others have had to move away, seeking new types of work in other locations. Hopefully, the economic problems in former mining areas will improve with time.

Our concern for the environment is great, but so is our need for mineral resources. As our population grows larger, we will need more of the metal ores, fossil fuels, and other products that the mining industry provides. Fortunately, reclamation efforts are constantly improving. In addition, efforts to recycle and reuse the resources that we already have are increasing across the country. Nonetheless, it will always be difficult to find a compromise between our great need for mineral resources and the environmental damage we cause by getting those resources.

Think About It

The next time you look at a parking lot, think about how much metal was used to build all of those cars and trucks. Almost all of that metal was refined from underground ores. If you wanted to put all of the metal in a parking lot back in the ground, how big of a hole would you have to dig?

Extension: Find out the top three mineral resources that your state produces. What part of the state do they come from? How are the minerals extracted from the ground and how are they used?

UNIT 4

Reshaping the Crust

Capitol Reef National Park, Utah (background);
Yellowstone Falls, Yellowstone National Park, Wyoming (inset)

Introducing Unit Four

As a park ranger, I know that protecting and preserving our natural environment isn't something done just for future generations. It's something we do for ourselves—for today—as well.

In addition to supporting the delicate balance of nature, parks also play a vital role in rejuvenating the visiting public. Somehow, being surrounded by unaltered nature puts the details of our busy lives into perspective. In these respects, what I and others are doing to preserve our environment for the future also has a direct, positive effect on those things that are important to us today.

This unit discusses the forces that have shaped and continue to shape the earth. Understanding these forces may help you find ways to help preserve our natural environment—for today as well as tomorrow.

Pat Buccello

Patricia Buccello
Park Ranger
Sequoia National Park

In this UNIT

Chapter 12

Weathering and Erosion

The face of our planet is constantly changing. A massive landslide may make dramatic changes in an instant. Other changes, such as forming soil or sculpting canyons, take place gradually over thousands of years. In this chapter, you will study how natural agents, such as gravity, water, wind, and ice, have helped reshape the surface of the earth.

Chapter Outline

These formations are found in Bryce Canyon National Park, Utah.

12.1 Weathering Processes

Section Objectives

- **Discuss the agents of mechanical weathering.**
- **Discuss the chemical reactions that decompose rock.**

Most rocks deep within the earth's crust were formed under high temperatures and pressures. When these rocks are uplifted to the surface, they are exposed to much lower temperatures and pressures and to the gases and water in the atmosphere. As a result, rocks on the earth's surface undergo changes in their appearance and composition. The change in the physical form or chemical composition of rock materials exposed at the earth's surface is called **weathering.**

There are two types of weathering processes. **Mechanical weathering** processes physically break rock into smaller pieces but do not change its chemical composition. **Chemical weathering** processes break down rock by changing its chemical composition.

Mechanical Weathering

Mechanical weathering is strictly a physical process. Common agents of mechanical weathering are ice, plants and animals, gravity, running water, and wind. Physical changes within the rock itself aid mechanical weathering. For example, granite is formed deep beneath the earth's surface. As overlying rocks are removed, the pressure on the granite decreases. As a result, the granite expands, and long, curved cracks parallel to the surface, or *joints,* develop in the rock. When joints develop on the surface of the rock, the rock breaks into curved sheets that peel away from the underlying rock in a process called **exfoliation.**

Ice Wedging

A common kind of mechanical weathering is called **ice wedging.** Ice wedging occurs when water seeps into cracks or joints in rock and freezes. When the water freezes, its volume expands by about 10 percent, creating pressure on the rock. Every time the ice thaws and refreezes, it wedges farther into the rock, and the crack widens and deepens. This process eventually will split the rock apart, as shown in Figure 12–1.

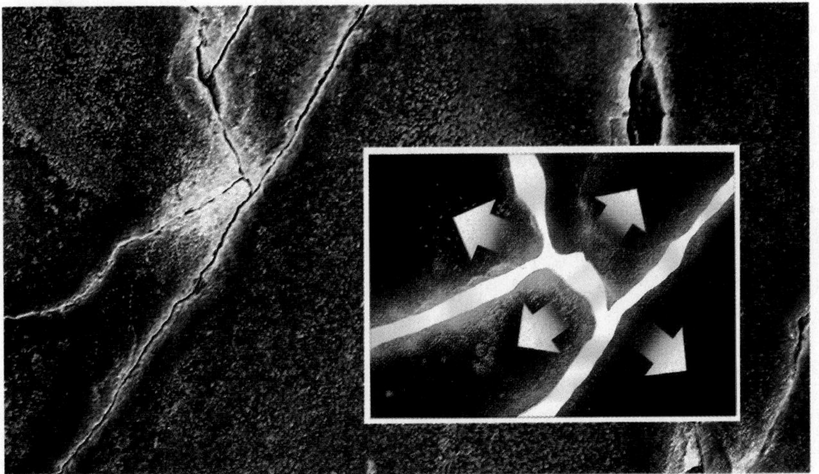

Figure 12-1. Ice acts as a wedge that breaks rock apart, as shown in the inset. Ice wedging is responsible for much of the shattering seen in the photograph.

Ice wedging commonly occurs at high elevations. It also occurs in climates where the temperature regularly varies above and below the freezing point of water, such as in the northern United States.

Organic Activity

Plants and animals are important agents of mechanical weathering. The roots of plants can work their way into cracks in rock. As the roots grow and expand, they create pressure that wedges the rock apart. The digging activities of burrowing animals, such as ground squirrels and prairie dogs, expose new rock surfaces to weathering. These plant and animal activities can be effective in weathering rocks over a long period of time.

Abrasion

The collision of rocks with one another, resulting in the breaking and wearing away of the rocks, is a form of mechanical weathering

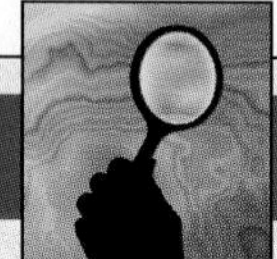

SMALL-SCALE INVESTIGATION

Mechanical Weathering

You can demonstrate the effects of mechanical weathering by abrasion by placing rocks in a container of water and shaking the container.

Materials

silicate rock chips; hand lens; 2 plastic containers (16 oz. or larger), one with a tight-fitting lid; water; strainer; clear glass jar or small beaker

Procedure

1. Examine the rocks with a hand lens, noting the shape and surface texture.
2. Fill the plastic container with the tight-fitting lid about half full of rocks. Add water to barely cover the rocks.
3. Tighten the lid, and shake the container 100 times.
4. Hold the strainer over the other container. Pour the water and rocks into the strainer.
5. Run your finger around the inside of the empty container. Write down what you feel.
6. Use the hand lens to observe the rocks.
7. Pour the water into the glass jar, and examine the water with the hand lens.
8. Put the rocks and water back into the container with the lid. Repeat Steps 3–7.
9. Repeat Step 8 two more times.

Analysis and Conclusions

1. Has the amount and particle size of the residue left in the container changed during the course of the investigation? Explain your answer.
2. How has the appearance of the rocks changed? How has the appearance of the water changed?
3. If the water from a small stream ran over a ledge of rock into a pool below, what would you expect to find at the bottom of the pool?

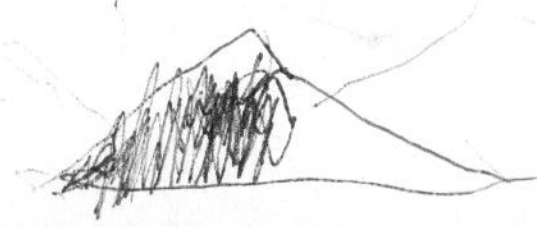

Figure 12-2. The roots of plants act like wedges to split rock.

called **abrasion.** Agents of abrasion are gravity, running water, and wind. Gravity causes loose soil and rocks to move down the slope of a hill or mountain. Rocks break into smaller pieces as they fall and collide. Running water or wind also can carry particles of sand or rock. These particles scrape against each other and against other stationary rocks, thus abrading the exposed surfaces. What evidence of abrasion have you seen at the beach?

Chemical Weathering

Chemical weathering, or decomposition, occurs when chemical reactions take place between the minerals in the rock and water, carbon dioxide, oxygen, and acids. These chemical reactions alter the internal structure of the original mineral and lead to the formation of new minerals. As a result, both the chemical composition and the physical appearance of the rock undergo changes.

Hydrolysis

Water plays a crucial role in chemical weathering. The change in the composition of minerals when they react chemically with water is called **hydrolysis.** For example, a type of feldspar combines with water and produces a common clay called *kaolin*. In this reaction, hydrogen ions from the water displace such elements as potassium or calcium in the feldspar. The feldspar thus is changed into clay by a chemical reaction.

The minerals affected by hydrolysis may dissolve in water. The water then carries the dissolved minerals to lower layers of rock in a process called **leaching.** Mineral ore deposits, such as bauxite the aluminum ore, may form when leaching causes a mineral to concentrate in a thin layer beneath the earth's surface.

Carbonation

When carbon dioxide (CO_2) from the air dissolves in water (H_2O), a weak acid solution called *carbonic acid* (H_2CO_3) is produced:

$$H_2O + CO_2 \longrightarrow H_2CO_3$$

Figure 12-3. Carbonic acid dissolved the calcite in limestone to produce this underground cavern.

The acid has a higher concentration of hydrogen ions than does pure water. These ions speed up the process of hydrolysis.

When some minerals come in contact with carbonic acid, they combine chemically with the H_2CO_3 and form a new product. This process is called **carbonation.** For example, carbonic acid reacts with calcite, which is a major component of limestone, and converts it to calcium bicarbonate. Calcium bicarbonate dissolves easily in water, so the limestone eventually is eaten away. The dissolving action of carbonic acid on limestone sometimes produces underground caverns such as that shown in Figure 12–3.

Oxidation

When metallic elements combine with oxygen, **oxidation** occurs. Oxidation often attacks rock with iron-bearing minerals. Iron (Fe) combines quickly with oxygen (O_2) dissolved in water and forms rust, or iron oxide (Fe_2O_3):

$$4Fe + 3O_2 \longrightarrow 2Fe_2O_3$$

Figure 12–4. The red tint to the soil in this Brazilian rain forest is due to oxidation.

The red color of much of the soil in the southeastern United States is due mainly to the presence of iron oxide produced by oxidation.

Acid Precipitation

Rainwater is naturally slightly acidic because it contains dissolved carbon dioxide. However, in industrial and heavily populated areas, waste gases containing oxides of nitrogen and sulfur are released into the air. These compounds combine with water in the atmosphere and produce *acid precipitation*. Acid rain, for example, has a greater ability to weather rock than does ordinary rain. Thus, the chemical weathering processes are greatly accelerated.

Plant Acids

Acids are produced naturally by some plants. Lichens and mosses grow on rocks and produce weak acids that can dissolve the surface of the rock. The acids also can seep into the rock and produce cracks that eventually cause it to break apart.

The Price of Corrosion

Corrosion is the natural chemical weathering of metals. When corrosion occurs in metals used in building or manufacturing, it poses a serious problem.

The cost of replacing or treating metals damaged by corrosion in the United States alone is estimated at $170 billion a year. However, the drain on natural resources also is enormous. For example, producing a single ton of steel requires 4 tons of coal.

Rust, an oxide that affects iron and steel, is by far the most common form of corrosion. Rust not only discolors but also weakens metal. It attacks the metal in buildings, bridges, culverts, and highways. Left unchecked, rust eventually destroys these structures.

Rust can also pose a threat within the home, where it attacks pipes and pipe fittings, furnace exhaust ducts, and electrical conduits.

Rust may ruin the appearance of an automobile, but it can also threaten the lives of drivers and passengers. Rust can damage the floor of the passenger compartment, allowing carbon monoxide fumes to enter the automobile. It can also weaken the suspension, steering, and brake systems.

Trying to stop rust is, at best, difficult. Steel can be galvanized, a process that covers the metal with a zinc coating, but this only delays the inevitable. Wherever moisture, heat, and steel exist together, you are likely to find rust.

What chemical reaction produces both corrosion and iron oxide in the soil?

Section 12.1 Review

1. In what kind of climate does ice wedging usually occur?
2. How do plants and animals help weather rocks?
3. What agents promote abrasion?
4. What chemical weathering process occurs when minerals in the rock react with carbon dioxide?
5. Automobile exhaust contains nitrogen oxides. How might these pollutants affect chemical weathering processes?

Section Objectives

- **Explain how rock composition affects the rate of weathering.**
- **Discuss how the amount of exposure determines the rate at which rock weathers.**
- **Describe the effects of climate on the rate of weathering.**

12.2 Rates of Weathering

The processes of weathering generally work very slowly. For example, carbonation dissolves limestone at an average rate of only about one twentieth of a centimeter every 100 years. At this rate, it could take up to 30 million years to dissolve a 150 m thick layer of limestone.

Rocks do not weather at the same rate. Different rates of weathering produce different formations, such as those shown in Figure 12–5. The rate at which rock weathers depends on a number of factors.

Rock Composition

The composition of rocks is a major factor in their rate of weathering. Often, igneous and metamorphic rocks on the earth's surface remain almost unchanged after all the surrounding sedimentary rock has weathered away. Of the minerals in igneous and metamorphic rock, quartz is the least affected by chemical weathering. Because

Figure 12-5. Different rock composition and structures account for the different rates of weathering in this rock formation.

quartz is one of the hardest minerals, it also resists mechanical weathering and retains its structure in tiny grains of sand.

Among sedimentary rocks, limestone and other rocks containing calcite are most rapidly weathered. Although these rocks resist mechanical weathering, they are easily weathered by carbonation.

Many other sedimentary rocks are affected mainly by mechanical weathering processes. The rate at which these rocks weather depends on the material that holds the fragments of the sediments together. For example, shales and sandstones that are not firmly cemented together gradually break up into clay and sand particles. However, conglomerates and those sandstones that are strongly cemented by silicates may resist weathering processes even longer than some igneous rocks.

Figure 12-6. When fractures bisect every edge of a smooth cube of rock, the surface area exposed to weathering doubles.

Amount of Exposure

The more exposure a rock receives to weathering processes, the faster it will weather. The amount of time the rock is exposed and the amount of surface area available for weathering are important factors.

Most rocks at the surface of the earth are broken by fractures and joints. Fractures also form natural channels through which water flows. Water penetrates the rock through these channels and breaks the rock by ice wedging. These processes usually split the rocks into a number of smaller blocks. Fractures and joints increase the surface area of a rock and allow weathering to take place more rapidly.

For example, picture a block of rock as a cube with six sides exposed. When the block is broken once, two more surfaces are created. When each of those blocks is broken, four more surfaces are created. Splitting the original block into eight smaller blocks, as shown in Figure 12–6, doubles the total surface area available for weathering.

Climate

In general, rainfall and the freezing and thawing produced by alternating hot and cold weather have the greatest effect on the rate of weathering. What chemical and mechanical weathering processes do you think are affected most by these conditions?

Climates in which variable weathering conditions exist can cause ice wedging fractures, which help to expose new rock surfaces. Chemical weathering then can attack the fractured rock more quickly. When temperatures rise daily or seasonally, the rate at which chemical reactions occur also accelerates.

In hot, dry climates, weathering takes place slowly. The lack of water limits the rate of the many chemical and mechanical weathering processes associated with water, such as carbonation and ice wedging. Weathering is also slow in very cold climates. In warm, humid climates, chemical weathering is fairly rapid. The constant moisture is highly destructive to exposed surfaces.

Figure 12–7. The photograph on the right shows Cleopatra's Needle after only one century in New York City. Co[...] fects of weathering [...] ings. The inset on th[...] the carvings just afte[...] was moved to New [...] set on the right show[...] ings after one century [...] York.

[...]athering rates is seen on Cleopatra's [...]le of granite. For 3,000 years, the [...] hot, dry climate scarcely changed [...]nited States government received [...] to Central Park in New York City. [...] to moisture and air pollution, Cleo- [...] severely. Ice wedging and chemi- [...]ation caused more harm in a [...]he preceding 30 centuries in the

[...]ation and slope of the surface [...] located, also influences the rate of weathering. Because it is colder at higher elevations, ice wedging often increases as the elevation increases. On steep slopes, such as mountainsides, weathered rock fragments are pulled downhill by gravity and washed along by heavy rains. New surfaces of the mountain are thus exposed more quickly to weathering.

Section 12.2 Review

1. Which mineral in igneous and metamorphic rocks is least affected by chemical weathering?
2. Why does fractured rock weather more rapidly than smooth rock?
3. What climatic factors influence weathering rates?
4. How does the topography of a region affect weathering rates?
5. How would Cleopatra's Needle probably have been affected if it had been in the cold, dry climate of Siberia for 100 years?

12.3 Weathering and Soil

Section Objectives

- **Explain how the composition of parent rock affects soil composition.**
- **Predict the type of soil produced in various climates.**

Weathering processes break and alter all rocks exposed at the surface of the earth. One result of weathering is the formation of **regolith,** the layer of weathered rock fragments covering much of the earth's surface. *Bedrock* is the solid, unweathered rock that lies beneath the regolith.

The lower regions of regolith are partly protected by those above and do not weather as rapidly as the upper regions. Eventually the uppermost rock fragments weather and form a layer of very fine particles. This layer of small rock particles provides the basic components of *soil*. Soil is a complex mixture of minerals, water, gases, and the remains of dead organisms. As plants and animals die, their remains decay and produce **humus,** a dark, organic material. Humus is part of any fully developed soil. The relation between bedrock, regolith, soil, and humus can be seen in Figure 12–8.

Soil Composition

The rock material in soil consists of three main types: clay, sand, and silt. These materials are classified by the size of their particles. Clay particles are less than 0.002 mm in diameter, silt particles range from 0.002 to 0.06 mm, and sand particles range from 0.06 to 2 mm. The proportion of these three materials in a particular soil depends mainly on the rock from which the soil was weathered, called its *parent rock.*

Soils containing large amounts of clay are weathered from parent rock that is rich in feldspar. When feldspar weathers, fine grains of silicate material containing aluminum and water are formed. The individual grains of clay are too small to be seen by the unaided eye. Other weathered minerals containing aluminum also may form clay.

Weathered granite and other rocks containing large amounts of quartz form sandy soils. Generally, sand consists of small quartz

INVESTIGATE!

To learn more about soil chemistry, try the In-Depth Investigation on pages 240–241.

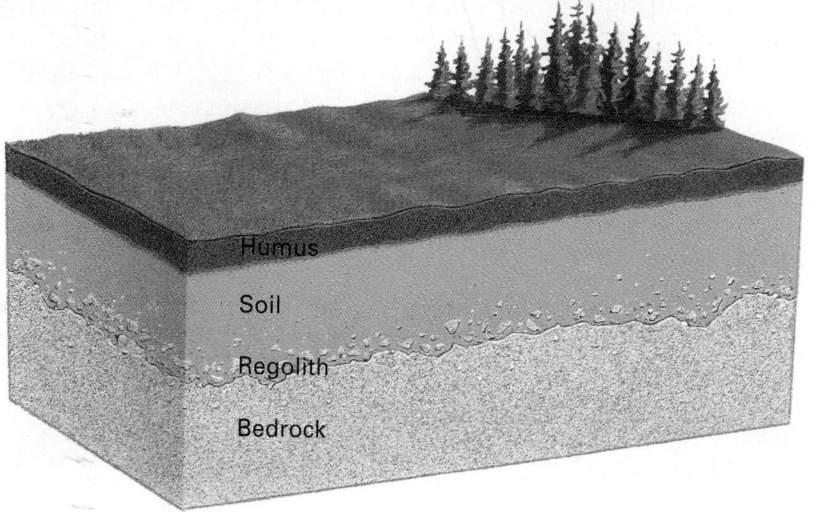

Figure 12–8. A blanket of regolith usually overlies the solid bedrock below. Soil and a layer of humus form over the regolith.

grains, which are large enough to be both felt and seen. Quartz rarely is weathered chemically, so quartz grains remain behind after other minerals in rock have broken down.

Silt particles are too small to be seen easily, but they give a gritty feel to the soil. Currents often carry silt along riverbeds and deposit it on soils near riverbanks.

The weathered mineral grains that form soil may be carried away from the location of the parent rock by water, wind, or glaciers. The soil resulting from the deposit of this material, called *transported soil,* may have a different composition than the bedrock on which it rests.

Soil Profile

Soil that rests on top of its parent rock, called *residual soil,* often develops distinct layers over a long period of time. To determine the composition of these soil layers, scientists study a **soil profile.** A soil profile is a cross section in which layers of the soil and the bedrock beneath the soil can be seen. Figure 12–9 shows the layers of a soil profile. The layers are called **horizons.** Fully developed

Figure 12-9. Notice the A horizon (black), B horizon (gray), and C horizon (brown) in this soil profile.

residual soils generally consist of three principal horizons: the A horizon, B horizon, and C horizon.

The A horizon of a residual soil is topsoil. Topsoil is a mixture of organic and small rock materials. Almost all the living things that live in soil inhabit the A horizon. Therefore, the A horizon contains humus and other organic materials. The A horizon is also the zone from which surface water leaches the minerals.

The B horizon is immediately beneath the A horizon. The B horizon is the subsoil, which contains the minerals leached from the topsoil, clay, and, sometimes, humus. In dry climates, the B horizon also may contain minerals that accumulate as the water in the soil evaporates rapidly.

The C horizon—the bottom layer of the soil—consists of bedrock that has been partially weathered. This bottom layer is in the first stages of mechanical and chemical change. In time, weathering will cause this horizon to develop into a B horizon and, finally, into an A horizon.

Transported soils do not have horizons. Instead, they are deposited in sorted layers or in unsorted masses by the action of water, wind, or ice.

Soil and Climate

Climate is one of the most important factors influencing soil formation. Climate determines the weathering processes that occur in a region. These processes partly determine the composition of the soil.

In humid tropical climates, where there is much rain and where temperatures are high, chemical weathering causes thick soils to develop rapidly. These soils are called **laterites** (LAT-UH-rites). This type of soil contains iron and aluminum minerals that do not dissolve easily in water. Rain leaches other minerals from the A horizon of the soil and the minerals sometimes collect in the B horizon. The heavy rains also wash away much of the topsoil, thus keeping the A horizon thin. However, because of the dense vegetation in humid tropical climates, organic material is continuously added to the soil. As a result, a thin layer of humus usually covers the B horizon. Why would tropical soil not be good for farming?

In desert climates, where rainfall is minimal, chemical weathering processes do not occur rapidly. As a result, the soil is thin and consists mostly of regolith—evidence that mechanical processes are occurring. The soil in arctic climates also is formed mainly by mechanical weathering. Because chemical weathering takes place slowly in arctic climates, the soil also is thin and is made up of rock fragments.

In temperate climates, where temperatures range between cool and warm and where rainfall is not excessive, both mechanical and chemical weathering processes take place. All three soil horizons may reach a thickness of several meters.

Two main soil types are found in temperate climates. In areas receiving more than 65 cm of rain a year, a type of soil called

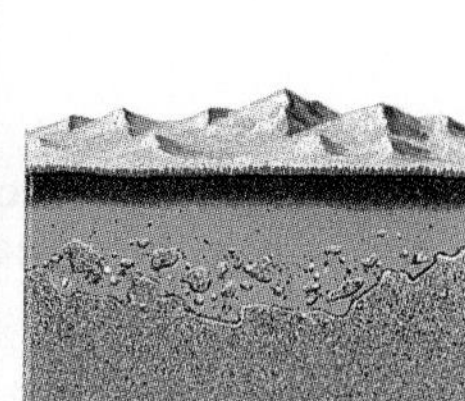

Arctic soil

Tropical soil

Figure 12-10. Tropical climates produce thick, infertile soils. Arctic climates produce thin soils.

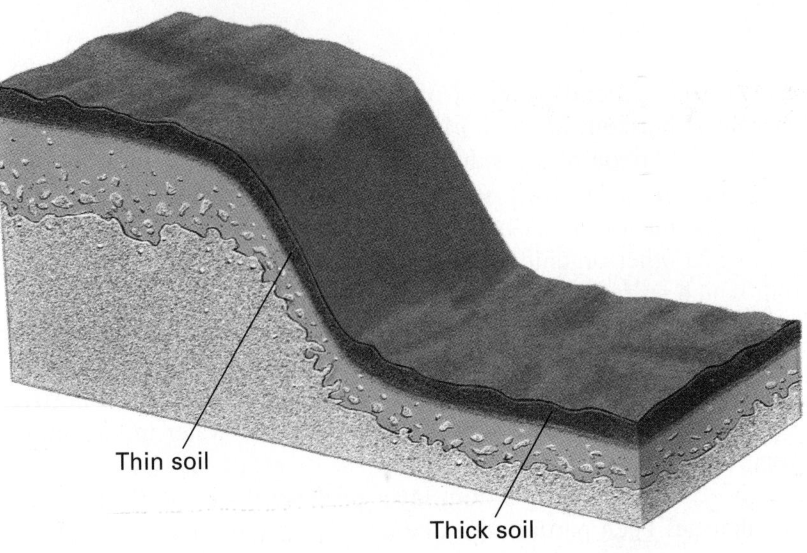

Figure 12-11. Soil is thick at the top and bottom of a slope; soil is thin along the slope itself.

pedalfer (PUH-DAL-fur) is found. Pedalfer soils contain clay, quartz, and iron compounds. The Gulf Coast states and states east of the Mississippi River have pedalfer soils. In areas receiving less than 65 cm of rain a year, a soil called *pedocal* (PED-uh-KAL) is found. The southwestern states and most states west of the Mississippi River have pedocal soils. Pedocal soils contain large amounts of calcium carbonate, which combines with excess hydrogen in the soil. This process makes the soil less acidic and very fertile.

Soil and Topography

Topography, or configuration of the land, also plays a role in soil formation. Because rainwater tends to run down steep slopes, much of the topsoil of the slope is washed away. Therefore, as shown in Figure 12–11, the soil at the top and bottom of a slope tends to be thicker than that on the slope itself. A study of soils in Manitoba, Canada, showed that the A horizon of soils on flat areas was more than twice as thick as that on 10° slopes.

The topsoil that does remain on a slope is often too dry and thin to support dense plant growth. The resulting lack of vegetation contributes to the development of a poor-quality soil. A thick humus layer cannot form because little organic matter is added to the soil. The soils on the sides of mountains are commonly thin and rocky, with few nutrients. Thus, cultivation of mountain soils is quite difficult. In contrast, lowlands that retain water tend to have thick, wet soils with a high concentration of organic matter, which forms humus. A fairly flat area with good drainage provides the best surface for formation of thick fertile layers of residual soil.

Section 12.3 Review

1. Why do some soils contain large amounts of clay?
2. Describe the differences between residual and transported soils.
3. Describe the three horizons of a mature residual soil.
4. Although desert and arctic climates are extremely different, their soils may be somewhat similar. Explain why.

12.4 Erosion

Various forces may move weathered fragments of rock away from where the weathering occurred. The process by which the products of weathering are transported is called **erosion.** The agents of erosion are gravity, wind, glaciers, and water in the form of ocean waves and currents, streams, and groundwater.

Section Objectives

- **Define *erosion* and list four agents of erosion.**
- **Identify various unwise farming methods that result in accelerated soil erosion and list four methods of soil conservation that prevent this damage.**
- **Discuss how mass movements contribute to erosion.**
- **Describe the major landforms.**

Soil Erosion

Soil erosion occurs worldwide and is normally a slow process. Ordinarily, new residual soil forms about as fast as the existing soil erodes. The eroded soil is often deposited elsewhere as transported soil. However, unwise use of the land and unusual climatic conditions can upset this natural balance. Once the balance is upset, soil erosion accelerates.

Accelerated Soil Erosion

Unwise farming and ranching methods increase soil erosion. For example, clearing trees and small plants and allowing animals to overgraze destroy the soil protection provided by plants. The upper soil layers are then exposed to the full effects of erosion.

Furrows in plowed land, especially those that are plowed up and down the slopes, allow water to run swiftly over the soil. As soil is washed away with each rainfall, a furrow becomes larger, forming a small gully. Eventually, the land is covered with deep gullies. This type of accelerated soil erosion is called *gullying*. The farmland shown in Figure 12–12 has been ruined by gullying. What two agents of erosion are at work in gullying?

Figure 12–12. This field is useless for farming because of heavy erosion called gullying.

Another type of soil erosion strips away parallel layers of topsoil and eventually exposes the surface of the subsoil or the partially weathered bedrock. This process is called **sheet erosion.**

Sheet erosion may occur when continuous rainfall evenly washes away the topsoil. Carried to streams by flowing water, the eroded topsoil may clog the stream or even change its course. Wind also can cause sheet erosion during unusually dry periods. The soil, made dry and loose by a lack of moisture, is carried away by the wind as clouds of dust and drifting sand. These wind-borne particles may move as dust storms that cover a wide area.

Constant erosion reduces the fertility of the soil by removing the A horizon, which contains the rich humus. The B horizon, which does not contain organic matter, is thereby exposed. The soil in this horizon is difficult to cultivate because it resists plant growth. Without plants, the B horizon has nothing to protect it from further erosion. As a result, within a few years, all of the soil layers could be removed by continuous erosion.

Soil Conservation

Rapid, destructive soil erosion can be prevented by soil conservation methods. Cultivating cover plants to hold the topsoil in place can protect topsoil in some areas. Farmers also can practice various crop planting methods designed to reduce erosion.

In a method called *contour plowing,* the soil is plowed in circular bands that follow the contour, or shape, of the land. This method of planting, shown in Figure 12–13, prevents water from flowing directly down slopes and so prevents gullying.

In *strip-cropping,* crops are planted in alternate bands. For example, a crop planted in rows, such as corn, may be planted in one band. Next to it may be planted a crop that fully covers the surface of the land, such as alfalfa. The cover crop protects the soil by absorbing and holding rainwater. Strip-cropping is often combined with contour plowing. The combination of these two methods can reduce soil erosion by 75 percent.

Figure 12-13. Contour plowing helps prevent gullying by causing water to flow along plowed furrows (left). Terracing builds up low ridges that slow movement of water downhill (right).

The construction of steplike ridges that follow the contours of a sloped field is called *terracing*. Terraces, especially those used for growing rice in Asia, as in Figure 12–13, prevent or slow the downslope movement of water, thus preventing rapid erosion.

In *crop rotation,* farmers plant one type of crop one year and a different type of crop the next. For example, crops that expose the soil to the full effects of erosion may be planted one year, and a cover crop the next year. Crop rotation stops erosion in its early stages, allowing small gullies formed during one growing season to fill with soil during the next.

In developing urban areas, clearing vegetation and removing soil to build housing and roads also contributes to erosion. It is important that people recognize the impact of these activities and that city planners consider soil conservation measures.

Figure 12-14. Talus accumulates at the base of a slope.

Gravity and Erosion

Gravity, through its downward pull, causes rock fragments to move down inclines. This movement of fragments down a slope is called **mass movement.** Mass movement can be either rapid or slow.

Rock breaks into fragments that move downslope. The rock fragments accumulate at the base of the slope in piles called **talus** (TAY-lus). As you can see in Figure 12–14, talus forms a slope that covers the base of a cliff.

Mechanical and chemical weathering may reduce talus to smaller fragments, which can move farther down the slope. Some fragments wash into gullies. Carried into successively larger waterways, the fragments eventually flow into rivers.

Rapid Mass Movements

The most dramatic mass movements occur rapidly. You have probably heard news reports of devastating, sudden slides of earth and rock. The fall of rock from a steep cliff is called a **rockfall.** A rockfall is the most rapid kind of mass movement. Rocks in rockfalls may range in size from tiny fragments to giant boulders.

The sudden movement of masses of loose rock and soil down the slope of a hill, mountain, or cliff is a **landslide.** Large landslides involving loosened blocks of bedrock generally occur on very steep slopes. You may be familiar with small landslides on cliffs and steep hills overlooking highways. Heavy rainfall, spring thaws, volcanic eruptions, and earthquakes can trigger landslides.

The rapid movement of a large mass of mud creates a **mudflow.** Mudflows occur in dry, mountainous regions during sudden, heavy rainfall or as a result of volcanic eruptions. Masses of mud move down slopes and through valleys, frequently spreading out in a large fan shape at the base of the slope. Many of the disastrous mass movements that occur in hillside residential communities of southern California are referred to as landslides but are actually mudflows.

Sometimes a large block of soil and rock becomes unstable and moves downhill under the influence of gravity. It then slides along the curved slope of the surface in one piece. This type of movement is called a **slump.** Slumping occurs along very steep slopes. Saturation by water and loss of friction with underlying rock causes loose soil to slip downhill over the more resistant rock.

Slow Mass Movements

Although most slopes appear to be stable, some slow movement actually is occurring. Catastrophic landslides are considered the most hazardous downslope movement. However, more material is ultimately moved by the greater number of slow mass movements.

CAREER FOCUS: *Natural Resources*

"I get a great sense of satisfaction from knowing the work I'm doing isn't just for today. It's going to matter in the future, too."

Lewis Nichols

Soil Conservationist

For Lewis Nichols, the career path to soil conservation began in a high school greenhouse.

"I had always been interested in plants and flowers. Then, when I was in high school, I became involved in a program in vocational agriculture. I worked in the greenhouse at the school during the year and had a job with the Soil Conservation Service during the summers," Nichols says.

Nichols, pictured at left, was determined to turn his interests into a profession. This East St. Louis, Illinois, native completed a bachelor's degree program in agronomy, the study of crop production and soil management, at Western Illinois University.

Like other soil conservationists, Nichols uses his technical training and on-the-job experience in many ways. He assists government agencies, industries, and private individuals in the areas of soil and water conservation, insect control, and land use. Though most of their work is done in the field, soil conservationists return to their offices to develop specific conservation practices for the problems being studied.

"Because your ability to provide practical assistance depends so much on work experience, I was somewhat frustrated in the beginning," he says. "Now that I have that experience, I find it very rewarding to know I'm able to help society by looking out for the future and providing a better world for our children—a world where the words *natural resources* still have meaning."

Hydrologists study water both in the field and in the lab. ▶

One form of slow mass movement is called **solifluction.** Solifluction, which means "soil flow," occurs in arctic and mountain climates where the subsoil is permanently frozen. In spring and summer, only the top layer of soil thaws. The moisture from this layer becomes trapped and cannot penetrate the frozen layers beneath. As a result, the surface layer becomes muddy and slowly flows downslope. Solifluction can also occur in warmer regions where the subsoil consists of hard clay. The clay layer acts like the permanently frozen subsoil in arctic climates, forming a waterproof barrier that prevents penetration of water deep into the soil.

Hydrologist

Hydrologists study the earth's surface and underground waters to locate adequate supplies of fresh water. They analyze the chemical and physical properties of water sources both on-site and in the laboratory.

A hydrologist, shown at lower left, works for government agencies and private industries in a variety of fields. Among these are soil and water conservation, pollution control, hazardous waste management, irrigation, development of hydroelectric power, and flood control. Irrigation projects such as the one shown above benefit from the hydrologist's knowledge of drainage and crop production.

Entry into the field of hydrology requires a bachelor's degree with course work in engineering, geology, geography, chemistry, mathematics, and physics.

Water Purification Technician

Every time you take a drink of water, your health depends on the work being done by water purification technicians at local water treatment plants.

Water purification technicians make sure that the water is fit to drink by monitoring the chemical processes and equipment involved in filtering and purifying water. Water purification technicians also know how to operate and maintain water purification equipment. This equipment can include systems for removing impurities in the water by various methods such as filtration and disinfection.

Technicians learn through on-the-job training or in programs available in community colleges.

For Further Information

For more information on careers in resource management, write to the Bureau of Land Management, Department of Interior, 1849 C St. NW, Mail Stop 1275 LS, Washington D.C. 20240. You might also explore the agency's Web site at www.blm.gov.

The extremely slow downhill movement of weathered rock material is known as **creep.** Soil creep is the most effective of all mass movements. However, it usually goes unnoticed unless buildings, fences, or other objects on the surface are moved along with it.

Many factors contribute to soil creep. Water separates and lubricates rock particles, allowing them to move freely. Growing plants produce a wedgelike pressure forcing particles apart. The burrowing of animals and repeated freezing and thawing also loosen rock particles. Gravity then slowly pulls the particles downhill.

Erosion and Landforms

Through weathering and erosion, the earth's surface is shaped into different physical features, or **landforms.** There are three major landforms—*mountains,* which are steep landforms of very high elevations, *plains,* which are flat or gently sloped surfaces generally not high above sea level, and *plateaus,* which are high-elevation flat surfaces. Minor landforms include hills, valleys, and dunes. The shape of landforms is influenced by rock composition and erosion agents.

All landforms are the result of two opposing processes. One process bends, breaks, and lifts the earth's crust, generally creating elevated landforms. The other process is the wearing action of weathering and erosion, reducing the land surface to sea level.

Erosion of Mountains

During the early stages in the history of a mountain, it undergoes uplift. As long as tectonic forces continue to uplift the mountain, it usually rises faster than it is eroded. Mountains that are being lifted are said to be youthful. Youthful mountains are rugged and have sharp peaks and deep, narrow valleys. During the intermediate, or mature, stage, a mountain is no longer rising. Weathering and erosion wear down the rugged peaks to rounded peaks and gentle slopes. The formations in Figure 12–15 illustrate the difference between youthful and mature mountains.

In its old stage, a mountain is reduced to a low, almost featureless surface near sea level. This surface is called a **peneplain** (PEEN-ih-PLANE), which means "almost flat." A peneplain usually has low, rolling hills, as seen in New England. Knobs of hard rock, such as granite, resist erosion and protrude above the peneplain. These knobs are called **monadnocks** (muh-NAD-NAHCKS).

Figure 12-15. The youthful mountains on the left contrast with the mature mountains on the right.

Figure 12–16. A mesa (left) is a small tablelike area. A butte (right) is a smaller, narrow-topped formation.

A peneplain can be mistaken for a depositional plain. However, the rocks beneath a peneplain have been folded and tilted by tectonic forces, while the rocks beneath a plain lie in horizontal layers.

Erosion of Plains and Plateaus

A plain is a flat landform generally not high above sea level. However, some plains such as the Great Plains of central North America have quite a range of elevation. They rise from near sea level in the east near the Mississippi valley to over 1,500 m at the foot of the Rockies. In fact, the plains along the foot of the Rockies are called the "High Plains."

A plateau is also a broad, flat landform. However, a plateau has a high elevation. Thus, a plateau is subject to much more erosion than a plain. Young plateaus, such as the Colorado Plateau in the southwestern United States, usually have deep stream valleys that separate broad, flat regions. Mature plateaus, such as those found in the Catskill region in New York State, are eroded into rugged hills and valleys.

The effect of weathering and erosion on a plateau depends on the climate and the composition and struture of the rock. In dry climates, resistant rock produces plateaus with nearly flat tops. As a plateau ages, erosion may dissect the plateau into smaller, tablelike areas called **mesas** (MAY-suhz). Mesas ultimately erode down to small, narrow-topped formations called **buttes** (BYOOTS). Mesas and buttes are shown in Figure 12–16. In dry regions, these landforms have steep walls with flat tops. In humid climates, the landforms are more rounded.

Section 12.4 Review

1. List several agents of erosion.
2. What natural phenomena can trigger a landslide?
3. What three farming methods increase the rate of soil erosion?
4. Could you determine if a landform is a peneplain or a depositional plain by flying over it in an airplane? Explain your answer.

Chapter 12 Review

Key Terms

- abrasion (221)
- butte (237)
- carbonation (222)
- chemical weathering (219)
- creep (236)
- erosion (231)
- exfoliation (219)
- horizon (228)
- humus (227)
- hydrolysis (221)
- ice wedging (219)
- landform (236)
- landslide (233)
- laterite (229)
- leaching (221)
- mass movement (233)
- mechanical weathering (219)
- mesa (237)
- monadnock (236)
- mudflow (233)
- oxidation (222)
- peneplain (236)
- regolith (227)
- rockfall (233)
- sheet erosion (232)
- slump (234)
- soil profile (228)
- solifluction (235)
- talus (233)
- weathering (219)

Key Concepts

Agents of mechanical weathering break rock into smaller pieces but do not change its chemical composition. **See page 219.**

Chemical weathering changes the composition of the minerals within a rock. **See page 221.**

Rock weathers at different rates depending partly on its mineral composition. **See page 224.**

The greater the amount of exposure, the faster a rock weathers. **See page 225.**

Rock weathers more rapidly in regions where rainfall is abundant and where alternating freezes and thaws occur. **See page 225.**

The parent rock from which soil is formed is the major factor determining the composition of the soil. **See page 227.**

Thick soils develop in tropical climates. Thin soils develop in climates where rainfall is minimal. **See page 229.**

The movement of the products of weathering away from where the weathering occurred is caused by natural agents. **See page 231.**

Farming methods that result in loss of topsoil accelerate soil erosion. Plowing fields and planting crops wisely help conserve soil. **See page 231.**

Slow and rapid mass movements of rock, soil, and mud are the major causes of erosion. **See page 233.**

Erosion wears away landforms as it levels the earth's surface. **See page 236.**

Review

On your own paper, write the letter of the term that best completes each of the following statements.

1. Physical stress on exposed surfaces causes rock to
a. decompose. b. break into smaller pieces.
c. melt. d. become buried.

2. A common kind of mechanical weathering is called
a. oxidation. b. carbonation.
c. ice wedging. d. leaching.

3. Oxides of sulfur and nitrogen that combine with water vapor cause
a. iron rain. b. acid rain.
c. mechanical weathering. d. carbonation.

4. The mineral in igneous rock that is least affected by chemical weathering is
a. calcite. b. quartz.
c. feldspar. d. mica.

5. The surface area of rocks exposed to weathering is increased by
a. burial. b. leaching. c. quartz. d. joints.

6. Chemical weathering is most rapid in
a. hot, dry climates. b. cold, dry climates.
c. cold, wet climates. d. hot, wet climates.

7. The chemical composition of soil depends to a large extent on
a. topography. b. its A horizon.
c. the parent material. d. its B horizon.

8. The soil in tropical climates is often
a. thick. b. dry. c. thin. d. fertile.

9. The transport of weathered materials by a moving natural agent is called
a. mass movement. b. weathering.
c. erosion. d. creep.

10. All of the following farming methods prevent gullying, except
a. terracing. b. contour plowing.
c. strip-cropping. d. irrigation.

11. The most effective of all mass movements is
a. a landslide. b. a rockfall.
c. a mudflow. d. creep.

12. When a mountain reaches old age, it is eroded to an almost featureless surface called a
a. peneplain. b. monadnock. c. talus slope. d. mesa.

Critical Thinking

On your own paper, write answers to the following questions.

1. Compare the weathering processes that affect a rock on top of a mountain and a rock buried beneath the ground.
2. Which do you think would weather faster, a sculptured marble statue or a smooth marble column? Explain your answer.
3. Compare the appearance of a 50-year-old highway that runs through a desert with that of a highway of the same age that runs through New York City.
4. If the A horizon and B horizon of a soil are relatively red, what do you think the underlying C horizon is composed of? Why?
5. Mudflows in the southern California hills are usually preceded by a dry summer and widespread fires, followed by torrential rainfall. Explain why.
6. How can you determine whether a plateau in an advanced stage of evolution is made of sandstone? Explain.
7. Suppose that a mountain has been wearing down at the rate of about 2 cm per year for 10 years. In the eleventh year, scientists find that the mountain is no longer losing elevation. What do you think has happened?

Application

1. You are consulted on whether a limestone building or a quartzite building would better withstand the effects of acid rain. What would be your response and why?
2. If you were a farmer and could choose the ideal climate in which to grow your crops of deep-rooted plants, what climate would you choose? Explain your reasons.
3. Suppose you wanted to cultivate grapevines on a hillside in Italy. What farming methods would you use? Explain why.

Extension

1. Rocks on the moon are subjected to alternating very high and very low temperatures. The wide variation in temperatures does not produce rain or snow, however. Find out how these alternating temperatures affect moon rocks. Summarize your findings.
2. The Dust Bowl created in the 1930's in the southwestern United States was partially due to poor farming practices. Find out in what ways farmers damaged the soil. Find out, too, what farmers in that region are doing now to prevent a recurrence of the Dust Bowl. Report your findings.
3. Find out the evolutionary stages of the Andes Mountains, the Rocky Mountains, and the Appalachian Mountains, and make a relief map that illustrates their differences.

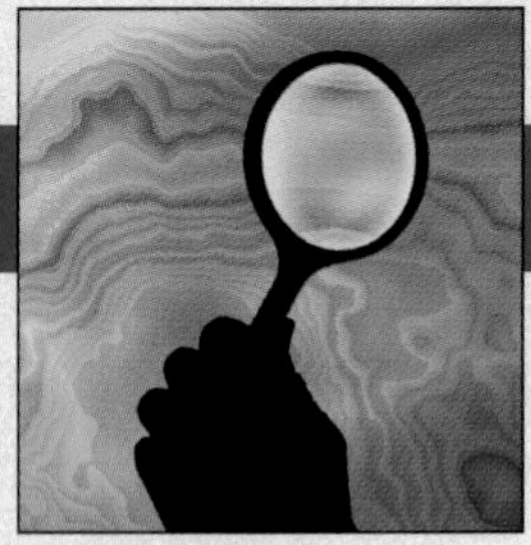

IN-DEPTH INVESTIGATION

Soil Chemistry

Materials

- ammonia solution
- medicine dropper
- subsoil sample (B/C horizons)
- 9 cork stoppers
- pH paper
- 9 test tubes
- topsoil sample (A horizon)
- water
- dilute hydrochloric acid
- safety goggles
- lab apron
- test-tube rack

Introduction

The kinds and amounts of minerals present in soil are important to life. To support plant life, the soil must have a proper balance of minerals. In order for plants to take in the minerals they need, the soil must also have the proper acidity. Plants use up some minerals, but allow others to accumulate. The minerals that are used up are replaced by the decay of dead plants or by the addition of fertilizers. The minerals that are most likely to be used up contain the elements nitrogen, phosphorus, and potassium. Minerals that tend to accumulate contain aluminum silicates, iron silicates, and calcium carbonate. Soils with an accumulation of aluminum and iron silicates are called *pedalfer* soils. The term *pedalfer* is derived from a combination of the Greek word πεδον (*ped*on) meaning "soil," *al*uminum, and the Latin word *ferrum* meaning "iron." Soils rich in calcium carbonate are called *pedocal* (*ped*on + *cal*cium) soils. These substances accumulate mostly in the B horizon of the soil. Hydrochloric acid has little or no effect on silicates, but it decomposes calcium carbonate, causing CO_2 gas to bubble out of solution.

pH Values for Some Common Substances

Substance	pH
Ammonia	11.5
Blood	7.5
Lemon juice	2.2
Milk	7.0
Milk of magnesia	10.5
Orange juice	3.5
Pure water	7.0
Sea water	8.5
Soft drink	3.0
Vinegar	2.8

Prelab Preparation

1. Review Section 12.3, pages 227–230.
2. Review the safety guidelines for handling caustic substances.
3. Define the terms acidic, alkaline, and neutral.
4. Acidity is measured on a scale called the pH scale. The pH scale ranges from 0 (acidic) to 14 (alkaline). A pH of 7 is neutral (neither acidic nor alkaline). To the left is a list of different substances and their pHs. Classify the substances as acidic, alkaline, or neutral.
5. The pH paper changes color in the presence of an acidic or alkaline substance. Take a strip of pH paper and wet it with tap water. Compare the color of the wet pH paper with the pH color scale. What is the pH of the tap water?

Procedure

1. **CAUTION:** *Wear your laboratory apron and safety goggles.* Place a small amount of the topsoil sample in a clean test tube until 1/8 full. Add water to the test tube until 3/4 full. Stopper

the test tube with a cork and shake. Set the soil and water mixture aside in the test-tube rack to settle. When the water is fairly clear, test it with a piece of pH paper. What is the pH of the soil sample? Is the soil acidic or alkaline?

2. Repeat Step 1 with the subsoil sample. What is the pH of this sample? Is the soil acidic or alkaline?
3. Pedalfer soils have a tendency to become acidic. Pedocal soils tend to become alkaline. Based on the pH results in Steps 1 and 2, predict whether your soil is pedalfer or pedocal.
4. To test your prediction, you will need to test the composition of your soil sample. Take out five rock particles from the subsoil sample. Place each particle in a separate test tube. Using the medicine dropper, add two drops of hydrochloric acid to the tubes. **CAUTION:** *If any of the acid gets on your skin or clothing, rinse immediately with cool water and alert your teacher.* Record your observations. How many of the rock particles were silicates? How many were calcium carbonate?
5. Place a small amount of the subsoil sample in a clean test tube until 1/8 full. Slowly add dilute hydrochloric acid to the test tube until about 2/3 full. Cork the tube and gently shake the mixture. **CAUTION:** *Always shake the test tube pointing away from yourself and other students. Make sure you are still wearing safety goggles.* After shaking, remove the stopper and set the test tube in the test-tube rack to allow the soil to settle. Note any reaction and record your observations. After the soil mixture has settled, draw a diagram of the test tube and its contents. Label each layer of material. If iron is present, the solution may look brown. What color is the clear liquid above the soil sample?
6. Using a medicine dropper, place 10 drops of the clear liquid in a clean test tube. Following the same precautions as in Step 4, add 12 drops of ammonia to the test tube. Test the pH of the resulting solution. If the pH is greater than 8, any iron present should settle out as a reddish-brown residue. The remaining solution will be colorless. If the pH is less than 8, add two more drops of ammonia and test the pH again. Continue until the pH reaches 8 or higher. Record your observations and draw a diagram of the contents of the test tube. Label each layer of material. Is iron present in your sample?

Analysis and Conclusions

1. Is your soil sample most likely pedalfer or pedocal? Explain your answer based on the results of the tests you performed in Steps 4–6 of the Procedure.
2. What type of soil, pedalfer or pedocal, would you treat with acidic substances such as phosphoric acid, sulfur, or ammonium sulfate to help plant growth? Explain why?
3. Explain why the acidic substances in the item above are usually spread on the surface of the soil.

Extension

Why has the use of phosphate and nitrate detergents been banned in some areas?

Chapter 13

Water and Erosion

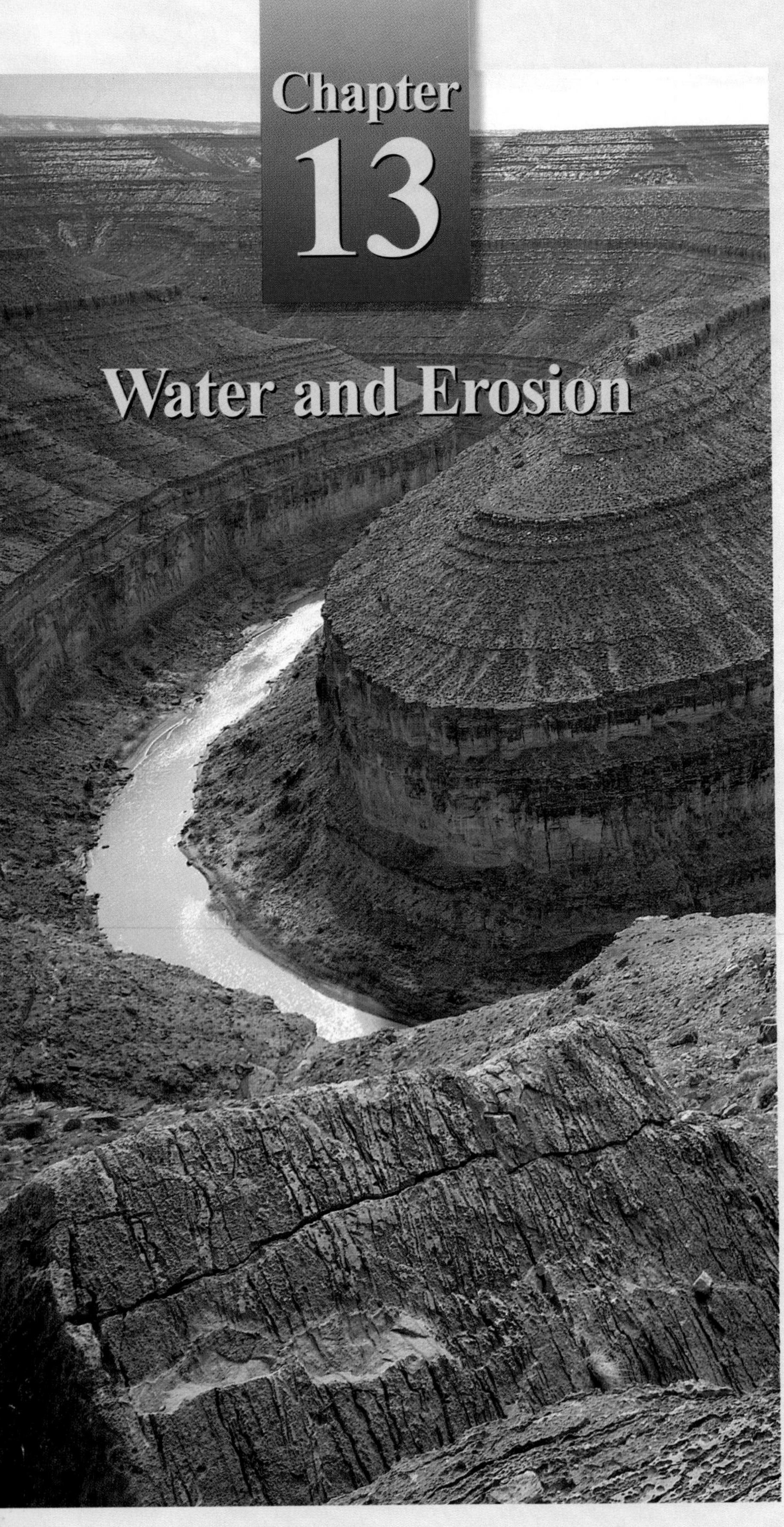

Thundering over a mighty falls or seeping silently through the ground, water is always at work. It carries away more soil and moves more of the rocks that make up the earth's crust than all the other forces of erosion combined. In this chapter, you will learn about the water supply of the earth and how surface water helps to reshape the earth's land.

Chapter Outline

◀ **The Goosenecks in Utah were formed by the San Juan River.**

13.1 The Water Cycle

Section Objectives

- **Outline the stages of the water cycle.**
- **Explain the components of a water budget.**
- **List two approaches to water conservation.**

The origin of the earth's water supply has puzzled people for centuries. Artistotle and other ancient Greek philosophers believed that rivers such as the Nile and the Danube could be supplied by rain and snow alone. It was not until the middle of the seventeenth century that scientists could accurately measure the amount of water received on the earth and the amount flowing in rivers. These measurements showed that the earth's surface actually receives up to five times more water than rivers carry off. So, a more puzzling question than ''Where does the earth's water come from?'' is ''Where does the water go?''

There is a continuous movement of water from the atmosphere to the earth's surface and back to the atmosphere again. This cycle of water movement is called the hydrologic cycle, or **water cycle,** and is illustrated in Figure 13–1.

The process by which liquid water changes into water vapor is called *evaporation.* Each year, about 500,000 km^3 of water evaporate into the atmosphere. About 86 percent of this water evaporates from the ocean. The remaining 14 percent of the water evaporates from lakes, streams, and the soil. Water vapor also enters the air by *transpiration,* a process by which plants give off water vapor into the atmosphere. Together, the processes of evaporation and transpiration are called **evapotranspiration.** The continents lose about 70,000 km^3 of water each year through evapotranspiration. Another 40,000 km^3 flows over the land into rivers as **runoff** or soaks deep into the soil and rock underground, becoming **groundwater.**

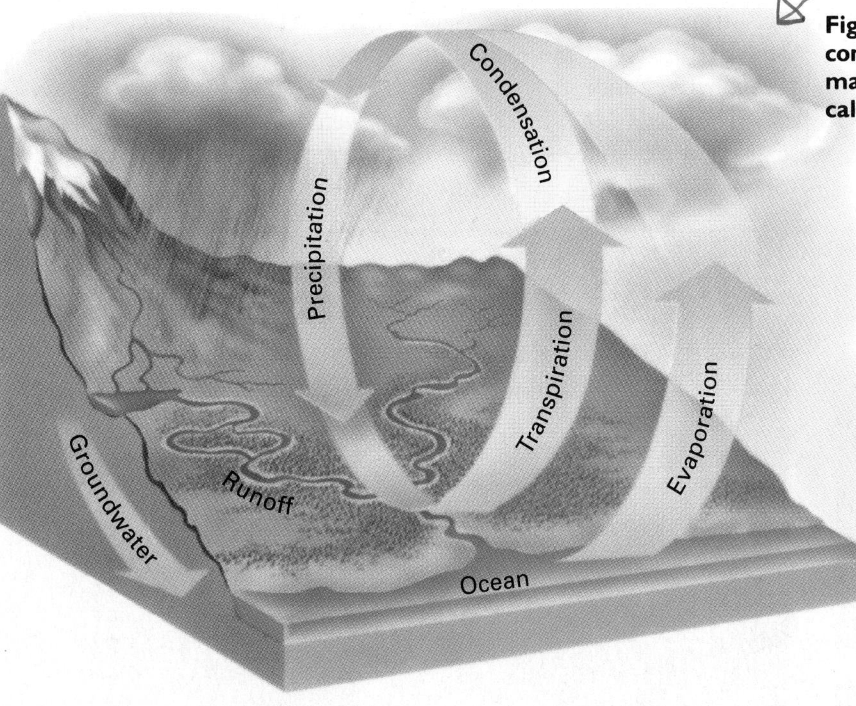

Figure 13–1. Evapotranspiration, condensation, and precipitation make up the continuous process called the *water cycle*.

Another process of the water cycle is **condensation.** When water vapor rises in the atmosphere, it expands and cools. As the vapor becomes cooler, some of it condenses, or changes into tiny liquid water droplets, and forms clouds.

The third process of the water cycle is **precipitation.** Precipitation is the process by which water falls from clouds to the earth's surface as rain, snow, sleet, and hail. About 75 percent of all precipitation falls on the earth's oceans. The remaining 25 percent falls on the land surface and becomes runoff or groundwater. Essentially all of this water eventually returns to the atmosphere by evapotranspiration, condenses, and falls back to earth to begin the cycle again.

Water Budget

The continuous cycle of evapotranspiration, condensation, and precipitation gives the earth its **water budget.** A financial budget is a statement of expected income—money coming in—and expenses—money going out. When the money coming in equals the money going out, the budget is balanced. In the earth's water budget, precipitation is the income. Evapotranspiration and runoff are the expenses. The water budget of the earth as a whole is balanced because the amount of precipitation is equal to the amount of evapotranspiration and runoff. However, the water budget of a particular area, called the *local water budget,* usually is not balanced.

Factors that affect the local water budget include temperature, the presence of vegetation, wind, and the amount and duration of rainfall. When the precipitation exceeds the evapotranspiration and runoff of an area, the result is moist soil and possible flooding. When evapotransiration exceeds precipitation, the soil becomes dry and irrigation may be necessary. Vegetation reduces runoff in an area but increases the evapotranspiration. Wind increases the rate of evapotranspiration.

The factors that affect the local water budget vary geographically. For example, the Sahara Desert receives very little precipitation compared with that received in the tropical rain forests of

Figure 13–2. Tropical rain forests (left) require large amounts of annual rainfall. Deserts (right) have low annual rainfalls.

EarthBeat

Life Cycle of a Lake

All precipitation does not either evaporate or immediately flow from the land to the ocean. Sometimes this water collects in a depression in the land and creates some of the most beautiful formations of water on the earth—lakes.

Lakes occur most frequently in high latitudes and in mountainous areas. However, they also are common in low areas near the ocean, such as in Florida, and near rivers with low gradients. Most of the water in lakes comes from precipitation and melting ice and snow. Springs, rivers, and runoff directly from the land also are sources of lake water.

Like rivers, lakes have a life cycle. Unlike rivers, most lakes are relatively short-lived. Many lakes eventually disappear because too much of their water drains away or evaporates. A common cause of excess drainage is an outflowing stream that cuts its bed below the lake basin. A lake may also lose its water if the climate becomes drier and evaporation exceeds precipitation.

Lake basins may be destroyed if they are filled with sediments. Streams that feed a lake carry and deposit sediments in the lake. Sediments also may be eroded from the lake basin itself by waves cutting into its sides. These sediments build up on the shores of the lake and gradually fill it in, as shown in the illustrations.

Organic deposits from vegetation also may accumulate in the basins of shallow lakes. As these deposits grow denser, a bog or swamp may form. The lake basin eventually may become dry land.

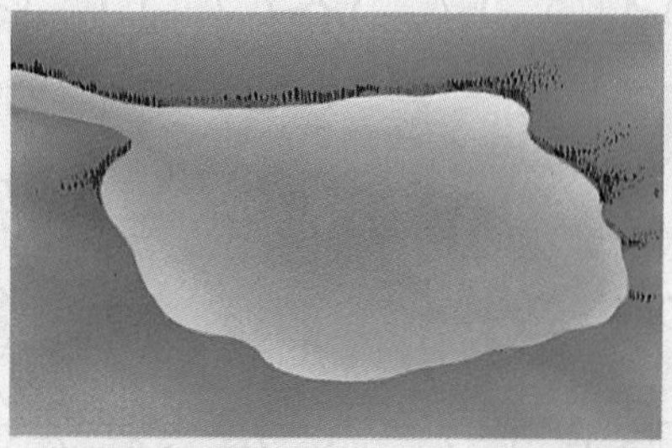

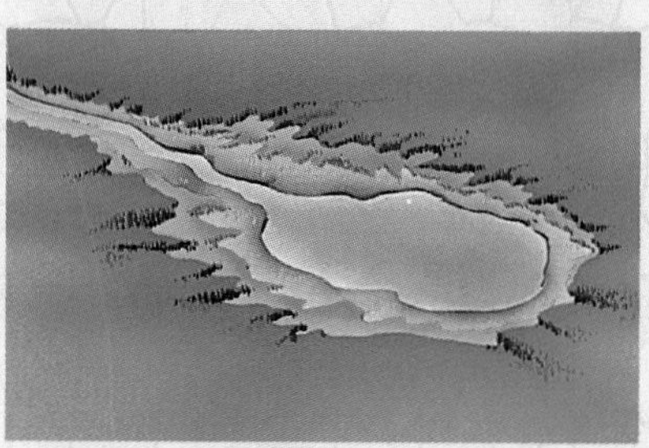

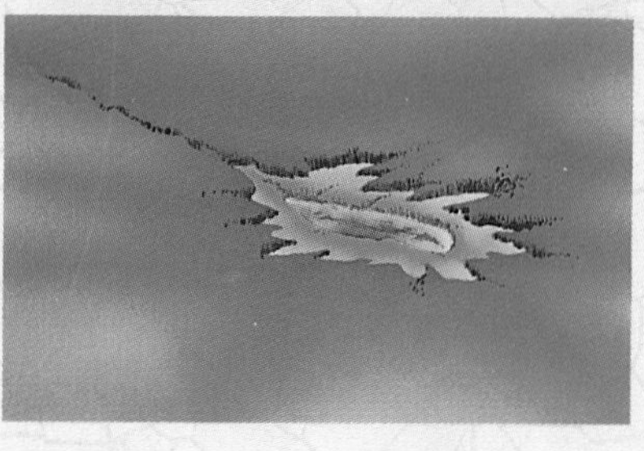

What process of stream deposition is similar to the filling of a lake basin with sediments?

Brazil. The rate of evapotranspiration in Kenya, which is at the equator, is generally much greater than the rate in Finland, which is much closer to the North Pole.

The local water budget also changes with the seasons in most areas of the earth. Generally, cooler temperatures slow the rate of evapotranspiration. During the warmer months, evapotranspiration increases. Would you expect that streams generally would transport more water in warmer or in cooler months?

Water Conservation

According to recent estimates, each person in the United States uses about 95 m^3 of water each year. Water is used for bathing, washing clothes and dishes, watering lawns, carrying away wastes, and, of course, drinking. Agriculture and industry use very large amounts of

Figure 13-3. As water becomes polluted, the supply available for human consumption is reduced.

water. In the United States, the greatest amount of water is used to irrigate crops. As the population of the United States increases, so, too, does the demand for water.

On the average, 90 percent of the water used by cities and industry is returned to rivers or to the oceans as waste water. Some of this waste water contains harmful materials, such as toxic chemicals and metals. These toxic materials destroy plants and animals in the water and pollute rivers, such as the one shown in Figure 13–3. Polluted rivers cannot support much life.

Scientists have identified two approaches that can be used to ensure that enough fresh water is available today and in the future. The first approach is conservation. Many people favor enacting and strictly enforcing antipollution laws. Many concerned people also support programs that educate the public about the limits of water resources and the danger of wasting them. The second approach is finding other supplies of fresh water. Scientists are experimenting with several methods of **desalination,** the process of removing salt from ocean water. Someday desalination may provide coastal cities with nearly all of their fresh water. The best way of maintaining an adequate supply of fresh water at present, however, is the wise use and conservation of the fresh water that is now available.

Section 13.1 Review

1. Name the major processes of the water cycle.
2. What is evapotranspiration?
3. Explain why many local water budgets vary.
4. What is desalination?
5. Regions that have much vegetation generally receive much rainfall. How do these factors affect the local water budget?

13.2 River Systems

Section Objectives

- **Describe the way in which a river develops.**
- **Explain how a stream causes erosion.**
- **Distinguish youthful, mature, and old river valleys.**

A river system is made up of a main stream and all the feeder streams, called **tributaries,** that flow into it. The land from which water runs off into these streams is called the drainage basin, or **watershed,** of the river system. The ridges or elevated regions of high ground that separate watersheds are called **divides.**

A river system begins to form when local precipitation exceeds evapotranspiration. The soil in the area soaks up as much water as it can hold. Gravity causes water that cannot soak into the soil to move downslope as runoff. When runoff moves across the land surface, it picks up and carries away weathered rock.

The force of the water running across the land surface may erode a narrow ditch, called a *gully.* The runoff from nearby slopes tends to flow into the gully. Each time rain falls, the water moving through the gully erodes the gully further, making it wider and deeper. Eventually, these processes—precipitation, downslope movement of water and rock particles, and erosion of the land—will form a fully developed valley with a permanent stream.

> **INVESTIGATE!**
>
> *To learn more about soil and runoff, try the In-Depth Investigation on pages 258–259.*

Stream Erosion

The path that a stream follows is called its **channel.** The main stream and its tributaries form a network of channels that drain the watershed of the river system. These channels lengthen and branch out at their upper ends, where runoff first enters the streams. The process of lengthening and branching of a stream is called **headward erosion.** Headward erosion carries away sediments from the slopes of a watershed at the upper end of a stream. Erosion of the watershed slopes extends a river system, which, in turn, extends the area of the drainage basin.

Headward erosion also can enlarge a river system through **stream piracy.** Stream piracy is the capture of a stream in one watershed by a stream with a higher rate of erosion in another

Figure 13-4. The tributaries feeding into this river are fed by runoff from slopes. Together the slopes make up the watershed of the river.

watershed. In this process, the captured stream drains into the other river system. Stream piracy extends the area of land drained by the capturing stream.

Channel Erosion

The edges of a stream channel that are above water level are its banks. The part of the stream channel that is below the water level is its bed. A stream gradually becomes wider and deeper as it erodes its banks and bed. Various factors affect the rate at which a stream erodes its channel.

Stream Loads

As a stream flows, it carries more than just water. The stream also transports soil, loose rock fragments, and dissolved minerals. The

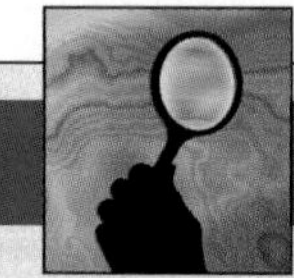

SMALL-SCALE INVESTIGATION

Soil Erosion

The uncontrolled runoff of surface water is a major cause of soil erosion. You can demonstrate soil erosion by making a model of a hillside.

Materials

rectangular pan about 23 × 33 cm, fine sand, sink lined with paper towels, brick or other support, water faucet, additional pan or container to catch water and sand, small ruler or straightedge, clock or watch

Procedure

1. Fill the pan about half full with moist sand as shown in the photo.
2. Place the pan in the sink so that the end resting on the brick is under the water faucet.
3. Place the second pan or container so it catches sand and water that flow out of the first pan.
4. Slowly open the faucet until a gentle trickle of water falls onto the sand in the raised end of the pan. Let the water run for 15–20 seconds.
5. Turn off the water and draw the pattern of water flow over the sand.
6. Press the sand back into place and carefully smooth the surface with a small ruler or straightedge. Repeat Step 4, but this time increase the rate of water flow slightly, but not enough to splash the sand. Draw a pattern of water flow as before.
7. Repeat Steps 4 and 5 two more times with different rates of water flow. For each trial, observe and draw the pattern of water flow.

Analysis and Conclusions

1. Compare the rate of erosion with the rate of water flow.
2. Compare the patterns of erosion with the rate of water flow.
3. How does the rate of water flow affect gullies?
4. Without changing the rate of water flow, how could the rate and effects of erosion be reduced on an actual hillside?

materials carried by a stream are called the **stream load.** The stream load has three forms—the *suspended load,* the *bed load,* and the *dissolved load.* The suspended load consists of particles of fine sand and silt. The velocity, or speed, of the water keeps these particles suspended, and they do not sink to the stream bed. The bed load is made up of larger, coarser materials, such as coarse sand, gravel, and pebbles. This material moves by sliding, by rolling, and by **saltation,** or short jumps. The dissolved load is mineral matter transported in liquid solution.

Although all three forms of stream load contribute to stream erosion, the bed load has the greatest effect. As rock fragments are carried along, they scrape the bottom and sides of the stream channel, wearing it away. Therefore, streams with large bed loads erode their channels more quickly than those with small bed loads. Large rocks are sometimes scraped over one area of the stream bed in a whirlpool motion. This scraping creates a bowl-shaped cavity called a *pothole,* like the one shown in Figure 13–5.

Figure 13-5. This photograph shows a pothole formed by the whirlpool action of rocks carried by a stream.

Discharge and Gradient

The **discharge** and velocity of a stream, as well as its load, affect how a stream cuts down and widens its channel. The volume of water moved by a stream in a given time is its discharge. The faster a stream flows, the higher is its discharge and the greater is the load it can carry. A swift stream carries more sediment and larger particles than a slow-moving stream carries. Hence, swift streams erode their channels more quickly than slow-moving streams do.

The velocity of a stream depends mainly on its **gradient,** or the steepness of its slope. Gradient is the change in elevation of the stream over a given horizontal distance. The gradient of a stream varies along the stream channel. Near the **headwaters,** or the beginning of a stream, the gradient is steep. This area of the stream generally has a high velocity, which causes rapid channel erosion. As the stream nears its mouth, where the stream enters a larger body of water, its gradient often becomes gentle. Its velocity and erosive power decrease. The stream channel eventually is eroded to a nearly flat gradient by the time it reaches the sea.

Water and Wind Gaps

Movements of the earth's crust can raise or lower the surface of the land. When the land surface is uplifted slowly by geologic forces, the existing stream channels may be eroded downward at the same rate as the land around them is elevated. When this happens, a deep notch is formed where the stream has eroded its channel through the raised mountains. This notch is called a **water gap.** An example is the Delaware Water Gap. In some cases, the land is uplifted faster than the stream can erode its channel. Because streams cannot flow uphill, the water gap is abandoned. The notch through which water no longer flows is called a **wind gap.** A water gap also is changed into a wind gap if the stream flowing through the gap is captured by another stream. The Cumberland Gap in the Appalachian Mountains is an example of a wind gap.

Stages of a River System

As a river erodes its banks and bed, it changes the landforms it passes through and alters its own course. The development of a river system is divided into three stages—youthful, mature, and old. These stages are not based on the actual age of the river but on its shape and how it erodes the land. The time required for a river to pass through each of these stages depends on the composition and structure of the rock through which the river flows.

Youthful Rivers

In its youthful, or early stage, a stream usually erodes its bed more rapidly than it erodes its banks. This produces a V-shaped valley with steep sides, like the one shown in Figure 13–6. Waterfalls and rapids are common features of youthful streams. These features are especially common in stream channels cut into hard rock, because the rock resists erosion.

Youthful rivers usually have relatively few tributaries. For this reason, a youthful river usually carries a small volume of water. Much of the precipitation falling on the watershed of a youthful river system does not reach the main stream because so few tributaries have developed. Instead, much of the precipitation may form lakes at high elevations.

Mature Rivers

A mature river, by comparison, has well-established tributaries. It drains its watershed effectively. Because of good drainage and many tributaries, a mature river can carry a larger volume of water than a youthful river can carry. A mature river, however, tends not to deepen its channel as much as a youthful stream does. Instead, erosion occurs mostly along the valley walls when the river overflows

Figure 13–6. A young river has a narrow valley and fast-moving water (left). A mature river forms a broader floodplain (right).

its banks and covers the valley floor. A mature-river channel usually occupies only a small part of the wide and relatively flat valley floor that it produces. Most of the waterfalls and rapids that existed during the youthful state of a mature river have disappeared. The gradient also has become less steep.

A mature stream with a low gradient tends to curve back and forth across the flat valley floor. A slight bend in the stream channel usually becomes a wider curve, because the water flows fastest around the outside edge of the curve. The faster-flowing water erodes the outside bank of the curve more quickly than the slower-moving water erodes the inner bank. The slower-moving water often deposits sediments along the inner bank. This process enlarges the curve and shifts the stream channel toward the outside bank. Generally, a series of these wide curves, called **meanders,** form across the valley floor.

Frequently, a meander becomes so curved that is almost forms a loop, separated by only a narrow neck of land. When the river eventually cuts across this neck, it deposits sediments at both ends of the meander and eventually abandons it. The meander is thus isolated from the river, as shown in Figure 13–7. If the water remains in the abandoned meander, an **oxbow lake** is formed.

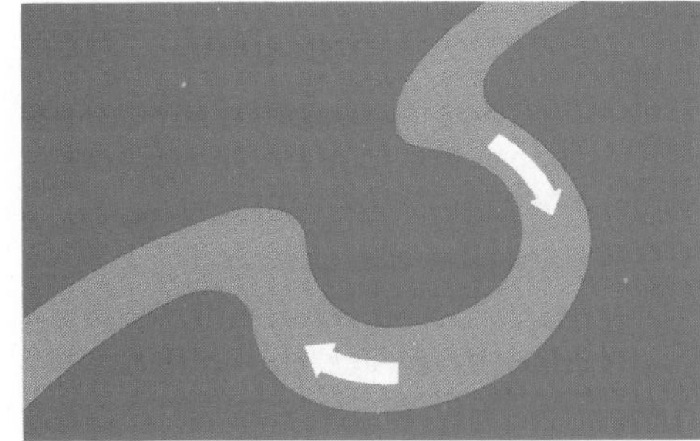

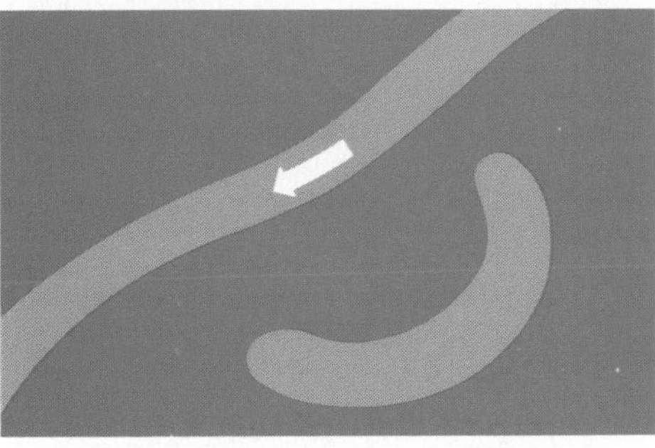

Figure 13–7. An oxbow lake results when part of a meander is cut off and abandoned by the river.

Old Rivers

As a river continues to age, its gradient and velocity decrease. The stream no longer erodes the land; instead, it begins to deposit its sediments in its own channel and on its banks. A broad, flat plain is formed. More meanders develop, and there are fewer tributaries, as smaller tributaries merge and become larger. How do you think the drainage of an old river compares with the drainage of a mature river?

Rejuvenated Rivers

Any movement of the earth's crust that increases the slope of the land will change the gradient of existing streams. A **rejuvenated** river is one whose gradient has become steeper in this way. The increased gradient of a rejuvenated river allows the river to cut more deeply into the valley floor. Rejuvenation often results in the formation of steplike terraces on both sides of a stream valley. These terraces provide evidence that the valley floor has been uplifted and a new floor has been cut through. There are many terraces along the Mississippi River.

Section 13.2 Review

1. How does a river system begin?
2. Define *watershed* and *divide*.
3. What process results in the lengthening and branching of a river channel?
4. Name the three types of stream loads.
5. What feature might help you distinguish a rejuvenated river from a young river?

Section Objectives

- **List two types of stream deposition and explain the differences between them.**
- **Describe the change in a stream that causes flooding.**
- **Identify direct and indirect methods of flood control.**

13.3 Stream Deposition

The total load a stream can carry is greatest when a large volume of water is flowing swiftly. When the velocity of the water decreases, the ability of the stream to carry its load decreases. As a result, part of the stream load is deposited as sediment. When the velocity increases again, some or all of the deposited sediment is carried away. In some cases, stream deposits remain in place for only a brief time. In other cases, these deposits can become permanent features of the land.

Deltas and Alluvial Fans

Most of the load carried by a stream is deposited when the stream reaches a large body of water. As a stream empties into a large body of water such as an ocean or a lake, the velocity of the stream decreases greatly. The load is usually deposited at the mouth of the stream in a fan shape with its tip facing upstream, as illustrated in Figure 13–8. This fan-shaped deposit at the mouth of a stream is called a **delta.** The exact shape and size of a delta are determined by waves, tides, off-shore depths, and sediment load of the stream.

When a stream descending a steep slope reaches a flat plain, the speed of the stream is suddenly reduced. As the stream suddenly slows down, it deposits some of its load on the level plain at the base of the slope. This deposit forms a fan-shaped heap with its tip pointing upstream. This deposit is called an **alluvial fan.** In desert and semidesert regions, temporary streams often form alluvial fans. An alluvial fan differs from a delta in the following three ways. First, the sediment that forms an alluvial fan is deposited on dry ground, whereas the sediment that forms a delta is deposited in water. Second, an alluvial fan is made up of coarse, angular sand and gravel. A delta, though, usually is made up of mud. Third, the surface of an alluvial fan is sloping, while the surface of a delta is relatively flat.

Figure 13–8. When a stream meets an ocean or lake, the sediment carried by the stream is deposited as a delta (below). Runoff from slopes in desert regions deposits sediment on the valley floor, building up an alluvial fan (inset).

A River and a City

At the beginning of the fifteenth century, Bruges, Belgium, was one of the great commercial and cultural centers of Europe. Waterways, such as the canal shown below, brought the city wealth through trade. Located at the base of the Zwin estuary, Bruges was easily accessible from the North Sea. Merchant ships from all over Europe and the Far East regularly docked at the harbor on the Zwin. In turn, traders from Bruges sailed to other parts of the world, carrying fine fabrics from the city's famous textile mills.

However, the prosperity of Bruges was short-lived. By the end of the fifteenth century, the last of the textile mills were closing, the shipping trade was virtually nonexistent, and the once powerful and populous city had been nicknamed *Bruges-la-Morte,* Bruges-the-Dead.

The decline of Bruges can be traced to stream deposition, the process in which a stream or river deposits its sediment as it enters a larger body of water. The Reie River, which flows through Bruges, broadened at the city of Damme, just northeast of Bruges, to form the Zwin estuary. The Zwin then emptied into the North Sea. As the Reie flowed into the Zwin, it deposited some sediment. In turn, as the Zwin flowed into the North Sea, it deposited its sediment. Thus, the Zwin estuary was becoming blocked at both ends. Soon, access to Bruges was cut off, and the golden age of the city ended.

What natural process in the development of a river system contributed to the stream deposition at Bruges?

Flood Deposits

The size of a stream channel is determined by the average volume of water that flows in the stream. Some channels are small enough to step across; others are several kilometers wide. If the volume of water in a stream remains constant, its channel changes very little. However, the volume of water in nearly all streams changes continually. A dramatic increase in volume can cause a stream to flood, or overflow its banks and wash over the valley floor. The part of the valley floor that may be covered with water during a flood is called a **floodplain.**

Spring floods are common near headwaters in areas where the winters are harsh. During the winter, little evapotranspiration occurs and snow and ice remain on the land. The water released by melting snow cannot be absorbed by the frozen ground and runs off into surface streams, increasing the volume of water in those streams.

Figure 13-9. The regularly flooded area along a stream is called the *floodplain*.

Ice jams, too, increase the chance of spring flooding when ice blocks the stream channels.

Human activity also contributes to the size and number of floods in many areas. Sometimes the natural ground cover of plants and trees, which protects the surface from rapid runoff, is removed. Forest fires, logging, and the clearing of land for cultivation or housing development can increase the volume and speed of runoff, leading to more frequent flooding.

When a stream overflows its banks and spreads out over the nearby land, the stream loses velocity and deposits its coarser sediment load along the banks of the channel. The accumulation of these deposits along the banks eventually produces raised banks, which are called **natural levees.** Natural levees may be quite high and prominent along old river channels, such as the lower Mississippi River, where repeated flooding has occurred.

Not all of the load deposited by a stream in a flood will form levees. Finer sediments are carried over the floodplain by the flood waters and are deposited there. A series of floods produces a thick layer of fine sediment, which becomes a source of rich floodplain soils. Swampy areas are common on floodplains because drainage is usually poor in the area between the levees and the outer walls of the valley. Since these areas are prone to flooding, why do people choose to live on floodplains?

Flood Control

Flooding is a natural stage in the development of a stream and will continue to occur as the river system matures. However, flooding can become a problem when cities and farms are located on the fertile floodplains of rivers. The best safety measure is to leave such

Figure 13-10. Artificial levees (left) and dams (right) are often built to control floods.

areas in their natural state. However, since cities, towns, industrial complexes, and other construction is done in these areas, flood control is often necessary to minimize the loss of life and property.

There are indirect and direct methods of flood control. Indirect methods include forest and soil conservation measures that prevent excess runoff during periods of heavy rainfall. The most common method of direct flood control is the building of a dam. The artificial lakes that form behind dams act as reservoirs for excess runoff. The stored water can be used to generate electric power, to supply fresh water to populated areas, to irrigate farmland during dry periods, and for recreation.

Another direct method of flood control is the building of artificial levees. However, artificial levees offer only temporary protection. As a river deposits sediment along its bed, the height of the levee must be raised. Artificial levees also must be protected against erosion by the river.

In some cases, a permanent overflow channel, or *floodway,* can be an effective means of flood control. When the volume of water in a river increases, a floodway carries away excess water and prevents the main stream from overflowing.

Section 13.3 Review

1. Describe the differences between a delta and an alluvial fan.
2. What is a floodplain?
3. Why are spring floods common where the headwaters of a river are in an area of harsh winters?
4. What advantage other then flood control does a dam provide?
5. If you were picking a material to make an artificial levee, what major characteristic would you look for?

Chapter 13 Review

Key Terms

- alluvial fan (252)
- channel (247)
- condensation (244)
- delta (252)
- desalination (246)
- discharge (249)
- divide (247)
- evapotranspiration (243)
- floodplain (253)
- gradient (249)
- groundwater (243)
- headward erosion (247)
- headwater (249)
- meander (251)
- natural levee (254)
- oxbow lake (251)
- precipitation (244)
- rejuvenated (251)
- runoff (243)
- saltation (249)
- stream load (249)
- stream piracy (247)
- tributary (247)
- water budget (244)
- water cycle (243)
- water gap (249)
- watershed (247)
- wind gap (249)

Key Concepts

The continuous cycle of water movement on earth is called the water cycle. **See page 243.**

An area's water budget is the amount of water received and the amount lost. **See page 244.**

Careful conservation of the earth's limited supply of fresh water is important. **See page 245.**

The development of river systems illustrates the erosive power of water. **See page 247.**

The velocity of its water and the abrasiveness of its loads cause a stream to erode its channel. **See page 248.**

Youthful river systems tend to erode the land much faster than do mature or old river systems. **See page 250.**

A stream can deposit its sediments on land or in the water. **See page 252.**

Flooding is often the result of a large and sudden increase in the volume of water carried by a stream. **See page 253.**

Floods can be controlled through natural conservation and by building various flood-control structures. **See page 254.**

Review

On your own paper, write the letter of the term that best completes each of the following statements.

1. The change of water vapor into liquid water is called
   a. runoff. b. evaporation.
   c. desalination. d. condensation.
2. Vegetation gives off water vapor into the atmosphere through a process called
   a. condensation. b. rejuvenation.
   c. saltation. d. transpiration.
3. In a water budget, the income is precipitation and the expense is
   a. evapotranspiration and runoff.
   b. condensation and saltation.
   c. erosion and conservation.
   d. rejuvenation and sedimentation.
4. The process that turns sea water into fresh water is
   a. desalination. b. transpiration.
   c. conservation. d. rejuvenation.
5. The land areas from which water runs off into a stream are called its
   a. tributaries. b. divides.
   c. watersheds. d. gullies.

6. Tributaries branch out and lengthen as a river system develops by
   a. headward erosion. b. condensation.
   c. saltation. d. runoff.
7. The stream load that includes gravel and large rocks is the
   a. suspended load. b. dissolved load.
   c. runoff load. d. bed load.
8. When a young river deepens its channel faster than it can cut into its sides, the result is
   a. a gradient. b. a V-shaped valley.
   c. a floodway. d. an oxbow lake.
9. A stream whose gradient has been increased by movement of the earth's crust is said to be
   a. rejuvenated. b. meandering.
   c. eroded. d. suspended.
10. The triangular formation that occurs when a stream deposits its sediment at the base of a steep slope is called
    a. a delta. b. a meander.
    c. an oxbow lake. d. an alluvial fan.
11. The part of a valley floor that may be covered during a flood becomes the
    a. floodway. b. groundwater. c. floodplain. d. artificial levee.
12. One indirect method of flood control is
    a. soil conservation. b. dams.
    c. floodways. d. artificial levees.

Critical Thinking

On your own paper, write answers to the following questions.

1. How would the earth's water cycle be affected if a significant portion of the sun's rays were blocked by dust or other contaminants in the atmosphere?
2. How might the local water budget of Calcutta differ from that of Stockholm? Use an atlas to determine the geographic location of these two cities before thinking about your answer.
3. Desalination may someday provide an almost endless supply of fresh water. What other problem must be solved before the desalinated water can be widely used?
4. In the desert areas of the southwestern United States, there are many shallow, narrow ditches that cut through the landscape. What do you suppose these ditches are? What was the most likely cause of their formation? If these ditches were located anywhere else, what might happen to them? Why does this not happen in the desert?
5. The Colorado River is usually grayish-brown as it flows through the Grand Canyon. What causes this color?
6. Why do you think the surface of an alluvial fan is sloping and that of a delta is flat?

Application

1. Assume that you decided to test the color of the water of the Colorado River over a period of years and found that it was becoming clearer. What would you conclude was happening to the river?
2. If you were trying to locate a mature river for geologic exploration, what characteristics would you look for?
3. Developers are planning to build retirement communities on the floodplain of a river, but away from the banks. Considering only the safety aspect, would you argue for or against building these communities? Support your argument.
4. What steps might the developers in Question 3 take to protect the people and property in the communities?

Extension

1. Do some research in your community to find how much precipitation your area receives, how much water is lost to evapotranspiration, how much runs off into streams and rivers, and how much becomes groundwater. How does the water budget for your community change with the seasons? Report your findings to the class.
2. Select a major river system in your area of the country. Prepare a map that shows the main stream channel, major tributaries, watershed, divides, waterfalls, deltas, and oxbow lakes.

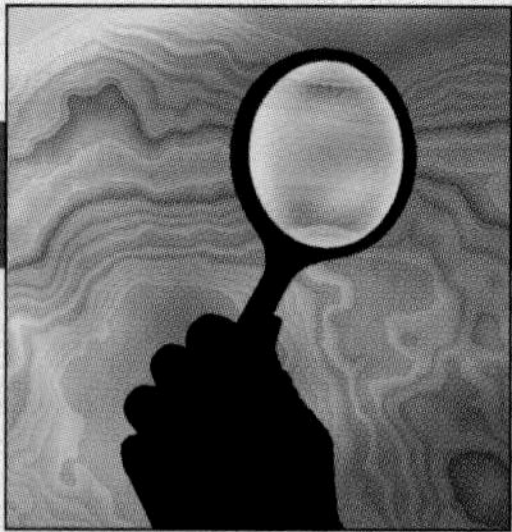

IN-DEPTH INVESTIGATION

Sediments and Water

Materials

- metric ruler
- water
- large nail
- graduated cylinder (100 mL)
- grease pencil
- timer
- dry, coarse sand
- dry, silty (unsorted) soil
- 2 cardboard juice containers (12-oz.)
- pan (23 × 33 × 5 cm or larger)

Introduction

Rain falling on a barren patch of sand or soil causes the surface to become wet. The rainwater then sinks deeper into the sand or soil. The amount of rainwater that can be held by any sediment depends on how much moisture is already in the sediment and the total amount of moisture the sediment can hold. If the water can move through the soil quickly enough, then the soil may never become filled with water. If a sediment is holding all the water it possibly can, then any additional water added to the sediment will cause the water to make puddles or flow downhill. Surface water flow, or runoff, is the primary cause of erosion of barren soil.

In this investigation, you will determine the erosional effect of water on different types of sediment.

Prelab Preparation

1. Review Section 13.2, pages 247–251.
2. Review the safety guidelines for hygienic care.
3. Pinch a few drops of water between your thumb and index finger. Hold your fingers at eye level and observe the water as you slowly move your thumb and finger apart. Describe and explain your observations.
4. Wet one of your index fingers. Place both of your index fingers on a lab table or desk and push your fingers away from you along the table's surface. Does the wet finger or the dry finger slide more easily? Explain why.

Procedure

1. Using the graduated cylinder, pour 300 mL of water into each of the juice containers.
2. Place the containers on a flat surface. With the grease pencil, draw a line around the inside of the containers marking the height of the water. On the outside of the containers, label one container *A*, and the other *B*. Empty and dry the containers.
3. Fill container *A* with the silty soil up to the line drawn inside the container. Tap the container gently to even out the surface of the sediment. Add more sediment if needed. Repeat this step, filling container *B* with sand.

4. Fill the graduated cylinder with 100 mL of water. Slowly pour the water into container *A*. Stop about every five seconds, allowing the water to be absorbed. Continue pouring until no more water is absorbed. There should be a thin film of water on the surface of the sediment. If more than 100 mL of water is needed, refill the graduated cylinder and continue this step.
5. Record the volume of water that you poured into the container.
6. Repeat Steps 4 and 5 using container *B*. Which type of sediment held more water?
7. Use the ruler to measure 1 cm above the surface of the sediments in both containers. Draw a line with the grease pencil to mark this height on the inside of both containers. Pour water from the graduated cylinder into container *A* until it reaches the 1-cm mark.
8. At the very bottom of the side of container *A*, push in a nail. Place the container inside the pan. At the same time, start the timer and pull the nail out of the container.
9. Observe the water level and record the amount of time it takes the water to drop to the sediment surface.
10. Repeat Steps 8 and 9 using container *B*. Which sediment was the water able to flow through faster?
11. Empty, clean, and dry the pan and the juice containers. Fill one container with fresh, dry sand and the other container with fresh, dry soil.
12. From a height of 30 cm, empty the container of sand in to the right side of the pan, forming a mound. From the same height, empty the container of soil into the left side. Measure and record the approximate heights of each mound in your notebook. Which mound is higher?
13. Slowly add water to each of the mounds until all the sediment is damp. Pack the sediments, making each mound as high as possible without toppling. Wash your hands. Measure and record the heights of the mounds. Which mound of wet sediment is higher? Is either mound higher than it was when dry? Explain.
14. Continue adding water to the mounds. What eventually happens? Explain.

Analysis and Conclusions

1. Based on your answer to the questions in Steps 6 and 10, would water erode an area of silty soil or an area of sand more quickly? Explain.
2. Would a hillside covered with sand or a hillside covered with silty soil be more easily eroded during moderate rainfall?
3. Would a hillside covered with sand resist erosion during an extended period of heavy rain? Would a hillside covered with silt resist the erosion?

Extension

Describe ways in which slopes covered with soil can be made more resistant to erosion.

Chapter 14

Groundwater and Erosion

Powerful agents of erosion are at work underground as well as on the earth's surface. Old Faithful Geyser in Yellowstone National Park in Wyoming and Mammoth Cave in Kentucky are evidence of those underground processes. In this chapter, you will learn how water moves beneath the ground and how it erodes the crust.

Chapter Outline

Old Faithful Geyser is located in Yellowstone National Park, Wyoming.

14.1 Water Beneath the Surface

Section Objectives

- **Distinguish between porosity and permeability.**
- **Identify the two moisture zones below the earth's surface.**
- **Relate the contour of the water table to the contour of the land.**
- **Describe how groundwater can be polluted.**

Precipitation that does not run off into streams and rivers usually seeps down through the soil. As the water seeps into the upper layers of the earth's crust, it occupies the air-filled openings, or pores, between rock particles. This underground water that fills almost all the pores in rock and sediment is called *groundwater.*

Groundwater is a plentiful source of fresh water on the earth. An estimated 90 percent of the earth's liquid freshwater supply is stored beneath the surface. In the United States alone, groundwater supplies about one-fifth of the freshwater needs. The amount of groundwater is possibly 50 times greater than the total amount of water in all the earth's rivers and lakes.

Rock Properties That Affect Groundwater

A body of rock through which large amounts of water can flow and in which much water can be stored is called an **aquifer.** Some aquifers are composed of porous layers of sediments, such as sand and gravel. Rocks with large pore spaces, such as sandstone, and highly fractured rock, such as limestone, also may be aquifers. Although a rock may be able to hold much water, if its pore spaces or fractures are not connected, the water cannot flow freely through it. Consequently, it is not an efficient aquifer.

Porosity

The amount of water a rock can hold is determined by the **porosity** of the rock. Porosity refers to the percentage of open spaces in a given volume of a rock or sediment. These spaces can be fractures, cavities caused by erosion, or pores between rock particles.

Porosity is influenced by several factors. One factor, **sorting,** is illustrated in Figure 14–1. Sorting is the amount of uniformity in the size of the particles, or grains, in rock or sediment. When a sediment is well sorted, its particles are all about the same size. This

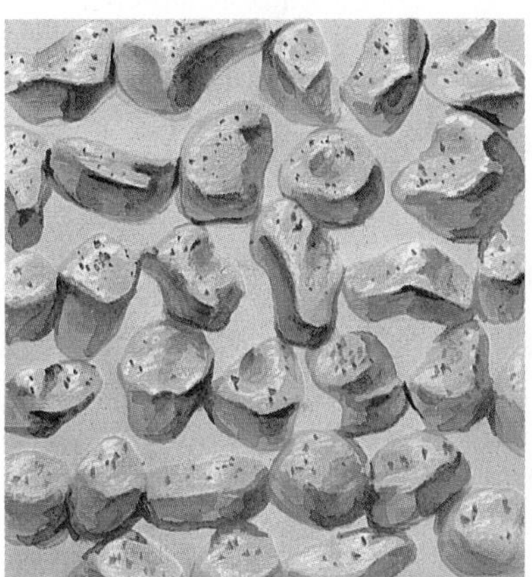
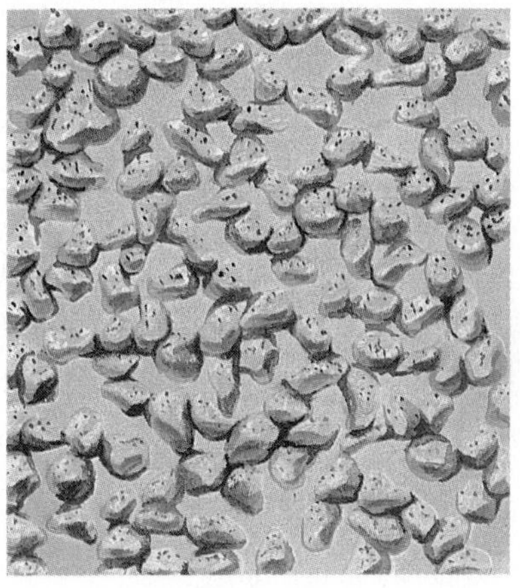

Figure 14–1. Well-sorted, large-grained rock (left) has high porosity. Well-sorted, small-grained rock (center) has equally high porosity. Poorly sorted rock with grains of many different sizes (right) has low porosity.

INVESTIGATE!

To learn more about porosity, try the In-Depth Investigation on pages 274–275.

uniformity means that few, if any, smaller particles fill the spaces between the larger grains. The size of the grains is unimportant. A sediment composed of all large particles can have the same porosity as a sediment composed of all tiny particles. A poorly sorted sediment contains particles of many sizes. Smaller particles fill the spaces between larger ones, making the rock less porous.

Another factor that influences porosity is the way particles are packed together. If they are packed loosely, there are many open spaces for water storage, and the rock has high porosity. Tightly packed rocks have few open spaces and, thus, low porosity.

Permeability

The **permeability** of a rock or sediment indicates how freely water passes through the open spaces in it. For a rock to be permeable, the open spaces must be connected. A rock can have high porosity, but

SMALL-SCALE INVESTIGATION

Permeability

You can demonstrate permeability and calculate the rate of drainage with a simple model.

Materials

sharpened pencil, 3 large paper or plastic cups (9 oz.), cheesecloth, rubber bands, ruler, sand, soil, gravel, 3 small thread spools or other supports, saucer or tray, measuring cup or graduated cylinder, water, clock or watch

Procedure

1. With the pencil point, make seven tiny holes in the bottom of each cup. Cover the holes with cheesecloth secured tightly by a rubber band.
2. Mark a line 2 cm from the top on each cup. Place the first cup on the spools in a saucer, as shown in the photo, and fill the cup to the line with sand.
3. Pour 120 mL of water into the cup. Time and record how long it takes for the water to drain through the cup.
4. Pour the water from the saucer into the measuring cup or graduated cylinder. Record the amount of water.
5. Repeat Steps 3 and 4 with the two other cups—one with soil, the other with gravel.
6. Calculate the rate of drainage for each cup by dividing the amount of water that drained by the time it took the water to drain.
7. Calculate the percentage of water retained in each cup by subtracting the amount of water that drained into the saucer from the original 120 mL. Divide the difference by 120.

Analysis and Conclusions

1. Which cup had the highest drainage rate?
2. Which cup retained the least water?
3. Consider the sample with the highest drainage rate and lowest percentage of water retained to be the most permeable. Which sample is the most permeable? the least permeable?

if the pores are not connected, the rock is not permeable. Permeability is affected by the size and sorting of the particles that make up a rock or sediment. The larger and more consistently sorted the particles are, the more permeable the rock or sediment tends to be. The most permeable rocks are those composed of coarse particles, such as sandstone. Other rocks, such as limestone, are permeable because they have interconnected cracks. Clay is composed of very fine-grained rock particles. Because of this, clay is **impermeable,** meaning water cannot flow through it.

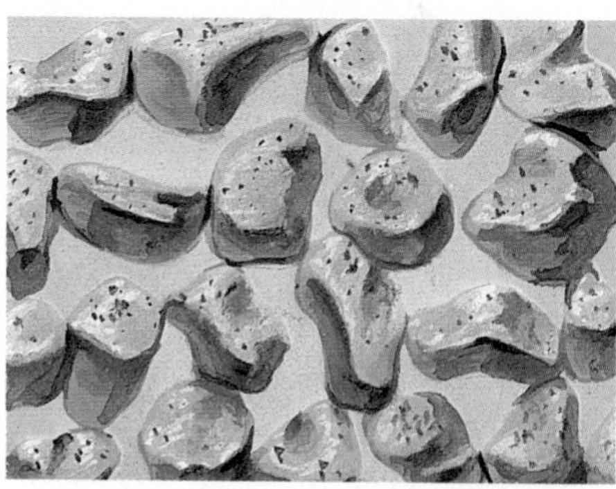

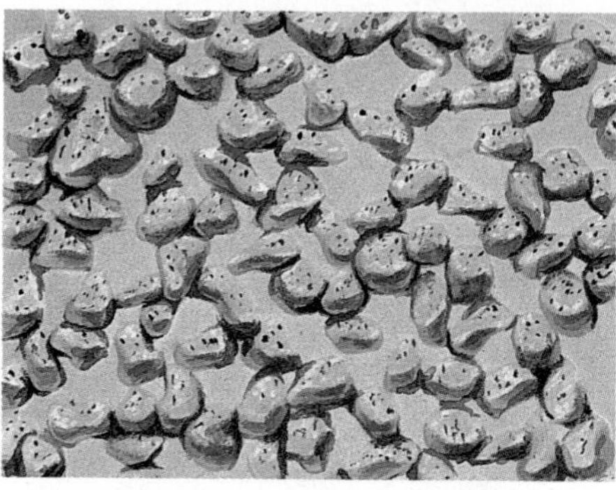

Figure 14–2. Friction slows the flow of water through rock. A large-grained rock (top) has less surface area to cause friction than does a small-grained rock (bottom).

Zones of Groundwater

Gravity pulls water down through soil and rock until it reaches impermeable rock. Water then begins to fill, or saturate, the spaces in the rock above the impermeable layer. As more water soaks into the ground, its level rises underground, creating the lower of the two zones of groundwater—the **zone of saturation.** The zone of saturation, as shown in Figure 14–3, is the layer of ground where all the pores are filled with water. The upper surface of the zone of saturation is called the **water table.**

The zone that lies between the water table and the earth's surface is called the **zone of aeration.** The zone of aeration is composed of three regions. The uppermost region holds soil water—water that forms a film around grains of topsoil. The bottom region, just above the water table, is the **capillary fringe.** Water is drawn up from the zone of saturation into the capillary fringe by capillary action. Capillary action is due to the attraction of water molecules to other materials, such as soil. When you use a paper towel to soak up a spill, you are relying on capillary action to draw moisture into the towel. Between the region of soil water and the capillary fringe is a middle region that normally remains dry except during rainfalls. If this middle region remains wet for an extended period of time, would you expect capillary action in the capillary fringe to increase or decrease?

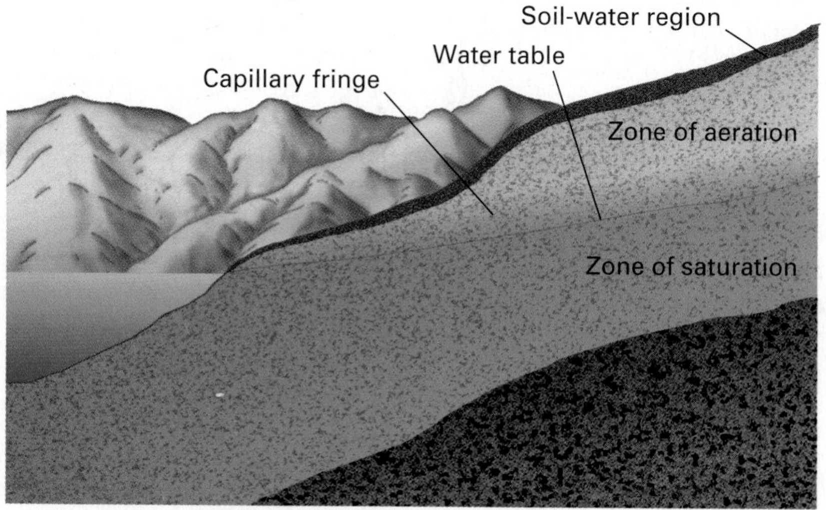

Figure 14–3. The diagram on the left illustrates the moisture zones beneath the earth's surface.

Movement of Groundwater

The permeability of the rocks in the zone of saturation influences the flow of groundwater. Like water on the earth's surface, groundwater flows downward in response to gravity. For example, it flows away from divides toward the lower-lying stream valleys.

The rate at which groundwater flows through the zone of saturation depends on the permeability of the aquifer and the *gradient* of its water table. Gradient is the steepness of any slope. Water passes quickly through highly permeable rock but slows as the permeability of the rock decreases. The velocity of a stream increases as its gradient increases. In like manner, the velocity at which groundwater moves increases as the gradient of the water table increases.

Topography and the Water Table

The water table is found at different depths below the ground surface. The depth depends on the topography of the land, the permeability of the rock, and the amount of rainfall. Generally, shallow water tables parallel the contours of the land, rising beneath hills and dipping under valleys. During periods of prolonged rainfall, the water table rises. During periods of drought, the water table deepens and flattens as water flows toward stream valleys and is not replaced by new rainfall.

Under most surface areas, there is only one water table. In some areas, however, a layer of impermeable rock lies near the surface above the main water table. This rock layer prevents water from seeping down into the zone of saturation. Water collects on top of this rock layer, creating a second zone of saturation and, thus, a second water table. These secondary water tables, called **perched water tables,** are illustrated in Figure 14–4.

Conserving Groundwater

Groundwater is a major source of water for irrigation in the United States. In many communities, it is the only source of fresh water. Although groundwater is renewable, the supply is limited.

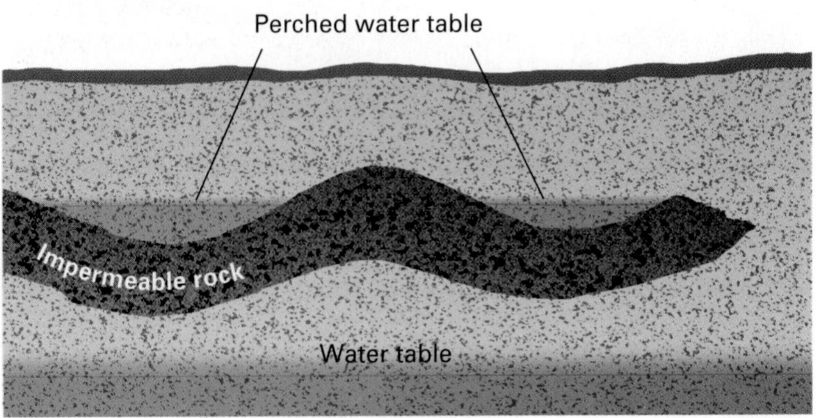

Figure 14–4. A perched water table lies above the main water table.

Figure 14–5. Polluted groundwater can be rid of toxic material in water purification plants such as this one in San José, California.

Groundwater collects and moves slowly, and the water taken from aquifers may not be replenished for thousands of years. Like other natural resources that are limited, groundwater must be conserved.

Groundwater can be polluted in many ways. Waste dumps and leaking underground storage tanks for toxic chemicals are common sources of groundwater pollution. Agricultural and lawn fertilizers and pesticides are other groundwater pollutants. Leaking sewage systems can introduce dangerous biological pollutants into groundwater sources. Also, if too much groundwater is pumped from an aquifer near the ocean, the water table may drop below sea level. The lowered water table allows salt water from the ocean to flow into the aquifer, contaminating the water supply.

Communities can regulate their use of groundwater to help conserve this valuable resource. They can monitor the level of the local water table and discourage excess pumping when the water table falls too low. After water is used, it generally drains into sewers or sewage canals. Because groundwater replenishes itself slowly, some communities recycle used water. Used water is purified and then is pumped back into the ground to replenish the groundwater supply.

Section 14.1 Review

1. What is the difference between the porosity and the permeability of rock?
2. Name and describe the two zones of groundwater.
3. How does the contour of a shallow water table compare with the local topography?
4. Pumping too much water from an aquifer can deplete the supply of groundwater. In what type of location might this practice lead to contamination of the groundwater supply? How?

Section Objectives

- **Describe artesian formations.**
- **Describe two land features produced when groundwater is heated beneath the surface.**

14.2 Wells and Springs

Two common ways through which groundwater comes to the earth's surface are wells and springs. A well is a hole that is dug below the water table and that fills with groundwater. A spring is a natural flow of groundwater to the earth's surface that is found where the ground dips below the water table. Wells and springs are classified into two groups: ordinary and artesian.

Ordinary Wells and Springs

For an **ordinary well** to function properly, it must penetrate into highly permeable sediment or rock below the water table. If the rock is not sufficiently permeable, groundwater cannot flow into the well quickly enough to replenish the water that is withdrawn. If the well is not deep enough, it will dry up when the water table falls below the bottom of the well.

Pumping water from a well creates a cone-shaped depression in the water table around the well, as illustrated in Figure 14–6. This lowered area of the water table is called the **cone of depression.** If a large amount of water is taken from a well, the cone of depression may drop to the bottom of the well, causing the well to go dry. The lowered water table can extend several kilometers away from the well, causing surrounding wells to also go dry.

Ordinary springs are commonly located in rugged terrain because the ground surface often drops below the water table. However, such springs usually do not flow continuously because the water table may drop below the ground surface during dry periods.

A spring that is formed where a perched water table intersects the ground surface is likely to flow continuously. Water filters down through permeable rock until it is stopped by the impermeable rock that forms the perch. At this point the water comes to the ground surface as a spring. Because of the impermeable rock beneath, the level of the zone of saturation can never fall below the ground

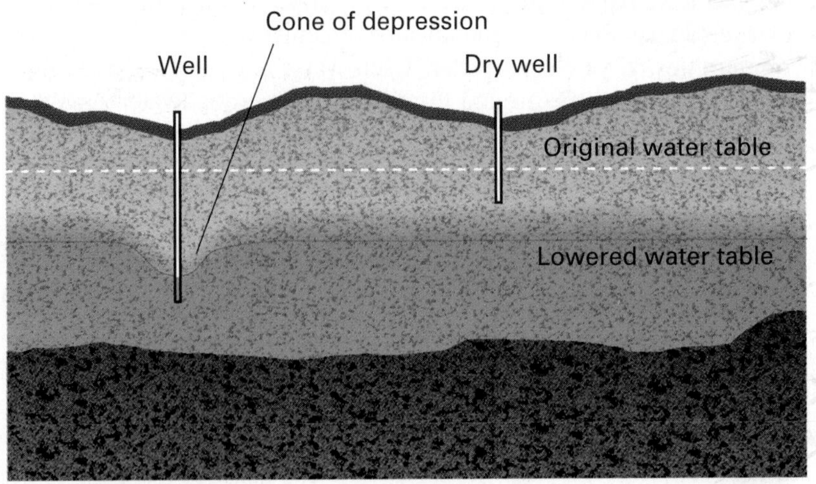

Figure 14–6. A cone of depression develops around a pumping well. Notice that the lowered water table has caused a neighboring well to go dry.

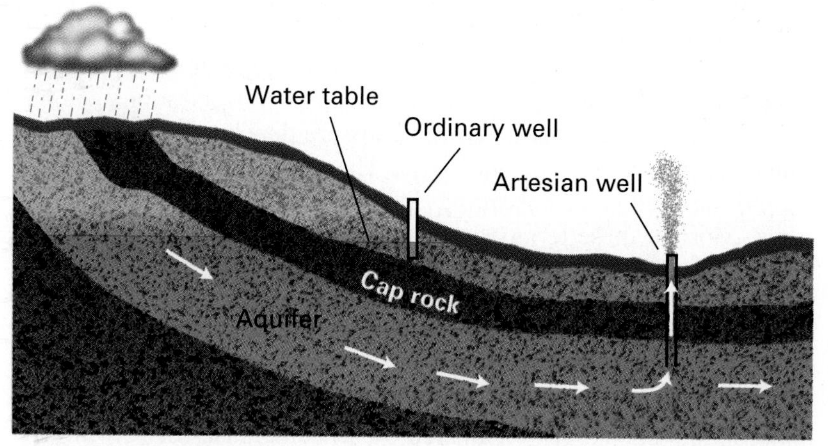

Figure 14–7. The aquifer in this artesian formation dips under the impermeable cap rock. An artesian well must penetrate the cap rock to reach the groundwater in the aquifer.

surface slope. Therefore, the spring is essentially permanent. The spring will disappear only if there is a severe drought that causes the perched water table to completely dry up.

Artesian Wells and Springs

The groundwater that supplies many wells comes from local precipitation. However, one type of well, called an **artesian well,** gets its water from far, possibly hundreds of kilometers, away. An artesian well is one through which water flows freely with no pumping necessary.

The source of water for an artesian well is an arrangement of permeable and impermeable rock called an **artesian formation.** An artesian formation is a sloping layer of permeable rock sandwiched between two layers of impermeable rock and exposed at the ground surface. The permeable rock is the aquifer, and the top layer of impermeable rock is called the **cap rock,** as illustrated in Figure 14–7. Precipitation drains down into the exposed area of the aquifer. As the water flows downward, pressure increases because of the weight of the overlying water. An artesian well is dug through the cap rock to reach the water in the aquifer. Because the water in the aquifer is under pressure, it quickly flows up through the well. The level of the water in an artesian well usually is much higher than the level of the surrounding water table.

In addition to supplying artesian wells, artesian formations are the source of water for some springs. When cracks occur naturally in the cap rock, water under pressure in the aquifer flows through the cracks. This flow forms **artesian springs,** also known as *fissure springs*. An artesian spring is often the source of water in a desert oasis. Why do you think ordinary springs are not the usual source of water in desert oases?

Hot Springs and Geysers

Groundwater is sometimes heated beneath the earth's surface. The water can be heated as it passes through areas where there has been recent volcanic activity or near pockets of molten rock. Hot groundwater that rises to the surface before cooling produces a **hot spring.**

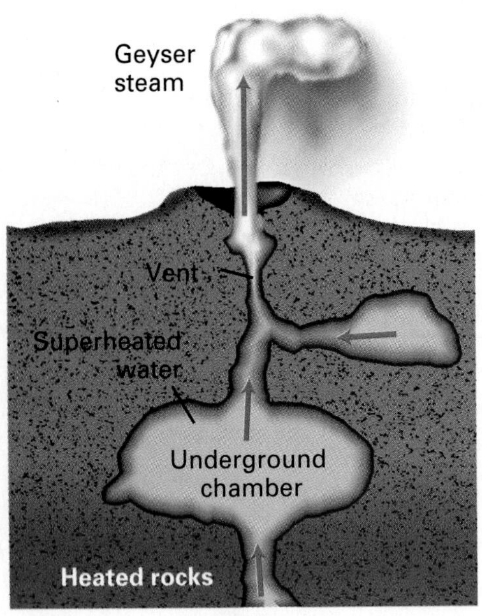

Figure 14–8. The underground chambers and vent of a geyser enable water to become superheated and eventually erupt to the surface.

To be called a *hot spring,* the water must be at least as warm as human body temperature—37°C.

Because greater quantities of minerals dissolve in hot water than in cold water, hot springs often contain large concentrations of minerals. Many of these dissolved minerals are deposited when the water reaches the surface and cools. They frequently accumulate around the mouths of hot springs, forming steplike layers, or terraces. Most of these terraces are made of **travertine,** a form of water-deposited calcite. Travertine is usually white when freshly deposited but turns gray as it weathers. It also may be red, brown, or yellow if iron compounds or other minerals are present in the spring.

Hot springs that form in areas of recent volcanic activity may become **mud pots.** A mud pot forms when the rock surrounding a hot spring is chemically weathered by volcanic gases dissolved in the water. The weathered rock mixes with the hot water to form a sticky, liquid clay that bubbles at the surface. Mud pots sometimes are called *paint pots* because the clay may be highly colored by the minerals or organic material they contain.

Hot springs that erupt periodically are called **geysers.** Some geysers erupt from surface pools and shoot up sheets of water and steam. Others erupt through small surface vents, sending narrow columns of water and stream high into the air. Both types of geysers tend to build deposits of the silicate mineral *geyserite*. The water from the geyser evaporates, leaving behind the mineral deposits.

The underground structure of a geyser consists of a narrow vent that leads to the surface and connects with one or more underground chambers, as illustrated in Figure 14–8. Heated groundwater fills both the vent and the chambers. The water at the bottom of the vent is superheated by the hot rocks around it and is under pressure from the water above. Because water boils at a higher temperature when it is under high pressure, the water at the bottom of the vent does not boil. Eventually, the water higher in the vent, which is under less pressure, begins to boil. The boiling water produces steam that pushes the water above it to the surface. Release of the water near the top of the vent relieves the pressure on the superheated water farther down. With the sudden release of pressure, the superheated water changes immediately into steam and explodes toward the surface. The eruption continues until most of the water and steam are emptied from the vent and storage chambers. After the eruption, groundwater begins to collect again and the process is repeated, often at regular intervals.

Section 14.2 Review

1. Why are ordinary springs commonly found in steep terrain?
2. Describe the rock layers in an artesian formation.
3. In what two ways can heated groundwater reach the surface of the earth?
4. What features of geysers cause them to erupt rather than flow to the surface like hot springs? Explain why.

14.3 Groundwater and Chemical Weathering

Section Objectives

- **Explain how caverns and sinkholes form.**
- **Identify the features of karst topography.**

As groundwater passes through permeable rock, it dissolves minerals in the rock. The warmer the rock is and the longer the water travels through the rock, the more minerals the water will dissolve. Water that contains relatively large amounts of dissolved minerals, especially ions of calcium, magnesium, and iron, is called **hard water.** Water that contains few dissolved minerals is called **soft water.**

Dissolved minerals make hard water unacceptable for many common uses. For example, soap added to hard water will not produce suds. Also, many people prefer not to drink hard water because of its metallic taste. What are some household appliances or fixtures that may be damaged by the buildup of mineral deposits from hard water?

Results of Weathering by Groundwater

One way that minerals become dissolved in groundwater is through chemical weathering. As water moves through organic materials and soil, it leaches carbon dioxide and combines with it to form carbonic acid. This weak acid chemically weathers the rock it passes through, breaking down and dissolving the minerals in the rock.

Caverns

Rocks rich in the mineral calcite, such as limestone, are especially vulnerable to chemical weathering. Although limestone is not a porous rock, its layers are usually fractured by vertical and horizontal cracks. As groundwater flows through these openings, carbonic acid slowly dissolves the limestone and enlarges the cracks. When this chemical weathering process enlarges a number of connected cracks and cavities, a cavern forms, as shown in Figure 14–9. A cavern is a large cave, often containing many smaller, connecting chambers. Caverns are common in areas with extensive limestone deposits. The Carlsbad Caverns in New Mexico and Mammoth Cave in Kentucky are excellent examples of large limestone caverns.

Figure 14–9. The formations in the Carlsbad Caverns in New Mexico are made of calcite, a mineral dissolved from limestone.

Sinkholes

During dry periods, the water table is low and caverns are not completely filled with water. When there is no longer support from the water beneath, the roof of a cavern close to the surface of the ground may collapse. This process forms a circular depression called a **sinkhole.** When such sinkholes later fill with water, they become sinkhole lakes.

Stalactites and Stalagmites

A cavern that lies above the water table does not fill with water; however, water still passes through the surrounding rock. When water containing dissolved calcite drips from the ceiling of a limestone cavern, some of the calcite solidifies on the ceiling. As this calcite builds up, it forms a suspended, cone-shaped deposit called a **stalactite** (stuh-LAK-TITE). When drops of water fall on the cavern

Disappearing Land

Hundreds of small ponds and lakes dot central Florida's landscape. Their formation is no mystery to residents of Winter Park, Florida, who witnessed the ground beneath a three-story house collapse, creating a hole 30 m deep. Within a few days, the hole expanded to 100 m across—the size of a football field. It swallowed up several houses, part of a four-lane highway, a swimming pool, a parking lot, five cars, and a truck. The cost of the damage from this cave-in reached about $2 million.

The catastrophe that struck Winter Park actually started thousands of years ago. Over the years, groundwater flowing beneath central Florida slowly dissolved the surface layer of limestone. So much limestone eroded away that giant underground caverns formed.

When groundwater fills a cavern, it takes the place of rock that has dissolved. Groundwater provides the pressure that helps to support the limestone above the cavern. However, if the water level drops, the ground above the cavern loses its support and caves in, forming a sinkhole.

This is what happened in Winter Park after a two-year drought. With no precipitation to replace the groundwater used by Winter Park residents, the caverns beneath the town remained empty and the surface collapsed.

During the same week, eight more sinkholes appeared in Winter Park.

What measures might Winter Park residents take to help prevent more sinkholes from forming?

floor, deposited calcite builds up to form a cone called a **stalagmite** (stuh-LAG-MITE). Often a stalactite will grow downward and a stalagmite will grow upward until they meet, forming a continuous column of calcite.

Natural Bridges

When the roof of a cavern collapses in several places, a relatively straight line of sinkholes is created. The uncollapsed rock between each pair of sinkholes forms an arch of rock called a **natural bridge.**

A natural bridge also can form when a surface river enters a crack in a rock formation, runs underground, then reemerges. As shown in Figure 14–10, the river erodes the rock it passes through, enlarging the opening, and eventually forms a natural bridge.

Figure 14–10. The eroding effect of an underground river produced this natural bridge in Utah.

Karst Topography

Regions where the effects of the chemical weathering due to groundwater are clearly visible at the surface are said to have **karst topography.** This topography is named after the scenic Karst region of Croatia, where this type of topography is well developed. The features of these areas include many closely spaced sinkholes, caverns, and streams that disappear into fissures in the rock. These streams then emerge in caves or through other fissures many kilometers away. In the United States, karst topography is found in Kentucky, Tennessee, southern Indiana, northern Florida, and Puerto Rico.

Perhaps the most spectacular karst topography on earth is on the Vogelkop Peninsula on the tropical island of New Guinea. Sinks more than 300 m deep have formed so close together that they are separated only by razor-sharp ridges. The terrain forms a pattern like that left in dough cut out by a biscuit cutter.

Features characteristic of karst topography also can be found in some arid regions. In such areas, the water table is far below the surface, and groundwater must pass through a thick zone of aeration. As it seeps downward, the groundwater may wash out fine grains from weakly cemented sediments. Small pipes form and extend outward from drainage gulleys. In some areas of the Badlands of South Dakota, for example, these pipes extend for many meters.

When the pipes collapse, sinkholes and caverns like those of areas with karst topography are formed. However, the pipes never extend below the water table, as do many sinkholes in limestone regions with true karst topography.

Section 14.3 Review

1. What is the effect of carbonic acid on rocks?
2. How are caverns formed?
3. What is the difference between a stalactite and a stalagmite?
4. Why might you expect to find springs in regions with karst topography?

Chapter 14 Review

Key Terms

- aquifer (261)
- artesian formation (267)
- artesian spring (267)
- artesian well (267)
- cap rock (267)
- capillary fringe (263)
- cone of depression (266)
- geyser (268)
- hard water (269)
- hot spring (267)
- impermeable (263)
- karst topography (271)
- mud pot (268)
- natural bridge (271)
- ordinary spring (266)
- ordinary well (266)
- perched water table (264)
- permeability (262)
- porosity (261)
- sinkhole (270)
- soft water (269)
- sorting (261)
- stalactite (270)
- stalagmite (271)
- travertine (268)
- water table (263)
- zone of aeration (263)
- zone of saturation (263)

Key Concepts

The porosity and permeability of a rock or sediment determine whether it is an efficient aquifer. **See page 261.**

There are two zones of moisture beneath the ground. **See page 263.**

The contour of shallow water tables is influenced by the local topography. **See page 264.**

Pollution can make the use of groundwater dangerous to living organisms. **See page 265.**

Artesian formations are the source of artesian wells and springs. **See page 267.**

Hot springs and geysers result when groundwater is heated. **See page 267.**

Caverns and sinkholes form as a result of chemical weathering. **See page 269.**

Karst topography demonstrates the effects of groundwater as a chemical weathering agent. **See page 271.**

Review

On your own paper, write the letter of the term that best completes each of the following statements.

1. Any body of rock through which water can flow and which can store enough water for domestic or industrial use is called
a. a well. b. an aquifer. c. a sinkhole. d. an artesian formation.

2. The total volume of the open spaces in a rock is called its
a. viscosity. b. permeability. c. capillary fringe. d. porosity.

3. When all the particles in a sediment are about the same size, the sediment is said to be
a. fractured. b. well sorted. c. permeable. d. poorly sorted.

4. The ease with which water can pass through a rock or sediment is called
a. permeability. b. porosity. c. carbonation. d. velocity.

5. The underground layer of rock where all of the open spaces are filled with water is called the
a. zone of aeration. b. cap rock. c. water table. d. zone of saturation.

6. The upper surface of the zone of saturation is called the
a. cap rock. b. water table. c. gradient. d. travertine.

7. The slope of a water table is its
a. gradient. b. permeability. c. porosity. d. aquifer.

8. It takes a long time for polluted groundwater to become pure again because
   a. groundwater can only be replaced artificially.
   b. groundwater is replenished slowly.
   c. groundwater travels very quickly.
   d. groundwater can never be replaced.
9. A natural flow of groundwater that has reached the surface is
   a. a spring. b. an aquifer. c. a well. d. a travertine.
10. Pumping water from a well causes a local lowering of the water table known as a
   a. cone of depression. b. horizontal fissure.
   c. hot spring. d. sinkhole.
11. Hot springs that are formed in areas of recent volcanic activity may become
   a. caverns. b. natural bridges.
   c. geysers. d. mud pots.
12. Travertine usually forms
   a. terraces. b. natural bridges.
   c. sinkholes. d. caverns.
13. When part of the roof of a cavern collapses, the result is a
   a. sinkhole. b. stalactite. c. horizontal fissure. d. geyser.
14. Calcite formations suspended from the ceiling of a cavern are called
   a. stalagmites. b. stalactites. c. sinks. d. aquifers.
15. Regions where the results of chemical weathering by groundwater are clearly visible are said to have
   a. sink topography. b. karst topography.
   c. limestone topography. d. artesian formations.

Critical Thinking

On your own paper, write answers to the following questions.

1. A rock can be porous, yet impermeable. Explain how.
2. In areas where the water table is at the surface of the land, what type of terrain would you expect to find?
3. Describe an artesian formation and explain how the water in an artesian well may have entered the ground many hundreds of kilometers away.
4. Explain how a mud pot forms. Why are mud pots sometimes called paint pots?
5. Explain the process that results in the formation of stalactites and stalagmites. Can you think of another process in nature that produces shapes similar to stalactites?
6. Do you think an area with karst topography would have many or few surface streams? Explain your answer.

Application

1. You are having an unusually dry summer. What effect would you expect this dry season to have on the capillary action in the soil?
2. Your city has been experiencing shortages in its water supply. What would be your advice on replenishing the groundwater artificially? What are your reasons for making this suggestion?
3. Construct a **concept map** using 10 of the new terms listed on the previous page by making connections that illustrate the relationship among the terms.

Extension

1. Research the source of the fresh water used in your community. How much of it is groundwater? Are there times of the year when the water table in your area drops low enough that compulsory water conservation measures are enacted? Write a report that summarizes your findings.
2. Garbage, trash, and other solid waste products often are buried in sanitary landfills. Water that seeps into the wastes can carry soluble pollutants into the groundwater. Find out how solid wastes are disposed of in your community. What measures are taken to protect the groundwater supply from contamination? How is the groundwater tested to verify its purity?

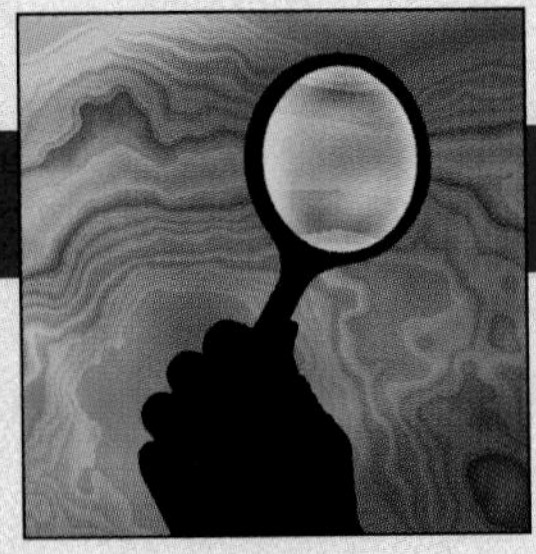

In-Depth Investigation

Porosity

Materials

- beaker (100 mL)
- plastic beads (4 mm)
- plastic beads (8 mm)
- graduated cylinder (100 mL)

Introduction

Whether soil is composed of coarse pieces of rock or very fine particles, there is always some empty space between the pieces of solid material. The empty space is called pore space. Porosity is calculated by dividing the volume of the pore space by the total volume of the soil sample. Thus, if 50 cm^3 of soil contains 5.0 cm^3 of pore space, then the porosity of the soil sample is $5.0\ cm^3/50\ cm^3 = 0.10 = 10\%$. The result is generally written as a percentage.

In this investigation, you will measure and compare the porosity of three samples that represent rock particles.

Prelab Preparation

1. Review Section 14.1, pages 261–262.
2. What is *meniscus?*

Procedure

1. Use a graduated cylinder and water to determine the volume of the beaker. Record the volume.
2. Dry the beaker and fill it to the top with large (8-mm) plastic beads. Gently tap the beaker to settle and compact the beads. Add more beads as needed to fill the beaker until the beads are level with the top. Do the beads represent well-sorted large rock particles, well-sorted small rock particles, or unsorted rock particles? What is the total volume of the beads, including the pore space?

3. Fill the graduated cylinder with water to the top mark and note the position of the water level. Carefully pour the water from the cylinder into the beaker with the large beads until the water level just reaches the top of the beads. Check the new water level in the graduated cylinder. How much water did you add? What is the volume of the pore space?
4. Calculate the porosity of the beads. Record the porosity as a decimal and as a percentage.
5. Repeat Steps 2–4 using the small (4-mm) plastic beads. Do the beads represent well-sorted large rock particles, well-sorted small rock particles, or unsorted rock particles? What is the volume of the pore space between the small plastic beads? What is the porosity of the small beads?
6. Drain and dry both sets of beads. Mix together equal volumes of the small and large beads. Repeat Steps 2–4 using the mixed-size beads. Do the beads represent well-sorted or unsorted rock particles?

Analysis and Conclusions

1. Compare the porosity of the large beads with the porosity of the small beads.
2. Does porosity depend upon particle size? Explain.
3. What effect did mixing the bead sizes have on the porosity? Explain the effect.

Extensions

1. Each of the three graphs shown below represents one of the following properties plotted against particle size: porosity, permeability, and capillarity. Particle size is plotted on the vertical axes, increasing from bottom to top. Porosity, permeability, and capillarity are plotted on the horizontal axis, increasing from left to right.
   a. Which graph represents porosity? Explain.
   b. Which graph represents permeability? Explain.
   c. Which graph represents capillarity? Explain.
2. What would be the effect on the porosity of coarse gravel if it were mixed with fine sand? Conduct an experiment to find out if your answer is accurate.

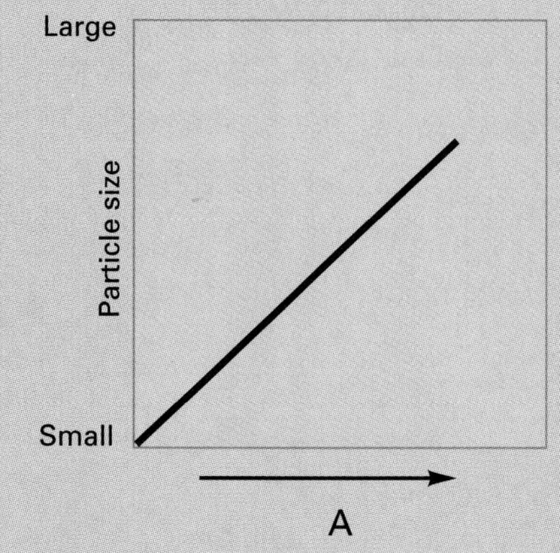

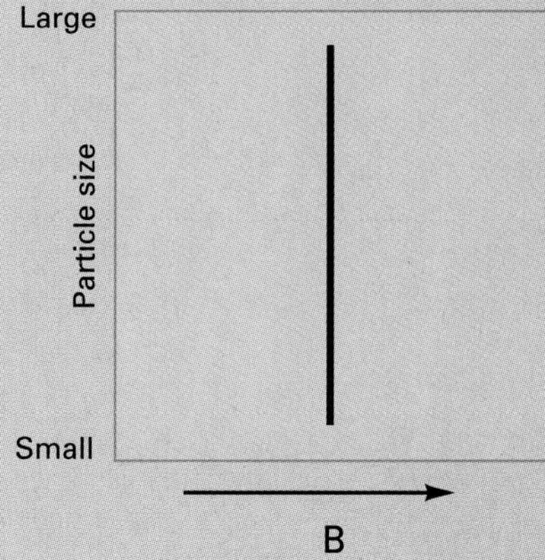

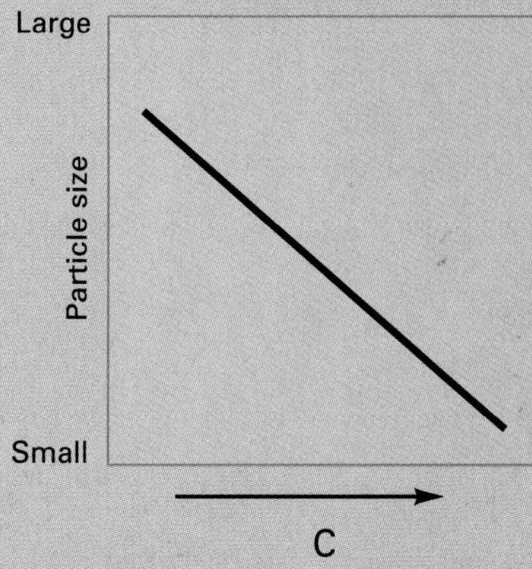

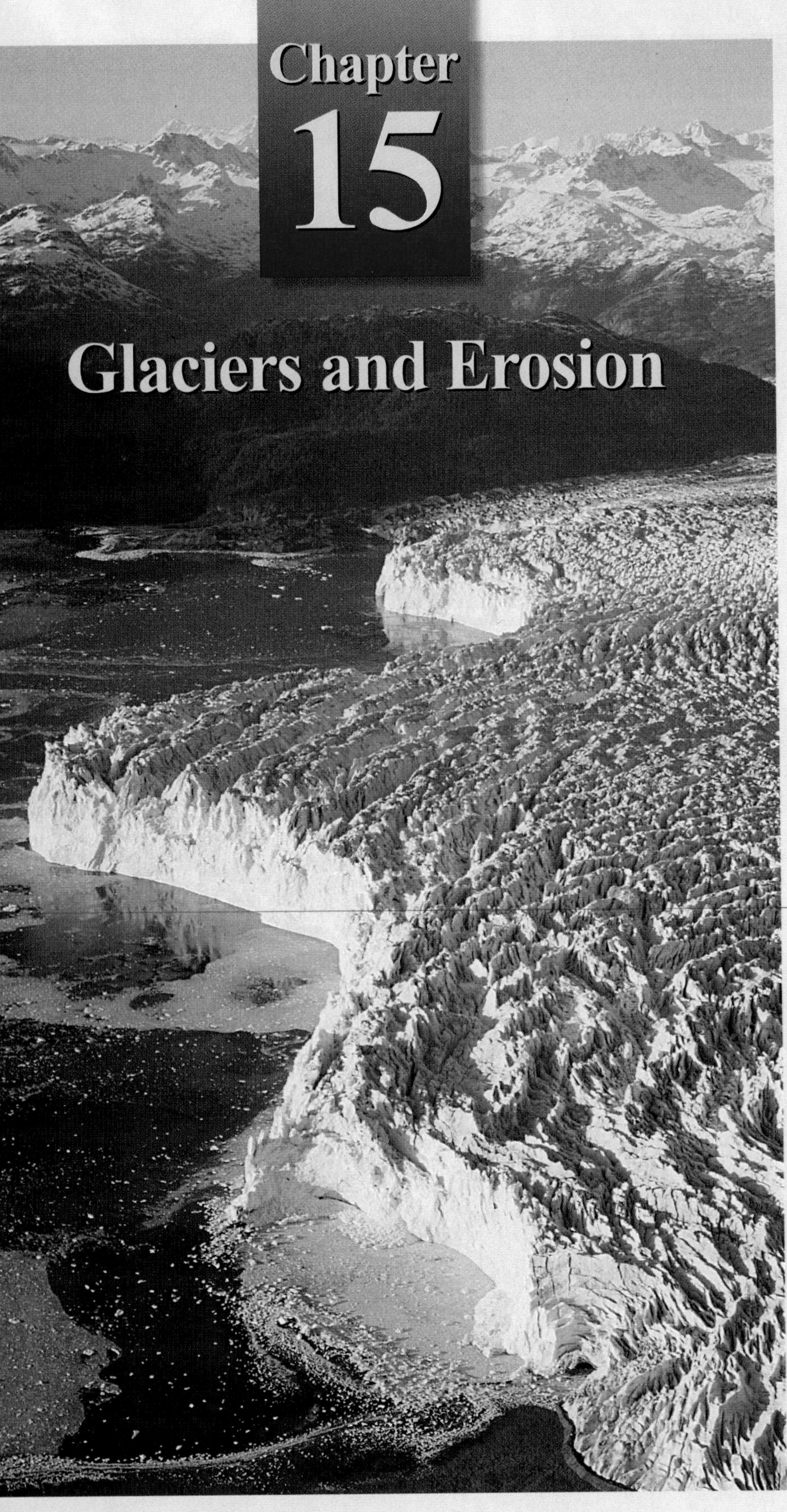

Chapter 15

Glaciers and Erosion

Massive sheets of moving ice once covered much of the earth's land area. The ice advanced and retreated many times during the earth's geologic history. Scientists are not sure why this occurred, but they have proposed several theories.

Glaciers still exist in some parts of the world, but worldwide evidence of glaciers tells us that they were once much more widespread.

In this chapter, you will learn to recognize the signatures left by these powerful reshapers of the earth's surface.

Chapter Outline

The Columbia Glacier flows into Prince William Sound, Alaska.

15.1 Glaciers: Moving Ice

Section Objectives

- **Describe how glaciers form.**
- **Compare two main kinds of glaciers.**
- **Explain two processes by which glaciers move.**

A single snowflake is lighter and more delicate than a feather. However, you can gather a handful of soft snow and squeeze it to make a firm snowball. In a process similar to the compacting of snow to make your snowball, natural forces compact snow to create enormous masses of moving ice. These masses of moving ice, called **glaciers,** are powerful agents of erosion. In fact, they have carved some of the most spectacular features on the earth's surface.

Formation of Glaciers

Water that reaches the earth's surface as rain quickly runs off and eventually flows to the ocean or seeps into the earth as groundwater. However, at high elevations and in polar regions, moisture often reaches the surface as snow and may remain on the ground all year. The **snowline** is the elevation above which ice and snow remain throughout the year. The elevation of the snowline varies from place to place. At the equator, it occurs at about 5,500 m above sea level; near the poles, the snowline is near sea level. Figure 15–1 shows the snowline in the Colorado Rocky Mountains.

A **snowfield,** or ice field, is an almost motionless mass of permanent snow and ice. Snowfields are formed by an accumulation of ice and snow above the snowline. Snowfields cover most of the land near the poles and the tops of some mountains at lower latitudes.

Because the average temperature at high elevations and in polar regions is always near or below the freezing point of water, the snow accumulates year after year. The partial melting and refreezing that periodically occur changes the snow crystals into small grains of ice. This grainy ice is called **firn,** from a German word that means "of last year."

In the deepest layers of accumulated snow, the pressure of the overlying layers becomes so great that the ice grains flatten and the air between the firn is squeezed out. The firn then loses its white

Figure 15–1. The snowline in Colorado's Rocky Mountains is at about 2,800 m.

Figure 15–2. A valley glacier descends from a snowfield on this Alaskan mountain (left). An enormous snowfield in Tongass National Forest, Alaska, covers most of the mountaintop (right).

color and becomes a bright steel-blue color that is characteristic of glacial ice. A glacier is formed when snow and ice accumulate to a great enough thickness that the ice starts to move downslope due to gravity. Glacial formation is greatest in regions where temperatures are low and snowfall amounts are high.

The growth of a glacier depends on the balance between snowfall received and ice lost by melting and evaporation. As long as new snow is added faster than it melts, evaporates, or breaks off at the sea as icebergs, the glacier will continue to increase in size. When the ice disappears faster than snow is added, the glacier will decrease in size. Very slight differences in average yearly temperatures and snowfall may upset the balance between new snow and ice loss. Thus, the increase or decrease in the size of a glacier may be an indicator of annual climatic change.

Types of Glaciers

There are two main types of glaciers; they are distinguished by their size and where they are formed. One type of glacier is formed in mountainous areas. As the ice moves down a valley, it produces a **valley glacier,** which is a long, narrow, wedge-shaped mass of ice. Valley glaciers are best developed in the high mountain regions of the world such as in coastal Alaska, the Himalayas, the Andes, the Alps, and New Zealand.

The other type of glacier covers large land areas. These masses of ice, called **continental ice sheets,** occupy millions of square kilometers. Today, continental ice sheets are found only in Greenland and Antarctica.

Antarctica is covered by the largest continental ice sheet in the world. This ice sheet is one and a half times as large as the area of the mainland United States. In some places, the Antarctic ice sheet is more than 4,000 m thick.

About 90 percent of Greenland also is buried under a continental ice sheet. The Greenland ice sheet is over 3,000 m at its maximum ice thickness. The amount of water contained in the two great ice

INVESTIGATE!

To learn more about glaciers, try the In-Depth Investigation on pages 294–295.

sheets of Greenland and Antarctica is enormous. Scientists estimate that if melted, the ice would release enough water to raise world-wide sea level by more than 60 m. What effect would this rise in sea level have on coastlines and low islands?

Movement of Glaciers

Glaciers are sometimes called *rivers of ice*. A glacier, however, moves very differently from the water in a river. Water in a river and ice in a glacier both move downward in response to gravity. However, the ice in glaciers cannot move rapidly or flow easily around barriers as can water in a river. On average, a glacier moves about 100 m per year. Some glaciers may travel only a few centimeters per year, while others may move a kilometer or more.

Glacial motion has been a subject of much scientific study. Most scientists agree that glaciers move by two basic processes—**basal slip** and **internal plastic flow.**

Basal Slip

The weight of the ice in a glacier eventually exerts enough pressure to melt the ice where it comes in contact with the ground. The water from the melted ice acts as a lubricant between the base of the ice and the underlying rock. This allows the glacier to move forward by slipping at its base.

Basal slip also allows a glacier to work its way over small barriers in its path by melting and then refreezing. For example, if the ice pushes against a rock barrier, the pressure causes some of the ice to melt. The water from the melted ice then flows around the barrier and freezes again as the pressure is removed.

Internal Plastic Flow

Glaciers also move by internal plastic flow, in which solid ice crystals slip over each other, causing a slow forward motion. In studying glacial motion, scientists have driven stakes across a valley glacier, as shown in Figure 15–3. They found that the speed of internal plastic flow is not the same at all parts of the glacier. The rate of motion

Figure 15–3. A line of blue stakes driven into a valley glacier moves to the position of the red stakes as the glacier flows. Notice that the center of the glacier moves faster than the sides.

at a given point is determined by the slope and by the thickness and temperature of the ice. The internal plastic flow is faster nearer to the surface of a glacier than near its base. Friction caused by contact of the glacier ice with the rock surface slows the flow rate. The center of the glacier moves faster than its sides. Why do you think this occurs?

Features of Glaciers

Due to the pressure from the weight of ice above, the interior of the glacier moves by internal plastic flow. Pressure also causes the ice to melt and refreeze, enabling the base of the glacier to slip. The surface ice, meanwhile, remains brittle. Because the glacier flows unevenly beneath the surface, points of tension and compression build on the brittle surface, which then buckles and cracks.

Large cracks, called **crevasses** (krih-VASS-uz), form in the brittle surface of a glacier. Crevasses may be more than 30 m deep. They may be hidden by a thin crust of snow that breaks under the slightest weight. Traveling over the top of a glacier is very dangerous and should be attempted only by experienced climbers.

SCIENCE & TECHNOLOGY

Monitoring Ice Streams in Antarctica

▲ Ice streams could one day cause the West Antarctic Ice Sheet to break apart if the recent warming trend continues.

Some climatologists warn that the earth's average surface temperature is rising at a dangerous rate. These scientists argue that human activities such as burning fossil fuels continue to raise the level of carbon dioxide in the atmosphere and could change the earth's climate drastically within the next century. Other climatologists see the recent warming trend as a natural fluctuation in the earth's climate.

Because the retreat of the world's glaciers is often cited as a sign of global warming, scientists on both sides of the debate want to understand more about glaciers. In the last century, thousands of small glaciers have shrunk, causing a significant rise in sea level. However, changes in the larger glaciers, or continental ice sheets, are much more difficult to interpret.

To explore the effects of climate on continental ice sheets, scientists studying the Antarctic ice sheet are testing a technique called *radar interferometry.* With this technique, the glaciologists hope to monitor ice streams, which carry ice from the interior of the ice sheet to the sea.

Radar interferometry involves comparing two radar images of the same area taken a few days apart. In the Antarctica study, images of an ice stream taken six days apart were slightly different because of the

A continental ice sheet usually has a center of snow and ice accumulation. Ice sheets move outward in all directions from the center. For example, the ice sheets that now cover Antarctica and Greenland move outward toward the shores of their landmasses. Along some parts of the coast of Antartica, the ice sheet has moved out over the ocean, forming ice shelves. The largest of these shelves, the Ross Ice Shelf, covers an ocean area the size of Texas. The rise and fall of the tides breaks off large pieces of the leading edge of this ice. These large blocks of ice, called *icebergs*, drift into the ocean. Because most of an iceberg is below the surface of the water, icebergs pose a hazard to ships. One of the largest icebergs ever observed in the Antarctic was twice the size of Connecticut.

Section 15.1 Review

1. In what regions does snow accumulate year after year?
2. What main type of glacier covers Greenland?
3. Explain the type of glacial movement known as *internal plastic flow*.
4. Compare and contrast valley glaciers and continental ice sheets.

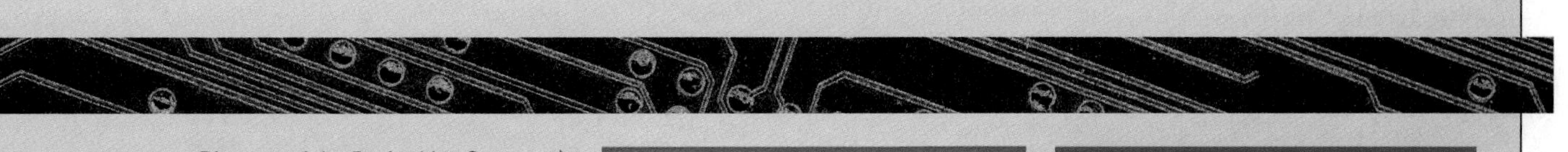

Diagram of the Rutford Ice Stream (left) and the interference pattern created by successive radar images (right)

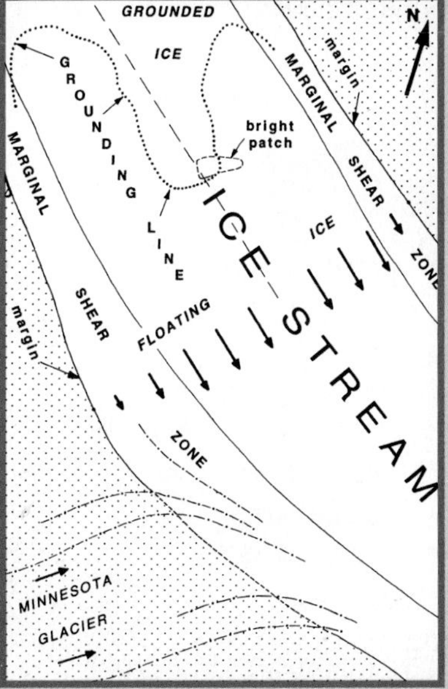

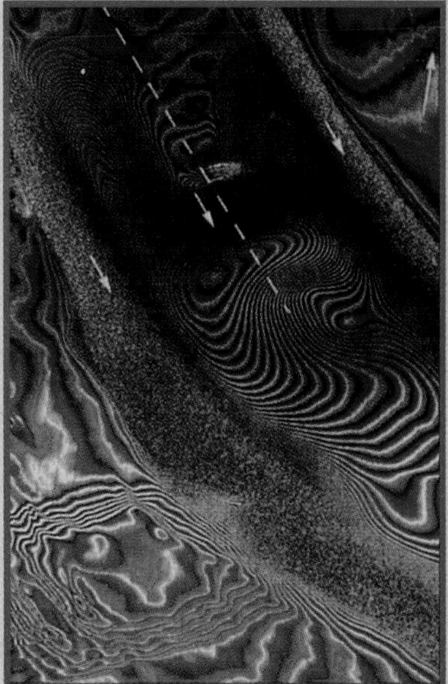

flow of ice in the stream. When combined, the two images create a pattern of interference where differences occur. This pattern is then color-coded in a sequence from blue to yellow to red. By counting the number of colored bands produced by the two Antarctic images, scientists were able to calculate the speed of the ice stream.

Periodically checking the flow rate of ice streams will help scientists judge whether large sections of ice are likely to break away from the main ice sheet. Rapid upstream movement of the ice streams' headwaters could even signal the collapse of the entire West Antarctic Ice Sheet. Interferometry provides glaciologists with a way to watch for such changes without actually working in the harsh, frigid environment of an ice sheet.

Why is it important to monitor ice streams in the Antarctic ice sheet?

15.2 Landforms Created by Glaciers

Section Objectives

- **Describe the landscape features that are produced by glacial erosion.**
- **Name and describe five features formed by glacial deposition.**
- **Explain how lakes are formed by glacial action.**

Both valley glaciers and continental ice sheets are powerful agents of erosion. Valley glaciers carve out rugged features in the mountains through which they flow. Massive continental ice sheets produce a smooth landscape as they plane off all existing land features except the highest mountains.

Glacial Erosion

The glacial processes that change the shape of mountains begin in the upper end of the valley where a valley glacier forms. As a glacier wedges its way through a narrow valley, it breaks off rock from the valley walls, causing the walls to become steeper. The moving glacier also pulls blocks of rock from the floor of the upper valley. These actions create a bowl-shaped depression called a **cirque** (SURK), illustrated in Figure 15–4. This French word, meaning "circus," refers to the resemblance of the depression to a round circus theater. Sharp and jagged ridges form between the cirques. These sawtoothed ridges are called **arêtes** (uh-RATES), which means "spines" in French. Sometimes, several arêtes are joined and form a sharp, pyramid-like peak called a **horn.**

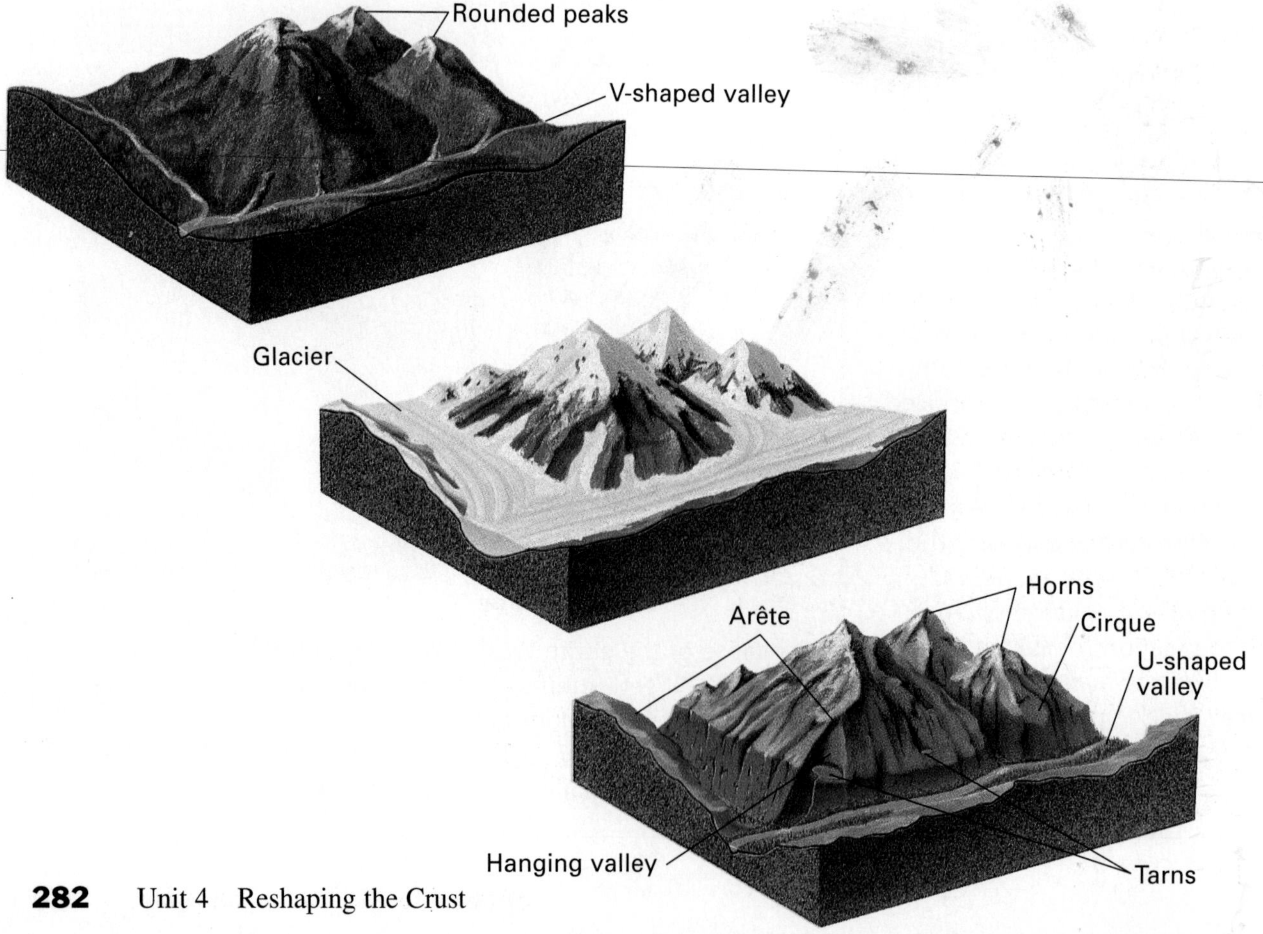

Figure 15–4. This diagram illustrates the features created by glacial erosion. The upper diagram, before glaciation, shows V-shaped valleys cut by young streams. Notice in the middle diagram that the glacier is eroding the mountains to form cirques, arêtes, and horns. In the lower diagram, small lakes, or tarns, fill the cirques.

Figure 15–5. Notice the rugged landscape of this typical U-shaped glaciated valley in the Mackenzie Mountains in Canada.

As the valley glacier flows down through an existing valley, it picks up large amounts of rock. These rock fragments, which range in size from microscopic pieces to large boulders, become embedded in the ice.

Solid rock over which a glacier has moved is often polished by the scraping action of the tiny rock particles embedded in the ice. Larger rocks carried by the ice may gouge out scratches or deeper grooves in the bedrock. Large rock projections may be rounded by a passing glacier. These rounded projections usually have a smooth, gently sloping side facing the direction from which the glacier came. The other side is steep and jagged because rock is pulled away as the ice passes over. The resulting rounded knobs of rock are called **roches moutonnées** (RAWSH-MOOT-un-AZE), which in French means "sheep rocks," because they resemble the backs of sheep.

As the valley walls and floor are scraped away by a glacier, the original stream-cut V-shape of the valley is changed into a glacial U-shape. Because glacial action is the only means by which a valley can acquire a U-shape, scientists can easily tell whether a valley has been glaciated.

Small tributary glaciers in adjacent valleys often flow into a main valley glacier. Because a smaller tributary glacier has less ice and cutting power than the main valley glacier, its U-shaped valley is not cut as deep into the mountains. When the ice melts, the smaller tributary glacial valley is left suspended on the mountain high above the main valley floor and is called a **hanging valley.** What natural feature do you think will appear when a stream flows from a hanging valley to the main glacial valley?

The landscape created by massive continental ice-sheet erosion is quite different from the sharp, rugged features shown in Figure 15–5, which were eroded by valley glaciers. Continental ice sheets erode by leveling landforms, producing relatively smooth, rounded landscapes.

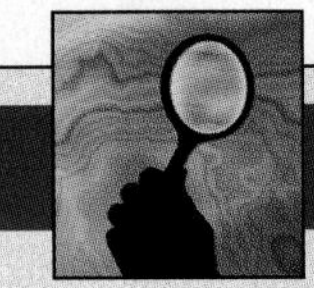

Small-Scale Investigation

Glacial Erosion

Many of the earth's natural features have been shaped by the movement of glaciers. You can demonstrate the effects of glacial erosion with a simple model.

Materials

plastic container (about 15 × 10 × 5 cm); mixture of sand, gravel, and small rocks; water; freezer; modeling clay (1–2 lb.); hand towel; sand (1–2 lb.); flat box (about 30 × 20 cm); soft wood board (about 30 cm long); rolling pin or large dowel

Procedure

1. Put the sand, gravel, and rock mixture in the bottom of the plastic container. Fill the container with water to a depth of about 4 cm. Freeze the container until the water is solid. Remove the ice block from the container.
2. Use a rolling pin or large dowel to flatten the modeling clay into a rectangle about 20 × 10 × 1 cm. Grasp the ice block firmly with the hand towel. Place the block with the gravel-and-rock side down at one end of the clay. Press down on the ice block and move it along the length of the flat clay surface. Sketch the pattern made in the clay by the ice block.
3. Next, press damp sand into the bottom of a shallow rectangular box. As in Step 2, move the ice block along the surface of the sand while pressing down lightly.

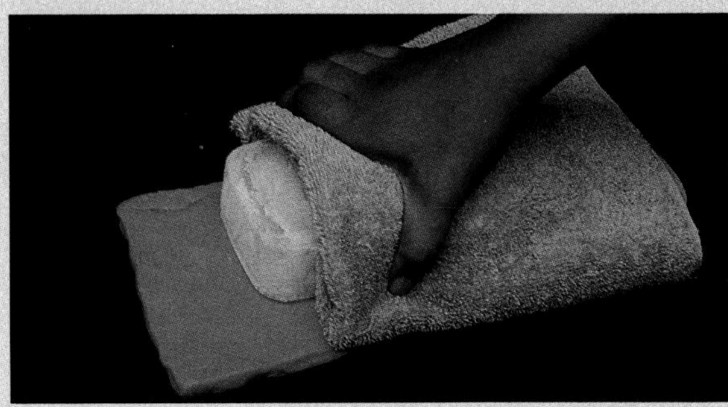

4. Repeat the process outlined in Steps 2 and 3 using a soft wooden board in place of the clay and sand.

Analysis and Conclusions

1. Describe the effects of moving the ice block over the clay, sand, and board.
2. Did any material from any surface become mixed with material from the ice block? Did the ice deposit material on any surface?
3. What glacial land features are represented by the features of your clay model? your sand model? your wood model?
4. Based on your observations, predict the results of glacial erosion on rock.

Existing valleys may be gouged out and deepened. Exposed rock surfaces are smoothed and rounded by continental ice sheets, just as bulldozers flatten landscapes. Rock surfaces also are scratched and grooved by rocks carried at the base of the ice sheet. These scratches and grooves run parallel to the direction of glacial movement.

Glacial Deposition

The ice of a glacier melts when a valley glacier reaches lower elevations or when a change in climate melts continental ice sheets. When a glacier melts, all of the material accumulated in the ice is deposited. Glacial deposits usually can be identified easily. For example, large boulders, called **erratics,** are transported by glaciers. The composition of erratics usually differs from that of the bedrock over which they now lie.

Various land features are formed by **glacial drift**—the general term for all sediments deposited by a glacier—or by the meltwaters from a glacier. One type of glacial drift, called **till,** is made of unsorted deposits of rock material. Till is deposited from sediments scraped off by the base of the glacier or is left behind when glacial ice melts. Another type of glacial drift is called **stratified drift.** Stratified drift is material that has been sorted and deposited in layers by streams flowing from the melted ice, or **meltwater.** Many transported soils originate as glacial drift, such as those found in Iowa and Illinois.

Figure 15–6. The dark line of unsorted rock material shown above is a medial moraine on Aletch Glacier in Switzerland.

Till Deposits

Landforms made from glacial till are called *moraines.* Moraines are ridges of unsorted rock material on the ground or on the glacier. There are several types of moraines, as illustrated in Figure 15–7. A **lateral moraine** is one that is deposited along the sides of a valley glacier, usually as a long ridge. When two or more valley glaciers join, their adjacent lateral moraines combine to form a **medial moraine.** Medial moraines form the dark stripes on the glacier surface shown in Figure 15–6.

The unsorted material left beneath the glacier when the ice melts makes up the **ground moraine.** Much of the landscape from Ohio west to the Montana Rockies and north into Canada is covered with ground moraine. The soil of ground moraine is often very rocky. An ice sheet may mold ground moraine into **drumlins.** Drumlins are long, low, tear-shaped mounds of till, often found in clusters. The long axes of the drumlins are parallel to the direction of glacial movement.

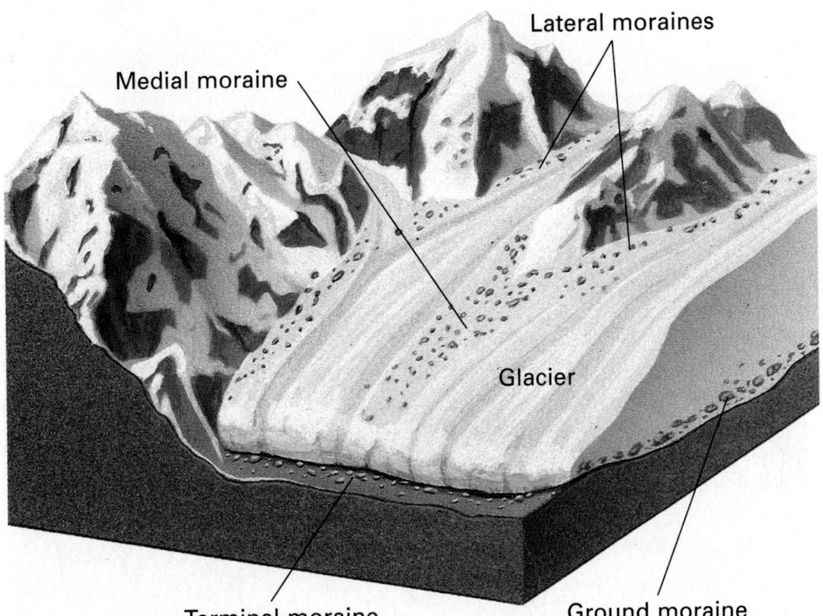

Figure 15–7. This diagram shows the types of moraines formed by glacial deposition.

Till deposited at the snout or front of a melting glacier forms a **terminal moraine.** Terminal moraines are belts of small ridges of till with many depressions that contain lakes or ponds. Large terminal moraines, some more than 100 km long, can be seen across the midwestern states from Minnesota to Ohio, especially south of the Great Lakes. A long terminal moraine runs the full length of New York's Long Island and another forms Cape Cod in Massachusetts.

Stratified Drift Deposition

Melting occurs in a glacier mainly during the summer months. Streams of meltwater flow from the edges, the surface, and beneath the glacier. The glacial meltwater usually appears milky in color due to the presence of very fine rock particles that were ground into fine powder by glacial erosion. Along with the small rock particles, the meltwater carries drift that it deposits as a large **outwash plain** in front of the glacier. An outwash plain is a deposit of stratified drift, which usually lies in front of a terminal moraine and is crossed by many meltwater streams.

Most outwash plains are pitted with depressions called **kettles.** A kettle forms when a portion of glacial ice is buried in drift. As the piece of ice melts, it leaves a cavity in the drift. The drift collapses into the cavity and produces a depression. Kettles often fill with water, forming kettle lakes.

When continental ice sheets recede, long, winding ridges of gravel and coarse sand may be left behind. These ridges, called **eskers** (ESS-kurz), consist of stratified drift deposited by streams of meltwater flowing in ice tunnels within the glaciers. Eskers may extend for tens of kilometers like raised, winding roadways.

Figure 15–8. This kettle lake in Saskatchewan, Canada, was formed by glacial deposition.

Glacial Lakes

Glaciers often form lake basins by eroding out surface areas, leaving depressions in the bedrock and deepening existing valleys. Thousands of lake basins in Canada, the upper midwestern states, New England, and New York State were gouged from solid rock by a continental ice sheet.

Thousands of other glacial lakes were formed by deposition rather than by erosion. Many lake basins were left in the uneven surface of ground moraine deposited by glaciers. Lake basins such as these are found in many glaciated areas of North America and Europe.

Other glacial lakes formed when terminal and lateral moraines blocked existing streams. The area south of the Great Lakes, from Minnesota to Ohio, has belts of moraines and associated lakes. Minnesota, the "Land of 10,000 Lakes," was completely glaciated and has evidence of all the different glacial lake types.

History of the Great Lakes

The Great Lakes of North America are the result of a combination of erosion and deposition by continental ice sheets. Glacial erosion widened and deepened broad river valleys covered by the ice sheets. Moraines to the south blocked off the ends of these valleys. As the ice sheets melted, the meltwater flowed into the valleys and was held there by the moraines to form lakes.

In their early stages, with the melting ice sheet to the north, the lakes had outlets to the south. Most drainage was through the Wabash and Illinois rivers, which flowed into the Mississippi River. Later, as the ice sheets melted, the lakes grew larger and also drained into the Atlantic Ocean through the Susquehanna, Mohawk, and Hudson River valleys.

After the ice age, the land slightly uplifted. The lake beds were also uplifted and slowly shrank to their present size. The upward tilt in the land also caused the lakes to drain to the northeast through the St. Lawrence River. As Lake Erie flowed northeast into Lake Ontario, Niagara Falls were created. Figure 15–9 illustrates the general geologic history of the Great Lakes.

Salt Lakes

Many of the basins found in the southwestern part of the United States contained large lakes during the periods when the continental ice sheets were present to the north. However, many of these lakes had no outlet streams, so water could leave them only by evaporation. When the water evaporated, salt that was dissolved in the water was left behind. This made the water increasingly salty. For example, the Great Salt Lake in Utah is a remnant of a much larger ice-age lake called Lake Bonneville. Today it is the site of a vast salt plain around a much smaller lake. Salt lakes most often form in dry climates, such as the southwestern United States, where evaporation is rapid and precipitation is low.

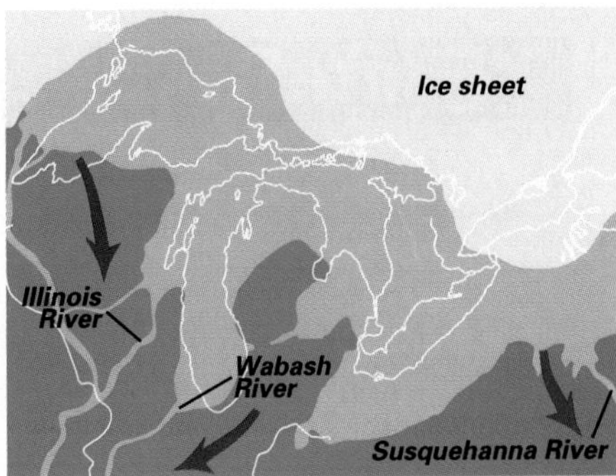

Early Ice Retreat

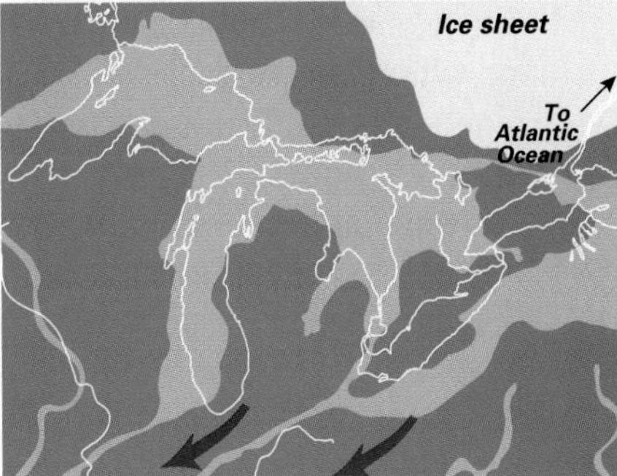

Late Ice Retreat

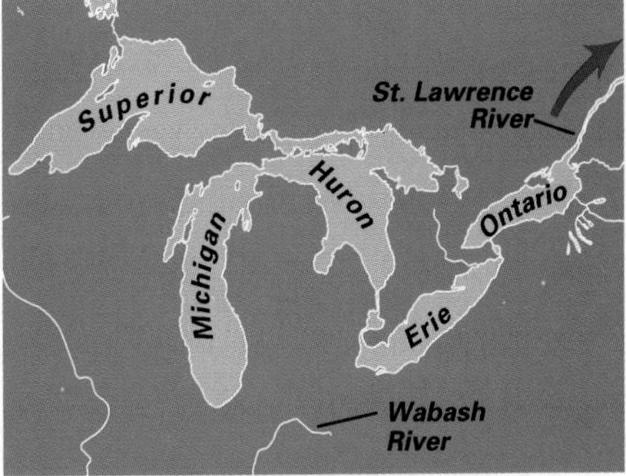

Today's Great Lakes

Figure 15–9. The ice sheet that covered northern North America formed enormous lakes that drained to the south. As the ice sheet retreated, the lakes became smaller and the drainage pattern changed (center). Uplifting of the land reduced the Great Lakes to their present size and established current drainage northeast through the St. Lawrence River (bottom).

EARTHBEAT

Lake Agassiz

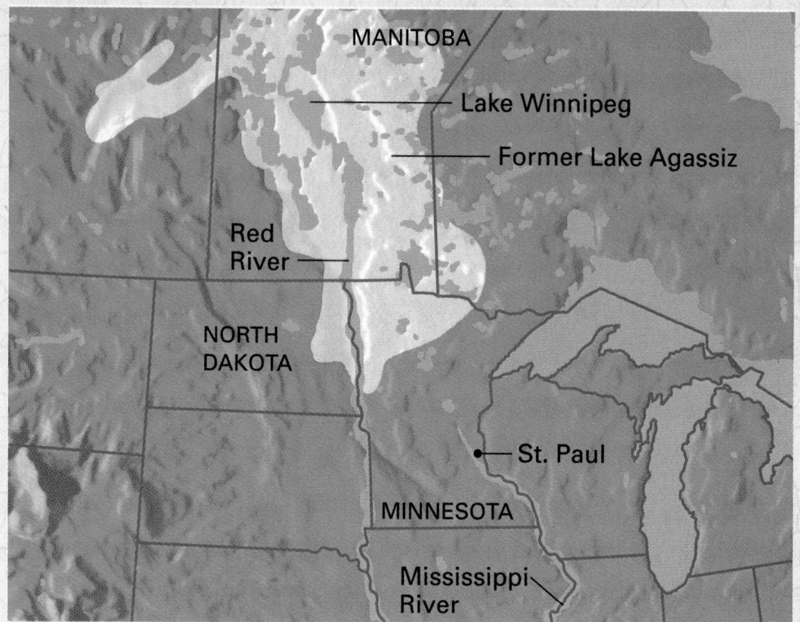

As the ice continued to retreat after the formation of the Great Lakes, a tremendous glacial lake was formed to the northwest of the Great Lakes. The lake was named Lake Agassiz, after Louis Agassiz, a nineteenth-century Swiss scientist. Agassiz became an early and active proponent of the idea of widespread glaciation, based on his extensive observations of land features in the Swiss alpine valleys.

Lake Agassiz developed when an ice dam blocked the northward flow of the glacial meltwater in the Red River Valley. The lake covered parts of North Dakota, Minnesota, and the Canadian province of Manitoba. The water of Lake Agassiz had to flow southward into the Mississippi River where St. Paul is now located.

When the ice block melted, the lake quickly drained, leaving a great fertile plain in its place. Depressions in the lake floor remained filled with water and formed remnant lakes. Lake Winnipeg in Canada is the largest remnant of the former vast Lake Agassiz.

Today a much smaller Red River flows for nearly 950 km through a fertile farmland of wheat fields that now cover this ancient lake bed.

What type of glacial deposit formed the lake bed of Lake Agassiz?

The composition of the water in a salt lake depends on the dissolved minerals it contains. The minerals either are brought in by streams or result from the chemical and biologic processes occurring in the lake. The salt deposits left in the dry beds of ancient salt lakes often contain valuable minerals. For example, borax, which is used as a cleaning material and in the manufacture of glass and steel, is mined from dry lake beds in the Mojave Desert of California.

Section 15.2 Review

1. How is a cirque produced?
2. How does a kettle form?
3. How can terminal and lateral moraines form glacial lakes?
4. Compare glacial sediments deposited directly by glacial ice with those deposited by glacial meltwater.

15.3 Ice Ages

A long period of climatic cooling during which continental ice sheets cover large areas of the earth's surface is known as an **ice age.** Scientists have found evidence of several major ice ages during the earth's geologic history. The earliest ice age occurred more than 600 million years ago. The most recent ice age began more than 2 million years ago. Its massive ice sheets started to retreat only 18 thousand years ago.

Section Objectives

- **Describe the climatic cycles that exist during an ice age.**
- **Identify and summarize the theory that best accounts for the ice ages.**

Climate During Ice Ages

An ice age probably begins with a long-term decrease in the earth's average temperatures. A drop in average temperatures of only about 5°C, combined with an increase in snowfall, sets the stage for an ice age.

Scientists know that climates were generally colder during the last ice age. However, studies indicate numerous cycles of advance and retreat of the continental ice sheets during this ice age. The ice sheets advanced during cold periods and retreated during warmer periods. The time when the ice sheets advance during an ice age is a *glacial period.* The times of warmer temperatures between the colder glacial periods are called *interglacial periods*.

Glacial Periods

As Figure 15–10 shows, glaciers covered nearly one-third of the earth's land surface during the last glacial period. Most glaciation was located in North America and Eurasia. In some parts of North

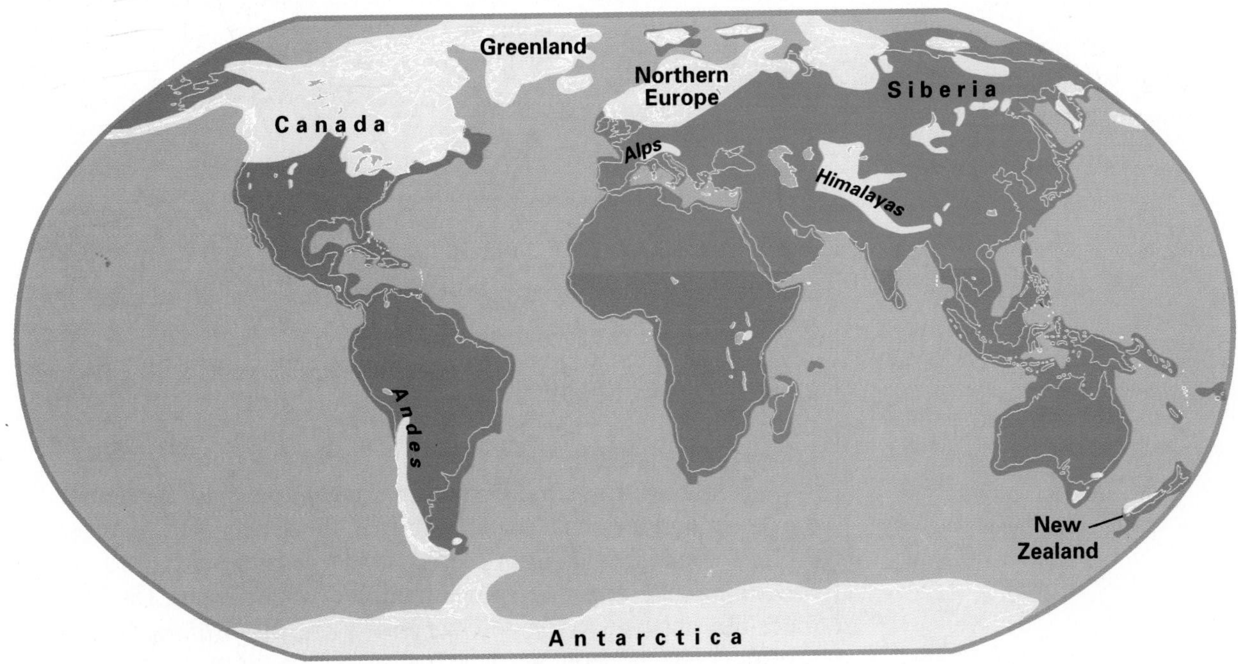

Figure 15–10. During the last ice age, over 30 percent of the earth's land area was covered with ice sheets, as shown here in light blue.

Figure 15–11. During the ice ages, many valley glaciers had a geologic impact on the land. In Alaska, valley glaciers such as this one in Glacier Bay extended much farther than they do today.

America, the ice was several kilometers thick. So much water was locked up in ice that sea level was about 140 m lower than it is today.

Canada and the mountainous regions of Alaska were buried under ice. In the mountains of the western United States, numerous small valley glaciers joined to form several large valley glaciers. They flowed outward from the Rockies, Cascades, and Sierra Nevada ranges. A great continental ice sheet with its center over the Hudson Bay region of Canada spread as far south as the Missouri and Ohio rivers. A large continental ice sheet, centered over the Baltic Sea, spread south over Germany, Belgium, and the Netherlands and west over most of Great Britain and Ireland. It flowed eastward over Poland and Russia, and long valley glaciers formed in the Alps and the Himalayas. In Siberia, another continental ice sheet formed. Large glaciers also formed in the southern hemisphere. The Andes Mountains of South America and most of New Zealand were covered by mountain ice fields and valley glaciers. Many land features that formed during the last glacial period are still easily recognizable. What features would you expect to see?

Causes of Ice Ages

Scientists have proposed a number of theories to account for ice ages. Each theory explains why the earth experienced the gradual cooling and the increase in precipitation that brought on the expansion of the glaciers. The theories also explain why the glaciers retreated during the interglacial periods.

The theory that most scientists now accept is called the **Milankovitch theory.** Milutin Milankovitch, a Serbian scientist, proposed that small, regular changes in the earth's orbit and in the tilt of the earth's axis caused the ice ages.

Three kinds of periodic changes occur in the way the earth moves around the sun. One change is in the shape of the earth's orbit. The orbit varies from nearly circular to more elongated and back to circular about every 100 thousand years. Figure 15–12 illustrates this change. A second change is in the tilt of the earth's axis. Over a period of about 41 thousand years, the tilt of the axis varies

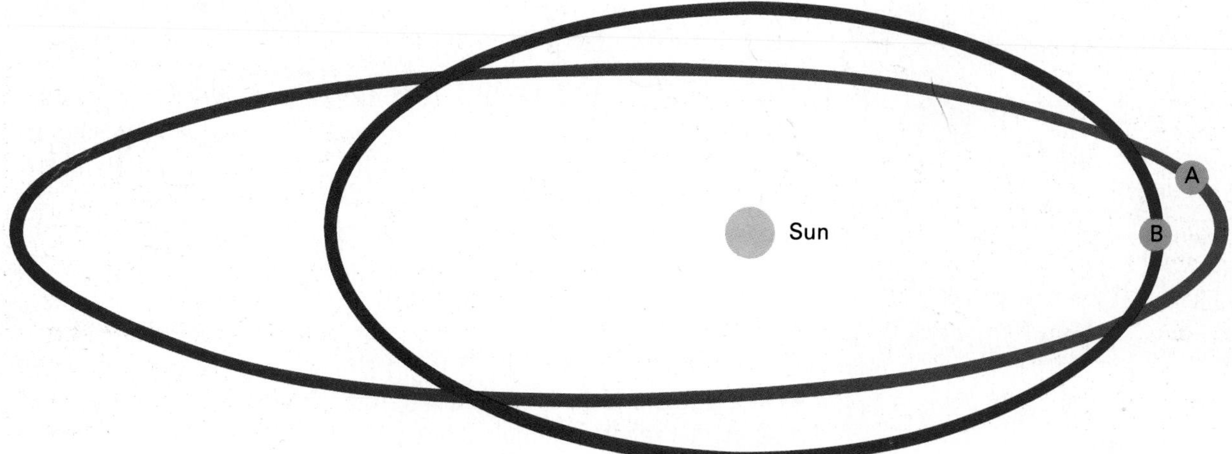

Figure 15–12. The distance of the earth from the sun changes as the earth's orbit varies between exaggerated pattern A and pattern B

between 21.5° and 24.5°. A third periodic change results from the circular motion, or precession, of the earth's axis. Precession causes the axis to change its position. The axis of the earth traces a complete circle in space every 26 thousand years. Milankovitch calculated how these three changes periodically change the distribution of solar energy reaching the earth's surface.

Evidence from the ocean floor—in the shells of dead marine animals—has provided support for the Milankovitch theory. Scientists have observed that the ratio of two isotopes of oxygen in the shells of these animals is related to the temperature of the sea water when the animals lived. The shells are found in layers of sediment on the ocean floor. They can show if a marine sediment formed during an ice age, when seawater temperatures were lower. Scientists have found that the record of ice ages found in marine sediments closely follows the cycle of cooling and warming predicted by the Milankovitch theory.

There also are other theories to explain the causes of ice ages. However, like the Milankovitch theory, most of these theories state that the ice ages resulted from a change in the amount of solar energy reaching the earth's surface. The explanations for the changes vary. Some theories propose that the changes in solar energy might have been caused by varying amounts of heat produced by the sun. Other theories propose that volcanic dust blocked the sun's rays. Another theory proposes that ice ages were caused by tectonic plate motion changing positions of the continents, which prevented warm ocean currents from reaching regions near the poles.

Section 15.3 Review

1. What happens to the sea level during an ice age?
2. What is a glacial period?
3. What must a theory of the ice ages explain?
4. Explain the theory of the ice ages that most scientists accept at the present time.
5. Briefly describe three other theories of the cause of ice ages.

Chapter 15 Review

Key Terms

- arête (282)
- basal slip (279)
- cirque (282)
- continental ice sheet (278)
- crevasse (280)
- drumlin (285)
- erratic (284)
- esker (286)
- firn (277)
- glacial drift (285)
- glacier (277)
- ground moraine (285)
- hanging valley (283)
- horn (282)
- ice age (289)
- internal plastic flow (279)
- kettle (286)
- lateral moraine (285)
- medial moraine (285)
- meltwater (285)
- Milankovitch theory (290)
- outwash plain (286)
- roche moutonnée (283)
- snowfield (277)
- snowline (277)
- stratified drift (285)
- terminal moraine (286)
- till (285)
- valley glacier (278)

Key Concepts

Formation of glaciers requires adequate snowfall and low temperatures. **See page 277.**

The two main types of glaciers are valley glaciers and continental ice sheets. **See page 278.**

Glaciers move by basal slip and by internal plastic flow. **See page 279.**

Glaciers erode the valleys through which they flow, producing characteristic landforms. **See page 282.**

When glaciers melt, they deposit their sediments, creating features that can be readily identified. **See page 284.**

Glaciers may form lake basins by eroding the land or by depositing sediments. **See page 287.**

During an ice age, glacial advances alternate with glacial retreats. **See page 289.**

The Milankovitch theory is the most commonly accepted theory of the cause of ice ages. **See page 290.**

Review

On your own paper, write the letter of the term that best completes each of the following statements.

1. Snow accumulates year after year in polar regions and
   a. in southern Canada. b. in the United States.
   c. below the snowline. d. at high elevations.
2. Glaciers formed in mountainous areas are called
   a. continental ice sheets. b. valley glaciers.
   c. icebergs. d. ice shelves.
3. Antarctica is covered by the earth's largest
   a. iceberg. b. valley glacier.
   c. continental ice sheet. d. outlet glacier.
4. A glacier will move by sliding when the base of the ice and rock are separated by a thin layer of
   a. water. b. snow. c. pebbles. d. drift.
5. When a glacier moves by internal plastic flow,
   a. its center moves fastest.
   b. its bottom moves fastest.
   c. its edges move fastest.
   d. the whole ice mass moves at the same speed.
6. Glacial erosion may produce a bowl-shaped depression known as
   a. a moraine. b. an esker. c. a cirque. d. a horn.
7. As a glacier moves through a valley, it carves out
   a. a U-shape. b. a V-shape. c. an esker. d. a moraine.

8. Unsorted glacial deposits are called
   a. stratified drift. b. outwash plains.
   c. eskers. d. till.
9. Long, winding ridges of gravel and sand deposited by a meltwater stream under the ice are called
   a. eskers. b. drumlins.
   c. outwash plains. d. medial moraines.
10. A kettle is a
   a. hill. b. depression. c. ridge. d. mound.
11. A deposit of stratified drift is called
   a. a drumlin. b. outwash.
   c. a ground moraine. d. a roche moutonnée.
12. Many glacial lakes are formed in
   a. outwash plains. b. kettles.
   c. eskers. d. arêtes.
13. During the last glacial period, the average temperature was about
   a. 5°C lower than today. b. 15°C lower than today.
   c. 35°C lower than today. d. 50°C lower than today.
14. One component of the Milankovitch theory is
   a. the circular motion of the earth's axis.
   b. continental drift.
   c. volcanic activity.
   d. landslide activity.
15. A proposed cause of the ice ages is decreased solar energy reaching the earth due to
   a. a lunar eclipse. b. blockage by volcanic dust.
   c. sinking of the land. d. increased storm activity.

Critical Thinking

On your own paper, write answers to the following questions.

1. Imagine that there is a village in Greenland located at the edge of the ice sheet. During the year there is an unusually large amount of snowfall. How might this snowfall affect the ice sheet? What danger might this pose for the inhabitants of the village?
2. Why is it important for scientists to monitor and study the continental ice sheets that cover Greenland and Antarctica?
3. Antarctic explorers need special training to travel safely over the ice sheet. Besides the cold, what structural aspects of the glaciers might be dangerous?
4. In addition to decreasing temperature and increasing snowfall, what other phenomenon might signal an impending ice age?

Application

1. List some features caused by glacial erosion and deposition that you might see on a car trip.
2. In what ways might past glacial action in New England and New York State affect tourism and recreation in those regions today?
3. A group of scientists is trying to find evidence to support Milankovitch's theory of the earth's ice ages. List some kinds of research they might need to do.

Extension

1. Many tourists visit Glacier National Park in Montana and Kettle Moraine Park in Wisconsin to see glaciated landscapes. Find out about other areas in the United States and Canada where you can see glacial features. Send for information about one of these places. You might try writing to the local tourist bureau. Once you have obtained your information, put together a travel brochure that would persuade people to visit the place.
2. Antarctica and Greenland are among the few places on earth where you can experience ice-age conditions. Imagine that you are a scientist studying the ice sheet on Antarctica. Write a diary of one day's activities, including observations of the scenery and the research projects you are working on. Consult reference materials for help with details.

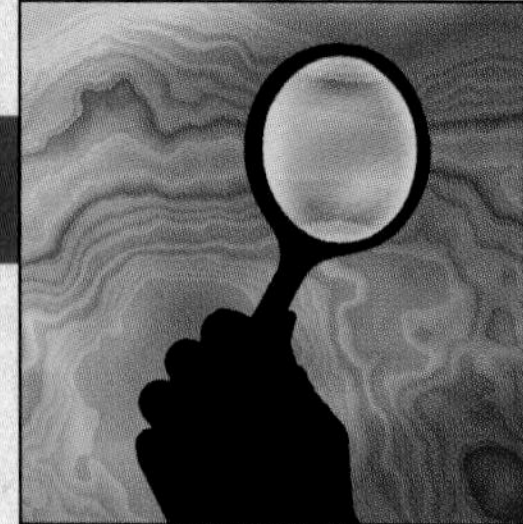

IN-DEPTH INVESTIGATION

Glaciers and Sea Level

Materials

- metric ruler
- sand and small pebbles (3–5 lb.)
- water
- milk carton (small)
- shallow pan (30 × 40 × 10 cm)
- wooden block (5 × 5 × 5 cm)
- freezer

Introduction

Today, glaciers hold only about 2.2% of the earth's water, but if the polar ice sheets melted, the coastal areas of many countries would be flooded. In the United States, many major cities, such as New York, New Orleans, Houston, and Los Angeles, would be flooded if the sea level rose only a few meters.

In this investigation, you will construct a model to simulate what would happen if the Antarctic ice sheet melted.

Prelab Preparation

1. Review Section 15.1, pages 277–281.
2. The day before the investigation, fill the milk carton with water and place it in a freezer.
3. Area (A) is calculated by multiplying length (l) times width (w) [$A = l \times w$]. Area is expressed in square units. Volume (V) is calculated by multiplying length (l) times width (w) times height (h). Volume is expressed in cubic units. Calculate the area of the two-dimensional shape and the volume of the three-dimensional shape shown below.

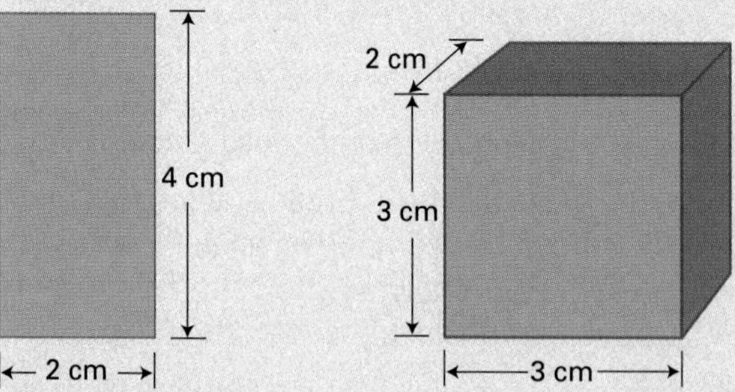

Procedure

1. Calculate and record the approximate surface area of the bottom of the pan.
2. Obtain the milk carton filled with frozen water. Remove the carton from the block of ice. Calculate and record the overall volume of the block and the area of one side of the ice block.
3. Add the sand and small pebbles to one end of the pan so that they cover about half the area of the pan, sloping toward the middle. Elevate this end of the pan using the wooden block as shown in the figure on the next page.

4. Slowly add water to the opposite end of the pan. Be sure that the water does not cover the sand, but only touches the edge of it. Measure and record the depth of the water at the deepest point. Then measure and record the distance from the end of the pan covered with sand to the point where the sand touches the water.
5. Place the block of ice lengthwise in the pan on top of the sand as shown in the figure above. What percentage of the total area of the pan calculated in Step 1 is covered by ice?
6. As the ice begins to melt, pick up the ice block. Note the appearance of the bottom of the ice block and of the sand under the ice. What is happening to the ice block? What is happening to the sand under the ice block? Place the ice block back on the sand.
7. While the ice is melting, calculate the expected rise in the pan's water level using the following formula:

$$\text{rise in water level (cm)} = \frac{\text{volume of water in ice block (cm}^3\text{)}}{\text{area of pan covered with water (cm}^2\text{)}}$$

8. When the ice is completely melted, measure and record the depth of water at the deepest point in the pan. What is the difference in the water level after the ice has melted in the pan? How does this compare with the value you calculated in Step 7? Explain any differences.
9. Remeasure and record the distance from the end of the pan covered with sand to the point where the sand touches the water. What is the difference in this distance after the ice has melted in the pan?

Analysis and Conclusions

1. How is the ice block model different from a real glacier on the earth?
2. How does this model represent what would happen on the earth if the Antarctic polar ice sheet melted?
3. In this investigation, you used a physical model to simulate an occurrence in nature. How else do scientists use models? What kinds of errors may occur when using models?

Extension

The total area of the earth is approximately 511,000,000 km^2. About 70% of the earth's surface is covered with water. The volume of water locked as ice in the Antarctic ice sheet is approximately 17,900,000 km^3. Calculate the area of the earth, in square kilometers, that is covered with water. Then find the average worldwide rise in sea level that would occur if the Antarctic ice sheet melted. Convert your answer to meters. How is the mathematical model for the rise in sea level similar to the physical model used in this investigation?

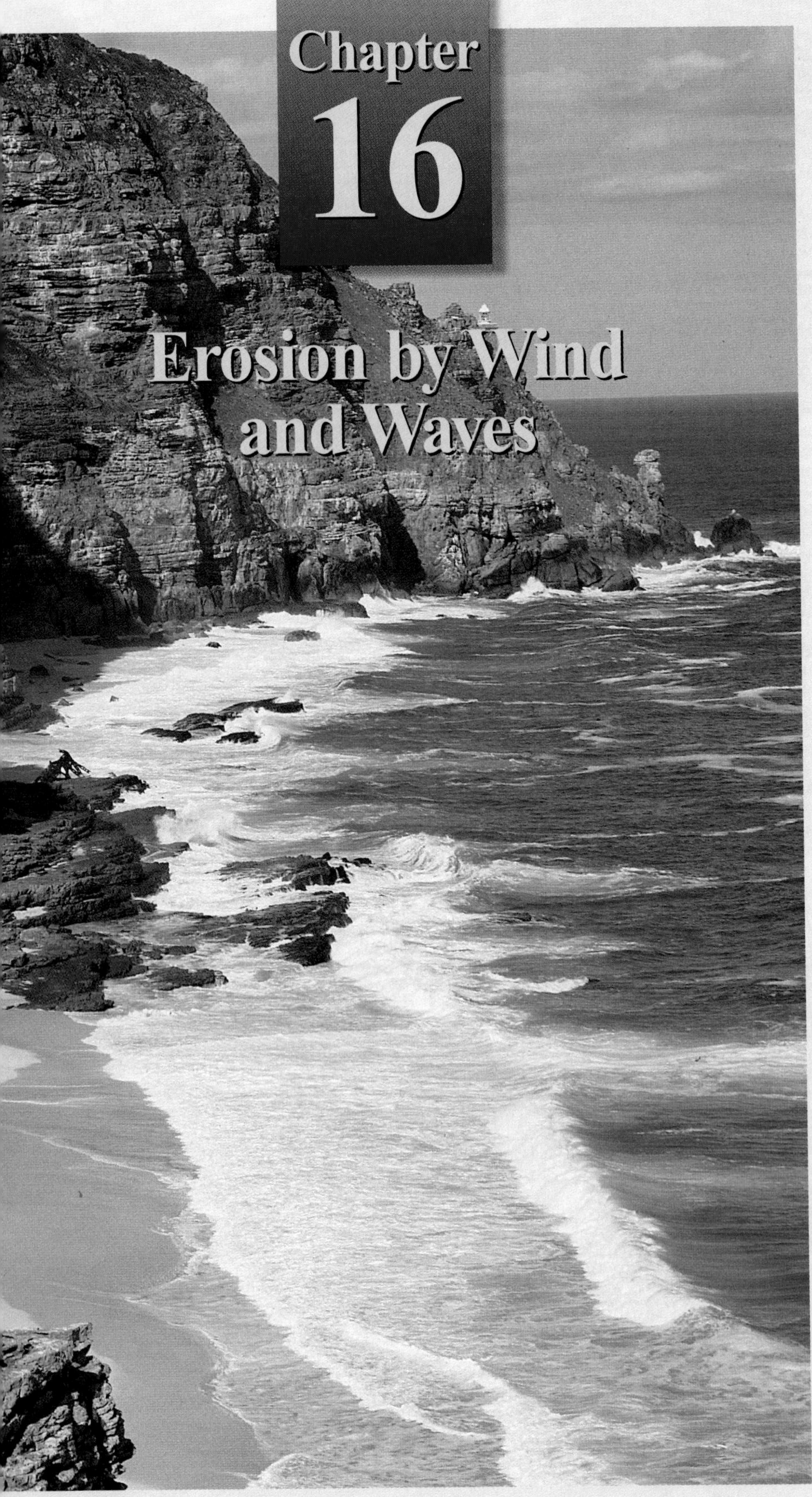

Imagine that you are standing on a beach. Suddenly you see swirling sheets of sand. All you can hear is the howl of the wind and the thundering of waves against the shore.

In reading this chapter, you will learn how the wind leaves its mark on the land. You will also learn how the wind and water work together to constantly change features of the earth where the land and oceans meet.

Chapter Outline

This rugged coastline is at the Cape of Good Hope, South Africa.

16.1 Wind Erosion

Section Objectives

- **Describe two ways that the wind erodes the land.**
- **Compare the two types of wind deposits.**

There is energy in the wind. Some of the energy can be used to move a sailboat or turn a wind turbine. However, this energy can also erode the land. The wind erodes dry land much more effectively than it erodes moist land. Moisture makes soil heavier and causes some soil and rock particles to stick together. Moist particles are more resistant to the force of the wind than are dry particles.

As the wind erodes the land, it carries along with it rock particles of various sizes. These particles are categorized as either sand or dust, usually silt and clay. Sand is loose fragments of weathered rocks and minerals. Most grains of sand are made up of quartz. Other common minerals in sand grains are mica, feldspar, and magnetite. Sand grains range in diameter from 0.06 mm to 2 mm. The smallest grain of sand is barely visible to the unaided eye; the largest grain of sand is about the size of a pinhead.

Dust consists of particles smaller than the smallest sand grain, or less than 0.06 mm in diameter. Most dust—silt and clay—is microscopic fragments of rock and minerals that come from the soil or from volcanic eruptions. Other sources of dust are plants, animals, and bacteria. Dust may also occur as a by-product of the burning of fuels and of certain manufacturing processes.

Wind moves sand and dust in different ways. Wind cannot keep aloft even the smallest particle of sand. Instead, sand grains are moved along by a series of jumps and bounces, much the way pebbles are moved by streams. Such movements are called *saltation*. Saltation occurs when the wind speed becomes great enough to roll sand grains along the ground. When rolling sand grains collide with one another, some sand grains bounce up, as illustrated in Figure 16–1. Once in the air, a sand grain moves ahead a short distance, then falls. As a sand grain falls, it strikes other sand grains. These sand grains also may be thrown into the air or rolled ahead by the impact. Saltating sand grains move in the same direction that the wind is blowing. However, the grains seldom rise more than 1 m above the ground, even in very strong winds.

In contrast to sand grains, dust particles can be lifted by the wind and carried high into the air. Even gentle air currents can keep dust particles suspended in the air. Dust from volcanic eruptions may

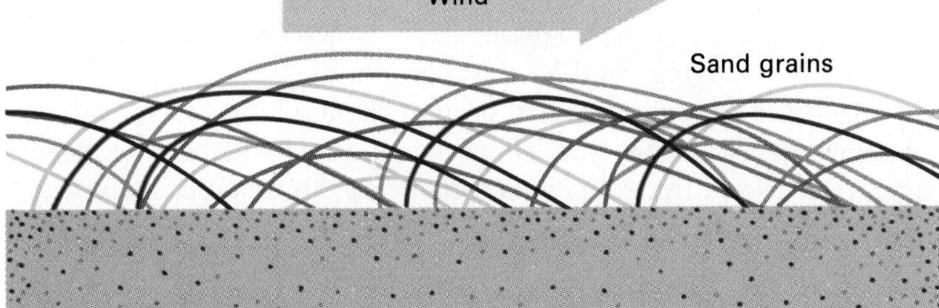

Figure 16–1. Sand grains move by saltation, making low, arcing jumps when blown by the wind.

Figure 16–2. The formation of desert pavement prevents the erosion of the material beneath it.

remain in the atmosphere for several years before falling to the ground. Strong winds may lift enormous amounts of dust into the atmosphere and create large dust storms. Some dust storms cover hundreds of square kilometers and darken the sky for several days.

Effects of Wind Erosion

The force of the wind erodes the land in a number of ways. The most common form of wind erosion is **deflation.** The term *deflation* is from the Latin word *deflare,* which means "to blow away." During the process of deflation, wind removes the top layer of fine, very dry rock or soil particles. Left behind are rock fragments that are too large to be lifted by the wind. These remaining gravel particles often form a surface of closely packed small rocks called **desert pavement,** which protects the underlying land from erosion. Desert pavement is shown in Figure 16–2.

Deflation is a serious problem for farmers because it blows away the best soil for growing crops. Deflation may cause shallow depressions to form in areas where the natural plant cover has been removed. As an area of dry, bare soil is exposed to wind, the wind strips off the topsoil. Once all the soil is gone, a shallow depression called a **deflation hollow** remains. A deflation hollow may be eroded further by wind and water, expanding to a width of several kilometers and a depth of 5 m to 20 m.

Desert pavements and deflation hollows are common in dry climates because wind erosion occurs more rapidly in deserts than it does in humid climates. In the desert, soil layers are thin and contain little moisture. Consequently, desert soils may be swept away by the wind.

Sand grains carried by the wind increase its erosive power, just as the stream load increases the erosive power of running water. Abrasion is the weathering of rock particles by the impact of other rock particles. Abrasion by wind-blown sand results in much erosion in areas where there are strong, steady winds, large amounts of loose sand, and relatively soft rocks.

Pebbles and small stones in deserts and along some beaches are exposed to wind abrasion. As a result, the surfaces of the rocks become flattened and polished on two or three sides. This polishing produces facets on the rocks. Rocks thus smoothed by wind abrasion are called **ventifacts,** from the Latin word *ventus,* meaning "wind." The direction of the wind that formed a ventifact can be determined by the appearance of the ventifact. How would you determine wind direction by studying ventifacts in a certain area?

Scientists once thought that large rock structures, such as desert basins, natural bridges, rock pinnacles, and rocks perched on pedestals, were caused by wind erosion. However, it is more likely that such large features were produced by erosion due to surface water and weathering. Erosion of large masses of rock by wind-blown sand takes place very slowly and only close to the ground, where saltation occurs.

Wind Deposition

All the material eroded by the wind is eventually deposited. The size of the particles carried by the wind depends on the wind speed. As its speed increases, the wind is able to carry larger particles. When the speed of the wind decreases, the heavier part of the load drops, forming a deposit. If the deposited fragments are not carried away by a later strong wind, they are covered by future deposits. Eventually the weight of the top layers of built-up wind deposits compresses the covered fragments together. Pressure and the cementing action of minerals carried in water can bind the fragments together, thus forming sedimentary rock.

Dunes

The best-known wind deposits are **dunes,** which are mounds of wind-blown sand. Dunes form where the soil is dry and unprotected and the wind is strong. Dunes are common in some deserts and along the shores of oceans and large lakes.

A dune begins to form when a barrier slows the speed of the wind. This reduction in wind speed causes sand to accumulate on the sheltered side of the barrier, as illustrated in Figure 16–3. Sand also accumulates at the base of the exposed side. As more sand is added, the deposit acts as a barrier itself, and the dune continues to grow, eventually burying the original barrier.

Usually the gentlest slope of a dune is the side facing the wind, or the windward side, because the wind is constantly reshaping that side. The sand that is blown over the crest, or peak, of the dune tumbles down the opposite side, called the *slipface*. The slipface has a steeper slope than the windward side. Wind sweeping around the ends of a dune often causes two long, pointed extensions to form.

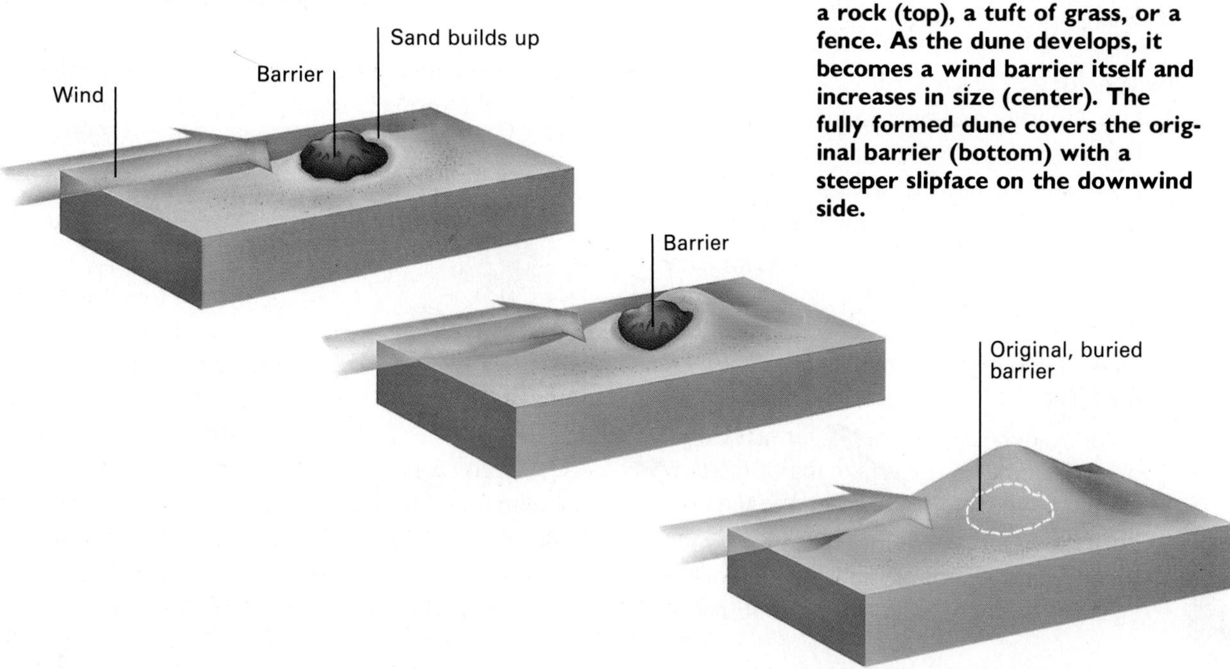

Figure 16–3. Wind-driven sand builds up around a barrier such as a rock (top), a tuft of grass, or a fence. As the dune develops, it becomes a wind barrier itself and increases in size (center). The fully formed dune covers the original barrier (bottom) with a steeper slipface on the downwind side.

These give the dune a crescent shape. A crescent-shaped dune is called a *barchan* (BAHR-kan). The open side of the crescent faces away from the wind. Barchans are a common type of dune and are found in most deserts. Another type of dune, called a *parabolic dune,* is also crescent-shaped. However, the open side of a parabolic dune faces into the wind. These dunes often form as sand is blown from a deflation hollow and collects around its rim.

In desert areas, where the amount of sand is great, a series of ridges of sand form in long, wavelike patterns. These ridges are called *transverse dunes*. This type of dune forms at a right angle to the wind direction. A similar type of dune, called a *longitudinal dune,* also forms in the shape of a ridge. However, longitudinal dunes lie parallel to the direction in which the wind blows.

The complicated shapes of most dunes result from the force and direction of the wind. If the wind usually blows from the same

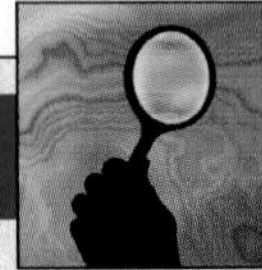

SMALL-SCALE INVESTIGATION

Dune Migration

You can demonstrate how dunes migrate by making a model dune and simulating natural events.

Materials

wax pencil or marker; ruler; shallow box or tray; paper bag, large enough to hold one end of the box; very fine, dry sand; safety goggles; filter mask; hair dryer; clock or watch

Procedure

1. Use a marker and ruler to mark the side of the box at 5-cm intervals, as shown in the photo.
2. Place the box halfway inside the paper bag so the bag will catch any blowing sand.
3. Fill the box about half full of fine, dry sand. Make a dune in the sand about 10–15 cm from the open end of the box. Look at the side of the box and record to the nearest centimeter the location of the peak of the dune.
4. Put the safety goggles and filter mask on. Hold the hair dryer level with the top of the dune about 10 cm from the open end of the box.
5. Turn the hair dryer to low speed for 1 minute. Identify and record the new location of the peak of the dune to the nearest centimeter.
6. Repeat Step 5 three times, first running the hair dryer for 2 minutes, then 3 minutes, and then 5 minutes. After each trial, record the location of the peak of the dune.
7. Flatten the sand. Place a barrier such as a rock in the sand. Position the hair dryer as in Step 4. Run the dryer for 3 minutes.

Analysis and Conclusions

1. In what direction did the dune migrate?
2. How far overall did the dune migrate?
3. What was the average distance the dune migrated per minute?
4. In Step 7, where does the dune form? What steps might be taken to slow down the process of dune migration? Explain your answer.

Figure 16–4. These loess deposits are located just west of Tarkio, Missouri.

direction, dunes will move downwind. Dune migration occurs as sand is blown off the windward side, over the crest, and is built up on the slipface. In fairly level areas, dune migration continues until a barrier is reached. To prevent dunes from drifting over highways and farmland, the planting of grasses, trees, and shrubs or the building of fences is often necessary.

Loess

The wind carries dust higher and much farther than it carries sand. Fine dust may be deposited in such thin layers that it is not noticed. However, thick deposits of yellowish, fine-grained sediment can be formed by the accumulation of windblown dust. This material is known by its German name, **loess** (LESS), meaning "loose." Although loess is soft and easily eroded, it can break into vertical slabs. Loess sometimes forms steep bluffs, such as those shown in Figure 16–4.

A large area in northern China is covered entirely with loess. The material in this deposit came from the Gobi Desert, in Mongolia. Large deposits of loess are also found in central Europe. In North America, loess is found in the midwestern states, along the eastern border of the Mississippi River valley, and in eastern Oregon and Washington State. These deposits probably were built up by dust from dried beds of glacial lakes and outwash. Loess deposits are extremely fertile and provide excellent soil for grain-growing regions of the United States.

Section 16.1 Review

1. Define *deflation*.
2. Describe how dunes form and migrate.
3. Describe how loess deposits are formed.
4. Compare the composition and shape of dunes and loess deposits.

16.2 Wave Erosion

Section Objectives

- **Compare the formation of six features produced by wave erosion.**
- **Define a beach and discuss the way in which it is formed.**
- **Describe the movement of sand along a shore and the features it produces.**

As wind moves over the ocean, it produces waves and currents that erode the bordering land. Wave erosion changes the shape of the earth's **shorelines,** the boundary where the ocean and the land meet. Shorelines are temporary and very unstable boundaries.

Shoreline Erosion

The power of waves striking rock along a shoreline may shake the ground like a small earthquake. Seismographs often record such vibrations. The great force of waves may break off pieces of rock and throw them back against the shore. The rock fragments grind together in the tumbling water. This abrasive action eventually reduces most of the rock fragments to small pebbles and sand grains.

Chemical weathering also attacks the rock along a shoreline. The waves force salt water and air into small cracks in the rock. Substances in the air and water produce a chemical action that may enlarge the cracks. The enlarged cracks, in turn, provide an increased surface area for physical and chemical weathering.

Much of the erosion along a shoreline takes place during storms. Large waves release tremendous amounts of energy against the shore. Huge blocks of rock, like those shown in Figure 16–5, can be broken off and eroded by such waves. A severe storm can noticeably change the appearance of a shoreline in a single day.

Sea Cliffs

In places where waves strike directly against rock, erosion usually produces a steep structure called a *sea cliff,* like the one shown in Figure 16–5. The waves slowly notch the base of the cliff. The notch cuts under the overhanging rock, until the rock eventually falls. The cliff is gradually worn back and made steeper. Many shorelines consist of such high, nearly vertical sea cliffs.

Figure 16–5. Sea cliffs develop where waves strike directly against rock along a shoreline. Point Bonita Lighthouse in Golden Gate, California, is built on a resistant rocky headland that extends from a sea cliff.

The rate at which sea cliffs are eroded by waves is dependent upon the resistance of the rock along the shoreline. Soft rock is eroded very rapidly. For example, sea cliffs made up of loose glacial deposits along the shore of Cape Cod are being worn away at the rate of about 1 m/yr. Old maps of England shows that sea cliffs made of soft sedimentary rock have been worn back several kilometers over the past few centuries. In contrast, shorelines made up of hard rock, such as the granite of the Seychelles Islands in the Indian Ocean, show little change over hundreds of years. Resistant rock areas that project out from shore are called **headlands.** Those areas with less-resistant rock form bays.

Sea Caves, Arches, and Stacks

A sea cliff is seldom eroded evenly. Fractured rock is more easily eroded than nonfractured rock. Waves often cut deeply into fractures and weak rock along the base of the cliff, forming a large hole, or a *sea cave*. Wave action can enlarge a sea cave, producing a *sea arch* when the waves cut completely through a headland. Offshore isolated columns of rock, which had once been connected to the sea cliffs or headland, are called *sea stacks*. In time, the sea stacks will be eroded so that they stand no higher than the water.

Terraces

Most erosion of a sea cliff takes place above water. However, as a sea cliff is worn back, a nearly level platform usually remains beneath the water at the base of the cliff. This platform is called a **wave-cut terrace.** As the waves cause the cliff to retreat, some of the rock from the base of the cliff scrapes the wave-cut terrace until it is almost flat. Other eroded material may be deposited some distance from the shore, creating an extension to the wave-cut terrace called a **wave-built terrace.**

Figure 16–6. Wave erosion of sea cliffs causes cliff retreat and forms isolated sea stacks, as shown below.

The water above a terrace is shallow. Waves lose much of their energy as they pass through shallow water. As the energy of the waves lessens over a terrace, the rate of erosion of the cliff is greatly reduced. However, if the terraces erode completely or if the sea level becomes higher, waves may begin to erode the cliff again.

Beaches

Waves create various features along shorelines by eroding the land and depositing the rock fragments. A **beach** is one such shoreline feature. A beach is a deposit of sand or larger rock fragments along an ocean shore or a lakefront. As waves wash up on the shore, they move sand and small rock fragments forward. In retreating, the water moves some rock fragments away from the shore. Beaches form where the amount of rock fragments moving toward the shore is greater than the amount moving away from the shore. After the beach has formed, the rates at which fragments move toward and away from shore tend to equalize.

Table 16–1 Beach Materials

Material	Diameter (mm)
Boulders	More than 256
Cobbles	64 to 256
Gravel	
Coarse	16 to 64
Fine	2 to 16
Sand	
Coarse	0.6 to 2
Medium	0.2 to 0.6
Fine	0.06 to 0.2
Silt	Less than 0.06

Composition of Beaches

Most people think of beaches as areas of light-colored, small-sized sand grains along the water. Actually, the sizes and kinds of materials found on beaches vary widely, as indicated in Table 16–1. Many beaches are covered with pebbles or large rock fragments and have no sand.

The composition of beach materials depends on the source rock. For example, granite yields light-colored fragments that are mostly quartz and feldspar. Beaches of this type of sand are common along the North American shorelines, where granite is abundant. Beaches of black sand, as shown in Figure 16–7, are found on Hawaii and on other volcanic islands. The black sand comes from the volcanic rocks common to these islands. Some beaches along the Oregon and Washington shorelines are also made up of dark volcanic sand.

Figure 16–7. This beach in Hawaii is made up of black sand eroded from the volcanic rock that makes up the island.

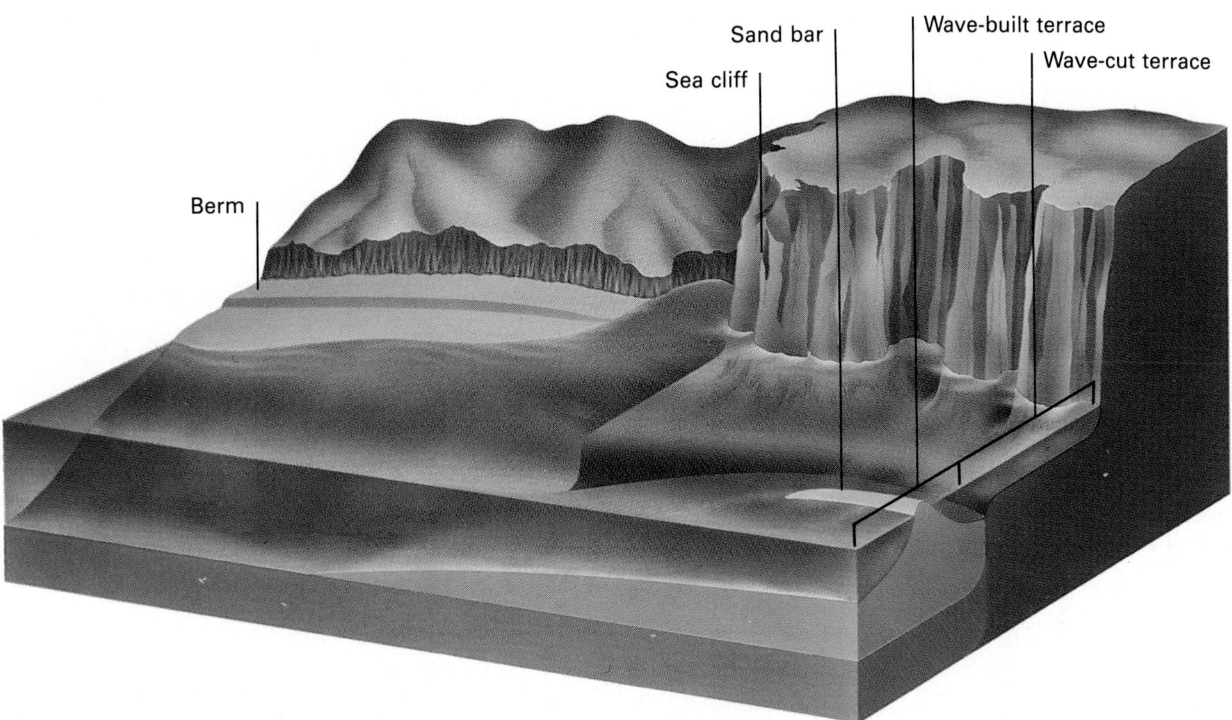

Figure 16–8. Notice that wave erosion creates a variety of coastal landform features.

In many locations sand is carried to the shore by rivers. Thus the source rock of much of the sand found along a beach may be in the valley of a nearby river. Some beaches, such as those in Florida, may consist of fragments of shells and coral that are washed ashore. In other locations, such as Cape Cod in Massachusetts, beach sand has been deposited by glaciers.

The Berm

Most rock fragments, from fine sand to large cobbles, which are rounded fragments larger than pebbles, are moved toward and away from the shore by waves. The smaller and lighter fragments are moved farthest and most easily. Each wave reaching the shore moves individual sand grains forward slightly. Although each advance is small, the total action of several thousand waves each day may move sand grains a great distance. The sand piles up on the shore, producing a sloping surface. During high tides or when large storm waves come in, sand is deposited at the back of this sloping beach. As a result, most beaches have a raised section called the **berm.** The area of beach above the berm is the part most often used by people for recreation.

The appearance of beaches changes seasonally. During winter, large storm waves remove sand from the beach on the seaward side of the berm. Thus the berm is higher and steeper during the winter. Waves erode and move the beach sand seaward. The sand carried away from the berm is deposited offshore. These sand deposits form a long underwater ridge called a **sand bar.** In summer, waves return the sand from offshore to widen the beach on the seaward side of the berm.

INVESTIGATE!

To learn more about beaches, try the In-Depth Investigation on pages 314–315.

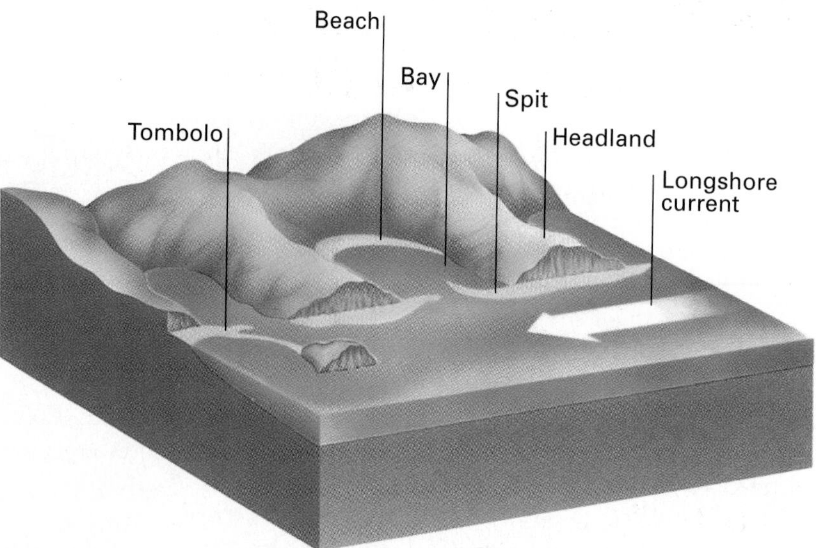

Figure 16–9. Spits and tombolos form as longshore currents deposit sand at headlands, openings to a bay, or offshore islands.

Longshore-Current Deposits

The direction in which a wave approaches the shore determines how it will move the sand grains and other beach materials. Most waves approach the beach at a slight angle and retreat in a direction that is more perpendicular to the shore. As a result, individual sand grains are moved by the waves in a zig-zag motion. The general movement of sand along the beach is in the direction in which the waves strike the shore.

Waves moving at an angle to the shoreline push water along the shore, creating *longshore currents.* A longshore current is a movement of water parallel to and near the shoreline. Longshore currents transport sand in a direction parallel to the shoreline. The effects of a longshore current are shown in Figure 16–9.

Sand moving along a relatively straight shore keeps moving until the shoreline changes direction. Shoreline direction changes at bays and headlands. The longshore current slows and sand is deposited at the far end of the headland. A long, narrow deposit of sand connected at one end to the shore is called a **spit.** Currents and waves may curve the end of a spit into a hook shape. Beach deposits may also connect an offshore island to the mainland. Such connecting ridges of sand are called **tombolos.** What process of stream deposition is similar to the way a spit forms?

Section 16.2 Review

1. Name several coastal landform features produced by wave erosion.
2. What determines the composition of beach materials?
3. Define *longshore current.* What coastal landform features do these currents produce?
4. What are the similarities in the way a barchan forms and the way a hooked spit forms?

16.3 Coastal Erosion and Deposition

Section Objectives

- **Explain how changes in sea level relative to the land affect coastlines.**
- **Describe the features of a barrier island.**
- **Compare the types of coral reef.**
- **Analyze the effect of human activity on coastal land.**

Coastlines, the boundaries between the land and the ocean, are among the most rapidly changing parts of the earth's surface. The coastal area extends from relatively shallow water to several kilometers inland. The study of a coastline can help scientists understand its hazards and predict future changes in it.

Processes That Affect Coastlines

Most coastlines are the result of various processes working together. However, scientists have found that many coastline features are formed by a change in sea level relative to the land. Coastlines are affected by the long term rising or falling of sea level and by the long term uplifting or sinking of the land that borders the water. These and other more rapid processes, including wave erosion and deposition, are constantly changing the appearance of coastlines.

Coastlines and Sea-Level Changes

A change in the amount of ocean water causes sea level to rise or fall, covering or exposing coastlines. During the last glacial period, which ended about 11,000 years ago, some of the water that is now in the ocean existed as continental ice sheets. Scientists estimate that the ice sheets then held about 70 million km^3 of ice. Now the ice sheets in Antarctica and Greenland hold only about 25 million km^3 of ice. During the last glacial period, the water that made up the additional 45 million km^3 of ice must have come from the oceans. As a result, sea level was probably about 140 m lower than it is today. Since the last glacial period, the ice sheets have been melting and sea level has been rising, as indicated in Figure 16–10. Sea level is now rising at an average rate of about 1 mm/yr. If the Antarctic and Greenland ice sheets were to melt, the oceans would rise about 60 m. Low-lying coastal regions would be submerged. Many large cities, including New York, Los Angeles, Miami, and Houston, are located in such coastal areas and may be endangered by a gradual rise in sea level.

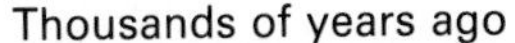

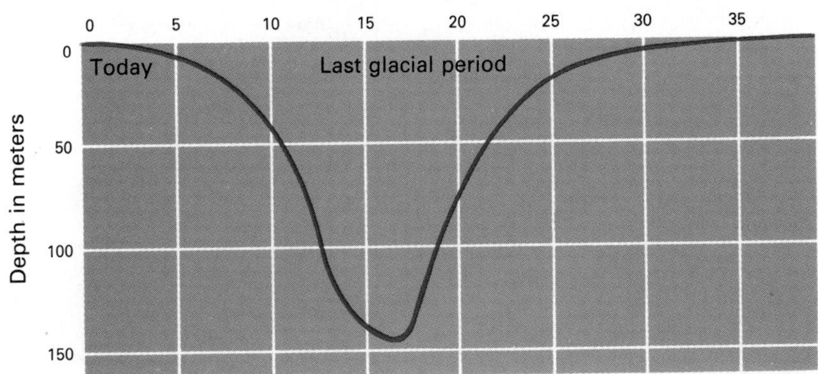

Figure 16–10. This graph shows how sea level has changed during the past 35,000 years. 0 on the graph represents present sea level.

Coastlines and Land Changes

An exposed coastline may become flooded if the land sinks. A coastline that is submerged may be exposed if the land rises. Sinking or rising of the land is sometimes part of the isostatic adjustment to weight changes of the earth's crust. Lithosphere that becomes heavier will sink lower into the asthenosphere; lithosphere that becomes lighter will rise.

In some parts of Europe, particularly in Scandinavia, the land is rising. Docks have actually been raised out of the water over the course of the past several centuries. Apparently, the heavy ice sheet that covered this region during the last glacial period caused the land to sink, or subside. As the ice sheet melted and retreated, this weight was removed and the land began to slowly rise. Parts of North America near Hudson Bay are also slowly rising for the same reason.

Large-scale movements of the earth's crust can cause coastlines to sink or to rise. For example, much of the east coast of the United States is slowly sinking. Some scientists think that the side of a continent near the trailing edge of a moving crustal plate tends to sink. This hypothesis may help explain the sinking of the east coast, since it is located toward the trailing edge of the westward moving North American plate.

Coastlines of continents near a plate boundary may also be exposed to forces created as the plates come together. Parts of such coasts may rise and others may sink. This seems to be the case on the west coast of the United States, which shows evidence of rising in most places but sinking in others. For example, much of the coast of California, Oregon, and Washington has risen over the past few million years. In contrast, the San Francisco bay area of California has sunk.

Submergent Coastlines

When sea level rises or the land sinks, a **submergent coastline** is formed. Divides between neighboring valleys become headlands separated by bays and inlets. Beaches are usually short, narrow, and rocky because they have just begun to form. The highest parts of the

Figure 16–11. Notice how the features of a submergent coastline are eroded from youth (left) to old age (right) as sea level rises relative to the land.

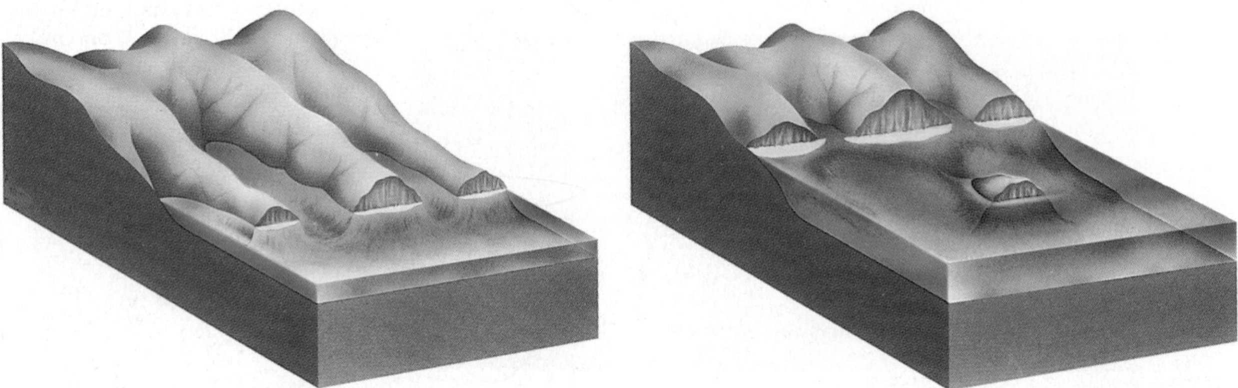

submerged land may form offshore islands. Irregular coastlines in many parts of the world, such as the northeast coast of the United States, seem to have been formed by submergence. However, submergent coastlines are eventually changed by wave action. The headlands are eroded, shorelines become straighter, and beaches become longer and wider.

The mouth of a river valley that is submerged by ocean water may become a wide, shallow bay that extends far inland. This type of bay, where salt water and fresh water mix, is called an **estuary** (ESH-chuh-WER-ee). The mouths of most rivers along the United States' east coast are estuaries. The slightly salt water of estuaries is an important source of such seafood as shrimp, crab, and oysters. Pollution of the rivers flowing into an estuary can destroy this valuable food source in such productive waters as Chesapeake Bay.

After the last ice age, the U-shaped glacial valleys along the coasts of Alaska, Chile, Greenland, Norway, and western Canada

The Flooding of Venice

The city of Venice, Italy, called "the Jewel of the Adriatic," is built on about 120 islands in the Adriatic Sea. Venice is famous not only for its canals but also for its priceless art treasures.

However, the waters of the Adriatic Sea pose a constant threat to the city. Every year, winter storms cause sea water to flood its public squares, walkways, and buildings. Venice's famous St. Mark's Square floods an average of 40 times a year. Overexposure to floodwaters is slowly eroding the foundations of its Byzantine and Renaissance architecture.

These floods are the result of three factors: erosion, rising sea level, and sinking land. Erosion of barrier beaches and sandbars allows high tides to reach farther into Venice. Also, like oceans worldwide, the Adriatic Sea is rising. Its average sea level has risen 11 cm this century while Venice has sunk almost 13 cm.

The weight of added sediments is causing the city to sink about 1 mm each year. Removal of too much groundwater and the weight of the buildings within the city have caused additional sinkage.

To prevent future and possibly permanent damage to Venice, the city restricts the pumping of groundwater. City planners are also considering building massive floodgates to block high tides that may threaten the city.

What will be the effect of the restriction on pumping of groundwater?

became flooded with ocean water as sea level rose. These flooded glacial valleys, called **fiords** (fee-ORDZ), form spectacular narrow, deep bays with steep walls.

Emergent Coastlines

When the land rises or sea level falls, an **emergent coastline** results. If an emergent coastline has a steep slope and is exposed rapidly, it will erode, forming sea cliffs, narrow inlets, and bays. A series of wave-cut terraces may be exposed as well. Much of the west coast of the United States is an emergent coastline.

Along some emergent coastlines, a gentle slope is formed when part of the continental slope is slowly lifted and exposed. A relatively smooth coastal plain with few bays or headlands and many long, wide beaches results. Much of the Florida coast is an emergent coastline.

Figure 16–12. South Padre Island, off the coast of Texas, is a long, narrow barrier island.

Barrier Islands

Most of the east coast of the United States from New York to Texas is a relatively flat coastal plain. As sea level has risen over the past 15,000 years, the shoreline has moved inland as much as 150 km across this gently sloping coast. Some geologists think that this movement isolated former dunes from the old shoreline, forming **barrier islands.** Barrier islands are long, narrow offshore ridges of sand, many over 100 km long. They lie nearly parallel to the shoreline 3 km to 30 km offshore. There are nearly 300 barrier islands between Long Island, New York, and Padre Island, Texas. Other barrier islands may have formed when sand spits were separated from the land by storms. Waves also may pile up ridges of sand scraped from the shallow offshore sea bottom. These deposits then may be moved toward the shore, or shoreward, by waves, currents, and winds. Winds blowing toward the land often create a line of dunes on the shoreward side of the islands. The dunes are usually 3 m to 6 m high, although they may reach a height of 30 m.

Between barrier islands and the shoreline is a narrow region of shallow water called a **lagoon.** Lagoons are often nearly filled with mud brought in by streams. At high tide, water may cut channels through a barrier island to form a passageway, or tidal inlet, into the lagoon. As a result, large areas of some lagoons become **tidal flats.** A tidal flat is a muddy or sandy part of the shoreline that is visible at low tide and is submerged at high tide. Most tidal flats are covered with plants that can grow in the salty water.

Barrier islands have become popular recreation and resort areas, but they can be hazardous places to live. Because of their low elevation, barrier islands are severely eroded by large waves from storms, especially hurricanes. During a storm, sand is washed from the ocean side toward the inland side of the island, causing most barrier islands to migrate toward the shoreline. Some barrier islands are being eroded at an average rate of 20 m/yr. Houses built along the beach have collapsed as beaches have disappeared during storms.

Coral Reefs

Another coastal feature is a **coral reef.** Corals are small marine animals that live in warm, shallow sea water. A coral extracts calcium carbonate from ocean water and uses it to build a hard outer skeleton. Corals attach to each other to form large colonies. New corals grows on top of dead ones, forming a coral reef, which is a submerged ridge made up of millions of coral skeletons.

Some coral reefs form around tropical volcanic islands. The coral colony grows in the shallow water near the shore. This type of coral reef around the coast of the island is called a **fringing reef.** As the ocean lithosphere bends under the weight of the volcano, both the volcano and reef sink. The coral reef, however, builds higher because the animals can only live near the surface of the water. The coral thus forms a **barrier reef** offshore around the remnant of the volcanic island. Finally, the island disappears completely under water, leaving a nearly circular coral reef, called an **atoll,** surrounding a shallow lagoon.

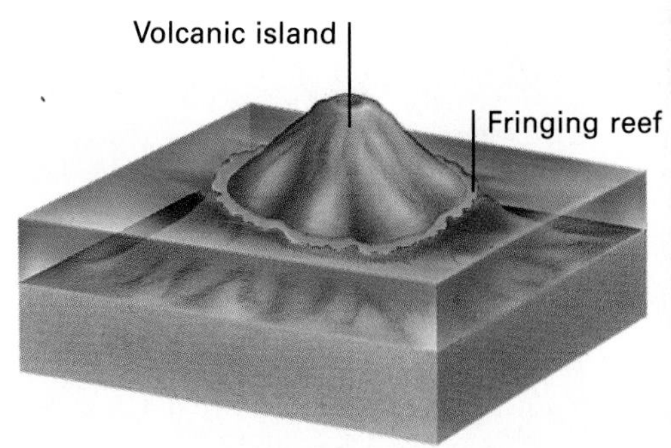

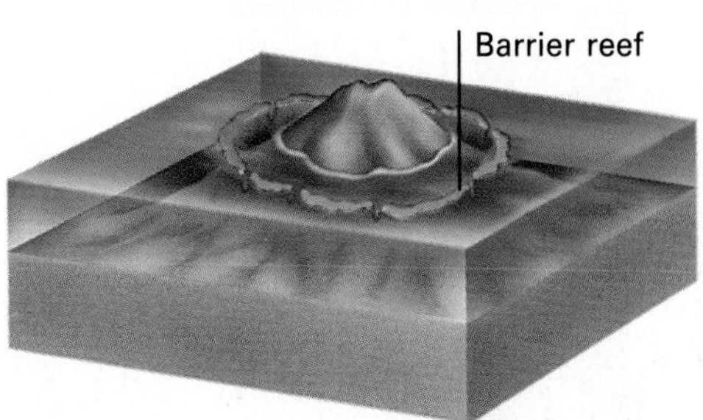

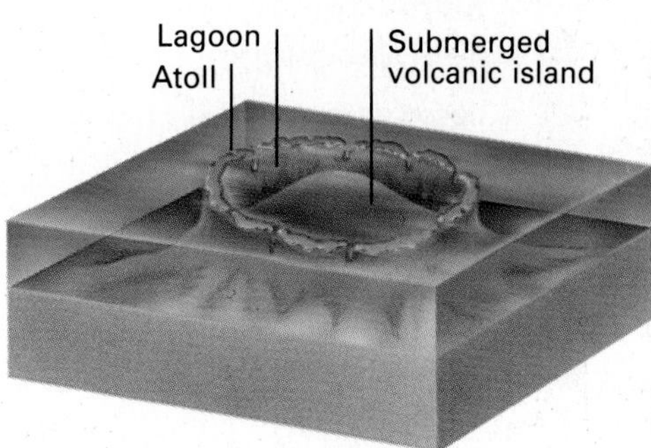

Figure 16–13. A coral reef changes from a fringing reef (top) to a barrier reef (center) to an atoll surrounding a lagoon (bottom). These changes occur as a volcanic island slowly sinks and the reef builds higher.

Preserving the Coastline

Coastal lands are used for commercial fishing, shipping, industrial and residential development, and recreation. However, coastlines are fragile, and they can be easily damaged.

As the earth's population increases, the use of shorelines will also increase. For example, about 5 percent of the land in the United States is coastal. But, population experts estimate, by the year 2000, one half of its people will live in coastal areas. New harbors will be dug along the Atlantic and Pacific oceans, the Gulf of Mexico, and the Great Lakes.

Pollution, which has already destroyed one tenth of the shellfish-producing waters of the country, will probably continue to increase. Oil spills are a threat as tankers travel near the shorelines and oil wells are drilled offshore. If shoreline use continues as in the past, much of the existing coastal land will be less desirable.

If the coastal zone is to be used at all, private owners and government agencies must work together to preserve it. Efforts have been made. For example, some New Jersey shoreline towns have brought sand from inland to rebuild beaches eroded by severe winter storms. The ocean floor just offshore also can be a source of sand for such beach preservation projects.

Section 16.3 Review

1. How does the formation of a submergent coastline differ from the formation of an emergent coastline?
2. Why are barrier islands particularly subject to erosion?
3. How does a fringing reef become an atoll?
4. Predict the effect that cutting into the shoreline to create a new harbor will have on the longshore current.

Chapter 16 Review

Key Terms

- atoll (311)
- barrier island (310)
- barrier reef (311)
- beach (304)
- berm (305)
- coral reef (311)
- deflation (298)
- deflation hollow (298)
- desert pavement (298)
- dune (299)
- emergent coastline (310)
- estuary (309)
- fiord (310)
- fringing reef (311)
- headland (303)
- lagoon (310)
- loess (301)
- sand bar (305)
- shoreline (302)
- spit (306)
- submergent coastline (308)
- tidal flat (310)
- tombolo (306)
- ventifact (298)
- wave-built terrace (303)
- wave-cut terrace (303)

Key Concepts

The wind erodes the land by weathering and by removing sediments. **See page 297.**

The two types of wind deposits are made up of particles of different sizes. **See page 299.**

Beaches result from the deposition of sediments by waves. **See page 304.**

Waves weather and erode the shoreline, producing characteristic features. **See page 305.**

Longshore currents move sediments along a shoreline. **See page 306.**

Coastlines are exposed or submerged as sea level changes. **See page 307.**

Barrier islands are unstable coastline features. **See page 310.**

Coral reefs that surround volcanic islands undergo changes as the islands sink. **See page 311.**

Human activities affect land along the coasts. **See page 311.**

Review

On your own paper, write the letter of the term that best completes each of the following statements.

1. Loose fragments of rock and minerals measuring between 0.06 mm and 2 mm are referred to as
a. pollen. b. dust.
c. sand. d. desert pavement.

2. Wind moves sand by
a. saltation. b. abrasion. c. emergence. d. depression.

3. The most common form of wind erosion is
a. migration. b. abrasion.
c. saltation. d. deflation.

4. Dunes move primarily by
a. abrasion. b. deflation.
c. migration. d. submergence.

5. Unlayered, yellowish, fine-grained deposits are called
a. beaches. b. loess.
c. dunes. d. desert pavement.

6. The most important erosion agent along shorelines is
a. waves. b. wind.
c. weathering. d. tides.

7. All of the following shoreline features are produced by wave erosion of sea cliffs except
a. spits. b. sea stacks.
c. wave-cut terraces. d. sea arches.

8. The composition of beach deposits depends on
   a. the climate. b. the source rock.
   c. the time of year. d. wave action.
9. Longshore-current deposition of sand at the end of a headland produces a
   a. sand bar. b. dune. c. spit. d. sea cliff.
10. Sea level is now
   a. stationary. b. falling about 1 mm/yr.
   c. rising about 1 cm/yr. d. rising about 1 mm/yr.
11. A coastline resulting from a rise in sea level or subsidence of the land is
   a. emergent. b. submergent. c. glaciated. d. volcanic.
12. Barrier islands tend to migrate
   a. seaward. b. along the shore.
   c. in the summer. d. toward the shore.
13. Coral reefs are produced by marine organisms that
   a. swim in circles.
   b. extract minerals from sea water to form hard skeletons.
   c. attach themselves to sand bars.
   d. build nests of sand.
14. If shoreline resources continue to be used as in the past,
   a. coastal land will be greatly improved.
   b. the damage can be repaired.
   c. sea level will fall.
   d. coastal land will become less desirable.

Critical Thinking

On your own paper, write answers to the following questions.

1. The deserts of the southwestern United States contain many tall, sculpted rock formations that are the result of weathering and erosion. Was wind or water erosion the more likely agent responsible for these formations? Explain your answer.
2. Beautifully colored sunsets and sunrises are the result of dust in the atmosphere. Such sunsets and sunrises were visible around the world for two years after Krakatau, a volcanic island in Indonesia, erupted in 1883. Explain how this is possible.
3. Suppose that once each month for a year a satellite orbiting the earth takes a photograph of the same 1-km^3 sandy area of the Sahara. Would the surface features shown in these 12 photographs remain essentially the same, or would they be very different? Explain your answer.
4. What effect does the development of a wave-built terrace have on erosion of the shoreline? What phenomenon might counteract this effect?
5. Describe two ways that the melting of continental glaciers affects the relative level of the land and the sea.

Application

1. Suppose that scientists have observed that the size of the sand bars in a particular area has been decreasing steadily over the past 20 years. What does this observation suggest about the climate of the area? Explain why.
2. As you approach a large landmass by ship, you notice an island connected to the shore by a tombolo. What type of shoreline do you predict the mainland will have? Explain your answer.
3. Concept map ? ? ? Construct a **concept map** using the new terms listed on the previous page to show the relationship between landforms and the agents of erosion that produce them.

Extension

1. Imagine that you are a newspaper reporter who has traveled to another planet. Scientists know that, at one time, both wind and water eroded the surface of the planet. Prepare a newscast describing the landscape you see and explaining the processes that produced it. Present the newscast to the class.
2. Suppose that sea level continuously rises at the rate of 1 mm/yr and that other factors affecting the coastlines do not change. Draw a map showing what the east coast of the United States might look like in 10,000 years.

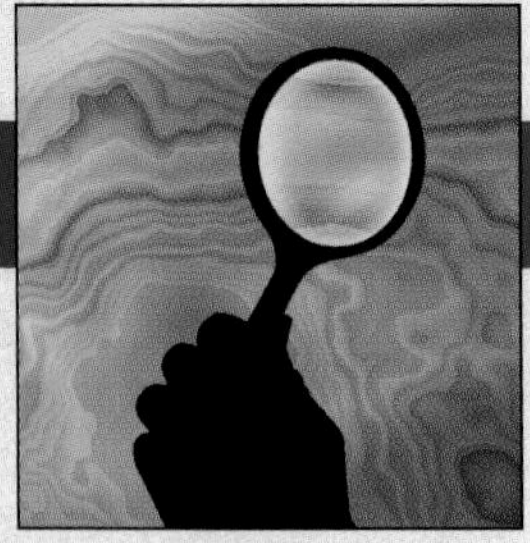

IN-DEPTH INVESTIGATION

Beaches

Materials

- masking tape
- 2 milk cartons (small)
- water
- sand (5–10 lb.)
- wooden block (large)
- pebbles
- rocks (small)
- plaster of Paris
- stream table or large plastic container
- metric ruler

Introduction

According to some estimates, 50 percent of the population in the United States lives within 50 miles of a shoreline. Beaches provide a source of income for many residents. Coastal management has become a growing concern as beaches are increasingly used for resources and recreation. The supply of sand for most beaches has been cut off by dams built on rivers and streams that would otherwise carry sand to the sea. Waves generated by storms also continuously wear away beaches.

Most waves that reach a beach move straight toward the beach. However, some waves in deeper water move at an angle to the beach. As these waves approach the beach at an angle, they set up a current called a longshore current. A longshore current moves parallel to the beach and in the same direction that the waves were moving in deep water.

In this investigation, you will examine a model showing how the forces generated by wave action build up, shape, and wear away beaches.

Prelab Preparation

1. Review Section 16.2, pages 302–306.
2. One day before you begin the investigation, make two plaster blocks. Mix a small amount of water with the plaster of Paris until the mixture is smooth. Before you pour the plaster mixture into the milk cartons, add five or six small rocks to the mixture for added weight. Let the plaster harden overnight. Carefully peel off the milk carton.

Procedure

1. Prepare a stream table or other similar large, shallow container. Make a beach by placing a mixture of sand and small pebbles at one end of the container. The beach should occupy about one fourth of the length of the container.
2. In front of the sand, add water to a depth of 2 to 3 cm. What happened to the beach when water was first poured into the container?
3. Using the large wooden block, generate several waves by moving the block up and down in the water at the end of the container opposite the beach. See Figure A.

Figure A

Continue this wave action until about half the beach has shown some movement. Describe the beach after this wave action has taken place. What happened to the particles of fine sand?

4. Predict what will happen to the beach if it has no source of additional sand.
5. Remove the sand and rebuild the beach.
6. In some places, breakwaters have been built offshore to protect beaches from washing away. Build a breakwater by placing two plaster blocks across the middle of the container. Leave a 4-cm space between the blocks. See Figure B.
7. Use a wooden block to generate waves as in Step 1. Record your observations.
8. Drain the water and make a new beach along one side of the container for about half its length. See Figure C. Predict what effect a longshore current would have on sand just offshore. How would this affect sand on the beach?
9. Using the wooden block, generate a series of waves from the same end of the container as the end of the beach. See Figure C. Record your observations. What happened to the beach? What happened to the shape of the waves along the beach?

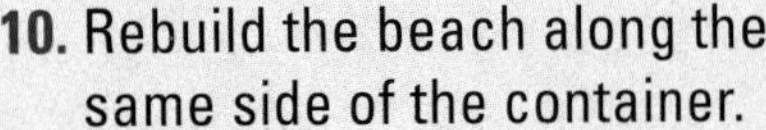

10. Rebuild the beach along the same side of the container.
11. A jetty or dike can be built out into the ocean to intercept and break up a longshore current. Place one of the small plaster blocks in the sand to act as a jetty. See Figure D.
12. As you did before, use the wooden block to generate waves. Describe the results.
13. Remove the wet sand and place it in a container. Dispose of the water. *NOTE: Follow your teacher's instructions for disposal of the sand and water. Never pour water containing sand into a sink.*

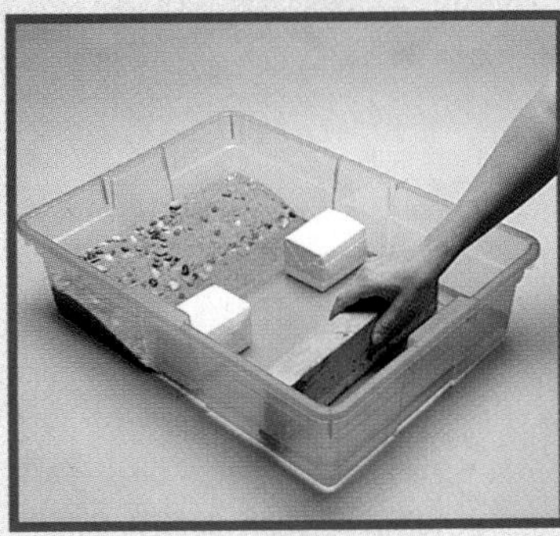

Figure B

Figure C

Figure D

Analysis and Conclusions

1. How does wave action build up a beach? How does wave action wear away a beach?
2. How do longshore currents change the shape of a beach?
3. What effect would a series of jetties have on a beach?

Extensions

1. What can be done to preserve a recreational beach area from erosion as a result of excessive use by people?
2. What can be done to preserve a recreational beach area from being washed away as a result of wave action and longshore currents?

What Comes Down Must Go . . . Where?

Imagine looking out your classroom window during a downpour. Billions of tiny raindrops splatter off everything in sight. Streams of water fall from your school's roof, forming dozens of puddles and streams on the ground. These miniature lakes and rivers swirl together, and tiny torrents carry away leaves, bits of trash, and other debris. A day or two later, the ground outside looks completely dry. Where does all that water go?

Materials

- red, blue, and orange pencils
- large reclosable storage bag (about 23 cm × 33 cm)
- large piece each of paper and cardboard
- scissors
- fine-tipped permanent marker
- umbrella, raincoat, or other foul-weather protection

In this activity, you will create a map of your school. Your map will show not only the buildings but also the paved and vegetated areas of the school's campus. By observing the type of ground cover and the slope of the terrain at various locations, you will predict whether rainwater will collect or run in tiny streams. You will also look for possible sources of pollution and places where erosion might occur. Later, you will go outside in the rain and find out whether your predictions were correct!

Part 1 (Outside on a fair-weather day)

1. Form a team with several of your classmates. Then, depending on the number of teams, divide your school's campus into equal portions. Each group will work on one part of the campus.

2. Cut out a piece of paper and a piece of cardboard that fit exactly into a large reclosable storage bag. This piece of paper will be your map.

3. Map one section of your school's campus. You should include buildings, paved areas (such as sidewalks, outdoor basketball or tennis courts, and parking lots), and vegetated areas (such as lawns, athletic fields, and natural areas).

4. a. Using a red pencil, mark the different areas on your map with arrows that point downhill. Use a narrower arrow to indicate a steeper slope, and a wider arrow to indicate a more gradual slope. Indicate flat areas with circles.

b. Using a blue pencil, mark the locations where you think surface water may collect or flow during a steady rain. These may include low-lying areas, the roofs of buildings, gutters, and drainage ditches.

c. Using an orange pencil, mark the locations that you think might contribute pollution to the runoff. (For instance, these might include oil stains in a parking lot or places where trash is usually found on the ground, such as near a dumpster.)

5. Seal your map and the piece of cardboard in the reclosable bag. When you are outside in the rain, you will use a permanent marker to write your observations on the outside of the bag.

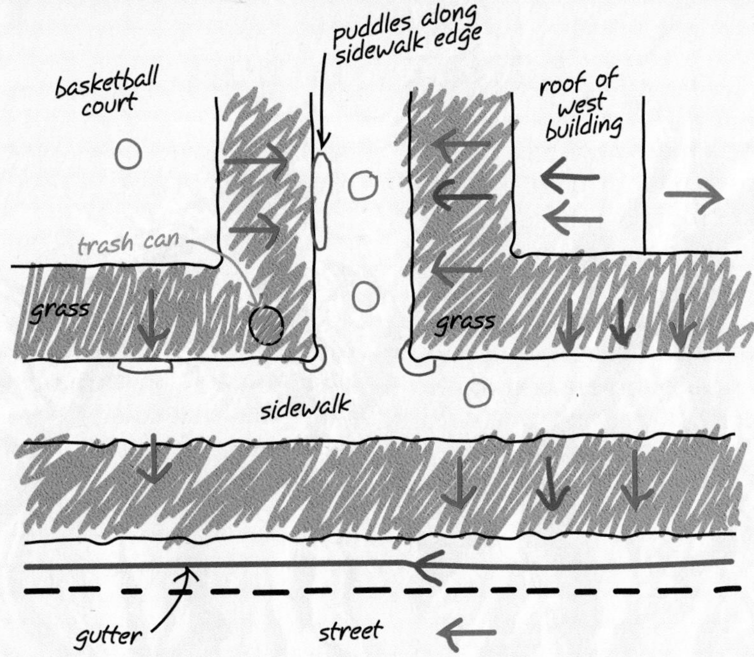

Part 2 (Outside during a steady rain)

6. Check your predictions. Dress appropriately and go outside to mark the actual places where water collects and runs. In places where water moves along the ground, use arrows to show its direction. Using *P*'s, mark the locations of pollutants that you observe in the water. Using *E*'s, mark the locations where erosion seems to be occurring. (Look for soil or natural debris that is being washed along by moving water.)

Part 3 (Back in the classroom)

7. Discuss some of the locations where your predictions were correct and incorrect. How do you explain the discrepancies between your predictions and observations?

8. Was the pollution that you observed suspended load, bed load, or dissolved load? Explain why there might have been pollution that you could not observe.

9. Explain how erosion on your school's campus could affect the erosion and deposition that occurs downstream.

10. Assemble the maps from your class into a single map of your school's campus. In your opinion, does most of the rainfall that hits your school's campus become groundwater or runoff? Where does runoff go when it leaves your school's campus? Your school's campus is probably part of a larger, local watershed. Find out what stream or other body of water the surface runoff in your area empties into.

Going Further

Following runoff from the top of a mountain through streams, lakes, and rivers all the way to the ocean is easy at the Gulf of Maine Aquarium's Web site. And that's just the beginning! Visit this site at octopus.gma.org.

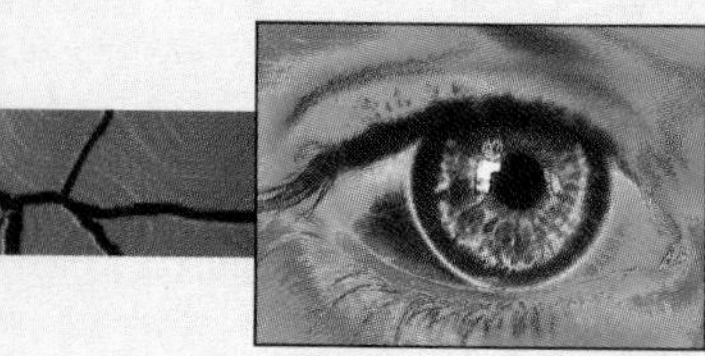

Eye on the Environment

Losing Ground

In the last 40 years, almost a third of the world's topsoil has been lost to erosion. Because of this, food production in some areas has been reduced to zero. At the same time, the world's population is growing by 250,000 people each day. With more people and less land to farm on, concerns about having enough food to go around are becoming more serious.

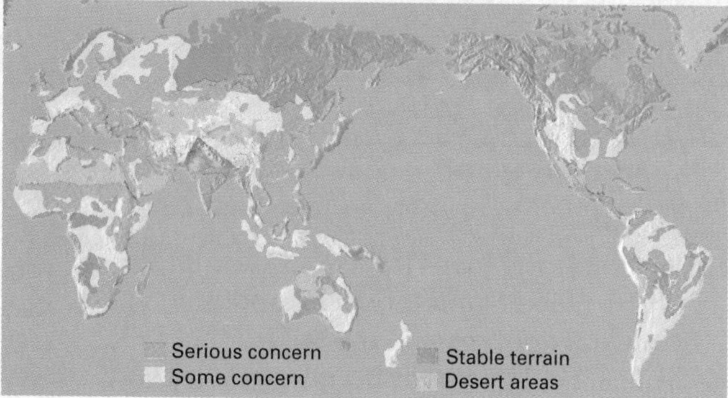

▲ **Top soil loss is affected by climate and population density.**

Understanding the Problem

Imagine a forest that is densely packed with trees. During a rainstorm, raindrops hit the tops of the trees and splatter down, eventually hitting the ground as a light but steady sprinkle. As rain collects on the ground, some of it soaks into the soil and the rest may trickle away, carrying tiny pebbles and a bit of soil along with it. For every hectare of forest like this, about 10–50 kg of soil is lost in an average year. (One hectare is a little smaller than two football fields.)

Now compare that forest with a recently plowed field. Remember that there is no cover of vegetation over the field—the soil is completely exposed to the full force of the weather. Not only that, but the soil is very loosely packed because it has been recently plowed. In addition, the plowing has torn up underground roots that might have helped to keep the soil in place.

As each raindrop hits the field, the force of the impact kicks up a tiny bit of soil. This flying soil may be carried away by the wind or by water. Most of it is lost to the tiny currents of water that form in the furrows of plowed fields. Either way, that little bit of soil is lost. It may not sound like a big deal, but it adds up fast. Because of wind and water erosion, one hectare of farmland can lose over 100,000 kg of soil in one year!

▲ **The soil that is dislodged as a raindrop hits the ground may be carried away by wind or water.**

Where Does It Go?

Wind can carry soil for hundreds of miles. Soil from Africa, for example, has been detected as far away as Brazil and Florida. If it is carried by water, the soil may settle onto the bed of a downstream river or lake. Either by wind or by water, much of it will be carried all the way to the sea. In fact, every year, thousands of tons of fertile topsoil end up at the bottom of the ocean.

The Little Colorado River sometimes takes on the color of its surrounding as it carries large amounts of eroded sediments. ▶

In a healthy ecosystem, topsoil that is lost through erosion is replaced by natural processes. These include the decomposition of organic matter by microorganisms and the breakdown of rocks by normal weathering. When a balance exists between the soil that is lost and the new soil that forms, the rate of topsoil loss is called *sustainable.* Currently, about 90 percent of the cropland in the United States is losing topsoil at faster than a sustainable rate.

Sustainable Farming

By changing their farming practices, farmers can drastically reduce the amount of soil that is lost from their fields. The critical step is that some vegetative cover must be left on the ground. This buffers the soil from the direct effects of wind and rain. For instance, in *no-till farming,* each new crop is planted among the remains of the previous crop. (Normally, a harvested field is tilled to overturn the soil and bury the remains of the old crop.) No-till farming ensures that there is always some vegetation left to cover the soil. Other practices that reduce erosion include ridge planting, contour plowing, and crop rotation.

Many farmers have already switched to methods of sustainable farming. Unfortunately, a more concerted effort will be necessary. As the world's population continues to increase, more food will be needed. Because of this, it will be more and more important to preserve the topsoil that we have left. The pressure to use sustainable-farming methods will be greater as time goes on and more soil is lost to erosion.

◀ **In this example of no-till farming, a new crop of soybeans grows up through the decaying remains of a corn crop.**

Think About It

Would erosion be greater on a level surface or a sloped surface (such as a field on the side of a hill)? Explain your answer.

Extension: Find out what is meant by the term *desertification.* How does it relate to topsoil erosion?

UNIT 5

The History of the Earth

Glen Canyon, Arizona (background);
Dawn Redwood forest leaf, Republic, Washington (inset)

Introducing Unit Five

Today's world emphasizes making immediate records of what is happening around us, but today's news is always a part of tomorrow's history. Knowing where we are in the field of earth science is important, but it's just as helpful to know how we got there—especially if we're looking at our current situation in terms of future implications.

The complex process of earth science is just that—a process that encompasses yesterday, today, and tomorrow. To understand that process, it is necessary to have the basic details recorded somewhere. Fortunately, even before there were science writers to help make those records, the earth recorded—and continues to "write"—its own detailed history.

In the following unit, you will study the earth's history through rock and fossil records that are as important to us today as they will be in the future.

Richard A. Kerr

Richard Kerr
Senior Writer, Research News Staff
Science

In this UNIT

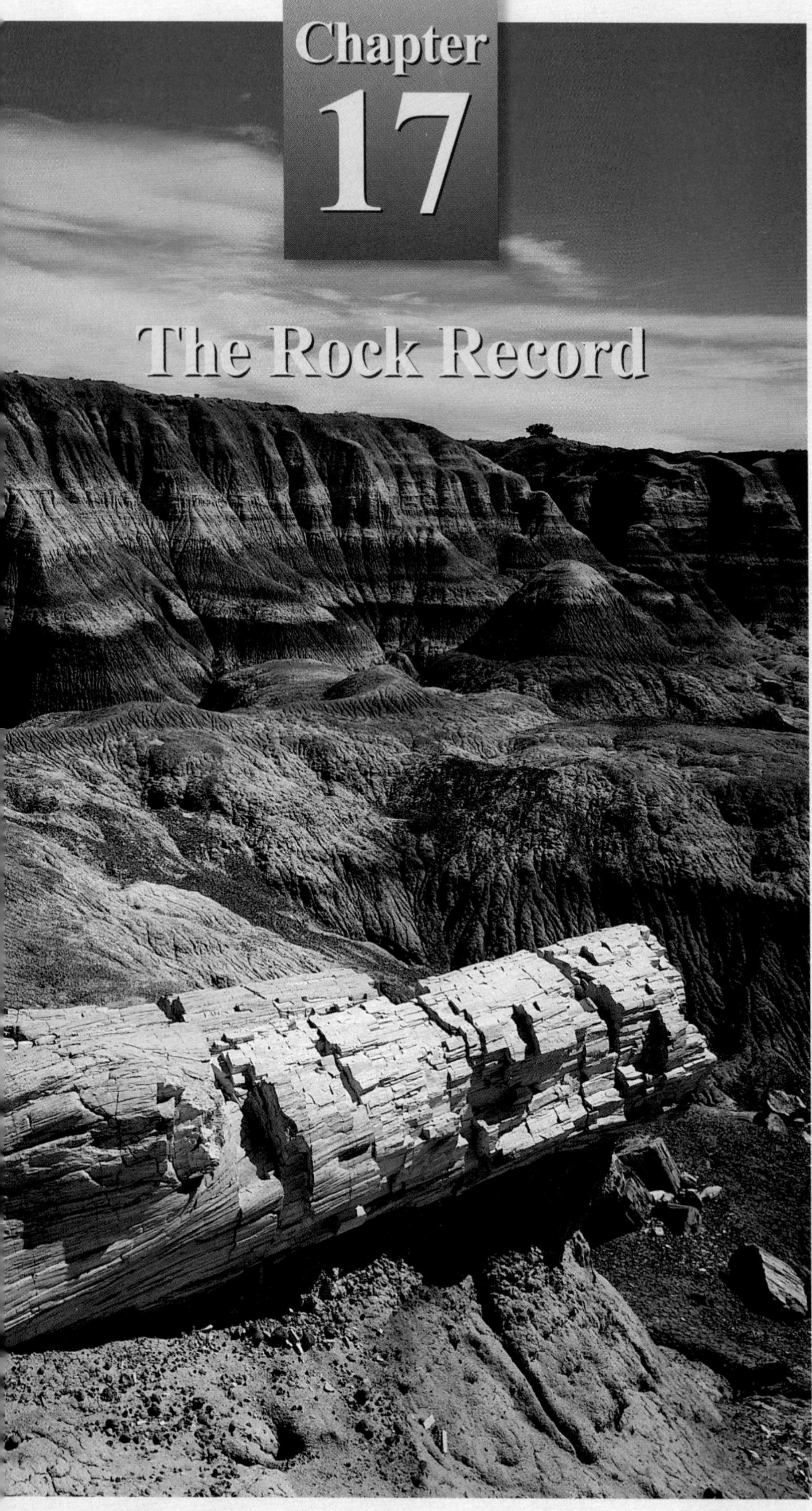

Rocks contain clues to the earth's past, including evidence of ancient life-forms. For example, the piece of rock shown here was once a living tree. Earth scientists can use the information found in the rock record to help them understand the geologic processes that have shaped the earth's crust.

In this chapter, you will read about the methods earth scientists use to uncover the geologic history of the earth.

Chapter Outline

A petrified log lies in a petrified forest at Blue Mesa, Arizona.

17.1 Determining Relative Age

Section Objectives

- **State the principle of uniformitarianism.**
- **Explain how the law of superposition can be used to determine the relative age of rocks.**
- **Compare three types of unconformity.**
- **Apply the law of crosscutting relationships to determine the relative age of rocks.**

Geologists estimate that the earth is about 4.6 billion years old. The idea that the earth is billions of years old originated with the work of James Hutton, an eighteenth-century Scottish physician and gentleman farmer. Hutton was a keen observer of the geologic changes taking place on his farm. Using scientific methods, Hutton drew conclusions based on his observations.

Hutton theorized that the forces changing the landscape of his farm were the same forces that had changed the earth's surface in the past. He thought that by studying the present, people could learn about the earth's past. Hutton's **principle of uniformitarianism** states that current geologic processes, such as volcanism and erosion, are the same processes that were at work in the past. This principle is one of the basic foundations of the science of geology. Geologists later refined Hutton's ideas by pointing out that although the processes are the same, the rates at which those processes occur may vary over time.

Before Hutton's work, most people thought that the earth was only about six thousand years old. They also thought that all geologic features had been formed at the same time. Hutton's principle of uniformitarianism, however, raised some serious questions about the earth's age. Hutton observed that the forces changing the land on his farm operated very slowly. He reasoned that millions of years must have been needed for those same forces to create the complicated rock structures found in the earth's crust. Hence, he concluded, the earth must be much older than previously thought.

Hutton's observations and conclusions about the age of the earth encouraged others to learn more about the earth's history. One way to learn about the earth's past is to determine the order in which rock layers and structures were formed.

Figure 17–1. The layers of sedimentary rock that make up Canyon de Chelly, in Arizona, were deposited over millions of years.

Layers of rocks, called *strata,* are like the pages in a history book, detailing the sequence of events that took place in the past. Using a few basic principles, scientists can determine the order in which rock layers were formed. Once the order of formation is known, a **relative age** can be determined for each rock layer. Relative age indicates that one layer is older or younger than another layer. It does not indicate the exact age of the rock.

Various kinds of rocks form layers. Igneous rocks form layers when successive lava flows stack up. In some metamorphic rock, such as marble, layers are also visible. To determine the relative age of rocks, however, scientists commonly study the layers in sedimentary rocks.

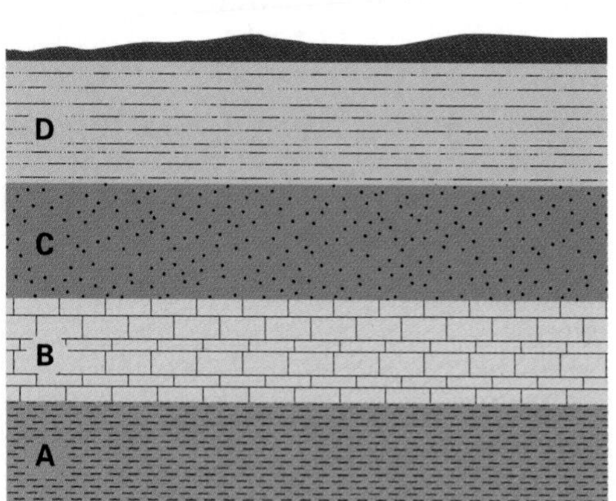

Figure 17–2. By applying the law of superposition, geologists are able to determine that layer *D* is the youngest layer in this rock bed.

Law of Superposition

The formation of sedimentary rock begins when sediments are deposited horizontally. As the sediments accumulate, they are compressed and harden into sedimentary rock layers, or beds. The boundary between two beds is called a **bedding plane.**

Scientists use a basic principle called the **law of superposition** in attempting to determine the relative age of a layer of sedimentary rock. The principle states that an undeformed sedimentary rock layer is older than the layers above it and younger than the layers below it. According to the law of superposition, layer *A* shown in Figure 17–2 was the first layer deposited, and thus it is the oldest. The last layer deposited was layer *D.* Is layer *B* older or younger than layer *C*?

Scientists also know that sedimentary rock generally forms in horizontal layers. Therefore, they can assume that most sedimentary rock layers that are not horizontal have been tilted or deformed by crustal movements after the layers were formed. Sometimes violent tectonic forces push older layers on top of younger ones, or movements of the crust may overturn a group of rock layers. In such cases, the law of superposition cannot be easily applied. Scientists must look for clues to the original arrangement of layers and then apply the law of superposition.

One clue to the original arrangement of rock layers lies in the size of the particles found in a layer of sedimentary rock. As you may recall from Chapter 10, the largest particles of sediment are usually deposited near the bottom of a layer. Therefore, the study of particle sizes in a sedimentary rock layer may reveal whether the layer has been overturned.

Another clue to the original position of rock layers is found in the shape of the bedding planes. In cross-bedded layers of sedimentary rocks, for example, sediments are deposited in curved sheets at an angle to the bedding plane. The tops of cross-bedded layers are often eroded before new layers are deposited. Therefore, the crossbeds are still curved at the bottom but flat across the top. By studying the shape of the bedding planes, scientists can determine the original position of cross-bedded layers.

Ripple marks can also be helpful in determining the order of rock layers. Ripple marks are small waves formed on the surface of sand by the action of water or wind. When the sand becomes sandstone, the ripple marks may be preserved. In undisturbed sedimentary rock layers, the peaks of the ripple marks point upwards. Thus, by examining ripple marks, scientists can establish the original arrangement of the rock layers. The relative ages of the rocks can then be determined using the law of superposition.

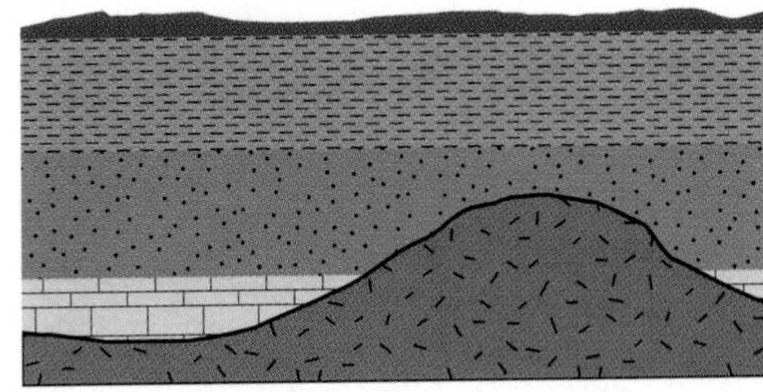

Noncomformity

Unconformities

Sometimes movements of the earth's crust lift up rock layers that were buried and expose these layers to erosion. Later, if the eroded surface is lowered or the sea level rises, sediments will again be deposited, forming new rock layers. The missing rock layers create a break in the geologic record, just like when pages are missing from a book. This break in the geologic record is called an **unconformity.** An unconformity indicates that for a period of time deposition stopped, rock was removed by erosion, and then deposition resumed. As illustrated in Figure 17–3, there are three types of unconformities.

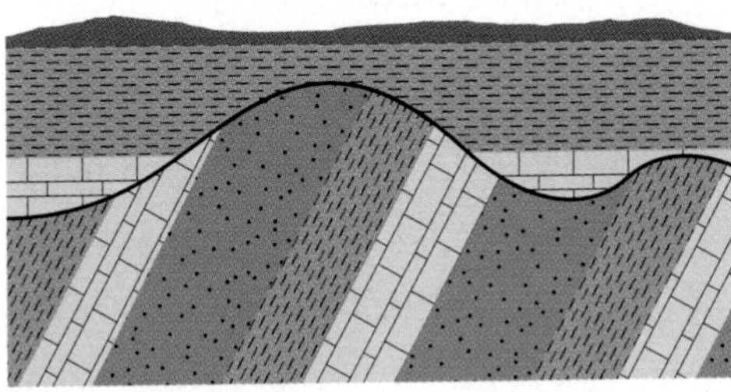

Angular unconformity

Nonconformity

Sedimentary rock is stratified, or deposited in layers. Metamorphic and igneous rocks are usually unstratified. An unconformity in which stratified rock rests upon unstratified rock is called a **nonconformity.** For example, unstratified rock such as granite forms deep within the earth. The granite may be lifted to the earth's surface by crustal movements. Once exposed, the granite begins eroding. Sediments may then be deposited on the eroded surface. The boundary between the sandstone and the granite layers is a nonconformity. It represents an unknown period of time during which the granite was eroded.

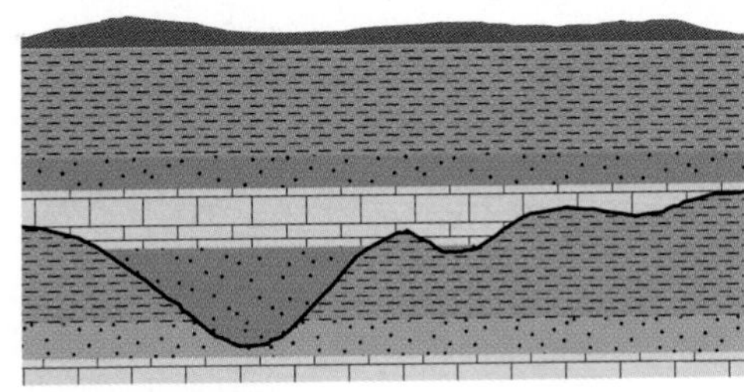

Disconformity

Figure 17–3. The three types of unconformity represent gaps in the geologic record.

Angular Unconformity

Another type of unconformity results when rocks deposited in horizontal layers are folded or tilted and then eroded. When erosion stops, a new horizontal layer is deposited on a tilted layer. The boundary between the tilted layers and the horizontal layer is called an **angular unconformity.** As Figure 17–3 shows, the bedding planes of the older rock layers are not parallel to those of the younger rock layers deposited above the angular unconformity.

Disconformity

Sometimes layers of sediments on the ocean floor are lifted above sea level without folding or tilting. Exposed to wind and running water, the surface layers are eroded. Eventually, the area again falls below sea level and deposition resumes. The boundary between the older, eroded surface and the younger, overlying layers is nearly horizontal and is called a **disconformity.** Although the rock layers look as if they were deposited continuously, a large time gap exists where the upper and lower layers meet at the unconformity.

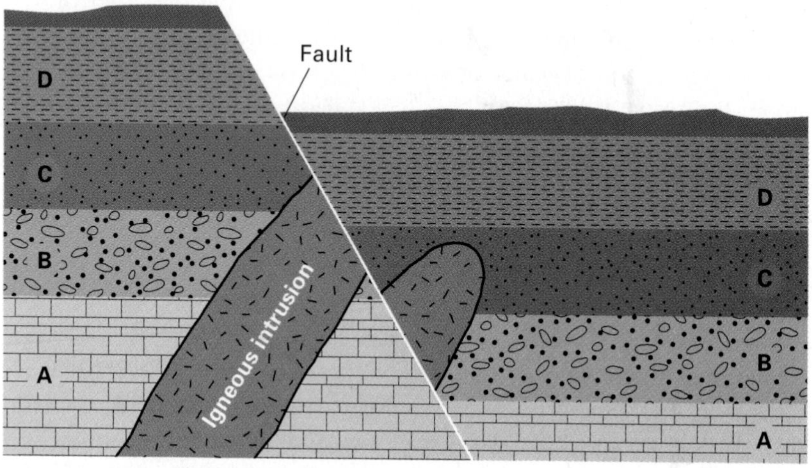

Figure 17–4. The law of crosscutting relationships can be used to determine the relative ages of rock layers and the faults and intrusions within them.

Crosscutting Relationships

When tectonic activity has disturbed rock layers, determining relative age using only the law of superposition may be difficult. In such cases, scientists may also apply the **law of crosscutting relationships.** The law of crosscutting relationships states that a fault or an intrusion is always younger than the rock layers it cuts through. A fault is a break or crack in the earth's crust along which rocks shift their position. An intrusion is a mass of igneous rock formed when magma is injected into overlying rock and then cools and solidifies. Crosscutting relationships can be extremely complex, and careful observation is necessary to accurately date rocks. Figure 17–4 shows a series of rock layers that contains both a fault and an igneous intrusion. As you can see, an intrusion cuts across layers *A*, *B*, and *C*. According to the law of crosscutting relationships, the intrusion is younger than layers *A*, *B*, and *C*. Look at the fault. What is the relative age of the fault compared with that of the rock layers?

Both the law of superposition and the law of crosscutting relationships can be applied to unconformities. According to the law of superposition, all rocks beneath an unconformity are older than those rocks above the unconformity. If a fault or intrusion cuts through the unconformity, the fault or intrusion is younger than all the rocks it cuts through above and below the unconformity.

Section 17.1 Review

1. What is the principle of uniformitarianism?
2. How can the law of superposition be used to determine the relative age of sedimentary rock?
3. Compare an angular unconformity with a disconformity. Which would be more difficult to recognize?
4. Suppose you find a series of rock layers in which a fault is cut off below an unconformity. Explain how you would apply the law of crosscutting relationships to determine the relative age of the fault and of the rock layers that were deposited above the unconformity.

17.2 Determining Absolute Age

Relative age indicates only that one rock layer is younger or older than another rock layer. In order to learn more about the earth's history, scientists often need to determine the actual age, or **absolute age,** of a rock layer.

Rates of Erosion and Deposition

One method of estimating absolute age involves studying rates of erosion. If scientists can measure the rate at which a stream erodes its bed, they can estimate the approximate age of the stream. For example, geologists have studied Niagara Falls and found that the edge of the falls is eroding at an average rate of 1.3 m per year. Based on the average rate of erosion, scientists then determined that the falls probably formed about 9,900 years ago, when the continental ice sheets of the last ice age where melting.

Determining absolute age using the rate of erosion is practical only with geologic features that formed within the past 10,000 to 20,000 years. For older surface features, like the Grand Canyon, the method is much less dependable. The Grand Canyon developed over millions of years. During that period, the rates of erosion have varied greatly.

Section Objectives

- **Summarize the limitations of using the rates of erosion and deposition to determine the absolute age of rocks.**
- **Describe the formation of varves.**
- **Explain how the process of radioactive decay can be used to determine the absolute age of rocks.**

Figure 17–5. The rocky ledge above Niagara Falls has been eroding for nearly 9,900 years.

Another method of estimating absolute age involves calculating the rate of sediment deposition. Using data collected over a long period of time, geologists can estimate the average rates of deposition for common sedimentary rocks such as limestone, shale, and sandstone. On the average, 30 cm of sedimentary rock are deposited over a period of 1,000 years. However, any specific sedimentary layer under consideration may not have been deposited at an average rate. For example, a flood can deposit many meters of sediment in just one day. In addition, the rate of deposition has probably changed over time. Therefore, this means of estimating absolute age is not very accurate.

Varve Count

You may have heard of people estimating the age of a tree by counting the growth rings in its trunk. Scientists have devised a similar method for estimating the age of certain sedimentary deposits. Some sedimentary deposits show definite annual layers. The layers, called **varves,** consist of a light-colored band of coarse particles and a darker band of fine particles.

Varves are usually formed in glacial lakes. During the summer, when snow and ice melt rapidly, a rush of water can carry large amounts of sediment into a lake. Most of the coarse particles settle quickly to form a layer on the bottom of the lake. With the coming of winter the surface of the lake begins to freeze. Fine clay particles, still suspended in the water, settle slowly to form a thin layer on top of the coarser sediments. A coarse summer layer and the overlying fine winter layer make up one varve. Thus, each varve represents one year of deposition. By counting the varves, scientists can estimate the age of the sediments.

Figure 17–6. Counting varves is one method of determining the absolute age of certain sediments.

Radioactive Decay

The discovery of radioactivity provided scientists with a more accurate method for finding the absolute age of rocks. As you may recall from Chapter 8, the nuclei of some elements emit particles and energy at a relatively constant rate. Elements that emit particles and energy are called *radioactive*. As radioactive elements emit particles and energy, they form new isotopes or elements. Because the rate of change is measurable and is not affected by external factors, radioactive elements function as natural clocks. Scientists determine the amounts of the old, radioactive element and the newly formed element in a rock. They then compare the relative amounts of the original and new elements to determine the absolute age of the rock.

Uranium is a radioactive element found in some rocks. One form of uranium, U-238, is particularly useful in establishing the absolute ages of rocks. U-238 has a mass number of 238 and an atomic number of 92. Remember, the atomic number is the number of protons in the nucleus. The mass number is the sum of the num-

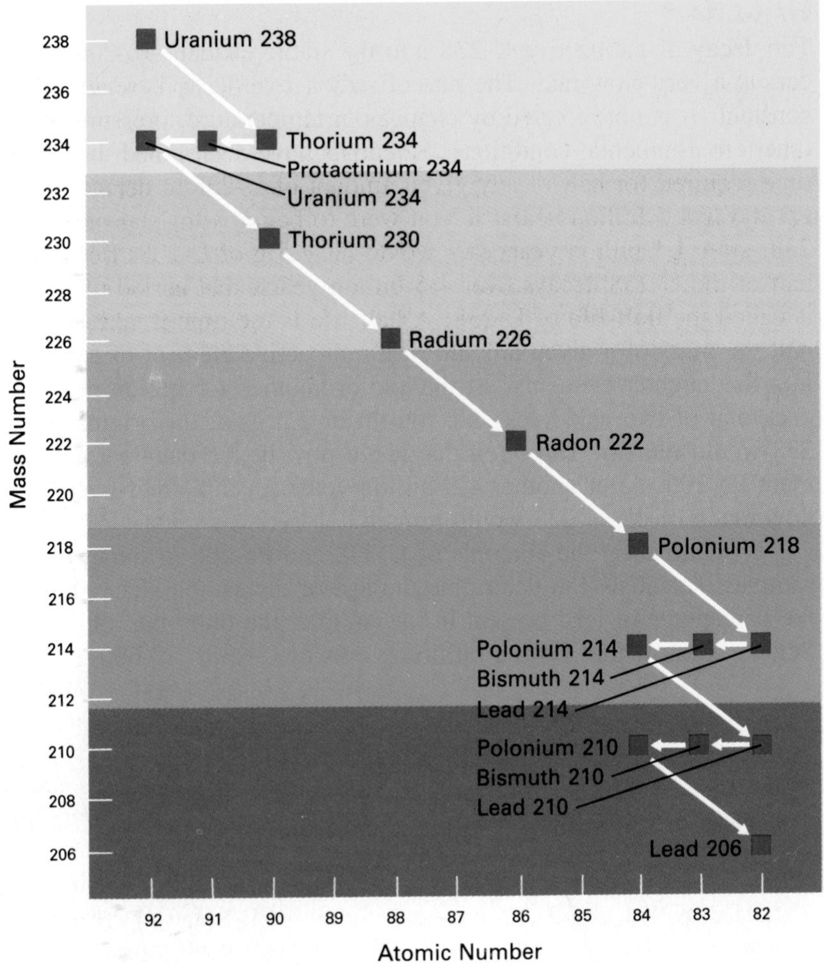

Figure 17–7. Through radioactive decay, the radioactive element U-238 eventually becomes the non-radioactive daughter element Pb-206.

ber of protons and neutrons in the nucleus. The highly radioactive nucleus of U-238 spontaneously emits two protons and two neutrons. This process, called *radioactive decay,* is illustrated in Figure 17–7. The loss of protons and neutrons from the nucleus decreases both the atomic number and the mass number. The result is a new element. In the first decay of U-238, the new nucleus has a mass number of 234 and an atomic number of 90. The new element is called *thorium* (Th-234).

The Th-234 nucleus is also radioactive, and one of its neutrons changes into a proton and an electron. The electron is emitted, leaving the proton in the Th-234 nucleus. This proton increases the atomic number by one, and thorium thus becomes the new element protactinium (Pa-234).

Protactinium is also radioactive, and the cycle of radioactive decay continues until a stable, or nonradioactive, form of the element lead (Pb) is produced. Pb-206, a form of lead with a mass number of 206, is the final product of this radioactive-decay chain. In this particular chain of decay, U-238 is known as the *parent element* and Pb-206 is known as a *daughter element*. Figure 17–7 shows the many daughter elements that form during the radioactive decay of U-238.

Half-Life

The decay of radioactive U-238 into the stable element Pb-206 occurs at a very slow rate. The rate of radioactive decay is relatively constant. It is not affected by changes in temperature, pressure, or other environmental conditions. Scientists have determined that the time required for half of any given amount of U-238 to decay into Pb-206 is 4.5 billion years. If you were to begin with 10 g of U-238, after 4.5 billion years you would have 5 g of U-238. Because half of the U-238 decays over 4.5 billion years, that period of time is called the **half-life** of U-238. A half-life is the time it takes for half the mass of a given amount of a radioactive element to decay into its daughter elements. At the end of another 4.5 billion years, or a total of two half-lives, one fourth, or 2.5 g, of the original U-238 would remain. Three fourths would now be the daughter element Pb-206. After another 4.5 billion years, or three half-lives, how much of the U-238 would remain?

By comparing the amounts of U-238 and Pb-206 in some rock samples, scientists can determine the age of the sample. The greater the percentage of lead present in the sample, the older the rock is. Scientists know that from a million grams of U-238, 1/7,600 g of

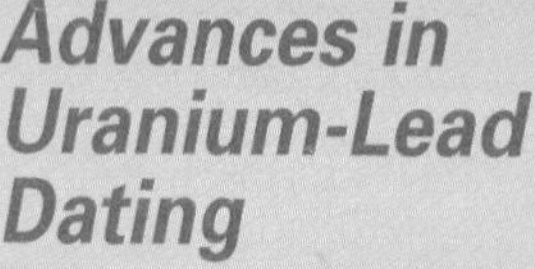

SCIENCE & TECHNOLOGY

Advances in Uranium-Lead Dating

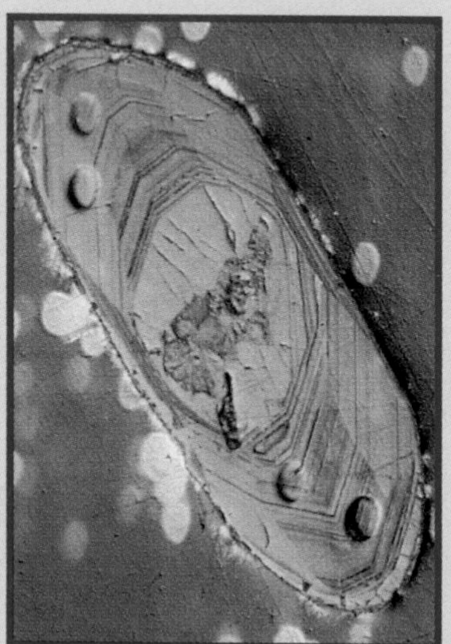

With modern technology, scientists can now date specific parts of a single mineral grain. The oblong pits on this magnified view of a zircon grain are places where samples have been removed for analysis.

Earth scientists theorize that the earth formed along with the rest of our solar system about 4.6 billion years ago. How long after its initial formation did the earth develop a core, mantle, and crust? When was the earth's surface cool enough to allow the first rocks and minerals to form? Scientists hope to answer these questions by finding and dating the earth's oldest rocks.

Dating ancient rocks is difficult, however. The earth's surface is constantly changing, and most ancient rocks have experienced high temperatures and pressures during their long existence. Over billions of years, tectonic activity and weathering have drastically altered these ancient rocks.

The mineral zircon, a type of silicate, has proven invaluable in dating old, complex rocks. Parts of some zircon grains can withstand extreme heat and pressure with their chemical structure intact. Zircon also contains significant amounts of the radioactive element uranium, which makes it a good candidate for uranium-lead dating.

With an instrument known as the *SHRIMP* (*S*uper *H*igh-

Pb-206 per year will be produced by decay. The U:Pb ratio can be used only when all the lead in the rock is known to have come from the decay of uranium. Because U-238 has an extremely long half-life of 4.5 billion years, it is most useful for dating geologic samples more than 10 million years old.

In addition to U-238, several other radioactive elements are used to date rock samples. Among these are potassium-40 (K-40), with a half-life of 1.3 billion years, and rubidium-87 (Rb-87), with a half-life of 48 billion years. Potassium is found in mica, clay, and feldspar. Potassium-40 is used to date rocks between 10 thousand and 4.6 billion years old. Dating with K-40 showed that the oldest Hawaiian islands are farthest from the hot spot where they originated. Often found in minerals with K-40, Rb-87 can be used to verify the age of rocks previously dated with K-40.

The amount of time that has passed since a rock was formed determines which radioactive element will give the more accurate age measurement. If too little time has passed since the radioactive decay process has begun, there may be too little of the daughter element for accurate dating. If too much time has passed, there may not be enough of the parent element left to obtain an accurate age measurement.

Using the *SHRIMP*, scientists can obtain uranium-lead dates from parts of microscopic mineral grains.

*R*esolution *I*on *M*icro*P*robe), scientists at Australia National University are able to date a tiny spot on a zircon grain. The zircon grain itself may be no larger than a speck of dust. By measuring the amount of uranium remaining and the amount of lead formed by the radioactive decay of the uranium, scientists are able to calculate when the parent rock crystallized.

The *SHRIMP* uses an ion beam to ionize the uranium and lead in a small part of the zircon grain. It separates the uranium and the lead by attracting the ions to opposite electrical charges. The flow of these ions produces an electrical current that the *SHRIMP* can measure precisely.

The *SHRIMP* has enabled scientists to date the oldest part of zircons, and thus the oldest rocks on earth. The oldest rock ever dated on earth is nearly 4.3 billion years old. Continued improvements in uranium-lead dating will undoubtedly add to our understanding of how the earth formed.

Why is it difficult to find and date the earth's original rocks?

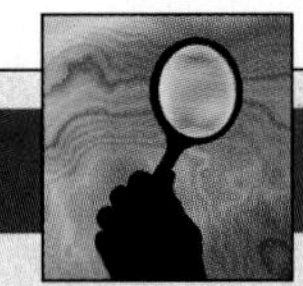

SMALL-SCALE INVESTIGATION

Radioactive Decay

Radioactive elements decay at a constant, measurable rate. The time it takes for half of any given amount of an original element to change into a new isotope or element is called a *half-life.* You can demonstrate the principle of radioactive decay with a simple model.

Materials

clock or watch with a second hand; sheet of ruled notebook paper, about 28 × 22 cm; scissors

Procedure

1. Record the time.
2. Wait 20 seconds, then carefully cut the sheet of paper in half. Select one piece, and set the other piece aside.
3. Wait 20 seconds, then cut the selected piece of paper in half. Select one piece, and set the other piece aside.
4. Repeat Step 3 until nine 20-second intervals have elapsed.

Analysis and Conclusions

1. In terms of radioactive decay, what does the whole piece of paper used in this investigation represent?

2. What do the pieces of paper that you set aside in each step represent?
3. What is the half-life of your "element"?
4. How much of your paper "element" was left after the first three intervals? after six intervals? after nine intervals? Express your answers as percentages.
5. What two factors in your model must remain constant for your model to be accurate? Explain your answer.

Carbon Dating

Some more recently formed geologic samples contain organic materials, the remains of once-living things. To determine the age of these samples, scientists use a form of radioactive carbon called *carbon-14* (C-14). Carbon-14 decays to form the daughter element nitrogen-14 (N-14).

Carbon-14 occurs naturally in the atmosphere combined with oxygen as carbon dioxide (CO_2). Most CO_2 in the atmosphere is formed with nonradioactive carbon-12 (C-12). Only a small amount of CO_2 is formed with C-14. The proportion of C-12 to C-14 in the atmosphere remains relatively constant.

All living plants and animals take in both C-12 and C-14. Plants absorb C-12 and C-14 in the form of CO_2 during photosynthesis. Carbon-12 and C-14 move through the food chain as animals eat plants. The C-12 and C-14 enter the tissues of plants and animals.

Figure 17–8. When organisms, such as this water beetle and grasshopper, die, the amount of C-14 in their tissues begins to decrease at a measurable average rate.

While those organisms are alive, the ratio of the two elements remains relatively constant. When a plant or an animal dies, however, the ratio begins to change. The organism no longer takes in C-12 and C-14. The amount of C-14 in an organism's tissues decreases as the radioactive carbon atoms decay to nonradioactive nitrogen-14 (N-14).

The half-life of C-14 is only about 5,730 years. To establish the age of a small amount of organic material, scientists first determine the proportion of C-14 to C-12 in the sample. They then compare that proportion with the proportion of C-14 to C-12 known to exist in a living organism.

C-14 is often used to establish the age of wood, bones, shells, and the organic remains of early humans. Improved techniques for detecting C-14 now make it possible to use carbon dating to establish the age of samples up to 50 thousand years old.

Section 17.2 Review

1. Why are calculations of absolute age based on rates of deposition not very accurate?
2. What are varves?
3. Explain how radioactive dating is used to estimate absolute age.
4. Suppose you have a shark's tooth that you suspect is about 15,000 years old. Would you use U-238 or C-14 to date the tooth? Explain your answer.

Section Objectives

- Describe four ways in which entire organisms can be preserved as fossils.
- List four examples of fossilized traces of organisms.
- Describe how index fossils can be used to determine the relative age of rocks.

17.3 The Fossil Record

Scientists called **paleontologists** (PAY-lee-on-TOL-uh-jists) study *fossils* to learn about the earth's past. Fossils are the remains or traces of animals or plants from a previous geologic time. Fossils are an important source of information for establishing both the relative and absolute ages of rocks. Fossils also provide clues to past geologic events, climates, and **evolution,** or the change of living things over time. The study of fossils is called **paleontology.**

Almost all fossils are found in sedimentary rock. Sediments, generally small and lightweight, cover but do not damage a dead organism. Sediments also protect dead organisms from being destroyed by other animals. The covering sediments may also slow down or stop the decaying process.

Fossils are almost never found in igneous rock or metamorphic rock. Igneous rocks are formed from hardened magma, or hot, molten rock. Therefore, living things covered by igneous rock usually burn up, leaving no remains to become fossilized. Metamorphic rocks undergo changes. Because the changes are caused by intense heat, pressure, and chemical reactions, any fossils in metamorphic rock usually are destroyed.

Kinds of Fossils

Fossils form in many different ways. Usually only the hard parts of organisms, such as bones, shells, and teeth, are preserved. In rare cases, an entire organism may be preserved. In some fossils, only a replica of the original organism remains. Other fossils merely provide evidence that life existed.

Figure 17–9. Fossils, such as these carbonized imprints of ferns, contain clues to the earth's past.

Preservation of Organisms

Normally, dead plants and animals are eaten by other animals or decomposed by bacteria. Left unprotected, even hard parts such as bones decay, leaving no trace of the organism. Only dead organisms that are buried quickly or protected from decay can become fossils.

Mummification One means through which an organism may be preserved is **mummification,** or drying. Mummified remains are often found in desert caves or buried beneath desert sand. Because most bacteria cannot survive without water, the mummified organism does not decay.

Figure 17–10. This spider is perfectly preserved in amber.

Amber Many insects have been found preserved in **amber**, or hardened tree sap. These insects became trapped in the sticky sap and were preserved when the sap hardened. In many cases, delicate features such as legs and antennae have been preserved, as shown in Figure 17–10. Even DNA has been recovered from amber.

Tar Beds The remains of animals have also been found preserved in tar beds. Tar beds are formed by thick petroleum oozing to the earth's surface. The tar beds were often covered by water. Animals that came to drink the water became trapped in the sticky tar. Other animals preyed upon the trapped animals and became trapped themselves. The remains of the animals were covered by the tar and preserved. For example, the bones of thousands of animals that lived more than 15,000 years ago have been found in the La Brea Tar Pits in southern California. In tar beds in Poland, entire woolly rhinoceroses have been found with their flesh and fur mostly intact.

Freezing The low temperatures of frozen soil and ice also protect and preserve organisms. Because most bacteria cannot survive freezing temperatures, organisms buried in frozen soil or ice do not decay. The frozen bodies of mastodons and woolly mammoths, two types of extinct elephantlike animals, have been found in Siberia and Alaska. Woolly rhinoceroses have also been found preserved in frozen arctic soil. These animals lived 5,000 to 40,000 years ago.

Petrification

Organisms usually are preserved through **petrification.** In this process, mineral solutions such as groundwater remove the original organic materials and replace them with minerals. Some common petrifying minerals are silica, calcite, and pyrite.

The replacement of the original materials is generally a very slow process, probably taking place molecule by molecule. The substitution of mineral for organic material often results in the formation of a nearly perfect mineral replica of the original organism. The petrified wood shown in Figure 17–11 is one example of this type of fossil. In the United States, petrified logs of cone-bearing trees can be seen in Petrified Forest National Park in Arizona. The minerals agate and chalcedony give these logs an overall gray color. However, some bright-colored streaks of other minerals are also visible. When some petrified fossils are viewed under a microscope, the de-

Figure 17–11. Look carefully at this photograph of petrified logs in Petrified Forest National Park, Arizona. Notice that the structure of the bark and the wood grain have been preserved.

tailed cell structure of the original tissues is clearly visible. In other petrified fossils, however, only the general outline of the organism remains.

Traces of Organisms

Sometimes no part of the original organism survives in fossil form. However, **trace fossils,** such as tracks, footprints, borings, and burrows, can still provide information about prehistoric life. The fossilized traces of ancient organisms are often remarkably detailed.

Trace Fossils A trace fossil, such as the footprint of an animal, is an important clue to the animal's appearance and activities. When scientists discover a footprint, such as that shown in Figure 17–12, they try to trace its history. Suppose a giant dinosaur left deep footprints in soft mud. Sand or silt may have blown or washed into the footprint so gently that the footprints remained intact. Then more sediment may have been deposited over the prints. As time passed, the mud containing the footprints hardened into sedimentary rock. The footprints were thus preserved. Scientists have discovered footprints of ancient reptiles, amphibians, birds, and mammals.

Figure 17–12. By studying the size and depth of these dinosaur footprints, scientists can learn about the size and weight of the dinosaur that made them.

Imprints, Molds, and Casts Carbonized *imprints* of leaves, stems, flowers, and fish made in soft mud or clay have been found preserved in sedimentary rock. An imprint displays the surface features of the organism. Shells of snails, parts of trees, and similar organic remains often leave empty cavities called *molds*. When an organism is buried in rock-forming sediments, its remains eventually decay or dissolve, leaving an empty space. A mold retains the shape

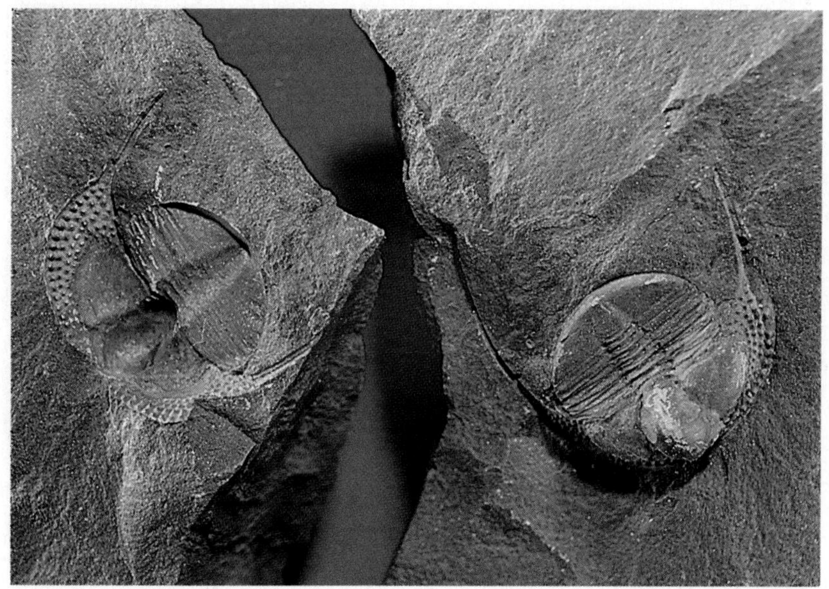

Figure 17–13. Over time, the mold of this trilobite (left) filled with sediments. The sediments then hardened and formed a cast of the original organism (right).

and surface markings of the original organism; however, it gives little information about the internal structure of the organism.

Sometimes sand or mud fills a mold and hardens, forming a natural *cast* such as that shown in Figure 17–13. A cast is a replica of the outer surface of the original organism. An artificial cast can be made by filling a mold with plaster.

Coprolites Some ancient worms, snails, and crabs are known to have existed because fossil remains of their waste material have been found. For instance, as ancient sea worms fed on small organisms on the ocean floor, they swallowed sand or mud as well. After the food was digested, the sand or mud was excreted as castings, or waste materials. These castings are found as fossils in some marine sediments.

Fossilized dung or waste materials from ancient animals are called **coprolites.** They can be cut into thin sections and observed through a microscope. The materials identified in these sections reveal the feeding habits of ancient animals such as dinosaurs.

Gastroliths Some dinosaurs had stones in their digestive systems to help grind their food. In many cases, these stones, called **gastroliths,** survive as fossils. Gastroliths can often be recognized by their smooth, rounded, and polished surfaces. However, identification of these stones is certain only if they are found within the remains of a dinosaur.

Interpreting the Fossil Record

The fossil record provides scientists with many clues to the geologic history of the earth. For instance, scientists can use fossils to find the relative ages of rocks. Fossils also reveal the ways in which environments have changed and how the changes have affected organisms throughout the geologic past of the earth.

INVESTIGATE!

To learn more about fossils, try the In-Depth Investigation on pages 342–343.

Figure 17–14. Trilobite fossils make good index fossils. If this Phacops trilobite fossil is found in a rock layer, the rock layer was formed about 400 million years ago in a marine environment.

Index Fossils

Certain fossils are found exclusively in rock layers of a particular geologic age. These fossils are called **index fossils.** To be considered index fossils, fossils must meet certain requirements. First, they must be present in rocks scattered over a wide area of the earth's surface. Second, index fossils must have features that clearly distinguish them from all other fossils. Third, the organisms from which the index fossils formed must have lived during a relatively short span of geologic time. Fourth, they must occur in fairly large numbers within the rock layers.

Paleontologists can use index fossils to establish the relative ages of the rock layers in which the fossils are found. For example, marine organisms called *ammonites* lived from 65 to 200 million years ago. Other marine creatures, called *trilobites,* lived from 245 to 570 million years ago. Therefore, in dating rock layers, scientists know that any rock layer containing trilobite fossils is older than a layer containing ammonite fossils. If a layer containing trilobite fossils was found above a layer containing ammonite fossils, what might scientists assume?

Scientists can use index fossils to date rock layers found in widely separated areas. An index fossil found in rock layers in different areas of the world indicates that the rock layers were probably formed during the same period.

Geologists also use index fossils to help locate oil and natural gas deposits. These deposits are formed from plant and animal remains that have been changed by chemical processes over millions of years. Index fossils are useful to petroleum geologists when they are drilling for oil and natural gas deposits.

Fossil Clues to the Past

Fossils also furnish scientists with important clues to the changes in climate and environment that occurred in the past. For example, fossils of alligator-like reptiles have been found in Canada, which today

EARTH BEAT

Clues to Climate

It is hard to imagine that the bitter cold climate of Antarctica was once much warmer and that the ice-covered landscape supported thick forests. However, studies of fossils found in ocean-floor sediments off the coasts of Antarctica indicate just that.

Scientists with the Ocean Drilling Project have brought us samples of sediments that ranged in age from 60 million years old to the present. These samples reveal much about the complex history of climatic changes in Antarctica.

Fossil spores and pollen over 39 million years old indicated that beech-tree forests once grew in Antarctica. Samples of sediment 37 to 60 million years old contained soil that is normally found in hot, humid climates. Thus scientists learned that prior to 37 million years ago the climate in the Antarctic was almost subtropical.

Fossils of freshwater organisms indicated that they were carried from Antarctic lakes to the ocean floor by rivers as recently as 20 million years ago. The fossils are evidence that at that time the climate was warm enough for unfrozen lakes to exist.

Scientists also found fossils of marine organisms that can live only in sunny coastal waters. Those found off the east coast were over 15 million years old, while those off the west coast were over 4.8 million years old.

What do these fossils indicate about the ice shelves that now cover the ocean off the east and west coasts of Antarctica?

has a relatively cold climate. Since alligators today live only in very warm climates, these fossils indicate that Canada once had a tropical climate. Tropical plant fossils also have been discovered in Antarctica just 400 miles from the South Pole. Fossils of marine animals and plants have been found in areas far from any ocean. These fossils indicate that such areas were once covered by an ocean.

Section 17.3 Review

1. Why are tar beds good sources of animal fossil remains?
2. List four types of fossils that can be used to provide indirect evidence of organisms.
3. How do geologists use fossils to date sedimentary rock layers?
4. Compare the process of mummification with the process of petrification.

Chapter 17 Review

Key Terms

- absolute age (327)
- amber (335)
- angular unconformity (325)
- bedding plane (324)
- coprolite (337)
- disconformity (325)
- evolution (334)
- gastrolith (337)
- half-life (330)
- index fossil (338)
- law of crosscutting relationships (326)
- law of superposition (324)
- mummification (335)
- nonconformity (325)
- paleontologist (334)
- paleontology (334)
- petrification (335)
- principle of uniformitarianism (323)
- relative age (324)
- trace fossil (336)
- unconformity (325)
- varve (328)

Key Concepts

According to the principle of uniformitarianism, the forces that are changing the earth's surface today are the same forces that changed the earth's surface in the past. **See page 323.**

Scientists use the law of superposition to determine the relative ages of rock layers. **See page 324.**

A nonconformity, an angular unconformity, and a disconformity represent interruptions in the sequence of rock layers. **See page 325.**

Scientists use the law of crosscutting relationships to determine the relative ages of rock layers. **See page 326.**

Because they change over time, the rate of erosion and the rate of deposition are not very reliable bases for determining absolute age of rocks. **See page 327.**

Varves are layers of sediment that form in glacial lakes. **See page 328.**

Radioactive elements decay at measurable average rates and can be used to determine absolute age. **See page 328.**

Entire organisms may be preserved through freezing, mummification, in amber, or in tar beds. **See page 335.**

Fossilized traces of organisms include trace fossils, imprints, molds, casts, coprolites, and gastroliths. **See page 336.**

Index fossils are found exclusively in rock layers of a particular geologic age. **See page 338.**

Review

On your own paper, write the letter of the term that best completes each of the following statements.

1. The concept that the present is the key to the past is part of the
   a. type of unconformity.
   b. law of superposition.
   c. law of crosscutting relationships.
   d. principle of uniformitarianism.
2. A sedimentary rock layer is older than the layers above it and younger than the layers below it, according to the
   a. type of unconformity.
   b. law of superposition.
   c. law of crosscutting relationships.
   d. principle of uniformitarianism.
3. A gap in the sequence of rock layers is
   a. a bedding plane.
   b. a varve.
   c. an unconformity.
   d. a uniformity.

4. An unconformity that results when new sediments are deposited on eroded horizontal layers is
   a. an angular unconformity. b. a disconformity.
   c. crosscut unconformity. d. a nonconformity.

5. A fault or intrusion is younger than the rock it cuts through, according to the
   a. type of unconformity.
   b. law of superposition.
   c. law of crosscutting relationships.
   d. principle of uniformitarianism.

6. The age of a rock in years is known as
   a. index age. b. relative age. c. half-life age. d. absolute age.

7. Varves are formed by layers of
   a. limestone mixed with coarse sediments.
   b. coarse sediments followed by fine sediments.
   c. shale followed by sandstone.
   d. sandstone followed by shale.

8. An atom with a mass number of 234 and an atomic number of 90 has
   a. 90 neutrons and 144 protons. b. 234 neutrons and 90 protons.
   c. 144 neutrons and 90 protons. d. 117 neutrons and 117 protons.

9. The process whereby the remains of an organism are preserved by drying is
   a. petrification. b. mummification.
   c. erosion. d. superposition.

10. Molds filled with sediments sometimes produce
   a. casts. b. gastroliths. c. coprolites. d. imprints.

11. Fossils that are found in many parts of the earth and that were formed by organisms that lived during a brief period of geologic time are known as
   a. coprolites. b. molds.
   c. gastroliths. d. index fossils.

Critical Thinking

On your own paper, write answers to the following questions.

1. Hutton developed the principle of uniformitarianism by observing geologic changes on his farm. What changes might he have observed?
2. How might a scientist determine the original positions of the sedimentary layers beneath an angular unconformity?
3. Of two intrusions, one cuts through all the rock layers. One other is eroded off and lies beneath several layers of sedimentary rock. Which intrusion is younger? Why?
4. What information about past climatic changes might scientists gather by studying varves?
5. How are the processes of mummification and freezing similar?

Application

1. A scientist is attempting to calculate the absolute age of sedimentary rock layers by using the average rate of deposition. However, between the third and fourth layers is a disconformity. What difficulties might this cause in determining the absolute ages of the layers below the disconformity?
2. In how many years would 16 g of U-238 turn into 0.5 g of U-238 and 15.5 g of daughter products?
3. Suppose that a certain type of fossil with unusual features is found in many areas of the earth. It represents a brief span of geologic time but occurs only in small numbers. Would the fossil make a good index fossil? Explain.

Extension

1. Examine an exposed area of sedimentary rock in your community. Sketch your findings. Write a report based on your conclusions about the rock bed's history.
2. Research two early methods for determining the age of the earth on the basis of the salinity of the oceans and the temperature of the earth. Report your findings to your class.
3. Find out more about trilobite fossils. When and where did trilobites live, and in what type of rock are trilobite fossils found? Write an article based on your findings.

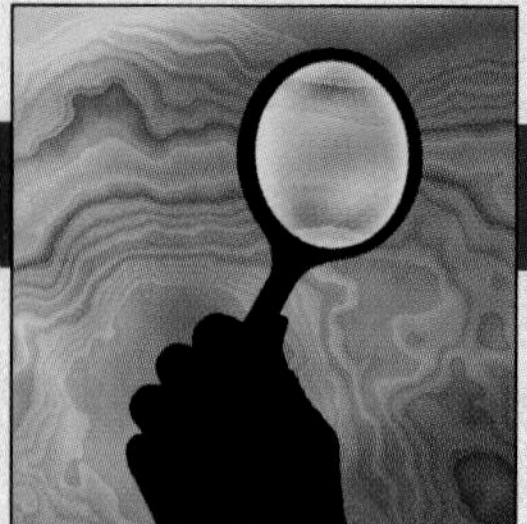

IN-DEPTH INVESTIGATION

Fossils

Materials

- hard objects (shell, key, paper clip, etc.)
- tweezers
- plastic container
- plastic spoon
- sheet of white paper
- modeling clay
- water
- newspaper
- soft carbon paper
- plaster of Paris
- leaf
- pencil or wooden dowel
- lab apron

Introduction

Paleontologists study fossils to find evidence of the kinds of life and conditions that existed on the earth in the geologic past. Fossils are the remains of ancient plants and animals or evidence of their presence. Some fossils are the preserved or altered bodies of organisms. Others, called trace fossils, such as footprints, tracks, burrows, and borings, provide indirect evidence of what an ancient animal looked like. A mold is formed when an animal or plant is buried in sediments that later harden into rock. In time, the body of the organism decays, leaving an empty space, or mold, in the rock with the same shape and surface markings as the organism. Under certain conditions, the mold may fill with minerals that harden to produce a replica of the outer surface of the original organism. This mineral replica is called a cast. Another type of fossil is an imprint, such as the thin impression made by a fish or a leaf. Carbon films on rock surfaces left by the decayed soft parts of an organism, such as a leaf, are examples of impressions. The shape and some surface features of the organism are visible in the carbon film.

In this investigation, you will use various methods to make models of trace fossils.

Prelab Preparation

1. Review Chapter 17, Section 17.3, pages 335–337.
2. Describe three ways in which fossils can be classified based on how they were formed.
3. Why are coal, oil, and gas called fossil fuels?

Procedure

1. Place a ball of modeling clay on a flat surface. Press the clay down to form a flat disk about 8 cm in diameter. Turn the clay over so that the smooth, flat surface is facing up.

2. Choose a small hard object. Press the object onto the clay carefully so that you do not disturb the indentation. Is the indentation left by the object a mold or a cast? What features of the object are best shown in the indentation? Sketch the indentation.
3. Fill a plastic container with water to a depth of 1 to 2 cm. Stir in enough plaster of Paris to make a paste with the consistency of whipped cream.
4. Spread several sheets of newspaper on a flat surface. Place the clay on the newspaper.
5. Using the plastic spoon, fill the indentation with plaster. Allow excess plaster to run over the edges of the imprint. Let the plaster set for about 15 minutes until it hardens.
6. On a second piece of smooth, flat clay, make a shallow imprint to represent the burrow or footprint of an animal. Fill the model trace fossil with plaster and let it harden.
7. After the plaster has hardened, remove both pieces of plaster from the clay. Do the pieces of hardened plaster represent molds or casts? Sketch your fossil imprint.
8. Place the carbon paper carbon side up on a flat surface. Gently place the leaf on the carbon paper and cover it with several sheets of newspaper. Roll the pencil or wooden dowel back and forth across the surface of the newspaper several times, pressing firmly to bring the leaf into solid contact with the carbon paper.
9. Remove the newspaper. Lift the leaf using the tweezers and place it carbon side down on a clean sheet of paper. Cover the leaf with clean newspaper and roll your pencil across the surface of the paper once again, as in Step 8.
10. Remove the newspaper and leaf. Observe and describe the carbon-print left by the leaf.

Analysis and Conclusions

1. Look at the molds and casts made by others in your class. Identify as many of the objects used as you can.
2. How does the carbon-print you made differ from an actual carbon-print trace fossil? Why is a carbon-print a trace fossil?
3. Why are carbon-prints, molds, and casts trace fossils?

Extension

Which of the following organisms—a rabbit, a housefly, an earthworm, a clam, a snail—would be most likely to form fossils? Which would leave trace fossils? Explain.

Chapter 18

A View of the Earth's Past

The rock record holds a fascinating story of the evolution of life on earth. Fossils, such as the one shown in the photo on this page, reveal a rich diversity of plant and animal life. They range from evidence of a single-celled marine organism at least 3.4 billion years old to the remains of the early ancestors of modern human beings. This chapter discusses the divisions of geologic time and the evidence of the evolution of living things throughout the earth's history.

Chapter Outline

◀ This fossil sea star lived about 400 million years ago.

18.1 The Geologic Time Scale

Section Objectives

- **Summarize the development of the geologic column.**
- **List the major units of geologic time.**

The surface of the earth is constantly changing. Mountains form and erode; oceans rise and recede. As conditions on the earth's surface change, various organisms flourish and then become extinct. Evidence of these changes is recorded in the rock layers of the earth's crust. To describe the sequence and length of these changes, earth scientists have developed a *geologic time scale*. This time scale is an outline of the development of the earth and of life on the earth.

The Geologic Column

Using the law of superposition and the study of index fossils, nineteenth-century scientists determined the relative ages of rock layers in areas throughout the world. No single area on earth contained a record of all geologic time. Therefore, scientists combined their observations from around the world to create a standard arrangement of rock layers. This ordered arrangement of rock layers, which is based upon the ages of the rocks, is called the **geologic column.** The geologic column represents a time line of the earth's history, with the oldest rocks at the bottom and the most recent rocks at the top.

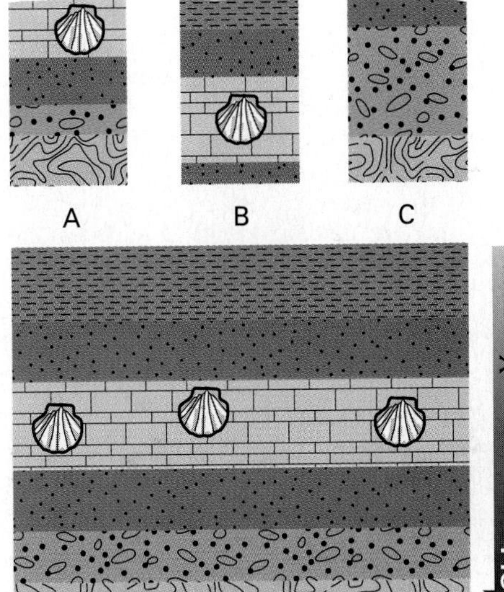

Figure 18–1. By combining observations of rock layers in areas *A*, *B*, and *C*, scientists can construct a geologic column.

Rock layers in the geologic column are distinguished from each other primarily by the kinds of fossils they contain and by the rock type. Fossils in the upper, more-recent layers resemble modern plants and animals. Most of the fossils in the lower, older layers are of plants and animals that are very different from those living today. In fact, many of the fossils found in older layers are from species that have long been extinct.

At the time the geologic column was developed, scientists based their estimates of the ages of rock layers on such things as average rates of sediment deposition. The development of radioactive dating methods, however, has enabled scientists to determine more accurately the absolute ages of rock layers in the geologic column.

Scientists can now use the geologic column to estimate the age of a rock layer even if the rock contains no radioactive minerals. The rock layer being studied is compared with a layer in the geologic column that has the same fossil content, relative position, and, sometimes, physical characteristics. If the two layers match, they were very likely formed at about the same time. Why is relative position important in determining the ages of rock layers?

Divisions of Geologic Time

Just as the calendar year is divided into months, weeks, and days, geologic time is divided into units. The geologic history of the earth is punctuated by major changes in the earth's surface or climate and by the extinction of various species. Geologists use these events as the basis for dividing the geologic time scale into smaller units.

65 million
245 million
570 million
4.6 billion

Cenozoic Era
Mesozoic Era
Paleozoic Era
Precambrian time

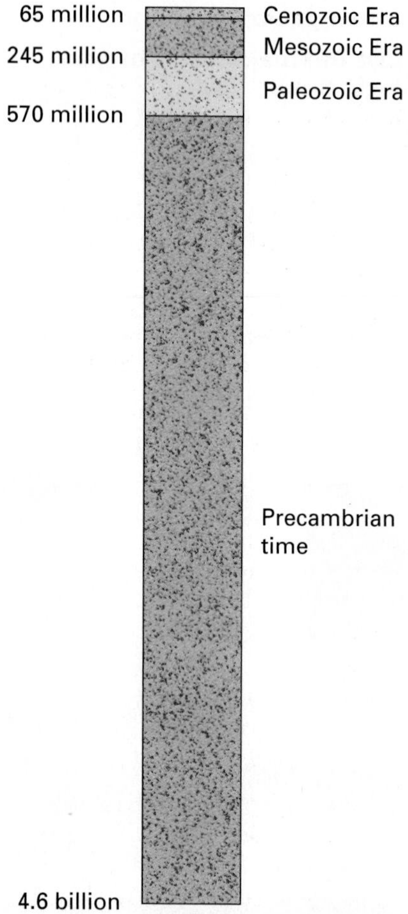

Figure 18–2. The present era, the Cenozoic, is the shortest era of geologic time, beginning 65 million years ago. Note its length in comparison to the other eras.

Rocks grouped within each unit contain a similar fossil record. In fact, a unit of geologic time is often characterized by fossils of a dominant life-form.

Eras

A very large unit of geologic time is an **era.** There are four geologic eras. The earliest is the Precambrian Era. This era is more commonly referred to as **Precambrian time** because it is much longer than the other geologic eras. Also, very few fossils exist in early Precambrian rocks, making it difficult to divide the 4 billion years of Precambrian time into smaller units. Precambrian rocks make up the oldest layers of the geologic column.

At the beginning of Precambrian time, the earth's crust was just beginning to solidify. Very early Precambrian rocks contain no evidence of life. However, a few fossils of bacteria and algae, thought to be the earth's first life-forms, have been found in rocks more than 3.5 billion years old. Rocks of late Precambrian time contain fossils of primitive worms, sponges, and corals. These fossils and other types of evidence have led most scientists to theorize that life began in the ocean.

The **Paleozoic Era** followed Precambrian time. The term *Paleozoic* comes from the Greek words meaning "ancient life." The Paleozoic Era lasted 325 million years. Rocks from this era contain fossils of a wide variety of both marine and land plants and animals.

The **Mesozoic Era,** a span of about 180 million years, followed the Paleozoic Era. *Mesozoic* means "middle life." Mesozoic rocks hold fossils of more-complex organisms such as reptiles and birds.

The present geologic era is the **Cenozoic Era.** *Cenozoic* means "recent life." Fossils of mammals are common in rocks of this era. The fossil sequence in the geologic column shows that earlier forms of life were generally followed by newer and more-complex forms.

Periods and Epochs

The eras have been divided into shorter time units called **periods.** Each period is characterized by specific fossils and is usually named for the location in which the rocks containing the identifying fossils were first found. For example, the Devonian Period of the Paleozoic Era was named for Devonshire, England. The term *Devonian* is applied to rock layers in the geologic column that developed at about the same time as the corresponding layers in Devonshire.

Rocks of the Cenozoic Era are the most recent and often the least deformed in the geologic column. They may contain an extremely detailed fossil record. This abundance of information has allowed scientists to divide the two periods of the Cenozoic Era into even shorter time units called **epochs.**

Section 18.1 Review

1. How was the geologic column developed?
2. What are the divisions of geologic time?
3. Why is the geologic column useful to earth scientists?

1-3

18.2 Geologic History

Section Objectives

- Identify the characteristics of Precambrian rock.
- Explain what scientists have learned from the geologic record about life during the Paleozoic era.
- Explain what scientists have learned from the geologic record about life during the Mesozoic era.
- Explain what scientists have learned from the geologic record about life during the Cenozoic era.

History is a written record of past events. Just as the history of different civilizations is recorded in books, the geologic history of the earth is recorded in rock layers. The type of rock that makes up each layer and the fossils that are found in each layer reveal much about the conditions that existed when the layer formed. For example, limestone forms in bodies of water. The presence of a limestone layer in a region indicates that the area was once covered by water. Cross-bedded sandstone, on the other hand, often forms from sand dunes. Thus, the presence of this sandstone may indicate that a desert or coastal dune area existed at that time.

Fossils also indicate the kinds of organisms that lived during that geologic time. By examining rock layers and fossils, geologists and paleontologists have found a large amount of evidence that supports the **theory of evolution.** Additional evidence that supports the theory of evolution comes from the examination of living organisms by biologists. The theory of evolution states that organisms change over time and that new kinds of organisms are derived from ancestral types. The theory of evolution by natural selection was proposed by Charles Darwin, an English naturalist, in 1859. Natural selection is often called the *survival of the fittest*.

One of the assumptions of evolution by natural selection is that organisms are adapted to their environment. Because of this adaptation, changes in the environment affect the organisms. Only those that can adapt to the environmental changes will survive. Organisms that cannot survive—in other words, those that are unfit to live and reproduce in the changing environment—become extinct.

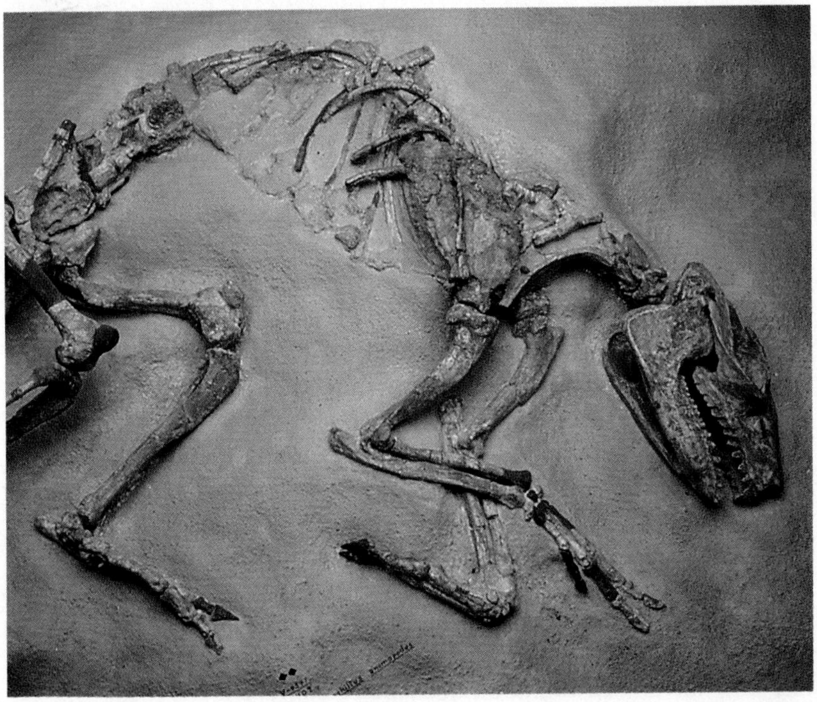

Figure 18–3. Many animals that existed in the past are now extinct. Shown here are the fossilized bones of an extinct South American plant eater.

Figure 18–4. Stromatolites, fossil cyanobacteria, are the most common Precambrian fossils.

Among the numerous types of environmental changes that affect the survival of organisms are major geologic and climatic changes. One example of a geologic change is a dramatic decrease in the amount of the earth's surface that is covered by water. At one time vast areas of the earth were covered by warm, shallow seas; these waters have since receded. An example of a major climatic change is a decrease in atmospheric temperature. Such a major decrease in temperature occurred at the time of the ice ages that have affected the earth periodically.

Based on evidence gathered from rocks and fossils, paleontologists try to determine how these changes affected the survival of various organisms. Earth scientists also attempt to learn what evolutionary changes may have taken place within organisms to enable them to survive environmental changes. The rocks and fossils also reveal which organisms became extinct, perhaps because they did not adapt to changes.

Precambrian Time

Precambrian time began with the formation of the earth nearly 4.6 billion years ago and ended about 570 million years ago. It makes up nearly 88 percent of earth's history. The Precambrian rock record is difficult to interpret. Most Precambrian rocks have been so severely deformed and altered by crustal activity that the original order of rock layers is rarely identifiable.

Large areas of exposed Precambrian rocks, called *shields,* are found on every continent. For example, a Precambrian shield covers much of eastern Canada. Precambrian shields are the result of several hundred million years of volcanic activity, mountain building, sediment formation, and metamorphism. Precambrian rocks were deformed and metamorphosed, causing minerals to collect near the surface of the earth as the rocks cooled. Nearly half of the deposits of valuable minerals in the world have been found in the rocks of Precambrian shields. Among these minerals are nickel, iron, gold, and copper.

Fossils are rare in Precambrian rocks, probably because Precambrian life-forms were soft-bodied. These early organisms lacked bones, shells, or other hard parts that commonly form fossils. Also, Precambrian rocks are extremely old. Some date back nearly 4.3 billion years. Extensive crustal movements such as folding and faulting, volcanic activity, and erosion over this long period of time probably destroyed most Precambrian fossils.

Of the few Precambrian fossils that do exist, the most common are *stromatolites,* reeflike deposits produced by cyanobacteria. Stromatolites still form today in warm, shallow waters. The presence of stromatolite fossils in Precambrian rocks indicates that shallow seas may have covered much of the earth during some periods of Precambrian time. Imprints of marine worms, jellyfish, and one-celled organisms were first found in the late Precambrian rocks of Australia and now have been found on the other continents.

The Paleozoic Era

The Paleozoic Era began about 570 million years ago and ended about 245 million years ago. Scientists theorize that at the beginning of the Paleozoic Era, the earth's landmasses were scattered in the world ocean. By the end of the Paleozoic Era, these landmasses had collided to form the supercontinent Pangaea. As you read in Chapter 4, this tectonic activity created new mountain ranges and lifted large areas of land above sea level.

Unlike Precambrian rocks, Paleozoic rocks contain an abundant fossil record. These fossils indicate that there was a dramatic increase in the number of plant and animal species on the earth at the beginning of the Paleozoic Era. As a result of this rich marine and land fossil record, geologists have been able to gather more facts about this era. Because of the wealth of information, they have divided the Paleozoic Era into seven periods.

The Cambrian Period

The Cambrian Period is the earliest period of the Paleozoic Era. From the fossils in Cambrian rocks, scientists have inferred that a variety of advanced forms of marine life first appeared during this period. These marine organisms, many with hard parts, quickly displaced the primitive Precambrian organisms as the dominant form of life. The explosion of Cambrian life may have been due in part to the warm, shallow seas that appear to have covered most of the continents during this period. Marine **invertebrates,** animals without backbones, thrived in these warm waters. The most common of the Cambrian invertebrates were the trilobites, hard-shelled animals that lived on the ocean floor. As you read in Chapter 17, scientists use trilobites as important index fossils for identifying layers of Cambrian rock throughout the world.

The second most common animals of the Cambrian Period were the brachiopods, a group of shelled animals. There is evidence that at least 15 different classes of brachiopods existed during this period. A few kinds of brachiopods still exist today, though they are

Figure 18–5. During the Paleozoic Era, various species of brachiopods, including this fossil brachiopod from the early Paleozoic Era, flourished in the warm, shallow seas.

rare. Other common Cambrian invertebrates include worms, jellyfish, snails, and sponges. No evidence of land-dwelling plants or animals, however, has been found among any of the Cambrian fossils discovered so far.

The Ordovician Period

During the Ordovician Period, the number of brachiopod species increased, and trilobites began to decline. Snails, clams, and other mollusks became the dominant life-forms. Large numbers of coral began to appear, although coral reefs did not occur until later. Colonies of tiny invertebrate animals called *graptolites* also flourished in the oceans during the Ordovician Period. Graptolite fossils are useful index fossils for this period.

An important development of the Ordovician Period was the appearance of the *ostracoderm,* a primitive fish. Fossils of ostracoderms are the oldest fossils of **vertebrates,** animals with backbones, yet discovered. Ostracoderms had no jaws or teeth. Unlike modern fish, their bodies were covered with bony plates. During the Ordovician Period, as during Cambrian and Precambrian times, there was no plant life on land.

Science & Technology

CAT Scanning Fossils

▲ Ultra-high-resolution CAT scanners provide unparalleled views of the internal structures of fossils.

Paleontologists studying dinosaur bones want to learn as much as possible from the fossils they find. Too often, however, this means destroying a fossil to look inside it. Recently, paleontologists have enlisted the help of *Computed Axial Tomography*, or *CAT* scanning, to examine dinosaur bones in a nondestructive way.

Without CAT scanning, paleontologists must examine the inside of a fossil by grinding away the specimen layer by layer. Unfortunately, by the time all of the fossil's internal structures have been revealed, the specimen is completely demolished.

Sectioning a fossil by layers also takes a lot of time. Scientists must carefully document each fresh surface because it will be destroyed later to uncover the next fresh surface. Scientists record their observations by measuring, drawing, photographing, and making an imprint of each new surface with a special chemical film.

Using CAT scans, paleontologists are able to document the interior of fossils more accurately and in much less time. More importantly, CAT scanning leaves the fossils intact. Scientists no longer have to destroy fossils in order to study the valuable information contained within.

CAT scanning was originally developed for medical diagnosis.

The Silurian Period

Marine life, both invertebrate and vertebrate, continued to thrive and evolve during the Silurian Period. Echinoderms, relatives of modern sea stars, and corals became more numerous. Scorpionlike sea creatures called *eurypterids* were abundant during this period. Fossils of giant eurypterids nearly 2.7 m in length have been found in western New York. Near the end of this period, the earliest land plants and animals such as spiders and millipedes evolved on land.

The Devonian Period

The Devonian Period is often called the *Age of Fishes* because rocks from this period contain fossils of many kinds of bony fishes. One type of fish, called a *lungfish,* had the ability to breathe air. Other air-breathing fish, called *rhipidistians,* had strong fins that probably enabled them to crawl out of the water and onto the land for short periods of time. The first true amphibian, *Ichthyostega,* probably evolved from these fish. Ichthyostega, which resembled a huge salamander, is thought to be the ancestor of modern amphibians such as frogs and toads. During the same period, land plants, such as giant horsetails, ferns, and cone-bearing plants, began to develop.

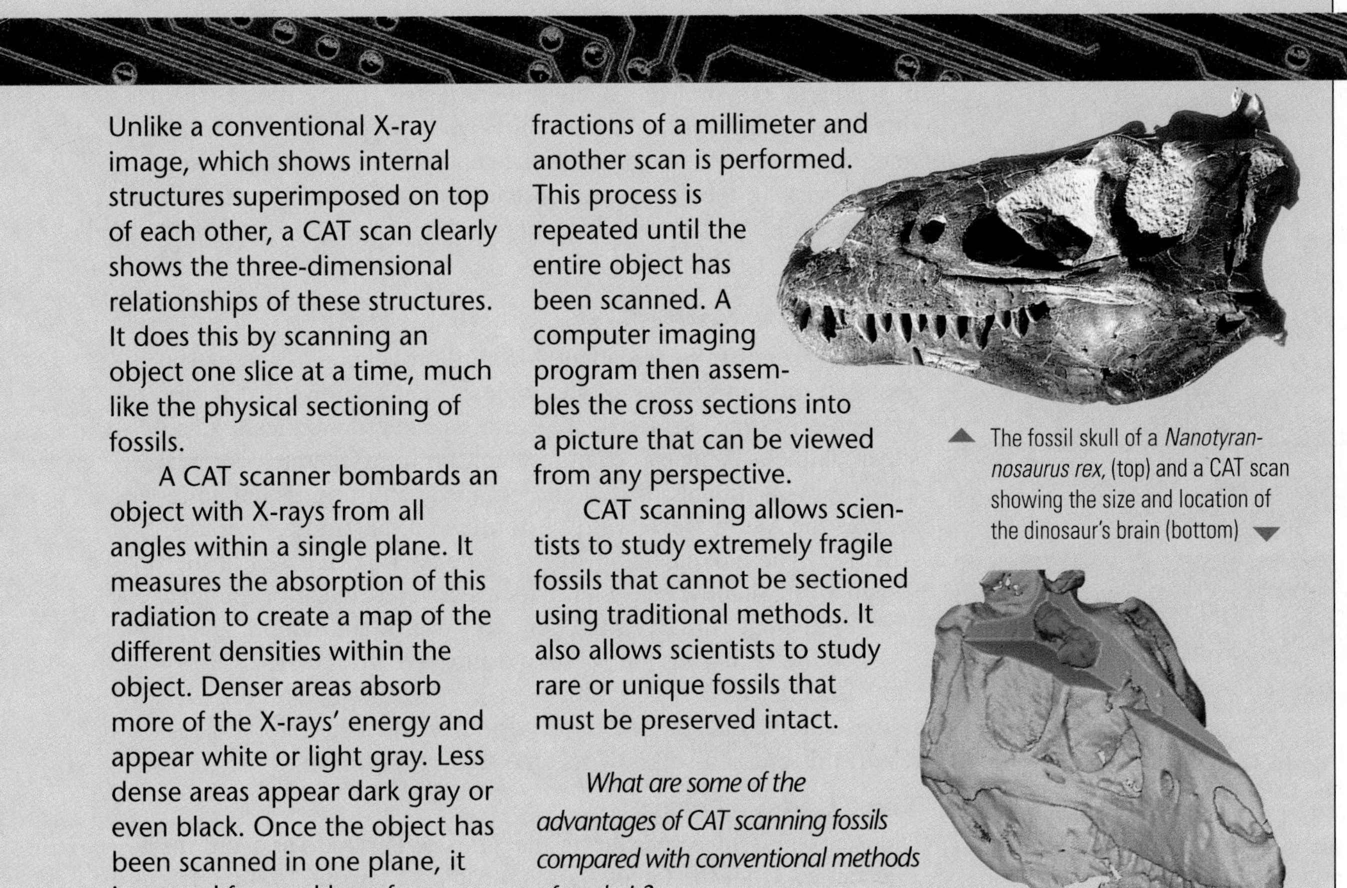

Unlike a conventional X-ray image, which shows internal structures superimposed on top of each other, a CAT scan clearly shows the three-dimensional relationships of these structures. It does this by scanning an object one slice at a time, much like the physical sectioning of fossils.

A CAT scanner bombards an object with X-rays from all angles within a single plane. It measures the absorption of this radiation to create a map of the different densities within the object. Denser areas absorb more of the X-rays' energy and appear white or light gray. Less dense areas appear dark gray or even black. Once the object has been scanned in one plane, it is moved forward by a few fractions of a millimeter and another scan is performed. This process is repeated until the entire object has been scanned. A computer imaging program then assembles the cross sections into a picture that can be viewed from any perspective.

CAT scanning allows scientists to study extremely fragile fossils that cannot be sectioned using traditional methods. It also allows scientists to study rare or unique fossils that must be preserved intact.

What are some of the advantages of CAT scanning fossils compared with conventional methods of analysis?

▲ The fossil skull of a *Nanotyrannosaurus rex,* (top) and a CAT scan showing the size and location of the dinosaur's brain (bottom) ▼

The Carboniferous Period

In North America, the Carboniferous Period is divided into the Mississippian and Pennsylvanian periods. During the Carboniferous Period, the climate was generally warm, and the humidity was extremely high all over the world. Forests and swamps covered much of the land. Scientists think that the coal deposits of Pennsylvania, Ohio, and West Virginia are the fossilized remains of these forests and swamps. During this period, the rock in which some major oil deposits are found was also formed. *Carboniferous* means "carbon bearing." A discussion of coal formation appears in Chapter 11.

Amphibians and fish continued to flourish during the Carboniferous Period. *Crinoids,* relatives of modern sea stars, were common in the oceans. Insects such as giant cockroaches and dragonflies were common on land. Toward the end of the Pennsylvanian Period, reptiles, the first vertebrates fully adapted to life on land, appeared. These early reptiles resembled large lizards.

The Permian Period

The Permian Period marks the end of the Paleozoic Era and a mass extinction of numerous Paleozoic life-forms. Nearly all the continents of the world had joined to form the supercontinent Pangaea. Collision of tectonic plates forced the Appalachian Mountains to rise so high that moist air could not rise over the mountaintops. On the northwest side of the mountains, areas of desert and dry savanna climates developed. The shallow inland seas that had covered much of the earth evaporated. As the seas retreated, many species of marine invertebrates, including trilobites and eurypterids, became extinct. However, fossils indicate that reptiles and amphibians were able to survive the changes in climate. In the following geologic era, the Mesozoic, reptiles would dominate the earth.

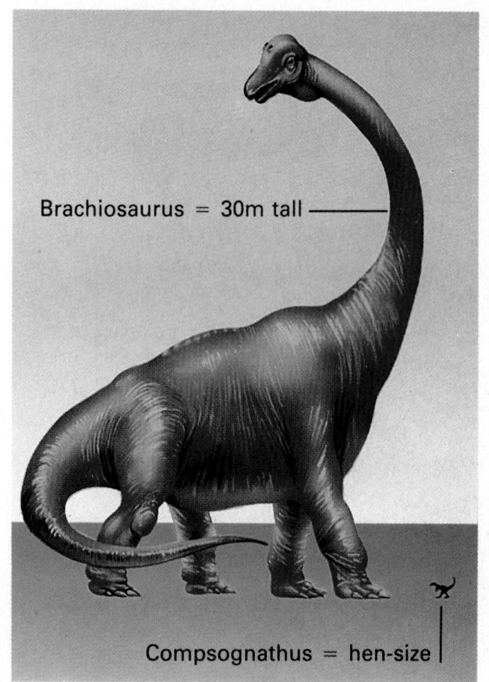

Figure 18–6. ***Brachiosaurus,*** **one of the largest known dinosaurs, stood 30 m tall. One of the smallest dinosaurs,** ***Compsognathus,*** **stood less than 1 m tall.**

The Mesozoic Era

The Mesozoic Era began about 245 million years ago and ended about 65 million years ago. Geologic evidence shows that during the Mesozoic Era the surface of the earth changed dramatically. The supercontinent Pangaea began to break up into separate continents. These new continents drifted and collided, uplifting mountain ranges such as the Sierra Nevada in California and the Andes in South America. The movements of the continents are discussed further in Chapter 19. Shallow seas and marshes covered much of the land. The continental climates were generally warm and humid.

Conditions during the Mesozoic Era favored the survival of reptiles. Lizards, turtles, crocodiles, snakes, and a great variety of dinosaurs flourished during the three periods of the Mesozoic Era. As a result, this era is also known as the *Age of Reptiles*.

The Triassic Period

Dinosaurs first appeared during the Triassic Period. The word *dinosaur* comes from the Greek words meaning "terrible lizard." Dinosaurs varied greatly in size. Some species of dinosaurs were no larger than squirrels. Other dinosaurs weighed as much as 10 bull elephants and some were nearly 30 m long. However, most of the dinosaurs of the Triassic Period were about 4 m to 5 m in length, and they could move quickly. These dinosaurs roamed through lush forests of cone-bearing trees and cycads—plants that resemble the palm trees of today.

Reptiles called *ichthyosaurs* inhabited the Triassic oceans. New forms of marine invertebrates also developed. The most common was the ammonite, a type of shellfish similar to the modern nautilus. Ammonite fossils are found throughout the world and serve as Mesozoic index fossils. The very earliest mammals, small rodent-like forest dwellers, made their appearance.

Figure 18–7. This fossil of a *pterosaur* is an example of the flying reptiles common during the Jurassic Period.

The Jurassic Period

Dinosaurs became the dominant life-form of the Jurassic Period. Fossil records indicate that two major groups of dinosaurs evolved. These dinosaurs are distinguished by their hip-bone structures. One group, called *saurischians* ("lizard-hipped"), included both herbivores, or plant eaters, and carnivores, or meat eaters. One of the largest saurischians was the herbivore *apatosaurus*. Apatosaurs weighed up to 50 tons and reached a length of 25 m.

The other major group of Jurassic dinosaurs, called *ornithischians* ("bird-hipped"), were herbivores. One of the best known of the ornithischians was *stegosaurus,* which had a single row of upright bony plates along its backbone. Stegosaurs were about 9 m long and stood about 3 m high at the hips.

Flying reptiles called *pterosaurs* were common during the Jurassic Period. Like modern bats, pterosaurs flew on skin-covered wings. Also found in Jurassic rocks are fossils of *archaeopteryx,* one of the first true birds with feathers.

Figure 18–8. *Archaeopteryx* and *apatosaurus* were animals that lived during the Jurassic Period.

Table 18–1

Era	Period	Epoch	Began (millions of years)	Characteristics from geologic and fossil evidence
Cenozoic	Quaternary	Holocene	0.011	End of last ice age; complex human societies develop.
		Pleistocene	2	Woolly mammoths, rhinos, and early humans appear.
	Tertiary	Pliocene	5	Large carnivores (bears, dogs, cats) appear.
		Miocene	24	Grazing herds abundant; raccoons, wolves appear.
		Oligocene	37	Deer, pigs, horses, camels, cats, and dogs appear.
		Eocene	58	Early horses, flying squirrels, bats, and whales appear.
		Paleocene	65	Age of mammals begins; lemeroids appear.
Mesozoic	Cretaceous		144	First flowering plants appear; mass extinctions (including all dinosaurs) mark end of Mesozoic Era.
	Jurassic		208	Dinosaurs are dominant life-form; flying reptiles and first birds appear.
	Triassic		245	Dinosaurs first appear; ammonites common; cycads and conifers abundant; first mammals appear.
Paleozoic	Permian		286	Pangaea comes together; mass extinctions mark the end of the Paleozoic Era.
	Carboniferous	Pennsylvanian	320	Giant cockroaches and dragonflies common; coal deposits form; first reptiles appear.
		Mississippian	360	Amphibians flourish; sea stars common in ocean; forests and swamps cover most of the land.
	Devonian		408	Age of fishes; first amphibians appear; giant horsetails, ferns and cone-bearing plants develop.
	Silurian		438	Echinoderms appear; eurypterids abundant; first land plants and animals appear.
	Ordovician		505	Brachiopods increase; trilobites decline; graptolites flourish; first vertebrates (fishes) appear.
	Cambrian		570	Advanced forms of marine life and first invertebrates appear; trilobites and brachiopods common.
Precambrian time			4,600	Formation of the earth; continental shields appear; fossils are rare; stromatolites are most common.

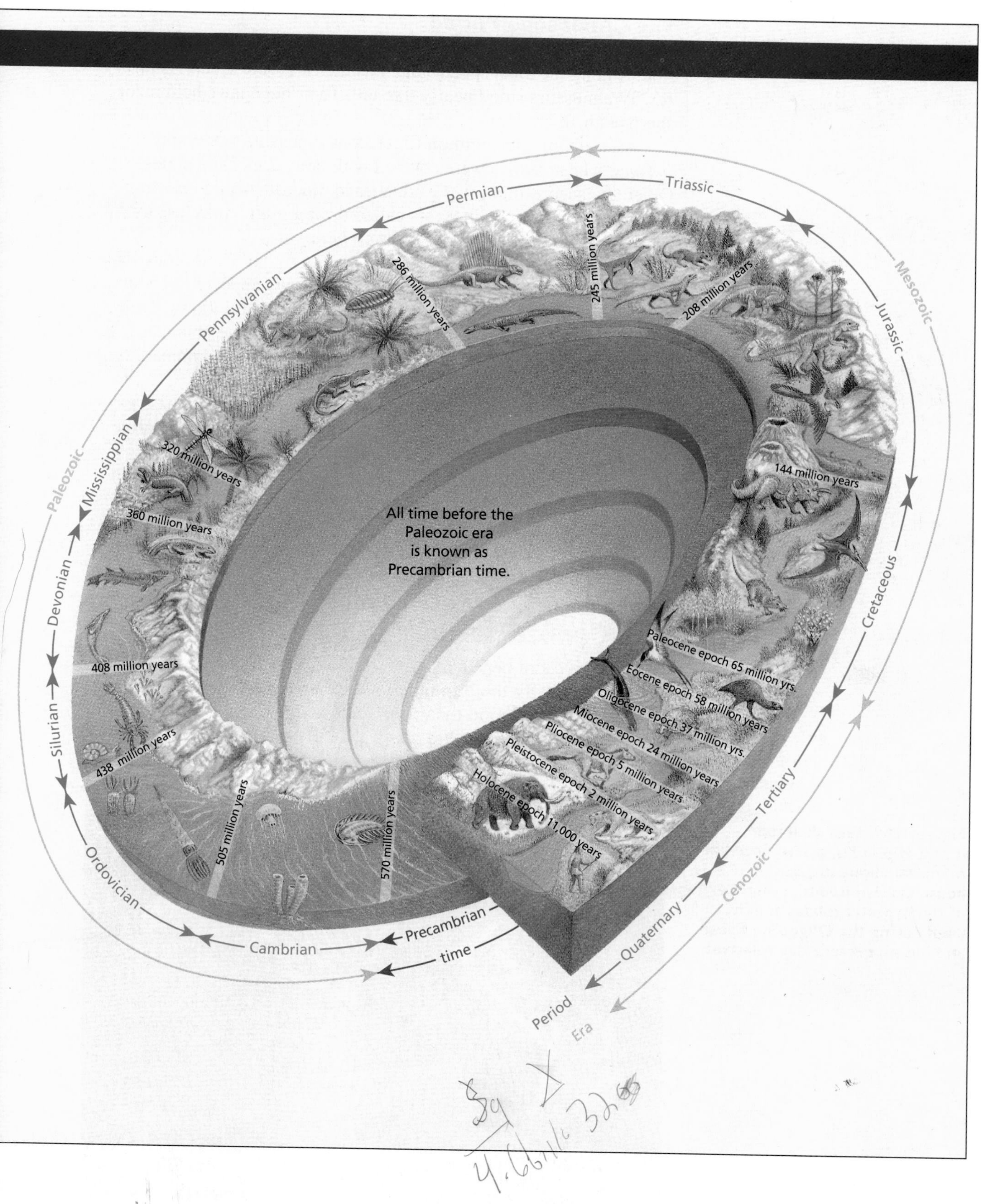

Permian
Triassic
Mesozoic
Jurassic
Pennsylvanian
286 million years
245 million years
208 million years
Paleozoic
Mississippian
320 million years
360 million years
144 million years
All time before the Paleozoic era is known as Precambrian time.
Devonian
Cretaceous
408 million years
Paleocene epoch 65 million yrs.
Eocene epoch 58 million years
Oligocene epoch 37 million yrs.
Miocene epoch 24 million years
Pliocene epoch 5 million years
Pleistocene epoch 2 million years
Holocene epoch 11,000 years
Silurian
438 million years
505 million years
570 million years
Tertiary
Cenozoic
Ordovician
Cambrian
Precambrian
time
Quaternary
Period
Era

The Cretaceous Period

Dinosaurs continued to dominate the earth during the Cretaceous Period. Among the most spectacular was the carnivore *Tyrannosaurus rex*. Tyrannosaurs stood nearly 6 m tall. Their huge jaws held razor-sharp teeth 15 cm long.

Also among the common Cretaceous dinosaurs were *ankylosaurs,* whose bodies were covered with bony armorlike plates. Other Cretaceous dinosaurs were horned dinosaurs called *ceratopsians* and duck-billed dinosaurs called *ornithopods.* Ankylosaurs, ceratopsians, and ornithopods were all herbivores.

Plant life seems to have evolved greatly by the Cretaceous Period. The first flowering plants, or *angiosperms,* appeared. The most common of these were trees such as magnolias and willows. Later, other trees such as maples, oaks, and walnuts became abundant. Angiosperms were so successful that they are still the dominant type of land plant.

The end of the Cretaceous Period was marked by the mass extinction of many species, including all dinosaurs and many marine animals, such as the ammonites. No dinosaur fossils have been found in rocks formed after the Cretaceous Period. As you read in Chapter 1, some scientists accept the meteorite-impact hypothesis as the explanation of this extinction. Other possible explanations include drastic changes in climate owing to the movement of continents and increased volcanic activity.

The Cenozoic Era

The Cenozoic Era began about 65 million years ago and includes the present period. By the beginning of the Cenozoic Era, the continents looked much as they do today, but they were located closer together. In general, the Cenozoic Era was a time of increased tectonic activity. As crustal plates collided, huge mountain ranges, such as the Alps and the Himalayas in Eurasia, were formed.

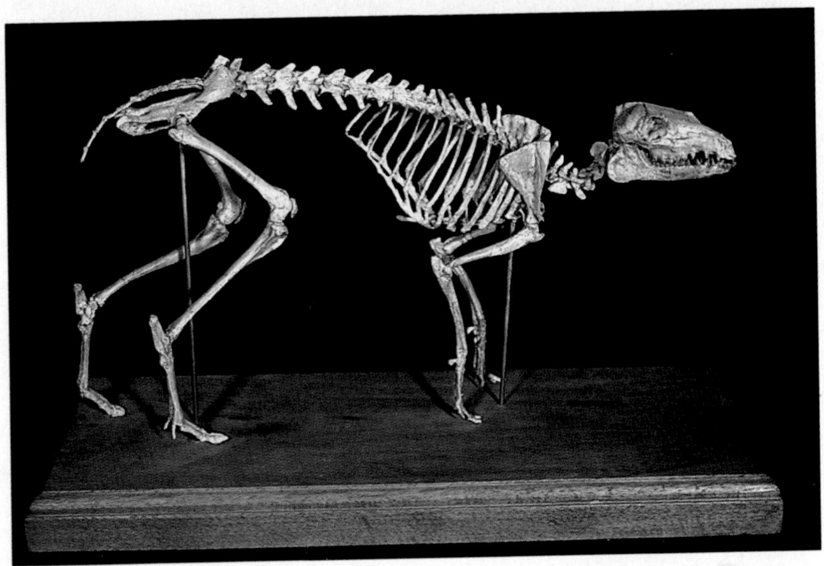

Figure 18–9. This skeleton is of a *Hipisodus,* a slim deer-like mammal about the size of a house cat. *Hipisodus,* a member of the hypertragulidae family, lived during the Oligocene Epoch and has no present day relatives.

During the Cenozoic Era, dramatic changes in climate appear to have occurred. At times, continental ice sheets covered nearly one third of the earth's land area. As temperatures fell during the ice ages, various species became extinct and other species appeared.

Although fossils of mammals first appear in rocks of the early Mesozoic Era, it was during the Cenozoic Era that mammals became the dominant life-form and underwent tremendous evolutionary change. The Cenozoic Era is thus often called the *Age of Mammals*. How do you think mammals survived during the ice ages?

The Cenozoic Era has been divided into two periods—the Tertiary and the Quaternary. The Tertiary Period includes the time before the last major ice age. The Quaternary Period began with the last ice age and includes the present. The Tertiary Period and the Quaternary Period have been further divided into seven epochs. The Paleocene, Eocene, Oligocene, Miocene, and Pliocene epochs make up the Tertiary Period. The Pleistocene and Holocene epochs are in the Quaternary Period.

Figure 18–10. The tarsier, a relative of the lemuroids, is the sole modern survivor of a group of primates common during the Cenozoic Era. Today, most lemurs exist only on the large island of Madagascar, off Africa's east coast.

The Paleocene and Eocene Epochs

The fossil record indicates that during the Paleocene Epoch many new mammals evolved. There were small rodents, about the size of modern squirrels or mice, and a few small carnivores. Fossils of the first primates, called *lemuroids,* also are found in rocks of the Paleocene Epoch. Lemuroids were small, tree-dwelling mammals with thick fur and long tails.

Other mammals, including *Hyracotherium,* which is the earliest known ancestor of the horse, developed during the Eocene Epoch. Fossil records indicate that the first flying squirrels, bats, and whales appeared during this epoch. Although the dinosaurs had become extinct, smaller reptiles continued to flourish. Worldwide temperatures dropped 4°C during the Eocene Epoch.

The Oligocene and Miocene Epochs

Geologic evidence indicates that during the Oligocene Epoch the climate continued to become cooler and drier. It became so dry that the Mediterranean Sea dried up, leaving only small, scattered salty lakes. Salt and other *evaporites* accumulated to a depth of 2,000 m on the floor of the sea. This change in climate favored the growth of grasses and cone-bearing and hardwood trees. Many of the earlier types of mammals became extinct. However, larger species of deer, pigs, horses, camels, cats, and dogs flourished.

The Miocene Epoch is often called the *Golden Age of Mammals*. The climate remained cool and dry. Great herds of primitive horses and camels roamed the plains, feeding on the abundant grasses. Fossil remains of members of the deer, rhinoceros, and pig families are commonly found in rocks of this epoch. *Baluchitherium,* a rhinoceroslike animal, lived during the Miocene Epoch. *Baluchitherium* (also called *Paraceratherium*) is the largest known land mammal ever to have existed. It was nearly twice as large as a modern elephant. Miocene rocks also contain fossils of raccoons, wolves, foxes, and the now extinct saber-toothed cat.

Figure 18–11. Saber-toothed cats were ferocious predators of the Pliocene and Pleistocene epochs. These remains were discovered in the La Brea tar pits in California.

The Pliocene Epoch

During the Pliocene Epoch, hunting animals—members of the bear, dog, and cat families—became fully evolved. These fierce carnivores, such as the saber-toothed cat in Figure 18–11, hunted the herds of grazing animals that inhabited the grassy plains. Fossils of the first modern horses are also found in rocks of this epoch.

Before the end of the Pliocene Epoch, great climatic changes occurred, and the continental ice sheets began to spread. With so much water locked in ice, sea level fell, and the Bering land bridge appeared between Eurasia and North America. The uplifting of the earth's crust between North and South America formed the central American land bridge. Various species migrated between the continents across these two major land bridges.

The Pleistocene and Holocene Epochs

During the Pleistocene Epoch, several periods of glaciation occurred over much of Eurasia and North America. Some animals had special characteristics that allowed them to endure the cold, such as the thick fur that covered woolly mammoths and rhinoceroses. Other species survived by moving to warmer regions. Less-adaptable species, such as giant ground sloths and dire wolves, became extinct. Fossils of the early ancestors of modern humans are found in rocks of the Pleistocene Epoch. Scientists have evidence that indicates these early humans were hunters. Their successful hunting may have been one factor that led to the extinction of many large mammals during the Pleistocene Epoch.

The Holocene Epoch, which includes the present, began about 11,000 years ago, as the last ice age ended. As the ice sheets melted, sea level rose an estimated 140 m, and the coastlines took on their present shapes. The Great Lakes were also formed as the ice sheets retreated. During the early Holocene Epoch, modern humans, or

INVESTIGATE!

To learn more about geologic time, try the In-Depth Investigation on pages 362–363.

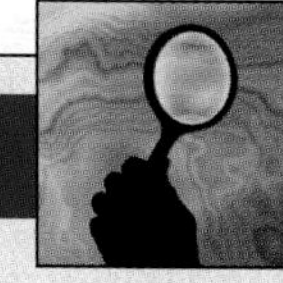

SMALL-SCALE INVESTIGATION

Geologic Time Scale

Geologic time spans some 4.6 billion years. Using a scale that represents time as distance, you can compare the length of eras and periods.

Materials

geologic time scale on pages 354–355; adding-machine paper, 5 m long; ruler; meter stick; 6 colored pencils

Era	Length of time (years)	Scale length
Precambrian	4,030,000,000	
Paleozoic	325,000,000	
Mesozoic	180,000,000	
Cenozoic	65,000,000 (to present)	

Procedure

1. Copy the table onto your own piece of paper.
2. Complete the table, using the scale 1 cm equals 10 million years.
3. Lay the adding-machine paper flat on a hard surface. Use the meter stick and a pencil to mark off the beginning and end of Precambrian time according to the time scale you calculated. Do the same for the other eras. Label the eras and color each era a different color.
4. Study the geologic time scale on pages 354–355. Use the scale 1 cm equals 10 million years and calculate the scale length for each period listed.
5. Mark the boundaries of each period by drawing a line across the width of the adding-machine paper. Label the periods on your scale.
6. Write in the major kinds of organisms that lived during each unit of time.

Analysis and Conclusions

1. Does the scale you made measure relative time, absolute time, or both?
2. When did humans first appear? What is the scale length from that period to the present?
3. Add the lengths of the Paleozoic, Mesozoic, and Cenozoic eras. What percentage of the total geologic time scale do these eras combined represent? What percentage of the total time scale does Precambrian time represent?

Homo sapiens, developed agriculture and began to make and use tools of bronze and iron.

Compared with the entire geologic time scale, human history is extremely brief. If you think of the entire history of the earth as one year, the first organisms would have appeared in early May. Early humans would have appeared on December 31 at 7 P.M.; and modern humans, not until 11:55 P.M. that night.

Section 18.2 Review

1. Why are fossils rare in Precambrian rocks?
2. How did the formation of Pangaea affect Paleozoic life-forms?
3. How did the ice ages affect animal life during the Cenozoic Era?
4. Compare the effects of the Permian extinction with those of the Cretaceous extinction.

1-4

Chapter 18 Review

Key Terms

- Cenozoic Era (346)
- epoch (346)
- era (346)
- geologic column (345)
- invertebrate (349)
- Mesozoic Era (346)
- Paleozoic Era (346)
- period (346)
- Precambrian time (346)
- theory of evolution (347)
- vertebrate (350)

Key Concepts

The geologic column is based on observations of the relative ages of rock layers throughout the world. **See page 345.**

Geologic time has been divided into eras, periods, and epochs. **See page 345.**

Precambrian rocks contain valuable minerals but few fossils. **See page 348.**

The rock record reveals the evolution of marine invertebrates and vertebrates during the Paleozoic Era. **See page 349.**

The rock record of the Mesozoic Era reveals an environment that favored the development of reptiles. **See page 352.**

The rock record of the Cenozoic Era includes the present period and reveals the rise of mammals as the predominant life-form. **See page 356.**

Review

On your own paper, write the letter of the term that best completes each of the following statements.

1. The geologic time scale is a
   a. scale for weighing rocks.
   b. timer used by geologists.
   c. rock record of the earth's past.
   d. collection of the same kind of rocks.

2. Scientists have been able to determine the absolute ages of most rock layers in the geologic column by using
   a. the law of superposition. b. radioactive dating.
   c. rates of deposition. d. rates of erosion.

3. An event that geologists would use in dividing the geologic time scale into smaller units is
   a. the eruption of a volcano. b. a change in rock color.
   c. a change in sediment type. d. the arrival of an ice age.

4. To determine the age of a specific rock, scientists might correlate it with a layer in the geologic column that has the same relative position, physical characteristics, and
   a. fossil content. b. weight.
   c. temperature. d. density.

5. *Paleozoic* means
   a. ancient life. b. middle life. c. recent life. d. primitive life.

6. Geologic periods may be divided into
   a. eras. b. epochs. c. days. d. months.

7. *Cenozoic* means
   a. ancient life. b. middle life. c. recent life. d. primitive life.

8. Precambrian time ended about
   a. 4.6 billion years ago. b. 570 million years ago.
   c. 65 million years ago. d. 25 thousand years ago.
9. The most common fossils found in Precambrian rocks are
   a. graptolites. b. trilobites. c. eurypterids. d. stromatolites.
10. An important index fossil for the Paleozoic Era is the
   a. ammonite. b. trilobite. c. ostracoderm. d. stromatolite.
11. The first vertebrates appeared during
   a. Precambrian time. b. the Paleozoic Era.
   c. the Mesozoic Era. d. the Cenozoic Era.
12. The *Age of Fishes* is the name commonly given to the
   a. Cambrian Period. b. Ordovician Period.
   c. Silurian Period. d. Devonian Period.
13. The *Age of Reptiles* is the name commonly given to
   a. Precambrian time. b. the Paleozoic Era.
   c. the Mesozoic Era. d. the Cenozoic Era.
14. Dinosaurs became the dominant life-form during the
   a. Permian Period. b. Silurian Period.
   c. Jurassic Period. d. Tertiary Period.
15. The first flowering plants made their appearance during the
   a. Cretaceous Period. b. Triassic Period.
   c. Carboniferous Period. d. Ordovician Period.
16. The *Age of Mammals* is the name commonly given to
   a. Precambrian time. b. the Paleozoic Era.
   c. the Mesozoic Era. d. the Cenozoic Era.

Critical Thinking

On your own paper, write answers to the following questions.

1. Explain how the law of superposition has aided scientists in the development of the geologic column.
2. Why is it difficult to divide Precambrian time into periods?
3. Paleontologists classify fossils by similarities in the structure of the hard parts of the organisms. Sometimes, however, scientists find that this is an unreliable method of classifying fossils. Explain.
4. There are rich deposits of coal in the Appalachian Mountains. During which geologic era were these deposits probably formed? Explain your answer.
5. What information in the geologic record might lead scientists to infer that shallow seas covered much of the earth during the Paleozoic Era?
6. Explain the basis scientists may have used for dividing the Cenozoic Era into the Tertiary and Quaternary periods.

Application

1. In 1966, two amateur fossil hunters found a fossil on the shore at Cliffwood, New Jersey. Scientists determined that the fossil was an insect intermediate between primitive ants and wasps. What types of information might they look for in this fossil discovery?
2. If you found a rock layer that contained dinosaur fossils, what assumptions could you make about the age of the rock layer?
3. Concept map ? ? ? Referring to the geologic time scale, construct a **concept map** that illustrates the relationship between the various time divisions and some of the organisms that first developed during those divisions.

Extension

1. Find out the locations of the Precambrian shields throughout the world. Shade in those areas on a map of the world.
2. Compare the evolution of life during the Paleozoic Era with that during the Mesozoic Era. How did environmental changes affect the development of organisms?
3. Research the discoveries made by British anthropologists Dr. Louis S. B. Leakey and Mary Leakey in Olduvai Gorge in Tanzania, Africa. Write a report based on your findings.

IN-DEPTH INVESTIGATION

Geologic Time

Materials

- meter stick
- adding-machine paper, 5 m long
- fine-point pen or sharpened pencil
- reference books with information on the geologic eras

Introduction

The vast time involved in geologic history is difficult to comprehend. You can imagine this difficulty when you try to compare a human's lifetime with the span of the earth's history. To understand the length of time it took for these earth-changing processes to occur and life on it to evolve, it is helpful to convert geologic time into a comprehensive, condensed scale. A conversion scale helps to unlock the real meaning of geologic time.

In this investigation, you will convert geologic time into a frame of reference that is easier to understand.

Prelab Preparation

1. Review Table 18.1, pages 354–355.
2. Review metric-scale conversions. For the purpose of this investigation, it is necessary to know that 1 m = 100 cm and 1 cm = 10 mm.

Procedure

1. By converting measures of geologic time to increments of distance, you can make a scale model of the earth's history. In order to translate time into equivalent distances, you must work out a scale. To do this, let 1 m equal 1 billion (1,000,000,000) years. Complete the following scale: 1 m = ? years; 1 cm = ? years; 1 mm = ? years.
2. Using your scale, convert each of the ages listed in the table on the next page to its equivalent distance in meters, centimeters, or millimeters. For example, an age of 9 million years becomes 0.009 m, 0.9 cm, or 9 mm.
3. Obtain a piece of adding-machine paper 5 m in length. At 1 cm from one end, draw a straight line across with a pencil or pen. Label this line with the word TODAY. Using the meter stick, measure 4.6 m from the TODAY line and draw another line. Mark this point as AGE OF THE EARTH. This point represents the earth's maximum age. Refer to the list of ages, and label the correct age.
4. Continue to mark off and label the distances you calculated for each geologic event. Be sure to label each geologic event and the age associated with the event.

Geologic Event	Age
Age of the earth	4.6 billion years
Oldest known plants	2 billion years
Oldest known animals	1.2 billion years
Paleozoic Era begins (Cambrian Period)	570 million years
Ordovician Period begins	505 million years
Silurian Period begins	438 million years
Devonian Period begins	408 million years
Mississippian Period begins	360 million years
Pennsylvanian Period begins	320 million years
Permian Period begins	286 million years
Mesozoic Era begins (Triassic Period)	245 million years
Jurassic Period begins	208 million years
Cretaceous Period begins	144 million years
Cenozoic Era begins (Paleocene Epoch)	65 million years
Eocene Epoch begins	58 million years
Oligocene Epoch begins	37 million years
Miocene Epoch begins	24 million years
Pliocene Epoch begins	5 million years
Pleistocene Epoch begins	2 million years
Holocene Epoch begins	11 thousand years

5. Use reference books to find the correct time in geologic history during which the following events occurred. Convert these times to their equivalent distances in meters. Place these events in the appropriate places on the adding-machine paper. For events occurring over long periods of time, mark the places on the paper that represent when the event begins and when the event ends. Connect these two points with a horizontal line.

a. trilobites appear, flourish, and become extinct
b. first flowering plants (angiosperms) appear
c. apes appear in Asia and Africa
d. dinosaurs appear, flourish, and become extinct
e. first birds appear
f. first reptiles appear
g. first mammals appear
h. early humans appear

Analysis and Conclusions

1. Explain why you cannot mark the position of important events in history such as the American Revolution on your paper scale.

2. Would it be possible to measure a distance on the paper that could represent the length of a person's lifetime? Explain your answer.

Extensions

1. What might be some reasons why we do not see the same animals and plants on the earth today that could be found thousands of years ago?

2. Why do some animals and plants, whose origins go back millions of years, still exist today relatively unchanged except for their size?

For billions of years, the surface of the earth has been tectonically active. Lithospheric plates have drifted slowly around the planet, colliding with each other to form supercontinents and then splitting apart into smaller landmasses. The existing continents resulted from this ongoing process.

This chapter discusses the global movement of the continents. It traces the evolution of North America and shows how the rock layers of the Grand Canyon reveal the history of a region.

Chapter Outline

◀ Grand Canyon National Park, Arizona, as seen from Yaki Point.

19.1 Movements of the Continents

Section Objectives

- **Identify the landmasses that made up Pangaea.**
- **Describe the breakup of the supercontinent Pangaea.**

Early one morning in May 1980, Mount St. Helens, a volcanic peak in the Cascade Range in southwest Washington, erupted violently. The eruption blew away one side of the mountain. The nearby forest was destroyed and instantly changed into a lifeless, barren landscape.

Sudden and dramatic events such as volcanic eruptions, earthquakes, and floods are reminders that the earth's surface is changing constantly. Changes such as these take place in minutes, hours, or days. However, most of the processes that shape the earth's crust occur very slowly. For example, changes in the locations of the continents and in the size and shape of the oceans have taken place over many millions of years. In fact, the earth's crust shows evidence of geological changes that have taken place over a period of about 4 billion years.

The rock record indicates that in the earth's long history there has never been a permanent or even typical arrangement of the continents and oceans. The location of the continents and oceans today is the result of plate tectonics, or the movement of crustal plates. The theory of plate tectonics states that the lithosphere is made up of many plates that ride on the asthenosphere. These plates drift slowly around the planet—usually no faster than a few centimeters each year. What are some consequences of this plate movement?

Formation of Pangaea

Geologists find it difficult to trace the movements of the continents that occurred very early in the earth's history. However, by combining evidence from many scientific fields, a general picture can be constructed. There is little doubt that continents and oceans existed during the earliest periods of the earth's history. All of the continents existing today contain large areas of deformed Precambrian rocks

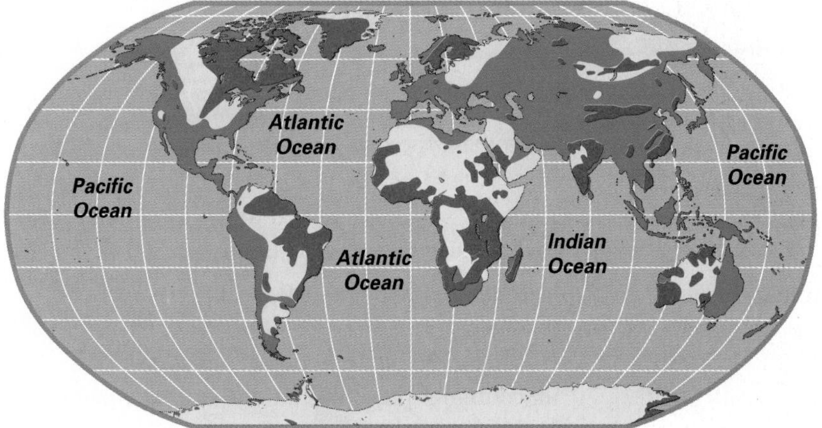

Figure 19–1. The major Precambrian cratons are shown in yellow. The exposed Precambrian rocks, or continental shields, are shown in red.

called **cratons.** Precambrian rocks within the cratons that have been exposed at the earth's surface are called *shields*. Cratons represent ancient cores around which the modern continents have formed. The exposed rocks of the continental shields supply geologists with evidence of mountain building, volcanic activity, and sediment deposition. However, the outlines of the continents and oceans as they existed during the Precambrian Era are not known.

During the Paleozoic Era, there were probably two or three supercontinents that formed, broke up, and re-formed as lithospheric plates drifted slowly around the earth. The supercontinent Pangaea, proposed by Alfred Wegener as the continental mass from which the modern continents split off, formed between 250 and 300 million years ago. The collisions of the various landmasses that resulted in the formation of Pangaea also caused much mountain building. The Appalachian Mountains of eastern North America and Russia's Ural

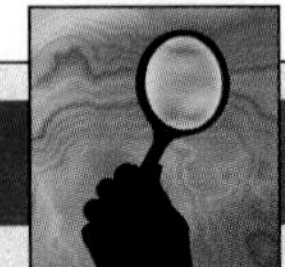

SMALL-SCALE INVESTIGATION

Plate Tectonics Theory

Although questions still remain about the movement of lithospheric plates, most scientists think that plate movement is caused by convection currents in the asthenosphere. You can demonstrate this theory by making a model of the asthenosphere and lithospheric plates.

Materials

water; shallow rectangular pan, about 23 cm × 33 cm; stove or 2 hot plates; dark-colored food coloring; 8 pieces of shirt-packaging cardboard, about 1 cm^2 each; marker; metric ruler

Procedure

1. Fill the pan with water until it is 3 cm from the top, and place it on the stove or hot plates.
2. Heat the water for 30 seconds over very low heat. Add a few drops of food coloring to the center of the pan. Observe and record what happens to the food coloring. Turn off the heat.
3. Label the cardboard pieces 1 through 8. Carefully place pieces 1 through 4 as close together as possible in the center of the pan. Place one of the four remaining cardboard pieces in each corner of the pan.

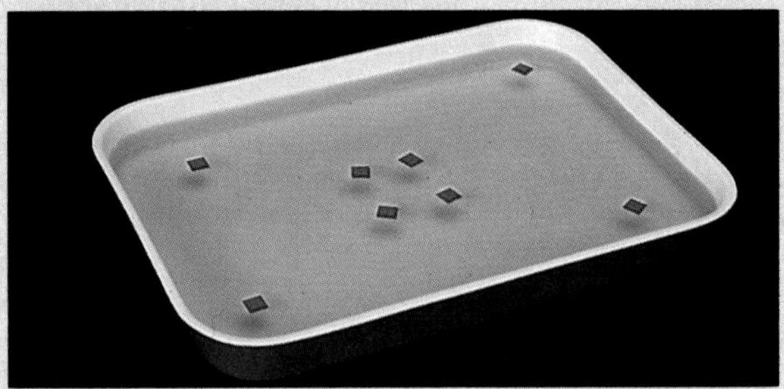

4. Turn on the heat. Sketch the pattern of movement for each cardboard piece. Turn off the heat. **Caution: Use a potholder when handling heated objects.**

Analysis and Conclusions

1. In the plate tectonics model, what do the water in the pan, the cardboard pieces, and the stove represent?
2. In Step 2, what happened to the food coloring in the water? Explain your answer.
3. Describe what happened to the cardboard pieces as the water was reheated in Step 4.
4. In what way is the model you made an inaccurate representation of the movement of lithospheric plates?

Mountains, which are the boundary between Europe and Asia, were formed during these Paleozoic Era collisions.

Pangaea, initially located close to the polar region of the Southern Hemisphere, covered about 40 percent of the earth's surface. The remaining 60 percent was covered by a single, large ocean called *Panthalassa*. A triangular body of water called the *Tethys Sea* started to cut into the eastern edge of Pangaea. North of the Tethys Sea lay a portion of Pangaea called **Laurasia.** This area included the land that would eventually form North America and Eurasia. To the south of the Tethys Sea lay the other part of Pangaea, which was called **Gondwanaland.** Gondwanaland eventually split to form South America, Africa, India, Australia, and Antarctica.

Because of its location near the southern polar region, the environment of Pangaea was harsh. Gondwanaland was partially covered by a continental ice sheet. Even the Sahara, one of the hottest places on the earth today, was once covered by a thick ice sheet. As Pangaea began to drift northward toward the equator, however, much of the ice sheet melted. This raised the level of the oceans and probably triggered global warming. As the climate grew warmer, the number of species of land plants and animals increased.

Breakup of Pangaea

About two hundred million years ago, Pangaea began to break into two landmasses. Just north of the equator a great east-west rift split the supercontinent, nearly separating Laurasia and Gondwanaland. They

Figure 19–2. Between 250 and 300 million years ago, the earth's landmasses had joined together in the supercontinent called *Pangaea* (left). About 200 million years ago, Pangaea began to break up, and by 135 million years ago, Laurasia and Gondwanaland were separated by the Tethys Sea (right).

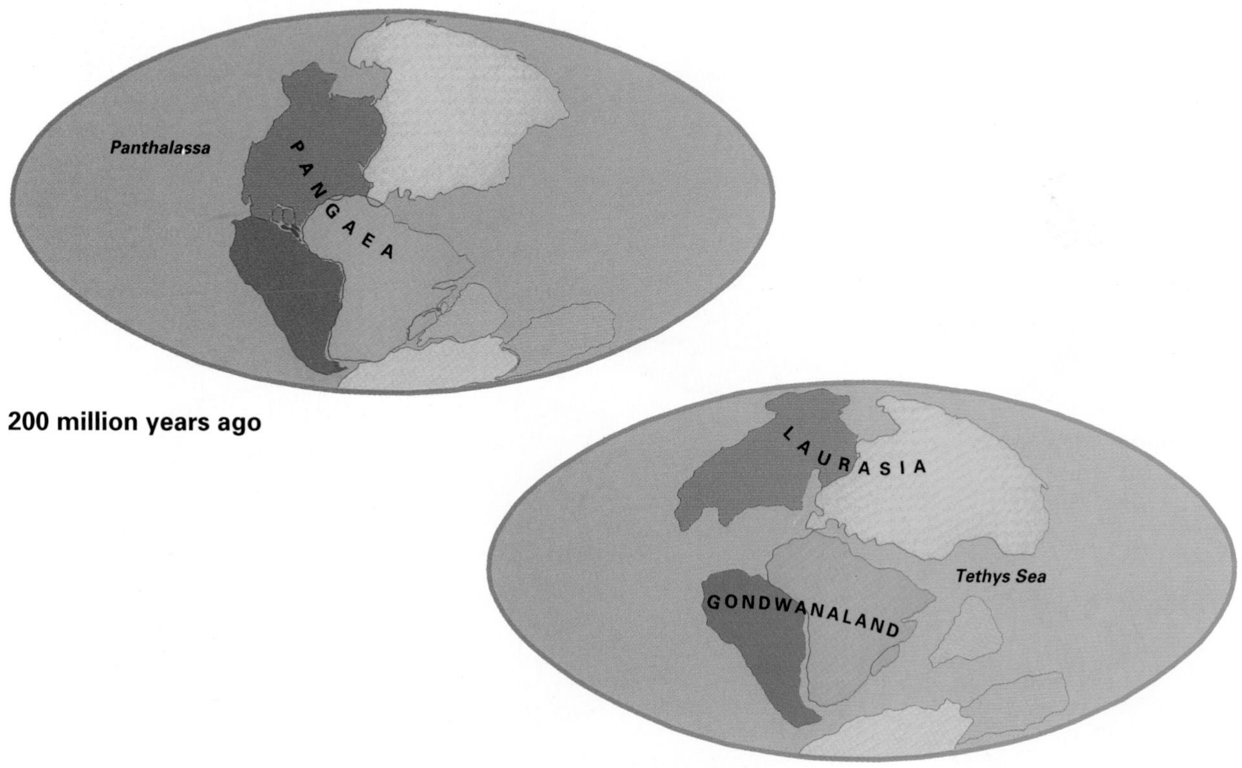

remained connected only where the southern tip of Spain touched the northwestern coast of Africa. East of this connection was the Tethys Sea; west of it the rift opened into Panthalassa. Ocean water poured into the rift, creating a narrow sea.

Geologic evidence indicates that as the rift opened, Laurasia began to rotate slowly in a clockwise direction. As this huge landmass moved, the sea located to its west began to widen. This body of water would become part of the Atlantic Ocean. The movement of Laurasia also resulted in the closing off of the Tethys Sea in the east. The Tethys Sea would become the much smaller Mediterranean Sea. The continued movement of Laurasia caused the east-west rift to diverge in a northward direction, separating North America from Eurasia and forming the North Atlantic Ocean.

As Laurasia began to break apart, Gondwanaland also began to split. A rift opened in Gondwanaland, splitting the landmass in two. The landmass that would become the continents of South America and Africa was located on one side of the rift. The landmass that consisted of India, Australia, and Antarctica was on the other side. India eventually broke away from Antarctica and began moving northward. Madagascar separated from Africa. About 100 million years ago, during the Cretaceous Period, a rift opened between Africa and South America, forming the South Atlantic Ocean. Over the last 65 million years Australia separated from Antarctica, and India, continuing its northward movement, collided with Eurasia, forming the Himalayas.

Slowly the continents moved into their present positions. As the continents drifted around the earth, they collided with smaller pieces

Figure 19–3. About 65 million years ago, the South Atlantic Ocean formed as South America split from Africa (left). Today the continents are still in motion (right).

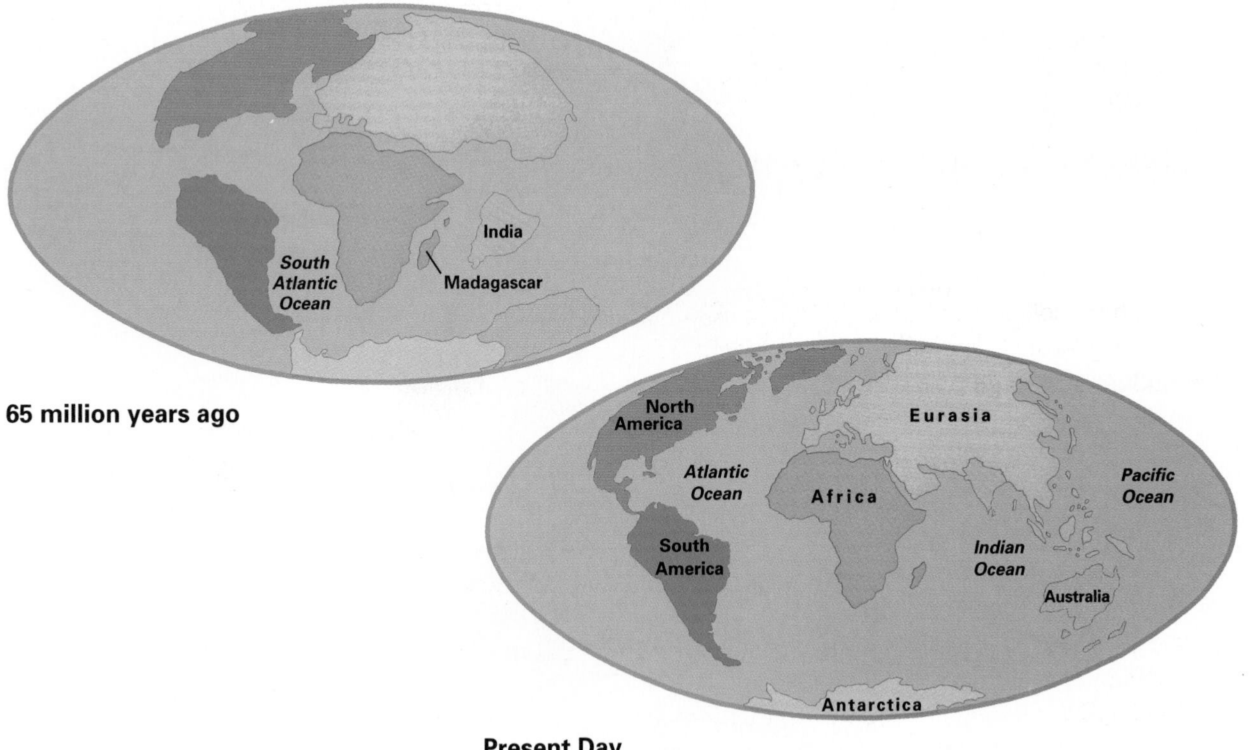

EarthBeat

Geography of the Future

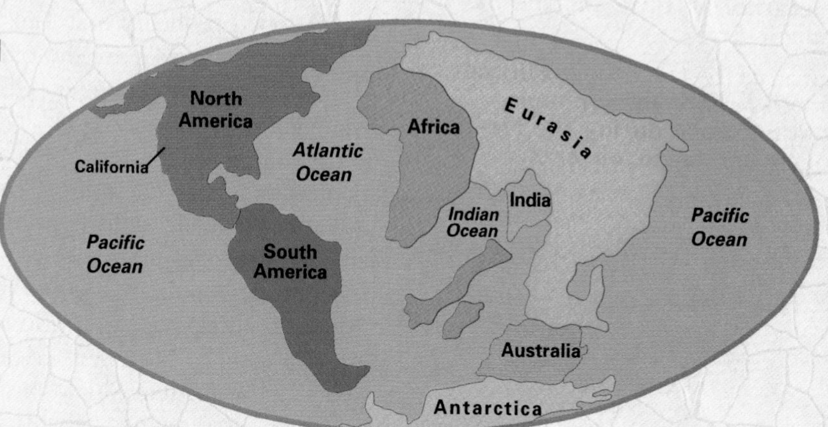

The movement of the crustal plates that change the oceans and the continents continues even today. For example, from actual measurements of plate movements, scientists now estimate that Hawaii and Japan are moving closer together by as much as 12.7 cm per year.

If the plates continue to move as they are now doing, in about 150 million years the face of the earth will be very different. Africa will collide with Eurasia, closing the Mediterranean Sea and probably forcing the Alps even higher. A new ocean will be formed as east Africa separates from the rest of the continent and moves eastward. As the North American and South American plates move westward and Eurasia and Africa move eastward, the Atlantic Ocean will become wider. Australia will continue to move north, eventually colliding with Eurasia.

In North America, Mexico's Baja Peninsula and the portion of California west of the San Andreas Fault will move to where Alaska is today. If this plate movement occurs as predicted, Los Angeles will one day be located north of where San Francisco is today.

Scientists predict that 250 million years from now all the continents may come together again to form a future supercontinent.

If the new supercontinent is located near the equator, how might the global climate and the level of water in the oceans be affected?

of continental crust and with volcanic islands formed on oceanic crust. These collisions welded new crust onto the continents and uplifted the land. Mountain ranges such as the Rockies, the Andes, and the Alps were formed. Tectonic plate motion also caused new oceans to open up and others to close.

Section 19.1 Review

1. What two major land areas made up the supercontinent Pangaea?
2. What processes that help to shape the earth's surface are evident in cratons?
3. Describe the formation of the North Atlantic Ocean and the South Atlantic Ocean.
4. Explain how the theory of plate tectonics relates to the formation and breakup of Pangaea.

Section Objectives

- Summarize the changes in the North American continent that occurred during both the Precambrian time and the Paleozoic Era.
- Summarize the changes in the North American continent that occurred during the Mesozoic and Cenozoic eras.

19.2 Growth of a Continent: North America

The ancient landmass around which the present North American continent developed probably first took shape during Precambrian time. Since Precambrian rocks are the oldest in the geologic column, they are usually covered by younger rock layers. In the Grand Canyon, for example, Precambrian rocks are exposed only at the very bottom of the canyon. This is where the Colorado River has cut through the sedimentary layers to a depth of over 1 km.

In eastern Canada and the northeast and upper midwest of the United States there is a large area where very old Precambrian rocks are exposed. This region is called the **Canadian Shield.** The Canadian Shield is the exposed portion of the craton around which the modern continent of North America has been built up. For 600 million years, the Canadian Shield has been the only part of North America that has remained mostly intact and not tectonically active.

To the south and west of the Canadian Shield is a central, stable region where the Precambrian rocks are buried under layers of sediments. The sediments are about 2.5 km thick in some places. They were deposited during the Paleozoic, Mesozoic, and Cenozoic eras when the land was covered by shallow seas. Surrounding the stable Canadian Shield and the platform around it are belts of mountains. These mountains exist where the crust has been folded or faulted by plate movements and by the invasion of magma from below. Figure 19–4 shows a cross section of the continent, showing the position of the stable central region and the mountains on either side.

Like similar rocks found all over the world, the Precambrian rocks of North America yield fossil evidence of only the simplest forms of life. Rocks of the Canadian Shield contain traces of microscopic bacteria. The fossils found in these rocks are thought to be more than 2 billion years old.

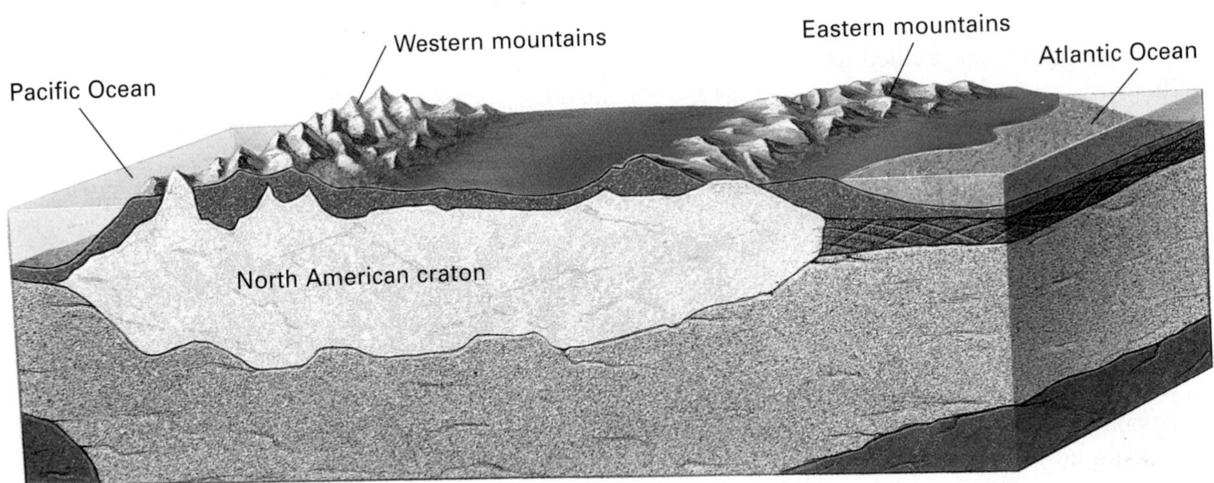

Figure 19–4. This cross section of the North American continent shows a central stable region bordered by two mountain belts.

Beginnings of a Continent

The rock structure of an existing continent can reveal a general outline of its geologic history. The rock structure of North America indicates that during the Paleozoic Era, much of what is now North America was part of a larger continent. This continent was located near the equator. A shallow sea covered much of what is now the United States.

Huge amounts of sediments were carried into the shallow water east and west of the present continent. The weight of the sediments caused the ocean floor to sink. More sediments were deposited and continued to accumulate on the sinking, shallow ocean floor. Eventually the sediments reached a thickness of over 12 km. Huge, Paleozoic, sediment-filled troughs called *geosynclines* developed. They now make up much of the older rock structures found along the eastern and western edges of North America. Thinner layers of Paleozoic sediments were also laid down over the stable region that is now the central United States.

As the Paleozoic Era ended 245 million years ago and the continents came together to form Pangaea, most of the warm, shallow seas disappeared. The marine invertebrates that had been so abundant for more than 400 million years vanished almost entirely. Their disappearance is thought to be related to the joining of the contenents into Pangaea. The large inland sea that covered much of what was to become North America dried up. Left behind were the huge deposits of salt now found buried deep in Permian layers from Kansas to New Mexico. It is likely that the earth's climate also changed. How might the change in climate have contributed to the disappearance of marine life?

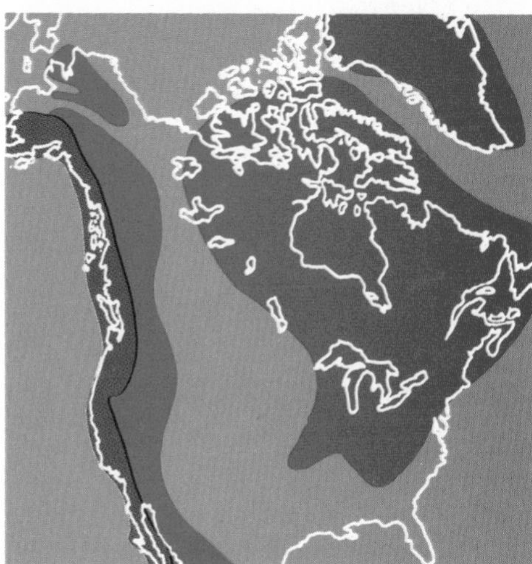

Figure 19–5. The top illustration shows that during the Paleozoic Era northeastern Canada was land (green area) and central North America was a wide sea (blue area). Sediment-filled troughs called *geosynclines* (brown areas) formed along the coasts. The bottom illustration shows that during the Cretaceous Period a shallow sea covered much of central North America.

Shaping the Continent

The burst of tectonic activity accompanying the breakup of Pangaea played a large role in shaping North America. The basic form of the continent was determined by the rifts that split Laurasia from Gondwanaland and North America from Eurasia. The North America that separated from Laurasia during the Mesozoic Era did not have the same shape as the Paleozoic continent that became part of Pangaea. Much of what is now the Atlantic coastal area was once attached to the African plate. Some of the ancient European plate became welded onto parts of present-day New England and Canada. In addition, new pieces of continental crust, called *terranes*, were added. The terranes were carried along on oceanic crust and were welded to North America as the oceanic crust subducted beneath the continental crust.

Early in the Mesozoic Era, a large mountain range, the ancient Appalachians, existed along what is now the eastern United States. The Appalachians were created during the collision that formed Pangaea. On the western side of the continent, a sea covered the geosynclinal trough that had been created during Paleozoic times. The central portion of the continent was periodically covered by a shallow sea. By the late Mesozoic Era, the Appalachian Mountains had been quite worn down. The Rocky Mountains began to rise east of the western geosyncline. This western uplift divided the continent into eastern and western land areas separated by an inland sea. The general landscape of North America, characterized by eastern and western highlands separated by a region of low elevation, began to take shape.

The Modern Continent

By the beginning of the Cenozoic Era, about 65 million years ago, North America had begun to take on the shape it has today. The inland sea that covered much of the central region during Mesozoic times had drained away. Only the present coastal plain along the Atlantic and Gulf coasts remained submerged. The tectonic activity that took place during the Cenozoic Era created most of the major landforms of western North America. Violent subduction and collision movements along plate boundaries in the western part of the continent caused the uplift of the Sierra Nevada in California. These mountains, which first appeared during the Jurassic Period, were eroded until they were almost flat, then uplifted again during the Cenozoic Era.

East of the Sierra Nevada, a series of tilted fault-blocks created a large area of alternating mountains and valleys. Uplift of a broad area around the Colorado River produced the Colorado Plateau. The Grand Canyon was created as the river cut into the rising land. Farther to the north, the Rocky Mountains continued to be pushed up. On the level platform to the east, sediments eroded from the Rockies were deposited on the platform, forming the Great Plains. Near the east coast of the continent, the ancient Appalachian Mountains continued to be worn down by weathering and erosion.

Volcanic activity has continued throughout the Cenozoic Era in the northwest. Washington, Oregon, and northern California are crossed by the volcanic Cascade mountain range. Lava plateaus created by volcanic eruptions formed to the east of the Cascades.

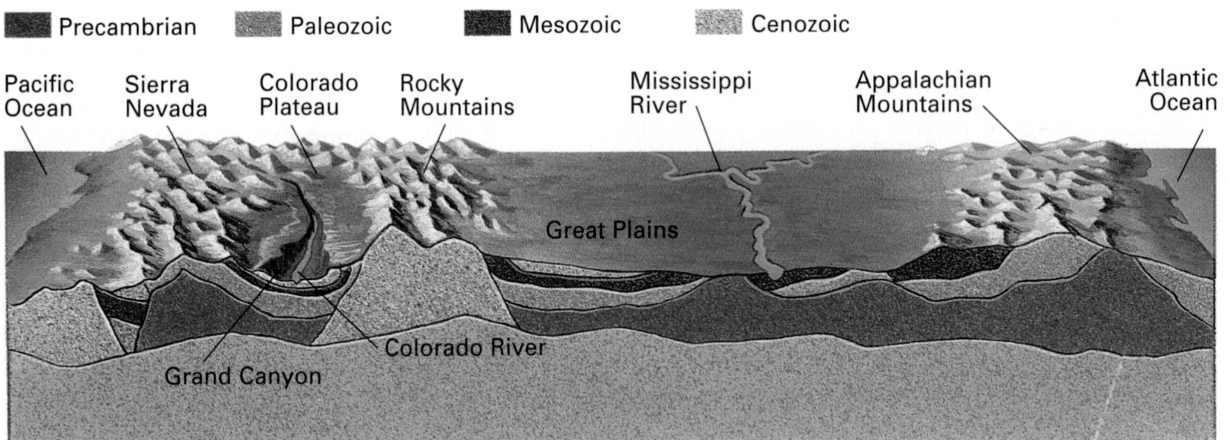

Figure 19–6. This east-west cross section of the United States, at about the latitude of Washington, D.C., shows the basic rock structures and the eras in which they were formed.

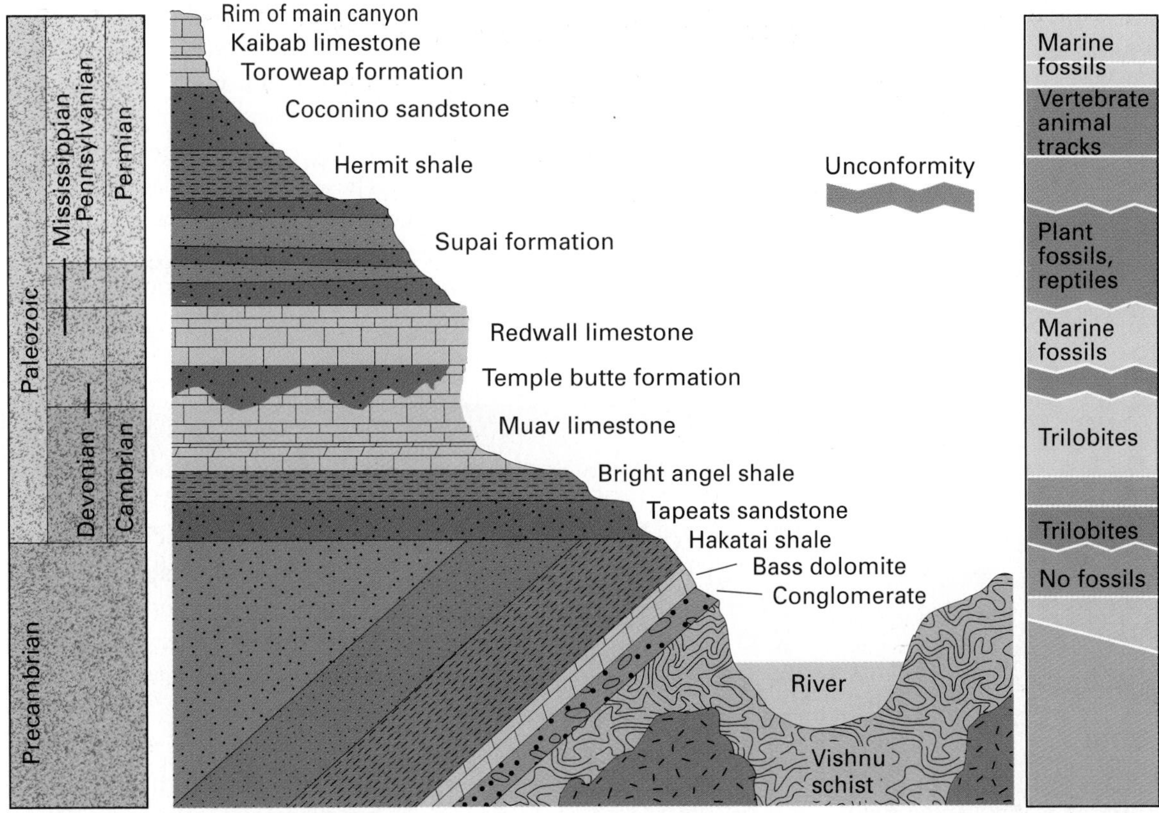

Figure 19–8. This cross section and geologic column of the Grand Canyon shows the positions of the Paleozoic strata deposited above Precambrian rocks and the associated fossils.

is represented by several deposits—the **Hermit shale,** the **Coconino sandstone,** the **Toroweap formation,** and the **Kaibab limestone.** These strata form the upper walls of the canyon, and the Kaibab limestone is the current surface rock of the Colorado Plateau.

The composition of these rock layers reveals more about the geologic history of the canyon region. For example, the light-colored layers you may have seen in many photographs of the Grand Canyon are composed of sandstone. Some of these sandstone layers show cross-bedding, which is characteristic of sand dunes. Therefore, it appears that these layers were laid down when the region was either an arid desert or a dune-covered beach. The layers of limestone and shale must have been deposited when the region was covered by a shallow sea. What do these facts tell you about the history of the Colorado Plateau?

History in Fossils

Through radioactive dating, geologists have estimated the absolute age of the fossils found in many of the rock layers of the Grand Canyon. The fossils are evidence of the kind of life that existed in this region during various periods. Limestone and shale, for example, often contain fossils of marine organisms such as clams, fishes, and corals. The Bright Angel shale contains many trilobites, marine invertebrates, and brachiopods.

INVESTIGATE!

To learn more about rock layers and history, try the In-Depth Investigation on pages 380–381.

The fossils show that a distinctive group of plants and animals lived during each period represented by the rock layers. Paleontologists can trace the changes in many of the organisms by examining the remains found in successive rock layers. Analysis of the fossils found in the older, lower rock layers through the younger, upper layers suggest a sequence of events. These fossils indicate that early, simple life-forms gradually either became extinct or changed into more-complex forms. The position, composition, and fossil content of the rocks of the Grand Canyon give a detailed and lengthy history of that part of the earth's surface.

CAREER FOCUS: *Fossils*

"You don't have to go far away to find fossils. I've seen some fine specimens in the rock walls of downtown Chicago department stores."

Mary Carman

Paleontologist

Mary Carman, former manager of the paleontology collection at the Field Museum of Natural History in Chicago, seems to have "inherited" her interest in fossils.

"My family always loved fossils," she says. "Whenever we got the chance, we'd make field trips to look for Indian artifacts and fossils of all kinds."

Following in the footsteps of her great-grandfather, an amateur geologist, and her grandmother, a chemistry teacher, Carman completed bachelor's and master's degree programs in geology at the University of Iowa. At the museum, her responsibilities ranged from handling all loans, acquisitions, and purchases of fossil specimens for the museum to coordinating displays and exhibits, and leading field trips and delivering lectures on a variety of topics associated with geologic history. In addition, Carman conducts her own research on a type of microscopic sea worm that existed some 200 to 600 million years ago. Fossils of these sea worms are valuable aids in locating oil deposits.

According to Carman, pictured at left, one of the rewards of her work is the chance to examine newly discovered fossils, such as one brought in to the museum by a Milwaukee collector.

"He had been fossil hunting in Iowa and found a fossil fish—a dipnoan—which is estimated to be some 375 million years old. We were excited not only because it was in excellent condition but also because it was the first one of its kind to be found in North

Stratigraphers do much of their work in the field.

Section 19.3 Review

1. List the four uppermost layers of sedimentary rock in the Grand Canyon. During what periods were they deposited?
2. What can the composition and fossil content of each layer of rock in the Grand Canyon reveal about the time when the layer was deposited?
3. If an outcrop shows alternating layers of sandstone and limestone, what can you assume about the geologic history of that region?

America. Although he wasn't sure what the fossil was, fortunately, he knew it belonged in a museum."

▲ Rock strata contain information on earth's history.

Stratigrapher

Stratigraphers study the earth's rock layers to learn more about geologic history. Specifically, these scientists examine rock layers to determine their age, origin, history, and fossil and mineral content.

An important tool used by stratigraphers is the seismic survey, in which sound waves are used to detect different strata. This information aids in locating rock layers where petroleum or natural gas might be found. The lower left photo shows stratigraphers at work.

Entry-level positions in stratigraphy require a bachelor's degree in geology, geophysics, or petroleum geology.

Paleontological Assistant

Paleontological assistants keep records of rock and fossil specimens that have been collected in the field. They clean, wash, and prepare the samples using special tools and cleaning solutions.

The paleontological assistants sort the specimens, compile background data on each item, and label the specimens. They then mount the specimens for study or experimental use. Most paleontological assistants work in research institutions or in large museums.

An entry-level position requires a bachelor's degree in geology or a related field such as biology.

For Further Information

For more information on careers in geologic history, write to the Paleontological Society, Office of the Secretary, Box 28200-16, Lakewood, CO 80228 or explore the University of California Museum of Paleontology's Web site at www.ucmp.berkeley.edu.

Chapter 19 Review

Key Terms

- Bright Angel shale (374)
- Canadian Shield (370)
- Coconino sandstone (375)
- craton (366)
- Gondwanaland (367)
- Hermit shale (375)
- Kaibab limestone (375)
- Laurasia (367)
- Muav limestone (374)
- outcrop (374)
- Redwall limestone (374)
- Supai formation (374)
- Tapeats sandstone (374)
- Toroweap formation (375)
- Vishnu schist (374)

Key Concepts

The supercontinent Pangaea formed during the late Paleozoic Era as lithospheric plates collided. Pangaea was made up of two major landmasses. **See page 365.**

The breakup of Pangaea began during the Mesozoic Era. **See page 367.**

The modern continent of North America is built up around a Precambrian craton. **See page 370.**

By the beginning of the Cenozoic Era 65 million years ago, the North American continent had begun to take on the shape it has today. **See page 372.**

The outcrops of the Grand Canyon reveal the geologic history of the region over millions of years. **See page 374.**

Scientists have used radioactive dating to estimate the absolute ages of fossils found in the rock layers of the Grand Canyon. **See page 375.**

Review

On your own paper, write the letter of the term that best completes each of the following statements.

1. Areas of exposed Precambrian rocks that may represent ancient continents are called
a. shields. b. fossils. c. outcrops. d. cratons.

2. During the Paleozoic Era, the continents were joined together in a huge landmass called
a. Panthalassa. b. Laurasia.
c. Pangaea. d. Gondwanaland.

3. North America and Eurasia were once part of a landmass called
a. Panthalassa. b. Laurasia.
c. Tethys. d. Gondwanaland.

4. South America, Africa, India, Australia, and Antarctica were once part of a landmass called
a. Panthalassa. b. Laurasia.
c. Tethys. d. Gondwanaland.

5. Pangaea began to break up about
a. 600 million years ago. b. 400 million years ago.
c. 200 million years ago. d. 100 million years ago.

6. The area of exposed Precambrian rocks found in North America is called the
a. New England Shield. b. Canadian Shield.
c. American Shield. d. Appalachian Shield.

7. During the Paleozoic Era, much of what is now the United States was covered by
a. a shallow sea. b. a desert.
c. an ice sheet. d. grasslands.

8. During the Mesozoic Era, new crust was added to western North America in the form of
a. shields. b. terranes. c. cratons. d. geosynclines.

9. North America had taken on its present shape by the beginning of
a. Precambrian time. b. the Paleozoic Era.
c. the Mesozoic Era. d. the Cenozoic Era.

10. The Colorado Plateau was uplifted during
a. Precambrian time. b. the Paleozoic Era.
c. the Mesozoic Era. d. the Cenozoic Era.

11. The absence of certain layers in the rock record of the Grand Canyon may be the result of
a. flooding. b. cross-bedding.
c. erosion. d. deposition.

12. The cross-bedded sandstone layers in the Grand Canyon indicate that the area may once have been covered by a
a. swamp. b. desert. c. glacier. d. forest.

13. The layers of limestone and shale in the Grand Canyon indicate that the area was once covered by a
a. shallow sea. b. desert. c. glacier. d. forest.

14. Fossils of marine organisms are often found in limestone and
a. sandstone. b. granite. c. lava. d. shale.

Critical Thinking

On your own paper, write answers to the following questions.

1. One hundred and fifty million years from now, the continents will all have drifted to new locations. How might these changes affect life on the earth?
2. Assume scientists know the rate at which North America and Eurasia are drifting farther apart on their respective continental plates. How might they determine when North America separated from Eurasia during the breakup of Pangaea?
3. If scientists discovered huge salt deposits in central Canada, what might they assume about the geologic history of that region?
4. If you found fossils embedded in glacial debris in Wisconsin, during which era might you assume the fossils were deposited? Why?
5. Geologists have found no deposits from the Pennsylvanian Period in the Grand Canyon. What might they assume about that period of the region's history?

Application

1. The Grand Canyon in Arizona and Zion and Bryce canyons in Utah are all located on the Colorado Plateau. However, the rocks of the Grand Canyon are the oldest, the rocks of Zion Canyon are next oldest, and those of Bryce Canyon are the youngest. What can you assume about the formation of these three canyons?
2. During an expedition to the Grand Canyon, you find traces of *Homo sapiens* dating to the Pleistocene Epoch. In which layer of the canyon, if any, would these traces most likely have been found? Explain your answer.

Extension

1. Write a report indicating how the breakup of Pangaea might have affected the distribution of animal species. Share your report with the class.
2. Shields provide the nucleus around which most continents have formed. Do research to find the name of each known shield and the continent on which it is located. Also find how the term *continental nucleus* applies to shields.
3. Describe in as much detail as possible what North America would look like if the Rocky Mountains had not been uplifted.

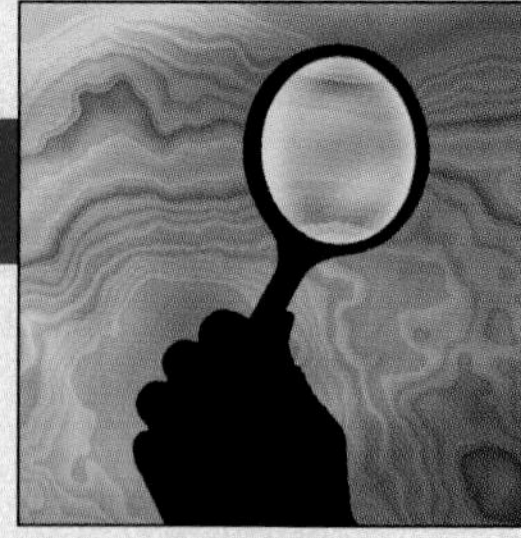

IN-DEPTH INVESTIGATION

History in the Rocks

Materials
- paper
- pencil

Introduction
Geologists have discovered much about the geologic history of North America by studying the arrangement of fossils in rock layers, as well as the arrangement of the rock layers themselves. Fossils provide clues about the environment during the time of their existence. Scientists can tell the age of the rocks in which the fossils are found because the ages of many fossils have been determined by radioactive dating. The information obtained by radioactive dating, fossil age, and rock arrangement helps to determine if any changes have occurred in the arrangement of the rock layers through geologic time.

In this investigation, you will discover how the geologic history of an area can be determined by examining the arrangement of fossils and rock layers.

Figure A

Geologic Periods	Brachiopoda	Echinodermata	Mollusca	Arthropoda	Chordata
	Name of Animal Group				
Quaternary			Pelecypod		Mammal
Tertiary			Pelecypod		Mammal
Cretaceous	Brachiopod	Echinoid	Gastropod Cephalopod		Shark
Triassic			Cephalopod		
Pennsylvanian	Brachiopod				
Mississippian	Brachiopod	Crinoid Blastoid	Cephalopod		
Devonian	Brachiopod			Trilobite	
Silurian	Brachiopod				
Ordovician			Cephalopod	Trilobite	

Prelab Preparation
1. Review Chapter 17, Section 17.1, pages 323–236, and Section 17.3, pages 334–339. Review Chapter 19, Section 19.3, pages 374–377.
2. Define the terms *index fossil* and *law of superposition* as they relate to the study of rock layers.

Procedure

1. Study the index fossils shown in Figure A on the previous page. Note their placement in related groups along with the geologic periods in which they lived.
2. Select one of the four fossil arrangements shown in Figure B below. This illustration shows how some of these fossils might be found in a series of rock layers. Record the number of the arrangement you are using.
3. Using Figure A, identify all the fossils in your arrangement and the geologic time in which they lived.

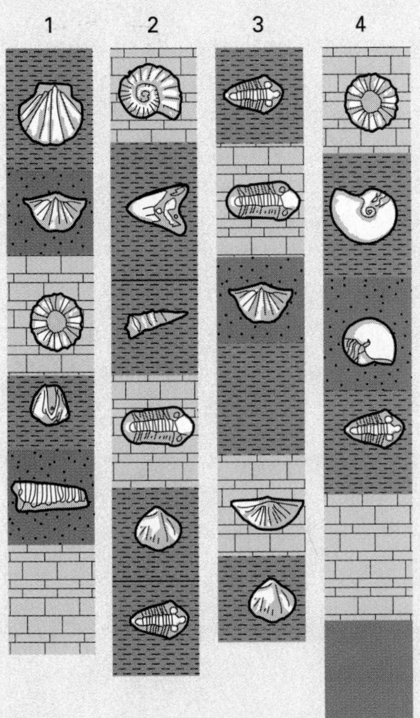

Figure B

4. List the fossil names in order from oldest to youngest.
5. Do the fossils in your arrangement appear in the correct order of their geologic times?
6. Do the fossils in your arrangement show a complete sequence of geologic periods with none missing? If not, which periods are missing?
7. Select another arrangement showing a different group of fossils. Repeat Steps 2–6 with this new group. Complete all four arrangements.

Analysis and Conclusions

1. What processes or events might explain the order in which each of the fossil arrangements were found?
2. Based on your observations in the procedure, why is it necessary that a fossil be found in a wide variety of geographic areas in order to be considered an index fossil?
3. Study Arrangement 3 in Figure B. Note that there is a rock layer containing no fossils between two rock layers that contain fossils. How might this have occurred?

Extensions

1. Collect fossils found in your area. Identify the fossils you have collected and describe what your area was like when the organisms existed.
2. Find out what types of sedimentary rock usually have the most fossils.
3. How are index fossils used to help petrologists locate oil deposits?

Where the Hippos Roam

Millions of years ago, ancestors of modern crocodiles lurked in the shallow waters of lakes and other bodies of water. They hunted fish and other animals, much as their descendants do today. If you could travel back in time to visit one of those lakes, you might see the ancestors of today's hippos there as well. Antelopes might browse along the edges of the lake, and rodents of various sizes might scurry back and forth.

When paleontologists examine the fossil of a prehistoric organism, they may discover clues about the organism's life. They may also answer questions about the environment it lived in: Was the area hot or cold? Was it humid or dry? Then, by putting all of these clues together, the paleontologists may be able to learn a little more about how organisms and environments change over time.

Unfortunately, studying a fossil site is no easy task! Discoveries such as those shown on pages 347 and 353 are rare. More often, a paleontologist may find a few teeth scattered over a very large area. In such cases, keeping track of where the fossils were found is very important. In this activity, you will use the data from a fossil site to create a map of fossil locations at that site. Then you will make some conclusions about the past environment, or *paleoenvironment,* at that location.

Materials

- colored pencils
- metric ruler

The animals that lived in lakes millions of years ago probably had similar lives to those that live in lakes today. ▼

◀ **Fossil crocodile teeth**

The table below shows the locations of fossils that were found spread out over 22,500 m^2. A team of paleontologists decided that this site, which measured 150 m × 150 m, was too large to work on all at once. Therefore, they decided to create a grid of 10 m squares. Starting in the northwest corner, they labeled the squares with the letters A–O from west to east. Then the team numbered the squares 1–15 from north to south. In this way, each fossil could be labeled with a letter and a number, depending on where it was found. For instance, the label A1 would signify the 10 m × 10 m square in the northwest corner of the site. Similarly, the label O15 would indicate the square in the southeast corner of the site.

Location of Fossil Teeth

Layer	Hippos	Rodents	Crocodiles	Bovids*
A	B11, C6, D3, I15, J10, L7, M6		C14, F7, G13, I3, L13, O2	
B	F2, J3, K1, K2	B10, B11, F13	H2, I7, K2, N5, N7	G14
C		B3, C10, D1, H8, M9, N4		A5, A6, E2, E4, E14, H7, H8, H12, K4, M1, N15

*Bovids are antelopes and other similar animals

1. Create a map of the fossil site. The scale should be 1 cm = 10 m. Label the grid of squares with letters and numbers. Using letters, show where the fossils were found. (Put an *H* in each square where a hippo tooth was found; use *R* to indicate a rodent tooth, and so on) Use a different color for each layer of sediment, and make a key to show which colors are which.
2. Based on the distribution of fossils in level C, what part of this site might have been covered by water? Devise a way to indicate that part on your map.
3. Describe how the environment at this site changed through time.
4. One member of the team wished to look for fossils of dry-climate plants at this site. Which layer or layers do you think would be most likely to yield fossils of this kind? Explain your answer.
5. When the paleontologists were analyzing these data, they proposed several hypotheses to explain the changing climate. One of them suggested that tectonic uplift had occurred, causing the area to gain elevation over time. A second paleontologist disagreed, stating that the area probably *lost* elevation over time. Whom do you agree with? Explain your answer.

Going Further

Did you know that western Wyoming was once covered by a huge body of water called Fossil Lake? Find out more by visiting Chicago's Field Museum of Natural History on the World Wide Web at www.bvis.uic.edu/museum.

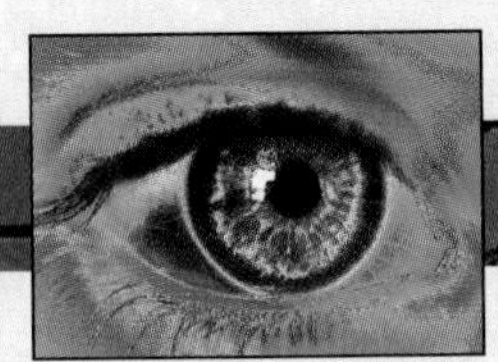

EYE ON THE ENVIRONMENT

The Mysterious Bone Bed

Recent work in eastern Africa has revealed one of the most amazing fossil sites in the world. It contains a rich variety of animals, and the fossils are perfectly preserved. An international team of scientists has been studying the site for over 10 years, but one of the deepest mysteries about the site remains to be solved.

Finding a Fossil Treasure Chest

More than 20 years ago, a geological survey team was working in Kipsaramon, a very remote area of western Kenya. They found some fossils on a hillside that looked promising for a paleontological study. They noted that the site deserved further research, but the area was largely untouched for many years. Researchers eventually returned to the site, and their work has revealed a bed of fossils that covers at least 2,000 m^2. The site is more densely packed with fossils than any other site in Kenya, and it is probably one of the most concentrated deposits of fossils in the world.

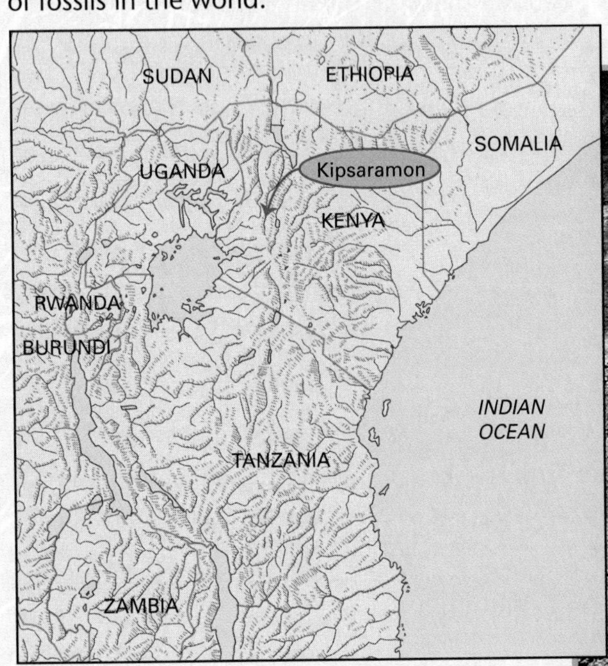

When an underground layer of sediment has a large number of fossil bones in it, scientists may refer to it as a *bone bed.* Bone beds are very exciting because a lot of data may be retrieved from just one site. In some cases, entire populations may be preserved there. This allows scientists to study the range of variation within a single species, and it may give information on how members of the species grew and interacted. In other cases, several different species may be present.

The Kipsaramon bone bed, which has been dated to about 15.5 million years, contains thousands of individual organisms. Dozens of different extinct species are present. These include rhinos, turtles, and apes, as well as crocodiles, antelopes, and squirrels. There are even some specimens of fossil wood. By putting together all of the information from the site, scientists have been able to get a picture of an entire prehistoric ecosystem.

The bone bed is only about 20 cm thick, but it is densely concentrated with fossils ▼

Jumble of Bones

▲ **In order to study the fossils at Kipsaramon, large sections of the bone bed are carefully removed from the ground and shipped to a lab in Nairobi. There, experts can slowly pick the fossils apart without having to leave them vulnerable to weather and curious passers-by.**

Almost all of the bones at the Kipsaramon bone bed are *disarticulated,* which means they are preserved as individual bones instead of as whole skeletons. In other words, teeth and hip bones are jumbled together with leg bones, skulls, and backbones.

In addition, the fossils are very closely packed together. The spaces between large bones are filled with smaller bones. Andrew Hill, who has led the international team of scientists working at the site, recalls an impressive experience: "I was cleaning bits of dirt and debris away from a beautifully preserved elephant tooth. Suddenly, I looked closer and realized that one of the bits wasn't dirt at all. It was the tooth of a primate! Moments later, as I was cleaning the new primate tooth, it happened *again.* I found a tiny rodent tooth stuck to the primate tooth."

Mysteries of the Past

Bone beds may form in a number of ways. A large number of animals may die at once, such as in a flood. A bone bed also may mark the location of a predator's favorite dining spot. When scientists study a bone bed, they look for clues to how it formed. For instance, fossils that were deposited during a flood usually show some wear and tear from the moving water. Similarly, fossils from a predator's den may show tooth marks, breakage, or other signs of the predator's work.

How did the Kipsaramon bone bed form? Scientists have developed several ideas about this question and tested them against the fossil evidence. So far, every hypothesis has turned out to be incorrect. The fact that the fossils are disarticulated shows that the bones were disturbed after they were deposited. The fossils show no signs of wear from movement or of damage from predators, however. How did they get jumbled up without getting damaged? How did they get so closely packed together? Scientists are seeking answers to these and other questions as they continue their investigations of the incredible Kipsaramon bone bed.

Think About It

What's your guess? Propose a hypothesis that explains how the Kipsaramon bone bed may have formed. How would you test your hypothesis?

Extension: Do some on-line research to find out about another bone bed. What fossils are present? How did it form?

Coast at Monterey, California (background);
Coral Reef, Bali, Indonesia (inset)

Introducing Unit Six

When I was young, I had the opportunity to sail up the coast of Maine on a yacht. Between my general interest in science and the romantic aspects of sailing the sea, that one-week trip was all it took to draw me to the fields of oceanography and marine geology.

Now, for the first time, we have the remote sensing tools necessary to see and map the details of things I could only wonder about back then—submarine canyons that look like the Badlands of South Dakota and channels that resemble the Mississippi River. These and other new perspectives have helped us realize how dynamic our earth really is.

The unit you are about to study discusses some of the important strides already made in understanding the geology of our oceans—an underwater world of discovery covering three quarters of the earth's surface.

Bonnie A. McGregor

Bonnie McGregor
Marine Geologist
U.S. Geological Survey

In this UNIT

Chapter 20

The Ocean Basins

Throughout much of recorded history, people regarded the ocean as a place of mystery and danger. They thought that fantastic monsters preyed on sailors and that the sea dropped off into the unknown.

Exploration of the oceans has replaced these fears with scientific curiosity and understanding. The oceans, which cover nearly 70 percent of the earth's surface, are no longer unfamiliar territory. In this chapter, you will study the features of the earth that lie beneath the ocean.

Chapter Outline

A crinoid living on brain coral near the British Virgin Islands

20.1 The Water Planet

Section Objectives

- **Name the major divisions of the global ocean.**
- **Describe the goals of oceanography.**

Nearly three quarters of the earth's surface lies submerged beneath a body of salt water called the global ocean. No other known planet has a similar covering of liquid water. Earth alone can be called the *water planet*.

The global ocean contains more than 97 percent of all the water on earth. Although the ocean is the most prominent feature of the earth's surface, it constitutes only a small part of the earth's total mass and volume. The mass of the ocean is only about 1/4,000 of the mass of the earth as a whole. The volume of the solid earth is about 800 times greater than the volume of water in the global ocean.

Divisions of the Global Ocean

The global ocean is divided into three major oceans—the Atlantic, Pacific, and Indian oceans. The waters of the polar regions are sometimes called oceans as well. However, Antarctic waters are really the southern parts of all three major oceans. Due to its small size, the Arctic Ocean is really a sea of the Atlantic Ocean. The term *sea* is applied to smaller areas of the ocean that are partially surrounded by land, such as the Mediterranean, Caribbean, and Coral seas.

The division of the global ocean into specific oceans and seas is useful because each of these parts has special characteristics. For

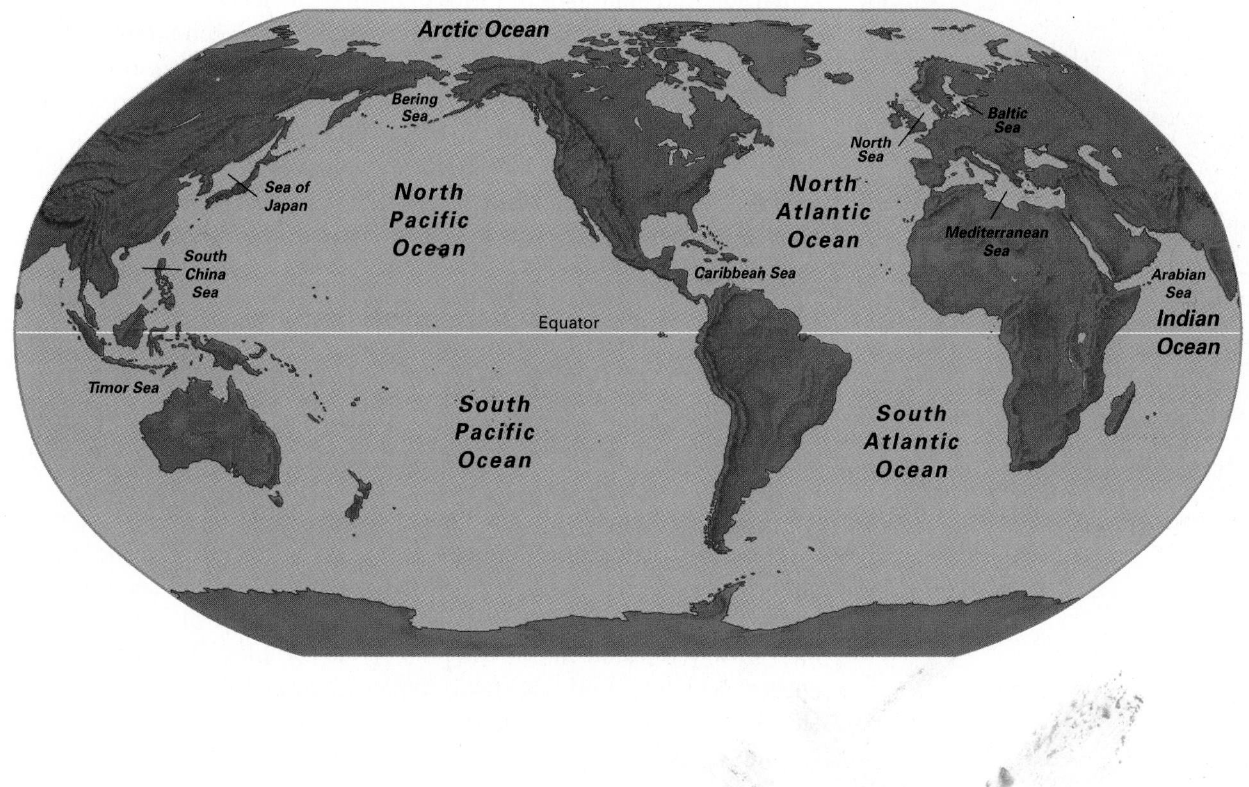

Figure 20–1. The global ocean is divided into specific oceans and seas.

example, in the Arctic Ocean the water near the surface is less salty than water in the other oceans.

The Pacific Ocean is the largest feature on the earth's surface. It contains more than one half of the ocean water on earth. With an average depth of 3.9 km, the Pacific is also the deepest ocean. The Sea of Japan and the Bering, Coral, and South China seas are part of the Pacific Ocean. The next largest in size is the Atlantic Ocean. The Mediterranean, Caribbean, and Baltic seas, as well as the Gulf of Mexico, are considered parts of the Atlantic Ocean. The Atlantic Ocean has an average depth of 3.6 km. The Indian Ocean is the third-largest ocean and has an average depth of 3.8 km. The Arabian and Red seas are part of the Indian Ocean.

Exploration of the Ocean

The branch of earth science known as *oceanography* is the study of the physical characteristics, chemical composition, and life forms of the ocean. Oceanography got its start in the 1850's. An American naval officer, Matthew F. Maury, used records from navy ships to derive data about ocean currents, winds, depths, and weather conditions. These observations, which he then published as the first textbook on the oceans, marked the beginning of the scientific study of the ocean.

In the late nineteenth century, the voyages of the British Navy ship H.M.S. *Challenger* laid the foundation for the modern science of oceanography. From 1872 to 1876, a team of scientists aboard the *Challenger* crossed the Atlantic, Indian, and Pacific oceans, measuring water temperatures at great depths and collecting samples of ocean water, sediments, and thousands of forms of marine life never before seen.

Today, many ships are equipped to perform oceanographic research. For example, the research ship *JOIDES Resolution,* the world's largest and most sophisticated scientific drilling ship, can drill as deep as 6.4 km into the ocean floor. The samples drilled by *JOIDES Resolution* provide scientists with valuable data about plate tectonics and the **ocean floor**. The ocean floor is made up of continental crust and oceanic crust that lie beneath the ocean waters.

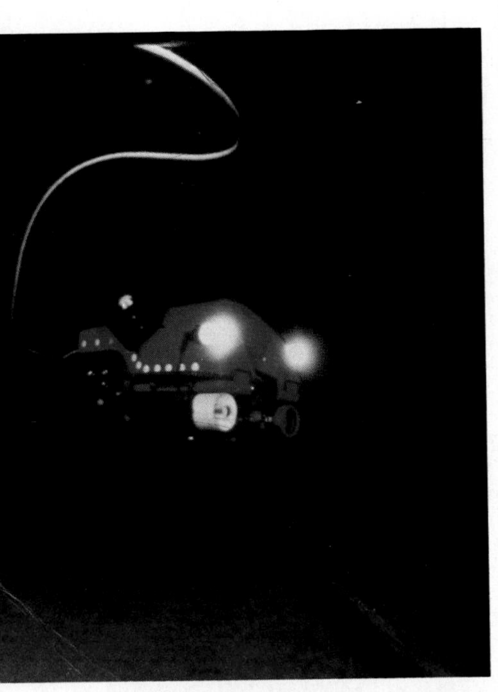

Figure 20–2. Robot submersibles like *Jason Jr.* send back detailed pictures of the ocean floor.

Submersibles

Underwater research vessels, called **submersibles,** also enable oceanographers to study the ocean depths. Some submersibles are piloted by people. The more advanced submersibles, however, are actually submarine robots similar to the *Jason Jr.*, shown in Figure 20–2. They can perform tasks ranging from photographing the ocean depths to collecting mineral samples from the ocean floor.

Several types of submersibles have been used for underwater research and exploration. One early type of submersible is the **bathysphere,** a spherical diving vessel that was first used for deep-ocean exploration. The bathysphere was carried to an area of the ocean, then slowly lowered into the depths on steel cables. The bathysphere, which carried scientists, remained connected to the research ship for communications and life support. Therefore, its movement and the tasks it could perform while under water were limited.

Figure 20–3. The bathyscaph *Alvin* has been used by oceanographers in deep-ocean research.

Another type of submersible, called a **bathyscaph,** is a self-propelled, free-moving submarine equipped for deep-ocean research. One well-known bathyscaph is the *Alvin,* a craft that holds one pilot and two scientists. The *Alvin,* shown in Figure 20–3, has made more than 2,000 dives and has helped scientists make exciting discoveries about the deep ocean. During one of the *Alvin*'s dives, startled oceanographers found communities of unusual marine life living at depths and temperatures where it was thought that almost no life could exist. Giant clams, blind white crabs, and giant tube worms were just three of the strange life forms discovered. They were found near the hydrothermal (hot water) vents along mid-ocean ridges. These animals had adapted to life in a hostile environment.

Robot submersibles also are making important discoveries about the ocean. These robot craft enable oceanographers to study the ocean at great depths and for long periods of time. Oceanographer Robert Ballard used *Argo,* a seeing-eye robot vessel, to locate the remains of the luxury liner *Titanic,* which sank in deep water off the coast of Newfoundland after striking an iceberg during its maiden voyage in 1912. The *Jason Jr.*, a highly maneuverable robot submersible equipped with underwater cameras, toured the inside of the sunken ship and returned pictures of barnacle-encrusted crystal chandeliers and delicate porcelain cups with the *Titanic* crest still visible. Today, robot submersibles tour the interior of undersea volcanoes and explore the depths of submarine canyons. Why are robot craft more practical for deep-ocean research than are craft designed to carry people?

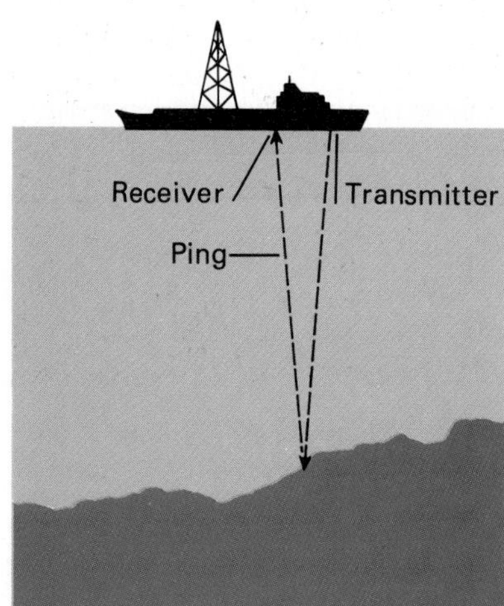

Figure 20–4. Active sonar sends out a pulse of sound. The pulse, called a *ping* because of the way it sounds, is reflected back when it strikes a solid object.

Sonar

Oceanographic research ships and submersibles are equipped with **sonar** to aid in mapping the ocean floor, as shown in Figure 20–4. *Sonar* is an acronym for ***so****und* ***n****avigation* ***a****nd* ***r****anging*. A sonar system consists of a transmitter and a receiver. The transmitter sends out a continuous series of sound waves from a ship to the ocean floor. The sound waves, traveling at the speed of about 1,500 m/sec through sea water, bounce off the solid ocean floor and are

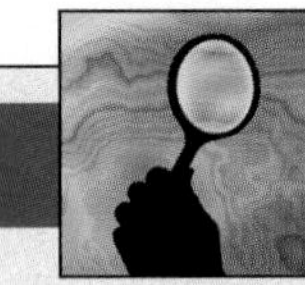

SMALL-SCALE INVESTIGATION

Sonar

You can demonstrate the principle of sonar by calculating the rate at which pulses travel along a spring to determine distance.

Materials

toy spring coil or flexible spring, heavy string, meter stick, stopwatch, adhesive tape

Procedure

1. Use heavy string to tie one end of the spring securely to a doorknob. Pull the free end of the spring taut and parallel to the floor. Keep the tension of the spring constant throughout the investigation.
2. Use tape to mark the floor directly beneath the hand holding the spring taut. Measure and record the distance from your hand to the doorknob.
3. Note the time. Hold the spring taut and hit the spring sharply with the side of your free hand. Strike the spring as close as possible to your hand holding the spring.
4. Check the time again to see how long it takes for the pulse to travel to the doorknob and back to your hand.
5. Repeat Steps 2 through 4 three times. Each time, hold the spring 60 cm closer to the doorknob than in the previous trial, keeping the tension constant by gathering coils into your hands as necessary.

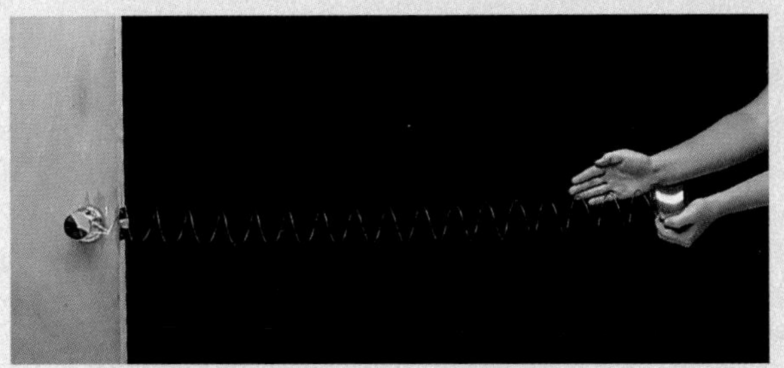

6. Calculate the rate of travel for each trial by multiplying the length of the spring between your hand and the doorknob by 2. Then divide by the number of seconds it took for the pulse to travel to the doorknob and back.

Analysis and Conclusions

1. Did the rate of travel of the pulse change during the course of the investigation?
2. If a pulse took 3 seconds to travel to the doorknob and back to your hand, what is the distance from the doorknob to your hand?
3. How is the apparatus you used like sonar? How is it different? Explain your answer.

reflected back up to the receiver. Scientists measure the amount of time it takes for the sound waves to complete this round trip. In this way, they can determine the depth of the ocean floor. With this information scientists can make maps and profiles of the ocean floor. One type of sonar can penetrate even the sediment layers and rock structures of the ocean floor. Its returning signal indicates the thickness of the sediments and the structure of the underlying rocks.

Section 20.1 Review

1. What are the three major divisions of the global ocean?
2. Define oceanography.
3. Most submarines use sonar as an aid in navigation. How would sonar enable an underwater vessel to steer its way through the ocean depths?

20.2 Features of the Ocean Floor

Section Objectives

- **Describe the main features of the continental margins.**
- **Describe the main features of the deep ocean basin.**

Exploration of the ocean depths at many locations has revealed the shape and makeup of the ocean floor. The ocean floor can be divided into two major areas, as illustrated in Figure 20–5. The **continental margins** are shallower portions of the ocean floor made up of continental crust. The **deep ocean basin** is the portion made up of oceanic crust.

Continental Margins

The line that divides the continental crust from the oceanic crust is not always obvious. Shorelines are not the true boundaries between the ocean floor and the continents. The real boundary lies some distance offshore, beneath the ocean itself.

Continental Shelf

Every continent is bounded in most places by a zone of shallow water, where the ocean covers the edge of the continent. This part of the continent that is covered by ocean water is called the **continental shelf.** The shelf usually slopes gently from the shoreline, dropping only about 1.2 m every 100 m. The water above the continental shelf is shallow. The average depth of the water covering a continental shelf is about 60 m. Although it is under water, a continental shelf is part of the continental margin, not the deep ocean basin.

The width of a continental shelf varies. On the west coast of South America, the continental shelf is only a few kilometers wide. On the west coast of Florida, however, the shelf extends 760 km into the Gulf of Mexico. The widest continental shelves extend out about 1,280 km from Siberia and Alaska into the Arctic Ocean. The continental shelf along the east coast of the United States has an average width of 170 km, while the shelf along the west coast is four times narrower.

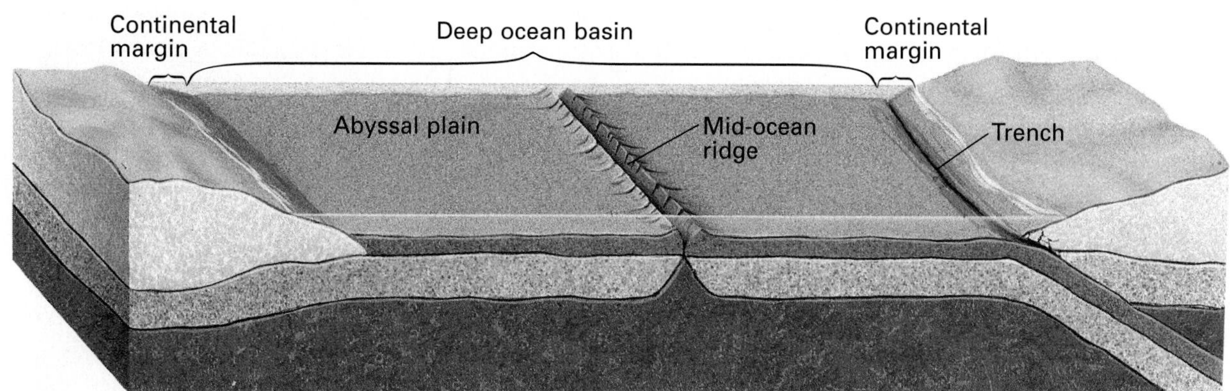

Figure 20–5. The ocean floor includes the continental margins and the deep ocean basin. Trenches, abyssal plains, and mid-ocean ridges are major features of the deep ocean basin.

The continental shelves are affected by changes in sea level. During the glacial periods, when continental ice sheets held great amounts of water, sea level fell. As the ice sheets melted and the water was added to the ocean, sea level rose. When sea level was lower than it is today, parts of the continental shelf were exposed to weathering and erosion. Today, with a higher sea level, the continental shelves are mainly surfaces of deposition. In general, the continental shelf is a smoother version of the land surface above the shoreline.

Continental Slope

At the seaward edge of a continental shelf is a steeper slope called a **continental slope.** The boundary between the continental crust and the oceanic crust is found at the base of the continental slope. Along the continental slope, the ocean depth increases to several thousand meters within a distance of a few kilometers.

Fishing on the Continental Shelves

Continental shelves provide the richest fishing grounds in the world. One of these is the Grand Banks, a stretch of shallow water 15 m to 300 m deep lying southeast of the Canadian province of Newfoundland.

Fish and shellfish are plentiful in the waters of this broad continental shelf because of the ample food supply. The warm Gulf Stream and the cold Labrador Current meet at the Grand Banks. These two currents mix the water, causing nutrients and microscopic sea organisms to become concentrated in that area. These organisms provide food for large numbers of small fish, which, in turn, provide food for larger fish.

Fishing is crucial to the economy of Canada's Atlantic provinces. Many people fish as a part-time occupation to help feed their families. Some farmers fertilize their land with calcium-rich crushed lobster shells. Grand Banks fishing has also helped develop industry in these provinces. Both Nova Scotia and Newfoundland have a number of seafood-processing plants. Ensuring a sustainable resource like the Grand Banks is of great importance.

Many nations with seacoasts, including the United States and Canada, now claim that their sovereignty extends into the ocean some 320 km (200 nautical miles) from shore in order to protect their fish and mineral resources. With management by these coastal nations, the valuable fishing grounds of the continental shelves, such as the Grand Banks, will not succumb to overfishing.

What is one possible consequence of claiming sovereignty so far from shore?

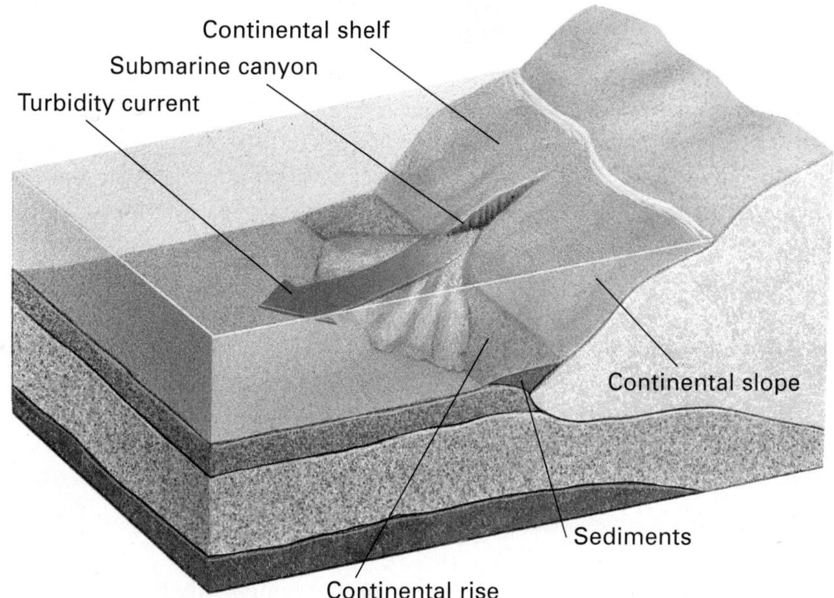

Figure 20–6. A continental margin is made up of distinct areas and features. Note the turbidity current on the continental slope. Turbidity currents carry tons of sediment from the continental shelves to the ocean floor.

The continental shelf and continental slope may be cut by deep V-shaped valleys. These deep valleys in the continental slope and shelf are called **submarine canyons.** The origin of these deep canyons is not completely known. However, they may have been caused by **turbidity currents.** These are currents that are very dense and that carry large amounts of sediment down the continental slopes. Turbidity currents form when landslides of various materials run down a slope. Most likely, these landslides are triggered by earthquakes or by the force of gravity. The speed and composition of the slides make them powerful agents of erosion—powerful enough, some scientists hypothesize, to deepen canyons in the continental slopes. Along the base of a continental slope may be sediments that have moved down the slope. These sediments form a raised wedge at the base of the continental slope called a **continental rise.**

Deep Ocean Basins

The deep ocean basin has its own distinctive features. These include broad, flat plains, submerged volcanic mountains, gigantic volcanic mountain ranges, and deep trenches. However, in the deep ocean basin the mountains are higher and the plains flatter than any found on the continents.

Trenches

Located in the deep ocean basin are the deepest features on the earth's surface. These long, narrow places are called **trenches.** The deepest place in the world is the Mariana Trench—over 11,000 m deep. The Mariana Trench is located in the North Pacific Ocean near the island of Guam. Most trenches are located along the Pacific Ring of Fire, where oceanic plates are subducting back into the mantle, below other plates. Trenches are associated with earthquakes, volcanic mountain ranges, and volcanic island arcs. Deep trenches in the Pacific Ocean occur offshore from Japan, Alaska's Aleutian Islands, the Philippines, and the west coast of South America.

Abyssal Plains

The parts of the ocean floor that are extremely flat and average 3 km to 4 km in depth are the **abyssal** (uh-BISS-uhl) **plains.** Abyssal plains cover about half of the deep-ocean basin. They are the flattest regions on earth. In some places, the depth changes less than 3 m over more than 1,300 km.

Sonar reveals that the abyssal plains are made up of thick layers of sediments deposited in the deep ocean basin. Currents probably carry the sediments from the continental margins. The trenches found along the edges of some continents trap the sediments being

CAREER FOCUS: *Oceans*

"I've had a number of unusual experiences as an underwater photographer. Swimming next to 4,100-kg whales was just one of them."

Chris Newbert

Underwater Photographer

Chris Newbert's travels as an underwater photographer have taken him to such exotic places as the Coral Sea and the Palauan reefs in the South Pacific. A sport diver, Newbert has explored and photographed the great oceans of the world.

A self-taught photographer as well, Newbert, pictured at left, has won numerous awards for his spectacular photos of marine life. To obtain these photos, Newbert has developed innovative techniques, including one that he calls "free-drift diving."

"I go far out to sea at night, dive down 30 to 60 meters, and just drift, photographing animals that migrate vertically," he said. When finished taking photos, he relies principally on a generator-powered light on his boat to guide him back to the surface.

In addition to having his photos published in books and magazines, Newbert has participated in photography exhibits throughout the United States. He also completed work as the photographer and author of his first book of underwater photography, *Within a Rainbowed Sea.* Newbert also conducts diving tours in many places around the world.

Not surprisingly, Newbert often encounters underwater companions. But he takes that in stride.

"Sharks, for example, will come after you from time to time. For the most part, though, I don't get too frightened."

Marine Surveyor

Marine surveyors gather data that can be used in mapping the ocean floor and shoreline. They

Scuba gear is basic equipment used by marine surveyors. ▶

carried down the continental slopes. Thus, abyssal plains are most widespread in the Atlantic Ocean, where there are no trenches along the continental margins. The Pacific Ocean, which has many trenches, has less-extensive abyssal plains.

Mid-Ocean Ridges

The most prominent features of the ocean basin are the *mid-ocean ridges,* shown in the undersea map in Figure 20–7. They are a continuous series of underwater mountain ranges that run along the floors of all the oceans. Only in a few places, such as in Iceland,

also calculate water depths at specific locations.

Marine surveyors must be able to use a variety of specialized equipment. Among this equipment are such things as wire drags and sonar to determine underwater depths. The surveyors' findings are important for navigation and for the construction of ports, docks, and bridges.

Marine surveyors, shown in the photograph to the left, do much of their fieldwork at sea. The data are then analyzed and charted in the surveyors' office.

Most marine surveyors begin their career preparation with a bachelor's degree in engineering.

Oceanographer

The field of oceanography includes four main areas of study. Marine geologists study undersea rock formations, coastlines, and the characteristics of the ocean floor. Marine chemists investigate the chemical composition of sea water and of the materials that make up the ocean floor. Physical oceanographers chart the movements of waves, currents, and tides. Marine biologists study the plant and animal life of the ocean.

Oceanographers doing research often spend months at a time at sea collecting data. They may use specialized equipment such as submersibles and underwater cameras.

A bachelor's degree in geology, chemistry, biology, or oceanography is a minimum requirement for entry into this field.

For Further Information

For more information on careers in marine science, write the National Oceanic and Atmospheric Administration, National Ocean Service, 1305 East–West Highway, Silver Spring, MD 20910, or visit their Web site at www.nos.noaa.gov.

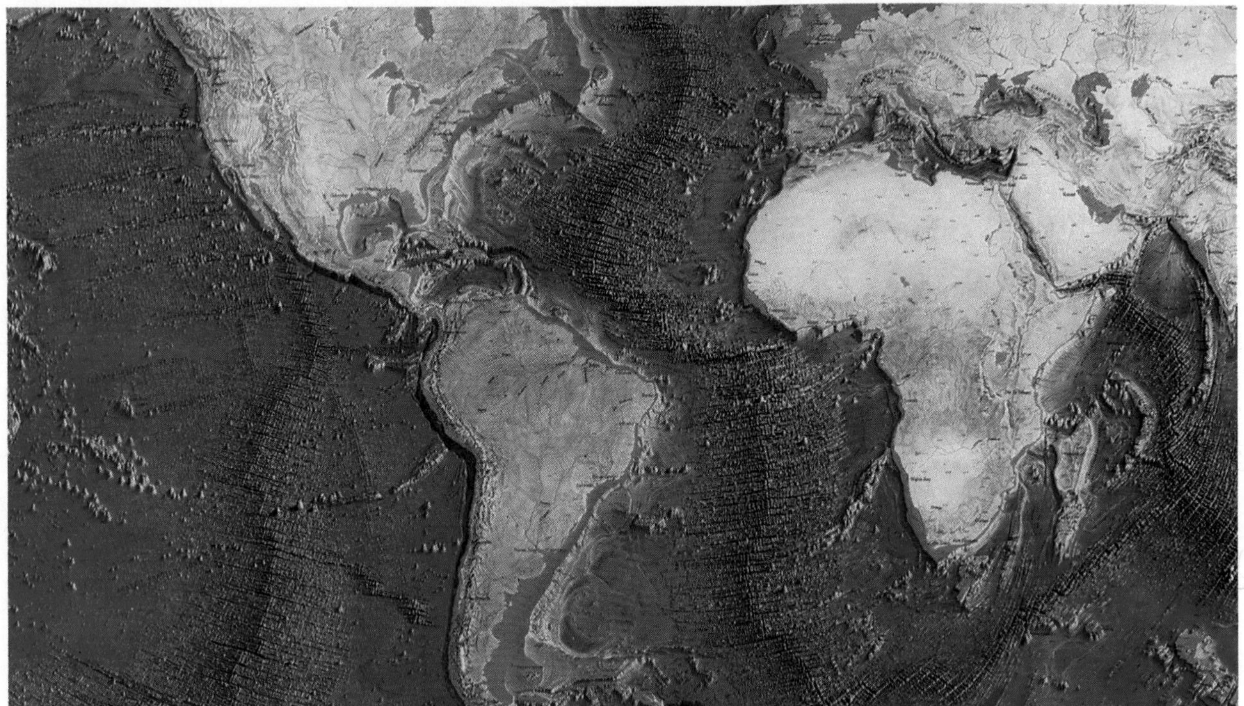

Figure 20–7. The Mid-Atlantic Ridge is clearly visible in this undersea map.

do the mid-ocean ridges rise above sea level. According to the theory of plate tectonics, magma reaches the sea floor and builds up a volcanic mid-ocean ridge, causing the ocean floor to be elevated. Mid-ocean ridges have a narrow depression, or *rift,* running along the center. This rift forms the earth's newest crustal rock and is evidence that the plates have moved away on each side of the ridge. Jagged peaks and valleys form when magma pushes up through the oceanic plate on both sides of the rift. Because different parts of the ridge separate at different rates, the new crust breaks into a series of faults, called **fracture zones,** that run perpendicular to the ridge.

Seamounts

Scattered along the ocean floor are isolated volcanic mountains called **seamounts.** Most of these are over 1,000 m high. Thousands of seamounts dot the Pacific Ocean floor; they are less common in the other oceans. Seamounts may rise above the surface of the ocean, forming oceanic islands such as Hawaii, Tahiti, and the Canary Islands. Some volcanic islands are flattened by waves. As the action of plate tectonics carries seamounts away from a mid-ocean ridge, the ocean crust sinks. As it sinks, the seamounts sink with it, beneath the ocean surface. These flat-topped, submerged seamounts are called **guyots** (GEE-oze).

(table mounts)

Section 20.2 Review

1. Name the major features of the ocean floor.
2. Why does the Pacific Ocean have less-extensive abyssal plains than the Atlantic does?
3. If sea level were to fall significantly, what would happen to the continental shelves?

20.3 Ocean-Floor Sediments

Section Objectives

- **Describe the formation of ocean-floor sediments.**
- **Explain how ocean-floor sediments are classified according to physical composition.**

Ocean-floor sediments differ greatly from one part of the ocean to another. The continental shelves and slopes are covered mainly with rock fragments. Most of these fragments have been carried into the ocean by rivers or have been washed away from the shoreline by wave erosion. The sediments are fairly well sorted according to size. Coarse gravel and sand are usually found close to shore because these heavier rock fragments are not easily moved offshore by waves and currents. Much finer particles are carried suspended in ocean water and are usually deposited at a greater distance from shore.

The bottom sediments in the deep ocean basin, beyond the continental slopes, are different from those found in shallower water. Samples of the sediments in the deep-ocean basins are obtained by taking **core samples** or by scooping up sediments. Core samples are taken by drilling into sediment layers. A core sample taken from the ocean floor is shown in Figure 20–8. What might scientists infer from a core sample containing a layer of sediment made up primarily of volcanic ash and dust?

Sources of Deep Ocean-Basin Sediments

The study of sediment samples shows that most of the sediments in the deep-ocean basin are made up of materials that settle slowly from the ocean water above. These materials may come from organic or inorganic sources.

Inorganic Sediments

Some of the ocean-basin sediments consist of rock particles that have been carried from the land to the oceans by rivers. When a river flows into the ocean, the river deposits its stream load as sediments. Most of these sediments are deposited along the shore and on the continental shelf. However, great quantities of these sediments occasionally slide down continental slopes to the ocean floor below. A single slide may move billions of kilograms of material down a slope. The tremendous force of the slide creates powerful turbidity currents that spread the sediments over the deep ocean basin.

Figure 20–8. Notice the layers of sediment in this core sample drilled from the ocean floor. Core samples reveal important clues about the geologic history of the earth.

Some deep ocean-basin sediments consist of fine particles of rock, including volcanic dust, that have been blown great distances out to sea by the wind. Once these particles land on the surface of the water, they begin to sink and gradually settle to the bottom of the ocean.

Icebergs provide sediments by carrying material from the land into the ocean. An iceberg is a large block of ice that has broken off from a glacier and floated out into the ocean. As a glacier moves slowly across the land, it picks up great quantities of rock. The rock becomes embedded in the glacier and moves with it. When the iceberg breaks off the glacier, drifts out to sea, and melts, the rock material sinks to the ocean floor.

Even meteorites contribute to ocean-basin sediments. Much of a meteorite vaporizes as it enters the earth's atmosphere. What cosmic dust remains falls to the earth's surface. Because so much of the earth's surface is ocean, most meteorite fragments fall into the ocean and become part of the sediments on the ocean floor.

Organic Sediments

The ocean basin is also covered with organic sediments. These are the remains of marine plants and animals. In many places on the ocean floor, almost all the sediments are organic.

The two most common substances in organic sediments are silica (SiO_2) and calcium carbonate ($CaCO_3$). Silica comes primarily from microscopic organisms, such as diatoms and radiolaria. Calcium carbonate comes mostly from the skeletons of tiny organisms called *foraminifera*. Animals, such as corals and clams, can also add calcium carbonate to ocean-basin sediments.

Chemical Deposits

Chemical deposits make up part of the ocean-floor sediments. Many chemical reactions take place in the ocean. During some of these reactions, solid materials are formed. These materials settle to the

Figure 20–9. The remains of diatoms (left) and radiolaria (right), both magnified hundreds of times, are important components of organic sediments on the ocean floor.

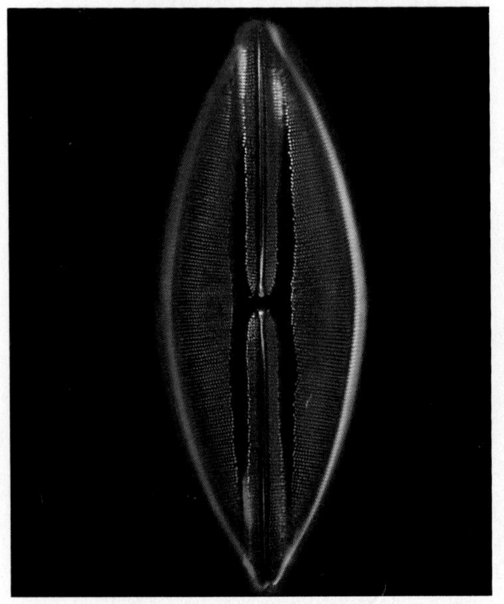

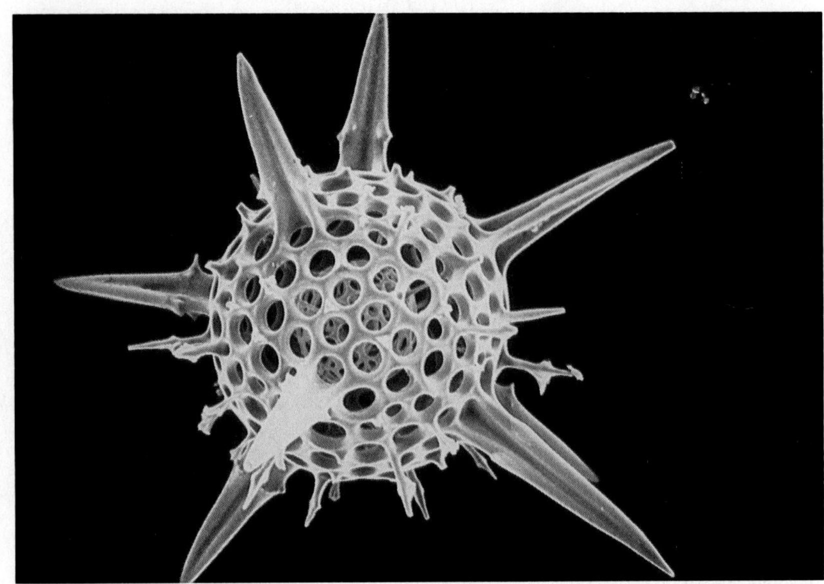

Figure 20–10. Undersea mining operations may allow industry to tap ocean-floor mineral resources someday. Nodules, such as these mined from the East Pacific Rise, are rich in minerals important to industry.

bottom. For example, potato-shaped lumps of minerals, or **nodules,** form on the abyssal plains. Nodules are composed mainly of oxides of manganese, nickel, and iron. They seem to have been formed by chemical changes affecting substances dissolved in ocean water. Other minerals found in ocean-floor sediments, such as the fertilizer phosphorite, are also formed directly in the ocean water.

Physical Classification of Sediments

Two general types of sediments are found on the ocean floor. **Muds** are very fine silt- and clay-sized particles of rock that have settled to the ocean floor. One common type of mud found on the abyssal plains is *red clay.* Red clay is at least 40 percent clay particles, mixed with silt, sand, and organic material. Actually, red clay can also be yellow-brown, blue, green, or gray.

About 40 percent of the ocean floor is covered with a soft, organic sediment called **ooze.** At least 30 percent of the ooze is made up of organic materials, such as microscopic particles of shells and the remains of diatoms, radiolaria, and foraminifera. The remaining material in ooze is made up of fine mud.

There are two types of ooze. One type, **calcareous ooze,** is mostly calcium carbonate. At depths below 4,500 m, calcium carbonate dissolves in the deep, cold ocean water. Therefore, calcareous ooze is never found below that depth. The second type of ooze, **siliceous ooze,** is mostly silicon dioxide, which comes from radiolaria and diatoms. Most siliceous ooze is found in the cool, rich ocean waters around the continent of Antarctica because of the abundance of these two marine organisms in that location.

INVESTIGATE!

To learn more about ocean-floor sediments, try the In-Depth Investigation on pages 404–405.

Section 20.3 Review

1. How do icebergs contribute to ocean-basin sediments?
2. Describe the composition of ooze.
3. There have been several recent attempts to develop an economical method of mining nodules from the ocean floor. Why is there such interest in ocean-floor nodules?

Chapter 20 Review

Key Terms

- abyssal plain (396)
- bathyscaph (391)
- bathysphere (390)
- calcareous ooze (401)
- continental margin (393)
- continental rise (395)
- continental shelf (393)
- continental slope (394)
- core sample (399)
- deep ocean basin (393)
- fracture zones (398)
- guyot (398)
- mud (401)
- nodule (401)
- ocean floor (390)
- ooze (401)
- seamount (398)
- siliceous ooze (401)
- sonar (391)
- submarine canyon (395)
- submersible (390)
- trench (395)
- turbidity current (395)

Key Concepts

The global ocean can be divided into three major oceans and numerous seas. **See page 389.**

Oceanography is the study of the oceans and seas. **See page 390.**

The continental margins include the continental shelf, the continental slope, and the continental rise. **See page 393.**

Features of the deep ocean basin include trenches, abyssal plains, mid-ocean ridges, and seamounts. **See page 395.**

Ocean-floor sediments form from inorganic and organic materials as well as chemical deposits. **See page 399.**

Based on physical characteristics, ocean-floor sediments are classified as mud or ooze. **See page 401.**

Review

On your own paper, write the letter of the term that best completes each of the following statements.

1. The largest of the oceans is the
a. Atlantic. b. Indian. c. Pacific. d. Arctic.

2. The continental and oceanic crust that lies beneath the ocean waters makes up the
a. ocean floor. b. continental margin.
c. continental slope. d. abyssal plain.

3. A self-propelled, free-moving submarine that is equipped for ocean research is a
a. turbidity. b. bathyscaph. c. bathysphere. d. guyot.

4. A system that is used for determining the depth of the ocean floor is
a. guyot. b. bathysphere. c. radiolaria. d. sonar.

5. Those portions of the ocean floor made up of continental crust are called
a. continental margins. b. abyssal plains.
c. mid-ocean ridges. d. trenches.

6. The submarine canyons were probably formed by
a. subduction zones. b. turbidity currents.
c. changes in sea level. d. the accumulation of sediments.

7. An accumulation of sediments at the base of a continental slope is called a
a. trench. b. turbidity current.
c. continental margin. d. continental rise.

8. The deepest parts of the ocean are called
a. trenches. b. submarine canyons.
c. abyssal plains. d. continental rises.

9. Extensive abyssal plains rarely form in those parts of the ocean basin that have
a. trenches. b. continental margins.
c. guyots. d. submarine canyons.

10. Volcanic mountains scattered along the ocean basin are
a. diatoms. b. seamounts.
c. mid-ocean ridges. d. foraminifera.

11. Submerged flat-topped seamounts are
a. guyots. b. diatoms. c. submersibles. d. bathyspheres.

12. Large quantities of the inorganic sediment that makes up the continental rise come from
a. turbidity currents. b. earthquakes.
c. diatoms. d. nodules.

13. Much of the silica on the ocean floor comes from
a. nodules. b. radiolaria and diatoms.
c. guyots. d. foraminifera.

14. Potato-shaped lumps of minerals on the ocean floor are called
a. guyots. b. foraminifera. c. nodules. d. diatoms.

15. Very fine particles of silt and clay that have settled to the ocean floor are called
a. muds. b. guyots. c. seamounts. d. nodules.

Critical Thinking

On your own paper, write answers to the following questions.

1. Suppose you are studying photographs of a newly discovered planet that has several large bodies of water. The three largest bodies of water—*A, B,* and *C*—are connected and occupy 70 percent of the surface of the planet. Two other bodies of water—*D* and *E*—are not quite so large and are partly surrounded by land. A sixth body of water—*F*—is completely surrounded by land. Which of these bodies of water resemble the earth's oceans? Which resemble seas? What type of body of water on earth does *F* most closely resemble?
2. The exploration of the ocean depths has been compared with the exploration of outer space. What similarities exist between these two environments and the attempts by people to explore them?
3. How might the continental shelves be affected if another ice age were to occur?
4. What might be the eventual fate of seamounts as they are carried along on the spreading oceanic crust?
5. A certain fish is known to exist only in one particular river in the central United States. Explain how the fossilized remains of this fish might become part of the sediments on the ocean floor.
6. Explain how it is possible that some red clays on the ocean floor have been found to contain material from outer space.

Application

1. Suppose you were searching off the east coast of the United States for the wreckage of an old Spanish galleon. Explain how sonar could aid your search.
2. Lava erupts from a volcano in the Cascade Mountains along the west coast of the United States. One component of the lava are traces of minerals similar to those found in nodules on the floor of the Pacific Ocean. How could these minerals have become part of the lava?
3. Concept map ? ? ? Construct a **concept map** starting with the term *ocean floor.* Add the new terms listed on the previous page and make connections to illustrate the relationships among them.

Extension

1. Prepare a brief report on the different types of submersibles. Your report should explain the special features of each type of submersible as well as how each has contributed to oceanographers' knowledge of the oceans.
2. Create an imaginary walking tour of the ocean basins. Your tour should begin at the edge of a continent—perhaps at a beach on the east coast of Florida. Explain exactly what tourists should look for along the continental margin and the ocean floor on their way to the western coast of Africa.

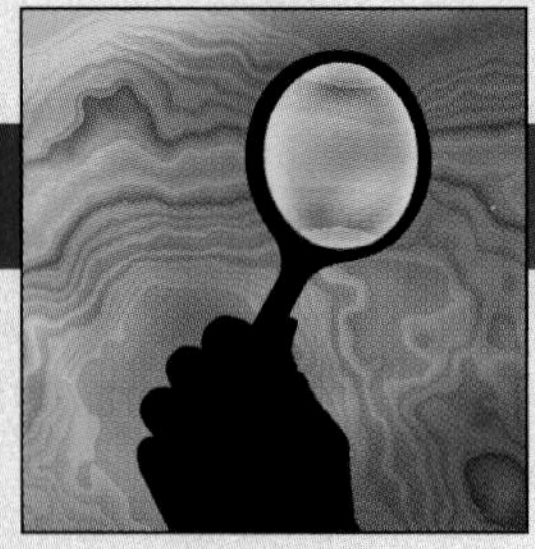

IN-DEPTH INVESTIGATION

Ocean-Floor Sediments

Materials

- 1 lb. dry mixed sand/soil
- paper cups
- stopwatch
- clear plastic column (32 in. × 1.5 in.)
- rubber stopper
- adhesive tape
- ring stand and clamp
- sieves (4 mm, 2 mm, 0.5 mm)
- water
- grease pencil
- balance
- teaspoon
- paper towels
- metric ruler

Introduction

Most of the ocean floor is covered with a layer of sediment that varies from 0.3 to 0.5 km in thickness. Much of this sediment is thought to have originated on land through the process of weathering. Through erosion, the sediment has made its way to the deep ocean basins.

Several factors determine where the sediment carried to the ocean will be deposited. One factor is the size of the particles that reach the ocean. By using soil samples of given particle size, you will determine the relationship between the size of particles and their settling rate in water. As you carry out this investigation, watch for any other factors that may affect settling rate.

Prelab Preparation

1. Review Section 20.3, pages 399–401.
2. On a surface protected by paper towels, prepare the mixed sand/soil samples. Stack the sieves on top of each other with the coarsest sieve on top and the finest on the bottom. Pour the mixed sand/soil sample into the top sieve, and sift it through to separate particles in the following size ranges: *Coarse*—the particles that remain on the top of the 4-mm sieve; *Medium*—the particles that pass through the 4-mm sieve but remain on the 2-mm sieve; *Medium-fine*—the particles that pass through the 2-mm sieve but remain on the 0.5-mm sieve; *Fine*—the particles that pass through the 0.5-mm sieve. Place each sample in a paper cup.

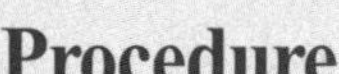

Procedure

1. Plug one end of the plastic column with a rubber stopper, and secure the stopper to the column with tape. Carefully fill the column with water to a level about 5 cm from the top, and place the column in a vertical position using the ring stand and clamp. Allow the water to stand until all large air bubbles have escaped. With the grease pencil, mark the water level on the column. This will be the starting line. Next, draw a line about 5 cm from the bottom of the column. This will be the finish line.

Soil samples	Trial 1	Trial 2	Trial 3	Average
Coarse	First time measurement:			
	Second time measurement:			
Medium	First time measurement:			
	Second time			

2. Have a member of your lab group dump one teaspoon of the coarse sample into the water column. The other group member should record two time measurements as follows:

   a. Using a stopwatch, start timing when the first particles hit the start line on the column, and stop timing when they reach the finish line. Repeat this procedure twice. Record the time for each trial in a table similar to the one shown above.

   b. Next, use the stopwatch to determine how long it takes the last particle in the sample to travel from the start line to the finish line at the base of the column. Repeat this procedure twice. Record the time for each trial in your table.

3. Determine the average time of the three trials for the first measurement. Do the same for the second measurement. Record the averages.

4. How long did it take for the first particles in the coarse sample to reach the finish line? How long did it take for all the coarse particles to reach the finish line?

5. Pour the sand/soil and water from the column into the container provided by your teacher. *Note: Do not pour sand or soil into the sink.*

6. Refill the plastic column with water up to the original level marked with the grease pencil.

7. Repeat Steps 1, 2, 3, and 5 for the remaining sample sizes. What were the settling times for the medium particles? the medium-fine particles? the fine particles? Record these measurements and the averages in your table.

8. Refill the plastic column with water. Pour 20 g of unsieved soil into the column and allow it to settle for five minutes. After five minutes, look at the column. Do layers of sediment appear in the column? Why does the water remain slightly cloudy even after most of the particles have settled?

Analysis and Conclusions

1. Compare the settling time of the medium particles with the settling time of the medium-fine particles.
2. Do similar-sized particles fall at the same rate?
3. Other than size, what factors would you expect to influence how rapidly particles fall in sea water?
4. How do the results in Step 8 help to explain why the deep ocean basins are covered with a very fine layer of sediment while areas near the shore are covered with coarse sediment?

Extension

What are some sources of the sediment that is found in the ocean?

Chapter 21 Ocean Water

Much of the earth's surface water exists in a liquid state. If the earth were much colder or hotter, the water would become a solid or a gas. The unique properties of liquid water make the oceans one of the most distinctive features of the earth. No other planet is known to have liquid water on its surface.

Chapter Outline

21.1 Properties of Ocean Water
- Composition of Ocean Water
- Salinity of Ocean Water
- Temperature of Ocean Water
- Density of Ocean Water
- Color of Ocean Water

21.2 Life in the Oceans
- Ocean Chemistry and Marine Life
- Sunlight and Marine Life
- Ocean Environments

21.3 Ocean Resources
- Fresh Water from the Ocean
- Minerals from the Ocean
- Food from the Ocean
- Ocean-Water Pollution

◀ A variety of ocean life flourishes in the coral reefs of the Indian Ocean.

21.1 Properties of Ocean Water

Section Objectives

- **Describe the chemical properties of ocean water.**
- **Describe the physical properties of ocean water.**

Pure liquid water is tasteless, odorless, and colorless. Water is the basic substance in which solids and gases are dissolved, forming the solution called *ocean water,* or *sea water.* Besides the dissolved substances, small particles of matter and tiny organisms may also be suspended in ocean water. Ocean water is a complex mixture that sustains a variety of plant and animal life.

Ocean water has both chemical and physical properties. Water's chemical properties are those characteristics that determine its composition and that enable it to dissolve other substances. Physical properties are those characteristics, such as temperature, density, and color, that can be used to describe ocean water. Scientists study all of these properties of ocean water in trying to understand the complex interactions among the oceans, the atmosphere, and the land.

Composition of Ocean Water

Each year, the earth's rivers carry about 400 billion kilograms of dissolved solids into the ocean. Most of these dissolved solids are salts. As water evaporates from the ocean, the salts and other minerals remain behind. Thus, salts and minerals are not returned to the land in the water that falls as rain and snow during the water cycle.

Gases also enter ocean water dissolved in rivers and streams or directly from the atmosphere. Some of the gases in ocean water come from volcanoes beneath the ocean. Most of the volcanic activity is from mid-ocean ridges and seamounts.

The relative amounts of the dissolved substances in ocean water have remained almost unchanged over millions of years. The processes that add new substances to the oceans seem to balance the processes that remove dissolved materials. For example, the deposition of sediments on the ocean floor adds billions of kilograms of material to the ocean each year. Marine plants and animals use up many of these substances to carry out their life processes.

Elements in Ocean Water

Ocean water is 96.5 percent pure water, or H_2O. Substances dissolved in ocean water contain about 75 chemical elements, mainly in ionic form. The six most abundant elements are chlorine, sodium, magnesium, sulfur, calcium, and potassium. See Table 21–1. There are also various trace elements—elements that exist in very small amounts—in ocean water. Among the trace elements are gold, zinc, and phosphorus.

Table 21–1
Elements in Ocean Water

Element	Percent of ocean water by mass
Chlorine (Cl)	1.94
Sodium (Na)	1.08
Magnesium (Mg)	0.13
Sulfur (S)	0.09
Calcium (Ca)	0.04
Potassium (K)	0.04

Dissolved Gases

The three principal gases in the atmosphere are nitrogen (N_2), oxygen (O_2), and carbon dioxide (CO_2). These same three gases are also the principal gases dissolved in ocean water. Of the three, carbon dioxide dissolves most easily in ocean water. Other atmospheric gases, which cannot dissolve as easily, are present in the ocean in smaller amounts. Usually the dissolved gases are in molecular form, just as they are in the atmosphere. Oxygen is mainly dissolved at the surface of the ocean, where it is in contact with the atmosphere and where marine plants can produce it.

Temperature affects the amount of gas that dissolves in water. Unlike solids, which dissolve better in warm water, gases dissolve better in cold water. In colder regions, water at the surface of the ocean usually dissolves larger amounts of gases than water in warm tropical regions does.

Dissolved gases can leave the ocean and return to the atmosphere. If, for example, the water temperature rises, less gas will remain dissolved, and the excess gas will be released into the atmosphere. Thus, the ocean and the atmosphere are continuously exchanging gases as conditions change.

Salinity of Ocean Water

Dissolved solids make up about 3.5 percent of the mass of ocean water. Sodium chloride (NaCl), or common table salt, makes up about 78 percent of the dissolved solids. The remaining portion consists of various other salts and minerals. The amount of dissolved solids present in ocean water is described as its **salinity** because most of the solids are in the form of salts. Salinity is the number of grams of dissolved solids in 1 kg of ocean water. For example, suppose that 1 kg of ocean water were evaporated and 35 g of solids remained. The salinity of this sample would be about 35 parts salt per 1,000 parts ocean water. This is written as *salinity = 35‰* (parts per thousand). Ocean water with a salinity of 35‰ has 3.5 percent of its total mass made up of dissolved salts. What percentage of dissolved salts would be present in water with a salinity of 40‰?

Both evaporation and freezing increase the salinity of ocean water. Only the water molecules are removed during evaporation and

Figure 21–1. This chart shows the percentages of dissolved solids in ocean water.

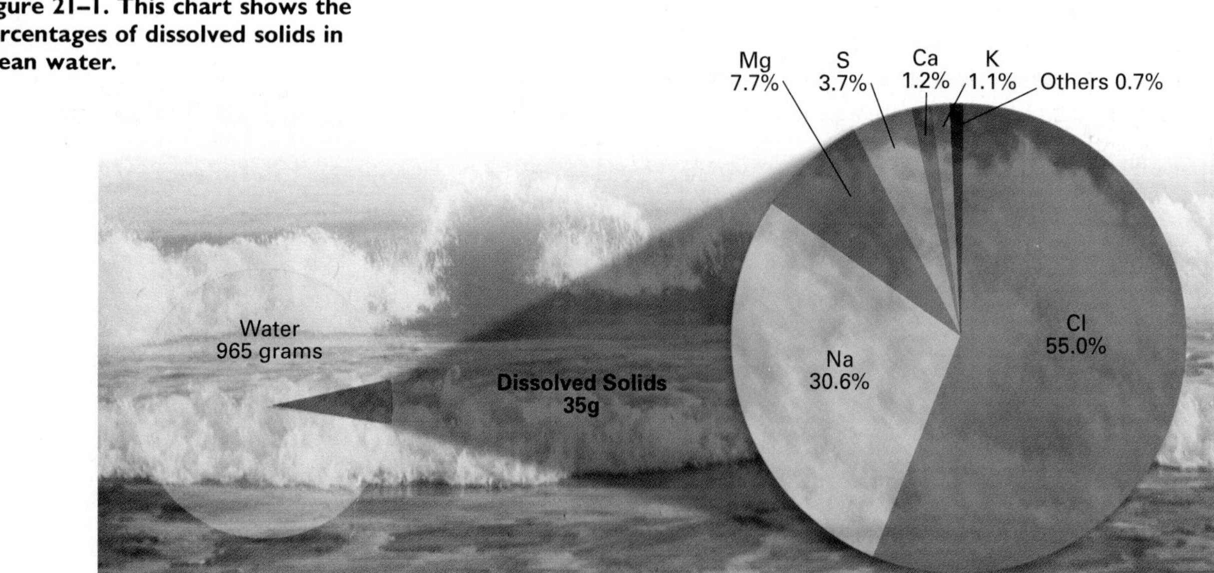

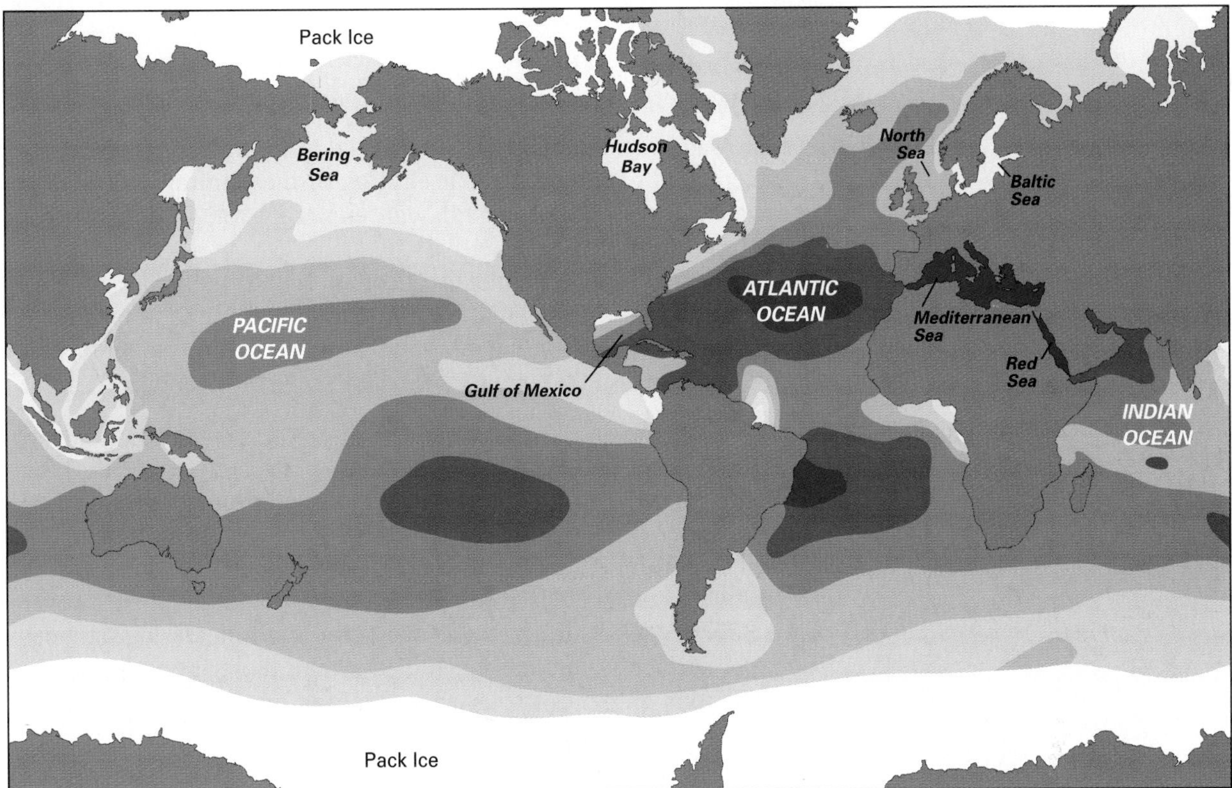

freezing. Dissolved salts and other solids remain in the ocean. When the rate of evaporation is high, the salinity of surface water increases. For this reason, tropical waters have a higher salinity at the surface than do polar waters. Salinity also varies with depth. Because water is constantly evaporating at the surface, surface waters usually have a higher salinity than do waters at great depths.

Over most of the surface of the ocean, salinity ranges from 33‰ to 36‰, with an average value of 34.7‰ for the global ocean. However, salinity at particular locations can vary greatly, as shown in Figure 21–2. The salinity of the Red Sea, for example, is over 40‰. The climate around the Red Sea is hot and dry. The high temperature of the sea leads to high levels of evaporation. In contrast, the salinity of the Baltic Sea is only about 20‰ because of the low temperatures and because the sea receives water draining from many rivers and runoff from melting snow. Because fresh water is constantly being added to the sea, the ratio of water to dissolved salts stays high and the salinity remains low. Salinity sometimes tends to be lower in surface waters near the equator because of the heavy tropical rains that fall there. In the Arctic Ocean, salinity is low because of the large amount of fresh water that enters the ocean from numerous rivers, melting ice, and because of the low rate of evaporation.

Although the salinity of ocean water may differ from one location to another, the relative amount of dissolved solids in ocean water does not change. This constant proportion is brought about by the continuous mixing of all parts of the ocean. Any dissolved substance deposited in the ocean is eventually spread uniformly throughout the water.

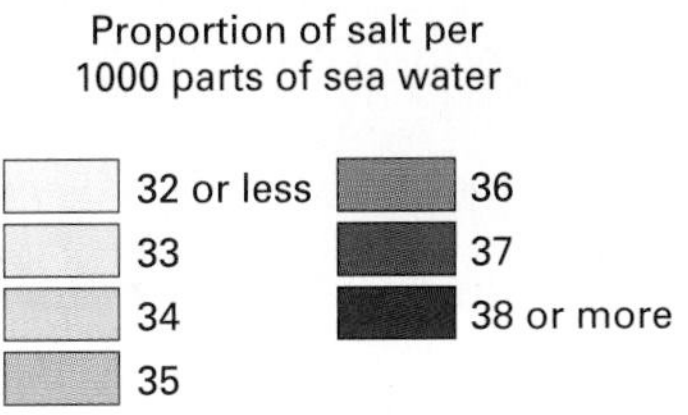

Figure 21–2. The average surface salinity of the global ocean varies from one location to another. Note areas of high and low salinity, for example, the high salinity of the Red and Mediterranean seas, and the low salinity of Hudson Bay and the Baltic Sea.

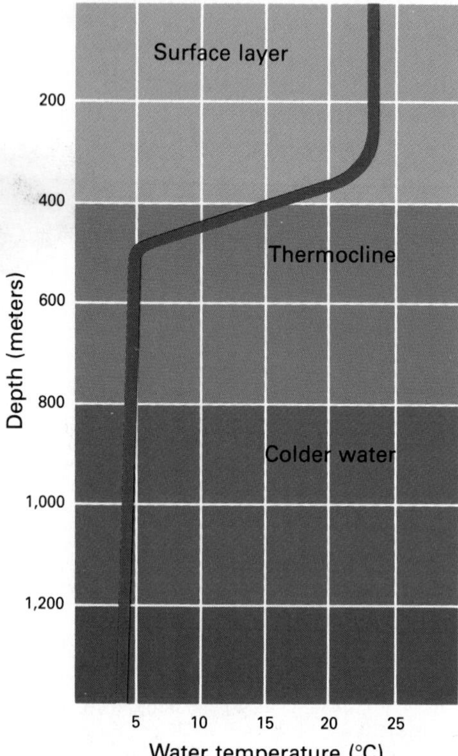

Figure 21–3. The temperature of ocean water decreases with increasing depth. Just below the surface is a thermocline, an area where the water temperature drops sharply.

Temperature of Ocean Water

Ocean water has the ability to absorb the long, invisible infrared wavelengths of sunlight. The absorption of these infrared rays heats the water. Thus infrared rays play an important role in determining the temperature of the ocean.

Infrared rays are completely absorbed within the upper zone of ocean water. This means that the sun can directly heat water near the surface of the ocean. In the deep zones of the ocean, the temperature of the water is usually about 2°C, which is just 4°C above the freezing point of salt water, which is –2°C.

Surface Temperature

The movement of water in the ocean thoroughly mixes the warmed surface water. This mixing distributes heat downward to a depth of 100 m to 300 m. Thus, the temperature of this zone of surface water is relatively constant, decreasing only slightly with depth. However, the temperature of surface water does drop with increasing latitude, so polar surface waters are much cooler than those in the tropics.

The total amount of solar energy falling on the surface of the ocean is much greater at the equator than at the poles. In tropical waters near the equator, average surface temperatures around 30°C are not unusual. Surface temperatures in polar oceans, however, usually drop below –2°C. Because ocean water freezes at about –2°C, vast areas of sea ice exist both in the Arctic Ocean and in Antarctic waters. A floating layer of ice that completely covers the ocean surface is called **pack ice.** Usually, pack ice is no more than 5 m thick because the bulk of the ice insulates the water below, and prevents it from freezing. The Arctic Ocean is covered by pack ice during most of the year. The Antarctic also has pack ice during its winter months.

In the middle latitudes, the surface temperature varies with the seasons. In some areas, the surface temperature may vary by as much as 10°C to 20°C between summer and winter.

The Thermocline

Because the sun cannot directly heat ocean water below the surface zone, the temperature of the water drops sharply as the depth increases. In most places in the ocean, this sudden temperature drop begins not far below the surface. This zone of rapid temperature change is called a **thermocline.**

The thermocline exists because the water near the surface becomes less dense as it is warmed by heat from the sun. This warm water cannot mix easily with the cold, dense water below. Thus, a thermocline marks the distinct separation between the warm surface water and the colder deep water. Below the thermocline, the temperature of the water continues to drop, but it does so very slowly, making the temperature of most deep ocean water just above freezing. Changing temperature or shifting currents may alter the depth of the thermocline or cause it to disappear completely. Nevertheless, a thermocline is usually present beneath much of the ocean surface. Thermoclines are also present in many lakes.

Density of Ocean Water

Density is the mass of a substance per unit volume. For example, 1 cubic centimeter of pure water has a mass of 1 g. Its density is 1.00 g/cm^3. Two factors affect the density of ocean water: salinity and the temperature of the water. Dissolved solids—mainly salts—add mass to the water in direct proportion to the amount of each one present. The large amount of dissolved solids in ocean water makes it denser than pure fresh water. Ocean water has a density between 1.026 and 1.028 g/cm^3.

Ocean water becomes more dense as it becomes colder and less dense as it becomes warmer. Temperature affects the density of ocean water more than salinity affects it. Thus, the densest ocean water is found in the polar regions, where the cold water is continually sinking to the ocean floor.

INVESTIGATE!

To learn more about influences on ocean water density, try the In-Depth Investigation on pages 424–425.

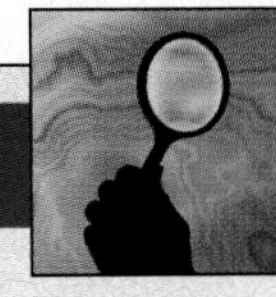

SMALL-SCALE INVESTIGATION

Density Factors

Temperature and salinity are two variables that influence the density of water. You can test these two variables using a simple straw float.

Materials

glass jar (1 L), distilled water, plastic soda straw, modeling clay, scissors, table salt, freezer, thermometer, metric ruler, marker, measuring cup

Procedure

1. Fill the jar with about 1 L of distilled water at room temperature.
2. Mark the length of the straw at 1-cm intervals. Fill 5 cm of one end of the straw with modeling clay.
3. Float the straw upright in the jar. If the straw does not float upright, cut off the open end at 1-cm intervals until it floats upright.
4. As the straw floats in the jar, use the 1-cm markings to estimate the length of the straw that remains below the water surface. Record your observations.
5. Put the jar in the freezer until the temperature of its contents drops to about 8°C. Remove the jar from the freezer and repeat Step 4. Record your observations.

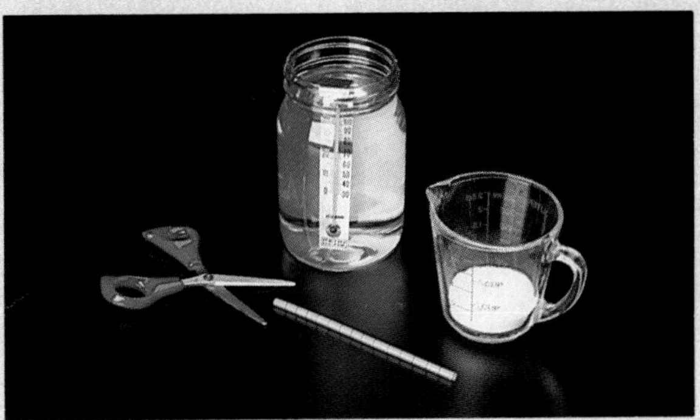

6. Wait until the water returns to room temperature. Add about 1/8 cup of salt to the water. Stir the water until the salt is completely dissolved. Repeat Step 4. Record your observations.

Analysis and Conclusions

1. In which trial was the water most dense? the least dense? Explain your answers.
2. Based on your observations, how might you make the distilled water in Step 5 less dense?
3. What effect would heating the water have on density?
4. Based on your observations, where would you expect the water in the ocean to be the least dense? the most dense?

Figure 21–4. In clear ocean water, blue light can penetrate up to 100 m before being absorbed. Until it is absorbed, blue light is reflected and makes the upper layers of ocean water appear blue. What can you infer about the depth of these vase and tube sponges?

Color of Ocean Water

The color of ocean water is determined by the way it absorbs or reflects sunlight. Because the ocean covers over 70 percent of the earth's surface, most of the sunlight that reaches the earth falls on the ocean. Much of this sunlight penetrates the surface of the ocean and is absorbed by the water.

Most of the various wavelengths, or colors, of visible light are absorbed by water. Only the blue wavelengths tend to be reflected. If you look almost straight down into clear water, the water appears blue because blue is the last color to be absorbed.

The blue color of the ocean may not always be apparent. When light rays strike water at an angle close to the horizontal, they are reflected rather than absorbed. This reflection causes the water to sparkle rather than to appear as a solid color. Also, many times, the natural blue color of the ocean is clouded by small rock particles and marine organisms suspended in the water, causing the water to appear green or brown. No light of any wavelength can penetrate ocean water at depths below a few hundred meters. Therefore, only the upper regions show color; the remainder is in total darkness.

Section 21.1 Review

1. Why does the water of the Arctic Ocean have relatively low salinity?
2. What is a thermocline?
3. How does temperature affect the density of ocean water?
4. Why would surface water in the North Sea be more likely to contain a higher percentage of dissolved gases than would surface water in the Caribbean Sea?

21.2 Life in the Oceans

Section Objectives

- **Explain how marine life alters the chemistry of ocean water.**
- **Explain why plankton can be called the foundation of life in the ocean.**
- **Describe the major zones of life in the ocean.**

Fossil evidence indicates that the first living organisms on the earth probably originated in the oceans more than 3 billion years ago. From that time to the present, marine organisms have changed, as have the physical and chemical properties of the ocean. Most marine organisms depend on two major factors for their survival: the essential nutrients available in ocean water and sunlight. Variations in the amount of either of these factors affect the ability of marine organisms to survive and flourish.

Ocean Chemistry and Marine Life

Animals and plants living in the oceans help maintain the chemical balance of ocean water. These organisms remove from the water all the nutrients and dissolved gases that they require for carrying out their life processes. At the same time, they return a tremendous variety of nutrients and gases to the water.

Nearly all life in the ocean is regulated by the life processes of plants. Plants absorb large amounts of substances containing carbon, hydrogen, oxygen, and sulfur. These elements are so highly concentrated in ocean water that the biological processes of marine organisms do not have any significant effect on their abundance. The elements nitrogen, phosphorus, and silicon are critical nutrients for the growth of plants. These substances are less abundant than carbon dioxide. Therefore, heavy plant growth can reduce the concentration of these essential elements to nearly zero in some areas of the ocean. When this occurs, the growth of plants in the depleted water is greatly slowed or even stopped. How does this process compare with the growth of vegetation on land?

Marine plants and animals absorb and store the substances they need for life. These substances are returned to the water when an

Figure 21–5. Like many marine organisms, this spotted puffer fish gets all it needs to maintain life, including food and oxygen, from the ocean water.

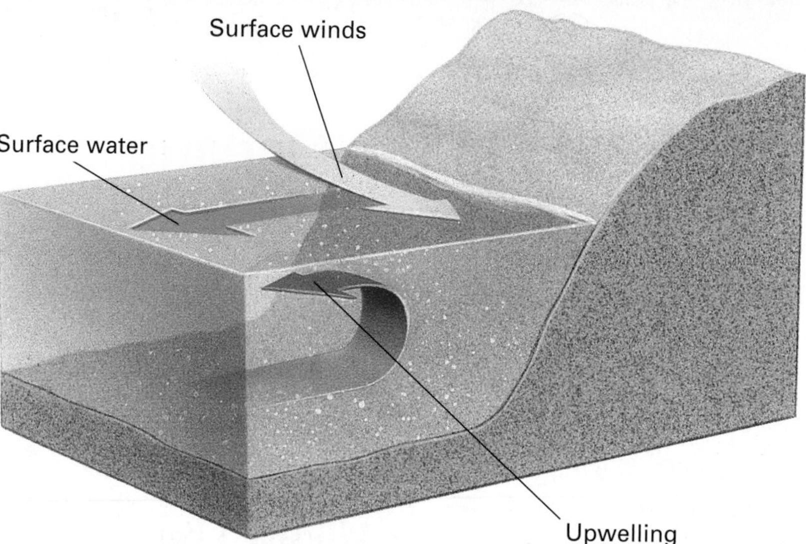

Figure 21–6. The illustration shows upwelling due to the offshore movement of surface water.

organism dies and its body decays. Bacteria in the water digest organic remains and release the essential nutrients. Gravity slowly pulls the organic remains away from the surface regions, where biological activity is greatest, and toward the ocean bottom.

In general, all of the elements necessary to marine life are consumed by marine plants and animals near the surface but released at great depths through decay. Thus, the deeper water becomes a storage area for the vital nutrients needed to support life. Deep-dwelling animals use the nutrients settling from the upper regions as a source of food. These important nutrients must be returned to the sunlit surface, however, before most organisms in the ocean can use them.

Several processes can cause deep ocean water to move upward, carrying with it the nutrients necessary for the survival of marine life. When wind blows steadily parallel to the shore along a coastline, surface ocean water is moved farther offshore. Deep water then moves upward to replace the surface water. This process, shown in Figure 21–6, is called **upwelling.** Upwelling occurs off the west coasts of North America, South America, and Africa. Areas of upwelling are rich fishing grounds.

In shallow water, wave action on the shore may be powerful enough to cause deep water to mix with surface water. Tides can also mix deep water with surface water. The distribution of life in the ocean depends, to a large extent, on the way life-supporting nutrients return to the surface from the ocean depths.

Sunlight and Marine Life

All marine plants and many microscopic marine organisms require sunlight as well as nutrients in the water. This means that plant growth in the ocean is restricted to the upper 100 m of water where light penetrates. Below about 100 m, there is not enough light to sustain photosynthesis. Within the zone where sufficient light exists, most regions of the ocean contain large amounts of free-floating, microscopic plants and animals called **plankton.** There are two main types of plankton: plants and other photosynthetic organisms known as **phytoplankton,** and animals called **zooplankton.** Phytoplankton remove dissolved gases and nutrients from the water and use the energy from sunlight to carry on photosynthesis. The phytoplankton then serve as the food source for the zooplankton.

Both forms of plankton are eaten by larger ocean animals called **nekton.** Nekton are forms of ocean life that swim, such as fish, dolphins, and squid. Because nekton can swim, they are able to search for food and avoid some predators.

In ocean food webs, plankton are consumed primarily by small fish and squid. These, in turn, become food for adult fish and other large marine animals. Some large animals, such as baleen whales, feed directly on plankton. Thus, phytoplankton form the base for the complex food webs that support life in the ocean.

Organisms that live on the ocean floor are called **benthos.** Benthos include marine plants that grow in sunlit shallow waters and animals such as oysters, sea stars, and crabs. Some types of benthos—sea anemones and corals, for example—attach themselves to the ocean floor.

Ocean Environments

The ocean can be divided into two general environments—bottom, or **benthic,** and water, or **pelagic.** The benthic environment is classified into five zones, and the pelagic environment is classified into two major zones. The amount of sunlight, temperature, and water pressure determine the distribution of marine life within these zones.

Benthic Zones

The shallowest benthic zone, the **intertidal zone,** lies between the low-tide and high-tide lines. Because of shifting tides and breaking waves, this zone is a relatively unstable environment for marine life. The waves scatter organisms, and the changing tides may leave the zone dry. Still, marine life flourishes here, attracted by sunlight, abundant nutrients, and well-mixed water. Crabs, clams, mussels, sea anemones, and seaweed populate this zone.

The largest number of benthic-zone-dwelling organisms live in the shallow **sublittoral zone.** This continuously submerged zone located on the continental shelves is dominated by sea stars, brittle stars, and sea lilies.

The **bathyal zone** begins at the continental slope and extends to a depth of 4,000 m. Because little or no sunlight reaches this zone,

Figure 21–7. A type of zooplankton, known as *Nauplius* is shown left. Dinoflagellate, shown right, is a type of phytoplankton that lives in the upper zone of the ocean where light is available for photosynthesis.

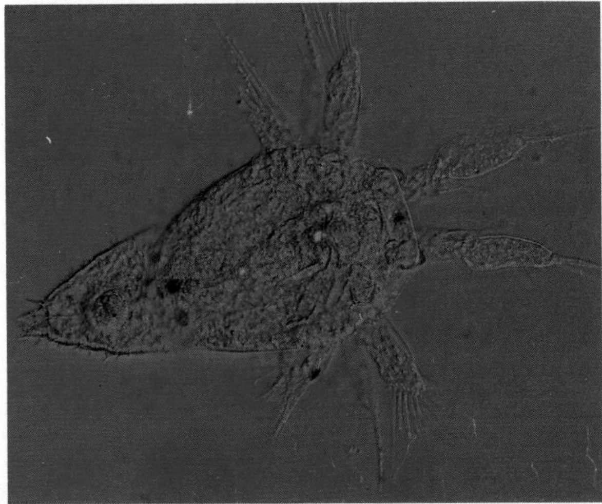

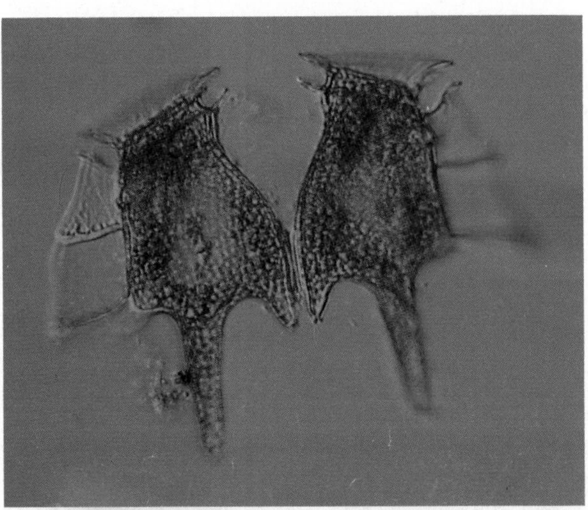

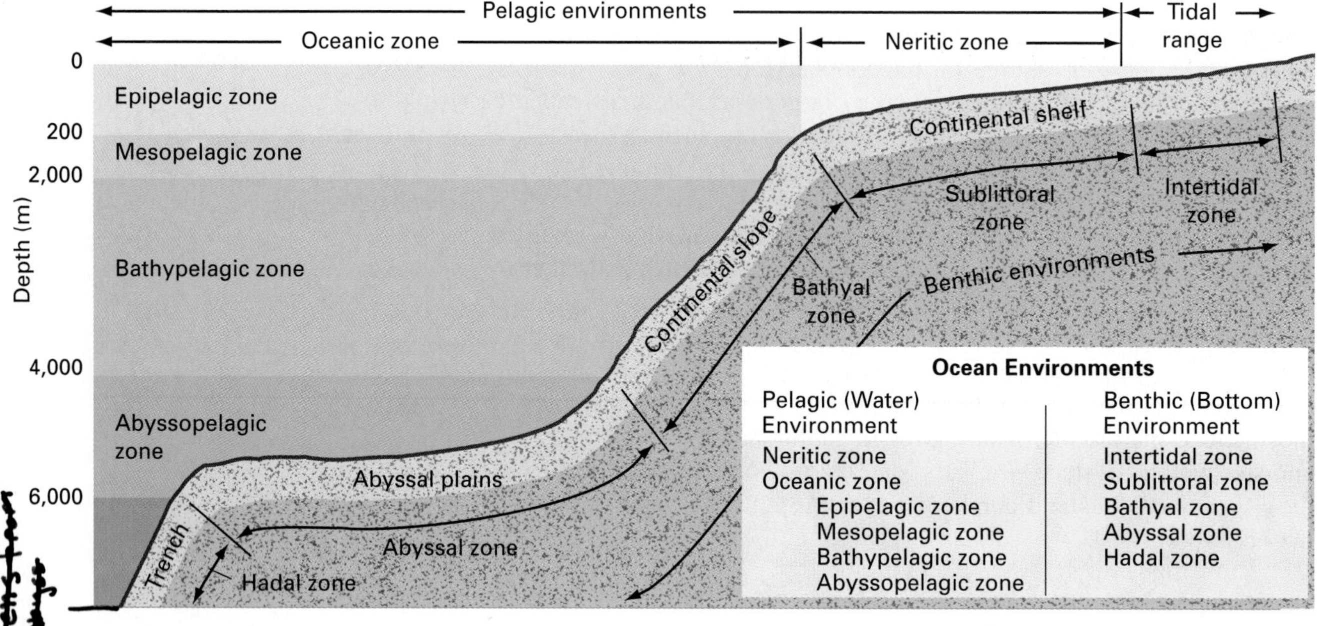

Figure 21–8. This diagram shows the classification and location of marine environments.

plant life is scarce. Among the animals that live in the bathyal zone are octopuses, sea stars, and brachiopods.

The **abyssal zone,** which has no light, begins where the bathyal zone ends and extends to a depth of 6,000 m. Organisms that live in the darkness of the abyssal zone include sponges, worms, and sea cucumbers.

The **hadal zone** is confined to the ocean trenches, which are deeper than 6,000 m. This zone is virtually unexplored, and life is sparse and dependent on food that falls from higher levels.

Pelagic Zones

The first pelagic zone is the **neritic zone.** This zone is located above the continental shelves. The neritic zone has abundant sunlight, generally moderate temperatures, and relatively low pressure. These conditions make it ideal for marine life. Plankton and nekton fill its waters, which are the source of much of the fish and seafood that people eat.

The **oceanic zone** extends into the deep and open ocean waters beyond the continental shelf. It is divided into four zones, based on depth. The *epipelagic zone* is the uppermost area. It is sunlit and populated by tuna, dolphin, and mats of floating sargassum weed. The dark *mesopelagic, bathypelagic,* and *abyssopelagic zones* occur at increasing depths. As a general rule, marine life decreases with depth in the pelagic environment.

Section 21.2 Review

1. In addition to the action of waves and tides, what other process causes deep ocean water to move upward?
2. Describe the two main types of plankton.
3. Describe the intertidal, neritic, bathyal, and abyssal zones and name some organisms found in each.

21.3 Ocean Resources

Section Objectives

- **Describe three important resources of the ocean.**
- **Explain the threat to ocean life posed by water pollution.**

The sea has long been a source of food and a means of transportation. Recently, however, people have begun to realize its importance as a source of natural resources needed to support modern civilization. Increased knowledge of the oceans, combined with the earth's expanding population, has caused new interest in the sea as a source of fresh water, minerals, and food.

Fresh Water from the Ocean

Throughout the world the need for fresh water is increasing rapidly. Developing countries need huge supplies of fresh water for industry and irrigation. Even some nations that formerly had abundant supplies of fresh water are facing shortages.

The increasing demand for water can be met in two ways. First, the fresh water now available can be conserved to avoid waste. Second, the amount of available fresh water can be increased. The water supply can be increased by finding a way to convert ocean water to fresh water at a reasonable cost.

One means of increasing the freshwater supply is through *desalination*. As you read in Chapter 13, desalination is the extraction of fresh water from salt water. One method of desalination now used is **distillation,** which involves heating ocean water to remove the salt. Heat causes the liquid water to evaporate, leaving dissolved salts behind. When the water vapor condenses, the result is pure fresh water. However, evaporating liquid water often requires a great deal of costly heat energy.

Another method for desalinating ocean water involves freezing. When the water freezes, the first ice crystals that form are free of salt. The salt remains in pockets of liquid water in the ice. The ice can be removed and melted to obtain fresh water. This process requires only about one-sixth the amount of energy needed for distillation.

A popular method for desalinating ocean water involves the use of special membranes that allow water under high pressure to pass through while blocking the dissolved salts. This method is known as *reverse osmosis* desalination. Large desalination plants are now in operation in Saudi Arabia, Kuwait, Mexico, and numerous islands and coastal desert communities where fresh water supplies from land are very limited.

Minerals from the Ocean

Mineral deposits are another natural resource from the ocean. On the first oceanographic research voyage, made by H.M.S. *Challenger* in 1872, strange black lumps of minerals, called *nodules,* were brought up from the ocean floor. More recent deep-sea exploration has shown that huge areas of the abyssal plains, particularly in the Pacific Ocean, are covered with these nodules.

Nodules are a valuable source of manganese, which is used in making certain types of steel. These nodules also contain iron, copper, nickel, and cobalt. Other nodules contain phosphates that are useful as fertilizers.

The way nodules are formed is not completely understood by scientists. However, measurement of their rate of growth suggests that they form very slowly, probably 1 to 10 mm per million years. The recovery of nodules is expensive and difficult because they are found in such deep water. If an economical method of mining them can be developed, the mineral-rich nodules could possibly replace some ores currently mined on land. However, the mining of nodules presents environmental concerns and raises questions about the ownership of the ocean floor and its minerals.

Some valuable minerals are now easily extracted from the oceans. For many centuries, salt has been obtained from ocean water. The ocean is the main source of magnesium and bromine. These two elements are refined from their dissolved salts. However, the concentration of most other useful minerals dissolved in the oceans is too small for extraction to be practical. For example, each cubic kilometer of ocean water contains about 6 kg of gold. However, obtaining just a few cents worth of gold would require

SCIENCE & TECHNOLOGY

Efficient Desalination

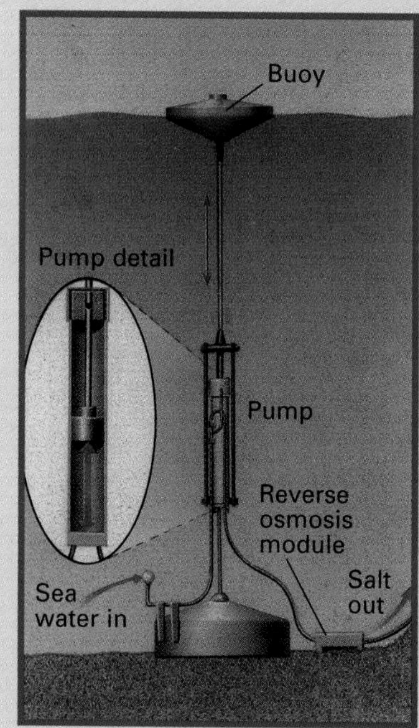

Although water covers more than 70 percent of the earth's surface, only 3 percent of it is fresh water. The other 97 percent is ocean water, too salty for drinking or irrigating crops. Desalination, however, can make the world's oceans a reliable source of fresh water for dry coastal regions.

Over the years, scientists have devised various ways of turning ocean water into fresh water and there are now thousands of desalination plants worldwide. Together these plants produce about 13 billion liters of fresh water every day. The cost, however, is about four times that of other sources.

Desalination is costly because desalination plants use a lot of energy. Distilling ocean water, freezing it, or pumping it through filters, as well as disposing of the resulting salt, all require energy. Approximately one third of the cost of desalted water comes from the fossil fuels needed to run land-based desalination plants.

To make desalination more efficient, chemist Mic Pleass and marine scientist Doug Hicks looked to the ocean itself for an alternative, cheaper source of power. Their invention, called *Delbuoy,* is a desalination system that is ocean-based and powered by waves. Simple in design, the Delbuoy system, illustrated at left, consists of a wave-driven pump and a

processing about 4 million liters of ocean water. Consequently, ocean water is not an economical source of gold.

The most valuable mineral resource taken from the ocean is the petroleum found beneath the sea floor. Deposits of offshore oil and natural gas exist along the continental margins in many parts of the world. As the worldwide need for energy has grown, these fossil-fuel resources have become very important. About one-fourth of the world's oil production is now obtained from offshore wells. As a result of new drilling techniques, oil and gas can be extracted far offshore and from depths over 1,000 m. However, the vast majority of offshore wells are in the shallow waters of the continental shelves. Productive offshore wells are located in the Gulf of Mexico, the North Sea, the Persian Gulf, and off the coasts of California, Alaska, and Australia.

Food from the Ocean

Of all the resources that the ocean is capable of supplying, the one in greatest demand is protein-rich food. At present a large part of the population of the world has a starchy diet. Such a diet can maintain life, but the lack of protein to build strong tissues decreases the

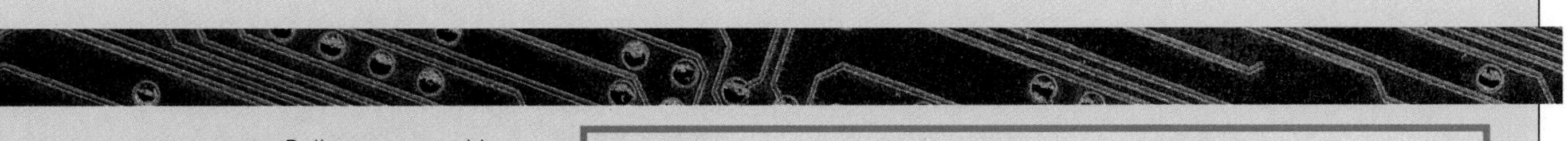

Delbuoy, a wave-driven desalination system, has been extensively tested in Puerto Rico.

reverse-osmosis module.

Delbuoy's pump is 3 m tall, anchored in 18 m of ocean water, and held upright by a buoy. The module contains a membrane through which water, but not salt, can pass. As the buoy rises and falls with the wave action, it operates the pump. The pump draws up ocean water, filters out the sand, and forces the water through the membrane in the reverse-osmosis module. The salt-free water is then pumped to storage tanks on land, and the salt is returned to the ocean.

The Delbuoy system is economical as well as efficient. While desalted water can cost as much as $8 per 1000 L in many Caribbean and Pacific islands, the Delbuoy system can deliver the same amount of water for about $5. If the Delbuoy system were mass-produced, it could someday deliver 1000 L of desalted water for as low as $1 to $2.

How might the Delbuoy system also reduce pollution of the environment?

Figure 21–9. Aquaculture establishments, such as this prawn farm in Hawaii, already provide a reliable, economical source of food.

ability of a person's body to fight disease. Perhaps half a billion people in the world suffer from some form of disease caused by a lack of protein in their diet. An important source of high-protein food is fish from the ocean. However, many ocean areas are already overfished and in decline.

In the future, **aquaculture,** the farming of the ocean, will become an important source of food production. Aquaculture involves developing and raising special breeds of marine animals and plants that yield large amounts of food. Aquaculture has already been used to successfully grow catfish, salmon, oysters, and shrimp on large aquatic farms. Similar methods might be used to breed other fish and shellfish in ocean farms. A major problem for aquaculturalists is that the ocean farms would be susceptible to pollution and may themselves be a local source of pollution.

Under the best conditions, an ocean farm could produce more valuable, protein-rich food than a land farm of the same size. Fish and shellfish change their food supply into protein more efficiently than do most land animals. No marine animals, except marine mammals, use energy to keep an even body temperature. Because of the buoyancy of water, they also use less energy supporting their weight against the force of gravity. In agriculture, only the top layers of soil can be used. In contrast, ocean farms may use a wide range of depths to produce food. Someday, the nutrient-rich bottom water may even be pumped to the surface as a way of fertilizing the ocean. How would this type of artificial upwelling benefit an ocean farm?

Ocean-Water Pollution

The oceans have been used as a dumping ground for many kinds of wastes, including garbage, sewage, and nuclear wastes. Until recently, most wastes were diluted, or destroyed as they spread throughout the ocean. But the growth of the world population and the increased use of more toxic substances have changed the situation. The ability of the ocean to absorb wastes and renew itself cannot match the increasing amount of waste that is being produced worldwide.

Figure 21–10. People have begun to recognize the importance of reducing ocean water pollution. This scientist is taking a water sample, which will be checked for impurities.

Productive coastal waters are in the greatest danger. Pollution has destroyed clam and oyster beds in some local areas. Sea birds have been found tangled in plastic products. Beaches have been closed because of sewage, medical wastes, and oil washed onto the sand.

Besides coastal waters and beaches, pollutants are now found in measurable amounts everywhere in the oceans. Traces of the insecticide DDT have been detected in the open ocean. Since lead was added to gasoline 60 years ago, the amount of lead in the oceans has increased. Both DDT and lead can cause problems in ocean food webs. As animals eat other animals, the amounts of DDT and lead build up in the bodies of the predators. In some areas of the world, high lead concentrations have made some fish inedible. Fortunately, DDT use in the United States has been banned, and leaded gasoline use is declining. Their levels are becoming reduced in the marine environment. Radioactivity from nuclear waste dumping, however, remains a potential threat. The global presence of marine pollutants indicates that the ocean cannot be used as a dumping ground forever.

Section 21.3 Review

1. Name and describe three methods for desalinating ocean water.
2. What ocean resource might eventually replace some minerals now mined on land?
3. What is aquaculture, and why is it important?
4. How does ocean pollution threaten the quality of life on the earth?

Chapter 21 Review

Key Terms

- abyssal zone (416)
- aquaculture (420)
- bathyal zone (415)
- benthic environments (415)
- benthos (415)
- distillation (417)
- hadal zone (416)
- intertidal zone (415)
- nekton (415)
- neritic zone (416)
- oceanic zone (416)
- pack ice (410)
- pelagic environments (415)
- phytoplankton (414)
- plankton (414)
- salinity (408)
- sublittoral zone (415)
- thermocline (410)
- upwelling (414)
- zooplankton (414)

Key Concepts

Ocean water dissolves solids and gases. **See page 407.**

The temperature, density, and color of ocean water vary around the earth. **See page 410.**

Marine life helps maintain the chemical balance of ocean water. **See page 413.**

Plant life regulates virtually all ocean life. **See page 413.**

There are two major environments of life in the ocean, each supporting certain types of organisms. **See page 415.**

The ocean is valuable as a source of fresh water, minerals, and food. **See page 417.**

Marine pollution threatens the ocean environments. **See page 420.**

Review

On your own paper, write the letter of the term that best completes each of the following statements.

1. The gas that dissolves most easily in ocean water is
a. carbon dioxide. b. argon.
c. nitrogen. d. oxygen.

2. The amount of dissolved solids in ocean water is called its
a. salinity. b. nekton. c. plankton. d. density.

3. A thin floating layer of ice that covers an ocean surface is called
a. an iceberg. b. a thermocline.
c. pack ice. d. benthos.

4. A zone of rapid water temperature change with increasing depth is called
a. a benthos. b. an abyssal zone.
c. a bathyal zone. d. a thermocline.

5. When liquid water is warmed, its density
a. increases. b. decreases.
c. remains the same. d. doubles.

6. Although most of the various wavelengths of visible light are absorbed by ocean water, the one wavelength that tends to be reflected is the color
a. violet. b. green. c. yellow. d. blue.

7. The process by which surface water is blown farther offshore and nutrient-rich deep water rises to take its place is called
a. desalination. b. distillation.
c. upwelling. d. aquaculture.

8. Drifting marine plants and animals are known as
a. plankton. b. benthos. c. nekton. d. tube worms.

9. Marine animals that can swim to search for food and avoid predators are called
a. phytoplankton. b. zooplankton.
c. nekton. d. benthos.

10. Sea anemones are an example of
a. zooplankton. b. nekton.
c. benthos. d. phytoplankton.

11. Perhaps the most unstable ocean environment is the
a. intertidal zone. b. abyssal zone.
c. bathyal zone. d. neritic zone.

12. The pelagic zone that has abundant sunlight, a fairly constant temperature, and relatively low pressure is the
a. bathyal zone. b. neritic zone.
c. abyssal zone. d. intertidal zone.

13. Which of the following is not a method for producing fresh water by desalinating ocean water?
a. distillation b. evaporation
c. reverse osmosis d. aquaculture

14. Lumps of minerals on the ocean floor are called
a. nekton. b. nodules. c. benthos. d. plankton.

15. Aquaculture is another name for
a. desalination. b. distillation.
c. ocean farming. d. rapid temperature changes.

16. In the future, scientists may be able to fertilize the ocean through
a. distillation. b. artificial upwelling.
c. mining nodules. d. dissolving gases.

17. The increased use of gasoline has led to an increase in the ocean of
a. plankton. b. upwelling. c. lead. d. salinity.

Critical Thinking

On your own paper, write answers to the following questions.

1. Suppose that climatic conditions over one of the earth's oceans changed dramatically. The changes resulted in a complete absence of upwelling and wave action. Explain what would happen to the marine life in this ocean and why.
2. How would a significant decrease in sunlight affect phytoplankton?
3. What impact would the decrease in sunlight discussed in Question 2 have on other forms of marine life?
4. When oceanographers first explored the deep ocean basin along mid-ocean ridges, they discovered a variety of marine life, including sightless crabs. Explain why these crabs are not handicapped by their sightlessness.

Application

1. Suppose you are contracted to build a desalination system for the state of Florida. In order to make your system as efficient as possible, you want to locate it in an area of the ocean where the salinity of the water is low. Where along the coast of Florida would you choose to build the site, near Miami or Tallahassee?
2. You have collected samples of ocean water from the three locations: a site just off the coast of Israel; a site on Prudhoe Bay, off the coast of northern Alaska; and a site off the coast of Sweden, on the Baltic Sea. All three samples have been brought to room temperature. From which location would the water have the greatest density? Why?
3. You have decided to start an aquatic farm. In which of the zones of ocean life should you locate your farm? Why?

Extension

1. Research the new foods being produced through aquaculture. What is the nutritional value of these foods? Draw a world map and color in the countries that are developing aquaculture foods.
2. Research the pollution of North American coastal waters. Make a chart showing how a marine food web is affected by lead pollution, or trace the impact of another pollutant on the marine food web.

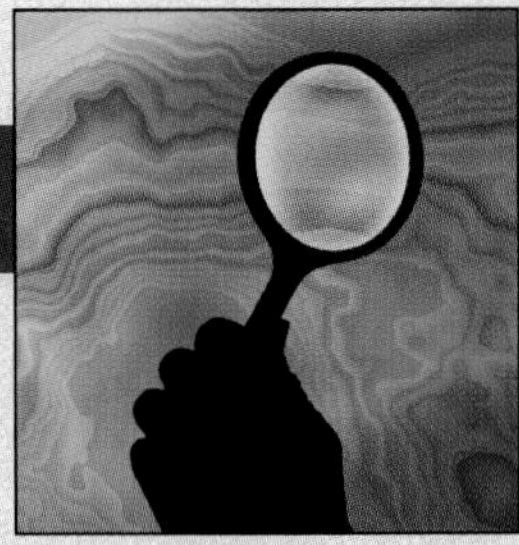

IN-DEPTH INVESTIGATION

Ocean Water Density

Materials

- plastic straw
- modeling clay
- beaker (250-mL)
- grease pencils (yellow and red)
- table salt
- Bunsen burner
- 50-mL graduated cylinder
- ring stand and clamp
- teaspoon
- heat-resistant gloves
- wire gauze
- safety goggles
- thermometer (Celsius)
- metric ruler
- scissors
- freezer
- water

Introduction

The density of ocean water varies in different areas of the ocean. Density differences are affected by the amount of dissolved solids, including salt, and the temperature of the water. Because salt is the most abundant dissolved solid in the ocean, a change in salinity will change the water's density. The salinity of an area of the ocean is also affected by the rate of evaporation or freezing and by the amount of fresh water and salts added by rivers and glacial runoff. The temperature of the ocean is determined by the amount of infrared radiation it receives.

In this investigation you will observe the effects of temperature and salinity on the density of salt water.

Prelab Preparation

1. Read Section 21.1, pages 407–412.
2. Review the safety guidelines for fire, heating, and eye and hand safety.
3. Define density.
4. Construct a hydrometer:
   a. Hold your index finger over one end of a plastic straw and slowly press the open end into a piece of clay until the straw is filled with about 5 cm of clay.
   b. Place the straw, clay end down, in a graduated cylinder filled with 50 mL of tap water. (If the straw does not float upright, cut off the open end at 1-cm intervals until it does.) Use a red grease pencil to mark the water level on the straw. Remove the straw from the water and draw a continuous line around the straw at the mark.
   c. Using a yellow grease pencil, draw lines around the straw at 1-cm intervals above and below the red line. The red line will be used as a reference point.

Procedure

1. Pour 100 mL of room-temperature tap water into a beaker. Stir in 2 teaspoons of salt until it is all dissolved. Measure and record the water temperature in a table of your own similar to the one shown under Step 5.

2. Place the hydrometer in the salt water. In your table, record the density by counting the marks above or below the red line to the water's surface.

3. Set up the ring-stand apparatus and Bunsen burner as shown in the photo at left. **CAUTION:** *Wear your safety goggles.*

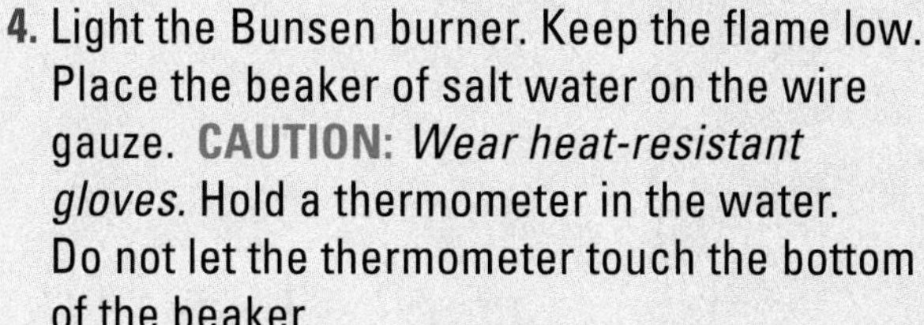

4. Light the Bunsen burner. Keep the flame low. Place the beaker of salt water on the wire gauze. **CAUTION:** *Wear heat-resistant gloves.* Hold a thermometer in the water. Do not let the thermometer touch the bottom of the beaker.

5. When the water's temperature reaches 25°C, turn off the Bunsen burner. Immediately place the hydrometer in the water. Record the relative density in your table.

Temperature (°C)	Density (cm above or below the red line)
25	
30	

6. Repeat Steps 4 and 5, heating the water until it reaches 30°C. As the temperature increases, does the water's density increase or decrease?

7. Heat the salt water until it begins to boil. Continue to boil the water for 5 minutes. Turn off the burner. Place the hydrometer in the water. Measure and record the water's density in another table as shown above.

Minutes of Boiling	Density (cm above or below the red line)
5	
10	
15	

8. Boil the water for another 5 minutes. Measure and record the water's density.

9. Repeat Step 8. Did the density of the water increase or decrease with each boiling?

10. Originally there was 100 mL of water in the beaker. How much water is in the beaker now?

Analysis and Conclusions

1. Explain how and why warming the water affected its density.
2. Based on your observations, infer what the density of polar ocean waters would be compared with the density of equally saline water near the equator. Explain your answer.
3. Why did the amount of water in the beaker change? Explain why boiling the water affected its density.

Extension

Place a beaker of salt water in a freezer until a crust of ice forms at the top. Break up and remove the ice from the water. Is the water denser or less dense than before the freezing? Explain why.

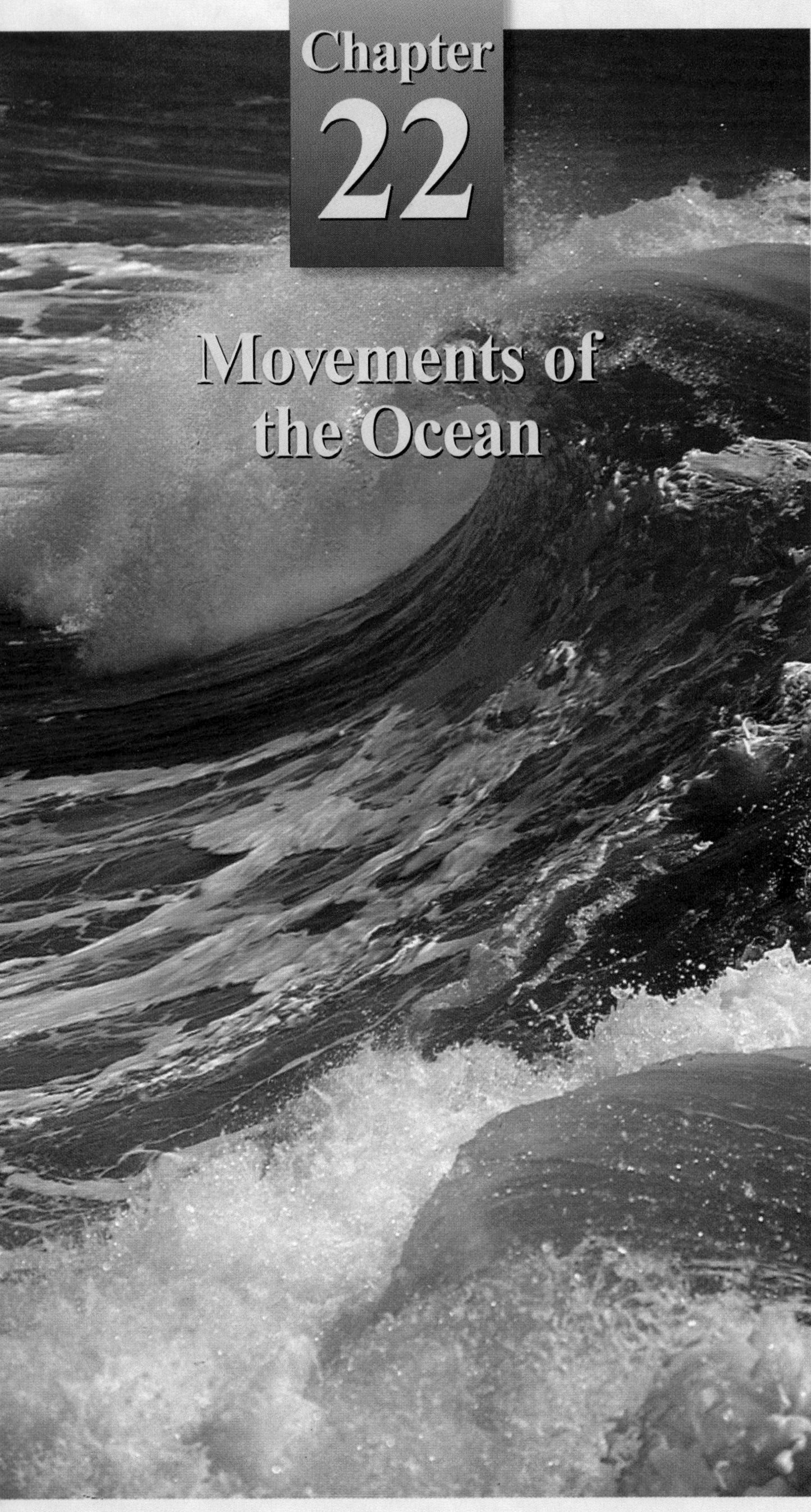

The waters of the ocean are constantly in motion. Waves rise and fall in varied rhythms. Ocean currents—some near the surface and some deep below—flow like rivers. The level of the ocean moves up and down with the tides. This chapter discusses the causes and effects of these powerful movements of the oceans.

Chapter Outline

The power of the ocean is visible in its waves.

22.1 Ocean Currents

Section Objectives

- **Discuss how wind patterns affect surface currents.**
- **Explain how differences in the density of ocean water affect the flow of deep currents.**

The waters of the ocean move in giant streams called **currents.** Some of these currents are strong enough to affect the speed and direction of ships. As a result, people have studied ocean currents since the early days of sailing.

Learning what causes ocean currents has not been easy. These complex movements of water are often difficult to trace. Modern-day oceanographers identify ocean currents by studying the physical and chemical characteristics of the water. They have even mapped the paths of athletic shoes from ships that sank in storms. From these data, they have determined a detailed pattern of ocean currents.

The movement of water in the ocean is still not completely understood. However, as oceanographers collect more-detailed data, they are better able to describe the movements and causes of ocean currents. Oceanographers know that there are two major types of ocean currents. **Surface currents** move on or near the surface of the ocean and are driven by winds. **Deep currents** move very slowly beneath the surface of the ocean and are caused by differences in the density of the water.

Surface Currents

Ocean water, like any substance, can be set into motion only if it receives energy. Steady winds provide this energy, moving large masses of water near the surface of the ocean. Thus, wind is the driving force of surface currents. Almost all of the surface currents of the ocean result from global wind patterns. Because the transfer of energy from the wind to the water occurs at the upper layer of the water, that part of the ocean moves fastest. The velocity of the water decreases with depth, and at a depth of approximately 100 m, most surface currents are almost undetectable.

Factors Affecting Ocean Surface Currents

Surface currents are controlled by three factors—the wind belts, the earth's rotational effects, and the location of the continents. The global wind belts that most directly affect the flow of surface currents are the *trade winds* and the *westerlies*. The trade winds are located just north and south of the equator. The westerlies are located in the middle latitudes, as shown in Figure 22–1.

In the Northern Hemisphere, the warm trade winds blow from the northeast, while in the Southern Hemisphere, they blow from the southeast. Both trade-wind belts push currents westward across the tropical latitudes of all three major oceans.

The temperate westerlies affect the surface currents in the middle latitudes. In the Northern Hemisphere, they blow from the southwest, pushing the currents eastward across the Atlantic and Pacific oceans. In the Southern Hemisphere, they blow from the northwest, pushing the earth's largest current, the West Wind Drift, completely around the world at the southern edges of all three major oceans.

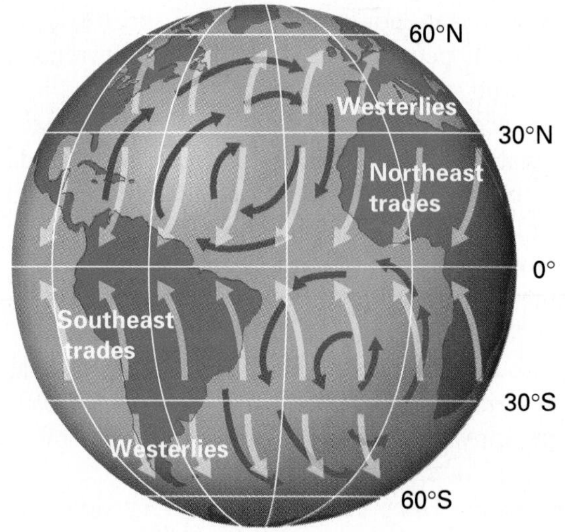

Figure 22–1. Global winds and the Coriolis effect together drive the surface currents of the oceans in great circular patterns. The white arrows represent global winds and the blue arrows represent ocean currents.

The **Coriolis effect** is also a major factor controlling surface currents. The Coriolis effect is the deflection of the earth's winds and ocean currents caused by the earth's rotation. As a result of the wind belts and the Coriolis effect, huge circles of moving water, called **gyres,** are formed. In the Northern Hemisphere, the flow is to the right, or clockwise. In the Southern Hemisphere, the flow is to the left, or counterclockwise. A gyre is usually divided into four currents based on direction of flow.

The third major factor involves the continents, which are obvious barriers to the surface currents. When an ocean surface current flows against a landmass, the current is deflected and divided.

Equatorial Currents

Warm equatorial currents are found in the Atlantic, Pacific, and Indian oceans. In each of these oceans, two warm-water currents, the *equatorial currents,* move in a westward direction. Between these westward-flowing currents lies a weaker, eastward-flowing current called the *Equatorial Countercurrent.*

Currents in the North Atlantic and Pacific

In the North Atlantic Ocean, the North Atlantic Equatorial Current pushes water through the Caribbean Sea and Gulf of Mexico and north along the east coast of North America. This swift, warm current is called the **Gulf Stream.** The Gulf Stream can move 100 million m^3 of water per second. By comparison, the Mississippi River moves 20,000 m^3 of water per second. The Gulf Stream moves from the Gulf of Mexico around the tip of Florida and north along the east coast of the United States. In the North Atlantic, the Gulf Stream meets the cold-water Labrador Current, which flows out of Baffin Bay and into the North Atlantic. When the warm air over the Gulf Stream blows over the cold water of the Labrador Current,

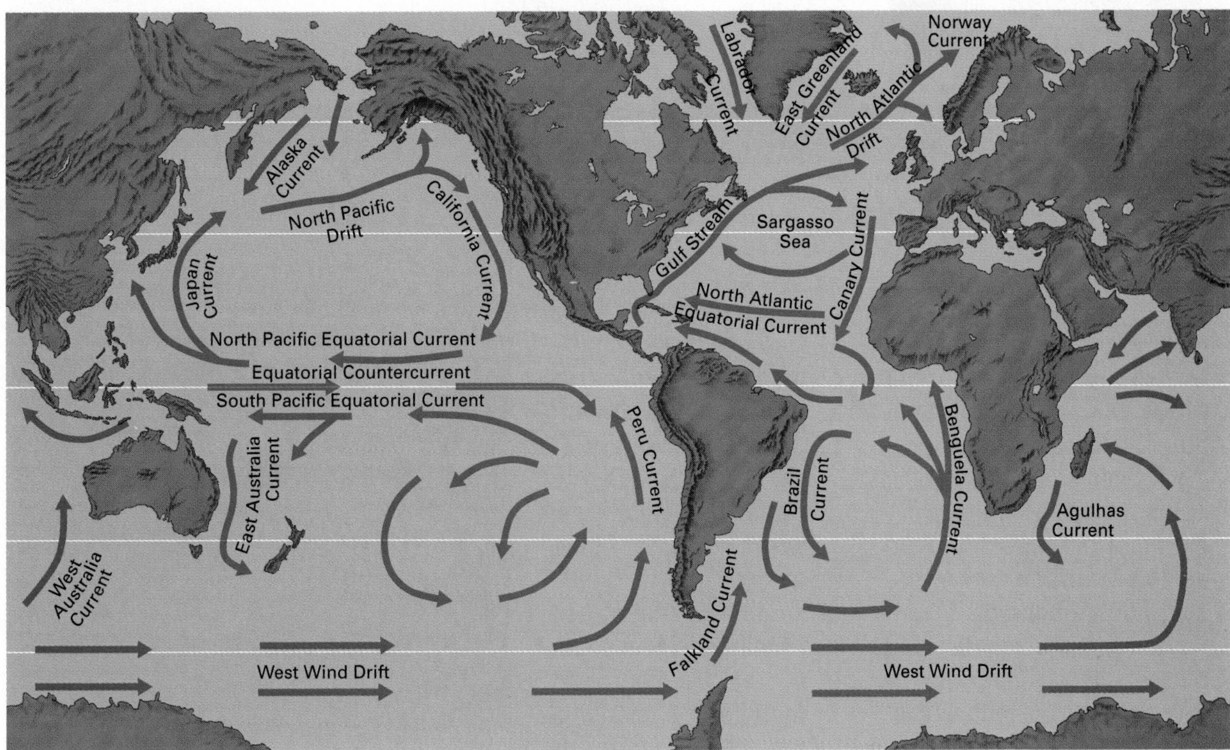

dense fog often results. South of Greenland, the Gulf Stream widens and decreases in speed until it becomes a vast, slow-moving warm current known as the *North Atlantic Drift*. A **drift** is a weak current. As the North Atlantic Drift approaches western Europe, it splits. One part becomes the Norway Current, which flows northward along the coast of Norway and keeps that coast ice-free all year. The other part is deflected southward and becomes the cool Canary Current, which eventually warms and rejoins the North Atlantic Equatorial Current.

Figure 22–2. This map shows the major surface currents of the oceans of the world. Warm-water currents are shown in red; cold-water currents are shown in blue.

As you can see in Figure 22–2, the Gulf Stream, the North Atlantic Drift, the Canary Current, and the North Atlantic Equatorial Current form the North Atlantic Gyre. At the center of this gyre lies an area called the *Sargasso Sea,* a vast area of relatively calm, warm water. Light winds blow over the sea, and great quantities of brown seaweed called *sargassum* float on its surface. The pattern of currents around the Sargasso Sea concentrates all kinds of floating debris. Even orange peels and plastic cups have been found there.

The pattern of currents in the North Pacific is similar to that in the North Atlantic. The warm Japan Current, the Pacific equivalent of the Gulf Stream, flows northward along the east coast of Asia. These waters then flow toward North America as the North Pacific Drift, eventually flowing southward as the cool California Current, which flows along the California coast.

Currents in the Southern Hemisphere

In the Southern Hemisphere, surface currents also move in gyres. However, the currents in southern oceans move counterclockwise.

In the most southerly regions of the Atlantic, Pacific, and Indian oceans, constant westward winds produce the *West Wind Drift*. It is the world's largest current. Since no landmasses interfere with the

EARTHBEAT

Shifting Currents

The surface currents in the Indian Ocean are divided into two circular patterns, or gyres. One gyre circulates north of the equator, and the other circlulates south of the equator. The southern gyre flows in a counterclockwise direction, as do all surface currents south of the equator.

North of the equator, major currents flow in clockwise gyres. However, the currents of the northern Indian Ocean change direction in response to the direction of strong seasonal winds called *monsoons.* The summer monsoon carries humid tropical ocean air into Asia and produces the heavy rains farmers there have come to depend on. The dry winter monsoon blows from the Eurasian continent and produces little precipitation.

From April to October, the wet monsoon blowing from the southwest drives the surface waters of the northern Indian Ocean in a clockwise gyre. The water flows eastward toward the southern coast of Asia, forming the Monsoon Current. This current then flows east along the coast of Indonesia, turns south, and joins the South Equatorial Current flowing westward.

From November to March, the southwest monsoon is replaced by a northeast wind. This dry monsoon changes the direction of the Monsoon Current, driving it westward along the Asian coast toward Africa.

At the equator, the current turns eastward, flowing parallel to, but in the opposite direction of, the South Equatorial Current. The eastward flow of water is called the Equatorial Countercurrent.

How do you think early explorers used these currents to travel back and forth to the Far East?

movement of this current, it completely circles Antarctica and flows across all three major oceans.

The surface currents in the Indian Ocean follow two patterns. In the southern part of the Indian Ocean, the currents follow a circular, counterclockwise gyre. In the northern part of the Indian Ocean, the currents are governed by winds called *monsoons,* which change direction with the seasons.

Deep Currents

In addition to the wind-driven surface currents, the ocean has cold, dense currents that flow very slowly, deep beneath its surface. These deep currents generally move much more slowly than do the surface currents. They are produced as cold, dense water of the polar regions sinks and flows beneath warmer ocean water toward the equator.

The movement of polar waters is a result of density differences. When water is cooled, it contracts, and the water molecules move closer together. This contraction makes the water more dense, and as a result, it sinks. When water is warmed, it expands and the water molecules move farther apart. Because the warm water is less dense, it remains above the cold water.

The higher density of polar waters is also a result of an increase in the salinity of the water. Salinity may increase in polar regions where water is frozen in icebergs and pack ice because when polar water freezes, most of the salt remains in the unfrozen water. Dense and more saline polar water sinks and forms a deep current that flows beneath less-dense ocean water. There is little mixing between the two layers. The deep-current layer rises only when winds blow surface water aside, causing the deep water to rise toward the surface as an upwelling.

Antarctic Bottom Water

The temperature of the water near Antarctica is close to the freezing point of ocean water, $-2°C$. The salinity is also high––35‰. These two factors make the water off the coast of Antarctica the densest and coldest ocean water in the world. This dense, cold water sinks to the ocean bottom and very, very slowly moves northward. It forms a deep-water current called *Antarctic Bottom Water.* The Antarctic Bottom Water moves along the ocean bottom for thousands of kilometers, reaching into the northern oceans to a latitude of approximately 40°N. This dense, cold water takes several hundred years to make the trip.

North Atlantic Deep Water

Oceanographers who study the deep currents in the Atlantic Ocean have discovered a general pattern. In the region of the North Atlantic just south of Greenland, the water is exceptionally cold and has a high salinity. This cold, salty water sinks and forms a deep current

Figure 22–3. The very dense and highly saline Antarctic Bottom Water travels beneath less-dense North Atlantic Deep Water.

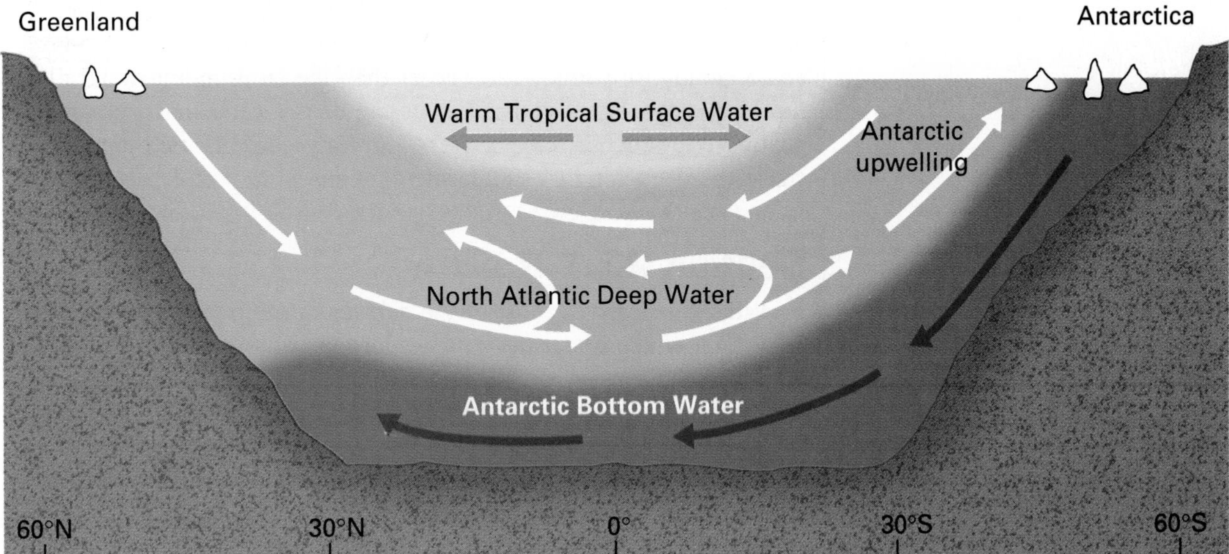

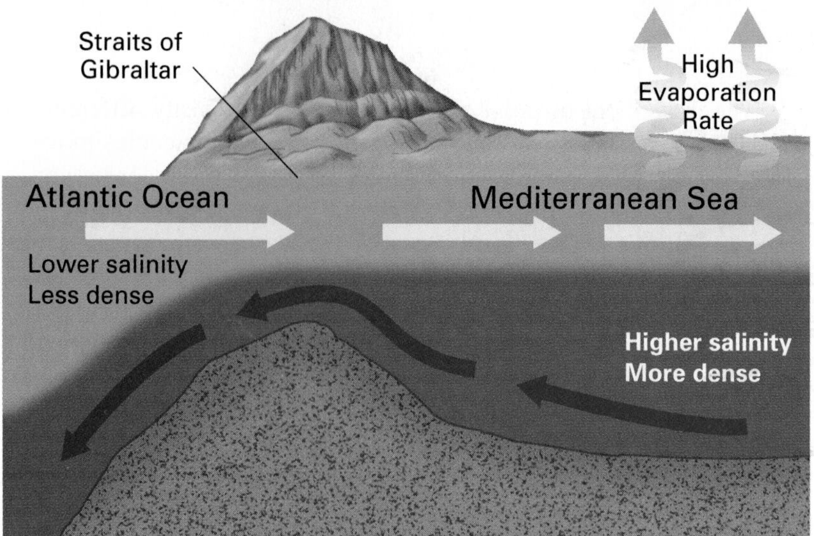

Figure 22–4. The highly saline dense water of the Mediterranean Sea forms a deep current as it flows over a high area in the straits of Gibraltar and into the less dense Atlantic Ocean.

that moves southward beneath the northward-flowing Gulf Stream. Most deep currents flow in a direction opposite that of the surface currents flowing above them. Near the equator this deep current divides. One part begins to rise, reverse direction, and flow northward again, while the rest of the current continues to flow southward toward the Antarctic. This portion of the North Atlantic Deep Water flows over the colder, denser Antarctic Bottom Water.

Deep Atlantic currents also exist near the Mediterranean Sea. Every summer, an increase in evaporation and a decrease in rainfall make the water of the Mediterranean more saline and thus denser. This denser water sinks and flows out through the straits of Gibraltar and into the Atlantic Ocean, creating a deep current. In turn, surface water from the Atlantic, which is less saline and thus less dense, flows into the Mediterranean Sea.

Turbidity Currents

A *turbidity current* is a strong current caused by an underwater landslide. Turbidity currents occur when large masses of sediment that have accumulated along a continental shelf or continental slope suddenly break loose and slide downward. The landslide mixes the nearby water with sediment. The sediment causes the water to become cloudy, or turbid, and denser than the surrounding water. The turbidity current gets under way as the dense water mass moves beneath the less dense, clear water. Turbidity currents are thought to cause the deepening of submarine canyons.

Section 22.1 Review

1. What force drives most of the surface currents of the oceans?
2. Which winds affect the surface currents on either side of the equator?
3. Which winds affect the surface currents in the middle latitudes of both hemispheres?
4. What two factors affect the density of ocean water and cause deep currents?
5. Explain the role of density in turbidity currents.

22.2 Ocean Waves

Section Objectives

- **Describe the formation of waves and the factors that affect wave size.**
- **Discuss the interaction of the shore and the waves.**

If you visit the seashore, you can observe the most obvious movement of the ocean—its **waves.** A wave is the periodic up-and-down movement of water. Waves are a way of transferring energy. As you can see in Figure 22–5, a wave has two basic parts—a **crest** and a **trough.** The crest is the highest point of a wave. The trough is the lowest point between two crests. Scientists study these two features to determine other wave characteristics.

The **wave height** is the vertical distance between the crest and the trough of a wave. The *wavelength* is the horizontal distance between two consecutive crests or between two consecutive troughs. The **wave period** is the time it takes for one complete wavelength to pass a given point. The period of most ocean waves ranges from 2 to 10 seconds.

The speed at which a wave moves is calculated by dividing the wave's wavelength by its period. This relationship is expressed by a simple formula:

$$\text{Wave speed} = \frac{\text{wavelength}}{\text{period}}$$

If a wave has a wavelength of 216 m and a period of 12 s, what is its speed?

Wave Energy

If there were no winds, the surface of the ocean would be almost as smooth as glass. Even the slightest breeze causes ripples in the ocean as a result of friction between the moving air and the water. The wind pushes directly against the side of a ripple. As the ripple receives more energy from the wind, it may grow into a wave. Although waves can be generated in different ways, the main source of wave energy is the wind. Therefore the longer the wind blows from the same direction, the more energy is transferred from the wind to the water. The more energy that is transferred, the larger the wave becomes. However, the wind seldom blows in one direction for very long.

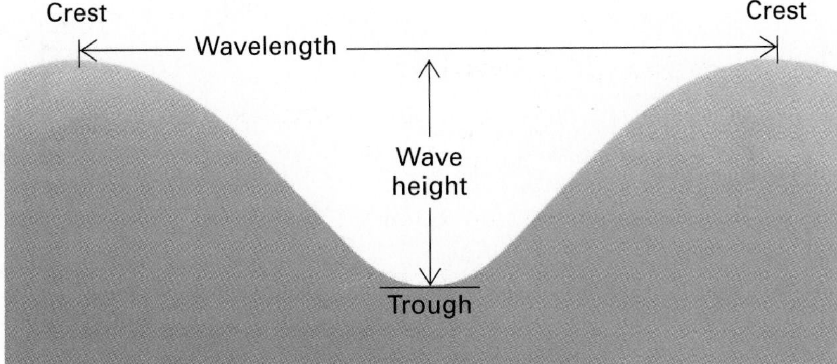

Figure 22–5. The vertical distance between the crest and the trough of a wave is the wave height. The distance from the crest of one wave to the crest of the following wave is the wavelength.

Usually the surface of the ocean is broken by many small waves moving in a number of directions in response to shifting winds. Larger waves, because of their large surface area, receive more energy from the wind than smaller waves do. Thus, larger waves tend to grow larger, and smaller waves quickly die out. One of a group of long, rolling waves that are all the same size is called a **swell.** Swells move in groups, one wave following the other. Swells that finally reach the shore may have been formed thousands of kilometers out in the ocean.

Water Movement in a Wave

Although the energy of a wave moves forward from water molecule to water molecule, the water itself moves very little. This surprising fact can be demonstrated by observing the movement of a cork floating on the water as a wave passes. The cork moves slightly forward as the wave approaches and then falls back an almost equal distance as the wave passes.

> **INVESTIGATE!**
>
> *To learn more about wave motion, try the In-Depth Investigation on pages 446–447.*

As a wave moves across the surface of the ocean, only the energy of the wave moves forward. The water does not move forward. The water particles within the wave move in a circular motion. During a single wave period, each water particle moves in one complete circle. At the end of the wave period, a circling water particle ends up almost exactly where it started.

As a wave passes a given point, the circle traced by a water particle on the ocean surface has a diameter that is equal to the height of the wave. Because waves receive their energy from wind pushing against the surface of the ocean, the energy received decreases as the depth of the water increases. As a result, water at various depths receives varying amounts of energy. Thus, the diameter of the circle that a particle moves in decreases with increasing depth, as shown in Figure 22–6. At a depth of about one half the wavelength, the circular motion is very slight. Below this point, there is almost no circular motion of water particles.

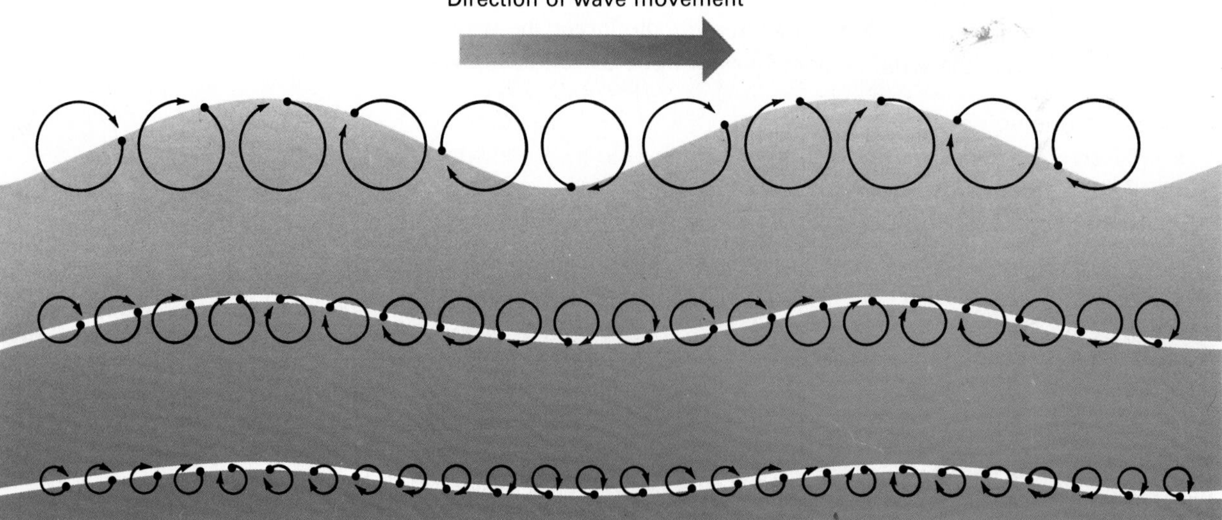

Figure 22–6. As a wave moves across the surface of the ocean, the water particles within the wave move in circular paths. The diameter of the circular path decreases with depth. Below a depth equal to one-half wavelength almost no circular motion occurs.

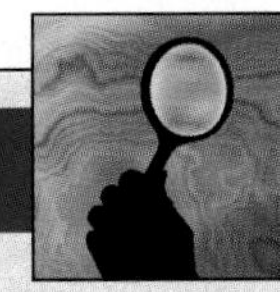

SMALL-SCALE INVESTIGATION

Waves

As a wave passes through water, the form of the wave moves across the surface. However, a wave does not carry water along with its motion. You can demonstrate this principle of wave movement with some common objects.

Materials

sink or rectangular pan (at least 40 cm × 30 cm × 10 cm), water, ruler, cork stopper, adhesive tape, teaspoon

Procedure

1. Fill the pan with water to a depth of 7 cm. Float the cork near the center of the pan. On each side of the pan, mark the location of the cork with a small piece of tape, as shown in the photo above.
2. Hold the spoon in the water, and slowly and carefully move it up and down in the water.
3. Continue to make waves in a slow, regular pattern. Observe the movement of the cork. Sketch what you have observed.
4. Remove the cork from the pan. Again use the spoon to make waves. This time, pay careful attention to creating a strong, steady series of waves. Remove the spoon from the pan.

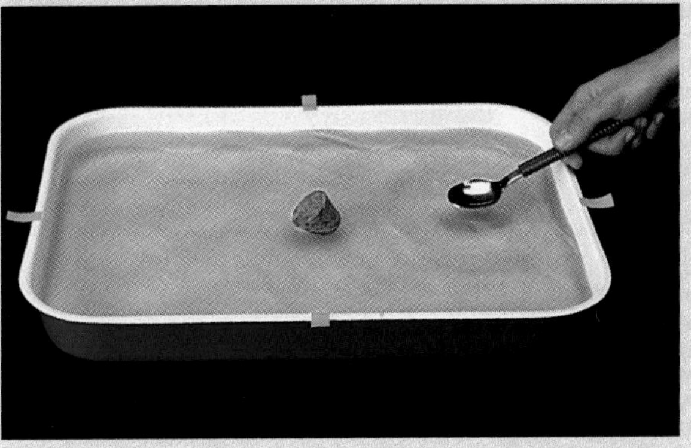

Observe what happens when the waves reach the edges of the pan. Write down or sketch what you observed.

Analysis and Conclusions

1. Describe the motion of the cork when a wave passes.
2. How does the cork move relative to the tape on the side of the pan? Explain your answer.
3. When a wave breaks on the shore of a beach, it is clear that the water is carried forward with the form of the wave. Based on your observations in Step 3, does this contradict your model? Explain.

Wave Size

Three factors determine the size of a wave. They are the speed of the wind, the length of time the wind blows, and the **fetch** of a wave. The fetch is the distance that the wind can blow across open water. Very large waves are produced by strong, steady winds blowing across a long fetch. Such conditions are most likely to occur during a storm. During a storm the steady high winds cause some waves to gather enough energy to reach great size. Strong, gusty winds produce choppy water with waves of various heights and lengths that may come from various directions. On calm days small, smooth waves move steadily across the surface.

The size of a wave will increase to only a certain height-to-length ratio before it collapses. The largest wave ever observed during a severe storm was 34 m high. Also, during storms with high wind speeds, the crest is blown off the wave, forming **whitecaps.**

Waves and the Shore

In shallow water near the shore, the bottom of the wave touches the ocean floor. A wave touches the ocean bottom where the depth of the water is about one-half the wavelength. Contact with the ocean floor creates friction, causing the wave to slow and eventually break.

Breakers

The height of a wave changes as the wave approaches the shore. The water involved in the up-and-down motion of a wave extends to a depth of one-half wavelength. As the wave moves into shallow water, the bottom of the wave is slowed by friction. The top of the wave, however, continues to move at its original speed. The top of the wave gets farther and farther ahead of the bottom. Finally, the top of the wave topples over and forms a **breaker,** a foamy mass of water that washes onto the shore. The height of the wave when it topples over is one to two times the height of the original wave. Figure 22–7 illustrates the formation of breakers.

Shells left on the shore by the retreating wave mark the farthest point reached by the wave. Breaking waves scrape sediments off the bottom and move them along the shoreline of sandy beaches. The waves also erode the cliffs of rocky shorelines.

The size and force of breakers are determined by the original wave height, wavelength, and the steepness of the ocean floor close to shore. If the ocean floor is quite steep, the height of the wave increases rapidly, and the wave breaks with great force. Breakers of this type are common along the Pacific Coast. If the shore slopes gently, the wave rises slowly. The wave spills forward with a rolling motion that continues as the wave advances up the shore. This type of breaker is common along the Atlantic Coast.

Figure 22–7. Breakers begin to form where the depth of the water is about equal to one-half wavelength. Notice in the diagram (below) that wave height increases and wavelength decreases as waves approach the shore.

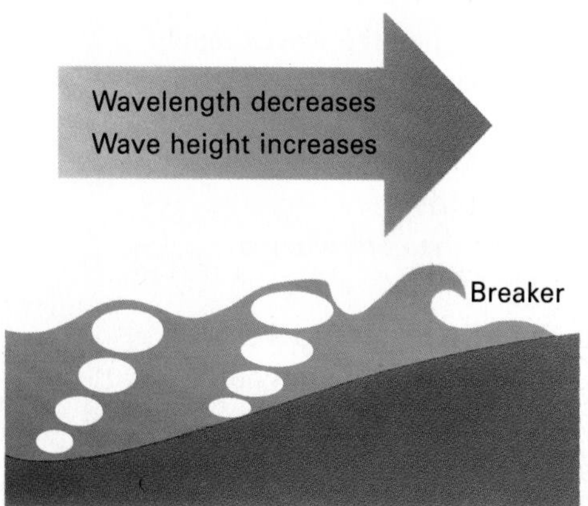

Figure 22–8. Waves strike the shore head-on as a result of refraction. Notice the waves approaching the shore at an angle. These waves bend as they draw closer to the shore.

Refraction

Most waves approach the shore at an angle. When a wave reaches shallow water, however, it bends. This bending is called **refraction.** Refraction occurs when the part of the wave nearest the shore strikes the shallow bottom and slows down. The end of the wave that is still in deeper water moves forward at its regular speed. The wave gradually bends toward the beach and strikes the shore head-on. The original direction of the wave does not affect this bending. Note the changing direction of the waves in Figure 22–8.

Undertows and Rip Currents

Water carried onto a beach by breaking waves is pulled back into deeper water by an irregular current called an **undertow.** An undertow is seldom very strong, and only along shorelines with steep drop-offs do they create problems for swimmers.

Some people confuse the generally weak undertow with the more dangerous **rip current.** Swift rip currents form when water from large breakers returns to the ocean through channels in underwater sandbars, which parallel the beach. Rip currents flow perpendicular to shore and may be strong enough to quickly carry a swimmer out into deep water.

Rip currents may be dangerous enough to force beaches to be closed to swimmers. Their presence can usually be detected by a gap in a line of breakers. A rip current may also be detected by turbid water, where sand has been stirred up by the current.

Longshore Currents

Sand bars are formed by another type of current called a *longshore current*. A longshore current forms as waves approach the beach at an angle. Longshore currents flow parallel to the shore. Great quantities of sand are carried by longshore currents. If there is a bay or inlet along the shoreline where waves refract, sand will be deposited as the energy of the waves decrease. The shoreline becomes straighter as the longshore current deposits sandbars, spits, tombolos, or barrier islands.

Tsunamis

The most destructive waves in the ocean, *tsunamis* are not powered by the wind. Tsunamis are seismic sea waves. Most are caused by earthquakes on the ocean floor, but some can be caused by volcanic eruptions and underwater landslides. Tsunamis are sometimes called *tidal waves,* which is misleading because tsunamis have no connection with tides.

Tsunamis have a long wavelength. In deep water, the wave height of a tsunami is usually less than 1 m, but the wavelength may be as long as 240 km. A tsunami commonly has a wave period of about 15 min and a speed of about 725 km/hr (as fast as a jet airliner). Because the wave height of a tsunami is so low in the open ocean, the tsunami cannot be felt by people aboard ships.

A tsunami has a tremendous amount of energy. The entire depth of the water is involved in the wave motion. All the destructive energy of this mass of water is delivered against the shore.

Near the shore the height of the tsunami greatly increases as its speed decreases. The wave may reach a height of 30 to 40 m as the water piles up. The arrival of a tsunami may be signaled by the sudden pulling back of the water along the shore. This occurs when the trough of the tsunami arrives before the crest. If the crest arrives first, a sudden, rapid rise in the water level occurs.

When the trough of a tsunami arrives first, the rapid withdrawal of water is followed by the arrival of the crest. The water then retreats with the arrival of the next trough. Many lives have been lost when people have misinterpreted such a retreat as the end of the tsunami. In Crescent City, California, following the Alaskan earthquake of 1964, the first wave of a tsunami crested 4 m above low tide. Three progressively smaller waves followed. People then returned to the shore, thinking the tsunami was over. However, a fifth wave struck in conjunction with the high tide, cresting at 6 m above low tide and killing 12 people.

The tsunami triggered by an earthquake in Chile in 1960 caused destruction all across the Pacific Ocean. It struck the coast of South America, then Hawaii, and crossed 17,000 km of ocean to strike Japan. Because tsunamis are most common in the seismically active Pacific Ocean, a tsunami warning center is located in Hawaii.

Section 22.2 Review

1. Explain how wave height, wavelength, and wave period are determined from the crest and trough of a wave.
2. What factors determine the size of a wave?
3. Why do incoming waves bend toward the beach until they strike the shore head-on?
4. What is the cause of most tsunamis?
5. Explain why waves slow down in shallow water.

22.3 Tides

Section Objectives

- **Describe the various forces that cause tides and that affect tidal patterns.**
- **Compare the power of currents in the open ocean with that of tidal currents in narrow bays.**

The daily changes in the level of the ocean surface are called **tides.** The force that causes the rise and fall of tides along coastlines was first identified in the late 1600's by Sir Isaac Newton. According to Newton's law of gravitation, the gravitational pull of the moon on the earth and its waters the major cause of tides.

The gravitational pull of the moon is strongest on the side of the earth nearest the moon. As a result, the ocean on the side of the earth facing the moon bulges slightly, causing a high tide within the area of the bulge. At the same time, another tidal bulge is created on the opposite side of the earth. To understand how this tide forms, you must realize that the earth and moon revolve around a common center of gravity, much like two skaters spinning in a circle while holding hands. As the earth and moon revolve, the outward force generated by their spinning causes the ocean on the side of the earth opposite the moon to bulge from the earth's surface, similar to the way a skater's scarf would swing out as the pair of skaters spins.

Low tides are formed halfway between the two high tides. Low tides form because ocean water flows away from the areas of low tide toward the areas of high tide.

Behavior of Tides

If the earth and the moon did not move, tides would always occur in the same locations. However, the earth rotates on its axis once every 24 hours. The moon moves through about 1/27 of its orbit around the earth in that same 24 hours. Thus, as the earth rotates, all areas of the ocean pass under the moon every 24 hr and 50 min.

The earth rotates from west to east. Therefore, the tidal bulges appear to move westward around the earth. Because there are two tidal bulges, most locations in the ocean have two high tides and two low tides daily. The difference between the levels of the high

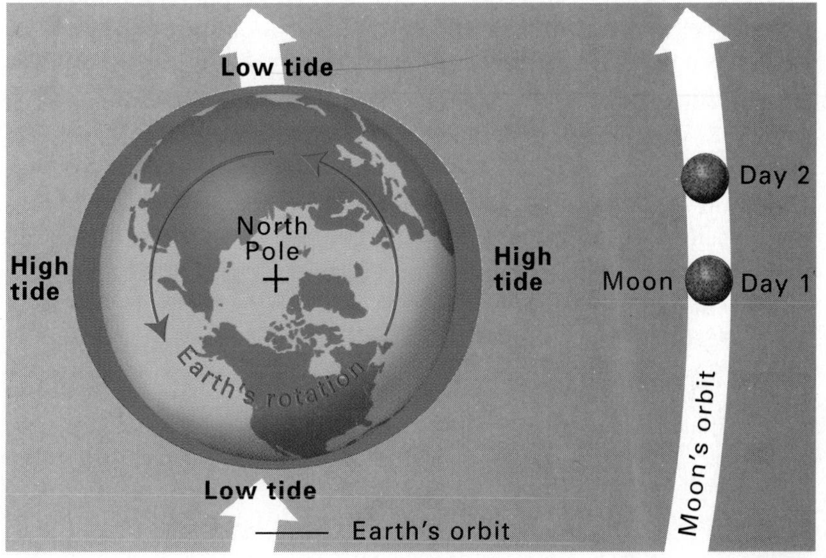

Figure 22–9. High tides form as the result of the gravitational pull of the moon and its motion around the earth. Most locations in the ocean have two high tides and two low tides daily.

tides and the low tides at a specific location is called the **tidal range.** The tidal range can vary widely from place to place. Because the moon rises about 50 min. later each day, the cycle of high and low tides also occurs 50 min. later.

The sun's gravitational pull also has an effect on the tidal range. However, the ability of the sun to affect tides is only one-half that of the moon. Nevertheless, the sun's gravitational pull can strengthen or weaken the gravitational pull of the moon on the tides.

During the periods of the new moon and the full moon, the earth, the sun, and the moon are all aligned. The combined gravitational pull of the sun and the moon acts on the earth, resulting in higher high tides and lower low tides. Therefore, the daily tidal range is largest at these times. During these periods, which occur twice a month, tides are called **spring tides.**

During the first- and third-quarter phases of the moon, the moon and the sun are at right angles to each other in relation to the earth. For that reason, the gravitational forces of the sun and moon do not act together; instead, they work against each other. As a result, the daily tidal range is small. Tides that occur during this time are called **neap tides.** How often do neap tides occur?

Science & Technology

Energy From the Ocean

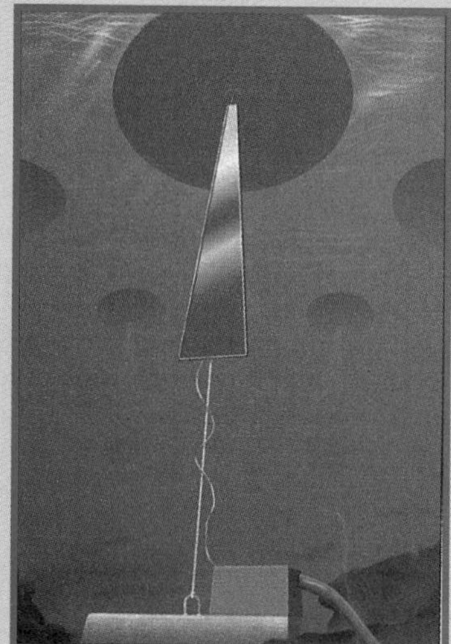

▲ This wave-powered system uses waves to stretch piezoelectric panels attached to buoys.

The search for renewable energy sources has led to some ingenious ways of tapping the ocean for energy. So far, scientists have found ways to generate electricity from three ocean features: waves, tides, and heat content.

Wave Energy

Scientists have designed a variety of machines for converting wave motion into electricity. Some wave-power systems use buoys and the up-and-down motion of waves to run submerged pumps that force water or air through turbine generators. Other systems consist of large, open-bottomed air chambers that act like floating pumps. Incoming waves raise the water level in the chamber and compress the air inside. The pressurized air then flows through the system's turbines. One innovative design, pictured at left, eliminates the need for turbine generators by using piezoelectric plastic, a material that produces electricity when it is stretched.

Tidal Energy

Generating electricity from tides is similar in principle to generating hydroelectricity. However, instead of using the flow of rivers to turn the turbines, tidal power plants rely on the flow of tides. A dam built across a bay or inlet captures water at high tide. As the tide ebbs, the stored

As the tidal bulges move around the earth, friction with the ocean floor slows the earth's rotation slightly. Scientists estimate that the average length of a day has increased by 10.8 min. since the dinosaurs became extinct 65 million years ago.

Tidal Variations

Although the global ocean is one body of water, the landmasses and irregularities in the ocean floor divide the ocean into several ocean basins. Tidal patterns are greatly influenced by the size, shape, depth, and location of the ocean basin in which the tides occur.

Along the Atlantic Coast of the United States, tides follow a semidiurnal, or twice daily, pattern. There are two high tides and two low tides each day, with a fairly regular tidal range. Along the shore of the Gulf of Mexico, however, tides follow a diurnal, or daily, pattern. There is only one high tide and one low tide each day. Along the Pacific Coast, the tides follow a mixed pattern, with an irregular tidal range. Pacific Coast tides often occur in the following sequence: a very high tide followed by a very low tide, then a lower high tide, followed by a higher low tide.

water is released through the dam's turbines to drive the electrical generators.

Thermal Energy

The process of producing electricity from the heat contained in ocean water is called Ocean Thermal Energy Conversion (*OTEC*). OTEC plants rely on steam to turn their turbine generators. The steam is created by pumping warm surface water through a vacuum chamber, which allows the water to vaporize well below its normal boiling point. The steam then turns special low-pressure turbines.

The Future of Ocean Power

Ocean power is not yet as cost-effective as producing electricity from fuel-burning or nuclear power plants. However, as ocean power technology becomes more reliable and more efficient, the ocean's renewable energy may become more affordable. Although scientists cannot predict all the environmental effects of ocean power, using ocean power would help to reduce the toxic waste and emissions associated with conventional power plants.

What do all three types of ocean power have in common?

▲ OTEC plant near Kailua Kona

Tidal patterns are also affected by **tidal oscillations** (AHSS-uh-LAY-shunz), slow, rocking motions of ocean water that occur as the tidal bulges move around the ocean basins. Along straight coastlines and in the open ocean, the effects of tidal oscillations are not very obvious. In some enclosed seas, such as the Baltic and the Mediterranean, tidal oscillations reduce the effects of the tidal bulges. As a result, these seas have a very small tidal range. In small basins and narrow bays off the major ocean basins, however, tidal oscillations may amplify the effects of the tidal bulges.

Tidal oscillations that produce the world's greatest tidal range are found in the narrow V-shaped Bay of Fundy, at the end of the Gulf of Maine in Nova Scotia. At times, the tidal range in the Bay of Fundy exceeds 15 m. The water oscillates and floods the bay at high tide. Then, as the water moves back to the other end of the Gulf of Maine, the resulting low tide almost completely drains the bay. See Figure 22–10 for two photographs that illustrate this effect. Other areas of the world with great tidal ranges include the English Channel, the waters off southern Alaska, and the Gulf of California in Mexico.

Figure 22-10. The difference between low tide (top) and high tide six hours later (bottom) in Canada's Bay of Fundy is nearly 15 m.

Figure 22–11. A tidal bore moves swiftly up Great Britain's River Severn. At times, the wall of water can reach 2 m in height, providing a wave on which surfers can ride.

Tidal Currents

As the ocean water rises and falls with the tides, it flows toward and away from the coast. This movement of the water is called a **tidal current.** When the tidal current flows toward the coast, it is called *flood tide*. When the tidal current flows toward the ocean, it is called *ebb tide*. The time period between flood tide and ebb tide with no tidal currents is called *slack water*.

Tidal currents are not found in the open ocean. Tidal currents are strongest between two adjacent coastal regions that have large differences in the height of the tides. In bays and along other narrow coastlines, tides may create rapid currents. Rugged coasts with large tidal ranges—such as the coasts of southeast Alaska, British Columbia, Canada, and Brittany, in northwestern France, for example—may have tidal currents that attain speeds of 20 km/hr. Ship captains who attempt to navigate in such waters must take these rapid currents into account.

When a river enters the ocean through a long bay, the tide may rush into the river. The surge of water that rushes upstream, called a **tidal bore,** is shown in Figure 22–11. The water can rise as high as 5 m. In some cases, the tidal bore rushes upstream in the form of a large wave, which eventually loses energy. The tidal bore that pours into the Amazon River resembles a small, moving waterfall as it moves rapidly upstream. The tidal bores in the River Severn in England travel almost 20 km/hr and reach up to 33 km inland.

Section 22.3 Review

1. What causes tides?
2. What are spring tides? What are neap tides?
3. Where might tidal currents be of concern to ships that are approaching the land?
4. What factors influence the tidal patterns in a particular location?

Chapter 22 Review

Key Terms

- breaker (436)
- Coriolis effect (428)
- crest (433)
- current (427)
- deep current (427)
- drift (429)
- fetch (435)
- Gulf Stream (428)
- gyre (428)
- neap tide (440)
- refraction (437)
- rip current (437)
- spring tide (440)
- surface current (427)
- swell (434)
- tidal bore (443)
- tidal current (443)
- tidal oscillation (442)
- tidal range (440)
- tide (439)
- trough (433)
- undertow (437)
- wave (433)
- wave height (433)
- wave period (433)
- whitecap (435)

Key Concepts

Most of the surface currents of the ocean are the result of global wind belts. The surface current patterns are, in turn, affected by the presence of landmasses and the rotation of the earth. **See page 427.**

Deep currents are produced as dense water near the North and South poles sinks and moves toward the equator beneath less-dense water. **See page 430**.

The wind is the primary source of wave energy. **See page 433.**

The size of a wave and the shape of the ocean floor close to shore determine the impact that a wave has on the shoreline. **See page 436.**

The gravitational effects of the moon and, to a lesser extent, the sun cause tides. **See page 439.**

Tidal currents are not found in the open ocean but may create rapid currents in narrow bays. **See page 443.**

Review

On your own paper, write the letter of the term that best completes each of the following statements.

1. The waters in the ocean move in giant streams called
   a. currents. b. westerlies. c. waves. d. tides.
2. The effect of the earth's rotation on winds and ocean currents is called the
   a. neap-tide effect. b. refraction effect.
   c. Coriolis effect. d. tsunami effect.
3. Two warm-water currents, the North and South Equatorial currents, flow
   a. northward. b. westward. c. southward. d. eastward.
4. The warm, swift current that flows up the east coast of the United States and then into the North Atlantic is the
   a. Japan Current. b. California Current.
   c. Labrador Current. d. Gulf Stream.
5. A weak current is known as a
   a. swell. b. drift. c. tide. d. crest.
6. The vast area of relatively still, warm water located in the middle of the North Atlantic is referred to as the
   a. Sargasso Sea. b. North Atlantic Drift.
   c. North Equatorial Current. d. Gulf Stream.
7. A strong, turbid current that is caused by an underwater landslide is called
   a. a surface current. b. an equatorial current.
   c. a turbidity current. d. a tsunami.

8. Deep currents are the result of
   a. the Coriolis effect.
   b. changes in the density of ocean water.
   c. the trade winds.
   d. neap tides.
9. The periodic up-and-down movement of water is a
   a. current. b. fetch. c. breaker. d. wave.
10. The highest point of a wave is its
   a. trough. b. crest. c. period. d. length.
11. The time it takes for one complete wavelength to pass a given point is called the
   a. wave speed. b. wave height. c. trough. d. wave period.
12. The length of open water across which the wind blows is the
   a. trough. b. sargassum. c. fetch. d. wave period.
13. When the faster-moving top part of a wave topples over the slower bottom part, the result is a foamy mass of water called a
   a. whitecap. b. rip current. c. breaker d. crest.
14. Sand bars may be formed by
   a. undertows. b. rip currents.
   c. tsunamis. d. longshore currents.
15. The daily changes in the elevation of the ocean surface are called
   a. waves. b. tides. c. swells. d. breakers.
16. The difference in the level between high tide and low tide is the
   a. neap tide. b. spring tide.
   c. tidal range. d. tidal oscillation.
17. The movement of water toward and away from the coasts due to tidal forces is called a
   a. tidal bore. b. tidal current.
   c. tidal range. d. tidal oscillation.

Critical Thinking

On your own paper, write answers to the following questions.

1. During winter in the northern Indian Ocean, winds called *monsoons* blow in a direction opposite to the direction that they blow during summer. What effect do these winds have on the surface currents?
2. Suppose that a retaining wall is built along a shoreline. What will happen to waves as they pass over the retaining wall?
3. Imagine that you are fishing from a small boat anchored off the shore of the Gulf of Mexico. You are lulled to sleep by the gently rocking boat but wake up to find your boat on wet sand. What happened?
4. Along the shoreline of a particular bay, high tide rises to about 20 m and low tide recedes to about 10 m. The bay is wide, with a large opening into the ocean. Should a tidal-power plant be built across the bay? Explain your answer.

Application

1. What could you do to create a deep current in a pan of slightly salty water? Explain how you would go about this.
2. From the deck of a stationary ship on the open ocean you notice that at least 15 minutes pass between the rise of one wave and the next. Can you assume that the sea is very calm? Explain why or why not.

Extension

1. Obtain a timetable of daily high and low tides for a shore near where you live. If the times are irregular, find out what causes the irregularities. Report your findings. If the times are regular, report on the factors that influence this regularity.
2. Find out how much electricity is provided by the La Rance, France, tidal power plant project. Find out what impact the project has had on the environment of the area. Report on your findings.
3. Concept map ? ? ? Construct a **concept map** using any of the new terms listed on the previous page and those in Chapter 16 to show how ocean currents, waves, and tides are related to the various features of coastlines.

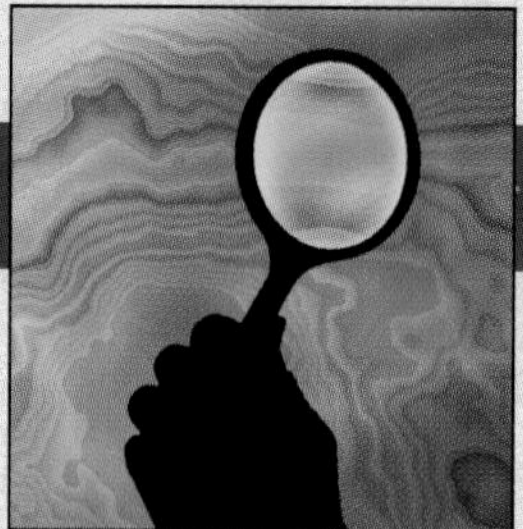

IN-DEPTH INVESTIGATION

Wave Motion

Materials

- graph paper
- sheet of paper (2 × 1 m)
- 3 colored pens or pencils
- marker
- thin rope (2.5 m)
- 2 cloth ties (about 50 cm in length)
- meter stick

Introduction

The source of wave movement in water is energy, which is generated primarily from wind. To a person standing on a beach and watching the waves come in to the shore, it may not be easy to understand that wind is the primary energy source of waves. Waves appear to move forward because of the actual movement of the water. However, only the energy of the waves moves forward; the water moves very little.

In this investigation, you will simulate wave motion to observe how energy generates wave motion in water. You will also observe the properties of waves.

Prelab Preparation

1. Review Section 22.2, pages 433–438.
2. Study the graph below. Identify the wave crests and wave troughs. The wavelength is the distance between two successive crests or troughs. What is the wavelength? What is the wave height?

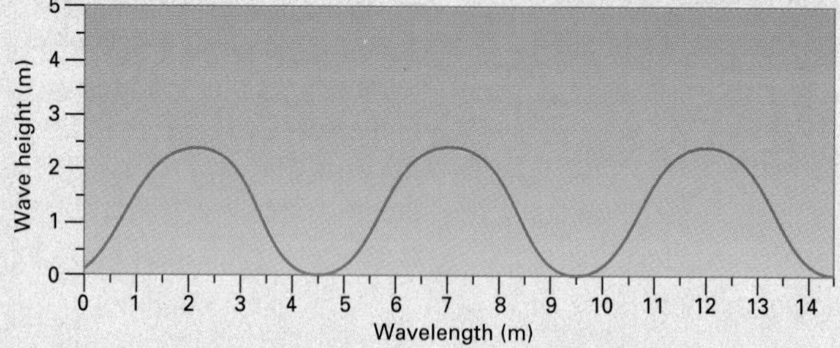

3. The wave period is the time required for two successive crests or troughs to pass a certain point. Wave speed can be calculated using the following formula:

$$\text{Wave speed} = \frac{\text{wavelength}}{\text{wave period}}$$

What is the speed of waves shown in the graph if the wave period is five seconds?

Procedure

1. Work with two partners for this investigation. Begin by tying one end of the rope to the leg of a chair or table. Be sure it is tied securely.
2. On the large sheet of paper, use the meter stick to draw a grid similar to the one shown in the illustration on the next page. Draw and label the grid using the measurements shown.
3. Place the sheet of paper on the floor and line up the rope along the 2-m line of the grid.

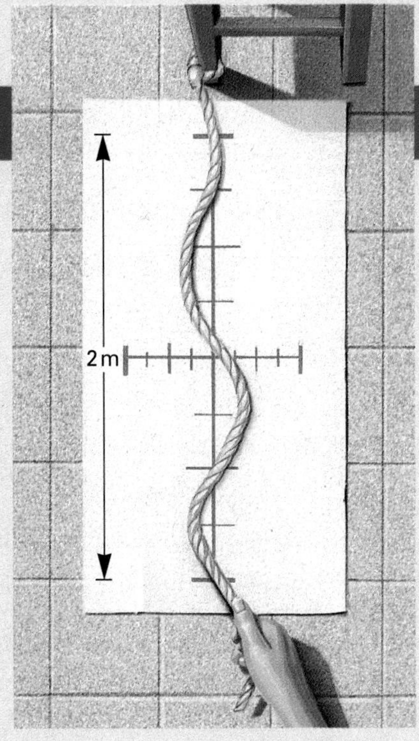

4. To make waves, you or one of your partners must move the free end of the rope from side to side as shown in the illustration. *Note: Be sure to maintain a constant motion with the rope.*
5. While one person is moving the rope, another person must mark on the paper where a crest of a wave hits. The third group member must mark on the paper where a trough of a wave hits.
6. On the graph paper, make a graph with wavelength (in meters) on the horizontal axis and wave height (in meters) on the vertical axis. Plot a wave that represents the wave that you observed in Step 5. Begin at 0 in the middle of the graph, and plot the height and the length. Indicate the direction of the wave's motion.
7. Continue the investigation by changing the motion of the rope. Move the rope at a fast speed. *Be sure to move the free end of the rope the same distance from side to side.*
8. As soon as a constant motion has been established, repeat Steps 5 and 6. Using a different colored pen or pencil, plot one of the waves on your graph.
9. Next, generate very small waves. Repeat Steps 5 and 6 using a third color.
10. On your graph, label a crest and a trough on each of the waves that you have plotted. What are the wavelengths of the three waves you plotted? What are their wave heights? Using the formula in the prelab preparation, calculate the wave speeds of the three waves represented on the graph if each wave period is six seconds.
11. Tie the two pieces of cloth around the middle of the rope about 15 cm apart. Set up another wave motion by moving the end of the rope. Observe and record the motion of the cloth ties compared with the motion of the waves.

Analysis and Conclusions

1. How do the wave motions differ on your graph? If these were real water waves, what might be the cause(s) of the different motions?
2. How is the action of the rope similar to wave movement in water?
3. How do the motions of the cloth ties differ from the wave's motion?
4. What do the motions of the cloth ties tell you about the wave movement in water?

Extension

Repeat the investigation using a 4-m rope. Construct a graph similar to your first graph, but extend the *x*-axis to plot a 4-m length. Observe and plot five waves of varying speeds and heights. Compare waves generated on a 2-m rope to those generated on a 4-m rope. Describe your results. Calculate wave speeds, using six seconds as the wave period for each wave that you plotted.

Mapping the Sea Floor From Space

In the late 1970's, oceanographers realized that underwater features such as seamounts and trenches affected the height of the water above them. Scientists observed that sea levels were higher above underwater mountain ranges, and lower above underwater lowlands and valleys. Such differences, which varied by several meters from place to place, were found to reflect differences in the gravitational pull of the ocean floor on the water above. And because military submarines are guided in part by changes in gravity, the United States Navy began a top-secret project to map these variations.

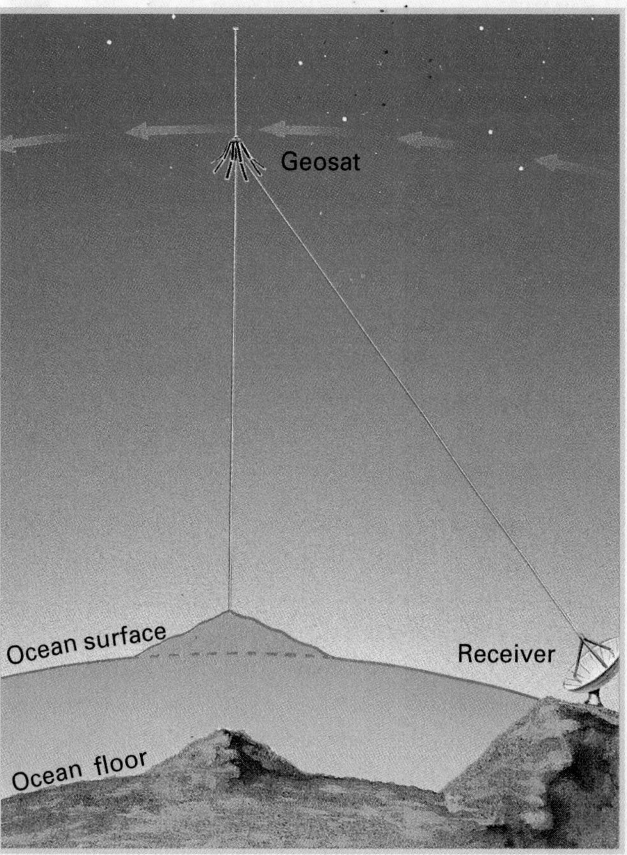

The Navy used a satellite, called *Geosat,* to chart the height of sea level all over the globe, generating huge amounts of data. The data were finally released to the public in 1995. Since that time, scientists have used this information to create what are probably the most important maps in the history of oceanography. These new maps of the ocean floor are so detailed that more than half of the seamounts they pinpoint were previously unknown.

1. Before these new maps became available, oceanographers used sonar to map the ocean floor. (See page 391.) Considering the fact that sonar equipment is carried aboard ships at sea, what major advantage is there to using a satellite to map the ocean floor?
2. Look at the map on the facing page. How does it support the theory of plate tectonics?
3. Compare the map on the facing page with the traditional map of the ocean floor on page 398. What features are similar? What features are different?
4. How do you think oceanographers will benefit from the new map?
5. How would you explain the deep valley that runs along the western edge of South America? (Look at the map on page 72 for a clue.)

Using radar pulses, a special satellite measures differences in the height of the ocean's surface, thereby indirectly determining the topography of the ocean floor.

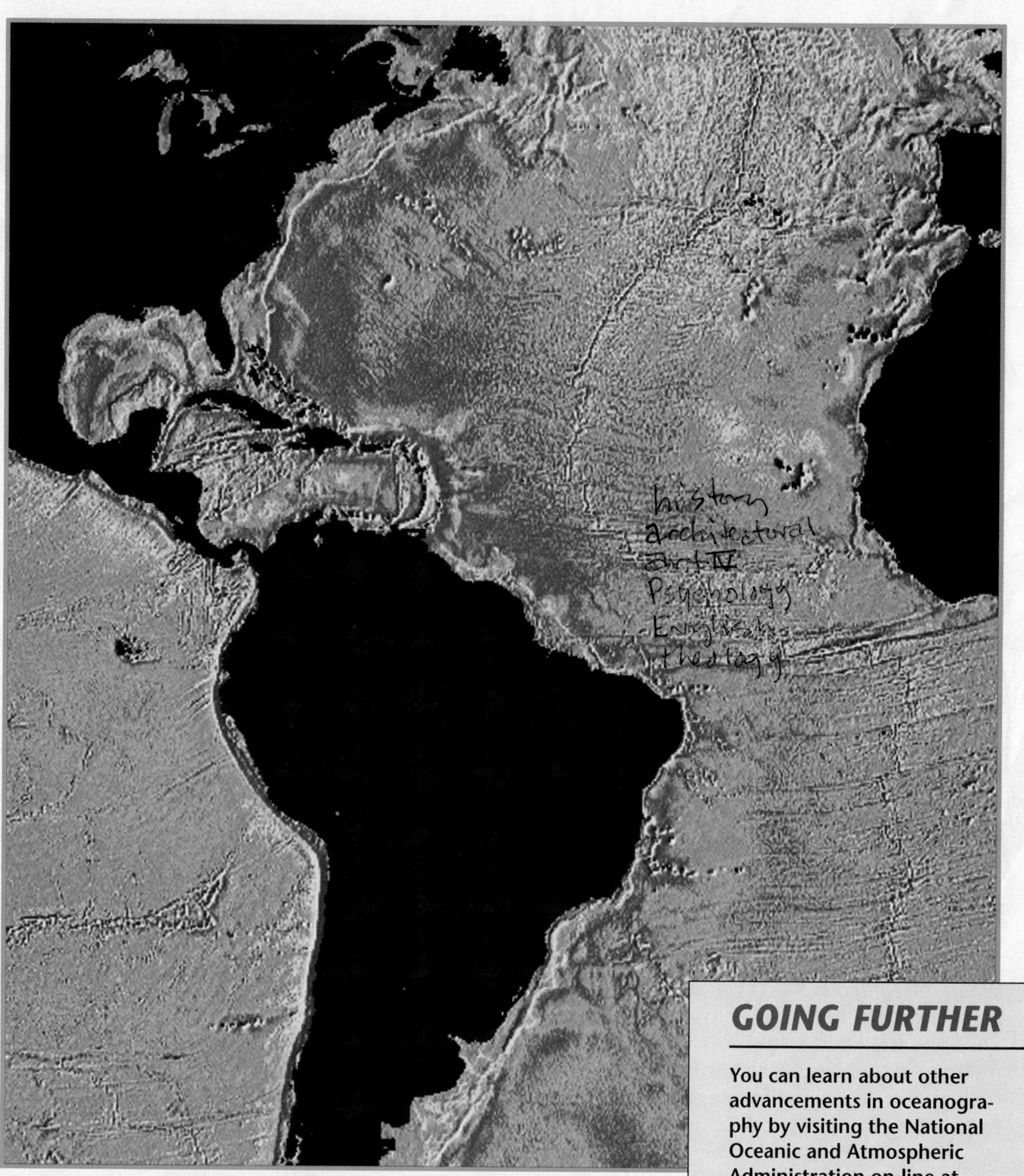

▲ **This map, generated by satellite measurements of the earth's surface gravity under the ocean, is much more detailed than conventional maps and shows parts of the ocean floor never before charted.**

GOING FURTHER

You can learn about other advancements in oceanography by visiting the National Oceanic and Atmospheric Administration on-line at www.ngdc.noaa.gov.

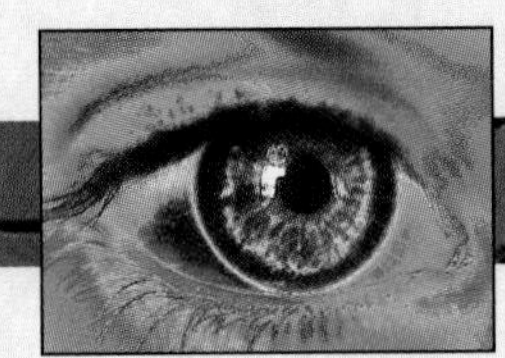

Eye on the Environment

A Harbor Makes a Comeback

Not so long ago, the harbor in Boston, Massachusetts, was known by many people as the most polluted harbor in the world. In recent years, however, citizens, scientists, and public officials have worked together to help Boston Harbor make one of the most impressive comebacks in history.

For many years, Boston Harbor was popularly known as the most polluted harbor in the world.

Polluted Since Colonial Times

When the United States declared its independence in 1776, the city of Boston was already a thriving metropolis. Residents dealt with their sewage as conveniently as possible—usually by dumping it directly into Boston Harbor. They hoped that the tides would carry the sewage out to sea, diluting it with enough water to render it harmless. Instead, much of the raw sewage remained in the harbor. For hundreds of years, it periodically washed up along the shore, sometimes causing outbreaks of cholera and typhoid among Boston residents. Over time, most of the species of fish and other marine life disappeared from the harbor.

In the late 1960's, a primary-treatment plant was built to remove some of the solid materials from the city's sewage before it was dumped into the harbor. Unfortunately, the plant could not meet the demands of the fast-growing city. About once a week, rainy weather caused the city's sewers to overflow. Even as late as the 1980's, millions of tons of sewage flowed directly into the harbor every year.

The state of the harbor became progressively worse. Beaches were closed, and the shell-fishing industry in Boston Harbor was shut down. Divers reported that a "gray, mayonnaise-like substance" covered the harbor's floor.

The Deer Island sewage-treatment facility is one of the largest of its kind in the world. The large, egglike tanks are used in the primary treatment of sewage.

Taking Action

Citizens of the Boston area had voiced concerns over pollution in the harbor for many years. The Clean Water Act of 1972 (CWA) required that sewage go through both primary and secondary treatment before being discharged into ocean waters. However, the facilities in Boston were not sufficient to do either one. Finally, in the mid 1980's, a federal judge found the state of Massachusetts guilty of violating the CWA. The state was placed under a court order to comply with the CWA by 1999. The Massachusetts Water Resources Authority was formed to oversee the massive cleanup. An 11-year, $6 billion plan was put into effect to drastically reduce the pollution in Boston Harbor.

Deer Island, which lies in the outskirts of the harbor, was chosen as the site for a new sewage-treatment complex. The complex includes both primary and secondary treatment plants that subject the sewage to a series of steps. Inorganic solids undergo a process in which they either settle to the bottom or float to the top of the sewage and are then removed. In addition, bacteria are used to remove organic material from the sewage. The solid "sludge" is removed, and, eventually, the remaining wastewater is disinfected before it is discharged into the ocean. Today, the primary and secondary plants that process the sewage from the Boston area are among the largest of their kind in the world.

A Success Story

The Deer Island complex is considered a grand success. By 1995, the beaches were reopened and were significantly safer. Plants, fish, and other organisms had returned to the harbor, and the ocean floor was much improved. Today, a variety of annual recreational events celebrate the harbor's amazing comeback.

The task of protecting the harbor is far from over, however. Officials continue to monitor the sewage-treatment process. They have conducted studies to make sure that the disinfected wastewater poses a minimal threat to local ecosystems. In addition, officials are aware that the long-term health of Boston Harbor will depend in part on the health of the surrounding watersheds. Eventually, the cleanup project will need to be expanded to include these areas.

Think About It

Why would the health of Boston Harbor depend on the health of the surrounding watersheds?

Extension: Find out how sewage from your area is treated and disposed of. How long has the current system been in effect?

UNIT 7

Atmospheric Forces

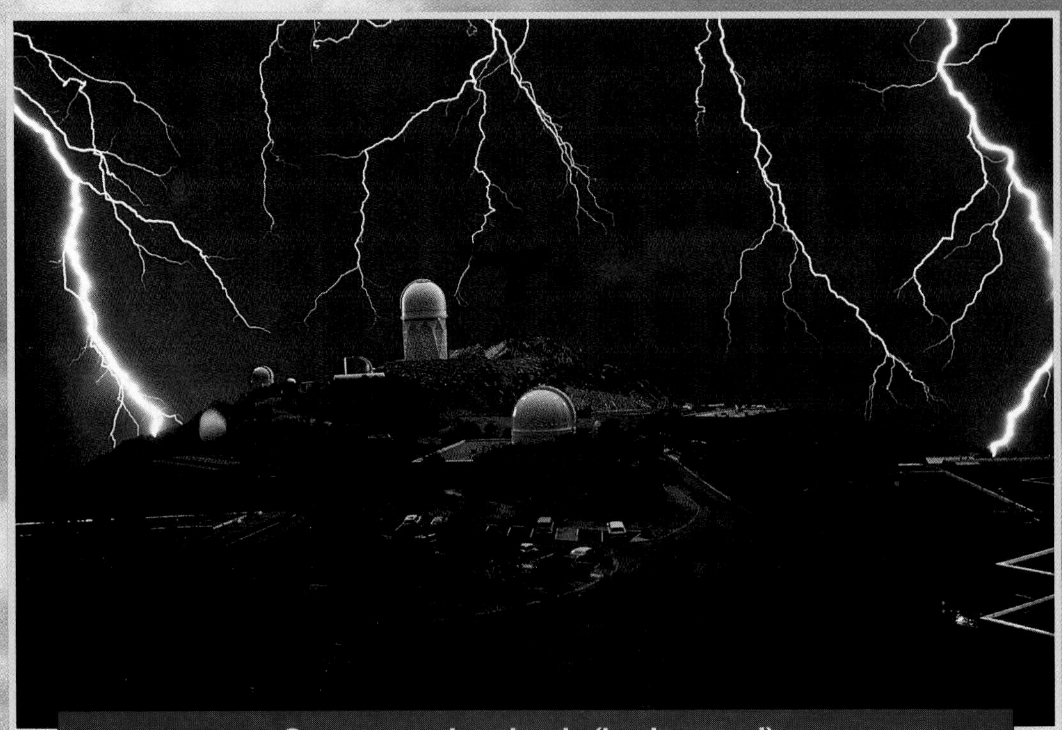

Stratocumulus clouds (background);
Lightning at Kit Peak National Observatory, Arizona (inset)

Introducing Unit Seven

Climate is not governed by political boundaries. Regardless of where you live, weather events happening halfway around the world today are likely to affect you within a week.

Fitting together the cause-and-effect patterns that influence the atmosphere of the entire globe is an important area of research and study today. From small matters of convenience—Shall I take along my umbrella?—to the largest concerns of human life and safety, weather is a significant, dramatic, and often violent factor in everyone's life.

In this unit, you will explore the global atmospheric forces that bring everyone in the world the same important phenomenon—their local weather.

Kenneth H Bergman

Kenneth Bergman
Climatologist
Climate Analysis Center
National Weather Service

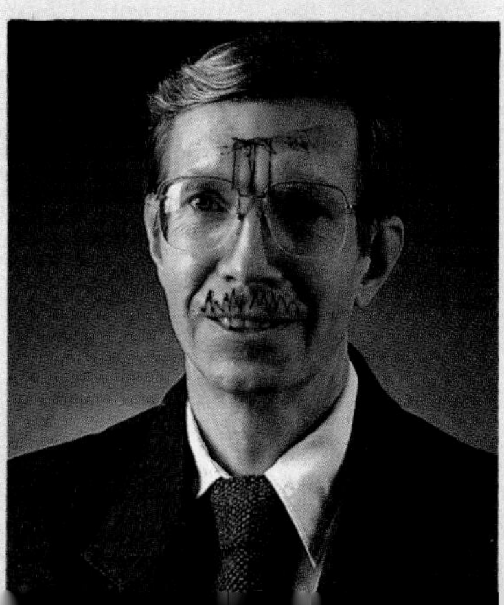

In this UNIT

Chapter 23

The Atmosphere

Have you ever thought of yourself as an organism that lives in a sea of gases, as a fish lives in water? People live within a narrow zone where the earth's atmosphere meets the land. The conditions within this thin layer have a major effect on peoples' lives.

In this chapter, you will learn how natural forces and processes act on this narrow zone to influence conditions such as air pressure, wind, and air pollution.

Chapter Outline

◀ **The wind provides energy to move this windsurfer over the water.**

23.1 Characteristics of the Atmosphere

Section Objectives

- **Discuss the composition of the earth's atmosphere.**
- **Explain how two types of barometers work.**
- **Describe the layers of the atmosphere.**
- **Identify the weather conditions that increase the effects of air pollution.**

The atmosphere is a layer of gases and tiny particles that surrounds the earth. The atmosphere influences almost every living thing. You breathe the gases of the atmosphere. The temperature of the atmosphere determines how you dress and what many of your daily activities will be.

As explained in Chapter 1, the study of the atmosphere is called *meteorology.* Meteorologists study all the characteristics of the atmosphere. They also study **weather** and **climate.** Weather is the general condition of the atmosphere at a particular time and place; it includes temperature, air movements, and moisture content. The general weather conditions over many years is climate. Meteorology, however, is not limited to weather and climate changes. Meteorologists may specialize in a particular area such as agriculture, aviation, forestry, or health.

Composition of the Atmosphere

The atmosphere, or air, is a mixture of chemical elements and compounds. As Figure 23–1 shows, the most abundant elements in air are the gases nitrogen, oxygen, and argon. The most abundant compounds in air are the gases carbon dioxide and water vapor. However, the graph does not include water vapor because the amount of water vapor varies greatly under different conditions.

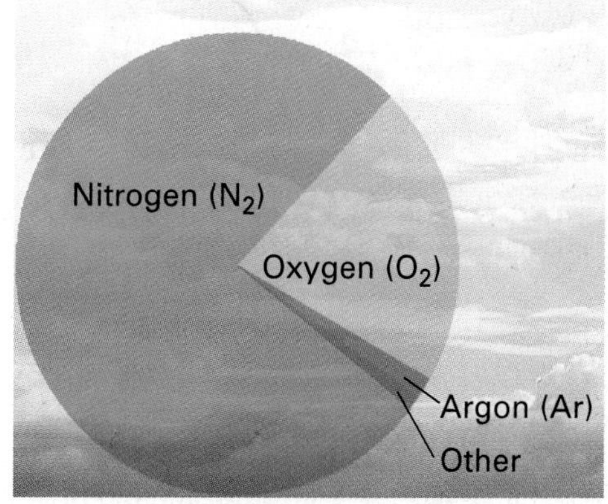

Figure 23–1. The pie chart shows the composition of dry air by volume. Other gases included in the 0.1% are neon (Ne), helium (He), xenon (Xe), and carbon dioxide (CO_2).

Water vapor is added to air by evaporation. Most water vapor comes from the oceans, but some also comes from lakes, ponds, streams, and the soil. Plants give off water vapor during transpiration, one of their life processes. At the same time that water vapor is being added to the atmosphere by evaporation, it is being removed by condensation and precipitation. The percentage of water vapor in air varies, depending on factors such as time of day, location, and season. Moist air may contain as much as 4 percent water vapor. Dry air has less than 1 percent water vapor. Because the amount of water vapor in air is variable, the composition of air is usually given for dry air.

Another important substance in the upper atmosphere is a form of oxygen called **ozone,** although it is present only in very small amounts. Oxygen (O_2) has two atoms per molecule, whereas ozone has three. What would the chemical formula for ozone be?

The ozone in the upper atmosphere is important because it protects the earth's inhabitants by absorbing harmful ultraviolet rays of the sun. Without ozone in the atmosphere, people would be severely sunburned by ultraviolet rays. Unfortunately, a number of human activities damage the ozone layer. Gases from aerosol spray cans, coolant used in refrigerators and air conditioning, and hydrocarbons from the burning of supersonic aircraft fuel break down ozone.

In addition to gases, the atmosphere usually contains various kinds of tiny solid particles, which are called *atmospheric dust*. Atmospheric dust includes mineral particles lifted from soil by winds, ash from fires, volcanic dust, and microscopic organisms. It may also include particles from meteors that have vaporized. When tiny drops of ocean water are tossed into the air as sea spray, the drops evaporate. Left behind in the air are tiny crystals of salt, another component of atmospheric dust. Large dust particles remain in the atmosphere only briefly because their weight causes them to fall. However, many particles that are so small they cannot be seen remain suspended in the atmosphere for months or years.

All over the earth and up to an altitude of about 100 km, the composition of dry air is nearly the same. Although nitrogen and oxygen are always being added to, as well as removed from, the atmosphere, the relative amounts of these gases do not change significantly.

Oxygen in the Atmosphere

The amount of oxygen in the atmosphere is the result of natural processes that maintain the chemical balance of the atmosphere. Animals, bacteria, and plants remove oxygen from the air as part of their life processes. Forest fires, the burning of fuels, and the weathering of some rocks also use up oxygen. Living things, burning, and weathering would quickly use up most atmospheric oxygen if it were not for various processes that add oxygen to air. Land and ocean plants produce large quantities of oxygen in daylight. During photosynthesis plants use sunlight, water, and carbon dioxide to produce their food. Oxygen is released as a product of photosynthesis. The amount of oxygen produced by plants each year equals that consumed by all processes. Thus, the oxygen content of the air is in a state of balance. It has not changed significantly over hundreds or even thousands of years.

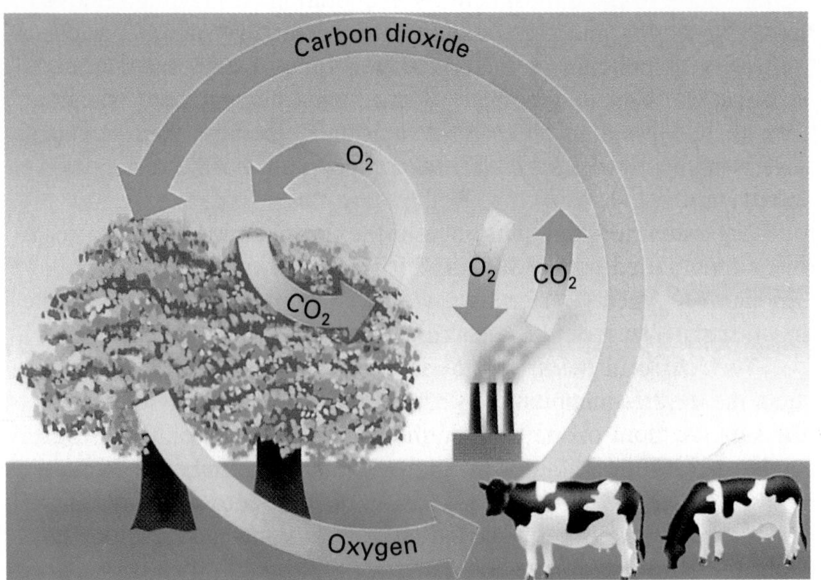

Figure 23–2. This illustration shows the consumption and production of oxygen (O_2) and carbon dioxide (CO_2) in the atmosphere. The oxygen–carbon dioxide cycle maintains a stable amount of oxygen and carbon dioxide in the atmosphere.

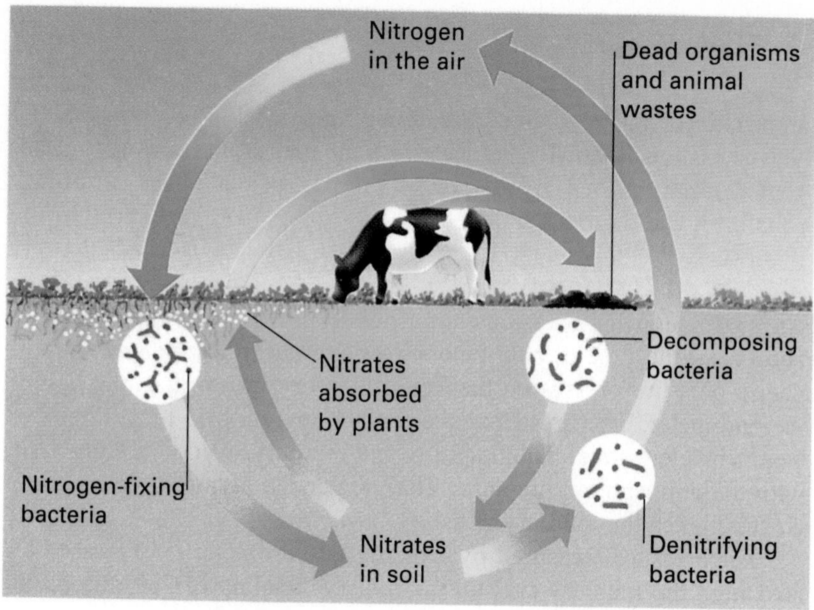

Figure 23–3. The nitrogen cycle maintains a stable amount of nitrogen in the air.

Nitrogen in the Atmosphere

The amount of nitrogen in the atmosphere is maintained through the **nitrogen cycle.** During the cycle, nitrogen moves from air to the soil, to plants and animals, and back again to the air. As Figure 23–3 shows, nitrogen is removed from the air mainly by the action of nitrogen-fixing bacteria. These microscopic organisms live in the soil and on the roots of certain plants. The bacteria chemically change nitrogen from the air into nitrogen compounds, which are vital to the growth of all plants. When animals eat plants, nitrogen compounds enter their bodies. These compounds are then returned to the soil through animal excretions or by the decay of dead organisms. In the soil, the processes involved in denitrification release nitrogen and return it to the atmosphere. A similar nitrogen cycle takes place among water-dwelling plants and animals.

Atmospheric Pressure

Gravity pulls the gases of the atmosphere toward the earth's surface and holds them there. Due to the pull of gravity, 99 percent of the total mass of atmospheric gases is found within 32 km of the earth's surface. The remaining 1 percent extends upward for hundreds of kilometers but gets increasingly thinner at high altitudes. In other words, there is less air at these higher altitudes.

A 1-cm^2 column of air that reaches from sea level to the top of the atmosphere has a mass of 1.03 kg and exerts a force of 10.1 newtons (N). In other words, at sea level, on every square centimeter of the earth's surface, the atmosphere presses down with a force of 10.1 N, which is about 2 lb. The ratio of the weight of the air to the area of the surface on which it presses is called **atmospheric pressure.** Since there is less air at higher altitudes, there is less weight pressing down on surfaces at those altitudes. Thus, the atmospheric pressure is lower at higher altitudes.

You have probably experienced the effects of changes in air pressure. If you have ever driven through the mountains or flown in an airplane, you may have had a popping sensation in your ears. This sensation is due to the decreased air pressure on the outside of your eardrum. When the air pressure on both sides of the eardrum is equalized, the popping stops.

Mercurial Barometer

An instrument that measures atmospheric pressure is called a **barometer.** One type of barometer is the *mercurial barometer.* Atmospheric pressure presses on the liquid mercury in a well at the base of the barometer. The pressure squeezes the mercury up to a certain height inside a tube. The height of the mercury inside the tube varies with the atmospheric pressure. The greater the atmospheric pressure is, the higher the mercury rises.

Air pressure measured by a mercurial barometer is expressed by how high the mercury rises in the tube. A reading of 760 mm on the

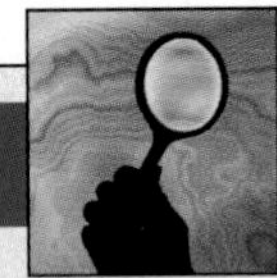

SMALL-SCALE INVESTIGATION

Barometric Pressure

A barometer measures changes in atmospheric pressure. You can construct a simple aneroid barometer with some common objects.

Materials

plastic wrap; coffee can with a diameter of about 10 cm, open at one end; rubber band; drinking straw, about 10 cm long; masking or adhesive tape; cardboard, 10 cm wide and at least 8 cm taller than the can; metric ruler

Procedure

1. Refer to the photo in making your barometer.
2. Secure plastic wrap tightly over the open end of the can with the rubber band.
3. Tape one end of the straw onto the plastic wrap near the center, as shown in the photo.
4. Fold the cardboard so that it stands upright and extends at least 3 cm above the top of the can.
5. Place the cardboard so that the free end of the straw just touches the front of the cardboard. Mark an *X* where the straw touches.
6. Draw three horizontal lines on the cardboard: level with the *X*, 2 cm above the *X*, and 2 cm below the *X*.

7. Position the cardboard so that the straw touches the *X* again. Tape the base of the cardboard in place so that it does not shift.
8. Observe the level of the straw at least once a day over a 5-day period. Record any changes.

Analysis and Conclusions

1. What factors affect how your model works? Explain your answer.
2. What does an upward movement of the straw indicate? a downward movement?
3. Compare your results with the barometric pressures listed in your local newspaper. What kind of weather was associated with high pressure? with low pressure?

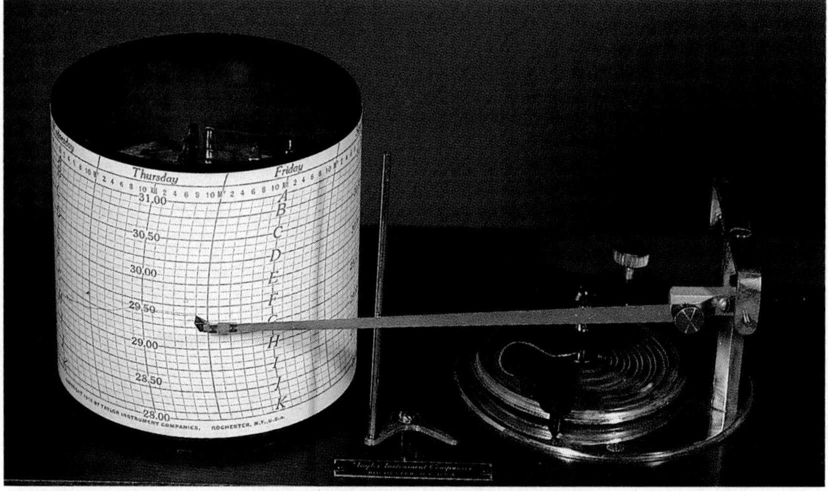

Figure 23–4. Aneroid barometers measure air pressure by the compression and expansion of a sealed metal box. When an aneroid barometer is constructed to keep a continuous record of air pressure, such as the one shown here, it is called a *barograph*.

barometer is called **standard atmospheric pressure** and indicates average atmospheric pressure at sea level. Standard atmospheric pressure, 760 mm or 29.92 inches of mercury, is sometimes referred to as *one atmosphere*.

Official weather maps use another measurement of air pressure, called *millibars* (mb). One millibar is equal to about 0.001 of standard atmospheric pressure.

Aneroid Barometer

The type of barometer most commonly used today does not contain mercury and is called an *aneroid barometer*. The word *aneroid* means "without liquid." Inside an aneroid barometer is a sealed metal container from which the air has been removed. When the atmospheric pressure increases, the sides of the container bend inward. When the pressure decreases, the sides bulge out again. These changes are indicated by a moving pointer on a scale. The movement of the pointer along the scale is controlled by the changing shape of the container. The scale is usually marked to show the pressure either in millimeters or inches of mercury, or in millibars. Aneroid barometers can be constructed to keep a continuous record of atmospheric pressure.

An aneroid barometer can also measure altitude above sea level. When used for this purpose, it is called an *altimeter*. The scale then registers altitude instead of pressure. At high altitudes, the atmosphere is less dense and exerts less pressure. Thus a lowered pressure reading can be interpreted as an increased altitude reading. To accurately measure altitude, an altimeter must be corrected for local weather conditions.

Layers of the Atmosphere

As altitude increases, air pressure decreases rapidly. There are no sharp pressure changes that separate the atmosphere into layers. The atmosphere does, however, show distinct differences in temperature with increasing altitude. The temperature differences mainly result from the way solar energy is absorbed as it moves downward through the atmosphere. Based on temperature differences, scientists identify four layers of the atmosphere.

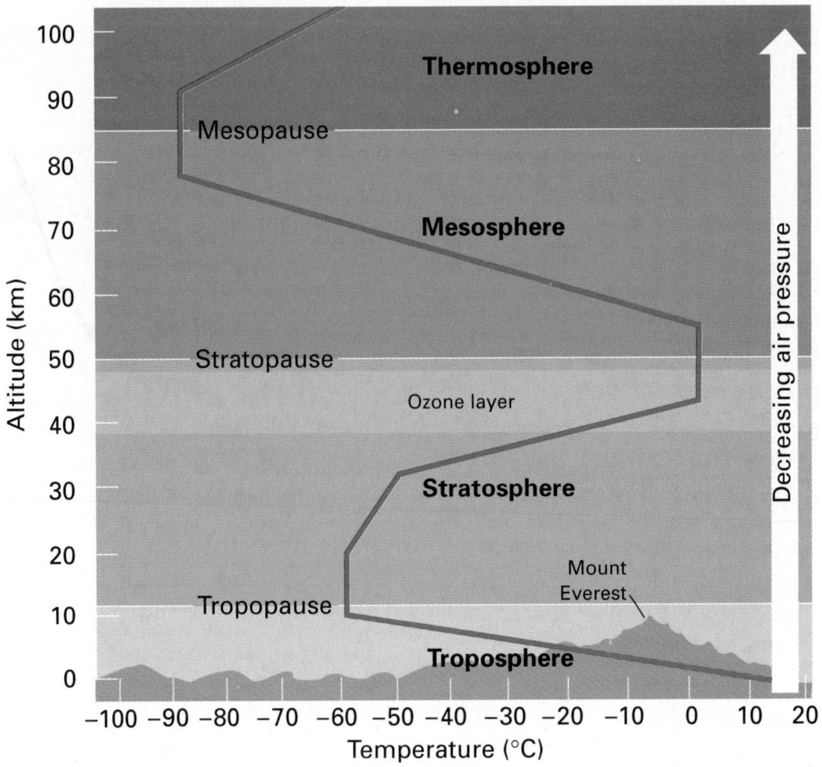

Figure 23–5. The red line indicates the temperature at various altitudes in the atmosphere.

The Troposphere

The atmospheric layer closest to the earth's surface is called the **troposphere.** The word *troposphere* comes from a Greek root meaning "change." The term aptly describes this atmospheric layer in which nearly all weather change occurs. Almost all of the water vapor and carbon dioxide in the atmosphere is found in the troposphere. Temperature within this layer decreases as altitude increases because there is an increase in distance from the warming effect of sunlight absorbed by the earth's surface. The temperature within the troposphere decreases at the average rate of 6.5°C/km. But at an average altitude of 10 km, the temperature stops decreasing. In this zone, called the **tropopause,** the temperature remains nearly constant. The tropopause is the upper boundary of the troposphere. The altitude of this boundary is not constant; it changes with latitude and with the season of the year. For example, at the equator the tropopause is found at an altitude of about 17 km. However, at the poles the altitude of the tropopause varies between 6 km and 8 km.

The Stratosphere

The layer of the atmosphere called the **stratosphere** extends upward from the tropopause to an altitude of 50 km. Almost all of the ozone in the atmosphere is concentrated in the stratosphere. At the base of the stratosphere, the temperature is about –60°C. In the upper stratosphere, the temperature begins to increase as the altitude increases. The air in the stratosphere gets warmer as a result of the direct absorption of solar energy by ozone.

The temperature of the ozone layer rises steadily to an altitude of about 50 km, where the stratosphere reaches its highest temperature. This high-temperature zone, called the **stratopause,** marks the upper boundary of the stratosphere.

The Mesosphere

Above the stratopause and extending to an altitude of about 80 km is the atmospheric layer called the **mesosphere.** In this layer, the temperature decreases as the altitude increases. In fact, the mesosphere is the coldest layer of the atmosphere, dropping to a temperature of −90°C. The upper boundary of the mesosphere, called the **mesopause,** is marked by a return to increasing temperatures.

The Thermosphere

In the atmospheric layer above the mesopause, called the **thermosphere,** the temperature increases steadily with altitude. In the thermosphere, nitrogen and oxygen atoms absorb solar energy. This process explains the high temperatures in the thermosphere. There are not enough data about temperature changes in the thermosphere to determine its upper boundary.

In the very thin air of the thermosphere, a thermometer cannot accurately measure the temperature. A thermometer measures the temperature of the particles, or the energy of the moving molecules, that strike it. Because the air in the thermosphere is so thin, the particles move rapidly but are very far apart. Therefore, they do not strike the thermometer often enough to produce an accurate temperature reading. Special instruments are needed to measure temperature accurately in the thermosphere. These instruments have recorded temperatures of more than 2,000°C in the thermosphere.

The lower region of the thermosphere, at an altitude of 80 km to 550 km, is often called the **ionosphere.** In the ionosphere solar rays absorbed by atmospheric gases cause the atoms of gas molecules to lose electrons and to produce ions and free electrons. The ionosphere gets its name from these ions. The ions and the free electrons are concentrated into four layers. The layers of free electrons can reflect radio waves back to the earth, as shown in Figure 23–6.

Above the ionosphere is the region where the earth's atmosphere blends into the almost-complete vacuum of interplanetary space. This zone of indefinite altitude, called the **exosphere,** extends for thousands of kilometers above the earth.

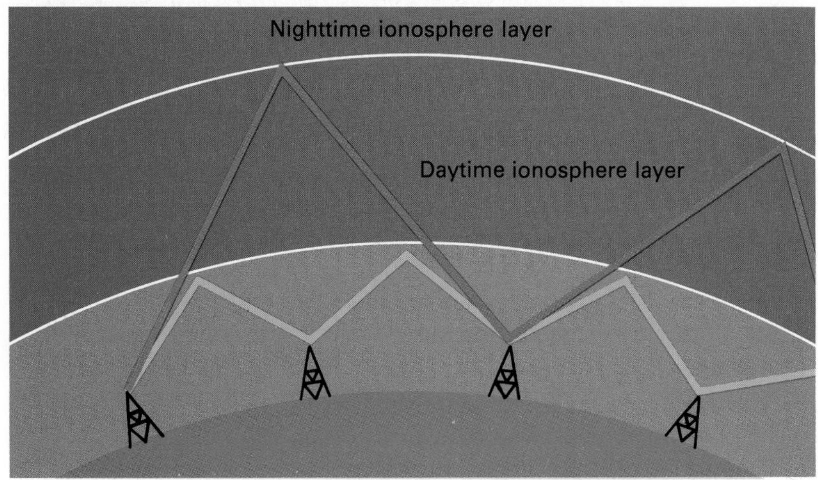

Figure 23–6. Radio waves can be transmitted around the world by reflecting them off of the ionosphere. At night the radio waves can travel farther because the lowest ion layer disappears and the waves are reflected off a higher ion layer.

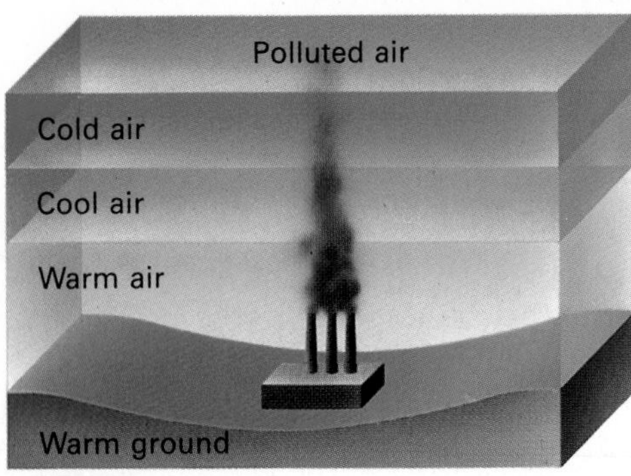

Normal

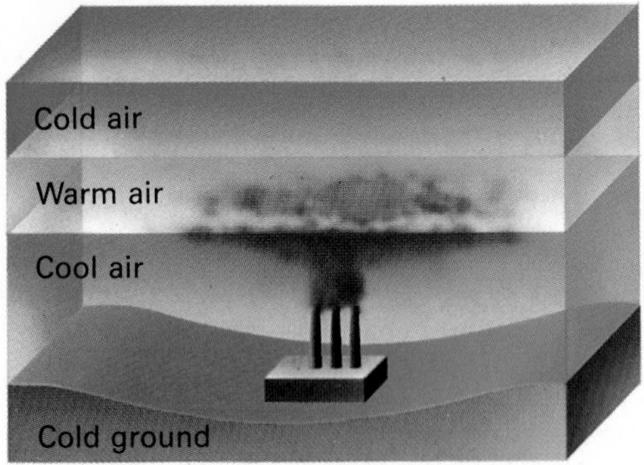

Inversion

Figure 32–7. Normal conditions are shown at top, and a temperature inversion is shown at bottom. During a temperature inversion, polluted cool air becomes trapped beneath a warm-air layer.

Air Pollution

Any substance in the atmosphere that is harmful to people, animals, plants, or property is an *air pollutant*. Many substances commonly found in air, such as sulfur dioxide, carbon monoxide, lead, and hydrocarbons, are known to be harmful when breathed by humans.

On a few occasions, severe air pollution has caused a large number of deaths. Less dramatic but perhaps more serious are the long-term effects on the health of people, especially children and the elderly, who breathe polluted air. Studies have shown that long-term exposure to pollution decreases one's ability to resist many illnesses.

The main source of air pollution is the burning of fossil fuels. Before automobiles became common, most air pollution resulted from the burning of coal by industry and in homes. Today most air pollution comes from the burning of coal and petroleum fuels. When these fuels burn, the sulfur that is released forms harmful sulfur dioxide gas. The operation of automobile engines produces several harmful substances, such as hydrocarbons, nitrogen oxides, carbon monoxide, and lead.

Acid precipitation is another harmful side effect of the burning of fossil fuels. Gases emitted by the burning of fossil fuels form acids when combined with water in the air. These acids fall to the earth as acid rain, mist, or snow. Over time, acid precipitation poisons fish, ruins soil, and kills crops and trees.

Air pollution can become an even more serious problem as a result of certain weather conditions. A common cause of periodic air pollution is the layering of warm air on top of cool air. Warm air, which is less dense, can trap the polluted cool air beneath it. Meteorologists refer to this condition as a **temperature inversion.** In some areas, topography may make air pollution even worse by keeping the polluted inversion layer from dispersing. Los Angeles, for example, is the site of frequent temperature inversions. Cool Pacific air becomes polluted, is trapped by a warm-air layer above it, and is prevented from moving by mountains that border the city. Under conditions in which air cannot circulate up and away from an area, trapped automobile exhausts produce *smog*. Smog is a general term for air pollution. It was named for a combination of smoke and fog.

Air pollution can be controlled only by preventing pollutants from being released into the atmosphere. International, federal, and local laws have been passed to reduce the amount of air pollutants produced by automobiles and industry.

Section 23.1 Review

1. What are the two most abundant elements in dry air?
2. What does a barometer measure?
3. In which layer of the atmosphere do weather changes occur?
4. Which industrial city—one on the Great Plains or one near the Rocky Mountains—would have fewer air-pollution incidents related to temperature inversions? Why?

23.2 Solar Energy and the Atmosphere

Section Objectives

- **Explain how radiant energy reaches the earth.**
- **Describe how visible light and infrared energy warm the earth.**
- **Summarize the processes of radiation, conduction, and convection.**

The earth's atmosphere is heated in several ways by the transfer of energy from the sun. Some of the heat in the atmosphere comes from the rays of the sun as they are absorbed by some gases in the atmosphere. Some heat enters the atmosphere indirectly as ocean and land surfaces absorb solar energy and then give it off as heat.

Radiation

All of the energy that the earth receives from the sun travels through space between the earth and the sun as *radiation*. Radiation includes all forms of energy that travel through space as waves. Light is the form of radiation that can be seen with our eyes. However, there are many other forms of radiation that we cannot see, such as X rays and radio waves.

Radiation travels through space in the form of waves at a very high speed—300,000 km/s. The distance from one wave crest to the next is called the *wavelength* of a wave. The various types of radiation differ in the length of their waves, as shown in Figure 23–8. Visible light, for example, consists of waves with various wavelengths that you see as different colors. Wavelengths shorter than those of visible light include ultraviolet rays, X rays, and gamma rays. Longer wavelengths include *infrared* (IN-fruh-RED) *waves* and radio waves. The waves that make up all forms of radiation are called *electromagnetic waves*. The complete range of wavelengths makes up the **electromagnetic spectrum.**

Figure 23–8. This illustration shows the electromagnetic spectrum from long radio waves to short gamma rays.

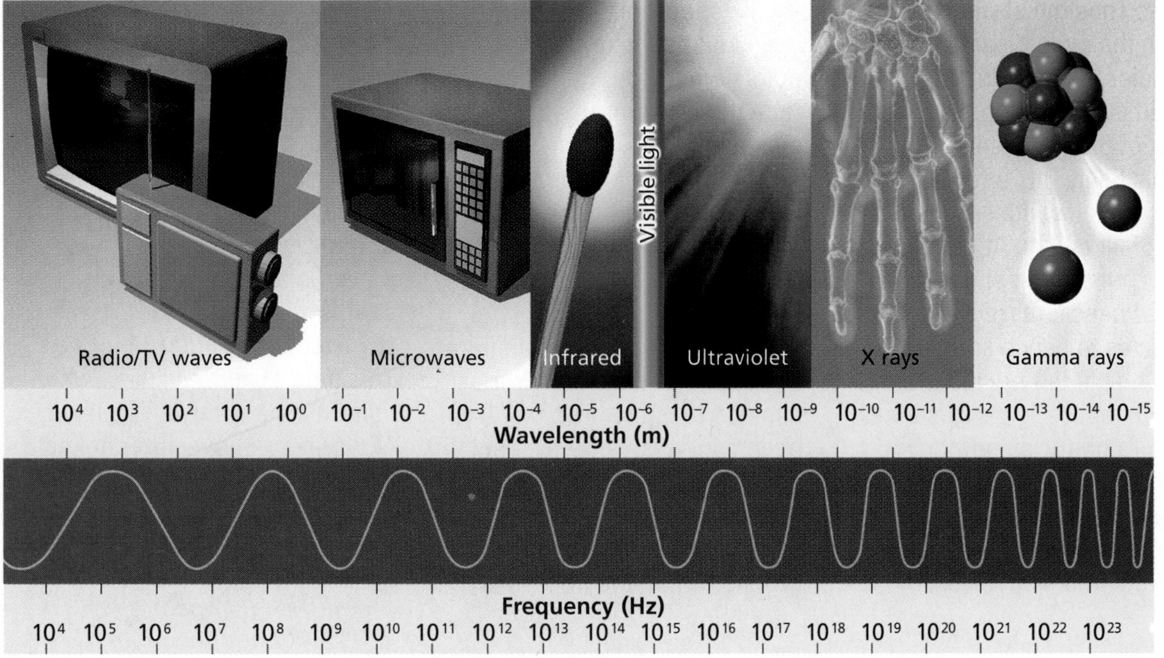

Almost all the energy reaching the earth from the sun is in the form of electromagnetic waves. A small amount of solar energy is carried to the earth by atomic particles emitted by the sun. Before the sun's radiation reaches the solid part of the earth, it passes through the earth's atmosphere. The atmosphere affects this radiation in several ways. First the molecules of nitrogen and oxygen in the upper atmosphere absorb the short wavelengths of X rays, gamma rays, and ultraviolet rays. This absorption occurs in the mesosphere and thermosphere. As solar energy is absorbed, molecules and atoms of nitrogen and oxygen lose electrons and become positively charged ions. In the stratosphere, ultraviolet rays are absorbed and act upon oxygen molecules to form ozone.

Thus, almost all of the shorter wavelengths are absorbed in the upper atmosphere. The small amount of ultraviolet radiation that does reach the earth's surface causes sunburn on skin that is exposed

EARTHBEAT

Visible Light and the Atmosphere

The atmosphere has a number of effects on the light that you see. The wavelengths of visible light that are most readily scattered by the fine dust in the air are the shorter ones. You see these wavelengths as blue and violet. Thus the sky looks blue when the air is clear. Larger particles, such as water droplets in clouds or fog, scatter most of the wavelengths of visible light. This effect makes the sky look white—a combination of all colors. At high altitudes, where the atmosphere is less dense, there are few particles to cause scattering. As a result, the sky looks dark blue. In space, the sky is black.

A rainbow is another example of the effects of the atmosphere on light. A rainbow is caused by the separation of sunlight into a spectrum of colors by raindrops. Refraction, or bending, of light rays in the raindrops separate white light into the entire visible spectrum. You can see these colors when you face the water droplets with the sun behind you.

A mirage is an optical illusion created by the atmosphere. Light rays are refracted as they pass through a boundary between hot air and cool air. You may have seen a mirage on a summer day when water seemed to appear on the dry pavement of a highway. This mirage occurs when a layer of cool air reflects the sky onto a layer of hot air close to the ground. When the light rays enter the hot air, they are bent upward, causing the sky to appear as a pool of water on the pavement. This is called an *inferior mirage.*

A distant mountain that seems to be suspended in the sky is called a *superior mirage.* This type of mirage forms when light rays pass through a layer of warm air above into cool air below.

Are the light rays in a superior mirage bent upward or downward?

to direct sunlight too long. Most of the solar rays that reach the lower atmosphere have longer wavelengths—those of visible and infrared waves. Most incoming infrared radiation is absorbed by carbon dioxide, water vapor, and other complex molecules in the troposphere. Only small amounts of visible light waves are absorbed as they pass through the atmosphere.

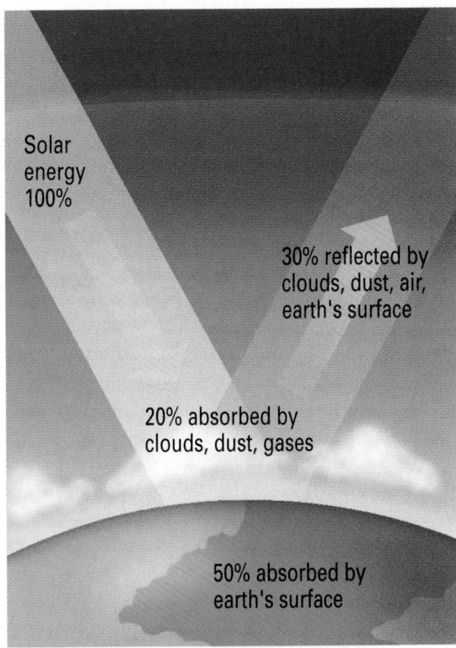

Figure 23–9. About 70% of the solar energy that reaches the earth is absorbed by the earth's land and ocean surfaces and by the atmosphere. The remainder, about 30%, is reflected back into space.

Scattering

Clouds, dust, and gas molecules in the atmosphere affect the path of radiation from the sun, causing scattering. Scattering means that water droplets and dust suspended in the atmosphere reflect and bend the rays. This bending sends the rays out in all directions without changing their wavelengths. In clear, cloudless air, scattering is also caused by the reflection of light off gas molecules. Scattering sends some of the radiation back into space. The remaining radiation continues downward toward the earth's surface. As a result of scattering, sunlight reaching the earth's surface comes from all directions. Scattering is what makes the sky appear blue and makes the sun appear red at sunrise and sunset. Short-wavelength rays, such as blue light, are more easily scattered than long-wavelength rays are. The sun appears red when it is low in the sky because more blue light rays are scattered by the atmosphere. Thus more of the longer-wavelength red light rays reach the earth's surface, giving the sun its red color.

Reflection

Of the total amount of solar energy reaching the earth's atmosphere, about 20 percent is absorbed by the atmosphere. About 30 percent is scattered back into space or is reflected from clouds or the earth's surface. The remaining 50 percent is absorbed by the surface.

When solar energy reaches the earth's surface, the surface either absorbs the energy or reflects the energy. The kind of surface on which the radiation falls determines to a large extent whether absorption or reflection occurs. Table 23–1 shows the amount of incoming solar radiation absorbed and reflected by various surfaces.

Table 23–1 Reflection and Absorption

	Percentage of solar radiation:	
Surface	Reflected	Absorbed
Soils	5–10	95–90
Desert	20–45	80–55
Grass	16–26	84–74
Forest	5–20	95–80
Snow	40–95	60–05
Water (high sun angle)	3–10	97–90
Water (low sun angle)	10–80	90–20

Because 30 percent of the total solar energy that reaches the earth's atmosphere is either reflected or scattered, the earth is said to have an average reflectivity of 0.3. The fraction of solar radiation reflected by a particular surface is called its **albedo.** The earth has an albedo of 0.3. The albedo of the moon is 0.07. What percent of the total solar energy reaching the moon is reflected?

Absorption and Infrared Energy

The solar radiation that is not reflected by the earth's surface is absorbed by the earth. Part of the absorbed radiation is composed of the infrared rays that have penetrated the atmosphere. When you feel the warmth of the sun, you are feeling infrared rays. The rocks, soil, water, and other earth materials are heated when they absorb infrared rays and visible light. The heated materials then produce their own infrared rays from the heat energy. These infrared rays have much longer wavelengths than the infrared rays that reach the earth's surface directly from the sun. The infrared rays with shorter wavelengths pass through the gases of the atmosphere. The infrared rays produced by the warmed materials of the earth's surface are mostly absorbed by water vapor and carbon dioxide in the atmosphere.

The Greenhouse Effect

The absorption of long-wavelength infrared rays from the earth's surface by gas molecules in the atmosphere traps heat energy and prevents it from escaping back into space. As a result, the lower atmosphere becomes warm. The warmed lower atmosphere keeps the earth's surface much warmer than it would be if there were no atmosphere. The process by which the atmosphere absorbs radiation has been compared to a greenhouse. The glass of a greenhouse allows the short wavelengths of visible light and infrared rays from the sun to pass through to the interior. The glass also prevents the long infrared rays emitted by the warmed surfaces within the greenhouse

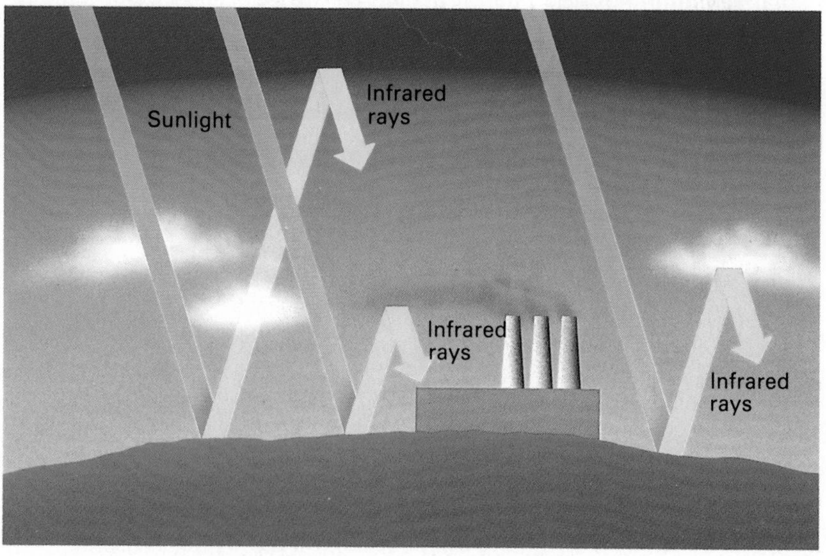

Figure 23–10. The visible and infrared rays of incoming sunlight pass through the water vapor and carbon dioxide of the atmosphere. Most of the longer infrared rays sent out by the warmed surfaces on the earth are trapped by these same substances.

from escaping to the outside. However, experiments have shown that a greenhouse is also heated because air warmed by the infrared rays is prevented from escaping. Thus, a greenhouse does not become heated in exactly the same way as does the atmosphere. Nevertheless, the process by which the atmosphere traps infrared rays over the earth's surface is called the **greenhouse effect.** By this process, the energy from the sun warms the air after having first been absorbed at the earth's surface. The atmosphere is heated mostly in the lower layer, the troposphere.

The average temperatures over all the earth's surface do not change much from year to year. Normally the amount of solar energy trapped is about equal to the amount that escapes into space. However, human activities may change this balance and cause the average temperature of the atmosphere to increase. For example, the burning of fossil fuels releases carbon dioxide into the air. Carbon dioxide absorbs infrared rays very effectively. Measurements have shown that the amount of carbon dioxide in the atmosphere has been increasing in recent years and seems likely to continue to increase in the future. Increases in the amount of carbon dioxide will intensify the greenhouse effect and may cause the earth to become warmer. Such general warming will probably cause climate changes in many parts of the world. Scientists are doing intensive research on the possible effects of global warming.

Variations in Temperature

Radiation from the sun does not heat the earth equally at all places at all times. How warm the atmosphere becomes in any region on the earth's surface depends on several factors. Latitude is the primary factor. Average temperatures are higher near the equator than near the poles. The direct rays of the sun striking near the equator are more effective in heating an area than the slanting rays striking the polar regions. Slanting rays spread their energy over a larger area than do direct rays.

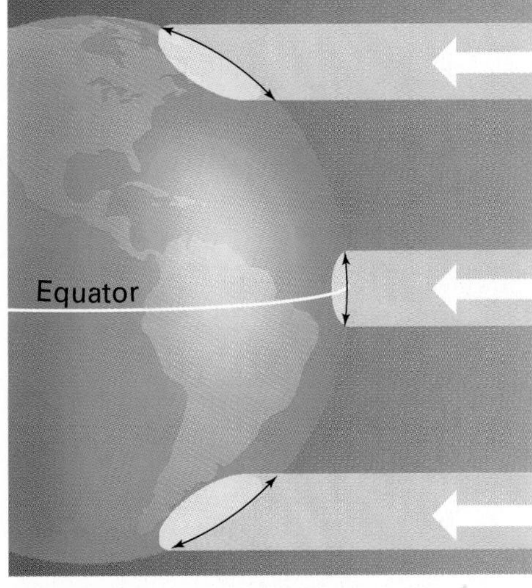

Figure 23–11. Temperatures are higher at the equator because solar energy is concentrated in a small surface area. Farther north and south, the same amount of solar energy is spread out over a larger surface area, and temperatures are lower.

Elevation is another factor in regional temperature variations. High elevations, such as mountaintops, become warm during the day, but they cool very quickly at night. The thinner air at high altitudes contains less water vapor and carbon dioxide to trap the heat. Desert temperatures usually show large changes between day and night because there is little water vapor present to hold the heat.

The temperature of water changes less than the temperature of land when solar energy is absorbed or when heat is given off. Water thus has a moderating effect. Under similar conditions, regions close to large bodies of water generally have more moderate temperatures. In other words, they will be cooler during daytime and warmer at nighttime than inland regions with the same general weather conditions. Also, the location of an area in relation to the wind patterns makes a difference. A region that receives winds off the ocean waters has more moderate temperatures than a similar region in which the winds blow from the land.

> **INVESTIGATE!**
>
> *To learn more about air density and temperature, try the In-Depth Investigation on pages 476–477.*

There is also a delay in the heating of the atmosphere both seasonally and daily. The sun is highest and its rays most direct in the Northern Hemisphere on June 21. You might expect that this would be the warmest time of the year. However, it is usually late July before enough heat has been absorbed and reradiated from the ground to really heat up the atmosphere. Only then does the warm, summer weather begin. For similar reasons, the warmest hours of the day are usually about 2:00 in the afternoon, although the sun is strongest at 12:00 noon.

Conduction and Convection

While most heating of the atmosphere comes from radiation, a small amount results from **conduction** and *convection*. In conduction, the molecules in a substance move faster as they become heated. These fast-moving molecules cause other molecules to also move faster. The motion makes the substance warm. Solid substances, in which the molecules are close together, make good conductors because heat can be transferred quickly from molecule to molecule. Air is warmed when it comes in contact with anything hotter than the air itself. However, the molecules of air are far apart. As a result, air is a poor conductor of heat. Some heating of the lower part of the atmosphere takes place through conduction when air comes into contact with the warmed surface of the earth.

Convection, as you learned in Chapter 4, involves the movement of gases or liquids when they are heated unevenly. The movement of air due to convection takes place when some of the air is heated by radiation or conduction. As that air is heated by radiation or conduction, it becomes less dense and rises. Nearby cooler air, which is more dense, sinks. As it sinks, the cooler air pushes the warm air up. The cold air is, in turn, warmed, and then it rises also. This continuous cycle of cold air sinking and warm air rising helps warm the earth's surface relatively evenly. Because heated air is less dense than cool air, it exerts less pressure on the earth than the same volume of cooler air does. Consequently, the atmospheric pressure is generally lower beneath a mass of warm air than it is under cool air. As dense, cool air moves into a low-pressure region, the less dense, warmer air is pushed upward. The general movement of cool air is always toward regions of lower pressure. These pressure differences, which are the result of the unequal heating that causes convection, create winds.

Section 23.2 Review

1. What type of solar radiation causes sunburn?
2. Why is the atmospheric pressure lower beneath a mass of warm air than beneath a mass of cold air?
3. You decide not to be outside during the warmest hours of a warm summer day. When will the warmest hours probably be? How do you know?

23.3 Winds

Section Objectives

- **Describe the global patterns of wind.**
- **Describe some factors that create local wind patterns.**

Because the earth receives more solar energy at the equator than at the poles, there is a belt of low air pressure at the equator. The heated air in the region of the equator is constantly rising. At the poles, the colder air is heavier and tends to sink. This sinking of cold air creates regions of high atmospheric pressure.

Pressure differences in the atmosphere at the equator and at the poles create a general movement of air worldwide. Air moves from high-pressure belts toward low-pressure belts. In very general terms, air near the earth's surface generally flows from the poles toward the equator, as shown in Figure 23–12. At higher altitudes, the rising warm air cools, and there is a general return flow of air from the equator toward the poles.

Global Winds

The circulation of the atmosphere as well as the oceans is affected by the rotation of the earth on its axis. The rotation causes surface winds in the Northern Hemisphere to be deflected to the right and those in the Southern Hemisphere to be deflected to the left. This motion is called the *Coriolis effect,* after the nineteenth-century French mathematician who first described it. A ball thrown on the earth's surface will curve only slightly due to the Coriolis effect. The ball travels far too short a distance to be affected. The ocean currents and winds of the world, however, are strongly deflected. Winds that would otherwise blow directly from a high-pressure area toward a lower-pressure area are deflected by the Coriolis effect.

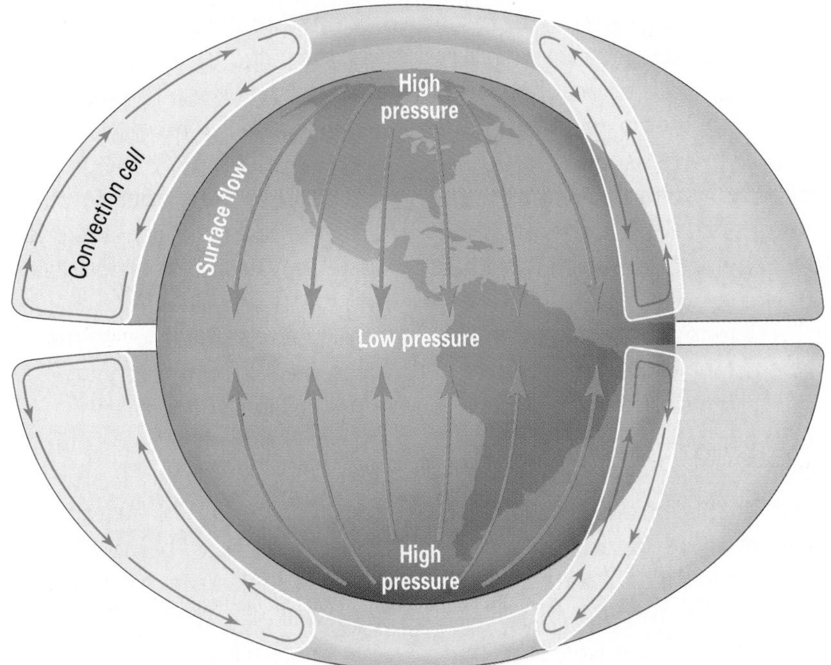

Figure 23–12. The blue arrows around the perimeter of the globe show the general circulation of air. Surface winds blow from polar high-pressure areas to equatorial low-pressure areas as shown by the red arrows. Wind movements are shown as if on a non-rotating earth, without the Coriolis effect.

A low-pressure belt exists at the equator because the heated air there tends to rise. As the warm air rises, it begins to move toward the poles. Around 30° latitude, some of this air sinks toward the earth's surface. At the surface, the descending air forms a high-pressure area from which air flows both north and south. At about 60° latitude, air flowing along the surface from the polar high and the high at 30° converges. This converging air rises, forming a low-pressure area at 60° latitude. The air flowing from the equator completes three looping patterns of flow called **convection cells.** The Northern and Southern hemispheres each have three convection cells.

Trade Winds

The winds in both hemispheres flowing toward the equator between 30° and 0° latitude are called **trade winds.** Like all winds, they are named according to the direction from which they flow. In the Northern Hemisphere, the trade winds flow from the northeast and are therefore called the *northeast trades*. The northeast trades are deflected to the right by the Coriolis effect. In the Southern Hemisphere, the trade winds are called the *southeast trades* and are deflected to the left.

SCIENCE & TECHNOLOGY

Energy from the Wind

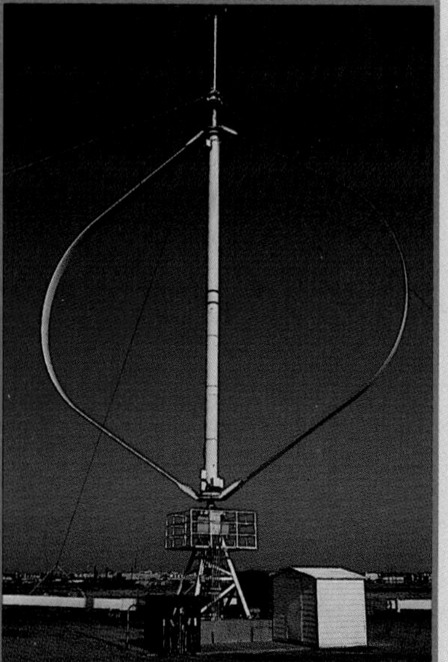

For more than a thousand years, people have used windmills for tasks such as grinding grain or pumping water. Today, more and more people are rediscovering the benefits of wind power as the number of wind turbines used to generate electricity continues to grow.

One result of the renewed interest in wind power is the reintroduction of the Darrieus turbine. Invented in 1920 by French engineer Georges Darrieus, the Darrieus turbine, shown at left, resembles an upside-down eggbeater. Because the turbine's blades spin on a vertical axis, the Darrieus turbine is mechanically simpler than propeller-type turbines, which spin on horizontal axes. The Darrieus turbine is designed to catch winds blowing from all directions, and it is much easier to maintain because most of its moving parts are located on the ground. The bend in the blades can be changed to maximize efficiency, while the turbine's vertical axis minimizes land use.

There also have been other, more recent, technological advances in wind turbines. These advances include both lighter and less costly materials, as well as the use of faster airfoil blades. Instead of being flat like earlier blades, an airfoil blade is shaped like an airplane wing. This shape allows air to flow

The trade wind systems of the Northern and Southern hemispheres meet at the equator in a narrow zone called the **doldrums.** The air in this warm zone moves mainly upward, and near the earth's surface the winds are weak and undependable. What problems might sailing ships have in the doldrums?

A belt of high pressure in the vicinity of 30° latitude is created by the descending air. This subtropical high-pressure belt is called the **horse latitudes.** The surface winds here are weak and changeable. The name *horse latitudes* was given to this region in the days when many sailing ships carried horses from Europe to the New World. Horses were often thrown overboard to save water for the sailors when the ships were becalmed and trapped in this zone because of the lack of wind.

Westerlies

In the subtropical highs, some of the descending air moving toward the poles is deflected by the Coriolis effect. This flow creates another wind belt in both the Northern Hemisphere and the Southern Hemisphere. These winds are known as the **westerlies.** In the Northern Hemisphere, the westerlies are southwest winds. In the Southern

smoothly over the blades, enabling the airfoil to use the lifting force of the wind more effectively.

The advantages to using the wind to generate electricity are that it consumes no fossil or nuclear fuel and produces no waste. Also, the wind itself costs nothing and is available in limitless supply. A disadvantage of wind power is the unpredictability of wind speed and direction. In addition, large wind farms can be noisy, visually unattractive, and deadly to migrating birds.

The photograph above shows some of the 13,000 wind turbines that have been installed in the Diablo Mountains of California. Home to some of the world's largest wind farms, California is now seeing its leadership in the multibillion-dollar market for wind-generated electricity being challenged. With steady growth in European countries such as the Netherlands, Denmark, Germany, and England, wind generation is quickly becoming the sustainable energy source of choice for utility companies worldwide.

What problems would there be in using wind turbines in areas along the equator and at 30° latitude?

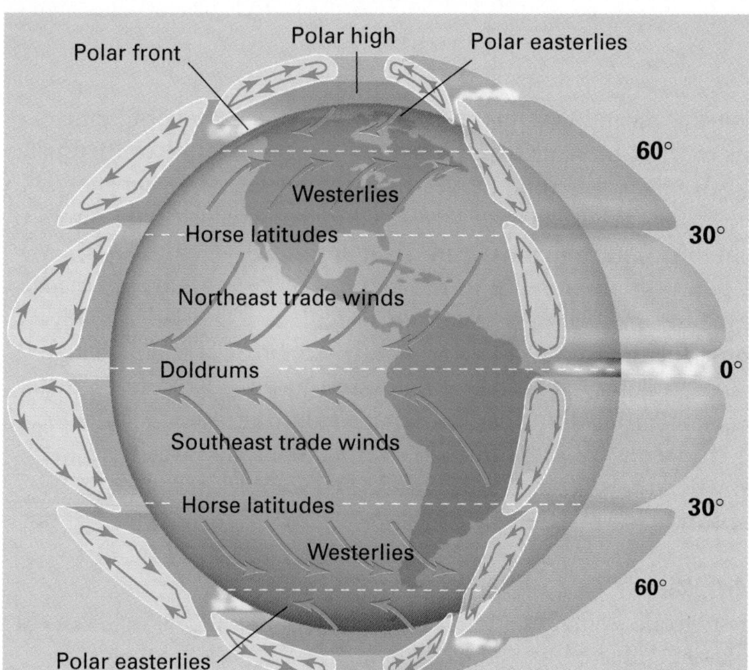

Figure 23–13. Both the Northern Hemisphere and the Southern Hemisphere have three wind belts. Wind belts are the result of pressure differences at the equator, the subtropics, the sub-polar regions, and the poles. The arrows are curved because of the Coriolis effect.

Hemisphere, they are northwest winds. The westerlies are located in a belt between 40° and 60° latitude. The contiguous United States is located within the westerlies, while Hawaii is located in the northeast trade winds, and parts of Alaska are in the polar easterlies.

Polar Easterlies

A third wind belt exists near each of the earth's poles, as shown in Figure 23–13. Poleward of the belt of westerlies, at about 60° latitude, is a belt of low pressure. These **subpolar lows** result when warm air moving poleward from the subtropical high is lifted by cold polar air moving toward the equator. Over the polar regions themselves, descending cold air creates areas of high pressure. The general surface movements of the cold polar air masses is toward the equator. Surface winds created by the polar high pressure are deflected by the Coriolis effect, becoming the **polar easterlies.** The polar easterlies are strongest where they flow off Antarctica. Where the polar easterlies meet warm air from the westerlies, a stormy region known as a polar front forms.

Wind and Pressure Shifts

As the sun's vertical rays shift northward and southward during the year, the positions of the pressure belts and wind belts also shift. Although the sun's rays move 47° in latitude, the average shift for the pressure and wind belts is only about 10° of latitude. However, even this small change means that some areas are in different wind belts during the year. In southern Florida, for example, westerlies prevail in the winter, but trade winds are dominant in the summer.

Jet Streams

Bands of high-speed winds exist in the upper troposphere and lower stratosphere over both the Northern and Southern hemispheres. These upper westerly winds are the **jet streams.** Because the temperature of polar air and middle-latitude air differs so greatly, the pressure also differs greatly. The cold polar air is much denser than the warmer air of the middle latitudes. The resulting pressure differences produce the polar jet streams. These bands of winds, found at an altitude of 10–15 km, are about 100 km wide and 2–3 km thick. The polar jet streams may reach maximum speeds of almost 500 km/hr. These winds do not blow steadily but instead change speed and position. The polar jet streams are important because they control the path of storms and have an effect on airline routes.

In the subtropical regions, very warm equatorial air meets the cooler air of the middle latitudes, creating the subtropical jet streams. Unlike the polar jet streams, the subtropical jet streams do not change much in speed or position.

Figure 23–14. In the daytime, air heats rapidly in valleys. The warm air rises up the slope as a valley breeze. At night, air at the mountaintop cools quickly. The heavy, cool air moves down the mountain slope as a mountain breeze.

Local Winds

The movements of air are also influenced by local conditions. A local feature that produces temperature differences often causes a local wind. Local winds are not part of the global wind belts. Gentle winds that extend over distances of less than 100 km are called *breezes*.

Land and Sea Breezes

Equal surface areas of land and water may receive the same total amount of energy from the sun. However, the land surfaces heat up faster than the water does. During daylight hours, therefore, a sharp temperature difference develops between a body of water and the land along its shore. This temperature difference is apparent in the air above the land and water. The warmer air above the land rises, and the cool air from above the water moves in to replace it. A cool wind moving from water to land, called a *sea breeze,* generally begins in the afternoon. Overnight the land cools more rapidly than the water does, and the sea breeze is replaced by a *land breeze*. A land breeze flows from the cooler land to the warmer water.

Mountain and Valley Breezes

In mountainous regions during the daylight hours, a gentle *valley breeze* blows up the slopes. The valley breeze is caused by warm air from the valleys moving up slope. At night the mountains cool more quickly than the valleys do. Then, cooler air descends from the mountain peaks, creating a *mountain breeze*. Campers in mountain areas may experience a warm afternoon quickly turn to a cold evening soon after the sun sets, as the cold air flows down slope and settles in the valleys.

Section 23.3 Review

1. What are the results of the Coriolis effect on wind flow?
2. What surface wind belt flows in the middle latitudes?
3. On a camping trip on the Oregon coast, you decide to hike to the ocean, but you are not sure of the direction. The time is 4:00 P.M. How might the breeze help you find the ocean? Why?

Chapter 23 Review

Key Terms

- albedo (466)
- atmospheric pressure (457)
- barometer (458)
- climate (455)
- conduction (468)
- convection cell (470)
- doldrums (471)
- electromagnetic spectrum (463)
- exosphere (461)
- greenhouse effect (467)
- horse latitudes (471)
- ionosphere (461)
- jet streams (472)
- mesopause (461)
- mesosphere (461)
- nitrogen cycle (457)
- ozone (455)
- polar easterlies (472)
- standard atmospheric pressure (459)
- stratopause (460)
- stratosphere (460)
- subpolar low (472)
- temperature inversion (462)
- thermosphere (461)
- trade winds (470)
- tropopause (460)
- troposphere (460)
- weather (455)
- westerlies (471)

Key Concepts

The earth's atmosphere is a mixture of gaseous elements and compounds and tiny suspended particles. **See page 455.**

Atmospheric pressure is measured by a barometer. **See page 458.**

The atmosphere is divided into four major layers that vary in temperature. **See page 459.**

Certain substances, when released into the air, can be harmful to people, animals, plants, and property. **See page 462.**

Most of the energy that reaches the earth from the sun comes in the form of electromagnetic waves. **See page 464.**

Visible light and some infrared rays from the sun penetrate the earth's atmosphere and heat materials such as rocks, soil, and water on the surface. **See page 466.**

Heat is transferred within the atmosphere by three processes—radiation, conduction, and convection. **See page 468.**

Air-pressure differences and the earth's rotation cause the various global wind belts. **See page 469.**

A surface feature, such as a body of water, a mountain, or a valley, can influence local wind patterns. **See page 473.**

Review

On your own paper, write the letter of the term that best completes each of the following statements.

1. During one part of the nitrogen cycle, nitrogen is removed from the air mainly by nitrogen-fixing
a. bacteria. b. waves. c. minerals. d. crystals.

2. Atmospheric pressure measured at sea level is
a. 99 percent. b. 1.03 kg. c. 32 km. d. 1.03 kg/cm^2.

3. A barometer measures
a. atmospheric pressure. b. wind speed.
c. ozone concentration. d. wavelengths.

4. Almost all of the water and carbon dioxide in the atmosphere is in the
a. exosphere. b. ionosphere.
c. troposphere. d. stratopause.

5. Radio stations can increase the distances they reach by bouncing radio waves off the
a. stratosphere. b. tropopause.
c. ionosphere. d. troposphere.

6. Around Los Angeles, frequent temperature inversions are the result of cool, polluted air being trapped by
   a. acid rain.
   b. a layer of colder air.
   c. mountains.
   d. the ocean.
7. Almost all of the energy reaching the earth from the sun is in the form of
   a. atomic particles.
   b. electromagnetic waves.
   c. ultraviolet rays.
   d. gamma rays.
8. Raindrops may separate sunlight into a range of colors, thereby causing
   a. a mirage.
   b. an inferior mirage.
   c. acid precipiation.
   d. a rainbow.
9. The process in which the atmosphere traps warming solar rays at the earth's surface is called the
   a. greenhouse effect.
   b. Coriolis effect.
   c. doldrums.
   d. convection cell.
10. Heat can be transferred within the atmosphere in three ways—radiation, conduction, and
    a. scattering.
    b. temperature inversion.
    c. weathering.
    d. convection.
11. A vertical looping pattern of air flow is known as
    a. the Coriolis effect.
    b. a convection cell.
    c. a trade wind.
    d. a westerly.
12. A gentle wind covering less than 100 km is called
    a. a jet stream. b. the doldrums. c. a breeze. d. a trade wind.

Critical Thinking

On your own paper, write answers to the following questions.

1. Explain how houseplants can increase the amount of oxygen in your home.
2. During a jet flight over the North Pole and toward a region in the middle latitudes, the pilot adjusts the altimeter. Why is this adjustment necessary?
3. Most aerosol sprays are banned in the United States. Which of the four layers of the atmosphere does this ban help protect? Explain your answer.
4. You hear a lecture about the earth's weather. The speaker says, "Infrared rays coming from the earth's surface heat the atmosphere much like a greenhouse is heated." Explain why that statement is incorrect.
5. What effect might jet streams have on airplane travel?
6. If there is a breeze blowing from the ocean to the land on the coast of Maine, about what time of day is it? How do you know?

Application

1. In a certain area of the country, many of the fish in a local lake have died. In addition, soils were found to be highly acidic, and nearby trees were losing their leaves. What kind of pollution may have caused these problems? What is the source of this pollution?
2. In a drive across the desert with your family, you see a distant sand dune that appears to be floating on air. How can you explain what you see?
3. In what ways would a knowledge of the global wind belts have helped a sixteenth-century explorer sailing between Spain and the northern part of South America?

Extension

1. Use encyclopedias or the Internet to find out more about rainbows. Draw diagrams showing how rainbows form and why a rainbow forms an arc. Present both diagrams to the class.
2. Do research on the Beaufort wind scale. Assign a Beaufort number to the morning and evening winds in your area each day for a week. Plot the results on a graph with the time and day on one axis, and the Beaufort number on the other axis. Share your graph with the class, and lead a discussion about any trends you notice.

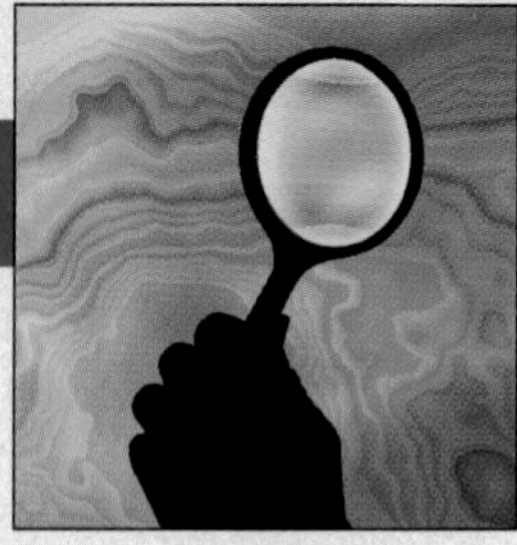

In-Depth Investigation

Air Density and Temperature

Materials

- 400-mL beaker
- hot water
- ice
- disposable syringe (60 cc)
- petroleum jelly
- glycerin
- Celsius thermometer
- cold water
- graph paper
- safety goggles

Introduction

Circulation in the earth's atmosphere is caused by the sinking of cold, dense air and the rising of warm, less dense air. The uneven heating of air causes this rising and sinking. The vertical movement of air caused by uneven heating is called convection. Convection is the main cause of air circulation in the earth's atmosphere.

In this investigation, you will observe how the density of air is affected by changes in its temperature.

Prelab Preparation

1. Review Section 23.1, pages 455–462, and Section 23.3, pages 469–473.
2. Density *(D)* is defined as the mass *(m)* per unit volume *(V)* of a substance, or $D = m/V$. A block of lead has a mass of 910 g and a volume of 80 cm^3. What is lead's density?
3. If a mass of a given substance remains the same while its volume increases, would the density of the substance increase or decrease?

Procedure

1. Remove the cap from the end of the syringe. Move the plunger until the cylinder is about two-thirds full of air. Add a dab of petroleum jelly to the end of the syringe and replace the cap.
2. Gently pull on the plunger and release. Read the volume of air in the syringe. Record your answer in a table of your own similar to that shown on the next page.

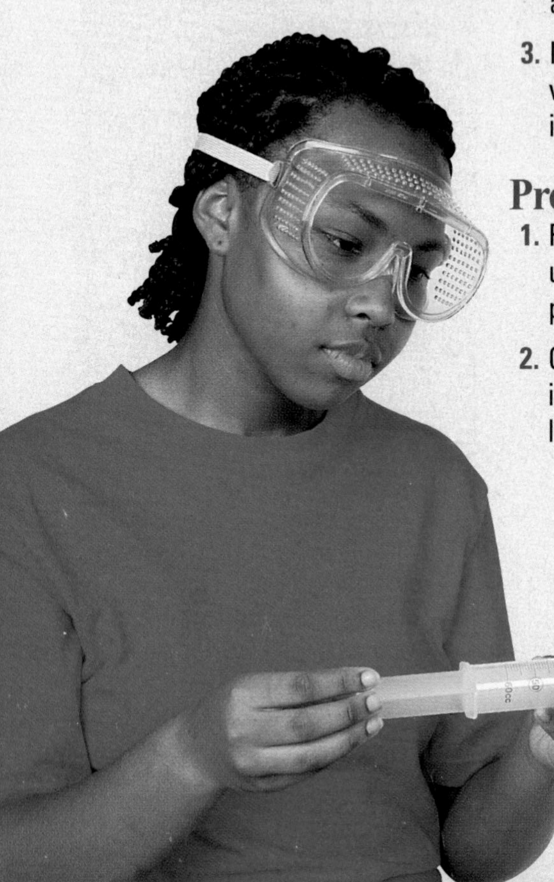

3. With your finger on the cap, gently push on the plunger and release. Read the volume of air in the syringe and record your answer in your table. **CAUTION:** *Wear your safety goggles.*
4. Take the average of the two volume readings. This number should be close to the volume at room pressure and temperature. Measure the room temperature, and record both the average volume and temperature in your table.
5. Add ice to a beaker of cold water until the temperature is 5 to 10°C. Do not let the temperature drop below 5°C.
6. Place the syringe in the water so that the air in the cylinder is completely covered with water. After five minutes, the temperature of the air in the syringe will be the same as the water temperature.
7. Repeat Steps 2–4 to find the volume of air at the new temperature. Record your observations in your table.
8. Repeat Steps 6 and 7 with water temperatures of about 15°C, 40°C, and 60°C. Use hot tap water for the higher temperatures. If the plunger sticks at higher temperatures, lubricate the cylinder wall with a few drops of glycerin.
9. Plot your data in a graph of your own similar to the one shown below. Draw a straight line through your data points. Some points may be slightly above or below this line. List reasons why all of the points do not fall on a straight line.

Temperature (°C)	Volume (cm^3)		
	Pull	Push	Average

Volume (cm^3)
40
20
0
0 10 20 30 40 50 60
Temperature (°C)

Analysis and Conclusions

1. According to your graph, what is the increase in volume, in cubic centimeters, as the temperature increases from 20°C to 40°C? As the temperature increases, does the density of the air in the syringe increase or decrease? Explain.
2. Based on your results, is hot air less dense or more dense than cold air? If the air is warmed, does it become less dense or more dense? Would warm air tend to rise or sink in a body of colder air?

Extensions

1. When a temperature inversion occurs, a layer of warm air is trapped beneath a layer of cool air. How might this condition affect air quality?
2. A temperature inversion may last several days. What might cause conditions to return to normal?

Chapter 24

Water in the Atmosphere

Water vapor is one of the most important substances in the earth's atmosphere. Although the amount of water vapor in the atmosphere is quite low, water vapor has major influences on weather conditions and climate. Suppose that all the water vapor in the atmosphere were to condense and fall to the earth as rain. The resulting worldwide layer of water would not be as thick as this book.

In this chapter, you will learn how water enters and leaves the atmosphere.

Chapter Outline

A large anvil-shaped cumulus cloud rises in the foreground with cirrus clouds above.

24.1 Atmospheric Moisture

Section Objectives

- **Explain how water vapor enters the air.**
- **Explain the meaning of humidity and describe how it is measured.**
- **Describe what happens when the temperature of air decreases at or below the dew point.**

Water exists in the atmosphere mainly in its gaseous form called *water vapor*. Water vapor is a tasteless, odorless, and invisible gas. However, you can detect the presence of water vapor by observing the spout of a tea kettle in which water is boiling. Notice the mist rising from the spout of the kettle. If you look carefully, you will see that the mist starts forming an inch or more above the spout. The space between the spout and the mist is filled with water vapor. The mist itself is condensed water vapor and is a liquid, not a gas.

Water in the atmosphere exists in two forms other than water vapor. Water can be a solid such as ice, or it can be a liquid in the form of water droplets.

Heat Energy and Water

Each water molecule of ice is strongly attached to others around it. Molecules of ice are held almost stationary in a definite crystalline arrangement. If you add enough heat to ice, the ice will change from a solid to a liquid to a gas. In other words, water changes from ice to liquid water to water vapor. When heat energy is added to ice, the molecules begin to move more rapidly. They break from their fixed positions, causing the ice to melt. In liquid water, the molecules move around each other, but they remain close together. When liquid water is heated, the random movement of molecules speeds up and causes the molecules to collide with one another. Such collisions can cause the molecules to move rapidly enough to evaporate and escape from the surface of the liquid. These evaporated molecules become the gas, water vapor.

Ice can change directly into water vapor without first becoming a liquid. The process by which a solid changes directly to a vapor is called **sublimation.** When the air is dry and the temperature is below freezing, ice and snow may sublimate into water vapor. Water vapor can also change directly into ice without becoming a liquid through a process called *deposition*. This is how frost is formed.

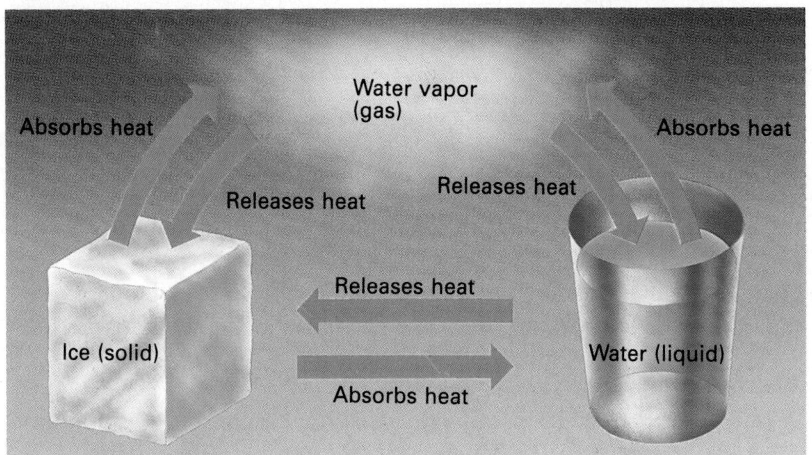

Figure 24–1. Water exists in three forms. Notice that water either absorbs heat or releases heat in changing from one form to another.

Although some water vapor enters the atmosphere through sublimation, most of the water vapor in the atmosphere comes from evaporation. Water absorbs heat energy from the sun. This heat causes molecules of water to evaporate and pass into the air. Most of this evaporation takes place in the regions near the equator, where the largest amounts of solar energy are received. The ocean is the principal source of atmospheric moisture. Billions of kilograms of water evaporate each day from the surface of the ocean. Evaporation from lakes, ponds, streams, and soil add water vapor to the atmosphere. Plants also give off water vapor, and a small amount of moisture comes from volcanoes and from the burning of fuels.

When water evaporates and enters the atmosphere, its molecules move much more rapidly than they did when they were in the liquid state of water. The energy absorbed is stored in the molecules and is called **latent heat.** *Latent* means "hidden." Latent heat will be released when water condenses and returns to the liquid state. Latent heat also is involved when water freezes or thaws.

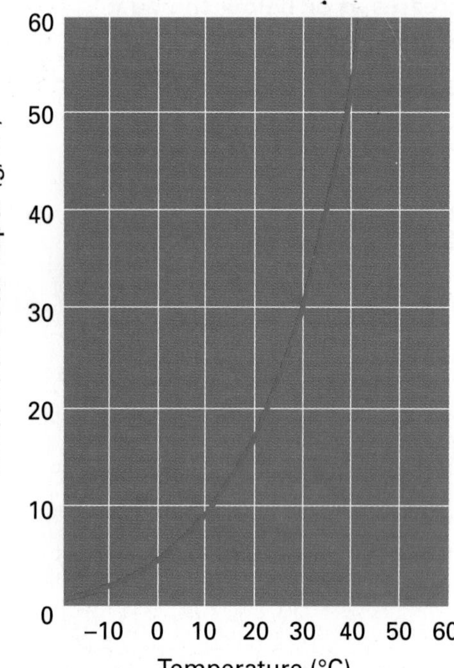

Figure 24–2. This graph shows the maximum amount of water vapor that a given volume of air can hold at various air temperatures.

Humidity

The amount of water vapor in the atmosphere is known as **humidity.** As water molecules evaporate into the air, the humidity of the air increases. When the air holds all the water vapor it can at a given temperature, the air is **saturated.** The amount of water vapor that a volume of air can hold increases as the temperature increases. As the temperature decreases, the capacity of a volume of air to hold water vapor also decreases. Figure 24–2 shows this relationship between air temperature and capacity to hold water vapor.

Relative Humidity

A common way to express the amount of water vapor in the atmosphere is by **relative humidity.** Relative humidity is a ratio. It compares the mass of water vapor in the air with the mass of water vapor that the air can hold at its saturation point. For example, suppose that 1 m^3 of air at a temperature of 20°C contains 13.9 g of water vapor. Meteorologists know that at the same temperature, 1 m^3 of air can contain 17.1 g of water vapor. They can thus calculate the relative humidity with this equation.

$$\frac{\text{(present) } 13.9 \text{ g/m}^3}{\text{(saturated) } 17.1 \text{ g/m}^3} \times 100 = \text{(relative humidity) } 81\%$$

For air at a fixed temperature, relative humidity changes as moisture enters or leaves the air. It also changes if the amount of moisture in the air remains the same but the temperature of the air changes. If the moisture remains the same, the relative humidity decreases as the temperature rises and increases as the temperature falls.

Measuring Relative Humidity

A **psychrometer** is an instrument used to measure relative humidity. A psychrometer, shown in Figure 24–3, consists of two identical thermometers. The bulb of one thermometer is covered with a damp wick while the bulb of the other thermometer remains dry. The person measuring relative humidity holds the psychrometer by a handle and whirls it. This whirling circulates the air around both thermometers. The water in the wick of the wet-bulb thermometer starts to evaporate. Evaporation requires heat, so heat is withdrawn from the thermometer. Consequently, the temperature reading of the wet-bulb thermometer is lower than that of the dry-bulb thermometer. When the air is dry, the rate of evaporation in the wick of the wet-bulb thermometer increases. The drier the air is, the more rapidly water evaporates from the bulb of the wet-bulb thermometer and the more quickly the wet bulb cools. The rate of evaporation will decrease as the air becomes more saturated.

Approximate relative humidity can be determined by using a table such as Table 24–1. Locate the column that represents the temperature difference of the thermometer readings and the row of the dry-bulb temperature. The value found where the column and row intersect is the relative humidity. If the difference between the thermometer readings is 5°C and the dry-bulb temperature is 12°C, what is the relative humidity?

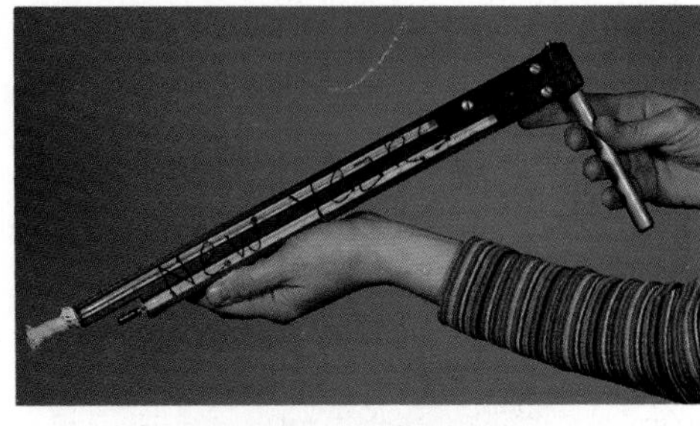

Figure 24–3. Meteorologists use a psychrometer and a relative humidity table, such as Table 24–1, to determine relative humidity. Relative humidity is always expressed as a percentage.

Table 24–1 Relative Humidity (in percentage)

Dry-bulb reading (°C)	Difference between wet-bulb reading and dry-bulb reading (°C)									
	1	2	3	4	5	6	7	8	9	10
0	81	64	46	29	13					
2	84	68	52	37	22	7				
4	85	71	57	43	29	16				
6	86	73	60	48	35	24	11			
8	87	75	63	51	40	29	19	8		
10	88	77	66	55	44	34	24	15	6	
12	89	78	68	58	48	39	29	21	12	
14	90	79	70	60	51	42	34	26	18	10
16	90	81	71	63	54	46	38	30	23	15
18	91	82	73	65	57	49	41	34	27	20
20	91	83	74	66	59	51	44	37	31	24
22	92	83	76	68	61	54	47	40	34	28
24	92	84	77	69	62	56	49	43	37	31
26	92	85	78	71	64	58	51	46	40	34
28	93	85	78	72	65	59	53	48	42	37
30	93	86	79	73	67	61	55	50	44	39

INVESTIGATE!

To learn more about relative humidity, try the In-Depth Investigation on pages 496–497.

Another instrument used to measure humidity is based on the fact that human hair stretches when the moisture in air increases. A piece of human hair will stretch about 2.5 percent when the relative humidity increases from 0 to 100 percent. The **hair hygrometer** is an instrument that records the changing length of a bundle of hairs when humidity changes. A variation in length causes a pointer to move on a scale or causes a pen to move on a graph, showing changes in relative humidity.

To measure humidity at high altitudes, an electric hygrometer is used. It may be carried up into the atmosphere in an instrument package known as a *radiosonde,* which is attached to a weather balloon. The electric hygrometer is triggered by passing an electrical current through a moisture-attracting chemical substance. The amount of moisture changes the electrical conductivity of the chemical substance. The change can then be measured and expressed as the relative humidity of the surrounding air.

SMALL-SCALE INVESTIGATION

Relative Humidity

Relative humidity is the ratio of the amount of water vapor in the air to the maximum amount of water vapor the air can hold at a given temperature. Using a relative humidity table, you can determine relative humidity.

Materials

thermometer; cheesecloth or light cotton material, about 5 × 5 cm; rubber band; water; small cardboard fan; relative humidity table on page 481

Procedure

1. Lay the thermometer on a table for a few minutes, until it adjusts to room temperature. Observe and record the dry-bulb temperature.
2. Secure the cheesecloth around the bulb of the thermometer with a rubber band. Moisten the cheesecloth with room-temperature water.
3. Hold the thermometer firmly at the top. Using your free hand, rapidly fan the bulb of the thermometer with the cardboard fan for 1 minute. Observe and record the wet-bulb temperature.
4. Repeat Steps 1–3 two more times. For each trial, subtract the temperature on the wet-bulb thermometer from the temperature on the dry-bulb thermometer. Use the table on page 481 to determine the relative humidity for each trial. Calculate the average relative humidity for the three trials. Record your results.

Analysis and Conclusions

1. How do you account for the difference between the readings of the dry-bulb and wet-bulb thermometers?
2. What is expressed by a relative humidity of 60 percent?
3. What might cause someone to feel hot and uncomfortable on a warm, humid day? Explain your answer.

Specific Humidity

Meteorologists use **specific humidity** to express the actual amount of moisture in air. Specific humidity is the number of grams of water vapor in 1 kg of air. For example, the very moist air in tropical regions might have a specific humidity of about 18 g/kg. The cold, dry air in polar regions, on the other hand, often has a specific humidity of less than 1 g/kg. Because specific humidity is measured only in units of mass, it is not affected by changes in temperature or pressure. Only a change in the amount of water vapor in the air changes the specific humidity.

Dew Point

You have learned that warm air can hold more water vapor than cold air can. The temperature to which air must be cooled to reach saturation is called the **dew point.** At any temperature lower than the dew point, water vapor begins to condense to a liquid or to change into a solid by sublimation. The dew point temperature of air depends upon the amount of water vapor already present in the air. When the air is nearly saturated with a relative humidity of nearly 100 percent, only a small temperature drop is needed for air to reach its dew point.

Air may cool to its dew point by *conduction* when the air contacts a cold surface. During the night, grass, leaves, and other objects near the ground lose heat. Their surface temperatures often drop to the dew point of the surrounding air. Air, which normally remains warmer than do surfaces near the ground, cools to the dew point when it contacts objects such as the cooler grass. The resulting form of condensation, which is shown in Figure 24–4, is called **dew.** You can see dew as tiny water droplets on many surfaces that have cooled. Dew is most likely to form on cool, clear nights when there is little wind.

If the dew point falls below the freezing temperature of water, water vapor will change directly into solid ice crystals, or **frost,** shown in Figure 24–4. Since frost forms from water vapor without first becoming liquid, it is not frozen dew. Frozen dew, which is relatively uncommon, forms as clear beads of ice. Frost often indicates that the temperature is low enough to damage growing plants.

Figure 24–4. The top picture shows dew, which formed on grass as the cold ground cooled the night air. The bottom picture shows frost, which formed on grass when the dew point dropped below freezing.

Section 24.1 Review

1. What is the principal source of water vapor in the atmosphere of the earth?
2. What term describes the temperature at which air reaches its saturation point?
3. How does a change in temperature affect relative humidity and specific humidity?
4. At a given temperature, 1 m^3 of air can hold 10 g of water vapor. What is the relative humidity at that temperature if 1 m^3 of air is holding 9 g of water?

Section Objectives

- **List the conditions that must exist for a cloud to form.**
- **Identify the types of clouds.**
- **Describe four ways fog may form.**

24.2 Clouds and Fog

Clouds and fog are visible masses of tiny water or ice particles suspended in the atmosphere. People commonly think of clouds as being high in the sky and fog as being close to the ground. However, clouds are not limited to high altitudes. Some clouds develop close to the surface of the earth. Both clouds and fog originate from water vapor in the air.

Cloud Formation

Clouds result from the condensation of water vapor throughout a large volume of air. For water vapor to condense, a solid surface must be available on which condensation can take place. Although the troposphere contains no large solid surfaces, it does contain millions of suspended particles of ice, salt, dust, and other solid matter. These suspended particles, called **condensation nuclei,** provide the necessary surfaces for cloud-forming condensation. Because the condensation nuclei are so small—less than 0.001 mm in diameter—they remain suspended in the atmosphere for a long time. Water molecules become attached to the particles, and as the molecules collect, water droplets are formed.

In order for clouds to form, the air must also be saturated. When air temperature reaches the dew point, the air is saturated, and water vapor begins to condense. Because the amount of water vapor that air can hold decreases as temperature goes down, cooling may lead to condensation. Several processes may bring about the cooling necessary for clouds to develop.

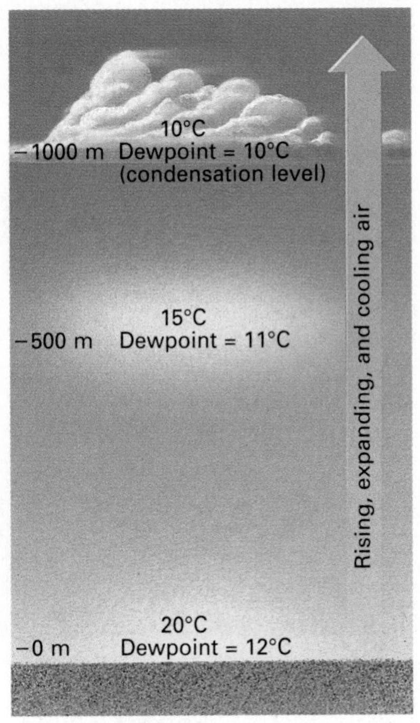

Figure 24–5. Notice that the rising air cools at the rate of 1°C for each 100 m. The dew point drops 0.2°C for each 100 m. In this illustration, temperature and dew point are the same at 1,000 m. Above that altitude, condensation begins and clouds form.

Convective Cooling

Because the higher layers of air compress the lower layers, the lower layers are more dense and have a higher pressure. When air rises into a region with a lower atmospheric pressure, its molecules move farther apart. This work of expanding the distance between molecules of air uses potential energy that is stored in compressed air. As this energy is used, the temperature of the air falls. The lowering of the temperature of a mass of air due to its rising and expanding is called **convective cooling,** as shown in Figure 24–5. Just as expansion causes air to cool, compression causes air to become warmer.

Adiabatic Temperature Changes Changes in temperature that result solely from the expansion or compression of air are called **adiabatic** (AD-ee-uh-BAT-ik) temperature changes. The adiabatic temperature changes associated with cloud formation take place at predictable rates. The adiabatic temperature of dry air decreases about 1°C for every 100 m the air rises. However, air usually contains moisture. Moisture may influence the rate of adiabatic change. Condensing water vapor in rising air gives off latent heat, which is absorbed by the air. This condensation slows the rate at which the air cools. The cooling rate of air varies with the amount of moisture it contains. The adiabatic rate for moist air varies from 0.5°C per 100 m for air with a low moisture content to 0.9°C per 100 m for air with a very low moisture content. The average rate is about 0.7°C per 100 m. Sinking air, whether moist or dry, heats at the rate of 1°C for every 100 m of descent.

Condensation Level Most clouds form by convective cooling. Figure 24–5 shows a typical sequence of cloud formation by convective cooling. On a sunny day, one area of the earth's surface may absorb more solar energy than do the areas around it. The air above this area of extra warmth becomes heated through conduction. The warmed air rises by convection and undergoes expansion and adiabatic cooling. When the air reaches a level where its temperature is lower than its dew point, the moisture in the air condenses to form a cloud. This level at which condensation forms, called the *condensation level,* is marked by the base of the clouds. Further condensation allows clouds to expand above the condensation level.

Forceful Lifting

Often air does not rise spontaneously; rather, an event occurs that forces the air to rise. The forced upward movement of air commonly results in the cooling of air and in cloud formation. Air can be forced upward when a moving mass of air meets sloping terrain, such as a mountain range. The forced rising air expands and cools, and clouds may form. As shown in Figure 24–6, entire mountaintops can be covered with clouds formed in this way. The large areas of clouds evident during storms are also formed by forceful lifting. These clouds are formed when a mass of warm air is pushed above a denser mass of cooler air. This process will be discussed further in Chapter 25.

Figure 24–6. Clouds can form as air is pushed up along a mountain slope and cooled below its dew point.

Figure 24–7. All clouds are of one of the three major forms. These forms, shown from top to bottom, are stratus clouds, cumulus clouds, and cirrus clouds.

Temperature Changes

Clouds will also form when one body of moist air mixes with another body of moist air with a different temperature. The mixing of the two bodies of air into one will cause the air's temperature to change. This temperature change may cool the combined air below its dew point. The result is condensation and cloud formation.

Advective Cooling

Another cooling process associated with cloud formation is **advective cooling.** Advective cooling is cooling produced when wind carries warm, moist air across a cold ocean or region of land. The cold water or land absorbs heat from the air, and the air cools. If the air is cooled below its dew point, condensation and fog or cloud formation takes place. Clouds formed by advective cooling are often very low in the atmosphere.

Classification of Clouds

Clouds are classified by their form and their altitude. The three forms are **stratus clouds, cumulus clouds,** and **cirrus clouds.** There are also three altitude groups: low clouds (up to 2,000 m), middle clouds (2,000–6,000 m), and high clouds (above 6,000 m).

Stratus Clouds

The most extensive clouds in the sky are stratus clouds. *Stratus* means "sheetlike" or "layered." The base of stratus clouds is low and may almost touch the earth's surface. Stratus clouds form where a layer of warm, moist air lies above a layer of cool air. When the overlying warm air cools below its dew point, wide clouds appear. Look at the stratus clouds shown in Figure 24–7. Stratus clouds cover large areas of sky, often blocking out the sun. The amount of rain that falls from stratus clouds is usually very little.

Two variations of stratus clouds, shown in Figure 24–8, are known as *nimbostratus* and *altostratus*. The terms *nimbo* and *nimbus* mean "rain." The dark *nimbostratus* clouds, unlike other stratus clouds, do bring heavy rains or snow. Altostratus clouds form at the middle altitudes. They are generally thinner than the lower stratus clouds and usually produce very little precipitation.

Cumulus Clouds

Puffy, vertical-growing clouds are known as cumulus clouds. *Cumulus* means "piled" or "heaped." Look at the cumulus clouds in Figure 24–7 and in Figure 24–8. Cumulus clouds are thick and have high, billowy tops. Cumulus clouds form when warm, moist air rises and cools. As the cooling air reaches its dew point, the clouds form. The flat base characteristic of most cumulus clouds shows where condensation began. The height of a cumulus cloud depends on the speed of the upward movement and the amount of moisture in the air. If the base of cumulus clouds begins at middle altitudes, they are called *altocumulus* clouds. Low clouds that are a combination of stratus and cumulus are called *stratocumulus* clouds.

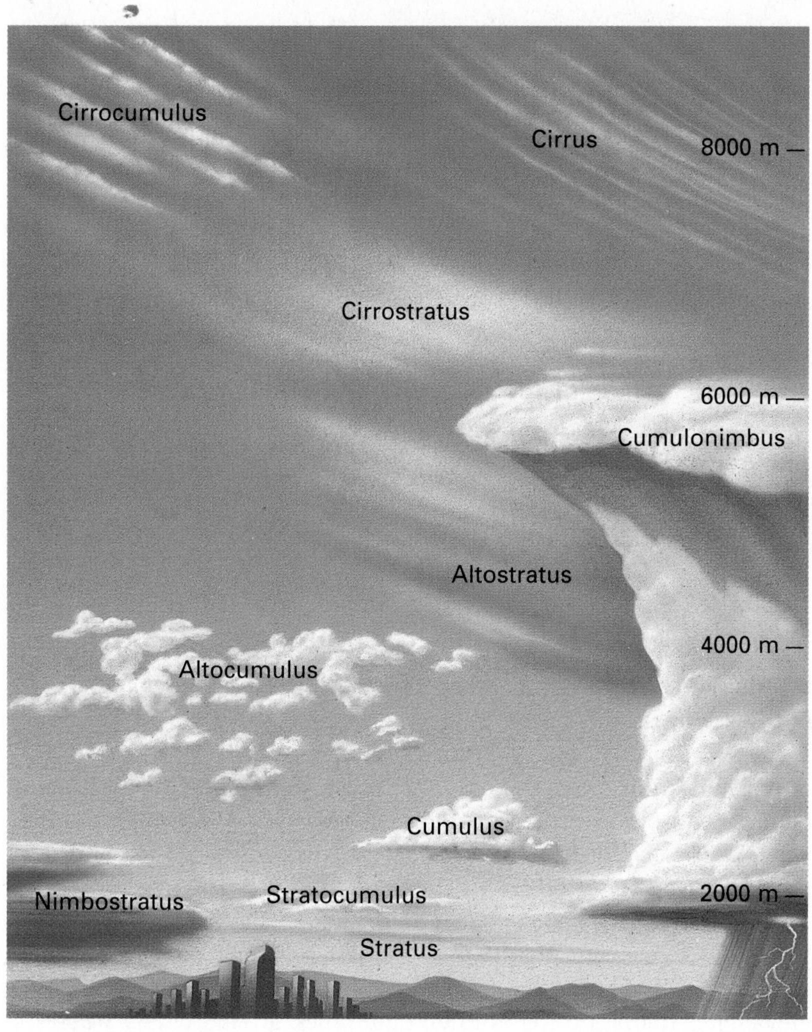

Figure 24–8. Meteorologists classify cloud types by form and altitude.

Cumulus clouds may form in fair weather and look like popcorn. On hot, humid days, cumulus clouds reach their greatest heights. The high, dark storm clouds produced are *cumulonimbus clouds,* or thunderheads, and they are often accompanied by rain, lightning, and thunder.

Cirrus Clouds

The highest clouds in the sky are wispy, feathery cirrus clouds, shown in Figure 24–7. *Cirro* and *cirrus* mean ''curly.'' Cirrus clouds form at altitudes above 6,000 m. The low temperatures at such high altitudes form clouds that are made up of ice crystals. Sunlight can easily pass through these thin clouds.

Locate the high-altitude, billowy clouds, called *cirrocumulus clouds,* in Figure 24–8. Cirrocumulus clouds are composed entirely of ice crystals. These clouds form quite rarely. Cirrocumulus clouds often appear just before a snowfall or a rainfall. Long, thin clouds called *cirrostratus clouds* form a high, transparent veil across the sky. A halo may appear around the sun or moon when either is viewed through a cirrostratus cloud. This halo effect is caused by the bending of light rays as they go through the ice crystals. Cirrostratus clouds may thicken and lower their base to form altostratus clouds.

Figure 24–9. Radiation fog, or ground fog—such as this fog in the Napa Valley in California—forms at night. The cold ground cools the air below its dew point, creating dense, low-lying fog.

Fog

Fog, like clouds, is the result of the condensation of water vapor in the air. The chief difference between the fog and clouds is that fog forms very near the surface of the earth when air close to the ground is cooled. For example, you may be familiar with one type of fog that results from the nightly cooling of the earth. The layer of air in contact with the earth becomes chilled below its dew point, and its water vapor condenses into droplets. This type of fog is called **radiation fog,** because it results from the loss of heat by radiation. This type of fog is also called *ground fog*. Radiation fog usually forms on calm, clear nights. It is thickest in valleys and low places because the dense, cold air in which it forms sinks to the lower elevations. Radiation fog is often quite thick around cities, where there are more sources of smoke and dust particles to act as condensation nuclei.

Another type of fog, called **advection fog,** forms when warm, moist air moves across a cold surface. Advection fog is common along coasts, where warm, moist air from above the water moves in over a cooler land surface. Also, dense fogs may form over the ocean when warm, moist air is carried over cold ocean currents.

Two other types of fog often form inland. An **upslope fog** is formed by the lifting and adiabatic cooling of air as it rises along land slopes. Upslope fog is really a kind of cloud formation at ground level. A type of fog known as **steam fog** usually forms over inland rivers and lakes. Steam fog is a shallow layer of fog formed when cool air moves over a warm body of water.

Section 24.2 Review

1. What is adiabatic cooling? How is it related to cloud formation?
2. What is the chief difference between fog and clouds?
3. On a hot and humid summer day, you notice enormous, billowy clouds with broad, flat bases. What can you predict about the weather? Explain how you know.

24.3 Precipitation

Section Objectives

- **Describe the various types of liquid and solid precipitation.**
- **Compare the two processes that cause precipitation.**
- **Describe how rain may be produced artificially.**
- **Describe how precipitation is measured.**

Any moisture that falls from the air to the earth's surface is called *precipitation*. Precipitation may occur in either liquid or solid form. The four major types of precipitation are rain, snow, sleet, and hail.

Forms of Precipitation

Rain is liquid precipitation. If the raindrops are smaller than 0.5 mm in diameter, it is called drizzle. Drizzle drops are close together, and, because they are so small, they fall very slowly. Drizzle results in only a small amount of total precipitation. Raindrops are between 0.5 mm and 5 mm in diameter. They may vary from a fine mist to large drops in a torrential rainstorm.

The most common form of solid precipitation is snow, which consists of ice particles. These particles may fall as small pellets, as individual crystals, or as crystals that combine into snowflakes. Most snow takes the form of snowflakes. Snowflakes may be as small as several millimeters in diameter or as large as several centimeters. Snowflakes tend to be small at low temperatures and become larger as the temperature nears 0°C. This size difference occurs because colder air is usually less moist. Consequently, less moisture is available for ice-crystal growth.

Extremely cold temperatures near the ground sometimes produce clear ice pellets called **sleet,** which forms when rain falls through a layer of freezing air. Occasionally rain does not freeze until it strikes a surface near the ground. In this event, it forms a thick layer of sheet ice, or **glaze ice,** as shown in Figure 24–10. The weight of ice often is great enough to break tree limbs and power lines. Conditions that produce glaze ice occur in what is called an ice storm.

Figure 24–10. Glaze ice forms as rain freezes on surfaces near the ground, such as on this plant.

Hail is solid precipitation in the form of lumps of ice. These lumps can be nearly spherical or irregularly shaped. Hail usually forms in cumulonimbus clouds. Convection currents within the clouds carry raindrops to high levels, where they freeze. As the frozen raindrops fall, they accumulate additional layers of liquid water on their surface. The hail may be carried up again into freezing air. Also, it may fall back through another layer of near-freezing moist air. Either way, the coating of water freezes, and another layer of ice forms. If the process is repeated a number of times, the hailstones may accumulate many layers of ice. Hailstones often consist of alternate layers of clear and cloudy ice. The clear ice forms as a hailstone passes through a layer of very moist air. The cloudy layer forms when water droplets with air bubbles between them freeze to the surface. Some hailstones become quite large. The impact of large hailstones can damage crops and property.

The Destructive Power of Hail

Hailstones start out as raindrops falling from clouds. Updraft convection currents carry the rain back up into freezing cloud regions, where layer after layer of supercooled water is added to the frozen drops. This water also freezes, making the hailstones larger. Some hailstones have grown from this process to the size of grapefruit.

The largest hailstone ever recorded had a circumference of over 44 cm and a mass of 758 g. Because of its size, this hailstone hit the ground at an estimated speed of 160 km/hr.

One hailstorm lasting only a few minutes can cause tremendous damage. For example, hail has been responsible for some airplane crashes. It has torn holes in the roofs of houses and automobiles and in the cabins of aircraft. Hail has killed people, livestock, and wildlife, and has stripped plants of their leaves.

The most consistent damage caused by hail is damage to crops. Not long ago, hail destroyed all the crops in 40 out of 53 villages in one region of Syria. This single storm left 60,000 people on the brink of starvation.

Each year in the United States, hail destroys over $200 million worth of wheat, corn, soybeans, and other crops. Hailstorms can have a devastating effect on individual farmers. Unlike drought, hail does not cause gradual damage. Instead, a family's entire crop, representing the work of an entire year, can be wiped out within minutes.

If you cut a hailstone in half, you will see layering. Explain why.

Causes of Precipitation

A cloud produces precipitation when its droplets or ice crystals become large enough to fall as rain or snow. Most cloud droplets have a diameter of about 20 micrometers. A micrometer equals 0.000001 m. A cloud droplet is thus smaller than the period at the end of this sentence. Droplets of this size easily remain suspended in the air. Even slight air movements prevent droplets from falling. Before a cloud droplet falls as precipitation, it must increase to about 100 times its normal diameter.

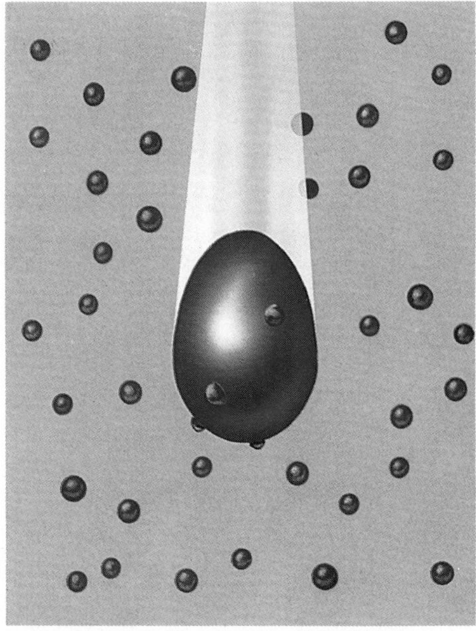

Figure 24–11. During coalescence, as cloud droplets fall, they collide and combine with small droplets. The resulting larger droplets fall as rain.

Coalescence

One way cloud droplets reach the precipitation stage is in a process called **coalescence** (KOE-uh-LESS-uns). Coalescence involves differences in size between cloud droplets. Generally, the original size of a cloud droplet depends on its condensation nucleus. Large nuclei tend to form large cloud droplets. Large droplets do not remain suspended in the cloud as long as small ones do. Instead, large droplets drift downward and, in doing so, collide and combine with smaller droplets. Each large droplet continues to coalesce until it contains at least a million times as much water as it did originally. By this time, the mass of each droplet is too great to be supported by the air, and the coalesced droplets fall as raindrops.

Supercooling

Precipitation may also form in clouds that contain water vapor, ice crystals, and water droplets that have gone through **supercooling.** Supercooled water droplets have a temperature of less than 0°C. The temperature of water droplets in these cold clouds may fall as low as −10°C. Yet even at this low temperature the water droplets do not freeze. They cannot freeze because there are too few **freezing nuclei**

Figure 24–12. The temperature at which ice crystals form determines their shape.

on which ice can form at these temperatures. Freezing nuclei are kinds of condensation nuclei with a crystalline structure similar to that of ice. Because only a few ice crystals will form, most of the water in the cloud remains as supercooled water droplets. Molecules evaporate more rapidly from the supercooled water droplets and condense on the ice crystals. The ice crystals increase in size until they gain enough mass to fall as rain or snow.

Meteorologists believe that almost all the rain and snow in the middle and high latitudes of the earth results from ice crystals in supercooled clouds. In the tropical regions, rain is commonly formed by the process of coalescence.

Cloud Seeding

Knowing how ice crystals form in clouds, meteorologists are able to take the first steps toward producing rain when it is needed. The method they use in attempting to cause or increase precipitation is called **cloud seeding.** In cloud seeding, freezing nuclei are added to supercooled clouds. In one method of cloud seeding, silver-iodide crystals, which resemble ice crystals, are used as freezing nuclei. Silver-iodide vapor is released into clouds from either burners on the ground or flares dropped from aircraft. In another method of cloud seeding, powdered dry ice is dropped from aircraft. The dry-ice particles cool the cloud droplets and cause ice crystals to form.

In some experiments with cloud seeding, seeded clouds produced more precipitation than unseeded clouds did. In other experiments, there was no significant difference in the amount of precipitation produced by seeded and unseeded clouds. In some cases, in fact, cloud seeding appeared to cause less precipitation. Meteorologists have thus concluded that cloud seeding may increase

Figure 24–13. Special equipment attached to the underside of cloud-seeding planes (inset) releases freezing nuclei into clouds. Meteorologists hope that cloud seeding will induce clouds to bring rainfall to drought-stricken areas.

precipitation under some conditions but decrease it under others. Research is under way to identify the conditions that produce increased precipitation. This information may help scientists improve their techniques for seeding clouds. Cloud seeding may eventually become a way to overcome the many problems associated with droughts. Successful cloud seeding may also help control severe storms. Meteorologists may be able to release precipitation from clouds before a storm can become too large.

Figure 24–14. A rain gauge is made up of a funnel and a cylindrical container.

Measuring Precipitation

A **rain gauge** is an instrument for measuring the amount of rainfall. One type of rain gauge consists of a wide-mouthed funnel placed over a cylindrical container. Rainwater passes through the funnel into the container below. In many rain gauges the mouth of the funnel is much larger than the container under it. Why do you think a rain gauge is constructed in this manner?

Another type of rain gauge consists of a funnel and a small divided bucket. Rain caught in the funnel fills one side of the bucket. The bucket then tips, dumping the water and allowing the other half to fill up. Each time one side of the bucket fills with 0.25 mm of rainwater, it tips and sets off an electrical device that records the amount. The rainwater dumped from the bucket is collected and weighed to check the accuracy of the record. A third type of gauge catches the water in a large bucket that is weighed continuously. The weight is recorded directly on a graph as centimeters of rain.

Rain gauges measure only the precipitation that falls in one spot. It is difficult to establish the total amount of rain that falls over a large area. Even over short distances, precipitation may vary by sizable amounts, especially in mountains and during thunderstorms.

Snow is measured by both the depth of accumulation and the water content. The depth of snow is measured with a measuring stick. The water content is determined simply by melting the snow and measuring the amount of water that results. The amount of water contained in any given volume of snow depends on the type of snow. For example, as much as 20 cm of dry snow may be needed to produce 1 cm of liquid water. But only 6 cm of wet snow is needed to produce 1 cm of liquid water. On average, 10 cm of snow will produce about 1 cm of liquid water.

Section 24.3 Review

1. Which form of liquid precipitation is made up of drops that are smaller than 0.5 mm in diameter?
2. In what parts of the world are raindrops commonly formed by the process of coalescence?
3. The usual purpose of cloud seeding is to create precipitation. How then might it help prevent a major storm?
4. If the bucket of a tipping-bucket rain gauge has emptied ten times, how much rain has fallen? Explain how you know.

Chapter 24 Review

Key Terms

- adiabatic (484)
- advection fog (488)
- advective cooling (486)
- cirrus cloud (486)
- cloud seeding (492)
- coalescence (491)
- condensation nuclei (484)
- convective cooling (484)
- cumulus cloud (486)
- dew (483)
- dew point (483)
- freezing nuclei (491)
- frost (483)
- glaze ice (489)
- hail (490)
- hair hygrometer (482)
- humidity (480)
- latent heat (480)
- psychrometer (480)
- radiation fog (488)
- rain gauge (493)
- relative humidity (480)
- saturated (480)
- sleet (489)
- specific humidity (483)
- steam fog (488)
- stratus cloud (486)
- sublimation (479)
- supercooling (491)
- upslope fog (488)

Key Concepts

Most moisture in the air comes from evaporation from the oceans. **See page 480.**

Humidity is the amount of water vapor in the air. **See page 480.**

When air reaches the dew-point temperature, water vapor condenses to a liquid or changes to a solid by sublimation. **See page 483.**

Clouds form when a body of air rises and cools below its dew point. **See page 484.**

The three major forms of clouds are stratus, cumulus, and cirrus. **See page 486.**

Radiation, advection, upslope, and steam fogs form in different ways. **See page 488.**

The four major forms of precipitation include rain, snow, sleet, and hail. **See page 489.**

Precipitation may be caused by coalescence or by supercooling. **See page 491.**

Cloud seeding is one way meteorologists attempt to induce clouds to drop rain. **See page 492.**

A rain gauge measures liquid precipitation. Snow is measured by its depth and water content. **See page 493.**

Review

On your own paper, write the letter of the term that best completes each of the following statements.

1. The process by which ice changes directly into water vapor is referred to as
   a. advection. b. conduction. c. sublimation. d. condensation.
2. Relative humidity is always expressed
   a. in g/m^3. b. as a percentage.
   c. in g/kg. d. in degrees Celsius.
3. To express the actual amount of moisture in the air, meteorologists use
   a. latent heat. b. relative humidity.
   c. specific humidity. d. dew point.
4. When air temperature drops, its capacity for holding water is
   a. slightly higher. b. much higher.
   c. about the same. d. lower.
5. The tiny water droplets that result when air is cooled by contact with a cold surface are called
   a. dew. b. frost. c. humidity. d. steam fog.

6. Changes in temperature that result solely from the expansion or compression of air are called
   a. adiabatic. b. supercooling. c. advection. d. latent cooling.
7. Clouds form when the water vapor in air condenses as
   a. the air is heated. b. the air is cooled.
   c. snow falls. d. the air is superheated.
8. Low, sheetlike clouds are called
   a. cirrus clouds. b. stratus clouds.
   c. cumulus clouds. d. cirrocumulus clouds.
9. The term *nimbo* or *nimbus* added to the name of any form of cloud means
   a. high. b. billowy. c. rain. d. layered.
10. The fog that results from the nightly cooling of the earth is called
    a. steam fog. b. upslope fog.
    c. radiation fog. d. advection fog.
11. Rain that freezes when it strikes the ground produces
    a. sleet. b. glaze ice. c. hail. d. frost.
12. Clouds in which the water droplets remain liquid below 0°C are said to be
    a. saturated. b. supersaturated.
    c. superheated. d. supercooled.
13. In one method of cloud seeding, silver-iodide crystals are used as
    a. freezing nuclei. b. dry ice.
    c. cloud droplets. d. a superheater.
14. A wide-mouthed funnel and a cylindrical container are used in making an instrument called a
    a. hygrometer. b. rain gauge. c. barometer. d. psychrometer.

Critical Thinking

On your own paper, write answers to the following questions.

1. Where would the air contain the most moisture—over Panama or over Antarctica? Explain your answer.
2. One body of air has a relative humidity of 97 percent. Another has a relative humidity of 44 percent. At the same temperature, which body of air is closer to its dew point? Explain your answer.
3. Why would polluted air more likely form clouds than clean air would?
4. In the tropical regions, raindrops are commonly formed by coalescence. Little precipitation there forms by supercooling. Why might this be true?
5. One day in January, 6 cm of very wet snow falls on your area. If all this snow melted quickly, how deep would the water be? Explain how you know.

Application

1. You are in school and you hear thunder outside. Describe the clouds that you would probably see if you looked out the classroom window.
2. You are camping in a valley on a calm, clear night. You awake and notice a thick fog. What type of fog is it, and how did it form?
3. You see lumps of ice the size of golf balls falling from the sky. Are you seeing sleet or hail? How do you know?

Extension

1. Look up the meanings and origins of the following terms used to describe clouds: *stratus, cumulus, cirrus.* Present your findings to the class.
2. At a convenient time of day, perhaps just after school or in the early evening, observe and sketch the cloud formations from a particular location. Do this at the same time and place every day for a week. Identify the cloud formations and share your drawings with other members of your class.
3. Do research to find out about a region that has little precipitation. Find out why the precipitation is so low. Find out how the dry climate influences plant and animal life and how people survive in the region. Report your findings to the class.

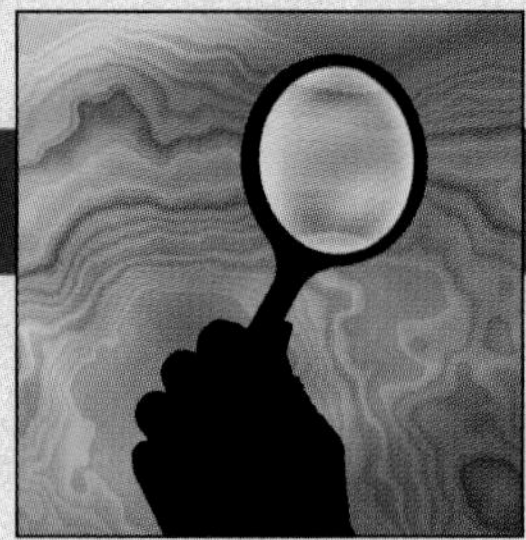

IN-DEPTH INVESTIGATION

Relative Humidity

Materials

- cotton cloth
- ring stand with ring
- Celsius thermometers (2)
- plastic container
- string
- water

Introduction

The earth's atmosphere acts as a storehouse for water that evaporates from the earth's surface. However, there is a limit to the amount of water vapor that the atmosphere can hold at a given temperature. When that limit is reached, the air is said to be saturated. Beyond this limit, no more water can be absorbed into the air. Air usually is not saturated with water vapor. Experiments show that air at a given temperature can never hold more than a certain amount of water vapor. For example, at normal room temperature (22°C) the air can hold about 20 grams of water per cubic meter. Relative humidity is a measure of how much water vapor is actually in the air compared with the maximum amount the air can hold at saturation. This comparison is expressed as a percentage.

When the air holds all the water it can, it is said to have a relative humidity of 100%. Assuming that 1 m^3 of air at 30°C could hold 30 g when saturated, you can express the relative humidity for the same air holding 20 g of water as 20 g/30 g = 0.67 = 67%.

If the air is not saturated, water can evaporate and enter the atmosphere. The energy needed to evaporate the water comes from the water itself. As the water evaporates, it loses heat energy. This loss of heat lowers the temperature of the water. A chart showing the cooling effect of evaporation can be used to determine relative humidity.

In this investigation, you will use wet-bulb and dry-bulb thermometer readings to determine relative humidity.

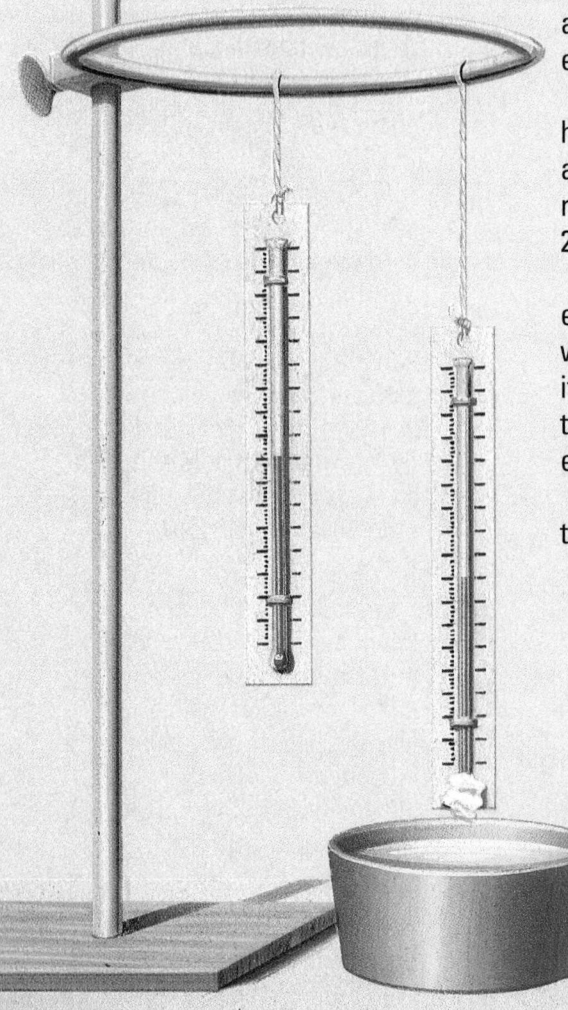

Prelab Preparation

1. Review Section 24.1, pages 480–483.
2. Study Figure 24-2 on page 480.
   a. How many grams of water can a cubic meter of air hold at a temperature of 20°C?
   b. How many grams of water can the air hold at 40°C?
3. If a cubic meter of air at 40°C holds 40 g of water, what is the relative humidity?

Procedure

1. Set up two thermometers as shown in the illustration on the previous page. Wrap a piece of cotton cloth around the bulb of one thermometer. Adjust the length of the string so that only the cloth and not the thermometer bulb is immersed in the water. With this setup you can measure both the air temperature and the cooling effect of evaporation.
2. Make a prediction. Will the two thermometers have the same reading? If not, which thermometer will have the lower reading?
3. Fan both thermometers rapidly with a piece of paper until the reading on the wet-bulb thermometer stops changing. Read the temperature on each thermometer.
   a. What is the temperature on the dry-bulb thermometer?
   b. What is the temperature on the wet-bulb thermometer?
   c. What is the difference in the two temperature readings?
4. Use the table on page 672 to find the relative humidity based on your temperature readings in Step 3. Look at the left-hand column labeled "Dry-Bulb Temperature." First, find the temperature you recorded in Step 3(a). Follow along to the right in the table until you come to the number that is directly below the "Difference in Temperature" (top row of the table) that you recorded in Step 3(c). This number, expressed as a percentage, is the relative humidity. What is the relative humidity at your laboratory station?

Analysis and Conclusions

1. Based on the relative humidity you found, can the air in your classroom hold more evaporated water?
2. If you wet the back of your hand, would the water evaporate and cool your skin?

Extensions

1. Suppose you exercise in a room in which the relative humidity is 100%.
   a. Would the moisture on your skin from perspiration evaporate easily?
   b. Would you be able to cool off readily? Explain.
2. Suppose you have just stepped out of a swimming pool. The relative humidity is low, about 30%. How would you feel—warm or cool? Explain.

Chapter 25 Weather

If you studied images of the earth taken from a satellite, you would see evidence that the atmosphere is far from quiet and peaceful.

In this chapter, you will learn how storms, lightning, and other atmospheric phenomena are produced and how meteorologists forecast the weather.

Chapter Outline

Lightning illuminates the night sky.

25.1 Air Masses

Section Objectives

- **Explain how an air mass forms.**
- **List and describe the types of air masses that usually affect the weather of North America.**

Differences in air pressure at different locations on the earth create wind patterns. The region along the equator receives much more solar energy than do the regions at the poles. Because warm air rises and cold air sinks, the heated equatorial air rises. As the air rises, it creates a low-pressure belt. Conversely, cold air near the poles sinks, creating high-pressure centers.

Air moves from areas of high pressure to areas of low pressure. Therefore, there is a general worldwide movement of surface air from the poles toward the equator. At higher altitudes the warmed air flows from the equator toward the poles. Temperature and pressure differences on the earth's surface alter this general wind pattern, however. These conditions create three *convection cells* in the Northern Hemisphere and three in the Southern Hemisphere. The earth's rotation also influences the wind pattern by causing the deflection of winds by the *Coriolis effect,* as described in Chapter 23. This deflection influences the direction of the prevailing wind belts.

In areas where air pressure differences are small, air remains relatively stationary. If the air remains stationary or moves slowly over a uniform surface, it takes on the characteristic temperature and *humidity* of that region. Such a large body of air—one with uniform temperature and moisture content—is called an **air mass.** An air mass is sometimes thousands of kilometers in diameter. A source region, where air masses form, must have a fairly uniform temperature and moisture content, such as occurs over polar or desert land areas or the ocean. Air masses that form over frozen polar regions are very cold and dry. Air masses that form over tropical oceans are warm and moist.

Figure 25–1. The movement of an air mass often is indicated by clouds.

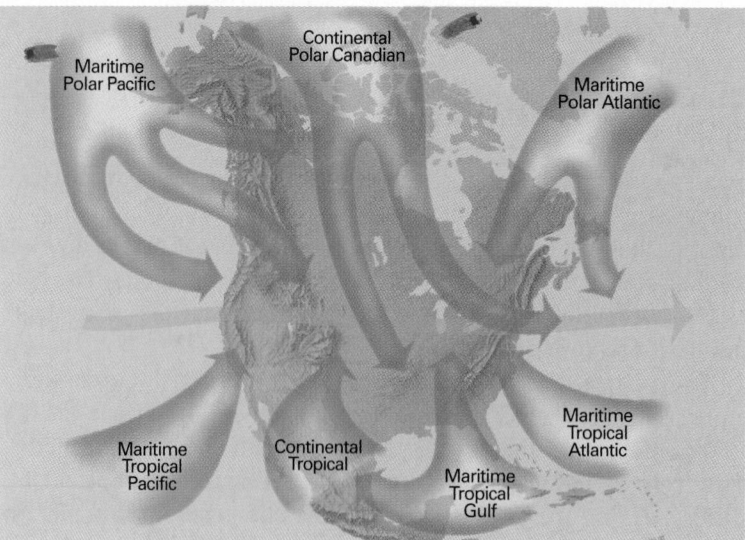

Figure 25–2. This map indicates the general direction of movement of the seven air masses that influence the weather in North America.

Types of Air Masses

Air masses are classified according to their source regions. The source regions also determine the temperature and the humidity of the air mass. The source regions for cold air masses are polar areas; these air masses are labeled *P*. The source regions for warm air masses are tropical areas; these air masses are labeled *T*. Air masses formed over the ocean are called *maritime* and are labeled *m*. Air masses formed over land are called *continental* and are labeled *c*. Maritime air masses are moist; and continental air masses are dry. The combination of tropical or polar air and continental or maritime air results in four main types of air masses. These air masses are **maritime polar** (mP), **maritime tropical** (mT), **continental polar** (cP), and **continental tropical** (cT).

An air mass may remain over its source region for days or weeks. Eventually it will move into other regions because of the worldwide wind patterns.

North American Air Masses

Air masses that strongly affect the weather of North America come from seven regions. These regions and the general directions of the air movements are shown in Figure 25–2. An air mass usually brings with it the weather of its source region, but it may change as it moves away. For example, cold, dry air may become warmer and more moist as it moves from land over the warm ocean. As the lower layers of air are warmed, the air rises; this warmed air can create heavy clouds and precipitation.

Polar Air Masses

Three polar air masses influence weather in North America. These air masses are called *continental polar Canadian, maritime polar Pacific,* and *maritime polar Atlantic*.

Continental polar Canadian air masses (cP) form in northern Canada over land that is covered by ice and snow. The cP air masses generally move southeastward across Canada and into the northern

United States. Occasionally these cool air masses reach as far south as the Gulf Coast of the United States. In the summer they usually bring cool, dry weather. In the winter they bring very cold weather to the northern United States.

Maritime polar Pacific air masses (mP) originate over the North Pacific Ocean, especially over the cool waters near Alaska's Aleutian Islands. These air masses are very moist but not extremely cold. The Pacific Coast is most affected by them. In the winter, maritime polar Pacific air masses bring rain and snow to the Pacific Coast. In the summer, they bring cool, often foggy weather. As they move inland and eastward, these air masses lose much of their moisture passing over the Cascades, the Sierra Nevada, and the Rocky Mountains. Thus, they may bring cool and dry weather by the time they reach the mid-United States.

Maritime polar Atlantic air masses (mP) form over the North Atlantic Ocean, especially over the cool waters between Greenland and Iceland. They move generally eastward toward Europe. However, they may pass over New England and eastern Canada. In the winter, they usually bring cold, cloudy weather and precipitation. In the summer, these air masses often produce cool weather with low clouds and fog.

Tropical Air Masses

Four tropical air masses influence the weather in North America. These air masses are called *continental tropical, maritime tropical gulf, maritime tropical Atlantic,* and *maritime tropical Pacific*.

Continental tropical (cT) air masses flow over North America only in the summer. They form over the deserts of northern Mexico and the southwestern United States and generally move northeastward, bringing clear, dry, and very hot weather.

Maritime tropical gulf and maritime tropical Atlantic air masses (mT) form over the warm waters of the Gulf of Mexico and the tropical North Atlantic Ocean. They move northward across the eastern United States, bringing mild, often cloudy, weather in the winter. In the summer, they bring hot, humid weather, thunderstorms, and hurricanes.

Maritime tropical Pacific air masses (mT) form over the warm areas of the North Pacific Ocean, but they rarely reach the Pacific Coast. However, in the winter, mT air masses may bring heavy precipitation and thunderstorms to the coast and the southwestern deserts.

Section 25.1 Review

1. Define *air mass*.
2. What is the name of the air mass that forms over the warm water of the Gulf of Mexico? What letters designate the source region of this air mass?
3. Suppose that snow is falling on the Pacific Coast area. What type of air mass is probably responsible for this weather? What letters designate the source region of this air mass?

Section Objectives

- **Compare the characteristic weather patterns of cold fronts with those of warm fronts.**
- **Describe how a wave cyclone forms.**
- **Describe the stages in the development of hurricanes, thunderstorms, and tornadoes.**

25.2 Fronts

When two unlike air masses meet, density differences usually keep the air masses separate. The air of a cool air mass is dense and does not mix with the less-dense air of a warm air mass. Thus, a boundary, called a **front,** forms between air masses. A typical front is several hundred kilometers long, but some fronts may be several thousand kilometers long. Changes in middle-latitude weather usually take place along the various types of fronts. Fronts do not exist in the tropics because there are no contrasting air masses.

Types of Fronts

In order for a front to form, one air mass must collide with another air mass. The kind of front that forms depends upon how the air masses are moving. When a cold air mass overtakes a warm air mass, a **cold front** is formed, as shown in Figure 25–3. The moving cold air lifts the warm air. If the warm air is moist, clouds will form. Large cumulus and cumulonimbus clouds typically form along a fast-moving cold front. Storms created along a cold front are usually short-lived and sometimes violent. A long line of heavy thunderstorms, called a **squall line,** may occur just ahead of a fast-moving cold front. A slow-moving cold front lifts the warm air ahead of it more slowly than does a fast-moving front. For that reason, a slow-moving cold front produces less cloudiness and lighter precipitation.

A warm air mass that overtakes a cooler air mass produces a **warm front.** The less dense warm air rises over the cooler air. The slope of a warm front is gradual, as you can see in Figure 25–3. Because of this gentle slope, clouds may extend far ahead of the surface location, or base, of the front. A distinct pattern of lowering clouds precedes the approaching base of a warm front. At the beginning of the pattern are high cirrus clouds. Behind the cirrus clouds

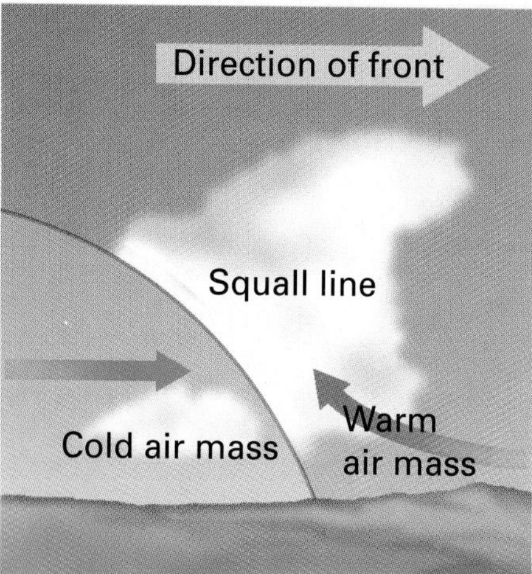

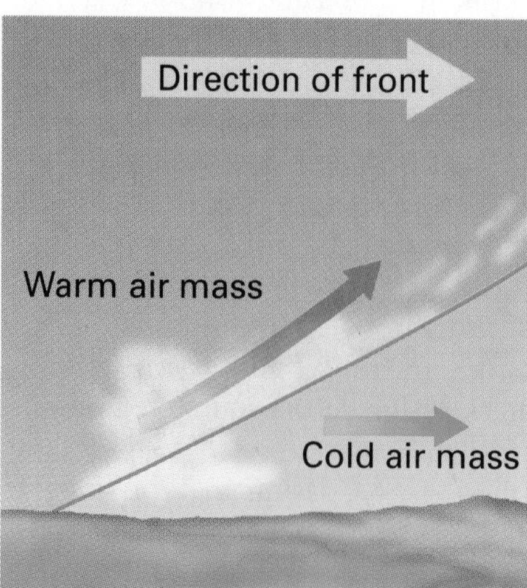

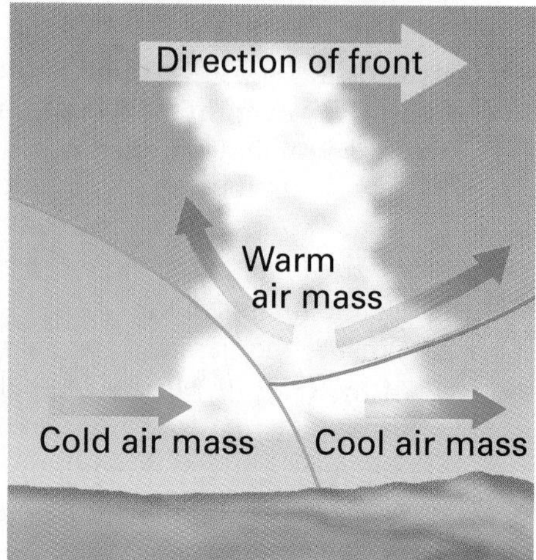

Figure 25–3. Cold and warm air masses collide to form a cold front (left), a warm front (center), and an occluded front (right).

are cirrostratus clouds, followed by altostratus, low stratus, and finally nimbostratus clouds at the base of the front. A warm front generally produces precipitation over a large area. A warm front may produce some violent weather if the body of warm air advancing over the cooler air is very moist.

Sometimes when two air masses meet, neither is displaced. The two air masses move parallel to the front between them. The front formed between the air masses is called a **stationary front,** because it does not move. The weather around a stationary front is similar to that produced by a warm front.

An **occluded front,** as shown in Figure 25–3, usually forms when a fast-moving cold front overtakes a warm front, lifting the warm air completely off the ground. The advancing cold front then comes into contact with cool air that develops beneath the lifted warm air. The warm front is completely cut off, or occluded, from the ground by colder air. The warm air is held above the ground high in the atmosphere. Figure 25–4 shows the symbols commonly used for the various types of fronts.

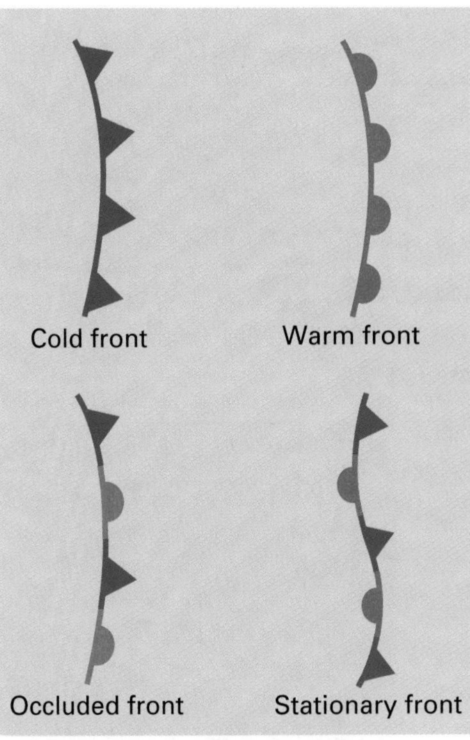

Figure 25–4. Above are the four frontal symbols meteorologists use to designate fronts.

Polar Fronts and Wave Cyclones

Over each of the earth's polar regions is a dome of cold air. The boundary at which the cold polar air meets the warmer air of the middle latitudes is called a **polar front.** A polar front circles the earth between 40° to 60° latitude in each hemisphere. The polar fronts move closer to the equator in the winter and back toward the poles in the summer. In the winter, the average position of the polar front in North America is across the middle of the United States. In the summer, it is north of the Great Lakes, across southern Canada.

Waves often develop along the polar fronts. A wave is a bend formed in a cold front or a stationary front. These waves are similar to the waves that moving air produces when passing over a body of water. However, they are much larger, often hundreds of kilometers in length. High-speed *jet-stream* winds, which you read about in Chapter 23, help develop a wave. As a wave develops in the polar front, meteorologists can detect shifts in the position and motion of the jet stream.

The waves along the boundary of a polar front are the beginnings of low-pressure storm centers called **wave cyclones.** Wave cyclones are large storms—up to 2,500 km in diameter. Their winds blow in circular paths spiraling upward around the low-pressure region at the center. Wave cyclones form not only along polar fronts but also along other cold or stationary fronts, and they strongly influence weather patterns in the middle latitudes.

Stages of a Wave Cyclone

In the first stage of a typical wave cyclone, there is a stationary front between a warm air mass and a cold air mass. At this stage the winds usually move parallel to the front. However, the winds on one side of the front blow in the opposite direction from the winds on

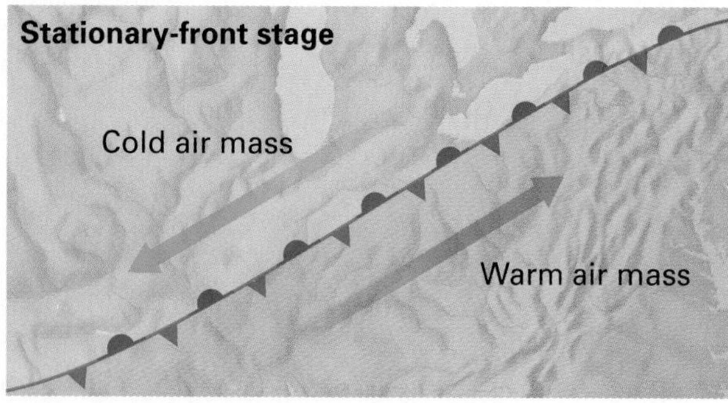

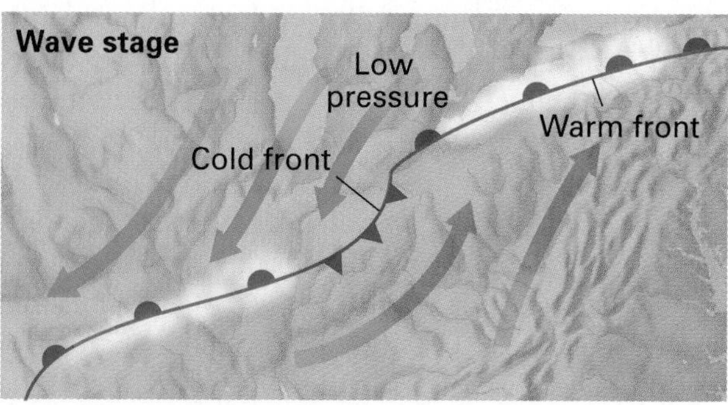

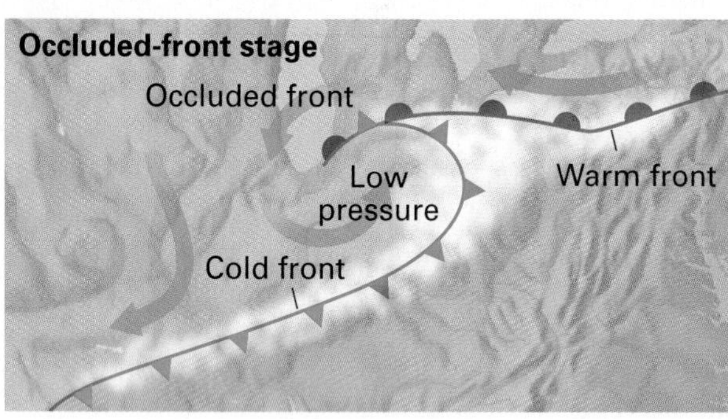

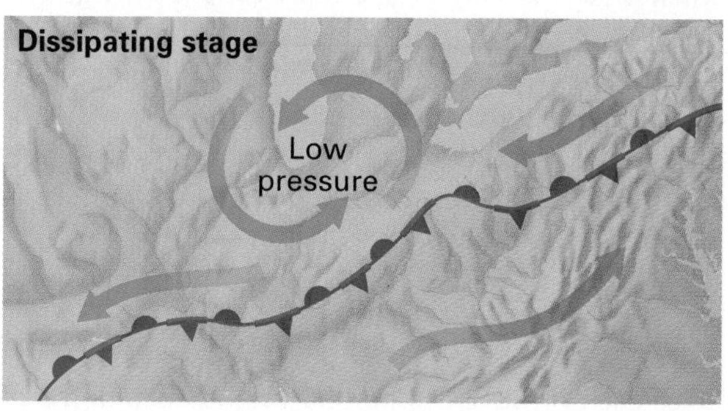

Figure 25–5. Wave cyclones occur along a cold or a stationary front (top, left). A wave cyclone begins when a bulge of cold air develops and advances slightly ahead of the rest of the front (top, right). As the fast-moving cold front overtakes the slow-moving warm front, an occlusion forms (bottom, left) and the storm reaches its highest intensity. Generally within 24 hours after the occlusion, the system loses all its energy and the wave cyclone disappears (bottom, right).

the other side. A wave develops along the stationary front as cold air that is moving toward the equator pushes into the warm air, forming a cold front. At the same time, warm air that is moving poleward pushes into the cold air, forming a warm front. The result of this air movement is a slow-moving warm front and a fast-moving cold front turning around a central low-pressure center. Clouds and precipitation spread along both the warm and cold fronts.

An intensified low-pressure region forms in the center, where the two fronts come together, as you can see in Figure 25–5. The less-dense warm air is lifted as it meets the cold air along the warm front. It is also pushed up by the advancing cold air behind the cold front. After a couple of days, the cold front overtakes the warm front and an occluded front develops. Generally within 24 hours after the occlusion, the wave cyclone dissipates.

A wave cyclone usually lasts several days. During this time, the air masses are in motion, and the disturbance moves with them. In North America, wave cyclones generally move in an easterly direction at speeds of about 32 to 64 km/hr, as they spin counterclockwise. North American wave cyclones follow several storm tracks. Most wave cyclones have a storm track from the Pacific coast to the Atlantic coast. As they pass over the western mountains, they lose their moisture and energy. Many, however, reintensify east of the Rockies as they receive energy from warm maritime tropical gulf air. Many enter in the Pacific Northwest and pass across the northern portions of the United States. Some enter through California and flow across to Texas and Florida. Some originate in the Rockies or Canada and affect only the midwestern and east-coast states.

Anticyclones

The air of an **anticyclone** sinks and flows outward from a center of high pressure, unlike a cyclone, which rises and flows inward toward a low-pressure center. Because of the Coriolis effect, the circulation of air around an anticyclone is clockwise in the Northern Hemisphere. Wave cyclones bring cloudy, stormy weather; anticyclones bring dry weather, since their sinking air does not promote cloud formation. If an anticyclone stagnates over a region for many days, it can cause droughts and air-pollution problems.

Hurricanes

Some tropical storms behave like middle-latitude wave cyclones. However, tropical storms are concentrated over a smaller area, lack warm and cold fronts, and are usually much more violent and destructive than are most wave cyclones. A severe tropical storm, with windspeeds starting at 120 km/hr, is called a **hurricane.** Hurricanes develop over warm, tropical oceans. Hurricane winds spiral rapidly in toward an intensely low-pressure storm center. Seldom more than 700 km in diameter, hurricanes are much smaller than wave cyclones but are much more powerful. These whirling storms are the most destructive storms that occur on the earth. The greatest number of hurricanes—an average of 20 per year—occurs in the western North Pacific Ocean, where they are called **typhoons.**

A hurricane begins when very warm, moist air over the evaporating ocean rises rapidly. When moisture in the rising warm air condenses, a large amount of energy in the form of latent heat is released. This heat increases the force of the rising air. Moist tropical air continues to be drawn into the column of rising air, releasing more heat and sustaining the process. An average hurricane has an energy content equal to the total amount of electricity used in the United States in six months.

A fully developed hurricane consists of a series of thick cumulonimbus cloud bands spiraling upward around the center of the storm, as shown in Figure 25–6. Rain is torrential. Winds increase

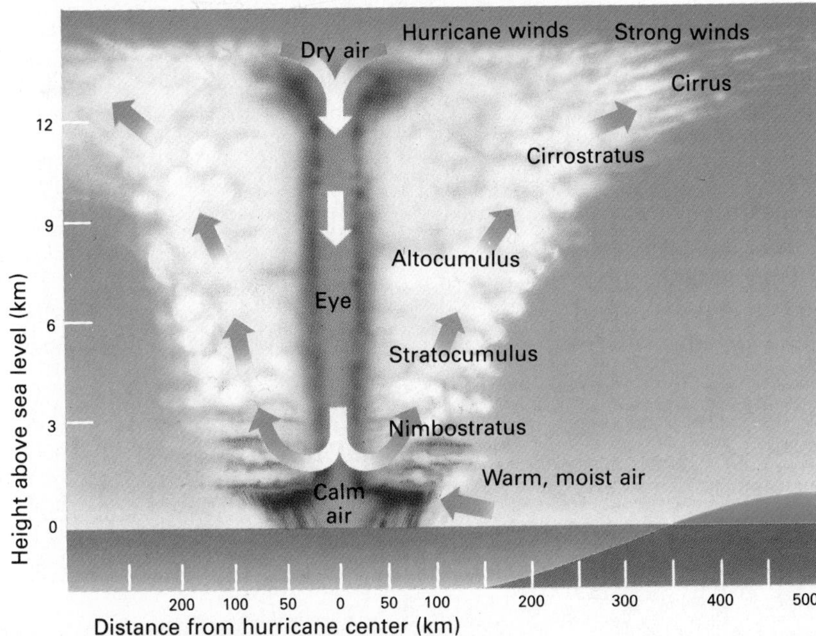

Figure 25–6. Thick clouds swirl upward around the center of a fully developed hurricane.

in velocity toward the center, or *eye*, of the storm, reaching speeds of up to 275 km/hr along the eyewall. The eye itself, however, is a region of calm, clear, sinking air. The most dangerous aspect of the hurricane is rising sea level and large waves, which can submerge vast areas of low-lying coastline. Most deaths during hurricanes are due to drowning.

Thunderstorms

A storm accompanied by thunder, lightning, and strong winds is called a **thunderstorm.** Thunderstorms often occur when a portion of air in a warm, moist mT air mass is heated and rises. High surface temperatures on the ground or in the ocean influence the rising of this warm, moist air. For that reason, thunderstorms occur most commonly in the late afternoon or early evening. Air rapidly rising over mountains or in a cold front also forms severe thunderstorms.

Weather Watch

In August 1969 Hurricane Camille slammed into the Gulf Coast of the United States. Winds up to 200 km/hr hammered coastal towns in Mississippi and Louisiana, ripping out power lines, flattening houses, and leaving 258 people dead.

In 1992 equally fierce Hurricane Andrew ripped across Florida and Louisiana, yet this time the death toll was lower. Thanks, in part, to improved weather forecasting, over a million people had evacuated safely.

Each year severe storms batter many parts of the United States, causing loss of life and an estimated $20 billion to $30 billion in damage. Hurricanes, tornadoes, blizzards, and floods all take their toll. The impact of these storms is evident in the wreckage they leave behind.

These disasters have had another, less obvious, impact. In response to the destructive potential of severe storms, people have looked for more accurate ways to predict the weather. This search has led to the development of the science of meteorology and to advancements in weather forecasting.

Recent technological advances have vastly improved the accuracy of weather forecasts. New systems of Doppler radar, atmospheric sensors, weather satellites, and supercomputers now make it possible for meteorologists to track killer storms and issue advance warnings.

What areas of the earth should meteorologists monitor to detect developing hurricanes? Explain why.

A thunderstorm develops in three distinct stages. The first stage of a thunderstorm is called the *cumulus stage*. At this stage, warm, moist air rises until the water vapor within the air condenses and forms a cumulus cloud. In the next stage, called the *mature stage,* violently rising warm, moist air swells higher and higher. The cumulus cloud grows until it becomes a dark cumulonimbus cloud. The cloud tops may reach heights of over 15 km, where the top spreads out in an anvil shape. Heavy, torrential rain showers and, occasionally, hailstones fall from the cloud. This precipitation produces a generalized cooling effect. While strong updrafts continue to rise, much of the air is dragged downward by the falling ice and raindrops, causing strong downdrafts. During the final stage, or *dissipating stage,* the strong downdrafts stop air currents from rising. The thunderstorm dies out as the supply of water vapor decreases. Why does the decrease in available water vapor cause the storm to subside?

During a thunderstorm, clouds discharge electricity in the form of lightning. The released electricity heats the air, causing it to expand rapidly. The rapid expansion and collapse of the air produces the loud noise known as *thunder*. For lightning to occur, the clouds must have areas with distinct electrical charges. The upper part of the cloud usually carries a positive electrical charge, while the lower part carries mainly negative charges. Lightning occurs as a huge spark that travels between the two parts of the cloud when the difference in their electrical charges becomes great. Scientists do not fully understand why clouds in thunderstorms have these differently charged regions. One theory states that large raindrops or ice particles carrying negative charges sink to the bottom of the cloud. Lighter, positively charged particles, such as cloud droplets and ice crystals, collect in the upper parts of the clouds.

Only about 10% to 20% of lightning takes place between the clouds and the ground. Yet each year lightning from thunderstorms kills several hundred Americans. A few precautions can reduce the risk of being struck by lightning. The first rule is to avoid any high location in open land. Lightning follows the shortest path between the cloud and the ground. Therefore, it is most likely to strike the highest object in a particular location, such as a tree, an isolated building, or a standing person. All metal objects and bodies of water should be avoided during a thunderstorm. However, the interiors of buildings, especially those with metal frames, and the interiors of automobiles are relatively safe. Metal structures are safe because the lightning is conducted to the ground along the metal exterior.

Tornadoes

The smallest, most violent, and shortest-lived severe storm is a **tornado.** A tornado is a whirling, funnel-shaped cyclone. A tornado forms when a thunderstorm meets high-altitude, horizontal winds. These winds cause the rising air in the thunderstorm to rotate. One of the storm clouds may develop a narrow, funnel-shaped, rapidly

Figure 25–7. Tornadoes, such as this one that occurred in Waukesha County, Wisconsin, can be recognized by their rapidly spinning funnel.

spinning extension. The extension reaches downward and may or may not actually touch the ground. If the tip of the funnel does touch the ground, it generally moves in a wandering, haphazard path. Frequently the funnel rises and touches down again a short distance away. The tornado generally covers a path not more than 100 m wide. Usually everything in that path is destroyed.

The destructive power of a tornado is due to the speed of the winds whirling in the funnel. The great destructive power demonstrated by these rapidly spinning winds indicates that they may reach speeds of up to 480 km/hr. Most of the injuries caused by tornadoes occur when people are trapped in collapsing buildings or are struck by objects flung by the wind. If you are outside during a tornado, you should immediately find a low place on the ground, lie down, and cover your head.

Tornadoes occur in many locations. However, they are most common in "Tornado Alley," in the south-central and midwestern United States, during the late spring or early summer. Tornadoes over the ocean are called **waterspouts.** Waterspouts are usually smaller and less powerful than tornadoes.

Section 25.2 Review

1. What kind of front forms when a cold air mass overtakes a warm air mass?
2. What is the general storm track commonly traveled by wave cyclones in North America?
3. If you are canoeing on a lake when a thunderstorm breaks out, are you in danger of being struck by lightning? Explain your answer.

25.3 Weather Instruments

Section Objectives

- **Describe the types of instruments used to measure air temperature and wind speed.**
- **Describe the instruments used to measure upper-atmospheric weather conditions.**

Weather observations are based, in part, on measurements of atmospheric pressure, humidity, and precipitation. In Chapter 23 you read about *barometers,* which are used to measure atmospheric pressure. In Chapter 24 you read about *psychrometers* and *hair hygrometers,* which are used to measure relative humidity, and *rain gauges,* which are used to measure precipitation. Weather observations are also based on measurements of temperature and winds. These measurements are made by special instruments.

Measuring Air Temperature

An instrument used to measure temperature is called a *thermometer.* Three types of thermometers are used by meteorologists. A common type of thermometer uses a liquid—usually mercury or alcohol—sealed in a glass tube. A rise in temperature causes the liquid to expand and move up the tube. A drop in temperature causes the liquid to contract and move down the tube. A scale marked on the glass tube indicates the temperature. Both the Celsius (C) and Fahrenheit (F) scales are commonly used in the United States.

A second type of thermometer is called a **bimetal thermometer.** This thermometer consists of a bar made up of two strips, each of a different metal. The metals, such as brass and iron, expand different amounts when heated. The bar will curve when heated and straighten again when cooled. An instrument called a **thermograph** measures temperature changes by recording the movement of the bar. In a thermograph a pen is attached to the bar. The pen point is placed against a chart on a rotating drum. As the drum rotates, the bending and straightening of the bar, and thus the temperature changes, are recorded on the chart.

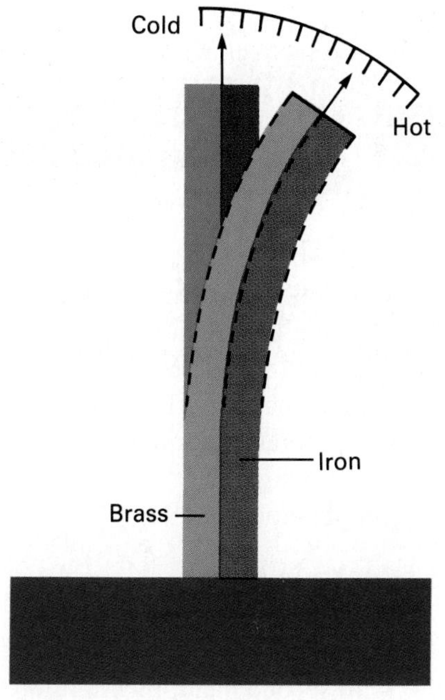

Figure 25–8. A bimetal thermometer curves when heated and straightens when cooled.

A third type of thermometer is an **electrical thermometer.** As the temperature rises, the electric current flowing through certain materials increases. The electric current flowing through these marials is translated into temperature readings. This type of thermometer is especially useful when an observer cannot be present.

Measuring Wind Speed and Direction

An instrument called an **anemometer** (AN-uh-MOM-uh-ter) measures wind speed. A typical anemometer consists of small cups attached by spokes to a shaft that rotates freely. The wind pushes against the cups, causing them to rotate. This rotation triggers an electrical signal that registers the wind speed—in m/sec, mph, or knots—on a dial. Some anemometers record the wind speed on a graph.

The direction of the wind is determined by a **wind vane.** Often a wind vane is shaped like an arrow with a large tail. The wind vane turns freely on a pole as the tail catches the wind. Thus, the arrow always points into the wind. A wind is described according to the direction from which it comes, so that *westerly* describes a wind that blows from the west. Wind direction may be described by one of 32 directions, or points, on the compass. Wind direction also may be recorded in degrees, beginning with 0° at north and moving clockwise to 360°. East is designated as 90°, and south, as 180°. How would a west wind be indicated in this system?

Measuring Upper-Atmospheric Conditions

Thermometers, anemometers, and wind vanes measure conditions of the atmosphere near the earth's surface. However, these conditions are only a part of the complete weather picture. An instrument package commonly used by meteorologists to investigate weather conditions in the upper atmosphere is the *radiosonde,* also discussed in

SMALL-SCALE INVESTIGATION

Wind Chill

Wind chill is a term used by meteorologists to describe the relationship between temperature and wind speed. While values are usually read from a chart, wind chill is based on the physics of evaporative cooling. You can demonstrate the cooling power of moving air with a simple model.

Materials

shallow pan, about 23 × 33 cm; water; thermometer; electric fan; clock or watch with second hand; metric ruler

Procedure

1. Place the pan on a level table, and fill it to a depth of about 1 cm with water at room temperature.
2. Lay the thermometer in the center of the pan with the bulb submerged. Be sure you can read the temperature without touching the thermometer.
3. Do not disturb the pan for 5 minutes, then read and record the water temperature.
4. Position an electric fan a few centimeters from the pan, and face the fan toward the pan. **Caution: Do not get the fan or cord wet.**

5. Turn on the fan at a low speed, and observe and record the water temperature every minute until the temperature remains constant.
6. If the fan has different speed settings, repeat Step 5 with the fan at a higher speed.

Analysis and Conclusions

1. How does the moving air affect the temperature of the water?
2. The moving air is the same temperature as the still air in the room. What causes the temperature to change?
3. On a cool, windy day, how would you dress to stay comfortable outdoors? Explain your answer.

Chapter 24. Carried aloft by a helium-filled weather balloon, a radiosonde measures relative humidity, air pressure, and air temperature. These measurements are sent back by radiowaves, where a receiver records the information. The path of the balloon is also tracked to determine the direction and speed of high-altitude winds. When the balloon reaches an extremely high altitude, it expands and bursts, and the radiosonde is parachuted back to earth.

Another instrument for determining weather conditions in the atmosphere is **radar.** Radar is an electronic device that transmits pulses of radio waves in the form of a beam. Objects that cross the beam reflect back to the transmitter, which receives the returning beams. Radar can detect objects that are too small or too far away for the human eye to see. For example, larger particles of water in the atmosphere reflect radar pulses. Thus, precipitation and storms, such as thunderstorms, tornadoes, and hurricanes, are visible on a radar screen. The newer Doppler radar can indicate the precise location, movement, and extent of a storm, as well as the intensity of precipitation and wind patterns within a storm. Radar data can then be analyzed on a computer screen. However, radar cannot detect cloud droplets and dust particles. Another device—the laser—emits a beam of light that meteorologists use to detect the presence of these fine particles and to measure the height of clouds.

Figure 25–9. Today meteorologists use a variety of instruments to gather and analyze weather data.

Instruments carried by weather satellites also provide important information about the atmosphere. For example, cameras on satellites take images of clouds. You have probably seen such images on a television weather report. Satellite images provide weather information for regions where observations cannot be made from the ground. The direction and speed of the wind at the level of the clouds can also be measured by examining a continuous sequence of cloud images. For night monitoring, satellite images made by using infrared energy reveal temperatures at the tops of clouds, at the surface of the land, and at the ocean surface. Satellite instruments can measure marine conditions such as the temperature and flow of ocean currents and the height of ocean waves.

Meteorologists also use supercomputers to understand the weather. Before computers were available, it was very difficult, and sometimes nearly impossible, to solve the mathematical equations describing the behavior of the atmosphere. In addition to solving many of these equations, computers can provide information that is useful in forecasting weather changes. Computers can also store weather records for quick retrieval. In the future, powerful computers may greatly improve weather forecasts and provide a much better understanding of the atmosphere.

Section 25.3 Review

1. How does a bimetal thermometer work?
2. What does a radiosonde measure?
3. If a wind vane is pointing toward the west, from what direction is the wind blowing? Explain your answer.

Section Objectives

- **Explain how a weather map is made.**
- **Describe the steps involved in preparing a weather forecast.**

25.4 Forecasting the Weather

Predicting the weather has challenged people for thousands of years. Early civilizations attributed control of weather conditions such as wind, rain, and thunder to the gods. More than 4,000 years ago, people attempted to forecast the weather using the position of the moon and stars as the basis for their forecasts.

Scientific weather forecasting began with the invention of basic weather instruments, such as the thermometer and the barometer. The invention of the telegraph in 1844 enabled meteorologists to share information about weather conditions quickly and led to the development of national weather services. In 1870 the United States formed a weather-forecasting agency as part of the Army Signal Corps. Twenty years later the agency was organized into the Weather Bureau. In 1970 it was renamed the National Weather Service. Because accurate weather forecasting is so important, all nations around the world cooperate in gathering and exchanging weather data.

Every six hours, weather observers at stations all over the world report weather conditions. These observers record the barometric pressure and whether it is rising or falling. They record the speed and direction of surface wind, and measure precipitation, temperature, and humidity. They note the type, amount, and height of cloud cover. Observers also record visibility and general weather conditions. Automated observing systems also collect similar data. In addition, many larger observation stations send up radiosondes to determine conditions high in the atmosphere. Each station puts the information in an international code and sends it to a central data-collection center. Weather centers around the world exchange the weather information they have collected.

The World Meteorological Organization (WMO) sponsors a program—World Weather Watch—to promote the rapid exchange of weather information. The organization helps developing countries establish or improve their meteorological services. It also offers advice on the effect of weather on natural resources and on human activities such as farming and transportation. WMO was founded in 1873 and is now part of the United Nations.

Making a Weather Map

Meteorologists prepare weather maps—the basic tools of weather forecasting—at the centers where the coded weather information is received. They revise these maps several times a day as new weather data are collected. Weather observations from the various stations reporting to the center are first translated into internationally recognized symbols. On the map, clusters of symbols are plotted around each reporting station, showing the conditions at that station. Such a cluster of symbols is called a **station model.**

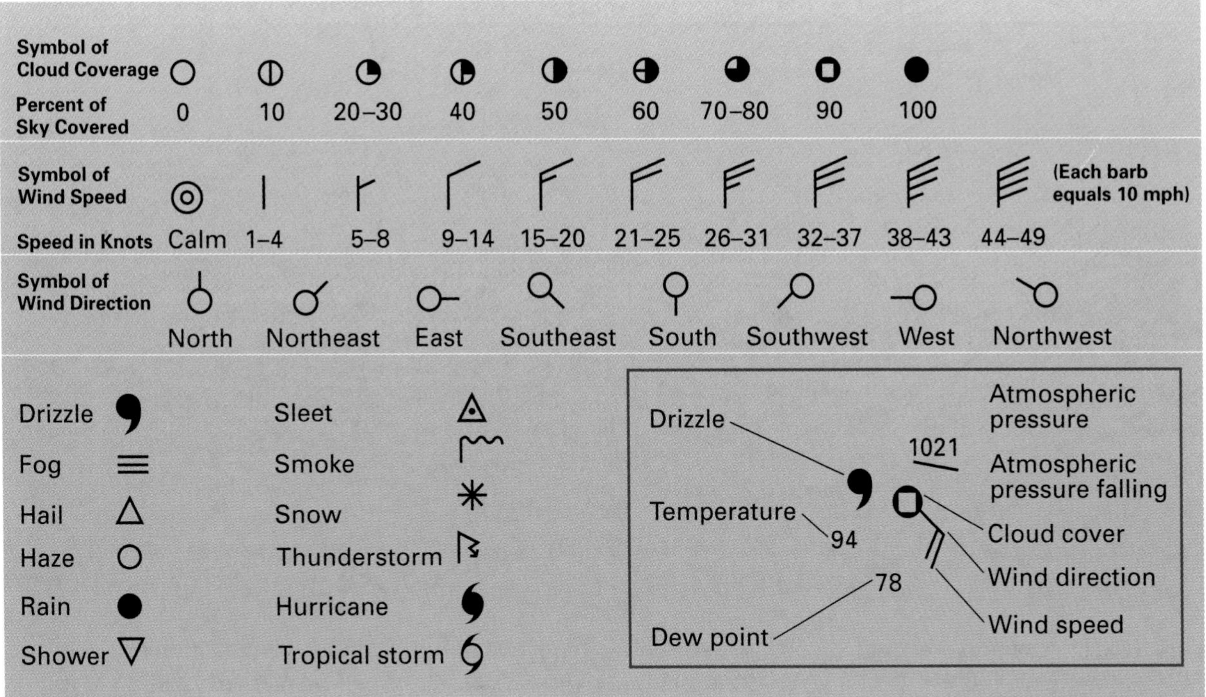

Common weather symbols describe cloud cover, wind speed, wind direction, and weather conditions such as type of precipitation and storm activity. These symbols and a station model incorporating them are shown in Figure 25–10. Notice that the symbols for cloud cover, wind speed, and wind direction are combined in one symbol.

Figure 25–10. Meteorologists use these symbols to indicate weather conditions, wind speed, cloud cover, and wind direction. A station model (inset) shows conditions around a weather station.

Other information included in the station model is the air temperature and the *dew point*. As explained in Chapter 24, the dew point is the temperature to which the air must be cooled to become saturated with water vapor. When the air is nearly saturated, there is little difference between its temperature and the dew point. Therefore, a comparison of these figures is an indication of the relative humidity at that time.

The station model also indicates the atmospheric pressure in millibars. The position of a straight line under this figure—horizontal or angled up or down—shows whether the atmospheric pressure is steady or is rising or falling.

On a weather map, lines are drawn to connect points of equal atmospheric pressure. These lines are called **isobars.** The spacing and shape of the isobars help meteorologists interpret their observations about the speed and direction of the wind. Closely spaced isobars indicate a rapid change in pressure and high wind speeds. Widely spaced isobars generally indicate a slow change in pressure and low wind speeds. Isobars that form circles indicate centers of high or low air pressure. Such centers are marked with an *H* for *high* or with an *L* for *low*. What configuration of isobars would represent a storm system?

The weather map is completed by marking in the location of fronts and areas of precipitation. Fronts are indicated by sharp

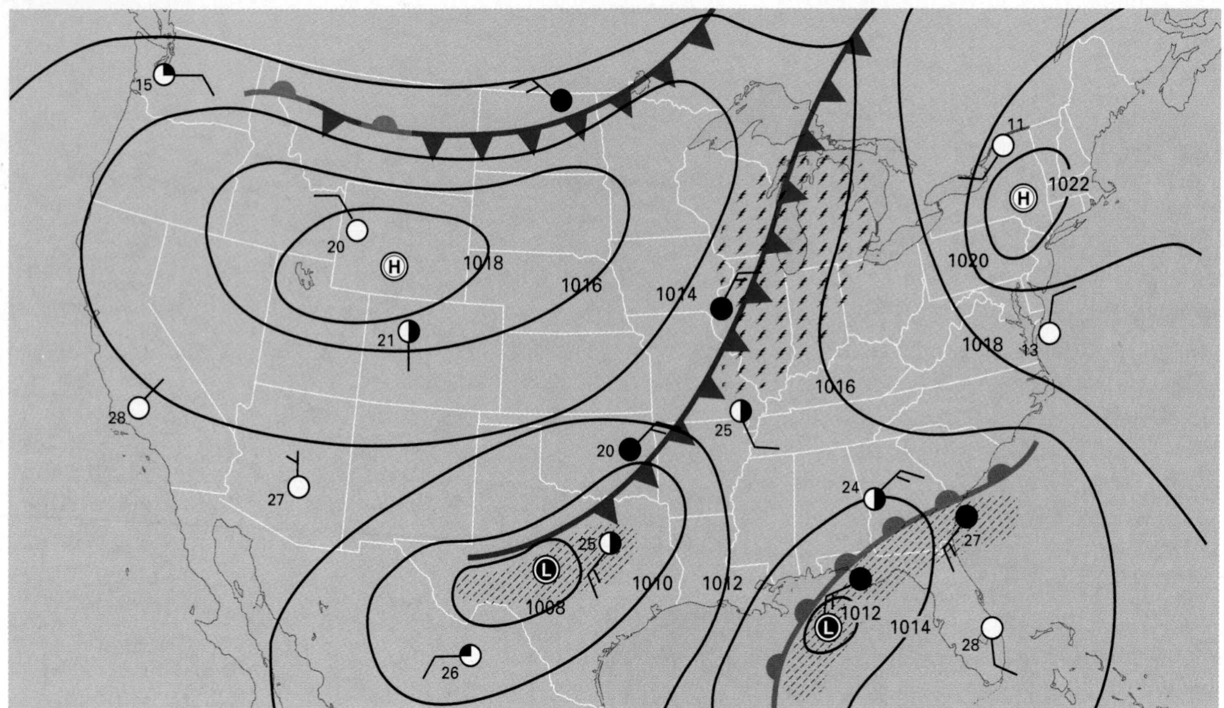

Figure 25–11. A typical weather map shows isobars, highs and lows, fronts, wind direction and speed, temperature, and precipitation.

changes in winds, temperature, or humidity. Areas between stations reporting precipitation are colored green or marked with a pattern of symbols. Use the weather map in Figure 25–11 to interpret the weather in your area the day the weather map was made.

Types of Forecasts

Meteorologists make two types of forecasts: daily and long-term. Daily forecasts predict weather conditions for a 48-hour period. Local daily extended forecasts look ahead five days. Long-range forecasts cover monthly and seasonal outlooks.

> **INVESTIGATE!**
>
> *To learn more about weather maps, try the In-Depth Investigation on pages 520–521.*

Daily and Long-Term Forecasts

To forecast the weather, meteorologists study the most recent weather map and compare it with maps for the previous 24 hours. This comparison allows them to follow the progress of large weather systems. The intensity and path of weather systems are plotted on these updated maps. Satellite images and radar reports of precipitation supply additional information. Meteorologists also use computers to prepare maps indicating conditions at many levels of the atmosphere. Although computers can make maps showing the possible weather conditions for the next several days, meteorologists must carefully interpret these maps because computer predictions are based on generalized descriptions. Using all the weather data available and their personal experience, meteorologists can then issue a forecast of the weather. Temperature, wind direction, wind speed, cloudiness, and precipitation can usually be forecast accurately, but it is often difficult to predict precisely when precipitation will occur or the exact amount that will fall.

Accurate weather forecasts can be made for three to five days, but accuracy decreases with each additional day. Extended forecasts

EARTHBEAT

Project VORTEX

The destructive power of tornadoes keeps meteorologists searching for better ways of predicting when tornadoes are likely to occur. In 1994 and again in 1995, more than 100 meteorologists participated in the *V*erification of the *O*rigins of *R*otation in *T*ornadoes *Ex*periment, or *VORTEX,* in hopes of improving tornado predictions. Armed with portable meteorological equipment, VORTEX scientists spent weeks chasing tornado-producing storms across Texas, Oklahoma, and Kansas at the peak of tornado season.

Earlier efforts by such storm chasers had led to the development of a new type of weather radar. This new radar could detect strong rotational winds within storms. The storm chasers then were able to provide the tornado sightings needed to verify which radar signatures led to tornado formation. These radars now form a nationwide network of *Nex*t Generation Weather *Rad*ars, or *NEXRAD,* which has extended tornado warnings by as much as half an hour.

VORTEX scientists used a mobile version of NEXRAD to capture the first three-dimensional radar images of a tornado from less than 3 km away. The images confirmed the theory that a tornado's fastest winds occur near the ground, and they revealed new features, such as powerful downdrafts in the tornado's vortex.

Understanding how tornadoes are structured and how they form should further improve tornado forecasts. The more timely and accurate these forecasts are, the greater the chances are of saving lives and property.

Why are tornadoes so destructive?

of 6 to 10 days are made by computer analysis of slowly changing large-scale movements of air. These changes help meteorologists predict the general weather pattern, but they still cannot forecast the long-term weather at a particular place. The distance between stations is too great to provide the necessary information.

Controlling the Weather

Some meteorologists are investigating methods of controlling rain, hail, and lightning. So far the most researched method for producing rain has been *cloud seeding*. In this process, freezing nuclei are added to supercooled clouds, causing rain to fall. Scientists in Russia have used cloud seeding with some success on potential hail clouds, causing rain to fall rather than hail. Cloud seeding is discussed further in Chapter 24.

Hurricanes have also been seeded with freezing nuclei in an effort to reduce the intensity of the storm. During *Project Stormfury,* which took place from 1962 to 1983, four hurricanes were seeded, with mixed results. Scientists have, for the most part, abandoned storm control and modification research, as it is not a feasible and attainable goal at the current level of technology.

Attempts have also been made to control lightning. The release of large quantities of ions near the ground can modify the electrical properties of small cumulus clouds. However, it is not known whether this method would affect the electrical properties of large

CAREER FOCUS: *Weather*

"One of the most interesting things about meteorology is that it's not an exact science. For researchers, there's always a challenge."

Mary Heffernan

Meteorologist

Not long ago, summaries of weather observation and information about forecast accuracy were available only 18 months after the fact—too late to be of practical use. Today, thanks to the work of meteorologists like Mary Heffernan of the National Weather Service, much summary and forecast data are available within 24 hours.

As a specialist in systems development, Heffernan has used her undergraduate training in computer science to design software needed for this more timely forecast-evaluation program.

"The system not only allows us to collect data more quickly, but it also helps us get feedback on forecast accuracy almost immediately, when it's most useful," she says.

This more immediate feedback allows weather forecasters to improve their reliability.

Heffernan, pictured at left, was also involved in developing another state-of-the-art support system for weather forecasters of the National Weather Service. Called AWIPS, for ***A**dvanced **W**eather **I**nteractive **P**rocessing **S**ystem,* this program makes it possible for computers to combine weather data from a variety of sources, including radar and satellites, more efficiently.

"Trying to understand what's happening in the atmosphere isn't always easy," says Heffernan, "but when you can see a need, define it, develop a solution, and finally see it implemented, the rewards are great."

Weather forecasters continually monitor changing weather patterns.

clouds. Seeding of potential lightning storms with silver-iodide nuclei has seemed to modify the occurrence of lightning, although no conclusive results have been obtained.

Section 25.4 Review

1. What is a station model?
2. Why are new and 24-hour-old weather maps compared?
3. Explain which region on a weather map—one with widely spaced isobars or one with closely spaced isobars—has stronger winds.

Severe-Storm Forecaster

Severe-storm forecasters work for the National Weather Service in stations such as those located in Kansas and in Florida. Kansas forecasters monitor severe thunderstorms and tornadoes. Florida forecasters monitor hurricanes that might threaten the East and Gulf coasts of the United States.

Severe-storm forecasters, such as the person shown at left, gather information from many sources, including satellites and radar. The collected data are then analyzed. If necessary, storm watches and warnings are issued.

A career in severe-storm forecasting requires a bachelor's degree in meteorology and experience in weather forecasting and analysis.

Weather Observer

Weather observers are also known as meteorological technicians. They observe, measure, and record weather conditions. Weather observers are also responsible for analyzing data, verifying the accuracy of the data, and supplying the results to meteorologists or other weather-information services.

Weather observers must be able to read instruments that measure temperature, wind velocity, air pressure, humidity, and rainfall.

Weather observers do not need a college degree, but a high school background in math and science is essential. Many people learn these skills in technical training programs.

For Further Information

For more information on careers in meteorology, write to the American Meteorological Society, 45 Beacon Street, Boston, MA 02108, or take a look at its home page at atm.geo.nsf.gov/AMS.

Chapter 25 Review

Key Terms

- air mass (499)
- anemometer (509)
- anticyclone (505)
- bimetal thermometer (509)
- cold front (502)
- continental polar (500)
- continental tropical (500)
- electrical thermometer (509)
- front (502)
- hurricane (505)
- isobar (513)
- maritime polar (500)
- maritime tropical (500)
- occluded front (503)
- polar front (503)
- radar (511)
- squall line (502)
- station model (512)
- stationary front (503)
- thermograph (509)
- thunderstorm (506)
- tornado (507)
- typhoon (505)
- warm front (502)
- waterspout (508)
- wave cyclone (503)
- wind vane (509)

Key Concepts

An air mass is a large body of air with uniform temperature and humidity. **See page 499.**

Polar air masses and tropical air masses influence the weather of North America. **See page 500.**

Cold and warm fronts are associated with characteristic weather conditions. **See page 502.**

A wave cyclone is a middle-latitude storm with a low-pressure center and fronts. **See page 503.**

Hurricanes, thunderstorms, and tornadoes are violent, destructive storms. **See page 505.**

Thermometers are used to measure air temperature. Anemometers are used to measure wind speed. **See page 509.**

Radiosondes, radar, satellite equipment, and computers are used to measure atmospheric weather conditions. **See page 510.**

Meteorologists prepare weather maps based on information from weather stations around the world. **See page 512.**

Meteorologists make daily and long-term forecasts of the weather. **See page 514.**

Review

On your own paper, write the letter of the term that best completes each of the following statements.

1. A region where air masses can form must be fairly
a. cold. b. warm. c. hilly. d. uniform.

2. In an air mass designated *cP*, the *c* stands for
a. continental. b. cold. c. coastal. d. cool.

3. Continental polar Canadian air masses generally move
a. southeasterly. b. northerly.
c. northeasterly. d. westerly.

4. The air masses that sometimes bring heavy rains to the deserts of the southwestern United States are called
a. continental polar Canadian. b. maritime polar Atlantic.
c. maritime tropical Pacific. d. maritime continental tropical.

5. The type of front formed when two air masses move parallel to the front between them is called
a. stationary. b. occluded. c. polar. d. warm.

6. A front that is completely lifted off the ground by cold air is called
a. cold. b. occluded. c. polar. d. warm.

7. The winds of a wave cyclone blow in circular paths around a
a. front. b. low-pressure center.
c. high-pressure center. d. jet stream.

8. The eye of a hurricane is a region of
a. hailstorms. b. torrential rainfall.
c. calm, clear air. d. strong winds.

9. In the mature stage of a thunderstorm, a cumulus cloud grows until it becomes a
a. stratocumulus cloud. b. altocumulus cloud.
c. cumulonimbus cloud. d. cirrocumulus cloud.

10. Tornadoes that occur over the ocean are called
a. waterspouts. b. typhoons. c. waves. d. hurricanes.

11. A wind with a direction designated as 90° is blowing from the
a. north. b. south. c. east. d. west.

12. An instrument package attached to a weather balloon is
a. an anemometer. b. a wind vane.
c. a radiosonde. d. a thermograph.

13. The lines on a weather map connecting points of equal atmospheric pressure are called
a. isobars. b. isotherms. c. highs. d. lows.

14. It is generally difficult to accurately predict
a. wind speed. b. amount of precipitation.
c. wind direction. d. temperature.

15. Concept map ? ? ? Of the following terms, which one would most likely be placed higher on a concept map designed for this chapter?
a. occluded front. b. wave cyclone.
c. tornado. d. air mass.

Critical Thinking

On your own paper, write answers to the following questions.

1. If the air in your region is warm and dry, what type of air mass could be responsible? What letters designate this air mass?
2. People on Vancouver Island, off the west coast of Canada, hear reports of a wave cyclone in the Gulf of Alaska. Is it likely that the wave cyclone will reach their area? Explain why.
3. Suppose a hurricane is passing over a Caribbean island. Suddenly, the rain and winds stop and the air becomes calm and clear. Is it safe to go outside? Explain your answer.
4. Is it safe to be on the street in an automobile during a tornado? Explain your answer.
5. In what direction would a wind of 315° make a wind vane point?
6. An air traffic controller is monitoring nearby airplanes by radar. The controller warns an incoming pilot of a storm a few miles away. How did radar help the controller detect the storm?

Application

1. Suppose you are traveling with friends through the desert in the southwestern United States and a thunderstorm occurs. You then tell them about the type of air mass that may have brought the storm. What did you say?
2. Suppose the air is warm and moist. You hear on the weather report, however, that a fast-moving cold front will reach your region the next day. What kind of weather conditions can you expect?
3. On a weather map, you see a station model with the circle half darkened and a straight vertical line extending upward. What can you say about the weather in that area for that day?

Extension

1. Research a recent, particularly destructive hurricane, thunderstorm, or tornado. Report your findings to the class. Use visual aids such as maps and charts in your report.
2. Watch or listen to the weather report for a week. Make a note each time the report mentions an approaching front and the weather it is expected to bring. Predict the weather for the next week and present your forecast to the class.

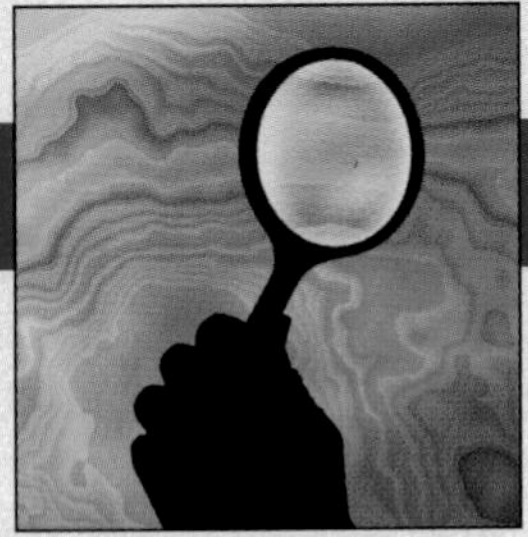

In-Depth Investigation

Weather Map Interpretation

Materials
- pencil
- paper
- colored pencils (red, blue)

Introduction

To read a weather map, you must be familiar with the meanings of various map symbols. Circles represent observation stations. A line pointing from the station circle indicates wind direction, extending in the direction from which the wind is blowing. A barb is used to indicate wind speed; each barb represents 10 mph, or about 11.5 knots. A knot (nautical mile/hour) is a unit of speed used in air and sea navigation. To the left of each station circle is the temperature in degrees Celsius. A line on a map connecting points of the same temperature is called an *isotherm.* Above each station circle is the atmospheric pressure in millibars. A millibar is a unit of pressure and is about 1 g/cm^2. A line joining points on a map that have the same pressure is called an *isobar.* Isobars help to identify fronts and give information about present and future wind movement. An *H* indicates a center of high pressure. An *L* indicates a center of low pressure. When an isobar crosses a front, the isobar forms a *V* with the point facing away from the low-pressure region. The *V* occurs where the isobar crosses a front because winds tend to blow along isobars and because there is a definite shift in wind direction from one side of a front to the other.

In this investigation, you will study the symbols used on a weather map to gain an understanding of the relationships among temperature, pressure, precipitation, and winds.

Prelab Preparation

1. Review Section 25.4, pages 512–517.
2. Make a copy of the weather map shown on the top of page 675.

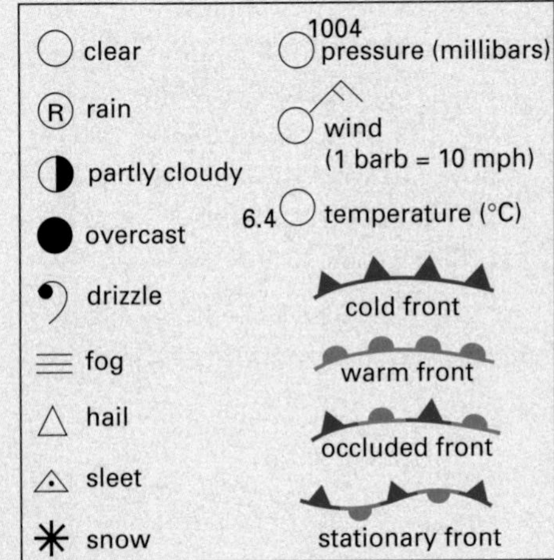

Procedure

1. Study the legend shown at left. Use the legend to interpret the weather map.
2. On your copy of the weather map, find a station whose temperature is 10.0°C. Draw a light pencil line through this station and through all other points that are also labeled 10.0°C.
3. If two stations have temperatures of 11.7°C and 9.5°C, there is an estimated point between them that is 10°C. Draw a light pencil line marking these points. Using a red pencil, draw a

continuous line connecting all the points of 10°C. The 10°C isotherm is a closed loop. However, some isotherms will be open curves or curved lines.

4. Using the same method as in Step 3, draw isotherms for every two degrees of temperature. Examples are 12°C, 16°C, and 22°C isotherms. Near cold and warm fronts, temperatures may change so rapidly over a short distance that it is best to leave out the 18°C isotherms. Label each isotherm with the temperature it represents.

5. Continuing on your copy of the weather map, find a station with a barometric pressure of 1004 millibars. Draw a light pencil line through this station and through all other points that are also 1004 millibars. Use the same method of estimation that you used in Step 3 to mark other points of 1004 millibars pressure. Connect all the points of 1004 millibars with a continuous blue pencil line.

6. Lightly draw isobars every 4 millibars, using the same method as in Step 5. Isobars for 992, 996, 1000, 1004, etc., should be drawn.

Analysis and Conclusions

1. What is the lowest temperature for which you have drawn an isotherm? What is the highest temperature for which you have drawn an isotherm?
2. Is either isotherm a closed loop? If so, which one?
3. Is the air mass identified by the closed isotherms a cold air mass or a warm air mass?
4. Is there a shift in wind direction associated with either front shown on your map? Describe the shift.
5. What is the value of the lowest-pressure isobar drawn? Is the isobar an open curve or a closed loop?
6. What is the value of the highest-pressure isobar drawn? Is the isobar an open curve or a closed loop?
7. Do the winds blow in a general clockwise or counterclockwise direction around a low-pressure region? Is the general direction clockwise or counterclockwise around a high-pressure region?
8. What are some weather conditions associated with fronts?

Extensions

1. Assume the following things will happen in the 24 hours after the observations were made for your map.
   a. The cold front, with its weather and low-pressure center, moves about 650 km in a southeasterly direction.
   b. The warm front and its weather move about 1000 km eastward.
   c. The high-pressure center and weather move about 800 km southeastward.

   Predict the weather conditions at Station A 24 hours after the observations for your map were made. Record your predictions in a table with columns listing pressure, wind direction, wind speed, temperature, and sky condition. Also make and record predictions for Station B and Station C.
2. Are you equally sure of all the predictions for Stations A, B, and C? Explain.

Chapter 26 Climate

On a December morning while residents of New York City are shoveling out after a snowstorm, residents of Los Angeles awake to another warm, sunny day. Daily weather conditions and seasonal weather patterns vary greatly around the world. For example, New York has a humid climate, warm summers, and cold winters, while Los Angeles has a dry climate and mild temperatures year-round.

In this chapter, you will learn about the factors that influence weather conditions and produce global climate patterns.

Chapter Outline

◀ **These varied landscapes in Kenya have several different climates.**

26.1 Factors That Affect Climate

Section Objectives

- **Explain how latitude determines the amount of solar energy received on earth.**
- **Describe how the different rates at which land and water are heated affect climate.**
- **Explain the effects of topography on climate.**

The average weather conditions of a region, or the weather patterns that occur over many years, are referred to as *climate*. Usually scientists describe climate in terms of the average monthly and yearly temperatures and the average amount of precipitation. Average temperatures are calculated by adding two or more temperature readings and dividing by the number of readings. For example, the average daily temperature is calculated by averaging the high and low temperatures of the day. The monthly average is determined by averaging the daily averages. The yearly average temperature can be calculated by averaging the monthly averages.

Another way scientists describe temperature is by indicating the **temperature range.** Temperature range is the difference between the highest and lowest temperatures of a day or month. The yearly temperature range is the difference between the highest and lowest monthly averages.

Using only average temperatures to describe climate can be misleading. For example, both St. Louis, Missouri, and San Francisco, California, have an average yearly temperature of about 13°C. However, St. Louis has a climate with cold winters and hot summers, while San Francisco has a generally mild climate all year.

Another major weather condition, precipitation, is described as the average precipitation a region receives in a year. However, average yearly precipitation alone does not accurately describe climate. For example, the average yearly precipitation for New York City and for Miami, Florida, is almost the same. In Miami, however, the rain falls mostly during the rainy season from May to October. In New York City, various forms of precipitation fall throughout the year. Clearly, the climates of these two cities differ.

An accurate description of climate cannot be given with only statistics; it must also include several factors that influence both temperature and precipitation. These factors include latitude, heat absorption and release, and topography.

Latitude

A major influence on the climate of a region is its latitude, or distance from the equator. Latitude determines the amount of solar energy received by, and the prevailing wind belts of, the region.

Solar Energy

The amount of solar energy that a location receives depends on two factors: the angle at which the rays of the sun strike the earth and the number of hours of daylight the location receives. The angle at which the sun's rays strike a region is determined by its latitude and by the tilt of the earth's axis. At the equator, the rays always strike the earth at a very high angle, nearly 90° for much of the year.

INVESTIGATE!

To learn more about climate, try the In-Depth Investigation on pages 538–539.

In equatorial regions, both day and night are about 12 hours long throughout the year. The result is steady high temperatures year-round and a yearly temperature range of only 3°C or 4°C in most areas. There are no summers or winters—only dry or rainy seasons.

At higher latitudes, the sun's rays strike the earth at an angle of less than 90°. The rays do not heat these areas as much because their energy is spread over a wider area. Thus, average yearly temperatures in these locations are lower than those at the equator. Also, the lengths of the days and the nights vary. For example, at 45° north or south of the equator, the hours of daylight vary from about 16 hours in the summer to about 8 hours in the winter. Therefore, the yearly temperature range is large—more than 30°C in some regions.

In the polar regions, the sun sets for only a few hours each day in the summer and rises for only a few hours each day in the winter. Thus, the annual temperature range is very large, but the daily temperature ranges are very small.

When the average daily temperatures in various parts of the world are plotted on a map, they form a series of temperature zones, as shown in Figure 26–1. The boundaries of these zones roughly follow the parallels of latitude. The average daily temperatures are

Figure 26–1. The zones of average daily sea-level temperature in January (top) and July (bottom) roughly follow the lines of latitude.

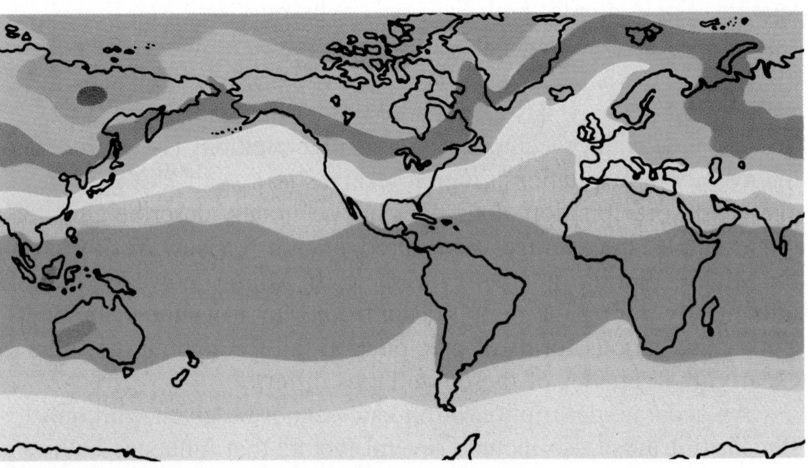

Average sea-level temperature in January

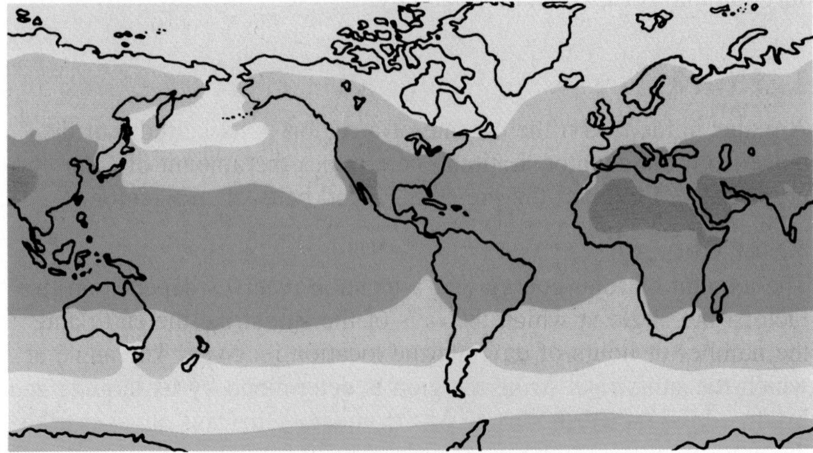

Average sea-level temperature in July

Degrees in Celsius (°C)

- Over 32
- 21 to 32
- 10 to 21
- −1 to 10
- −12 to −1
- −23 to −12
- −34 to −23
- −46 to −34
- Below −46

El Niño

From late 1982 to mid-1983, unusually violent weather wreaked havoc around the world. Typhoons and cyclones battered the Pacific United States, Hawaii, and Tahiti. Floods ravaged the southeastern United States, Bolivia, Ecuador, and Peru. Droughts struck Australia, southern Africa, Central America, Indonesia, and the Philippines. More than 1,000 people died, and property and crop losses soared to billions of dollars.

The cause of this unusual weather was discovered to be a warm Pacific current known as *El Niño.* It appears around Christmas about every three to ten years and lasts for about a year. El Niño begins with a weakening of the trade winds that usually push warm water toward the western Pacific Ocean. The weakened trade winds allow the warm water to flow eastward instead. This warm current heats the cool waters of South America's west coast to more than 27°C, upsetting the local marine ecology.

El Niño's effects are not limited to the equatorial Pacific region, however. Sea surface temperatures strongly affect atmospheric pressure and, thus, overall weather patterns. For example, the warming of the Pacific Ocean near South America shifts the subtropical jet stream and can cause unexpected weather around the world.

The severe consequences of the 1982–1983 El Niño prompted scientists to study the relationship between the ocean and the atmosphere more intensively. Using historical data, satellite measurements of sea surface topography, computer-aided models, and a network of meteorological buoys, scientists can now predict El Niño a year in advance. Improved forecasting significantly reduced losses from a 1986 El Niño and from three El Niño events that occurred between 1991 and 1995.

How might El Niño change the marine ecology on the west coast of South America?

highest near the equator and decrease with distance from the equator. However, the irregular shape of the temperature zones indicates that the amount of solar energy is not the only factor affecting climate.

Wind Patterns

Latitude also determines global wind belts that affect a region and, thus, the general direction of the wind in any particular location. The latitude ranges of these global wind belts are described in Chapter 23. Winds affect many weather conditions, such as humidity, precipitation, temperature, and cloud cover. Hence, regions with different prevailing winds often have different climates. The global wind pattern is also influenced by ocean currents and major mountain ranges.

Within the different global wind belts are various regions of low and high pressure. In the equatorial belt of low pressure—the doldrums—the air rises and cools, and water vapor condenses. As a

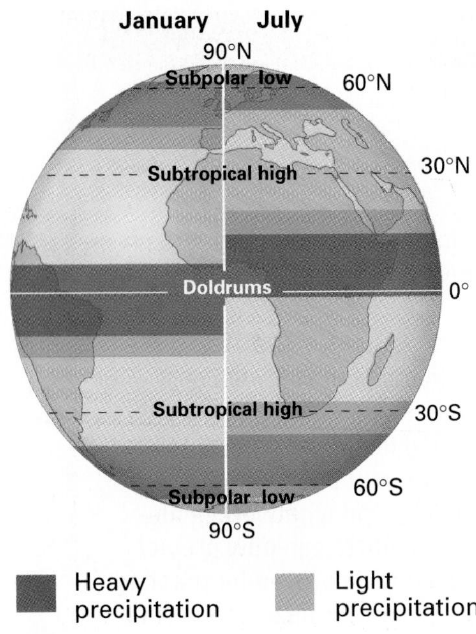

Figure 26–2. During winter in the Northern Hemisphere, global wind and precipitation belts shift to the south.

result, the equatorial regions of low pressure generally receive heavy precipitation. The amount of rainfall is most abundant in a belt around the equator and decreases steadily with increasing latitude. In the areas around 20° to 30° north and south latitude—the subtropical highs—the air is sinking, drying, and warming. Thus little precipitation occurs, causing most of the world's deserts to be located there.

Closer to the poles, at around 45° to 60° latitude, is a belt of higher precipitation. In these regions—the westerlies and subpolar lows—warm tropical air meets cold polar air, and *wave cyclones* frequently develop. At latitudes above 60°, average precipitation decreases in the cold, dry polar high-pressure air masses.

With the changing seasons, the global wind belts shift in a north-south direction, as shown in Figure 26–2. As the wind and pressure belts shift, the belts of precipitation associated with them also shift.

Heat Absorption and Release

The way solar energy strikes the earth and is absorbed or reflected also influences the surface temperature. Land heats faster and to a higher temperature than water does. One reason for this difference is that the land surface is solid and basically unmoving, while the water surface is liquid and continuously changing. Waves, currents, and other movements continuously replace warm surface water with cooler water from the ocean depths. This action prevents the surface temperature of the water from increasing rapidly. The surface temperature of the land, on the other hand, can continue to increase as more solar energy is received.

Land and water also absorb and release heat at different rates. The **specific heat** of water is higher than that of land. Specific heat is the amount of heat needed to raise the temperature of 1 g of a substance 1°C. A given mass of water requires more heat than does the same mass of land to increase its temperature the same number of degrees. Even if not in motion, water warms more slowly than land does. Water also releases heat more slowly than land does.

The average temperatures of land and water at the same latitude vary also because of differences in the loss of heat through evaporation. Evaporation affects water surfaces much more than it does land surfaces.

Ocean Currents

The amount of heat absorbed or released by the air is influenced by the temperature of ocean currents with which the air comes in contact. If winds consistently blow toward the shore, currents have a stronger effect on air masses. For example, the combination of a warm Atlantic current and steady westerly winds gives northwestern Europe an unusually high average temperature for its latitude. On the other hand, the warm Gulf Stream has less effect on the east coast of the United States because westerly winds usually blow the Gulf Stream and its warm maritime tropical air away from the coast.

Seasonal Winds

Heat differences between the land and the oceans sometimes cause winds to shift seasonally in certain regions. During the summer, the land heats more quickly than the ocean does. The warm air rises and is replaced by cool air from the ocean. Thus the wind moves landward. During the winter, the land loses heat more quickly than the ocean does, and the cool air flows away from the land. Thus the wind moves seaward. Such seasonal winds are called **monsoons.** They are strongest over the large landmass of Eurasia. For example, monsoons in southern Asia result from the heating and cooling of the northern Indian peninsula. In the summer, winds carry moisture to the land from the ocean, bringing heavy rainfall. In the winter, continental winds bring dry weather, sometimes even drought. Monsoon conditions also occur in eastern Asia and affect the tropical regions of Australia and East Africa.

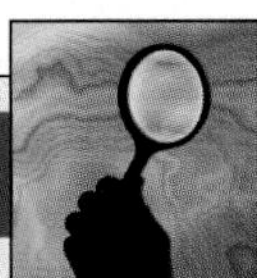

SMALL-SCALE INVESTIGATION

Evaporation

Evaporation affects temperature. The converse is also true. You can demonstrate how temperature affects the rate of evaporation of water.

Materials

portable clamp lamp (or flexible-neck lamp) with an incandescent bulb; 3 small Petri dishes or watch glasses; 3 thermometers; 50-mL graduated cylinder; ring stand with 2 rings; clock or watch; meter stick; water

Procedure

1. On your own piece of paper, make a data table similar to the one shown here.
2. Assemble the ring stand on a table, with support rings at heights of 25 cm and 50 cm above the base. Position the lamp directly over the rings at a height of 75 cm.
3. Place a Petri dish or watch glass on the base of the stand, and place one on each of the two rings.
4. Lay a thermometer across each dish, and turn on the lamp. Wait 3 minutes, then read and record the temperature shown on each thermometer.

Dish	Temperature	Amount of water evaporated
1		
2		
3		

5. Remove the thermometers, and add 30 mL of water to each of the three dishes.
6. Make sure the dishes are lined up directly beneath the lamp, and keep the light on for 24 hours.
7. Turn off the lamp, and carefully pour the water from the first dish into the graduated cylinder and record any change in volume. Repeat this process for the other two dishes.

Analysis and Conclusions

1. At what distance from the lamp did the greatest amount of water evaporate? the least?
2. Explain the relationship between temperature and the rate of evaporation.
3. Explain why puddles of water dry out much more quickly in summer than they do in fall or winter.

Figure 26–3. Air passing over a mountain range loses its moisture as it rises, expands, and cools. The dry air is compressed and warmed 1°C per 100 m as it descends on the other side of the mountain.

Topography

The topography, or shape of the land, also influences climate. The elevation, or height of landforms above sea level, produces distinct temperature changes. The average temperature decreases as altitude increases in the troposphere. For every 100 m increase in altitude, the average temperature decreases 0.7°C. Even along the equator, the peaks of high mountains are cold enough to be covered with snow.

Mountains influence the temperature and moisture content of passing air masses. When a moving air mass encounters a mountain range, it rises, adiabatically cools, and loses most of its moisture through precipitation. As the air descends on the other side of the range, it adiabatically warms (1°C every 100 m), compresses, and dries. Air flowing down mountain slopes, therefore, is usually warm and dry, as shown in Figure 26–3. One such wind is the **foehn** (FUHN)—a warm, dry wind that flows down the slopes of the Alps. Similar dry, warm winds that flow down the eastern slopes of the Rocky Mountains are called **chinooks.** A chinook wind can raise the air temperature very rapidly in a short period of time. In 1900, a chinook raised the temperature in a small town in Montana 17°C in three minutes. Would you expect to find more vegetation on the side of a mountain facing toward or away from the prevailing winds? Why?

Some winds that blow down mountain slopes are not warm. These winds have a source so high and cold that they remain cold even after heating. The **mistral,** which blows down from the snow-capped Alps to the Mediterranean Sea, is a strong, cold wind. Another cold mountain wind, the **bora,** blows from the mountains of Greece and the Balkan nations to the Adriatic Sea.

26.1 Section Review

1. List two factors that determine the amount of solar energy that an area receives. On what do these factors depend?
2. Does land or water heat more quickly? Why?
3. Compare monsoons and chinooks.

26.2 Climate Zones

Section Objectives

- **Name and describe the three types of tropical climates.**
- **Compare subarctic and tundra climates.**
- **List the various types of middle-latitude climates, and name the regions in which they are found.**
- **Explain why city climates may differ from rural climates.**

The earth has three major climate zones: tropical, polar, and middle-latitude. Each has distinct temperature characteristics. The warm zone immediately around the equator is the zone of **tropical climates.** Tropical climates have an average monthly temperature of at least 18°C, even during the coldest month of the year. Tropical climates are influenced by the continental and maritime tropical air masses, which develop close to the equator.

At the other extreme are the **polar climates.** In these regions, the average monthly temperature is never higher than 10°C. Continental and maritime polar air masses originate in these areas.

Between the tropical and polar climate zones is the zone of temperate climates, or **middle-latitude climates.** The average monthly temperature of these climates is no warmer than 18°C in the coldest month and no cooler than 10°C in the warmest month. In middle-latitude climate zones, the weather changes often because both tropical and polar air masses move across these regions. The middle-latitude climate zones are also frequently exposed to wave cyclones, with strong winds and heavy rains, that are produced along the polar front. The general boundaries of the three major climate zones are shown in Figure 26–4.

Tropical Climates

Each principal climate zone has a specific range of temperatures. However, there are several different types of climate within each of these zones because of differences in the amount of precipitation that occurs. For example, within the tropical zone there are three types of tropical climates: **tropical rain forest, tropical desert,** and **tropical savanna.**

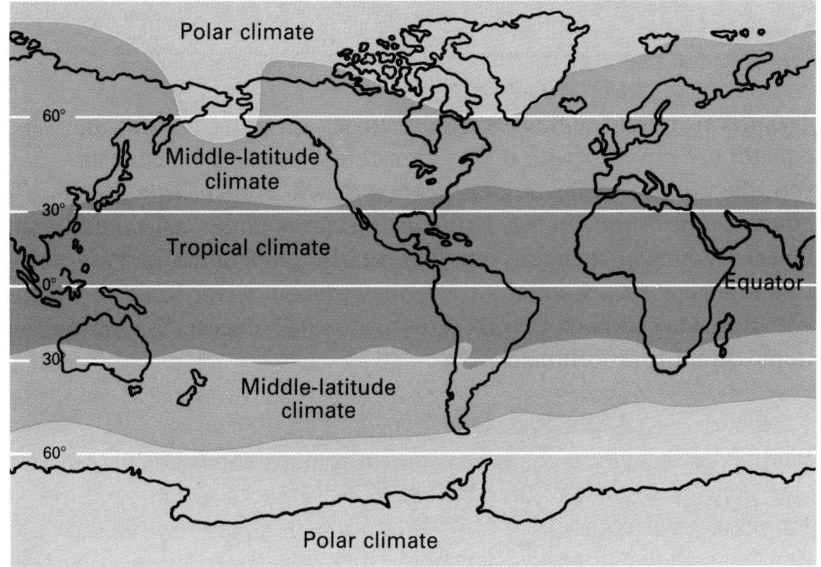

Figure 26–4. The general boundaries of the major climate zones are shown on this map.

EARTHBEAT

An Endangered Habitat

Tropical rain forests are one of the earth's most complex and varied natural habitats. Approximately one third of all the plant and animal species on earth live in rain forests. Scientists have identified 1,100 species of trees, shrubs, and grasses on less than 2.6 km^2 of land in the Colombian rain forest. In contrast, all of Great Britain and Ireland, which are located in a middle-latitude climate, support only about 1,450 species.

The plants in rain forests are sources for many medicines and for products used in agriculture and industry. For example, the copaiba tree produces a liquid similar to diesel fuel. The wild plants of the rain forests also may become an important source of new crops if cultivated varieties lose their resistance to diseases and insects.

Destruction of the earth's rain forests, however, is occurring at an alarming rate. Scientists predict that if this destruction continues at its present rate, all tropical rain forests will be destroyed over the next several decades. Huge tracts have been and continue to be burned to provide land for agriculture and urban development. In addition to destroying the forests, the burning of the trees adds carbon dioxide to the atmosphere. This carbon dioxide contributes to the greenhouse effect, which may be causing a rise in global temperatures. Also common are slash-and-burn farming practices: Forests are cut and burned down, crops are planted, then the land is abandoned when the minerals in the soil have been used up.

Approaches to reclaiming the rain forests include cultivating rain-forest products, using chemical fertilizers, and instituting crop rotation to restore the fertility of the soil.

How does destruction of the rain forests affect the tropical climate?

Tropical Rain Forest Climate

The warm, humid regions within 5° to 10° on either side of the equator are covered with dense, rain-forest vegetation. For that reason, the climate of this region is known as a *tropical rain forest climate*. The warm, moist, rising air produces an annual rainfall that is usually greater than 250 cm. The yearly temperature range is very small—about 3°C. Central Africa, the Amazon River basin of South America, and parts of Central America and Southeast Asia have tropical rain forest climates.

Tropical Desert Climate

Warm, dry weather conditions occur in regions about 20° to 30° latitude north and south of the equator. The Tropic of Cancer and the Tropic of Capricorn (23.5° N/S)—the limit to which the sun advances on the summer and winter solstices—fall within this zone.

These areas have a *tropical desert climate*. Tropical desert climates are influenced by the sinking, warming, and dry air masses of the subtropical highs. They include some of the earth's driest deserts. Annual rainfall in a tropical desert climate is less than 25 cm. The largest belt of tropical deserts extends across North Africa and southwestern Asia and includes the Sahara, Arabian, and Thar deserts. Smaller tropical deserts in the Northern Hemisphere are located in the Sonoran region of northern Mexico and in the southwestern part of the United States. The Kalahari Desert in southwest Africa and most of the interior of Australia also are tropical deserts.

Tropical Savanna Climate

A third type of tropical climate, the savanna climate, occurs in areas located between the tropical rain forest climate and the tropical desert climate. During different seasons, the precipitation belts shift toward the poles, producing very wet summers and very dry winters in these savanna regions.

The weather conditions favor the growth of the type of plants common to a *savanna*. As shown in Figure 26–5, a savanna consists mainly of open areas of coarse grasses that generally grow in clumps. Widely scattered on the grassland are drought-resistant trees and shrubs. Savanna climates are found in the areas that border the rain forests of South America and Africa. Parts of southeast Asia and northern Australia also have savanna climates. In southeast Asia the differences in the seasons are extreme because of the alternating monsoon rains and dry periods.

The Hawaiian Islands have both rain forest and savanna climates. These islands are located in the belt of trade winds that blow almost continuously from the northeast toward the equator. The trade winds bring heavy precipitation to the eastern windward slopes of the islands, causing a tropical rain forest climate. However, the western leeward slopes of the islands have sinking, downslope air and receive less rainfall, causing a savanna climate.

Figure 26–5. The vegetation in a savanna climate consists of drought-resistant grasses, shrubs, and trees, such as those in northern Australia.

Polar Climates

Climates that are mainly influenced by polar air masses—polar climates—occur in the regions located between 55° latitude and the poles. There are two types of polar climates: the **subarctic climate** and the **tundra climate.**

Subarctic Climate

All the land across North America, Europe, and Asia that lies between 55° and 65° north latitude, including most of Alaska, has a subarctic climate. Dry continental polar air masses control this climate. In the subarctic climate, yearly precipitation is only 25 cm to 50 cm. But due to low evaporation, forests of pine, spruce, and other cone-bearing trees can grow. Winters are severe, and summers are short. The subarctic climate has an unusually large annual temperature range. In fact, the largest yearly temperature range on earth—61°C—was recorded in subarctic Yakutsk, Siberia.

There is no subarctic climate in the Southern Hemisphere due to a lack of land between 55° and 65° south latitude.

Tundra Climate

The northern part of Alaska and other land areas north of the Arctic Circle have a tundra climate. This climate is named for the vegetation common to the region, the *tundra*. A tundra has no trees, but the ground is covered with mosses, lichens, and small flowering plants. It also has large expanses of rocky land with no vegetation. The yearly temperature range is not as great as that of subarctic climates. This is because areas with tundra climates are near the ocean, which holds some heat during the winter. The warmest months—July and August—have average temperatures of only 4°C. About 25 cm of precipitation are received in a year, mostly as snow.

In the Southern Hemisphere, only a few ice-free areas of Antarctica and some wind-swept desolate islands have a tundra climate. Poleward of the tundra climate are the ice caps. These areas, which are covered by pack ice or ice sheets, have no vegetation. Polar areas have very dry, cold air and receive little precipitation.

Figure 26–6. Although the Alaskan tundra has a cold average yearly temperature, many small plants bloom during the brief summer.

Middle-Latitude Climates

North America stretches from Central America, in the tropical climate zone, to northern Canada and Alaska, in the polar climate zone. However, the contiguous United States and southern Canada lie in the middle latitudes.

The climates of the various parts of the United States differ greatly. Wave cyclones bring most of the precipitation that falls on the United States and Canada. Because this precipitation falls unevenly across these two countries, they have several different middle-latitude climate types.

Marine West-Coast Climate

Between 40° and 60° north latitude, the Pacific Northwest of the United States has a **marine west-coast climate.** This land is in the belt of prevailing westerly winds of cool, moist maritime polar air. As these winds move east from the Pacific Ocean, the west coast receives precipitation. The coasts of Washington, Oregon, and northern California, where mountains block the movement of moist air toward the east, receive a great deal of moisture. The average yearly precipitation is 60 cm to 150 cm. The average temperature is a relatively cool 20°C in summer and a relatively mild 7°C in winter. The yearly temperature range is only 13°C. Most regions with a marine west-coast climate are covered with dense forests of cone-bearing trees.

Figure 26–7. The heavy rains of a marine west-coast climate help to maintain the growth of dense forests, such as this one in Prairie Creek Redwoods State Park in northern California.

Mediterranean Climate

Regions along the central and southern California coast have a climate like that of the coast of the Mediterranean Sea. This climate is thus called a **Mediterranean climate.** Regions with a Mediterranean climate are located between regions with a tropical-desert climate and those with a marine west-coast climate. They generally lie between 30° and 40° latitude. These regions have dry summers and wet winters. In the summer, the dry air of the subtropical highs blows over the region. In the winter, this pressure belt shifts southward, bringing wave cyclones to the region. Almost all of the yearly rainfall—an average of 40 cm—falls during the mild winter months. The yearly temperature range is small—only 7°C—with average summer temperatures of 21°C and average winter temperatures of 14°C.

Middle-Latitude Desert and Steppe Climates

Two types of middle-latitude dry climates are found between 35° and 50° north latitude in the interiors of Asia and North America. They are the **middle-latitude desert** and the **middle-latitude steppe climates.** Much of the land in the western United States, other than the coasts and mountains, has a middle-latitude desert climate. Little precipitation—less than 25 cm—falls annually in these deserts. Unlike tropical deserts, middle-latitude deserts have a winter season that may be quite cold and a summer season that ranges from warm to very hot. The vegetation of North American middle-latitude deserts consists of widely scattered drought-resistant shrubs and cacti.

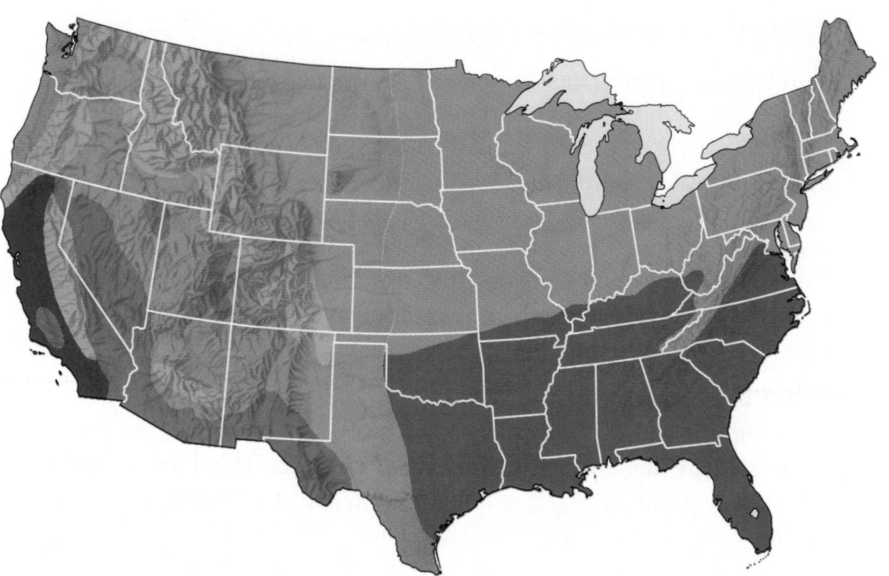

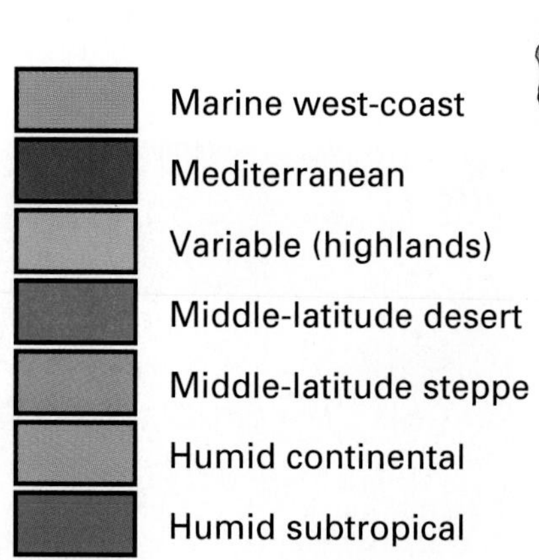

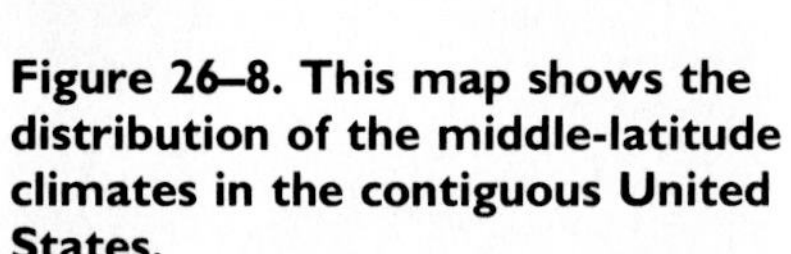

Figure 26–8. This map shows the distribution of the middle-latitude climates in the contiguous United States.

From the western United States, the dry air moves eastward. When it reaches the central part of the continent, it begins to pick up moisture from the maritime tropical gulf air moving northward.

At elevations of 1,200 m to 2,100 m, steppes gradually replace western deserts. Steppes receive 25 cm to 50 cm of rain a year, which supports a dense growth of grasses. The yearly temperature range is high—24°C. The average summer temperature is 23°C, and the average winter temperature is −1°C. As shown in Figure 26–8, the Great Plains, located to the east of the Rocky Mountains, have a steppe climate.

Humid Continental Climate

The wetter regions east of the steppe climate and extending to the east coast in North America and Asia have a **humid continental climate.** The location of the North American humid continental climate is shown in Figure 26–8. Areas with this type of climate are subject to both cold, dry continental polar air masses and warm, moist maritime tropical air masses. Summers are usually warm and humid as maritime tropical air masses move north. Winters are commonly very cold as continental polar air masses move south. When these air masses meet, weather conditions may change rapidly and violently. Seasonal changes are great, with yearly temperature ranges as great as 30°C. Average summer temperatures are as high as 25°C, and average winter temperatures as low as –5°C. Yearly average precipitation, mostly from wave cyclones and summer thunderstorms, is at least 75 cm. Forests of hardwood and softwood trees are found in a humid continental climate.

Humid Subtropical Climate

The southeastern coasts of continents located between 30° and 40° north and south latitude have a **humid subtropical climate.** The southeastern coast of the United States, shown in Figure 26–8, is one of these areas. In the summer, moist maritime tropical air masses move north across this region. These air masses bring warm, humid weather, heavy rains, and occasional hurricanes. In the winter, continental polar air masses moving from inland regions may bring brief but intense cold. For example, Charleston, South Carolina, has an average summer temperature of 27°C and an average winter temperature of 10°C. However, the temperature occasionally has plunged to −15°C in Charleston. The yearly temperature range in a humid subtropical climate is a relatively small 17°C. Annual precipitation in a humid subtropical climate is between 75 cm and 165 cm. The land in a humid subtropical climate is usually covered with dense forest.

Local Climates

The climate in any particular place may be influenced by local conditions as well as by the major factors that have been discussed. The elevation of the land, especially high mountain ranges and plateaus, is the most important factor affecting local weather conditions.

Large lakes influence local climates, and like oceans, they moderate temperatures. They can also cause an increase in precipitation on the shore farthest from the prevailing wind. For example, the eastern shore of Lake Michigan generally has more moderate temperatures, more cloudiness, and higher precipitation than the western shore does. Forests affect local climates by reducing the speed of the wind and by increasing the humidity.

Cities are *microclimates*. That is, they are small regions with their own local climate characteristics. In a city, the average temperature is 1°C to 2°C higher than that in surrounding rural areas. There are several reasons for this phenomenon. During the day, vegetation absorbs solar energy and gives off water vapor by transpiration. Because cities contain far less vegetation than rural areas do, less transpiration occurs, and therefore, more solar energy is available to heat the air. At night the air over cities is warmed by radiation from the materials in streets and buildings that have been heated during the day. Heavy traffic and the energy used for heating, lighting, and industry may also raise the air temperature in cities. More precipitation falls within cities and in areas downwind of cities than in rural areas. Dust, smoke, and other pollutants, carried into clouds by rising warm city air, form nuclei around which raindrops condense.

26.2 Section Review

1. Where do tropical desert climates occur?
2. In which type of climate does the largest yearly temperature range occur?
3. List the types of middle-latitude climates that occur in the contiguous United States and Canada.
4. A city in a middle-latitude desert climate might have the weather conditions of a steppe climate. Why?

Chapter 26 Review

Key Terms

- bora (528)
- chinook (528)
- foehn (528)
- humid continental climate (534)
- humid subtropical climate (534)
- marine west-coast climate (533)
- Mediterranean climate (533)
- middle-latitude climates (529)
- middle-latitude desert climate (533)
- middle-latitude steppe climate (533)
- mistral (528)
- monsoon (527)
- polar climates (529)
- specific heat (526)
- subarctic climate (532)
- temperature range (523)
- tropical climates (529)
- tropical desert climate (529)
- tropical rain forest climate (529)
- tropical savanna climate (529)
- tundra climate (532)

Key Concepts

Latitude determines the angle at which the sun's rays strike the earth. **See page 523.**

The different rates at which land and water are heated, the temperature of ocean currents, and seasonal winds all affect climate. **See page 526.**

The temperature and moisture content of the air are influenced by altitude and by the presence of mountains. **See page 528.**

The three types of tropical climates are based on different amounts of precipitation. **See page 529.**

Regions with subarctic and tundra climates receive little precipitation. **See page 532.**

Several types of middle-latitude climates occur in the contiguous United States and southern Canada. **See page 533.**

Industries, radiation from building materials, and lack of vegetation modify the climates of cities. **See page 535.**

Review

On your own paper, write the letter of the term that best completes each of the following statements.

1. At the equator the sun's rays always strike the earth
   a. at a low angle. b. at nearly a 90° angle.
   c. 18 hours each day. d. no more than 8 hours each day.
2. Nights are longest in the winter and shortest in the summer
   a. at the equator. b. at high altitudes.
   c. in the middle of the ocean. d. at the poles.
3. Water cools
   a. more slowly than land does.
   b. more quickly than land does.
   c. through the process of transpiration.
   d. because of waves and currents.
4. Ocean currents influence temperature by
   a. eroding shorelines.
   b. heating or cooling the air.
   c. washing warm, dry sediments out to sea.
   d. dispersing the rays of the sun.
5. Winds that blow in opposite directions in different seasons because of the differential heating of the land and the oceans are called
   a. chinooks. b. mistrals.
   c. monsoons. d. wave cyclones.

6. When a moving air mass encounters a mountain range, it
   a. stops moving.
   b. slows and sinks.
   c. rises and cools.
   d. reverses its direction.
7. Tropical deserts exhibit all of the following characteristics except
   a. location between 20° and 30° latitude.
   b. dense plant growth.
   c. influence of the subtropical highs.
   d. extremely dry conditions.
8. A tropical climate that is characterized by very wet summers and very dry winters is called
   a. a Mediterranean climate.
   b. a savanna climate.
   c. a trade-wind climate.
   d. an equatorial climate.
9. In regions with a Mediterranean climate, almost all the yearly precipitation falls
   a. during monsoons.
   b. in the summer.
   c. in the winter.
   d. during hurricanes.
10. Weather conditions tend to fluctuate rapidly throughout the year in a
    a. subarctic climate.
    b. middle-latitude desert climate.
    c. Mediterranean climate.
    d. humid continental climate.
11. The various structures and activities in cities affect the local climate by
    a. decreasing the average temperature.
    b. increasing both the average temperature and precipitation.
    c. increasing the average temperature and decreasing the precipitation.
    d. decreasing the precipitation.

Critical Thinking

On your own paper, write answers to the following questions.

1. The Milankovitch theory states that a periodic change in the tilt of the earth's axis was one factor in the onset of the ice ages. Use what you know about the factors affecting climate to explain how an ice age might have occurred.
2. If you enjoy a warm, dry climate all year but do not like the sparse vegetation in deserts, in what climate or climates would you live? Assume that you would not mind moving one or more times a year.
3. Explain why the vegetation in areas with a tundra climate is sparse, even though these areas receive precipitation that is adequate to support plant life.
4. Why do weather conditions change rapidly in a humid continental climate and remain relatively constant in the other middle-latitude climates?
5. Explain why the classification of climates often fails when you think only in terms of a specific location.

Application

1. Imagine you are going to build a vacation house near a coast with a warm offshore current. What must you investigate to determine how the current will affect the temperature on land? Explain your answer.
2. Suppose your family was moving to the mountains, but you do not like humid weather. Should you encourage them to find a house on the side of the mountains facing toward or away from the prevailing winds? Explain why.
3. Concept map ? ? ? Construct a **concept map** that correlates the various climates you learned about in this chapter with their geographical locations on the earth.

Extension

1. At the library, locate old copies of the local newspaper. Record the daily high and low temperatures for a summer month and a winter month during the past year. Graph the information and calculate the yearly temperature range. Report your findings to the class.
2. Use the information you have learned about climate zones to draw a map showing variation in vegetation around the world. Include areas with little or no vegetation as well as those with various types of vegetation. Share your map with the class.

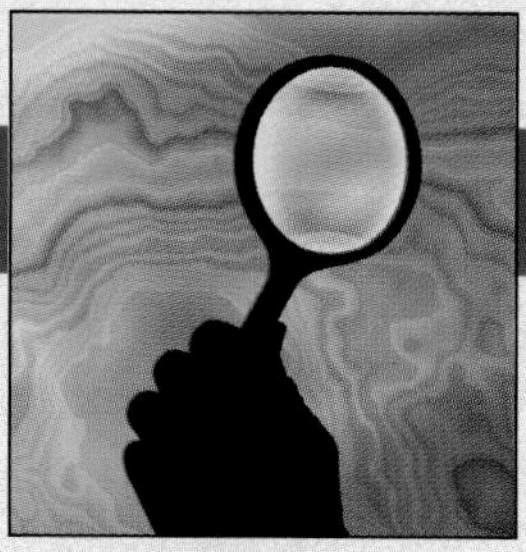

In-Depth Investigation

Factors That Affect Climate

Materials

- masking tape
- graph paper
- Celsius thermometers (2)
- black construction paper
- heat lamp
- soil
- two containers
- meter stick
- flashlight
- water

Introduction

The climate of a region is the summary of the region's weather conditions and changes throughout the year. The importance of climate as a geographical factor cannot be overestimated. In addition to its effects on human life, climate determines the type of soil and vegetation found in a given area. That can then determine how the land will be used to support the animal population. Although science and engineering projects are being designed to accommodate the weather, the distribution of world populations is still primarily determined by favorable natural climate.

In this investigation, you will explore how the angle of the sun's rays and the distribution of land and water affect climate.

Prelab Preparation

Review Section 26.1, pages 523–528.

Procedure

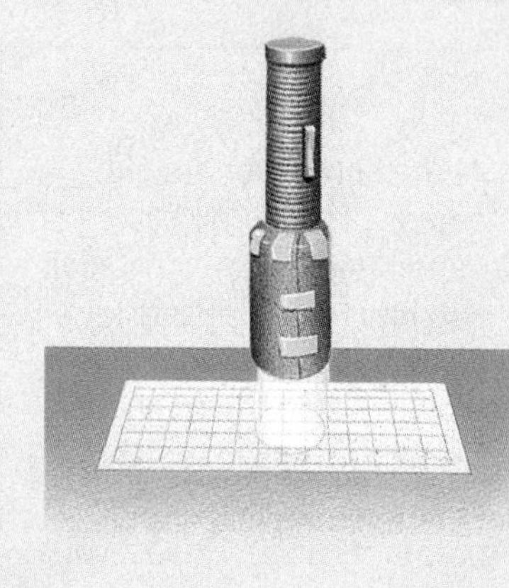

1. Roll the construction paper into a tube so that it is approximately the same size as the glass of the flashlight. Tape the edges of the paper together so that it does not unroll.
2. Attach the tube to the flashlight with the tape, as shown in the illustration at left. Turn on the flashlight and check to see that no light is coming from the seams you taped together.
3. Place the graph paper on a flat surface and hold the flashlight so that the edge of the paper tube is 15 cm from the graph paper and pointing straight down. Assume that the light is a sun ray and that the graph paper is the ground.
4. Draw a line around the lighted area on the graph paper. Count and record the number of squares that are completely illuminated by the light. You may need to darken the room.
5. Keep the flashlight 15 cm from another piece of graph paper, but hold the flashlight at a 45° angle to the graph paper. Repeat Step 4. Record the number of squares that are completely illuminated by the light.
6. Fill one of the containers with soil and the other with tap water. Place both containers on a flat surface next to each other.

7. Place the thermometer in the water and record the temperature.
8. Place the second thermometer in the container of soil, as shown in the illustration at right. The bulbs of both thermometers should be placed so that they are covered by no more than 0.5 cm of water or soil.
9. Place the heat lamp 25 cm above both containers. Record the temperature of each sample at 1, 3, 5, and 10 minute intervals after the lamp is turned on.
10. Disconnect the lamp and record the temperature of the soil and water after 5 minutes.

CAUTION: *Be sure to let the heat lamp cool before storing it.*

Analysis and Conclusions

1. Compare the area of the lighted squares obtained from the two trials in Steps 4 and 5. Which area is greater?
2. What area(s) of the earth is represented by the light striking the graph paper at a 90° angle, as in Step 4? What area(s) of the earth is represented by the light striking the graph paper at a 45° angle, as in Step 5?
3. Why do the sun's rays from overhead cause more warming than low-angle rays?
4. What conclusion can you draw about the angle of the sun's rays during the different seasons in the United States?
5. What substance, water or soil, absorbed more heat in Step 9?
6. What substance, water or soil, lost heat faster when the heat source was turned off in Step 10?
7. What conclusion can you draw about how land and water on the earth are heated by the sun?
8. Explain how the light striking the graph paper in Steps 4 and 5 is related to the effect of heat absorption in water and soil on the earth, as simulated in Steps 9 and 10.

Extensions

1. Continue this investigation using the flashlight and graph paper. Hold the flashlight at various angles to the paper, such as at a 20° angle, 0° angle, and so on. What areas on the earth are represented on the paper at the different angles you use?
2. Considering the earth's varied climates, what conclusions can you draw concerning the angle of the sun's rays on different locations on the earth?

Maps in Action

Snapshots of the Weather

Materials
- tracing paper
- pencil

From looking at a weather map, you might get the impression that the clouds, fronts, and other features it shows are standing still. Actually, the opposite is true. Weather patterns are constantly changing, and a weather map can only show what is happening at one particular instant. For this reason, meteorologists must rely on a sequence of maps to make predictions about local weather.

In this activity, you will have the opportunity to analyze a sequence of weather maps. The maps were taken from a daily newspaper and show weather patterns that occurred in the United States during a four-day period. You will make observations about what the maps show and do not show, and you will make a few predictions based on what you observe.

1. Look carefully at the maps shown on the facing page. What information do they show? Now look back at the weather symbols on page 513. What information is not included on these maps? Why do you think a newspaper might not include certain types of information when publishing daily weather maps?
2. Why do you think a newspaper that serves only a specific geographic region would publish the weather for the entire continental United States?
3. Describe how the weather patterns in your hometown changed during the four-day period shown.
4. During what season do you think this four-day period occurred? Explain your reasoning.
5. Trace the outline of one weather map on a separate piece of paper, but do not include any information on it. Make a prediction about the locations of the fronts on the day following this four-day period, and note the locations of the fronts on your new map.
6. Predict the temperature and precipitation patterns that occur in your hometown on the day following this four-day period.

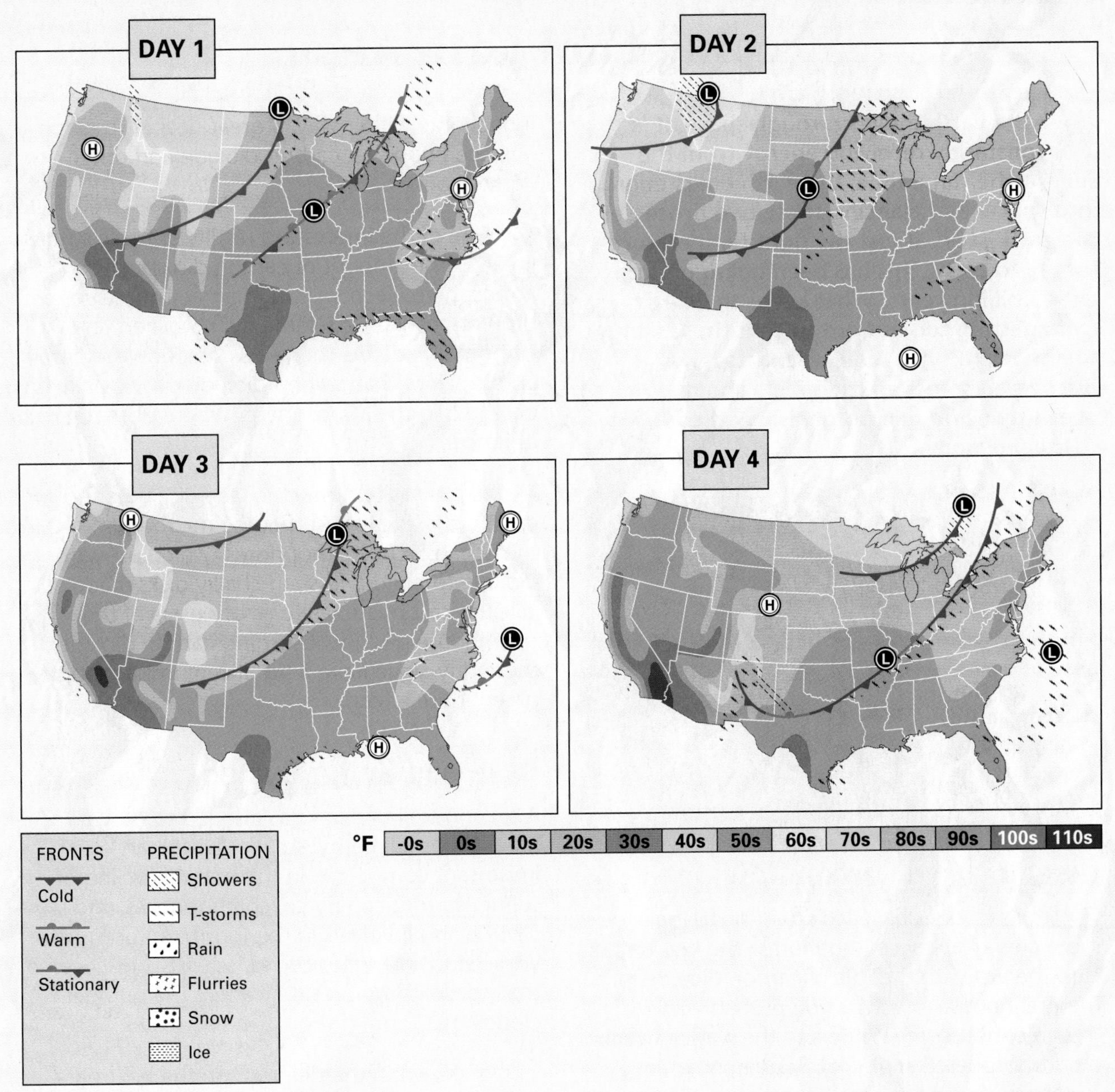

Going Further

Find out the latest information about weather patterns across the United States by visiting the National Climatic Data Center online at www.ncdc.noaa.gov.

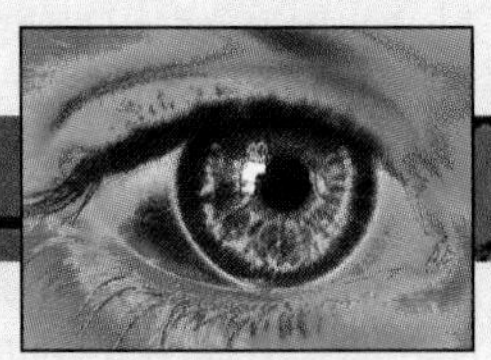

EYE ON THE ENVIRONMENT

Keeping Cool with Algae

As the earth moves through space, it is constantly receiving energy from the sun. At the same time, the earth emits energy into space. By balancing these two processes, the earth stays at just the right temperature for life to exist. This has been going on for over 3 billion years. What keeps the earth from getting too hot or too cold? In seeking to answer this question, scientists have discovered some amazing facts about ocean-dwelling microorganisms called *coccolithophores.*

This tiny microorganism may help to regulate the temperature of our planet.

Disappearing Carbon Dioxide

Two trails of clues have led scientists to look more closely at coccolithophores. The first trail of clues begins with the role of carbon dioxide in the greenhouse effect. As you probably know, the greenhouse effect refers to the way in which our atmosphere keeps heat trapped near the surface of the earth. The gases that create the greenhouse effect come from a variety of sources, including both natural processes and human activities. Some scientists are concerned that the carbon dioxide released by human activities may increase the greenhouse effect, resulting in global warming. Other scientists contend that natural changes in the earth's temperature (such as the occurrence of ice ages) may pose more of a threat to our planet's current inhabitants. Either way, a great deal of research has been done on carbon dioxide and other "greenhouse gases."

One of the interesting results of these studies involves the amount of carbon dioxide in the atmosphere. Each year, human activities such as burning fossil fuels release over 6 billion tons of carbon dioxide. As the years pass, however, only about half of that carbon dioxide can actually be detected in the atmosphere. Where does the rest of it go?

As it turns out, much of the disappearing carbon dioxide is absorbed by microorganisms in the oceans. Coccolithophores, which use carbon dioxide to build and shed chalky disks made of calcium carbonate, are primary users of the greenhouse gas. By absorbing carbon dioxide, these tiny algae have a significant impact on the greenhouse effect.

DMS and Sulfur Cycles

A second trail of clues has also led to coccolithophores. Dimethyl sulfide (DMS), an important compound in the natural cycling of sulfur, is produced by coccolithophores, and DMS is also involved in the formation of clouds. While studying sulfur cycles, scientists discovered that coccolithophores may affect the number of clouds that form in the sky.

As shown in the diagram on the next page, coccolithophores absorb sulfur compounds from ocean water. They use these compounds to produce DMS, which is then released into the air. Once in the atmosphere, DMS causes water vapor to condense and form clouds. Thus, when a patch of ocean has more coccolithophores, more clouds form in that area.

As more clouds form, they block sunlight from reaching the ocean. This keeps the

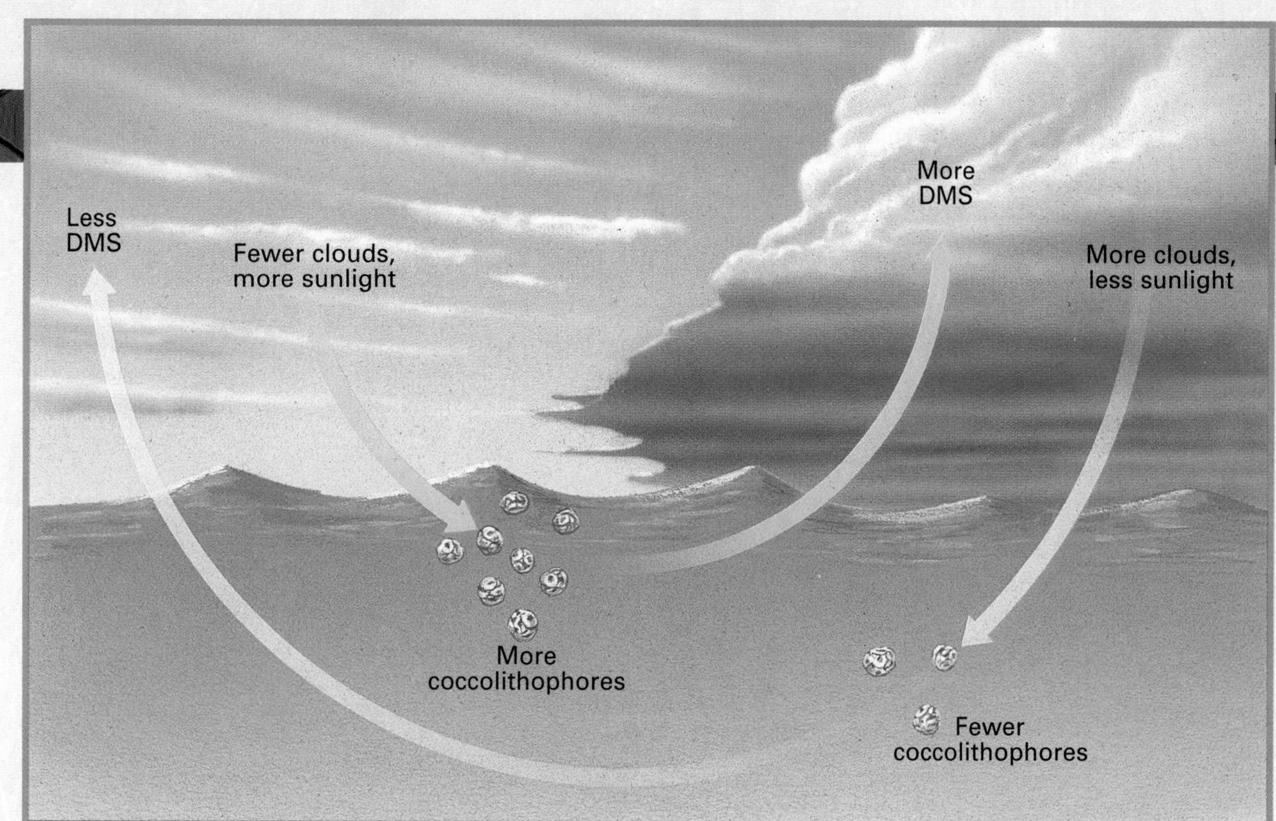

▲ **DMS-cloud-formation cycle**

coccolithophores from undergoing photosynthesis, and their numbers begin to decline. With fewer coccolithophores, less DMS is produced and fewer clouds form. Eventually, as more sunlight hits the ocean, more coccolithophores grow, more clouds form, and the cycle begins again. Scientists think that this process may have served as a natural thermostat for our entire planet for millions of years.

Tough Puzzles to Solve

Scientists have made great progress toward understanding how coccolithophores affect the atmosphere. However, whether we can use this knowledge to help diminish the threat of changes in our atmosphere's climate remains to be seen. For instance, to avoid the risk of global warming, some scientists have suggested ways to promote coccolithophore growth in the ocean. This would serve to reduce atmospheric carbon dioxide and create more clouds, both of which would cool the planet. Unfortunately, this would be risky because scientists do not know how marine ecosystems would be affected.

Research into coccolithophores and other marine microorganisms will continue. As it progresses, scientists will probably learn more about these life-forms. Our ability to apply this research to changes in our atmosphere may prove to be limited, but one thing is almost certain—you will be hearing more about these amazing microorganisms.

Think About It

Why would it be hard to predict the effects of artificially promoting coccolithophore growth?

Extension: Find out about other marine microorganisms. What do they do? How are they important to other organisms in marine ecosystems?

UNIT 8

Studying Space

Trifed Nebula in Sagittarius (background);
Shuttle astronaut retrieving satellite for repair (inset)

Introducing Unit Eight

We are small creatures on a small planet in a small corner of the universe. So far, we are unique—at least within this solar system.

Exploring and understanding the circumstances on earth and in space that allow us to exist and continue existing are especially important today. In the last century of our planet's 4.5-billion-year history, technology has given us the opportunity to affect the dynamic systems of the earth. For the first time, we have the ability to affect our protective environment and to alter or maintain the atmosphere that sustains life on our planet.

In this unit, you will study the moon, sun, planets, and stars. Although they may seem remote, they are important to our lives on earth. None of us is untouched by their presence.

Kathryn D Sullivan

Kathryn Sullivan
Astronaut
National Aeronautical and
Space Administration

In this UNIT

Chapter 27

Stars and Galaxies

Of all the stars in the sky, none appear as bright as the sun. Certainly, to people on the earth, the sun is the most important star: the earth could not exist without it. However, the sun has about the same size, temperature, and brightness as billions of other stars. The sun is an ordinary star with one important difference—it is orbited by an inhabited planet. In this chapter, you will learn about the stars and galaxies of the universe.

Chapter Outline

Only one star is present in this image taken by the Hubble Space Telescope. The rest of the objects are entire galaxies.

27.1 Characteristics of Stars

Section Objectives

- **Describe how astronomers determine the composition and surface temperature of a star.**
- **Explain why stars appear to move to an observer on the earth.**
- **Name and describe the way astronomers measure the distance from the earth to the stars.**
- **Explain the difference between absolute magnitude and apparent magnitude.**

Life on the earth is dependent upon the sun, the star nearest to the earth. A **star** is a body of gases that gives off a tremendous amount of radiant energy in the form of light and heat. Except for its relationship to the earth, the sun is similar to billions of other stars.

From the earth, most stars in the night sky appear to be tiny specks of white light. However, if you look closely at the stars, you will notice that they vary in color, as listed in Table 27–1. For example, the star Antares shines red, the star Rigel shines blue-white, and the star Arcturus shines orange.

Stars vary in size and mass as well as in color. Some stars are less than 20 km in diameter, far smaller than the earth. Other stars have a diameter 1,000 times that of the sun. The sun, a medium-sized star, has a diameter of about 1,392,000 km. Most stars that are visible in the night sky are medium-sized stars.

Many stars also have about the same mass as the sun, which is about 330,000 times more massive than the earth. Some small stars have only 1/50 of the sun's mass. Large stars may have more than 50 times the sun's mass. Stars also differ in composition, temperature, distance from the earth, and brightness.

Composition and Temperature

Before the Hubble Space Telescope was placed in orbit, it was very difficult to study stars from the earth. Consequently, astronomers learned about stars primarily by analyzing the light that they emitted. Astronomers direct starlight through a *spectrometer,* a device that separates light into different colors, or wavelengths. Starlight passing through a spectrometer produces a display of colors and lines called a *spectrum*. There are three types of spectra: emission, or bright-line; absorption, or dark-line; and continuous.

Of particular use to scientists are the dark-line spectra of stars. Analysis of dark-line spectra reveals certain characteristics, such

Table 27–1 Classification of Stars

Color	Surface temperature (°C)	Examples
Blue	Above 30,000	10 Lacetae
Blue-white	10,000–30,000	Rigel, Spica
Blue-white	7,500–10,000	Vega, Sirius
Yellow-white	6,000–7,500	Canopus, Procyon
Yellow	5,000–6,000	Sun, Capella
Orange	3,500–5,000	Arcturus, Aldebaran
Red	Less than 3,500	Betelgeuse, Antares

as composition and temperature. Every chemical element has a characteristic spectrum. The colors and lines in the spectrum of a star indicate which elements make up the star. Through spectrum analysis, scientists have learned that stars are composed of the same elements that are found on the earth. However, unlike the earth, hydrogen is the most common element in most stars, and helium is the second most common. Elements such as iron, sodium, and calcium make up the remaining mass of the stars.

The surface temperature of a star is indicated by its color, as shown in Table 27–1. Most stars are in the range of 2,800°C to 24,000°C, although a few are hotter. Usually a star shining with a predominantly blue light has an average surface temperature of 35,000°C. However, the surface temperatures of some blue stars are as high as 50,000°C. Red stars are the coolest, with average surface temperatures of 3,000°C. Yellow stars, such as the sun, have surface temperatures of about 5,500°C.

Motion

Two kinds of motion are associated with stars—actual motion and apparent motion. Because stars are so far away, their actual motion can be measured only with high-powered telescopes and other instruments. The apparent motion of stars, motion visible to the unaided eye in a dark sky, is due to movement of the earth. You can photograph the apparent motion of the stars from the earth. Aim a camera at the northern sky on a clear evening, and leave the shutter open for a few hours. The result will be a photograph similar to the one shown in Figure 27–1.

Figure 27–1. The stars appear as curved trails in this long-exposure photograph. The trails result from the rotation of the earth on its axis.

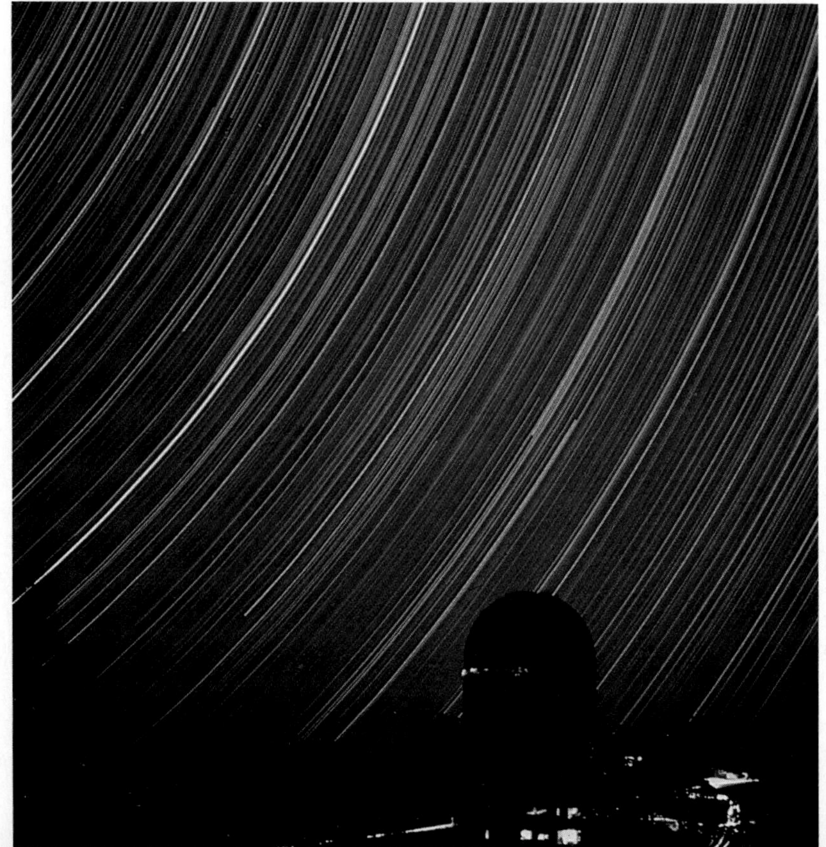

Figure 27–2. This photograph shows the telescope domes at Kitt Peak National Observatory in Arizona. Astronomers check the instrument at the focus of the 2.1-m telescope at the observatory (inset).

Almost all of the curves of light on the photograph are recordings of the apparent motion of stars. The circular trails left by most stars suggest they are moving westward around a central star called Polaris, the North Star. The circular pattern actually results from the rotation of the earth on its axis. Polaris is located almost directly above the North Pole and does not appear to move.

The earth's revolution around the sun causes the stars to appear to move in a second way. Stars located on the side of the sun opposite the earth are obscured by the sun. As the earth orbits the sun, different stars become visible during different seasons. The visible stars appear to shift slightly to the west every night. Over the period of a year, most stars appear to move across the sky a little each night and finally disappear below the western horizon.

Some stars are always visible in the night sky. In the Northern Hemisphere, they lie so close to Polaris that neither their apparent nightly nor annual orbits carry them below the horizon. These stars can always be seen circling the North Star. The circling stars are called **circumpolar** stars. The stars of the Little Dipper are circumpolar for most observers in the Northern Hemisphere.

Most stars have at least three actual motions. First, they rotate on an axis. Second, they may revolve around another star. Third, they move away from or toward the earth.

From the spectrum of a star, astronomers can learn more about the motion of that star. As you read in Chapter 1, the spectrum of a star moving toward or away from the earth appears to shift. The apparent shift in the wavelength of light emitted by a light source moving toward or away from an observer is called the *Doppler effect*. The colors in the spectrum of a star moving toward the earth are shifted toward blue. This shift, called *blue shift,* occurs because the light waves from a star appear to have shorter wavelengths as the star moves toward the earth.

A star moving away from the earth has a spectrum that is shifted toward red. This shift, called **red shift,** occurs because the wavelengths of light appear to be longer. Most distant galaxies, or large groups of stars, have red-shifted spectra, indicating that these galaxies are moving away from the earth.

Distance to the Stars

Because space is so vast, distances between the stars and the earth are measured in **light-years.** A light-year is the distance that light travels in one year. Since the speed of light is 300,000 km/s, light travels about 9.5 trillion km in one year. Light from the sun takes about 8 minutes to reach the earth, so the sun is 8 light-minutes from earth. By contrast, the star system nearest the earth is the Alpha Centauri system. The closest star in this system is Proxima Centauri. It is 4.3 light-years away, nearly 300,000 times the distance from the earth to the sun. Sirius, the brightest star seen from the earth, is 9 light-years away. Polaris is 700 light-years away.

Astronomers use many different methods to determine distances between the stars and the earth. One method is called **parallax.** As the earth circles the sun, observers are able to study the stars from slightly different angles. During a six-month period, a nearby star will appear to shift slightly, relative to stars that are farther away. The closer a nearby star is, the greater will be the amount of shift. From the amount of shift, astronomers can calculate the distance to any star within 1,000 light-years. Because the angle of shift is very small and extremely difficult to measure, scientists usually use photography to measure the shift. The star is photographed at the beginning and end of a six-month period, and its position in relation to other stars is studied each time.

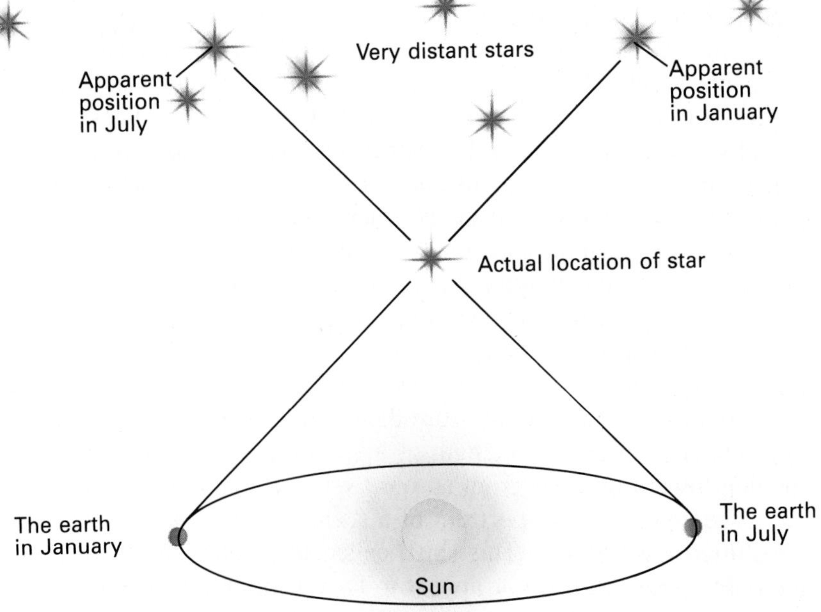

Figure 27–3. Observers on the earth see nearby stars against those in the distant background. The movement of the earth causes nearby stars to appear to move.

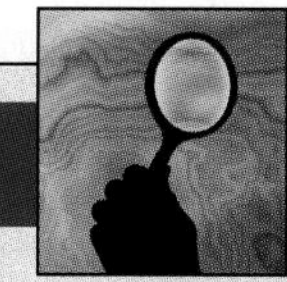

SMALL-SCALE INVESTIGATION

Parallax

You can demonstrate the principle of parallax by viewing an object from several locations.

Materials

5 paper or plastic plates, about 15 cm in diameter (1 red, 4 blue); thread; scissors; masking tape; meter stick; ladder

Procedure

1. Cut five 1-m lengths of thread. Tape one end of each piece of thread to the edge of a paper plate.
2. Tape the free end of each piece of thread to the ceiling at various vertical heights and randomly spaced about 30 cm apart, in a pattern similar to the one shown in the photo above.
3. Stand directly in front of and directly facing the red plate at a distance of several meters.
4. Close one eye and sketch the position of the red plate relative to the blue plates in the background.
5. Take several steps back and to the right of your original position. Repeat Step 4.
6. Take several more steps directly back and make another sketch.

7. Repeat Step 6 once again.

Analysis and Conclusions

1. Compare your drawings. Did the red plate change position as you viewed it from different locations? Explain your answer.
2. What kind of results would you expect if you continued to repeat Step 6 at greater and greater distances? Explain your answer.
3. If you noted the positions of several stars with a powerful telescope, what would you expect to observe about their positions if you sighted the same stars several months later? Explain.

You can demonstrate how parallax works by extending your arm in front of you and holding up your thumb. Close one eye and note the position of your thumb against the background. Then open that eye and close the other one. Now note the position of your thumb. Your thumb will seem to have moved across the background by about 3°. Because your eyes are a few centimeters apart, each eye sees your thumb from a different angle. Try repeating this exercise with your thumb close to your face. Your thumb will appear to jump from one side of your face to the other. The closer your thumb is to your face, the greater is its apparent motion. Similarly, the closer a star is to the earth, the greater is its apparent motion.

Astronomers estimate how bright a more distant star is by studying its spectrum. They compare this estimate of true brightness with the apparent brightness of the star. From these two measurements, astronomers can calculate the distance of a star from the earth.

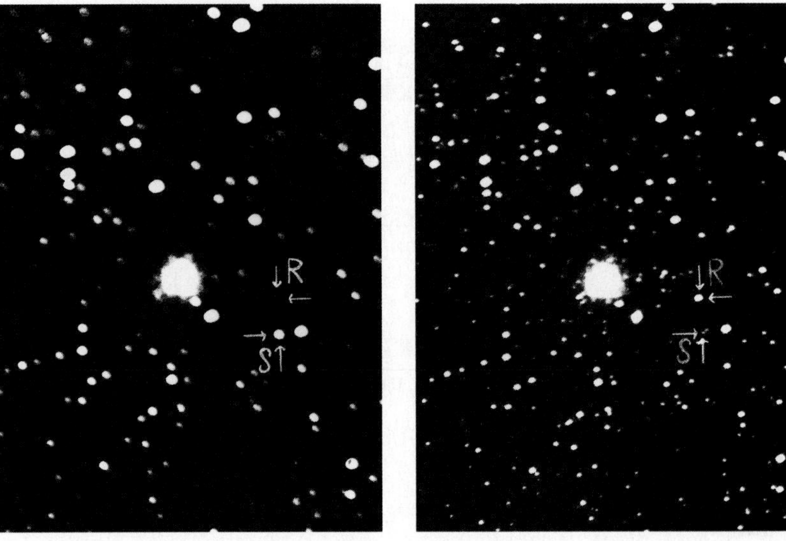

Figure 27–4. These two photographs show the variable stars R Scorpii and S Scorpii. In the photograph on the left, R Scorpii is faint and S Scorpii is bright. In the photograph on the right, R Scorpii is bright and S Scorpii is faint.

Some special stars serve as distance indicators. For example, a *Cepheid* (Sef-EE-id) *variable* star brightens and fades in a regular pattern. The brightness of some Cepheids varies by as much as 600 to 700 percent. This change in brightness is caused by a rhythmic swelling and shrinking of the star. Most Cepheids have regular cycles ranging from 1 day to 100 days. Cepheids with longer cycles are brighter. Astronomers can measure the cycle of a Cepheid and then estimate its true brightness. By comparing its apparent brightness and true brightness, astronomers can calculate the distance to the Cepheid variable. This, in turn, tells them the distance to the galaxy in which the Cepheid is located.

Stellar Magnitudes

Over 3 billion stars can be seen through ground-based telescopes. Of these, only about 6,000 are ever visible to unaided observers on the earth. Over a trillion stars can be observed from earth-orbiting telescopes, such as the Hubble Space Telescope. The visibility of a star depends on its brightness and its distance from the earth. Astronomers use two scales to describe the brightness of a star. One scale is based on how bright the star appears from the earth. The other scale is based on how bright the star would be if all stars were the same distance from the earth.

Apparent Magnitude

The brightness of a star as it appears from the earth is called its **apparent magnitude.** Astronomers use special instruments that are attached to telescopes to measure the light from a star. The measurement is then assigned a number on the scale shown in Figure 27–5. The brightest stars have the lowest numbers; the dimmest stars have the highest numbers. The most powerful telescopes can detect stars with an apparent magnitude of about 29. These stars are about 1.5 billion times fainter than the faintest star seen without a telescope.

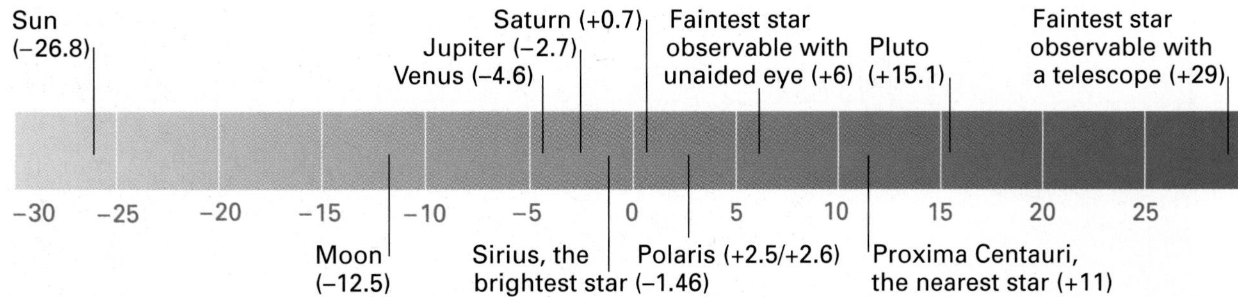

Figure 27–5. The sun, with an apparent magnitude of −26.8, is the brightest object in the sky. All other objects are dimmer, so their apparent magnitudes are higher.

The faintest star that can be seen by the unaided eye has an apparent magnitude of 6. A star of this brightness is called a sixth-magnitude star. A first-magnitude star is one of the brightest stars in the sky.

The apparent magnitudes of a few stars, some planets, the moon, and the sun are negative numbers because they are brighter than first-magnitude stars. For example, the brightest star in the night sky, Sirius, has an apparent magnitude of –1.46. The sun has an apparent magnitude of –26.8.

Absolute Magnitude

The apparent magnitude of a star depends on how much light the star emits and how far the star is from the earth. The true brightness, or **absolute magnitude,** of a star is how bright the star would appear if it were seen from a distance of 32.6 light-years.

Assuming the sun were 32.6 light-years away, it would be a fifth magnitude star. Therefore, the absolute magnitude of the sun is +5. Most stars have absolute magnitudes between –5 and +15. Thus, the sun is in the middle of the range of absolute magnitudes.

Each star has both an apparent magnitude and an absolute magnitude. The relationship between these two measures of brightness depends on the distance between the earth and the star. Stars that are less than 32.6 light-years from the earth appear brighter than they would if they were 32.6 light-years away. Consequently, these stars have apparent magnitudes that are lower than their absolute magnitudes. For example, the sun is only a fraction of one light-year from the earth. It has an apparent magnitude of –26.8 and an absolute magnitude of +5.

Stars that are more than 32.6 light-years from the earth appear dimmer than they would if they were only 32.6 light-years away. Consequently, these stars have apparent magnitudes that are higher than their absolute magnitudes. How far away is a star with an apparent magnitude of +7 and an absolute magnitude of +7?

> **INVESTIGATE!**
>
> *To learn more about star magnitudes, try the In-Depth Investigation on pages 568–569.*

Classification of Stars

Plotting the surface temperatures of stars against their absolute magnitudes reveals an interesting pattern. The graph illustrating this pattern is a Hertzsprung-Russell diagram, or **H-R diagram.** The graph is named for Ejnar Hertzsprung and Henry Russell, the astronomers

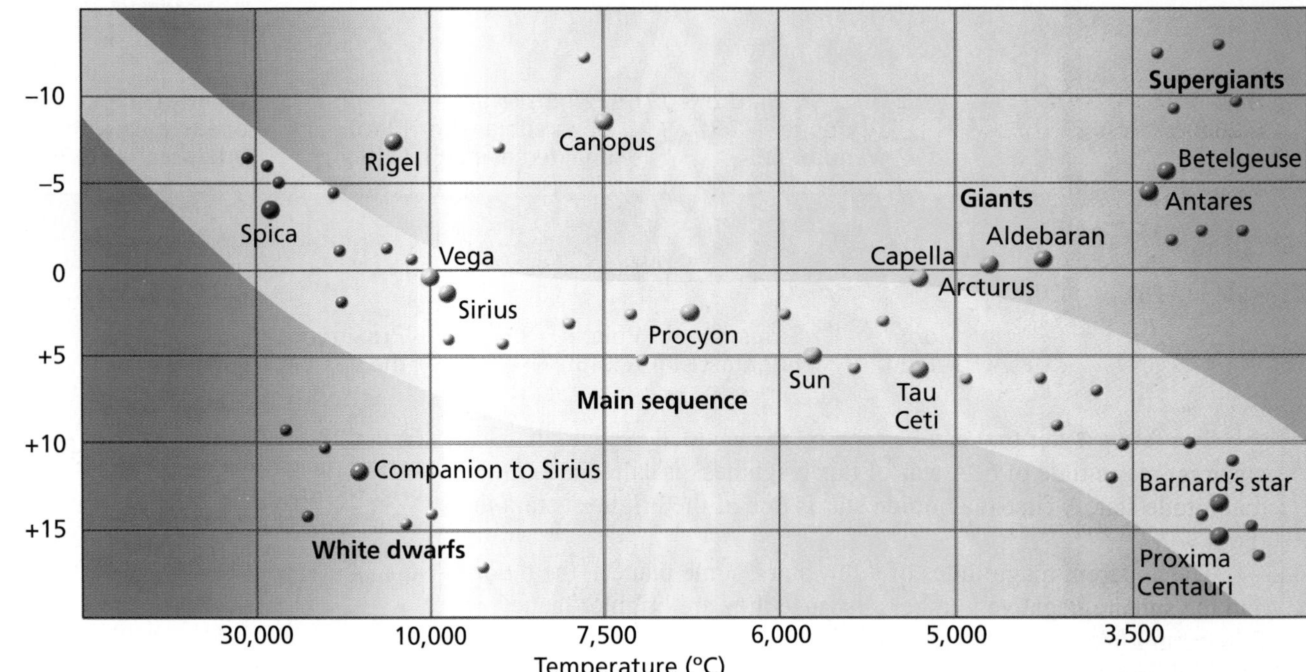

Figure 27–6. The Hertzsprung-Russell diagram shows the relationship between the surface temperatures and the absolute magnitudes of stars.

who discovered the pattern. As you can see in Figure 27–6, the brightness of most stars increases as their surface temperature increases.

The majority of stars fall within a band running through the middle of the H-R diagram. The band extends from cool, dim, red stars at the lower right to hot, bright, blue stars at the upper left of the H-R diagram. Stars within this band are called **main-sequence stars.** The sun and many stars that are visible in the night sky are main-sequence stars.

A group of cool, bright stars appears near the upper right corner of the H-R diagram. These stars are very large. A cool star can be very bright only if it has a large surface area to give off a large amount of light. These huge stars, such as Arcturus, are called **giants.** Some are so large that they are known as **supergiants.** Antares, with a diameter of more than 2.7 million km, is an example of a supergiant.

The stars found in the lower left of the H-R diagram are hot but dim because of their small size. They are known as **white dwarfs.** A typical white dwarf star is about the size of the earth.

Section 27.1 Review

1. What do astronomers analyze to determine the composition and surface temperature of a star?
2. As you observe the night sky, why do stars appear to move westward across the sky?
3. How do astronomers measure the distance to stars that are less than 1,000 light-years from the earth?
4. Assume that a star has an apparent magnitude of +2 and an absolute magnitude of +4. What do you know about the distance of that star from the earth?

27.2 Stellar Evolution

Section Objectives

- Describe how a protostar develops into a star.
- Explain how a main-sequence star generates energy.
- Describe the possible evolution of a star during and after the giant stage.

Since a typical star exists for billions of years, astronomers will never be able to observe one star throughout its entire life. Instead, astronomers have developed theories about the evolution of stars by studying stars in different stages of development.

A star begins as a **nebula** (NEB-yuh-luh; pl: nebulae), a cloud of gas and dust, such as the one shown in Figure 27–7. A nebula is usually composed of about 70 percent hydrogen, 28 percent helium, and 2 percent heavier elements. The particles of material in a nebula have a very weak gravitational attraction for one another. When a force, such as from the explosion of a nearby star, compresses some of the particles, the nebula begins to contract.

According to Newton's law of gravitation, gravitational force increases as distance decreases. Therefore, as the density of the particles increases, their gravitational attraction for one another increases. As these particles come together, a sphere of matter builds up within the cloud.

Gravitational forces cause the nebula to continue to shrink. As the nebula becomes smaller, it begins to spin more rapidly. You may have seen the effect of a decreasing diameter on the speed of a spinning object, such as an ice skater. As a spinning skater pulls his or her arms in closer to the body, the rate of spin increases.

The shrinking, spinning nebula begins to flatten into a disk of matter with a central concentration called a **protostar.** The increase in temperature in the center of a protostar has two causes. One cause

Figure 27–7. The Orion Nebula is a region in which star formation is currently taking place.

is collision. As the particles move toward the center of the nebula, they collide. Whenever solid objects collide, some of the energy of their motion is converted into heat energy. You can demonstrate this principle by rubbing your hands together. The motion of your hands against each other warms them.

Pressure also causes the temperature in the nebula to increase. As the nebula shrinks and the force of gravity pulls matter toward its center, the pressure in the core of the nebula increases. All materials become warmer when compressed. You can observe this principle by again watching an ice skater. The pressure of the skate blade on the ice heats the ice beneath the blade and causes it to melt.

The contracting and heating up of the nebula continues for several million years. As the collisions and pressure raise the temperature in the core of the protostar to over 10,000,000°C, *nuclear fusion* begins. Nuclear fusion is the process through which small atomic nuclei combine to form larger atomic nuclei, releasing energy. When fusion takes place, a protostar begins to generate energy and is considered a star.

A nebula may produce more than a single star. Often two or more stars created out of the same nebula revolve around each other,

SCIENCE & TECHNOLOGY

The Hubble Space Telescope Eyes New Stars

▲ The Hubble Space Telescope has a resolution high enough that it could distinguish two fireflies only 3 m apart, as well as their individual flashes, at a distance comparable to that between Washington, D.C., and Tokyo.

From its orbit around the earth, the Hubble Space Telescope has photographed newborn stars emerging from huge pillars of dense gas and dust. Such events are by no means local. The Hubble Space Telescope recorded the events occurring some 7,000 light-years away, in the Eagle Nebula. The largest of the pillars photographed by the Hubble telescope is shown at right. It is estimated to be 10 trillion kilometers high.

These pillars are thought to result from a process called *photoevaporation.* In this process, high-energy photons from intensely hot stars nearby bombard the nebula. This bombardment strips away the outer layers of gas and dust, leaving pillars of dense material that contain *evaporating gas globules,* or *EGGs.* The gas and dust in these globules is so dense that the globules may further condense to form new stars.

By watching stars form in the Eagle Nebula, astronomers may be able to answer the question, What determines a star's size? After studying Hubble's images, astronomers already have a better idea of how mature stars can affect the growth and size of a developing star. Photoevaporation caused by a mature star may strip away enough gas and dust from around an EGG that the

the force of gravity keeping them close together. A contracting nebula may also produce planets that revolve around a central star. Astronomers think that the sun, the earth, and the other planets in our solar system all formed at about the same time from the same nebula.

Main-Sequence Stars

The second and the longest stage in the life of a star is the main-sequence stage. During this stage, energy is generated in the core of the star as hydrogen atoms fuse to become helium atoms. Fusion releases enormous amounts of radiant energy. For example, when 1 g of hydrogen is converted into helium, the energy released is enough to keep a 100-W light bulb burning for 3,000 years. This energy moves outward in much the same way that energy rises upward through boiling water. The star does not expand, however, because the force of gravity pulls the matter inward. The energy from fusion balances the force of gravity, making the star stable in size. A main-sequence star maintains a stable size as long as it has an ample supply of hydrogen to fuse into helium.

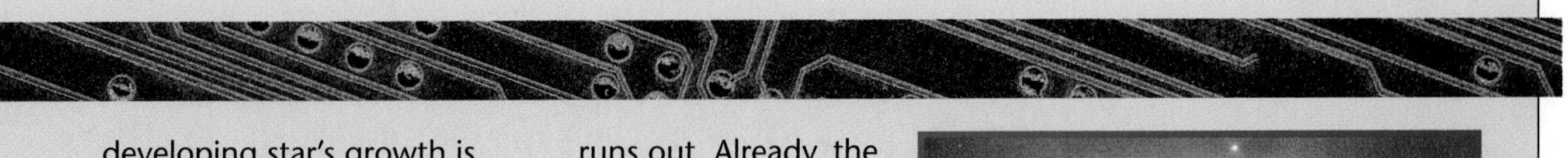

developing star's growth is stunted. If enough gas and dust is stripped away, the star may be prevented from forming in the first place. Photoevaporation may also prevent planets from forming around newborn stars.

However, photoevaporation is not a factor in the growth of all stars. Stars that form in isolation continue to condense, pulling in additional material until nuclear fusion begins. The newborn star's own radiant energy then produces a stellar wind that sweeps away any remaining material.

The Hubble Space Telescope will continue to advance our knowledge of the universe at least until the year 2005, when its guaranteed funding runs out. Already, the telescope has seen farther into space than any preceding it and has provided images in unprecedented detail of almost every type of celestial body known to astronomers.

Why would planets be unlikely to form around the stars in the Eagle Nebula?

The EGGs in the photograph lie at the tips of fingerlike protrusions extending from this pillar. Each EGG is slightly larger in diameter than our solar system.

Giants and Supergiants

A star enters its third stage when almost all of the hydrogen atoms within its core have fused into helium atoms. Without hydrogen as a source of fuel, the core of a star contracts under the force of the star's gravity. This contraction increases the temperature in the core of the star. The higher temperature causes the helium atoms in the core to fuse into carbon atoms. Hydrogen fusion continues to take place in a shell surrounding the helium core.

The combined hydrogen fusion and helium fusion release energy, which causes the outer shell of the star to expand greatly. The star's shell of gases grows cooler as it expands. The star is no longer a main-sequence star. Instead, it has become a giant or a supergiant. Giants are 10 or more times bigger than the sun. Supergiants are at least 100 times bigger than the sun.

The stages in the life of a star cover an enormous period of time. Scientists estimate that over a period of 5 billion years, the sun, a main-sequence star, has converted only 5 percent of its original hydrogen.

White Dwarf Stars

The end of helium fusion marks the end of the giant stage in the evolution of a medium-sized star. With energy no longer available from fusion, the star enters its final stages. The star loses its outer gases, revealing a core, which for a time heats and illuminates the expanding gases, which appear as a **planetary nebula.** Gravity causes the last of the matter in the star to collapse inward. What is left is a hot, dense core of matter—a white dwarf. White dwarfs shine for billions of years before they cool completely.

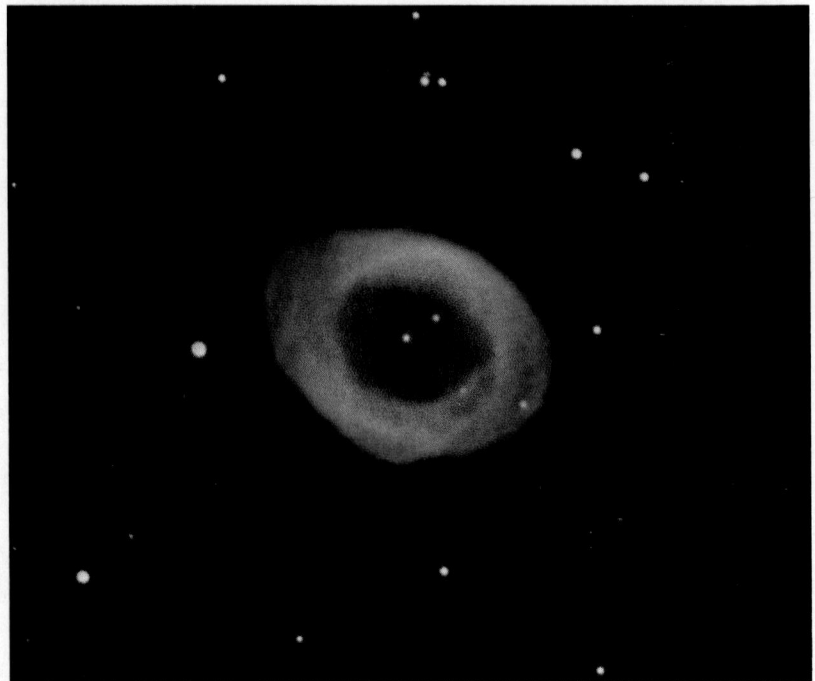

Figure 27–8. The outer shell of expanding gas around a dying star is a planetary nebula. The ring nebula in Lyra is shown here.

As white dwarfs cool, they become fainter and fainter. When a white dwarf no longer emits energy, it may become a dead star or *black dwarf*. Black dwarfs probably do not exist yet, as the universe is not old enough to have produced them. This may be the final stage for many stars.

Novas Some white dwarfs do not just cool and die. During the process of cooling, one or more large explosions may occur that release energy, gas, and dust into space. A white dwarf that has such an explosion is called a **nova.** The explosion may cause the star to become many thousands of times brighter. A nova may appear up to 1 million times brighter than the sun. Then, sometimes within only a few days, the nova begins to fade back to its normal brightness. A white dwarf may become a nova several times.

Astronomers think that a nova is most likely to occur in a white dwarf that revolves around a main-sequence star or a giant star. White dwarfs are denser than main-sequence stars and giants. Therefore, the white dwarf would have a greater surface gravity than its companion. As gases from the companion star accumulate on the white dwarf, the pressure builds until the white dwarf explodes as a nova.

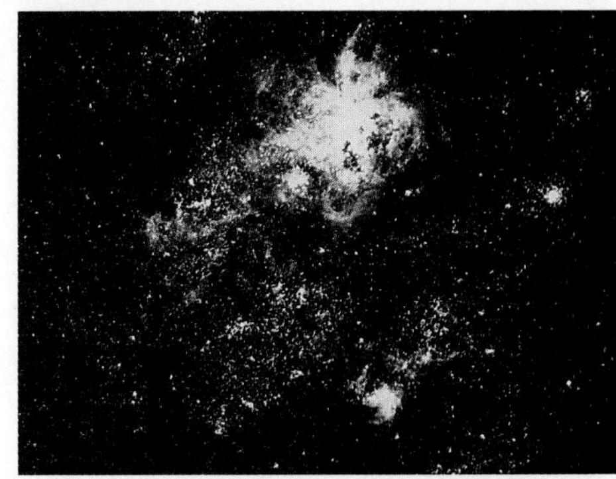

Figure 27–9. Supernova 1987A, shown here before and after, resulted from the explosion of a blue supergiant star. Visible only from the Southern Hemisphere, it was the brightest supernova in nearly 4 centuries.

Supernovas

Stars with masses 10 to 100 times that of the sun may produce explosions up to 100 times brighter than novas. In 1054, Chinese astronomers saw an explosion in the sky that was so bright they could see it during the day for three weeks. The amount of energy radiated during that time was equal to the energy produced by the sun over a period of 500 million years. What the Chinese astronomers saw was a **supernova,** a star that has such a tremendous explosion that it blows itself apart.

Supernovas occur in stars much larger than those that produce novas. After the supergiant stage, these larger stars contract with a gravitational force much greater than that of smaller stars. The collapse produces such high pressures and temperatures that nuclear fusion begins again. This time carbon atoms in the core of the star fuse into heavier elements, such as magnesium. These heavier elements then fuse into iron.

Fusion continues until the core is almost entirely iron. At this point, nuclear fusion stops. The iron begins to absorb huge amounts of energy from gravitational attraction. The iron core collapses, causing the outer part of the star to explode. During the explosion, the energy released approximately equals the amount of energy radiated by an ordinary star over its lifetime.

Neutron Stars After an explosion, the core of a supernova may contract into a very small but incredibly dense ball of neutrons, called a **neutron star.** A spoonful of matter from a neutron star would weigh 100 million tons on the earth. A neutron star with more mass than the sun may have a diameter of only about 30 km. Neutron stars rotate very rapidly.

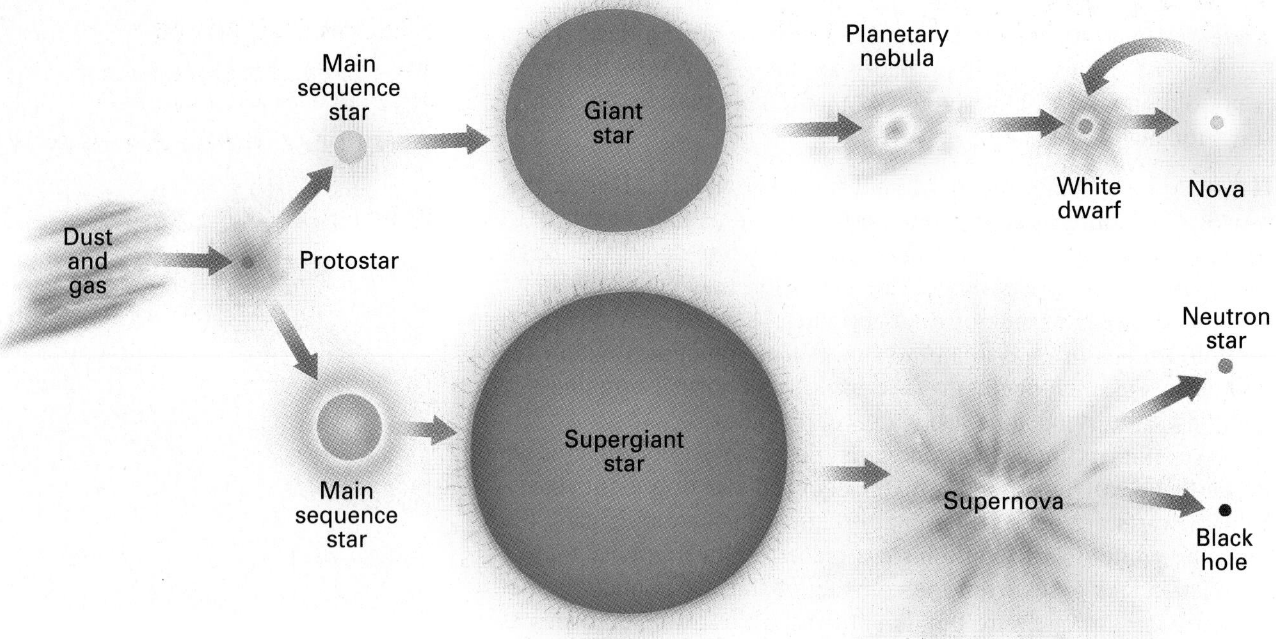

Figure 27–10. A star the size of the sun becomes a white dwarf near the end of its life cycle. A larger star may become a neutron star or a black hole.

Some neutron stars emit two beams of radiation that sweep across space like the light beams from a lighthouse. These neutron stars are called **pulsars.** Astronomers detect the radiation from most pulsars as radio waves.

Black Holes Some massive stars produce leftovers that are too massive to become neutron stars. These stars contract with even greater force. The force of the contraction crushes even the dense core of the star, leaving what astronomers think is a hole in space, or a **black hole.** The gravity of a black hole is so great that not even light can escape from it.

Since black holes do not give off light, locating them is difficult. However, astronomers theorize that a black hole can be observed by its effect on a companion star. Matter from the companion star is pulled into the black hole, disappearing forever from the universe. Just before the matter is pulled in, X rays are given off. Astronomers try to locate black holes by detecting these X rays. In the 1970's, astronomers identified what they think is a black hole within the constellation Cygnus. Today, astronomers speculate that massive black holes may be at the cores of many galaxies.

Section 27.2 Review

1. List two reasons why the temperature of a protostar increases.
2. What is the process that generates energy in the core of a main-sequence star?
3. What form of fusion occurs in a giant star?
4. What causes a nova explosion?
5. Explain why only very large stars can form black holes.

27.3 Star Groups

Section Objectives

- **Describe the characteristics that identify a constellation.**
- **Describe the three main types of galaxies.**
- **Explain the big bang theory.**

When you look into the sky on a clear night, you see what appear to be individual stars. These visible stars are only some of the billions of stars that make up the universe. However, only one in four of these stars is actually a single star. Astronomers estimate that about one third of the stars are double stars and the rest are groups of three or more stars. In addition to these double-star and triple-star systems, there also are groups and clusters of stars.

Constellations

Using a star chart and observing carefully, you can identify some of the star groups that form star patterns or regions. Although the stars that make up a pattern seem to be close together, they are not all the same distance from the earth. In fact, they may be very distant from one another.

If you look at the same area for several nights, the positions of the stars in relation to each other do not appear to change. Because of the tremendous distance from which the stars are viewed, they appear fixed in their patterns. However, if you observe each pattern at exactly the same time on successive nights, they will appear to shift a bit to the west from night to night. For more than 3,000 years, people have observed and recorded these shifting but seemingly fixed patterns, which are called **constellations.**

Astronomers recognize 88 constellations. Some constellations are named for real or imaginary animals, such as Ursa Major, the great bear, and Draco, the dragon. Other constellations are named for ancient gods or legendary heroes, such as Hercules and Orion. Most

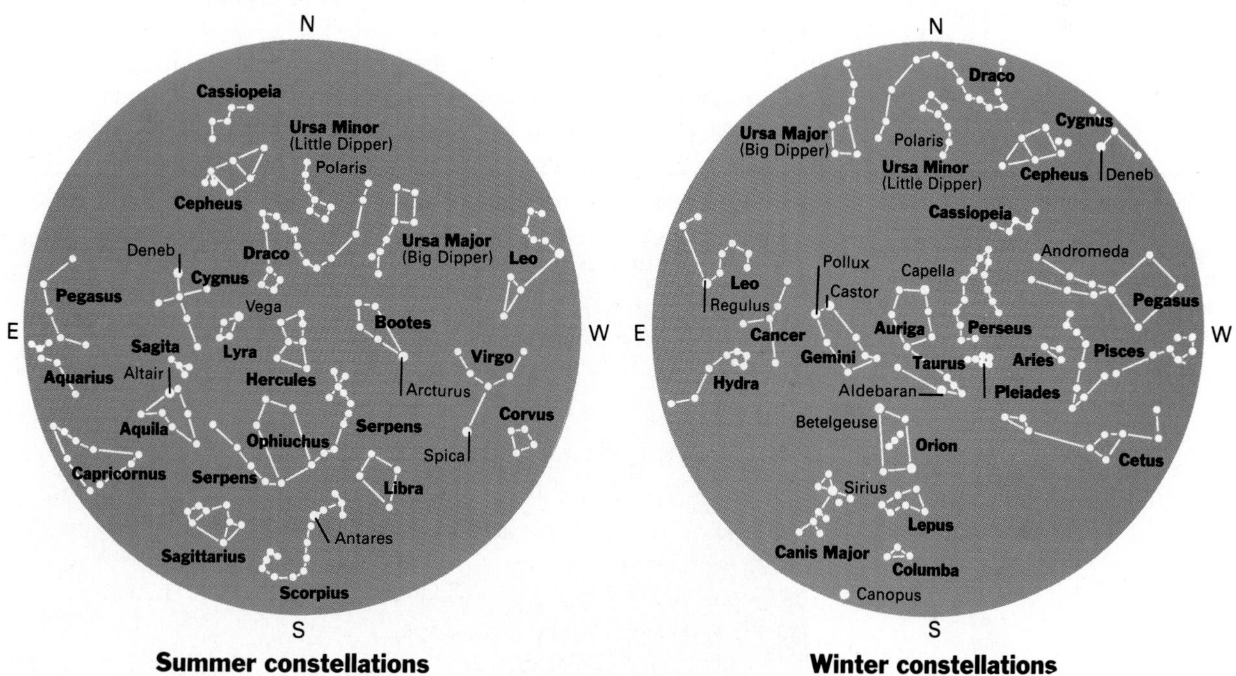

Figure 27–11. These charts show the constellations visible in the Northern Hemisphere. Different constellations are visible in the Southern Hemisphere.

constellations, however, do not actually look like the figures for which they are named.

Astronomers have long used the 88 constellations to divide the entire sky into sectors. Just as you can use a state map to locate a particular town, you can use a map of the constellations to locate a particular star.

Astronomers label the stars within each constellation according to apparent magnitude. The brightest star in a constellation is labeled *alpha,* or α; the second brightest is called *beta,* or β; and so on. Thus, the brightest star in Scorpius is called *Alpha Scorpii,* or α *Scorpii;* the second brightest is called *Beta Scorpii.*

Some bright stars in the sky also have individual names that are not connected with the constellation in which they are found. For example, Alpha Scorpii, which has a red glow similar to that of the planet Mars, is also called Antares, meaning "rival of Mars."

Galaxies

The major components of the universe are **galaxies.** These large-scale groups of stars are bound together by gravitational attraction. A typical galaxy is about 100,000 light-years in diameter and contains 100 billion stars.

Besides billions of stars, a galaxy also contains gas and dust clouds, or nebulae. Astronomers have discovered that some nebulae are bright. Through spectroscopic studies, astronomers have learned that there are two kinds of bright nebulae: those that glow from the hot gases within and those that shine by reflecting the light of nearby stars. Other nebulae are dark, such as the Horsehead Nebula shown in Figure 27–12. Dark nebulae are visible as dark areas amidst the stars. Dark nebulae absorb the light of more distant stars behind them.

Astronomers estimate that there are between 50 billion and 1 trillion galaxies in the known part of the universe. Two galaxies, the Large Magellanic Cloud and Small Magellanic Cloud, are the closest neighbors to the earth's galaxy, which is the *Milky Way Galaxy.* Even so, the Large and Small Magellanic clouds are 150,000 light-years

Figure 27–12. The distinctive shape of this dark nebula, the Horsehead Nebula, is reflected in its name.

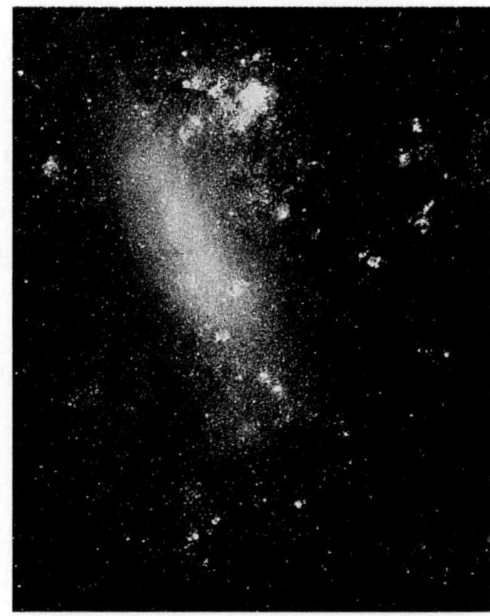

Figure 27–13. The three main types of galaxies are spiral (left), elliptical (center), and irregular (right).

away. Within 3 million light-years of the Milky Way Galaxy are about 17 other galaxies. These galaxies and the Milky Way Galaxy collectively are called the *Local Group*. If each of these galaxies is of typical size, how many stars are in the Local Group?

Types of Galaxies

In studying the vast number of galaxies, astronomers found that galaxies could be classified by shape into the three main types shown in Figure 27–13. One type, called a **spiral galaxy,** has a nucleus, or center, of bright stars and flattened arms of stars that spiral around the nucleus. The spiral arms contain millions of young stars, gas, and dust. Some spiral galaxies have a bar of stars that runs through the center. These galaxies are called **barred spiral galaxies.**

Galaxies of the second type vary in shape from nearly spherical to flattened disks. These galaxies are called **elliptical galaxies.** They are very bright in the center and do not have spiral arms. Elliptical galaxies have no young stars and contain very little dust and gas.

The third type of galaxy, called an **irregular galaxy,** has no particular shape. Irregular galaxies tend to be smaller and fainter than other types of galaxies. Some astronomers think that either the stars in irregular galaxies are of low mass and cannot organize into a regular pattern or they are the result of two galaxies colliding. Thus, the stars in an irregular galaxy are unevenly distributed throughout the galaxy.

The Milky Way

If you look into the night sky, you will see a cloudlike band of stars that stretches across the sky. Because of its milky appearance, this part of the sky is called the *Milky Way.* The Milky Way is the disk of the Milky Way Galaxy. The Milky Way Galaxy is a spiral galaxy in which the sun is but one of billions of stars. Each star seems to have its own motion. Some stars seem to be moving toward the sun, while others seem to be moving away from it.

The Milky Way Galaxy has a diameter of about 100,000 light-years. At its nucleus, the galaxy is 2,000 light-years thick, with the sun located about 30,000 light-years from the center. The Milky

Way Galaxy, like all spiral galaxies, rotates. The sun, which is located in one of the spiral arms, revolves around the center of the galaxy at a speed of about 250 km/s. At this speed, it completes one rotation in about 200 million years.

Star Clusters Besides single stars, the Milky Way Galaxy contains star clusters. These groups of hundreds of stars may be **open clusters** or **globular clusters.** A globular cluster has a spherical shape, while an open cluster is more loosely shaped. Globular clusters contain more stars than do open clusters. Also, globular clusters are distributed around the central core of the galaxy, while open clusters are located in the arms of the disk.

Binary Stars Most stars in the galaxy are **binary stars** or *multiple-star systems*. Binary stars are pairs of stars that revolve around each other. Multiple-star systems have more than two stars. In a multiple-star system, two stars may revolve rapidly around a common center of gravity, while a third star revolves more slowly at a greater distance from the pair. Such star systems are important to astronomy because they are used to determine stellar masses.

Formation of the Universe

The *big bang theory* is the most widely accepted theory explaining the formation of the universe. According to the big bang theory, all of the matter and energy in the universe was once concentrated in an extremely small volume. Then, about 17 billion years ago, the so-called "big bang" occurred. Matter and energy were propelled outward in all directions as the universe began to expand. As the matter and energy moved outward from the center, the force of gravity began to have an effect. Matter began to condense, forming the galaxies. The galaxies continued to move outward from the center. They are still moving outward today, as shown in Figure 27–14.

In 1960, astronomers discovered objects in the universe that are about 12 billion light-years from the earth. Light from these objects must travel for 12 billion years to reach the earth. Therefore, we see

Figure 27–14. Before the big bang, all the matter and energy in the universe was concentrated in an extremely small volume. When the big bang occurred, matter was sent outward in all directions. Galaxies began to form. The universe continues to expand today.

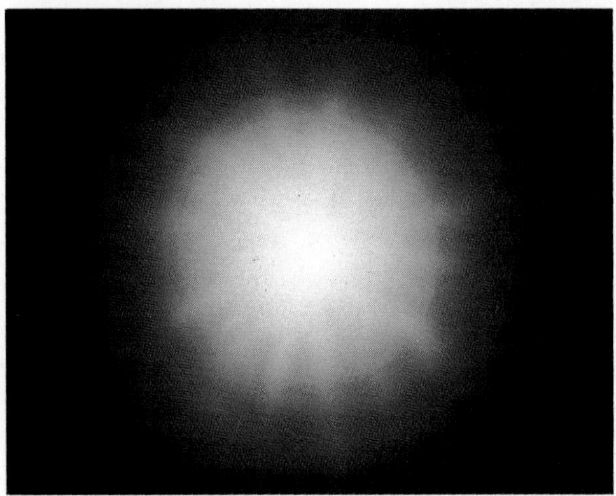

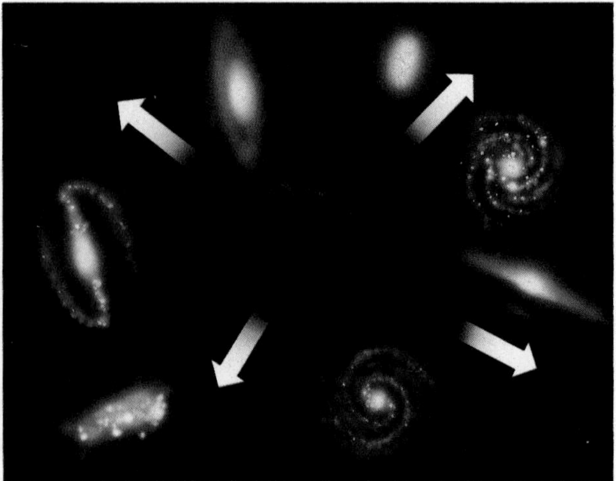

EARTHBEAT

Quasars

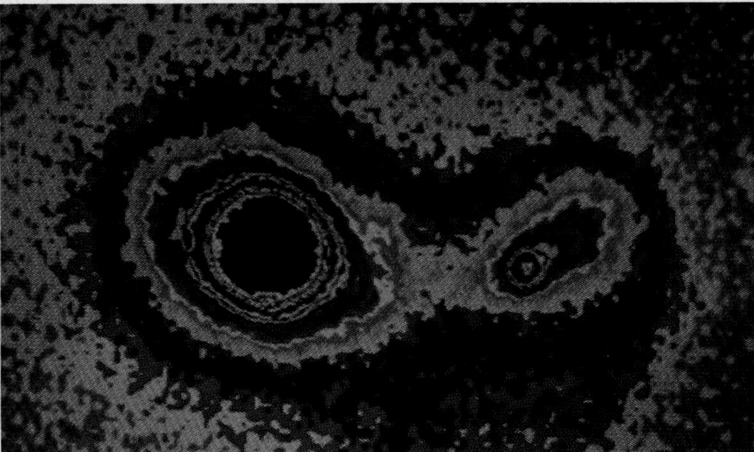

Quasars have been one of the most puzzling objects in the sky. Viewed through an optical telescope, a quasar appears as a small, faint star. The word *quasar* is a shortened term for *quasi-stellar radio source*. The prefix *quasi* means "similar to," and the word *stellar* means "star."

Quasars are the most distant objects that have been observed from the earth. According to its red shift, one quasar is moving away from the earth at a speed of more than 270 million m/s as the universe expands. That speed is 90 percent of the speed of light. In the computer-enhanced color photograph shown here, the red shift of the quasar (large object) and the faint galaxy (small object) are similar.

Many quasars are hundreds of times brighter than the brightest galaxy. In addition, quasars appear to be pinpoints of light, unlike galaxies, which include billions of bright objects. Scientists do not yet understand how such relatively tiny objects can emit so much energy. One hypothesis is that a quasar has a giant black hole in its center, which pulls mass from the surrounding space. The energy that is observed on earth is emitted as the matter falls into the black hole.

Astronomers have found one quasar that may be 15 billion light years away.

If the big bang occurred 17 billion years ago, how long after the big bang did the light now reaching the earth leave that quasar?

these objects as they were 12 billion years ago. These starlike objects, called **quasars,** give off radio waves and X rays that can be detected from the earth. Quasars were among the first objects formed after the big bang and seem to have evolved into galaxies.

Section 27.3 Review

1. Why do the patterns of the stars appear to shift westward slightly from night to night?
2. List the three basic types of galaxies.
3. What type of galaxy is the Milky Way Galaxy?
4. Describe a quasar.
5. Imagine that astronomers found a galaxy that was 20 billion light-years from the earth. How would they have to revise the big bang theory to account for such a discovery?

Chapter 27 Review

Key Terms

- absolute magnitude (553)
- apparent magnitude (552)
- barred spiral galaxy (563)
- binary star (564)
- black hole (560)
- circumpolar (549)
- constellation (561)
- elliptical galaxy (563)
- galaxy (562)
- giant (554)
- globular cluster (564)
- H-R diagram (553)
- irregular galaxy (563)
- light-year (550)
- main-sequence star (554)
- nebula (555)
- neutron star (559)
- nova (559)
- open cluster (564)
- parallax (550)
- planetary nebula (558)
- protostar (555)
- pulsar (560)
- quasar (565)
- red shift (550)
- spiral galaxy (563)
- star (547)
- supergiant (554)
- supernova (559)
- white dwarf (554)

Key Concepts

To determine the composition and surface temperature of a star, astronomers study the spectrum of the star. **See page 547.**

The stars appear to move westward around a central point and to move westward on successive nights across the sky. **See page 548.**

To measure the distance to a star from the earth, astronomers use direct and indirect methods. **See pages 550–553.**

Apparent magnitude and absolute magnitude are two ways of characterizing the brightness of a star. **See page 552.**

A protostar is considered a star when it begins to generate energy. **See page 555.**

A main-sequence star generates energy through hydrogen fusion. **See page 557.**

A giant is a large, cool star with a core in which helium fusion is occurring. **See page 558.**

A constellation is a visible star group that was identified by past cultures and is used by astronomers to divide the sky into sectors. **See page 561.**

Astronomers have identified three main types of galaxies. **See page 563.**

The big bang theory is the most widely accepted explanation of the formation of the universe. **See page 564.**

Review

On your own paper, write the letter of the term that best completes each of the following statements.

1. In the majority of stars, the most common element is
a. oxygen. b. helium. c. hydrogen. d. sodium.

2. The color of the hottest stars is
a. red. b. yellow. c. green. d. blue.

3. Stars appear to move in circular paths around Polaris because
a. the earth rotates on its axis.
b. the earth orbits the sun.
c. the stars revolve around Polaris.
d. Polaris is the center of the Milky Way Galaxy.

4. The change in position of a nearby star compared with the position of a faraway star is called
a. parallax. b. red shift.
c. blue shift. d. a Cepheid variable.

5. The brightest stars have apparent magnitudes that are
a. over +20. b. between +10 and +19.
c. between +1 and +9. d. negative numbers.

6. The absolute magnitude of a star is
   a. the relative brightness of the star.
   b. the true brightness of the star.
   c. the comparative brightness of the star.
   d. the apparent brightness of the star.
7. A protostar becomes a star when it begins to
   a. develop a red shift. b. generate energy.
   c. shrink and spin. d. explode as a nova.
8. A main-sequence star generates energy by fusing
   a. nitrogen into iron. b. helium into carbon.
   c. hydrogen into helium. d. nitrogen into carbon.
9. A dying star can shed some of its gases as a
   a. planetary nebula. b. white dwarf.
   c. globular cluster. d. supernova.
10. Black holes are difficult to locate because they
   a. move very quickly. b. do not give off light.
   c. have very low gravity. d. are far away from any stars.
11. A pattern of stars is called a
   a. galaxy. b. nebula. c. pulsar. d. constellation.
12. Stars appear in fixed locations in the sky because they
   a. are so far from the earth. b. do not move.
   c. are all moving toward the earth. d. are all in the same galaxy.
13. The basic types of galaxies are
   a. spiral, elliptical, and irregular. b. barred, elliptical, and open.
   c. spiral, quasar, and pulsar. d. open, binary, and globular.
14. Quasar formation is associated with
   a. nuclear fusion. b. main-sequence stars.
   c. the explosion of a supernova. d. the big bang.

Critical Thinking

On your own paper, write answers to the following questions.

1. If the spectrum of a star indicates that the star shines with a red light, approximately what is the surface temperature of the star?
2. Why are different constellations visible during different seasons of the year?
3. Explain why Polaris is considered to be a very bright star even though it is not a bright star in the earth's sky.
4. Why does heat build up more rapidly in a massive protostar than in a less massive one?
5. Explain why an old main-sequence star will be composed of a higher percentage of helium than will a young main-sequence star.
6. If all galaxies began to show blue shifts, what would this indicate about the size of the universe?

Application

1. If you determined that a certain star displayed a large parallax, what could you say about its distance from the earth?
2. Suppose that a scientist has discovered a red-dwarf star. Describe the likely size and surface temperature of such a star.
3. When looking through a nearby university's telescope, you observed a galaxy that has no young stars and contains little dust or gas. What kind of galaxy were you probably looking at?

Extension

1. The modern scale of star magnitudes is based on a system established in 129 B.C. by Hipparchus of Nicea. Conduct library research to find out how the modern scale differs from the one introduced by Hipparchus.
2. In February 1987, astronomers saw a supernova explosion in the Large Magellanic Cloud. Because Ian Shelton was the first astronomer to notice it, the explosion is known as Shelton's Supernova. Write a report on Shelton and his historic discovery.
3. Make a storyboard showing the development of the universe according to the widely accepted big bang theory.

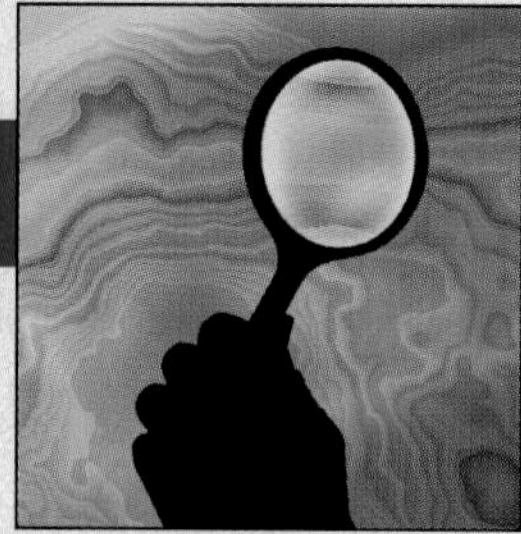

IN-DEPTH INVESTIGATION

Star Magnitudes

Materials

- 3-volt flashlight bulbs (2)
- masking tape
- large rubber band
- 3 AA batteries
- 2 paraffin bricks (12 × 6 cm)
- ruler
- 2 plastic-coated wires with stripped ends (15 cm and 20 cm)
- aluminum foil (12 × 12 cm)
- desk lamp with incandescent bulb

Introduction

Astronomers study the brightness, or magnitude, of stars. Except for the sun, stars are very faint and visible only at night. Thus, their brightness must be measured with a device called a photometer. An astronomical photometer consists of a surface that is sensitive to light and a device that measures the amount of light that reaches the surface. The brightest part of a star's spectrum can occur at any wavelength of optical light. Some stars look bluish while others look reddish. Color is an indication of a star's surface temperature. There is also a relation between the color of a star and its brightness. Photometers can be used to compare the colors of different light sources.

In this investigation, you will determine the effect of distance on brightness and the relation between temperature and color.

Prelab Preparation

1. Review Section 27.1, pages 547–548 and 550–553.
2. Review the safety guidelines for glassware safety. 
3. Constructing a flashlight:
   a. Arrange the bulbs and batteries as shown in the figure below. Using masking tape, attach the wires to the batteries and bulbs. The bulb should be on. If it is not, study the wiring diagram again and make adjustments.
   b. Tape the flashlight arrangement together so that it can be moved. Be sure to leave the wire loose at the end of the battery so that you can turn your flashlight on and off.
4. Constructing a photometer: Fold the aluminum foil in half with the shiny side facing out, and place it between the paraffin bricks. Hold the pieces together with a rubber band.

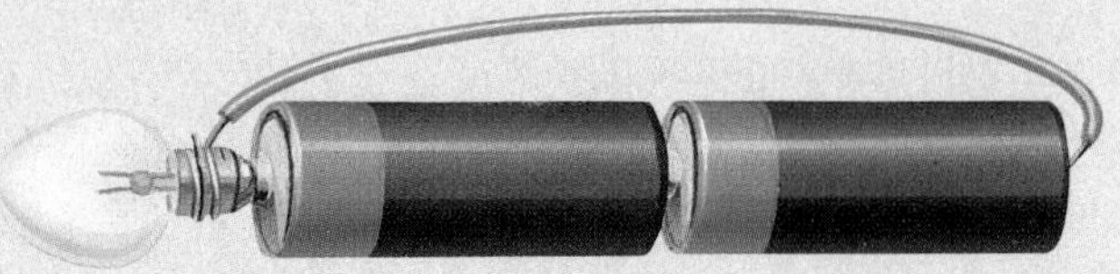

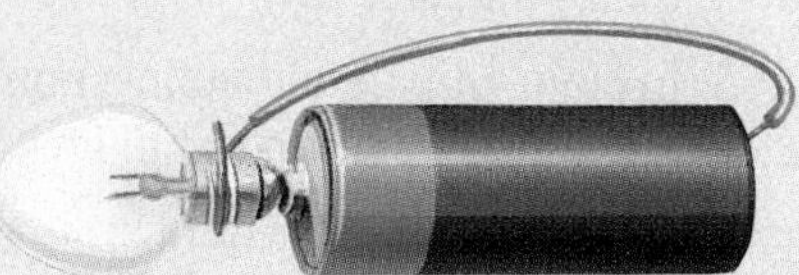

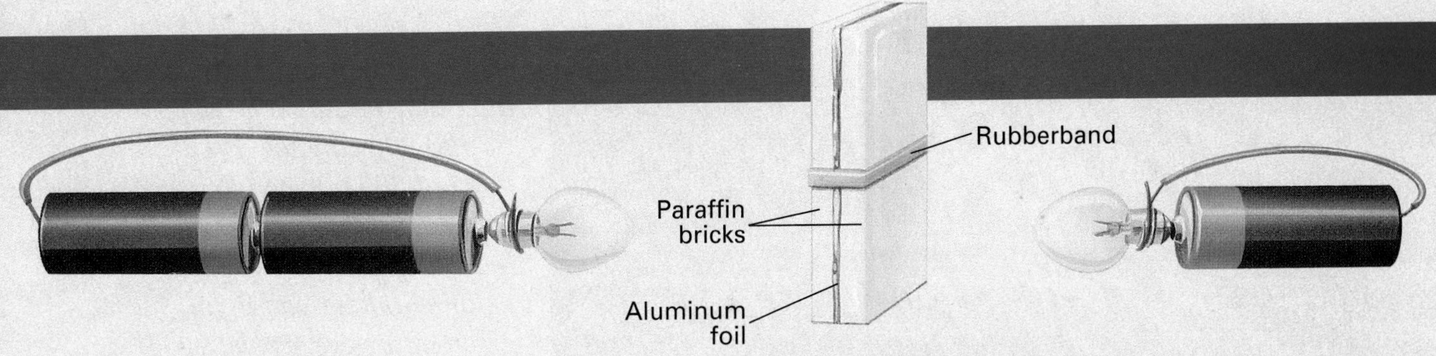

Procedure

1. Place the two flashlights on a table about 2 m apart. Place the photometer between them with the largest sides of the bricks facing each flashlight bulb, as shown in the figure above.
2. Turn on both flashlights, and turn off all room lights.
3. Move the paraffin photometer until both sides are equally bright. Measure the distance, in centimeters, from each flashlight bulb to the center of the photometer. Record these measurements in your lab report. Which way did you move the photometer to make it record equal brightness on both sides?
4. Square the distances you recorded in Step 3. Record these distances in your lab report.
5. Incandescent light bulbs have filaments inside that glow at a temperature much cooler than the sun's surface. Place the photometer between the desk lamp and a window on a bright day. Sunlight coming through a window will be the same color as the sunlight outdoors. Turn off any fluorescent ceiling lighting, and turn on the lamp.
6. Compare the color differences between the paraffin sides of your photometer. Which light, the light bulb or sunlight, is more yellow? Which light is more white?
7. Darken the room once again, and compare the colors of the bulb powered by one battery with the colors of the bulb powered by two batteries.

Analysis and Conclusions

1. As you move away from a light source, the brightness decreases in relation to the square of the distance. The ratio of the square of the distances you calculated in Step 4 is equal to the ratio of the brightnesses of the bulbs. What is the ratio of the square of the distance of the two-battery flashlight to that of the one-battery flashlight? What does this tell you about the relation of the brightnesses of the two flashlights?
2. An astronomer knows that two stars have the same spectra and should be the same brightness. Yet one is four times fainter than the other. How much farther away is the fainter star?
3. Based on the results of the investigation, would you expect a white star or a yellow star to be hotter?
4. Using your knowledge of the spectrum, would you expect a white star to be hotter or cooler than an orange star? Predict whether a blue star is hotter or cooler than a white star. Also predict whether a red star is hotter or cooler than an orange star.

Extensions

1. Find an incandescent bulb controlled by a dimmer. Watch the color of the light as it fades. Does it get more yellow or more white? Explain why.
2. If a red star is very cool with a low absolute magnitude, why are bright red stars visible in the sky?

Chapter 28

The Sun

Try to imagine the heat and light that would be produced by 300,000 candles. Then think of 300,000 candles squeezed into a single square inch of space. That is the amount of heat and light given off by every square inch of the sun's surface—a surface that is almost 12,000 times as great as that of the earth. Energy from the sun makes life on the earth possible. In this chapter, you will learn about the structure of the sun and how it produces such huge amounts of energy.

Chapter Outline

The sun's light energy is blocked from the earth during an eclipse.

28.1 Structure of the Sun

Section Objectives

- **Explain how the sun converts matter into energy in its core.**
- **Compare the radiative and convective zones of the sun.**
- **Describe the three layers of the sun's atmosphere.**

Throughout most of human history, people thought that the sun's energy came from fire. People knew that burning a piece of coal or wood produced heat and light. They assumed that the sun, too, burned some type of fuel to produce its energy. Not until this century did scientists discover the source of the sun's energy.

The tremendous heat and light of the sun make it appear to be a dazzling, brilliant ball with no distinct features. Because the sun's brightness can damage the eyes, astronomers use special scientific instruments to study the sun. Spectroscopic analysis of the sun's rays shows that most of the known elements are found in the sun. The rest are probably present as well, but in amounts too small to be detected. Hydrogen makes up 92 percent of the sun's content, and hydrogen and helium together make up almost 100 percent. Studies indicate that the sun has several regions and distinct features.

The sun has three basic regions: the core, the inner zones, and the atmosphere. The inner zones and the atmosphere are each further divided into smaller layers. Each layer of the sun has specific characteristics. However, the boundaries between layers are not distinct. Each layer blends gradually into the next. If you could travel through the sun, you could not tell when you left one layer and entered the next.

The Core

At the center of the sun is the core. The core makes up 10 percent of the sun's total diameter of 1,300,000 km. The temperature of the sun's core is about 15,000,000°C. No liquid or solid can exist at that high a temperature. The core, like the rest of the sun, is made up entirely of gas.

The sun's mass is 300,000 times greater than the earth's mass. Consequently, the force of gravity is much greater on the sun than on the earth. The sun's gravity is so strong that the center of the sun is 10 times denser than iron.

The enormous pressure and heat of the sun change the structure of the atoms within the core. On the earth, atoms normally consist of a nucleus surrounded by one or more electrons. The nucleus, composed of protons and neutrons, remains unchanged even if the number of electrons changes. Within the core of the sun, however, the heat and pressure strip electrons away from the atomic nuclei. The exposed nuclei can then be changed by nuclear reactions. The most common nuclear reaction occurring inside the sun is the fusion of hydrogen into helium, or nuclear fusion. In nuclear fusion, atomic nuclei are combined to form larger atomic nuclei.

Hydrogen Fusion

The nuclei of hydrogen atoms are the primary elements in the fusion occurring in the sun. The sun contains a greater number of hydrogen atoms than it does of any other kind of atom. A hydrogen atom, the

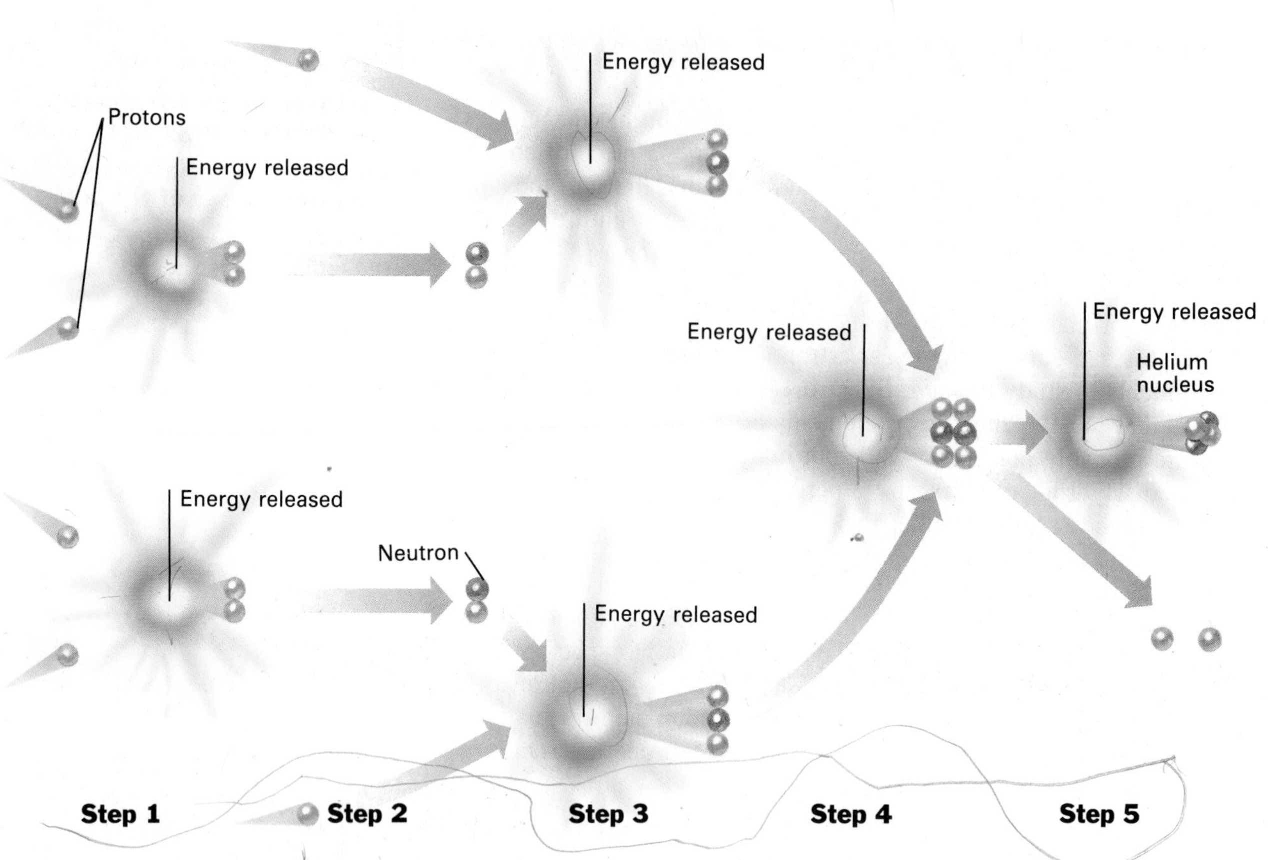

Figure 28–1. In the core of the sun, the nuclei of hydrogen atoms fuse into helium. The process converts mass into energy.

simplest of all atoms, usually consists of one electron and a nucleus of one proton. In the sun's core, hydrogen atoms have lost their electrons. Thus, only the protons remain.

The process of nuclear fusion that produces most of the sun's energy consists of five steps, as shown in Figure 28–1. First, two hydrogen nuclei, or protons, collide and fuse. Second, one of these protons changes into a neutron. Third, another proton combines with the proton-neutron pair, producing a nucleus made up of two protons and one neutron. Fourth, two of these nuclei collide and fuse. Finally, the resulting cluster throws off two protons. The remaining combination of two protons and two neutrons is the nucleus of a helium atom. During each step of the reaction, energy is given off.

The fusion of hydrogen nuclei may take place in a slightly different series of steps, but one of the final products is always a helium nucleus. The helium nucleus has about 0.7 percent less mass than the hydrogen nuclei that combined to form it. The lost mass is converted to energy during the hydrogen-to-helium reaction. This second product of fusion is the energy that causes the sun to shine.

Mass into Energy

In 1905, the famous physicist Albert Einstein proposed that a large amount of energy could be produced from a very small amount of matter. At the time he proposed this, the existence of nuclear fusion was unknown. In fact, scientists had not yet discovered the nucleus of the atom. Einstein's work led to his theory of relativity, a revolutionary view of the physics of matter, energy, and time. Einstein's proposal involved the equation $E = mc^2$. E represents the energy produced; m represents the mass, or the amount of matter, changed; and c represents the speed of light, which is 300,000 km/s.

Einstein's equation can be used to calculate the amount of energy produced from a given amount of matter. With this equation, astronomers are able to explain the vast quantities of energy that the sun produces, using only a very small amount of fuel. The sun changes more than 600 million tons of hydrogen into helium every second. Yet this is a small amount compared to the total mass of hydrogen in the sun. At this rate how much hydrogen is changed into helium each day?

Reactions other than hydrogen fusion also take place in the sun's core. For example, complex fusion reactions among nuclei of carbon, nitrogen, and oxygen atoms also produce energy.

INVESTIGATE!

To learn more about the sun, try the In-Depth Investigation on pages 588–589.

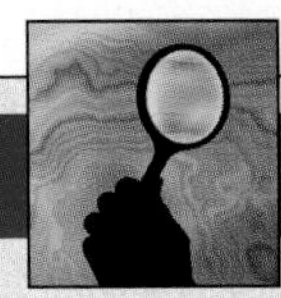

SMALL-SCALE INVESTIGATION

Solar Viewer

Solar telescopes enable astronomers to study the sun without looking directly at it. Using some common objects, you can make a solar viewer that functions somewhat like a solar telescope.

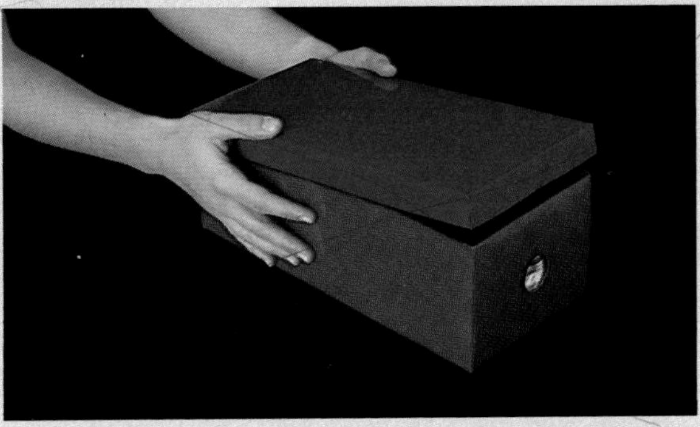

Materials

shoe box with lid; ruler; scissors; masking tape; 2 index cards; safety pin; piece of aluminum foil, about 4 cm × 4 cm

Procedure

1. Cut a hole with a diameter of about 3 cm in the center of one end of the shoe box. Tape an index card to the inside wall at the end of the box opposite the hole.
2. Tape the foil over the hole in the shoe box. Use the safety pin to make a tiny hole in the center of the foil. Put the lid on the shoe box.
3. Hold the box with the pinhole toward the sun. With your back to the sun, lift the lid and observe the image of the sun projected on the index card. Write down what you observe.

   Caution: Never look directly at the sun. Direct sunlight can damage your eyes.
4. Place a second index card inside the box at various distances from the pinhole. Write down what you observe at each distance.
5. Repeat Step 3 several times, making the pinhole slightly larger each time. Write down what you observe as the diameter grows larger.

Analysis and Conclusions

1. Does the image of the sun change in size or brightness as the distance between the pinhole and the index card is changed?
2. What happens to the image as you make the pinhole larger?
3. How is your solar viewer like a solar telescope? How is it different?

The Inner Zones

Before reaching the sun's atmosphere, the energy produced in the core moves through two inner zones. The zone surrounding the core is called the **radiative zone.** The temperature in this zone is about 2,500,000°C. In the radiative zone, energy moves from atom to atom in the form of electromagnetic waves, or radiation. This electromagnetic radiation transfers energy through space.

Around the radiative zone is the **convective zone,** which has a temperature of about 1,000,000°C. Energy produced in the core moves through this zone by *convection,* the transfer of energy by moving liquids or gases. Hot gases carry heat energy to the sun's surface. As the atoms of the hot gases move outward and expand, they radiate and lose heat. The cooling gases become denser than the other gases and sink to the bottom of the convective zone. There, the cooled gases are heated by the energy from the radiative zone and rise again. Thus, heat is transferred to the sun's surface as the gases continuously rise and sink.

The Sun's Atmosphere

Surrounding the convective zone is the sun's atmosphere. Although the entire sun is made of gases, the term *atmosphere* refers to the uppermost region of solar gases. This region has three layers.

The innermost layer of the solar atmosphere—the **photosphere,** or light sphere—is made of gases bubbling up from the convective zone. The temperature in the photosphere is about 6,000°C. This layer has a grainy appearance, called *granulation,* which results as the gases rise from and sink into the convective zone. Much of the energy given off from the photosphere is in the form of visible light. The visible light is what is seen from the earth. Therefore, the photosphere is considered the surface of the sun.

Figure 28–2. False colors have been added to this photograph of the sun to show temperature differences. The red area is actually cooler than the green and blue areas.

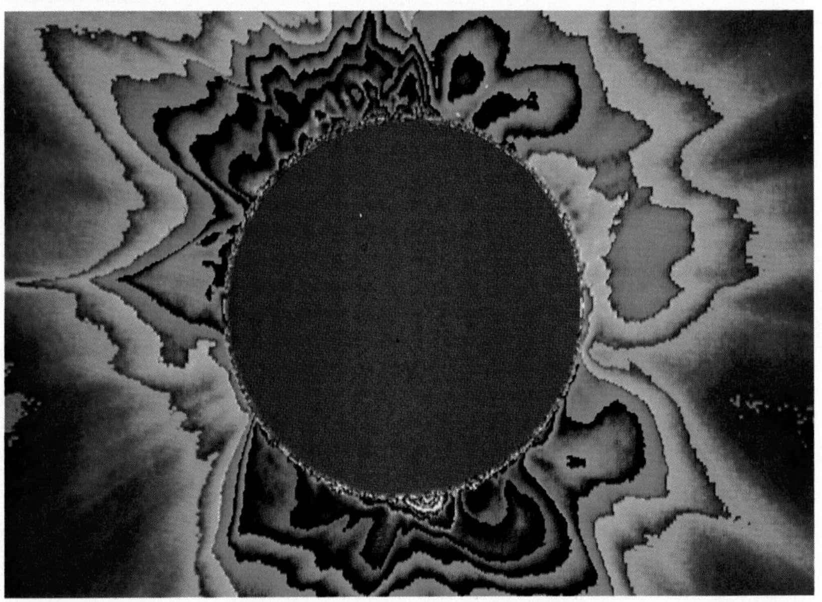

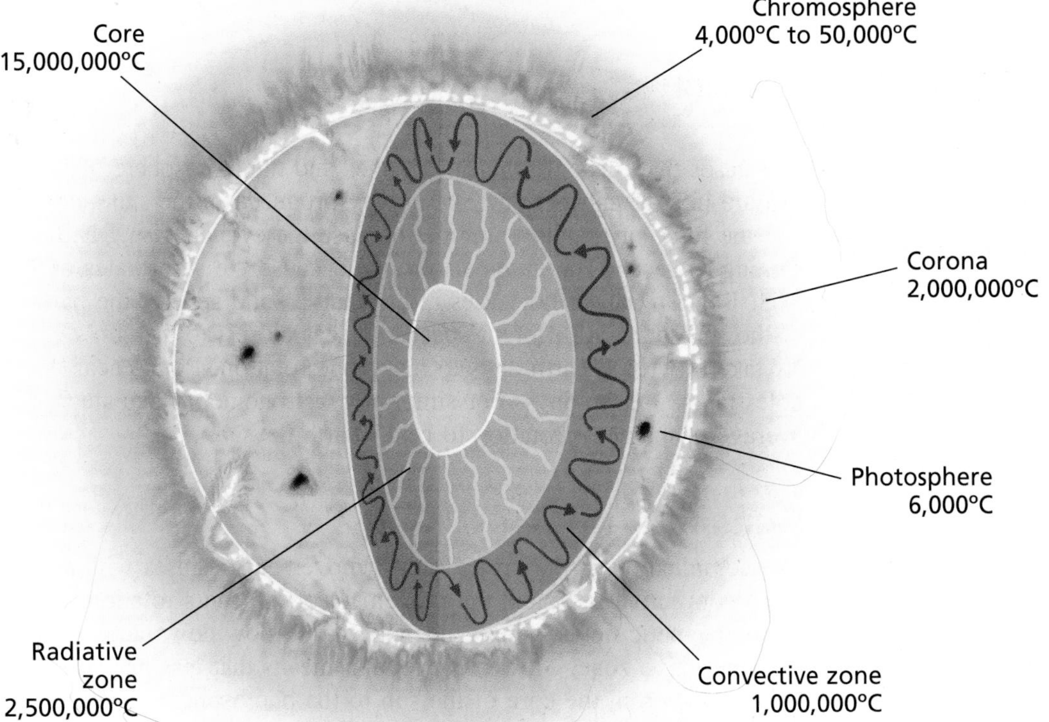

Figure 28–3. The boundaries of the sun's regions and layers that surround the sun's core are not distinct. Each layer blends gradually into the next.

Above the photosphere lies the **chromosphere,** or color sphere, a thin layer of gases that seems to glow with a reddish light. Its temperature ranges from 4,000°C to 50,000°C. The gases of the chromosphere move away from and toward the photosphere. In an upward movement, they occasionally form narrow jets of hot gas that shoot outward from the chromosphere and then fade away within a few minutes. Some of these jets reach heights of 16,000 km.

The outermost layer of the sun's atmosphere—the **corona** (kuh-ROE-nuh), or crown—blends into space. This layer is a huge cloud of gas heated by the sun's magnetic field to a temperature of about 2,000,000°C. Although the corona is relatively thin, it prevents most of the atomic particles from the sun's surface from escaping into space. However, some electrically charged atomic particles, or ions, stream out into space through holes in the corona. These particles, called the **solar wind,** flow to distant parts of the solar system.

The chromosphere and the corona are normally not seen from the earth because of the brightness of the blue sky during the day. Occasionally, however, the moon moves between the earth and the sun and blocks out the light of the photosphere. The corona then becomes visible.

Section 28.1 Review

1. Describe the most common type of nuclear fusion in the sun.
2. What are the two end products of fusion in the sun?
3. Will the amount of hydrogen and helium in the sun increase or decrease over the next few million years?
4. What layer of the sun's atmosphere can normally be seen?
5. How does the transfer of energy in the radiative zone differ from the transfer of energy in the convective zone?

Section Objectives

- **Explain how sunspots are related to powerful magnetic fields on the sun.**
- **Compare prominences and solar flares.**
- **Describe how the solar wind can cause auroras on the earth.**

28.2 Solar Activity

The gases that make up the inner zones and the atmosphere of the sun are in constant motion. The energy produced in the sun's core and the force of gravity combine to cause the cyclical rising and sinking of gases. The gases also move because the sun rotates on its axis. Being a ball of hot gases rather than a solid sphere, the parts of the sun rotate at different speeds. Places closest to the sun's equator take only 25.3 earth days to make one rotation. Points near the poles take 33 earth days. For simplicity, astronomers calculate the average of the sun's rotation to be 27 earth days.

Sunspots

The combination of the up-and-down movement of gases within the sun's convective zone and the movement of the sun's rotation produces magnetic fields. These magnetic fields slow down activity in the convective zone. Slower convection means that less gas is transferring heat from the core of the sun to the photosphere. Therefore, regions of the photosphere near strong magnetic fields are up to 3,000°C cooler than surrounding areas.

The cooler areas of the sun appear darker than the areas surrounding them. Cool, dark areas of gas within the photosphere that are caused by powerful magnetic fields are called **sunspots.** Large sunspots can be more than 100,000 km in diameter, which is several times the size of the earth. Photographs of sunspots are shown in Figure 28–4.

Astronomers have carefully observed sunspot activity. They have found that sunspots initially appear in groups about midway between

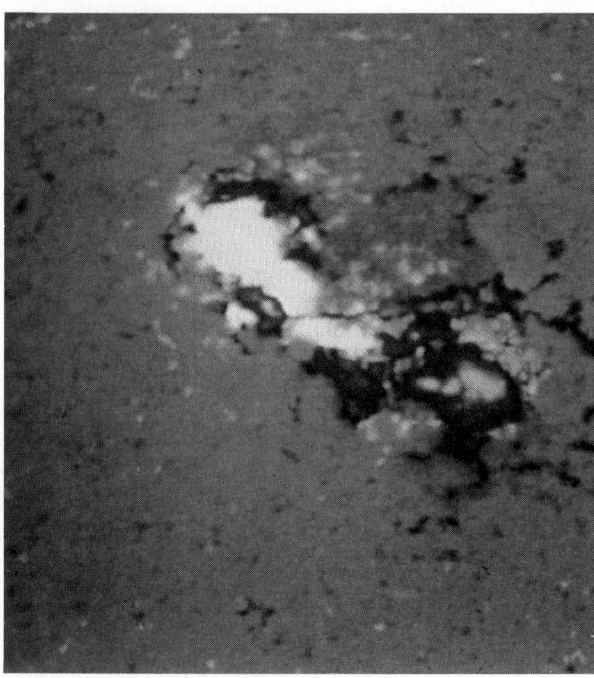

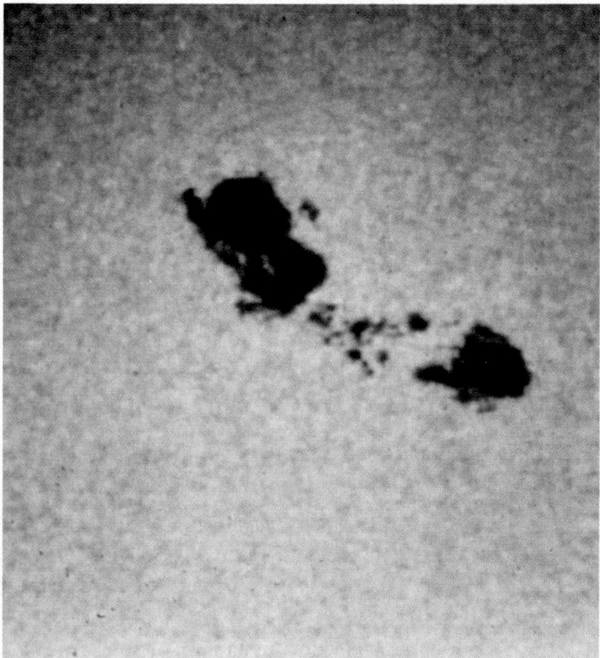

Figure 28–4. Sunspots usually appear in pairs—one magnetic north and one magnetic south. The photograph on the left has been artificially colored to show this polarity. The photograph on the right is of the same sunspots in visible light.

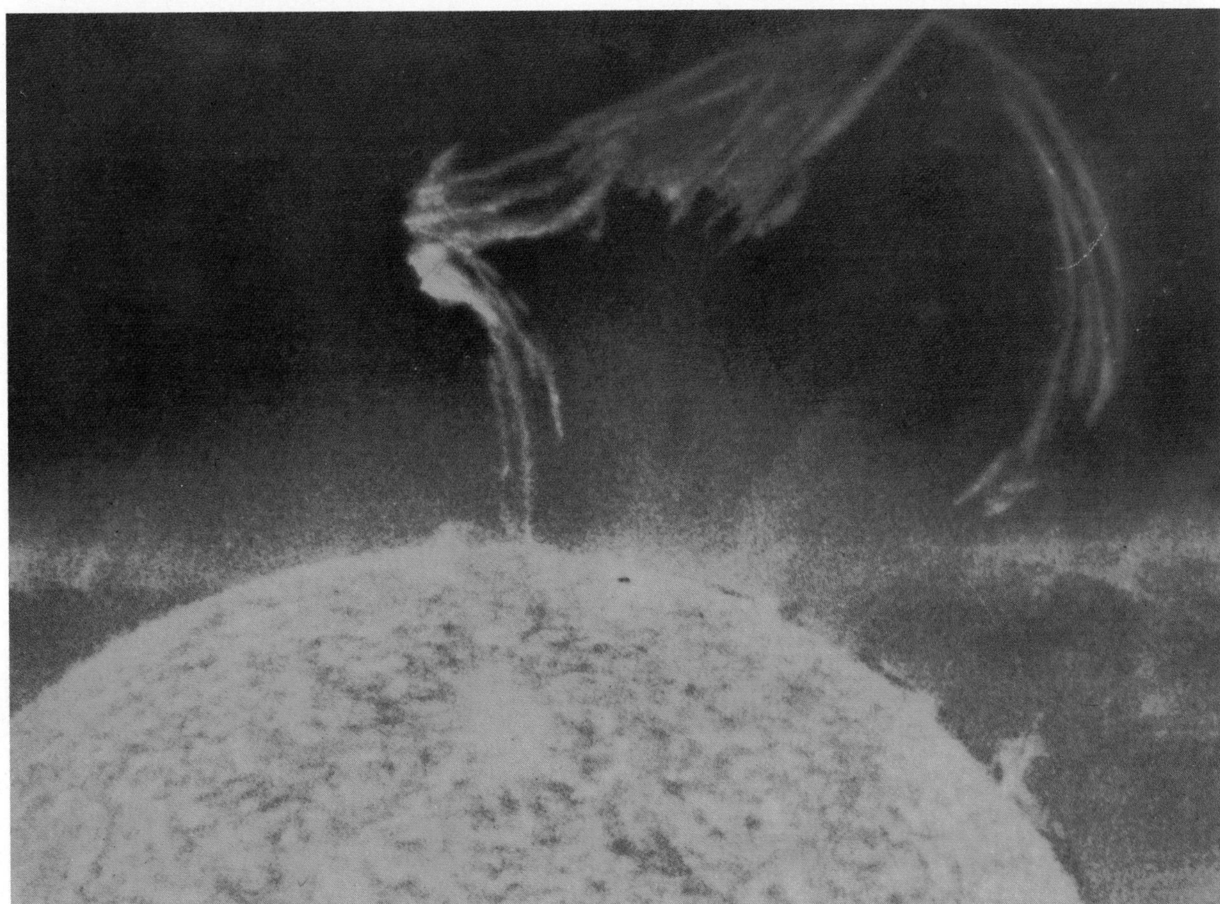

Figure 28–5. A solar prominence can arch half a million kilometers above the sun's surface.

the sun's equator and poles. As sunspots slowly disappear, new ones seem to appear near the sun's equator. The shift of sunspots was one of the first indications to astronomers that the sun rotates on its axis. The number of sunspots also varies according to an average 11-year cycle called the **sunspot cycle.**

A sunspot cycle begins when the number of sunspots is very low but is beginning to increase. Astronomers may not see any sunspots for several weeks. Then gradually they see more and more. The number of sunspots increases over the next few years until it reaches a peak. At a peak of sunspot activity, 100 or more sunspots may be visible. After the peak, the number of sunspots begins to decrease until it reaches a minimum. Another 11-year cycle begins when the number of sunspots begins to increase again. If the number of sunspots was very low in 1988, when will the next low point in the cycle occur?

Prominences and Solar Flares

The magnetic fields that cause sunspots also create other disturbances in the solar atmosphere. Great clouds of glowing gases, called **prominences,** form huge arches that reach high above the sun's surface, as shown in Figure 28–5. Each solar prominence follows curved lines of magnetic force from one sunspot area to another. Some prominences may last for several weeks, while others may exist for up to a year.

One of the most violent of all solar disturbances is a **solar flare,** a sudden outward eruption of electrically charged atomic particles. Solar flares may extend upward several thousand kilometers within minutes. Few eruptions last more than an hour. Solar flares usually occur near sunspots. During a peak in the sunspot cycle, five to ten solar flares may be visible each day.

Some of the particles from a solar flare are flung out so forcefully that they escape into space. These particles increase the strength of the solar wind. As the gusts of solar-wind particles enter the atmosphere of the earth, they can generate a sudden disturbance in the earth's magnetic field. Such disturbances are called *magnetic storms.* Although several small magnetic storms may occur each month, the average number of severe storms is less than one per year. Magnetic storms have been known to interfere with radio communications on earth.

Coronal Mass Ejections

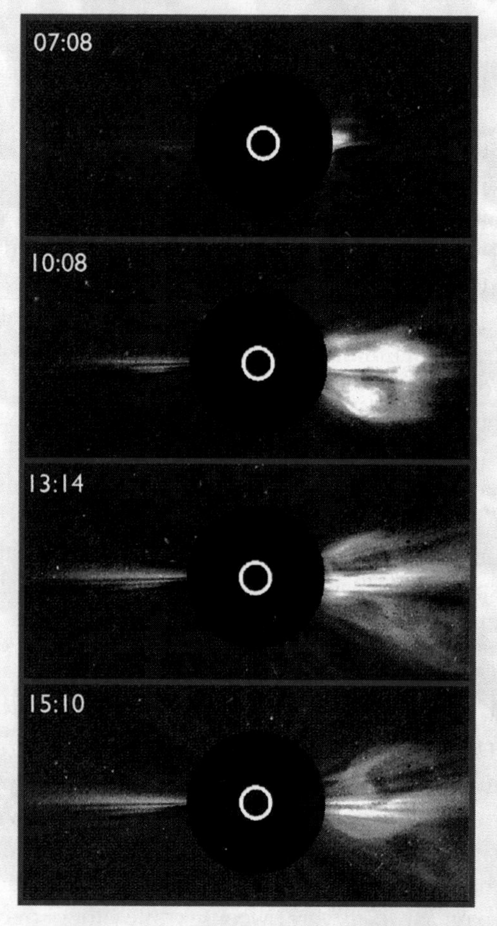

During an aurora, the night sky fills with colored lights and the air crackles as electrically charged atomic particles from the sun strike the earth's atmosphere. Scientists estimate that one night's display contains enough electricity to supply the entire United States for a year.

Astronomers once linked all auroras to solar-flare activity. However, recent findings by projects such as the Solar Maximum Mission (SMM) and the Solar and Heliospheric Observatory (SOHO) suggest that auroras result from solar events called *coronal mass ejections.* Caused by periodic shifts in the sun's magnetic fields, coronal mass ejections can release billions of kilograms of gaseous ions at speeds of up to 2,000 km/s. As you can see in the time-lapse photographs shown here, these ejections can grow larger than the sun itself (indicated by the white circle) in a matter of hours, and they could be the major driving force behind the solar wind.

Coronal mass ejections could also be the solar wind's largest source of ions. If enough of these charged particles reach the earth, they disrupt the ionosphere and interfere with radio transmissions. Like solar flares, coronal mass ejections pose radiation hazards, especially to satellites and to astronauts who are unprotected by the earth's atmosphere.

How do electrically charged particles from the sun cause auroras?

Figure 28–6. Auroras, such as these over Alaska, can fill the entire sky with a colorful curtain of light.

Auroras

On the earth the most spectacular effect of a magnetic storm is the appearance in the sky of bands of light called **auroras** (uh-RORE-uhz). Figure 28–6 shows an example of an aurora. When the electrically charged particles of the solar wind approach the earth, they are guided toward the earth's magnetic poles by the earth's *magnetosphere*. The magnetosphere is the space around the earth that contains a magnetic field. The electrically charged particles strike the gas molecules in the upper atmosphere, thereby producing green, red, blue, or violet sheets of light. Because of the effect of the magnetosphere, auroras are usually seen close to the magnetic poles. Depending upon which pole they are near, auroras are also called *northern lights* or *southern lights*.

Auroras usually occur between 100 km and 1,000 km above the earth's surface. They are most frequent just after a peak in the sunspot cycle, especially after solar flares. Across the northern United States, auroras are visible about five times a year.

Although auroras are most often visible in polar regions, they are sometimes visible near the equator. In 1921, the residents of some South Pacific islands were able to see the southern lights.

Section 28.2 Review

1. Why are sunspots cooler than surrounding areas?
2. How long is the sunspot cycle?
3. How are prominences different from solar flares?
4. What causes auroras?
5. If the strength of the sun's magnetic field decreased from its current level, why would the temperature of sunspots increase?

Section Objectives

- **Explain the nebular theory of the origin of the solar system.**
- **Describe how the planets developed.**
- **Describe the formation of the land, the atmosphere, and the oceans of the earth.**

28.3 Formation of the Solar System

Scientists have long debated the origins of the **solar system.** The solar system includes the sun and the bodies revolving around the sun. In the 1600's and 1700's, many scientists thought that the sun formed first and threw off the materials that later formed the **planets.** Planets are the nine major bodies orbiting the sun. In 1796, however, the French mathematician Marquis Pierre Simon de Laplace advanced a new hypothesis. It stated that the sun and the planets condensed out of the same spinning *nebula,* or cloud of gas and dust. The hypothesis also stated that the entire solar system formed at approximately the same time. Laplace's hypothesis developed into what became known as the **nebular theory.**

Formation of the Sun

According to current theory, the big bang spread matter throughout the expanding universe. Some of this matter then gathered into clouds of dust and gas. The cloud of gas and dust that eventually developed into our solar system is called the **solar nebula.** The solar nebula was larger than our solar system is now.

About 4 billion to 5 billion years ago, shock waves from a nearby supernova or some other force caused a cloud of dust and gas to contract, forming the solar nebula. A star—the sun—began to form in the center of the solar nebula. In Chapter 27, you read about the origin of a typical star. The sun is thought to have developed by this same process. Heat from collisions and pressure from the force of gravity caused the center of the solar nebula to become hotter and denser. When the temperature at the center of the nebula became great enough, hydrogen fusion began, and the sun was formed. About 99 percent of the matter in the solar nebula became part of the sun.

Formation of the Planets

While the sun was forming in the center of the solar nebula, planets were forming in the outer regions, as shown in Figure 28–7. The small bodies of matter in the solar nebula are called **planetesimals** (PLAN-uh-TESS-uh-mulz). Some planetesimals joined together through collisions and through the force of gravity to form much larger bodies called **protoplanets.** The protoplanets' gravity acted like giant magnets, pulling in other planetesimals from the solar nebula.

Eventually the protoplanets condensed into the existing planets and **moons.** Moons are the smaller bodies that orbit the planets. Planets and moons are smaller and denser than the protoplanets.

The distance between a protoplanet and the developing sun influenced the composition of the planet that formed from the protoplanet.

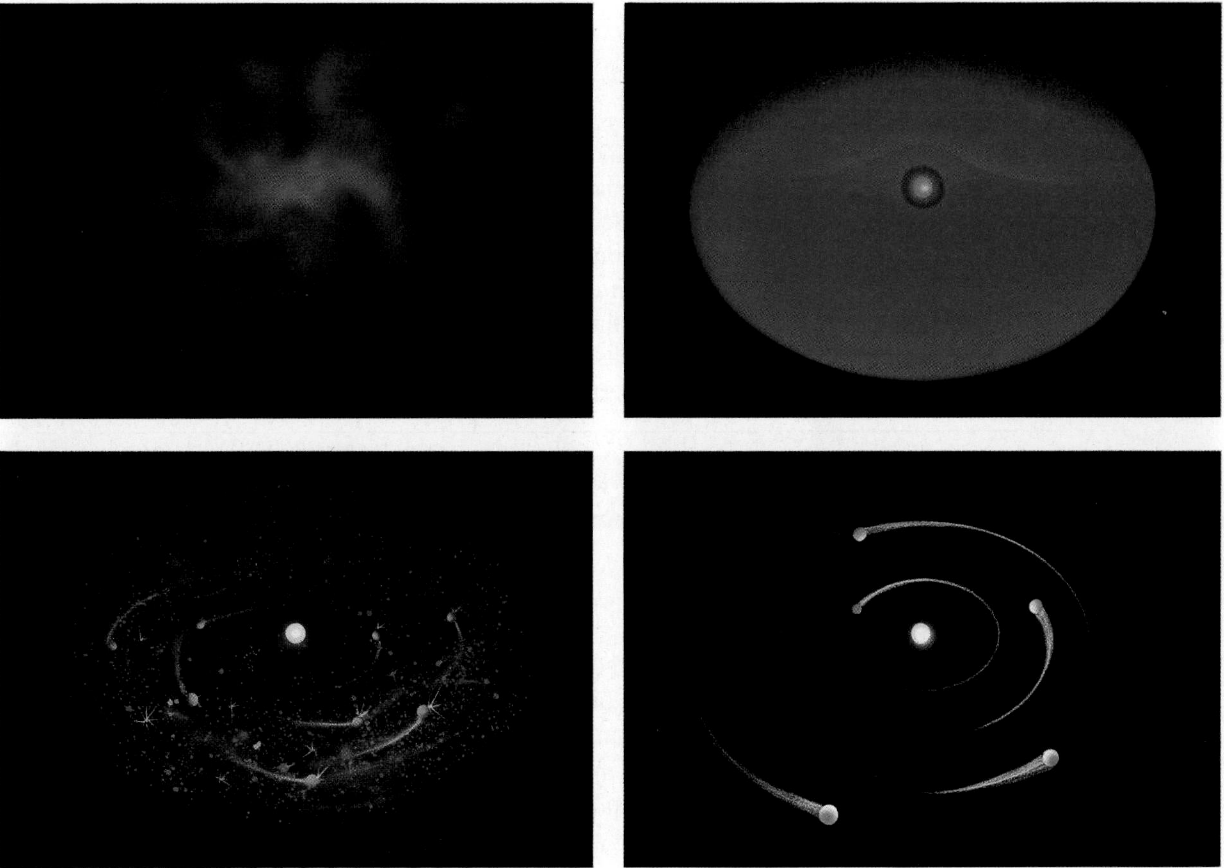

Figure 28–7. The matter that forms into planets begins as dust that forms into planetesimals in the nebula. The planetesimals combine to form protoplanets, which condense into planets.

The four protoplanets closest to the sun became Mercury, Venus, Earth, and Mars. They contained large amounts of the heavier elements, such as iron.

The next four protoplanets became Jupiter, Saturn, Uranus, and Neptune. These four protoplanets formed in the cold regions of the solar nebula. The icy material of the outer protoplanets consisted of helium and hydrogen and frozen gases, such as water, methane, and ammonia. Far from the heat of the sun, the outer protoplanets developed into huge planets. Thick layers of ice surrounded small cores of heavy elements.

The inner planets probably could not accumulate lighter gases because the temperature was too high so close to the sun. The energy and matter from an intense solar wind may also have stripped away the early atmosphere of lighter elements from the inner planets. Because of their greater distance from the sun and their tremendous gravitational forces, the outer planets were able to retain much of their original gases. Today these planets are referred to as the *gas giants*. These planets are called gas giants because they are composed mostly of gases and are such huge planets.

Pluto is farthest from the sun and is the smallest of the known planets. It is very cold, so cold that most of the elements that are normally gases are permanently solidified. Pluto may be best described as an ice ball of frozen gases and rock.

Formation of the Earth

When the earth first formed, it was very hot. Three sources of heat contributed to the high temperature on the new planet. First, the earth retained much of the heat produced when it collided with planetesimals. Second, the increasing weight of the outer layers compressed its inner layers, generating more heat. Third, *radioactive* materials, which emit high-energy particles, were very abundant when the earth first formed. When these particles were absorbed by surrounding rocks, the energy of motion of the particles was converted into heat energy.

CAREER FOCUS: *Space*

"Having the chance to work on questions I've wondered about my whole life is what I find extremely rewarding."

Sandra Faber

Astronomer

"I was one of those kids consumed with science. Rocks, biology, fossils, space—you name it. If it was science, I was interested," says Sandra Faber, professor of astronomy at the University of California Santa Cruz, and staff member at the Lick Observatory in Santa Cruz, California.

Faber, pictured at left, eventually chose astronomy as her field of study for a very specific reason.

"I was interested in the big picture," she says. "Somewhere along the line, it occurred to me that the questions about the universe posed in astronomy were the most all-encompassing."

Faber has a bachelor's degree in physics from Swarthmore College and a doctorate in astronomy from Harvard University. She is now attempting to answer those questions about the nature of the stars, planets, galaxies, and the universe itself. To gather the necessary data, Faber uses various kinds of telescopes, including telescopes mounted on satellites.

"One of my most exciting experiences came . . . at the end of a long project with six other astronomers to study the expansion of the universe to find out how uniform and regular it is," she says. "After analyzing the data, we were shocked to discover enormous motions over a huge volume of space. This work could have a major impact on the theories for the origin of structure in the universe."

Planetarium technicians have the universe at their fingertips. ▶

The Solid Earth

The temperature on the young earth was high enough to melt iron, the most common of the existing heavy materials. Gravity pulled the molten iron toward the center. Thus, as the earth developed, denser materials flowed to its center, and less-dense materials were forced to the outer layers.

The earth eventually separated into the three distinct layers. At the center is a dense *core,* composed mostly of iron and nickel. Around the core is the very thick layer of rock called the *mantle*. The outermost layer is a thin *crust* of less-dense solid materials.

Planetarium Technician

A planetarium technician, pictured at lower left, installs, operates, and maintains sound and projection equipment for planetariums. Many technicians help to develop special effects and audiovisual displays. Technicians may also construct and install public education displays in the exhibit area of planetariums.

Most planetarium technicians have vocational training in electronics, electromechanics, and construction.

▲ This telescope dome at Mauna Kea, Hawaii, houses the 3.8 m UK infrared telescope.

Solar Energy Systems Designer

Solar energy systems designers develop solar heating systems for new and existing structures. Solar energy systems designers must have a knowledge of energy requirements, local weather conditions, thermodynamics, and solar technology. In developing a particular system, a designer must determine the location, type, and size of the solar components to be used. Solar components include solar panels for energy collection, pumps to circulate water, and water storage tanks. The choice of solar components is based on engineering principles, energy needs, and the angle at which the sun's rays strike the structure. The designer often inspects the system during construction to make sure it is built correctly.

The minimum requirement for entry into the field of solar energy systems design is a bachelor's degree in engineering.

For Further Information

For more information on careers in astronomy, write the American Astronomical Society, 2000 Florida Avenue NW, Suite 400, Washington, D.C. 20009. You may also want to check out their Web site at www.aas.org.

The Atmosphere

The protoplanet that became the earth could not hold gases because its gravitational force was too weak. As collisions added more mass to the protoplanet, its gravity increased. Eventually, the protoplanet captured some of the hydrogen and helium that were abundant in the solar nebula. By the time the protoplanet had evolved into the earth, its atmosphere consisted primarily of hydrogen and helium. Today these two elements, which were common in the earth's first atmosphere, are mainly found in the upper atmosphere.

Much of the earth's first atmosphere was probably lost as a result of a solar explosion or the solar wind. The earth's second atmosphere resulted from explosions within the earth over 3 billion years ago. During the earth's early history, the heat in its interior caused volcanoes to form. The volcanic eruptions released large amounts of gases, such as water vapor, carbon dioxide, ammonia, and methane. These gases formed a new atmosphere. What would have happened to these gases if the earth had been much smaller?

The action of sunlight probably caused the ammonia and some of the water vapor in the atmosphere to change into nitrogen, hydrogen, and a little oxygen. Most of the hydrogen slowly escaped into space because it was too light to be permanently held by the earth's force of gravity. When cyanobacteria and early green plants appeared, the amount of oxygen increased. Both of these life-forms use carbon dioxide and release oxygen during photosynthesis. Slowly the amount of oxygen in the atmosphere increased to current levels. Some of the oxygen formed ozone, which collected in a higher atmospheric layer around the earth. This ozone layer shielded the earth and its inhabitants from the harmful ultraviolet radiation of the sun. The events that resulted in the formation of the earth's atmosphere are illustrated in Figure 28–8.

Figure 28–8. During the earth's early history, volcanic eruptions formed an atmosphere of water vapor, ammonia, carbon dioxide, and methane (left). The action of sunlight on this primitive atmosphere is thought to have triggered the chemical changes that began to form the earth's highly oxygenated atmosphere and its ozone layer (right).

The Oceans

As the atmosphere was developing, the earth was cooling enough for liquid water to form. Between 3 billion and 3.5 billion years ago, water vapor began to condense. It fell to earth as rain and

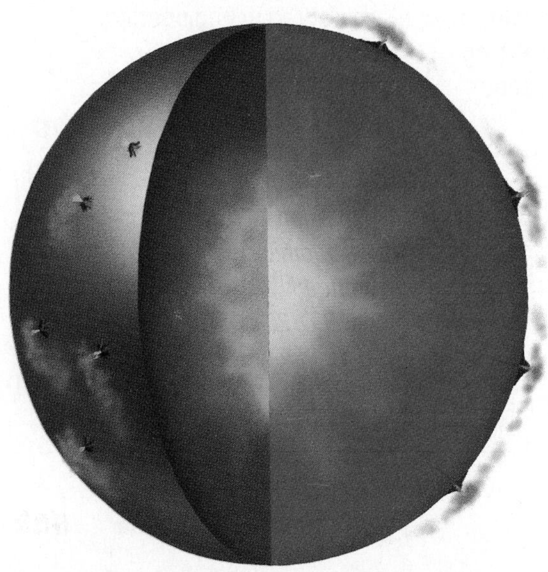

EARTHBEAT

The Solar Constant

Astronomers once thought that the sun's energy output remained steady. They called the amount of solar energy that a square centimeter of the earth receives every second the *solar constant.* To calculate the solar constant, scientists use the average distance between the sun and the earth and assume that the earth has no atmosphere.

Based on data from sensors sent outside the earth's atmosphere, however, astronomers now know that the solar "constant" is not constant. Records of solar energy output from programs such as the Solar Maximum Mission (Solar Max) and the Atmospheric Laboratory for Applications and Science (ATLAS) show that the solar constant can vary as much as 0.15 percent.

The sensor data also show that changes in the solar constant may be related to the same 11-year cycle that governs sunspots. Sunspots are cooler regions of the photosphere and can cause short-term reductions in solar output. Solar output varies the most when sunspots are most active. However, as sunspot activity peaks at the height of the solar cycle, the sun's average energy output also peaks.

The Solar and Heliospheric Observatory (SOHO), shown above, continuously monitors solar output. It is also equipped to study the sun's dynamic structure. Interactions among features such as the sun's magnetic fields, its seismic waves, and its convective zone may reveal why solar output fluctuates.

Solar constant studies will help explain how solar activity affects earth's atmosphere and climate and may enable scientists to detect climate changes that result from human activity.

If the sun's energy output stayed at a slightly lower level than usual for 1,000 years, what climate pattern might emerge?

formed oceans in the lower surface areas. The ocean water absorbed much of the carbon dioxide from the atmosphere. Water naturally absorbs carbon dioxide whenever its concentration in the atmosphere is high enough. By 1.5 billion years ago, the chemical composition of the atmosphere was similar to what it is today.

Section 28.3 Review

1. What two forces caused the solar nebula to develop into the sun?
2. How were planetesimals different from protoplanets?
3. List three reasons why the earth was hot when it first formed.
4. What gas in the atmosphere was absorbed by the oceans as they formed?
5. Explain the following statement: Volcanic eruptions aid in the creation of an atmosphere.

Chapter 28 Review

Key Terms

- aurora (579)
- chromosphere (575)
- convective zone (574)
- corona (575)
- moon (580)
- nebular theory (580)
- photosphere (574)
- planet (580)
- planetesimal (580)
- prominence (577)
- protoplanet (580)
- radiative zone (574)
- solar flare (578)
- solar nebula (580)
- solar system (580)
- solar wind (575)
- sunspot (576)
- sunspot cycle (577)

Key Concepts

The enormous pressure and heat at the sun's core converts matter into energy through fusion. **See page 571.**

Energy from the sun's core moves through the radiative zone and the convective zone before it enters the sun's atmosphere. **See page 574.**

The sun's atmosphere is composed of the photosphere, the chromosphere, and the corona. **See page 574.**

Sunspots are caused by powerful magnetic fields on the sun. **See page 576.**

Prominences and solar flares are two types of solar activity caused by disturbances in the solar atmosphere. **See page 577.**

The solar wind, composed of electrically charged particles, can cause auroras when the particles enter the earth's atmosphere. **See page 579.**

According to the nebular theory, the sun and the planets formed at the same time. **See page 580.**

Planets formed from small bodies of matter in the solar nebula. **See page 580.**

As the earth cooled down after its formation, the current atmosphere and oceans developed. **See page 584.**

Review

On your own paper, write the letter of the term that best completes each of the following statements.

1. According to Einstein's theory of relativity, in the formula $E = mc^2$, the *c* stands for
a. corona. b. core.
c. the speed of light. d. the length of time.

2. A nuclear reaction in which two atomic nuclei combine is called
a. fission. b. fusion. c. magnetism. d. granulation.

3. The portion of the sun in which energy moves from atom to atom in the form of waves is called the
a. radiative zone. b. convective zone.
c. solar wind. d. chromosphere.

4. The portion of the sun normally visible from the earth is the
a. core. b. photosphere.
c. corona. d. solar nebula.

5. The sunspot cycle lasts
a. 2 years. b. 5 years. c. 11 years. d. 19 years.

6. Sudden outward eruptions of electrically charged atomic particles from the sun are called
   a. planetesimals. b. coronas.
   c. sunspots. d. solar flares.
7. Gusts of solar wind can cause
   a. protoplanets. b. magnetic storms.
   c. nuclear fission. d. nuclear fusion.
8. Northern lights and southern lights are other names for
   a. prominences. b. auroras.
   c. granulations. d. protoplanets.
9. The hypothesis that the sun and the planets developed out of the same cloud of gas and dust is called the
   a. nebular theory. b. theory of relativity.
   c. nuclear theory. d. theory of convection.
10. The small bodies of matter that filled the solar nebula are called
   a. protoplanets. b. planetesimals.
   c. auroras. d. proton nuclei.
11. Compared with the size of the present-day planets, the protoplanets were
   a. much smaller. b. slightly smaller.
   c. similar. d. larger.
12. The first atmosphere of the earth had a high percentage of
   a. helium. b. oxygen. c. nitrogen. d. water vapor.
13. In the process of photosynthesis, cyanobacteria and green plants give off
   a. oxygen. b. carbon dioxide.
   c. hydrogen. d. helium.
14. Water vapor began to condense into oceans about
   a. 1 million to 1.5 million years ago.
   b. 1 billion to 1.5 billion years ago.
   c. 3 billion to 3.5 billion years ago.
   d. 15 billion to 15.5 billion years ago.

Critical Thinking

On your own paper, write answers to the following questions.

1. Explain how the transfer of energy in a pan of hot water is similar to the transfer of energy in the sun's convective zone.
2. Predict what would happen to the number of sunspots if the sun's magnetic field suddenly increased in strength.
3. If the earth's magnetosphere shifted, what would happen to the area where auroras are most often visible?
4. How would the layers of the earth be different if the planet had never been hotter than it is today?
5. How would the atmosphere of the earth be different if the earth had formed from a much larger protoplanet?

Application

1. You are asked to explain the following statement to an elementary school class: Solar hydrogen is responsible for life existing on the earth. What would you say?
2. A group of senior citizens has decided to head north to view the northern lights. When would you advise them to go in order to see the most frequent displays?
3. Concept map ? ? ? Using the new terms listed on the previous page, construct a **concept map** that, among other things, illustrates the path of energy from the core of the sun to the far side of the earth.

Extension

1. Write a short story describing an imaginary trip to the center of the sun. Describe each layer and zone through which you would pass.
2. In the early 1900's, two scientists proposed an alternative to Laplace's theory about the origin of the solar system. These scientists, Thomas Chamberlin and Forest Moulton, proposed a dualistic theory. Find out what caused them to question Laplace's theory. Describe the theory they proposed.
3. The earth's atmosphere had very little oxygen before photosynthetic organisms existed. Do research to find out how important green plants, particularly in the rain forests, are for the earth's oxygen supply today.

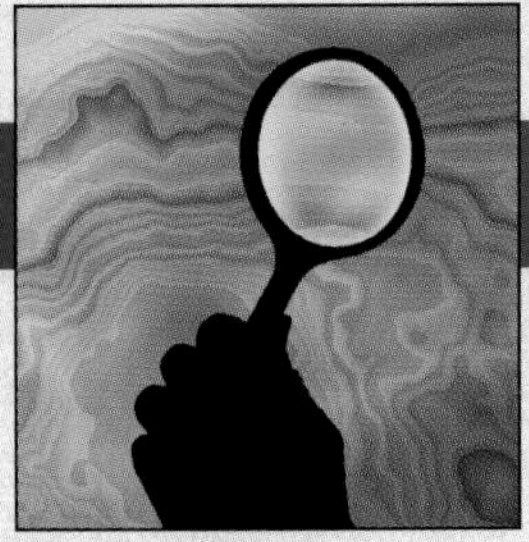

In-Depth Investigation

Size and Energy of the Sun

Materials

- shoe box with lid
- piece of very thin sheet metal (2 × 8 cm)
- glass jar with lid
- aluminum foil
- black paint, flat finish
- modeling clay
- safety pin
- Celsius thermometer
- index card
- desk lamp with 100-watt bulb
- masking tape
- metric ruler
- scissors
- pencil

Part I Introduction

The sun is an average of 150 million kilometers away from the earth. Yet scientists have been able to measure the sun's size and distance away from the earth. Scientists use complicated astronomical instruments to make these measurements. However, it is possible to measure the sun's size with a simple instrument called a solar viewer and the knowledge of the relation between the size of an object and the size of the object's image as seen through the viewer.

In this investigation, you will perform a simple experiment from which you can calculate the sun's diameter. Then you will collect energy from sunlight and estimate the amount of energy produced by the sun.

Prelab Preparation

1. Review Section 28.1, pages 571–573.
2. Review the guidelines for glassware safety. 
3. Describe how a pinhole image of the sun is formed.
4. Construct a pinhole solar viewer. Refer to the Small-Scale Investigation on page 573 for directions on making a solar viewer.
5. Construct a solar collector.
   a. Punch a hole in the jar lid or use one already prepared by your teacher. **CAUTION:** *When using hole-punching equipment you must be supervised by an adult.*
   b. Gently bend the sheet-metal piece around a pencil to shape it. Then gently place the metal piece around the thermometer bulb so it fits snugly. Be careful not to press too hard. Next bend the remaining metal outward to collect as much sunlight as possible. See the illustration at left.
   c. Paint the sheet metal black.
   d. Slip the top of the thermometer through the hole in the lid. Mold the clay around the thermometer on the top and bottom of the lid to hold the thermometer steady. Then secure the thermometer and clay to the lid with masking tape. Place the lid on the jar so that the thermometer bulb is in the middle of the jar, as shown in the figure at left.

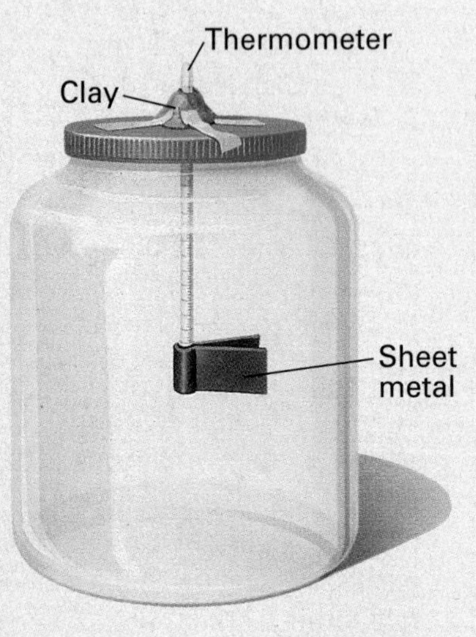

Procedure

1. Aim the pinhole in the solar viewer toward the sun. **CAUTION:** *You should never look directly at the sun.* Adjust the box until the sunlight forms a small image on the index card at the other end of the viewer.

2. Carefully measure and record the diameter of the image. Measure and record the distance between the image and the pinhole.
3. The ratio of the diameter of the sun's image to its distance to the pinhole is equal to the ratio of the sun's diameter divided by the distance to the sun, as shown by the following equation:

$$\frac{\text{sun's diameter}}{\text{sun's distance to pinhole}} = \frac{\text{image diameter}}{\text{image's distance to pinhole}}$$

If the distance between the sun and the earth is 150 million kilometers, calculate the diameter of the sun using your data.

Part II Introduction

Energy, whether from a light bulb or from the sun, spreads out in all directions. The amount of energy received decreases with the square of the distance between the energy source and the point that receives the energy. In other words, if a light is moved three times farther away than its original distance from a point, the energy received at that point is nine times less.

Procedure

1. Place the solar collector so that the black metal is angled toward the sun. If it is cool outside, use sunlight shining through an open or untinted window.
2. Watch the temperature reading on the thermometer until it reaches a maximum value. Record this value.
3. Allow the collector to cool in the shade until it reaches room temperature.
4. Place the lamp at the end of a table. Remove any reflector or shade from the lamp.
5. Turn the collector toward the lamp and put the collector about 30 cm from the lamp.
6. Turn on the lamp and wait one minute. Gradually move the collector toward the lamp. Watch the temperature carefully. Stop when the temperature reaches the maximum temperature that was achieved in sunlight. If the temperature rises above that level, move the collector back and let it cool to room temperature. Repeat the experiment and watch the temperature's rise more carefully.
7. Once the temperature has stabilized at the same level reached in sunlight, record the distance between the center of the lamp and the thermometer bulb.

Analysis and Conclusions

1. What is the sun's diameter as given in the textbook?
2. Explain any difference in your answer and the textbook value.
3. The collector absorbed as much energy from the sun at a distance of 150 million km as it did from the 100-W bulb at the distance you measured. Using the distance to the sun as 1.5×10^{13} cm, calculate the power of the sun in watts as follows.

$$\frac{\text{power}_{\text{sun}}}{(\text{distance}_{\text{sun}})^2} = \frac{\text{power}_{\text{lamp}}}{(\text{distance}_{\text{lamp}})^2}$$

4. The sun's power is quoted as 3.7×10^{26} watts. Compare your answer with this value. Describe any sources of measuring inaccuracy with your method.

Extension

Would the experiment have worked with a fluorescent bulb? Explain.

Visible in the night sky are objects that shine with a steady light and wander across the sky. Ancient astronomers called these moving objects planetae, *the Greek word for "wanderers." They are now called planets. In this chapter, you will learn about the planets and other bodies in the solar system.*

Chapter Outline

◀ Jupiter, shown here with four of its moons, is the largest planet in the solar system.

29.1 Models of the Solar System

Section Objectives

- **Compare the models of the universe developed by Ptolemy and Copernicus.**
- **Summarize Kepler's three laws of planetary motion.**

Two thousand years ago, many philosophers developed ideas about the universe based on what they saw. One of these was Aristotle, a Greek philosopher who lived from 384 to 322 B.C. Through observations and reasoning, Aristotle promoted an earth-centered, or **geocentric,** model of the solar system. This model stated that the sun, the stars, and the planets revolved around the earth.

This early Greek model, however, did not explain why some planets appear to reverse direction occasionally, moving from east to west instead of the usual west to east. This backward motion is called **retrograde motion.**

About 500 years after Aristotle, another Greek astronomer, Claudius Ptolemy, proposed a model of the universe in which each planet had two motions. One motion was the revolution of the planet around the earth. As the planet revolved, it also moved in a series of small circles, like the links of a chain. Ptolemy called these small circles **epicycles.** According to Ptolemy, the motion of the planet in its epicycle would make it appear to move backward at times.

Copernicus's Model

In the 1500's, Nicolaus Copernicus, a Polish astronomer, challenged the geocentric ideas of Aristotle and Ptolemy. Copernicus proposed a **heliocentric,** or sun-centered, model of the solar system. According to this model, the earth and the other planets revolve around the sun. Copernicus also proposed that all planets orbit in the same direction but that each moves at a different speed and distance from the sun. The faster planets, therefore, pass the slower planets from time to time.

Confirmation of Copernicus's model finally came in the early 1600's. At that time, the Italian scientist Galileo Galilei was able to observe the motions of the planets with the newly invented telescope. He collected evidence that proved that the heliocentric model was the correct one.

Kepler's Laws

While Galileo was making his observations, other scientists were collecting data that also supported a heliocentric solar system. Prior to the invention of the telescope, Tycho Brahe, a Danish astronomer, devoted his life to making detailed observations of the positions of the stars and planets. Near the end of his life, Brahe hired a German astronomer, Johannes Kepler (1571–1630), as his assistant. Kepler was able to explain Brahe's precise observations in mathematical terms. Kepler developed three laws that explained most aspects of planetary motion.

Law of Ellipses

Kepler's first law states that each planet orbits the sun in a path called an **ellipse.** An ellipse is an oval whose shape is determined by two points within the figure, as shown in Figure 29–1. Each of these points is called a **focus** (pl: foci). The sun is at one focus of the orbit of a planet. If you draw a line from any point on the ellipse to each of the two foci, the total length of the lines will always be the same. Some ellipses look almost like circles. In fact, a circle is a special kind of ellipse in which the two foci are at the same point. Other ellipses are more elongated ovals.

Because the orbits of the planets are ellipses, a planet is not always the same distance from the sun. The point where an orbit is closest to the sun is the *perihelion*; the point where it is farthest from the sun is the *aphelion*. Why would a circular orbit have neither a perihelion nor an aphelion?

The distance of a planet from the sun is usually defined as the average of the distances from the sun at its perihelion and its aphelion. For example, the aphelion of the earth's orbit is about 152 million km from the sun; the perihelion is about 147 million km from the sun. The average of 147 million and 152 million is 149.5 million. This average distance between the earth and the sun is known as one **astronomical unit,** or AU. The distance between the sun and other planets is usually measured in astronomical units.

Law of Equal Areas

Kepler's second law describes the speed at which planets travel at different points in their orbits. By studying Brahe's data, Kepler found that the orbit of the earth was a nearly perfect circle, with the sun off-center. He found that the earth moves fastest when it is closest to the sun. He calculated that a line from the center of the sun to the center of the planet sweeps through equal areas in equal periods of time.

Imagine a line that connects the center of the sun to the center of a planet. When the planet is near the sun, the imaginary line is relatively short. The planet is moving rapidly and in ten days, for example, the imaginary line sweeps through a short, wide triangular

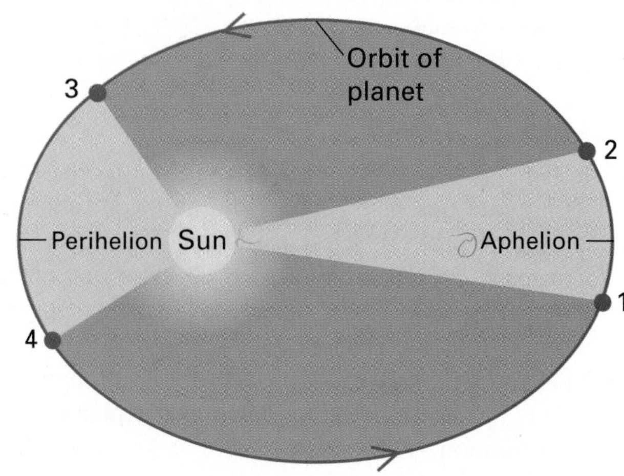

Figure 29–1. You can draw an ellipse using two pins as the foci and a string (left). Kepler's second law states that the areas a planet sweeps out in a given period of time—the shaded areas between points *1* and *2* and points *3* and *4*—are equal.

sector. When the planet is farther from the sun, the imaginary line is longer. However, the planet is moving more slowly and the imaginary line sweeps through a long, thin triangular sector in ten days. Kepler's second law states that the area of the long, thin sector is the same as the area of the short, wide sector.

Law of Periods

Kepler's third law describes the relationship between the average distance of a planet from the sun and the **orbit period** of the planet. The orbit period is the time required for the planet to make one revolution around the sun. According to Kepler's third law, the cube of the average distance of a planet from the sun (r) is always proportional to the square of the period (p). The mathematical formula that describes this relationship is $K \times r^3 = p^2$, where K is a mathematical constant. When distance is measured in AU's and the period is in earth-years, $K = 1$ and $r^3 = p^2$.

For example, the radius of the earth's orbit, or its distance from the sun, is 1 AU, and its period is 1 year. Putting these numbers into the formula yields $1 \times 1^3 = 1^2$. This simplifies to $1 = 1$. Jupiter is 5.2 AU's from the sun, and its period is 11.9 years. The cube of 5.2 is 140.6. The square of 11.9 is 141.6. The two results, 140.6 and 141.6, are approximately equal. Apparent errors in the law of periods are caused by rounding off the distance or periods. When the distance and period are calculated with enough accuracy, the period squared always equals the distance cubed.

Newton's Application of Kepler's Laws

Kepler's laws explained how the planets orbit the sun. Isaac Newton asked why the planets move in this way. The explanation that Newton eventually gave described both the motion of objects on the earth and the motion of the planets in space. He hypothesized that a moving body will change its motion only if an outside force causes it to do so. For example, a ball rolling on a smooth surface will continue to move in a straight line unless something causes it to change direction. The tendency of a moving body to move in a straight line at a constant speed until an outside force acts on it is called **inertia.** Inertia also refers to the tendency of an object to remain at rest until an outside force acts on it.

Newton compared a planet to a rolling ball. Because a planet does not follow a straight path, an outside force must cause it to curve. Newton identified this force as *gravity,* the attractive force that exists between all objects in the universe. The gravitational pull of the sun keeps the planets in orbit around the sun.

Section 29.1 Review

1. What is the major difference between Ptolemy's and Copernicus's models of the universe?
2. According to Kepler, what is the shape of planetary orbits?
3. Upon what observation did Kepler base his second law?
4. If an asteroid is 4 AU's from the sun, what is its orbit period?

Section Objectives

- **Identify the basic characteristics of Mercury and Venus.**
- **Identify the basic characteristics of Earth and Mars.**

29.2 The Inner Planets

The four planets closest to the sun are called inner planets. These planets are Mercury, Venus, Earth, and Mars. The inner planets are also called **terrestrial planets,** because they are similar to the earth. Because these planets formed close to the heat of the sun, materials with low boiling points were driven off. The planets consist mostly of solid rock, with a metal core. Inner planets do not have rings. The number of moons per planet varies from zero to a maximum of two. The inner planets also have bowl-shaped depressions on their surfaces called *impact craters* that resulted from collisions of the planets with objects made primarily of rock. These collisions took place during the later stages of the formation of the solar system.

Mercury

Mercury is the planet that is closest to the sun. Consequently, Mercury has a shorter orbit period than any other planet—88 days. In fact, the ancient Romans named the planet *Mercurius* after the swift messenger of the gods. Mercury rotates very slowly on its axis—only once every 59 days. Mercury is so close to the sun that light from the sun usually obscures the planet from view. Even the best photographs taken from the earth show Mercury as only a fuzzy ball. Mercury has no moons.

In 1974 and 1979 *Mariner 10* visited Mercury. The spacecraft transmitted photographs to the earth that revealed a surface that was heavily cratered, like that shown in Figure 29–2. The large number of craters suggests that Mercury has changed little since the formation of the solar system. Some craters appear to be filled with hardened lava. If this is the case, Mercury was once volcanic. The photographs also showed a line of cliffs hundreds of kilometers long. These cliffs may be wrinkles in the crust, which developed when the once-molten core cooled and shrank.

INVESTIGATE!

To learn more about craters, try the In-Depth Investigation on pages 612–613.

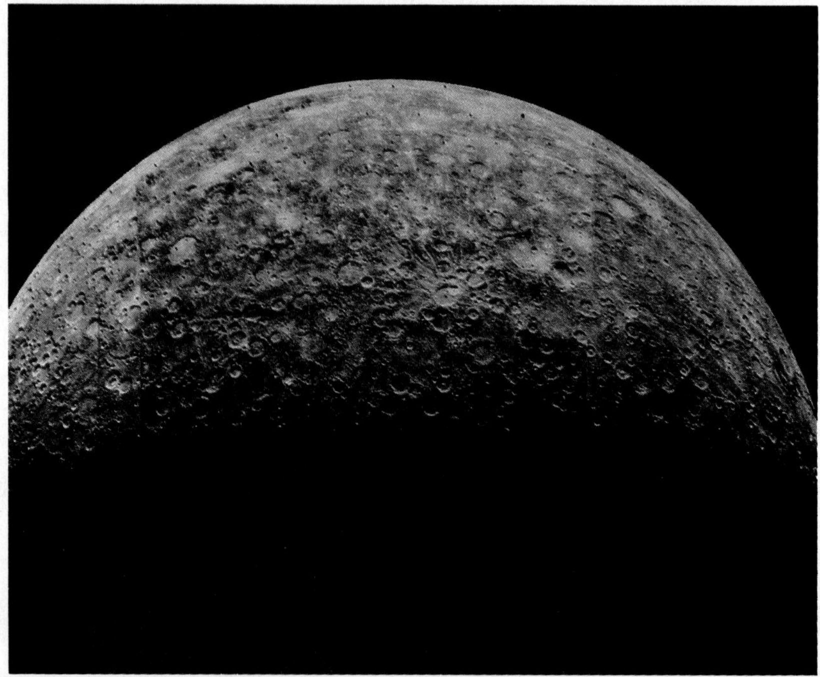

Figure 29–2. The surface of Mercury, shown in this photograph taken by *Mariner 10,* probably looks very much like it did shortly after the solar system was formed.

Mercury has a thin atmosphere for two reasons: its closeness to the sun and its size. Because Mercury is so close to the sun, solar heat causes the gas molecules near the surface of the planet to move rapidly. Because the planet is so small, its gravitational pull is too weak to hold enough rapidly moving gas molecules to form a dense atmosphere. The absence of a dense atmosphere and its slow rotation contributes to the huge daily temperature range on Mercury. During the day, the temperature may reach as high as 427°C. At night the temperature may plunge to –173°C.

Before the *Mariner 10* expedition, astronomers thought Mercury was too small and rotated too slowly to have a magnetic field. However, the instruments on *Mariner 10* did detect a weak magnetic field, which suggests that Mercury has a core of molten iron.

Venus

The second planet from the sun is Venus. The orbit period of Venus is 225 days. Like Mercury, Venus rotates very slowly on its axis—once every 243 days. The direction of rotation is opposite that of the other planets. For this reason, the sun, viewed from Venus, rises in the west and sets in the east. Like Mercury, Venus has no moons.

In some ways Venus is the earth's twin. The two planets are of almost the same size, mass, and density. However, Venus is much hotter than the earth because it is closer to the sun and its dense atmosphere provides an insulating effect. The average surface temperature on Venus is approximately 453°C, and its atmospheric pressure is almost 100 times the atmospheric pressure on the earth.

The high temperature and dense atmosphere of Venus are closely related. Astronomers think that when Venus was formed, its temperature was lower and its atmosphere was less dense than it is today. Astronomers also think that oceans may have formed on the surface of Venus at that time, and that there may have been much volcanic activity.

Figure 29–3. Venus has approximately the same size, mass, and density as the earth. Its surface (left) is composed of basalt and granite rocks. However, its dense atmosphere (right), composed primarily of carbon dioxide, is quite different.

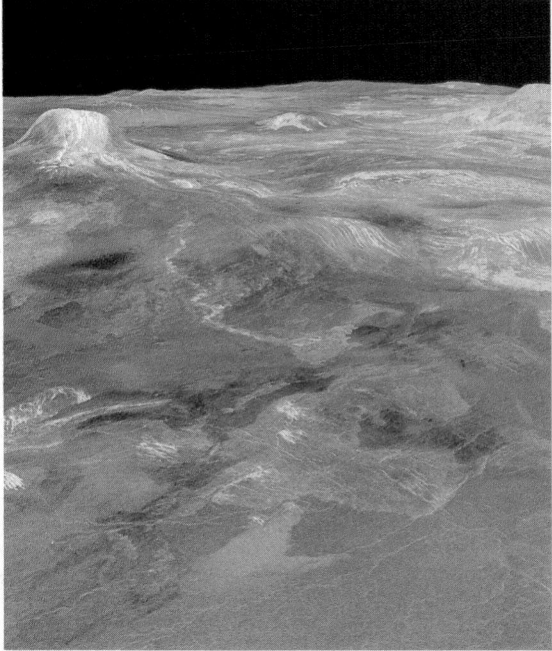

Astronomers theorize that, as the sun became hotter, the oceans on Venus evaporated. Because there was no water to combine with the carbon dioxide released by the erupting volcanoes, the level of carbon dioxide in the atmosphere steadily increased. As a result, the atmospheric pressure increased. Carbon dioxide still accounts for about 96 percent of the atmosphere of Venus. This blanket of carbon dioxide allows orange wavelengths of solar energy to penetrate the atmosphere of the planet but blocks the escape of heat. Consequently, the planet stays too hot to support life. High above the surface of the planet, the temperature decreases dramatically. There is very little water vapor in the atmosphere of Venus. Therefore, instead of containing water vapor, the clouds of Venus are composed of droplets of sulfuric acid.

In the 1970's, the Soviet Union sent six *Venera* probes to explore the surface of Venus. The probes survived the heat and pressure of the atmosphere long enough to transmit surface photographs of the rocky landscape of Venus. The photographs showed a smooth plain, with some mountains and valleys. Other instruments carried by the probes indicated that the surface of Venus is composed of basalt and granite. These are two types of rock commonly found on the earth. Between 1990 and 1994, the *Magellan* orbiter produced radar images of most of the Venusian surface. Domes and mountains, volcanoes and vast lava plains are the most common features on Venus.

Figure 29–4. The earth is covered with oceans of water and an atmosphere that can support life.

Earth

The third planet from the sun is Earth. The orbit period of the earth is 365.24 days, and the planet completes one rotation in 23 hours and 56 minutes. The earth is the fifth largest planet. It has one moon.

The earth has had an extremely active geologic history. Geologic records indicate that over the last 200 million years, the earth's continents separated from a singe landmass and drifted to their present positions. Weathering and erosion have changed and continue to change the surface of the earth.

Life on the earth is possible because of the distance of the planet from the sun. The temperature is warm enough for water to exist as a liquid. Mercury and Venus are so close to the sun that they are too hot to retain liquid water. Mars and the outer planets are so far from the sun that water cannot exist in liquid form. Most of the water on these planets is in the form of ice. The earth is the only planet in the solar system that has oceans of liquid water.

Geologists theorize that as oceans formed on the earth, water combined with carbon dioxide in the atmosphere. Since carbon dioxide did not build up in the atmosphere, solar heat was able to escape. Consequently, the earth maintained the moderate temperatures needed to support life, an average of 14°C. Plants and cyanobacteria contributed additional oxygen to the atmosphere. The earth is the only known planet with the proper combination of water, temperature, and oxygen to support plant and animal life as we know it.

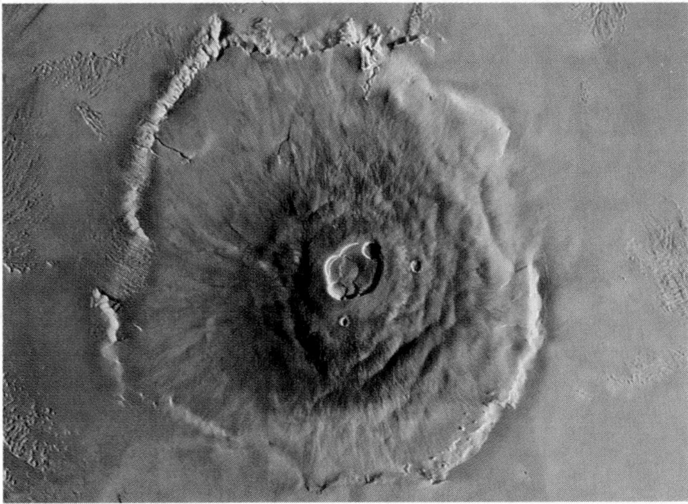

Figure 29–5. Evidence of the active volcanism on Mars (left) is found on the surface of the planet in the form of volcanoes such as Olympus Mons (right), the largest known volcano in the solar system.

Mars

Mars is the fourth planet from the sun. Its orbit period is 687 days; and its rate of rotation, 24 hours and 37 minutes. Therefore, the lengths of a day on Mars and on the earth are almost the same. Mars also has seasons much like the earth's because the tilt of its axis is nearly the same as that of the earth's axis. Mars has two moons.

Mars is geologically active, with large surface areas covered with lava. The volcanoes on Mars are the largest in the solar system. The largest of these, *Olympus Mons,* is three times higher than Mount Everest, and its base is about the size of Nebraska.

A system of deep canyons also covers part of the surface. The largest, the *Valles Marineris,* is as long as the United States—4,500 km. Huge fault zones in the crust of the planet may have formed these canyons.

The atmospheric pressure and temperature on Mars are presently too low for water to exist as a liquid. However, two U.S. spacecraft, *Viking 1* and *Viking 2,* found evidence of erosion by water. Therefore, astronomers can infer that Mars once had a warmer and wetter climate. The temperature near the equator of Mars now approaches 20°C during the summer. Near the poles it drops as low as –130°C during the winter. The little water that remains on Mars is trapped in polar ice caps or is possibly frozen beneath the surface.

One of the tasks of the *Viking* probes was to search for signs of life on Mars. The probes found no convincing evidence that life has ever existed on the planet. However, recent evidence from Martian meteorites discovered in the Antarctic ice may indicate that primitive life may have existed on Mars. Final proof may not come until future spacecraft reach Mars in the next several years.

Section 29.2 Review

1. What evidence suggests that Mercury has changed little since it was formed?
2. What aspect of the earth makes its temperature favorable for life?
3. Describe two ways in which Mars is similar to the earth.
4. What is unusual about the rotation of Venus?

Section Objectives

- **Identify the basic characteristics of Jupiter and Saturn.**
- **Identify the basic characteristics of Uranus, Neptune, and Pluto.**

29.3 The Outer Planets

The five planets farthest from the sun are called the *outer planets*. They are Jupiter, Saturn, Uranus, Neptune, and Pluto. The first four are called the *giant planets* because they are the largest planets in the solar system. They are also called the **Jovian planets** because they are similar to Jupiter. The smallest and usually the most distant planet in the solar system is Pluto. Pluto is different from all the other planets and may not have formed in the same way.

Although the Jovian planets are much larger and more massive than the inner planets, they are far less dense. Each of these planets has a thick atmosphere made up mostly of hydrogen and helium gases. Also, each probably has a core of rock, metals, and water.

Jupiter

The first of the outer planets, Jupiter, is the fifth planet from the sun. Jupiter, shown in Figure 29–6, is by far the largest planet in the solar system. Its mass is twice that of the other eight planets combined. Its orbit period is almost 12 years. Jupiter rotates faster than any other planet—once every 9 hours and 50 minutes. Jupiter has at least 16 moons and one ring made up of millions of particles.

Jupiter seems to have a liquid metallic core surrounded by lighter elements. Jupiter's large mass causes the temperature and pressure in its interior to be much greater than those within the

Figure 29–6. The Great Red Spot of Jupiter, in the lower left of the image, is similar to a giant storm on the earth. The colored cloud bands of Jupiter, however, are unlike any earthly phenomenon.

Figure 29–7. Jupiter's most distinctive feature is its Great Red Spot, an on-going, massive storm.

earth. Temperatures in the interior of Jupiter rise as high as 30,000°C. The intense pressure has changed most of the interior of the planet into a sea of liquid metallic hydrogen. Electrical currents in this hot liquid may be the source of Jupiter's enormous magnetic field.

Jupiter is mostly made up of gases; 92 percent of the planet is hydrogen and helium, very much like the composition of the sun. Due to the high temperature and great pressure, the surface of the planet is not solid, but rather a mixture of hot gases and liquids. Astronomers theorize that when Jupiter was formed about 4.6 billion years ago, it did not have enough mass to enable nuclear fusion to begin. As a result, Jupiter never became a star.

The surface of Jupiter is very unique because of its alternating light- and dark-colored cloud bands. The orange, gray, blue, and white bands spread out parallel to the equator. The colors suggest the presence of ammonia, methane, and water vapor. Scientists think that the rapid rotation of Jupiter causes these gases to swirl around the planet, forming the bands. The average temperature of the atmospheric layers is –160°C. The lower layers are about 20°C, closer to the temperature of the earth's atmosphere. Jupiter also has lightning and thunderstorms. A combination of lightning, ammonia, methane, and water vapor is thought to be necessary for the development of life; yet there is no evidence for the possibility of life on Jupiter.

To an observer on the earth, the most distinctive feature of Jupiter is the Great Red Spot. Astronomers think that heated material rising to the surface from the interior of the planet causes this feature. The Great Red Spot is a giant rotating storm, somewhat like a hurricane on the earth, that has been raging for several hundred years or more. New observations from the *Galileo* spacecraft, which reached Jupiter in 1995 and dropped an entry probe into its atmosphere, indicate that Jupiter has wind speeds of over 600 km/hr. The huge amount of data received during *Galileo's* visit promises new insights on the mysterious, hidden interior of Jupiter.

Figure 29–8. Saturn has at least 20 moons and a complex system of rings.

Saturn

The sixth planet from the sun, Saturn, is over half a billion kilometers farther from the sun than Jupiter is. Saturn is the second largest planet in the solar system. Its average temperature is −176°C. It has at least 20 moons and several rings.

Saturn spins very rapidly, rotating on its axis every 10 hours and 40 minutes. The rapid rotation of both Saturn and Jupiter causes each to bulge out at its equator and to flatten at its poles. Saturn also has bands of colored clouds that run parallel to its equator. Both planets probably have small, rocky cores, interiors of liquid metallic hydrogen, and dense atmospheres of hydrogen and helium gas.

Saturn differs from Jupiter in three ways. One difference is that Saturn is much less dense than Jupiter. In fact, Saturn is the least

Table 29–1 Planetary Data

	Average distance from sun (10^6 km)	Diameter (km)	Orbit period (earth time)
Mercury	57.9	4,878	88 d
Venus	108.2	12,104	225 d
Earth	149.6	12,756	365.24 d
Mars	227.9	6,794	687 d
Jupiter	778.3	142,796	11.9 yr
Saturn	1,427	120,660	29.5 yr
Uranus	2,871	52,400	84.1 yr
Neptune	4,497	50,450	164.8 yr
Pluto	5,914	2,445	248.6 yr

*at least

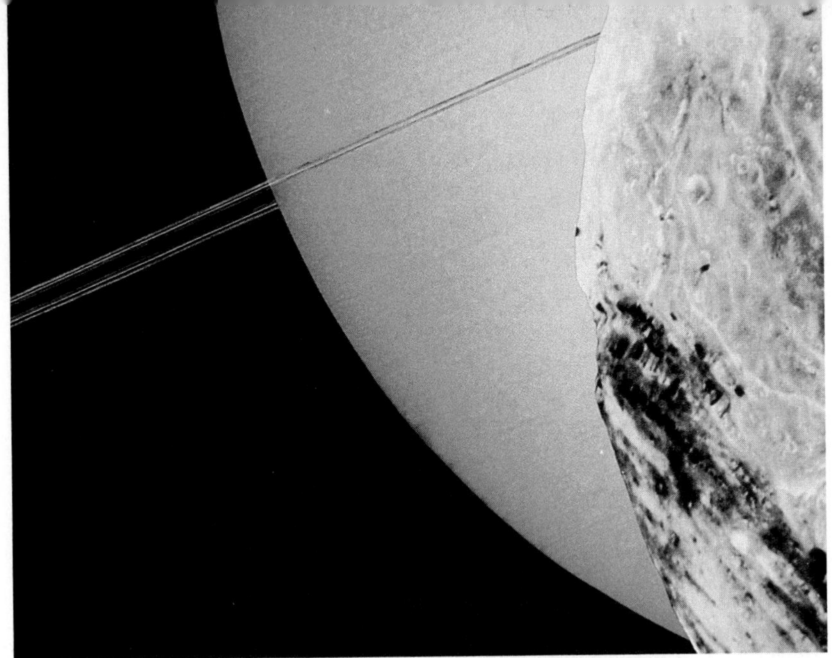

Figure 29–9. Uranus, with its thin set of rings, is shown here as seen from its moon Miranda.

dense planet in the solar system, with a density less than that of water. Another difference is that the period of Saturn, which is 29.5 years, is nearly 20 years longer than that of Jupiter. Saturn also differs from Jupiter and from all the other planets in the solar system because it has a much more complex system of rings. These rings are discussed in Chapter 30.

Uranus

Uranus is the seventh planet from the sun and the third largest planet in the solar system. Uranus was discovered in 1781. It was the first planet to have been discovered since ancient times. Because Uranus is nearly 3 billion km from the sun, astronomers have great difficulty

Rate of rotation (earth time)	Number of satellites	Surface temperature (°C)	Dominant atmospheric gases
59 d	0	–173 to 427	He
–243 d	0	453	CO_2
23 hr 56 min	1	–13 to 87	N_2, O_2
24 hr 37 min	2	–130 to 20	CO_2
9 hr 50 min	16*	–160	H_2, He
10 hr 40 min	20*	–176	H_2, He
17 hr 14 min	15*	–215	H_2, He, CH_4
16 hr 06 min	8*	–225	H_2, He, CH_4
6.4 d	1	–208 to –233	CH_4, NH_3

in trying to study the planet. Uranus has at least 15 moons and at least 11 small rings. Its orbit period is 84 years.

The most distinctive feature of Uranus is its unusual rotation. Most planets, including the earth, rotate with their axes perpendicular to their orbital planes as they revolve around the sun. Uranus, however, rotates like a rolling ball. The axis of Uranus is almost horizontal to the plane of its orbit. The exact rotation rate of Uranus was not discovered until 1986, when *Voyager 2* passed by Uranus. Astronomers were then able to determine that Uranus rotated about once every 17 hours.

The greenish color of Uranus indicates that its atmosphere contains methane. Like the atmospheres of the other outer planets, the atmosphere of Uranus contains mainly hydrogen and helium. The average cloud-top temperature of Uranus is −215°C; however, astronomers believe that the temperature of the planet is much higher below the clouds. There may be a mixture of liquid water and methane beneath the atmosphere. Scientists also think that the center of Uranus is a core of rock and metals, with a temperature of about 7,000°C.

SCIENCE & TECHNOLOGY

Galileo Probes Jupiter

▲ The first step in *Galileo's* long journey was aboard the space shuttle *Atlantis.*

On December 7, 1995, after a six-year voyage of more than 3.7 billion kilometers, the spacecraft *Galileo* arrived at Jupiter. The 5.3-meter-long spacecraft flew what is called a *gravity-assist trajectory,* passing by Venus once and by the earth twice to gain enough speed to reach Jupiter.

Galileo's first task was to monitor an atmospheric probe that it launched while still five months and 51 million kilometers from Jupiter. The 336-kg probe plunged into Jupiter's atmosphere just as *Galileo* neared the planet. For 57 minutes after deploying its parachute, the probe transmitted data about the composition and meteorology of Jupiter's atmosphere.

The probe data that *Galileo* relayed to earth surprised mission scientists in many ways. For example, the probe detected wind speeds much higher than expected. Rather than decreasing as the probe fell, wind speeds increased to more than 640 km/hr. This wind pattern suggests that Jupiter's internal heat affects the planet's weather more than heat from the sun does.

Perhaps more surprising was the probe's chemical analysis of Jupiter. Scientists had thought that Jupiter formed from a solar nebula and later pulled in planetesimals rich in

Neptune

The eighth planet from the sun, Neptune, is similar to Uranus in size and mass. Its orbit period is 165 years, and it rotates about every 16 hours. Neptune has eight moons and possibly four rings.

The existence of Neptune was predicted before it was discovered. After Uranus was discovered, astronomers noted variations from its expected orbit. They suspected that the gravity of an unknown planet beyond Uranus might be responsible. In the mid-1800's John Couch Adams, an English mathematician, and Urbain Leverrier, a French astromomer, independently calculated the position of such a planet. Three years later, a German astronomer, Johann Galle, discovered a bluish-green disk where Leverrier had predicted it would be—4.5 billion km from the sun. Astronomers named the planet *Neptune* after the Roman god of the sea.

Data from the *Voyager* spacecraft indicate that Neptune's atmosphere is made up largely of hydrogen, helium, and methane. The upper atmosphere of Neptune is composed of white clouds of frozen methane. These clouds appear as continually changing bands between the equator and the poles of Neptune.

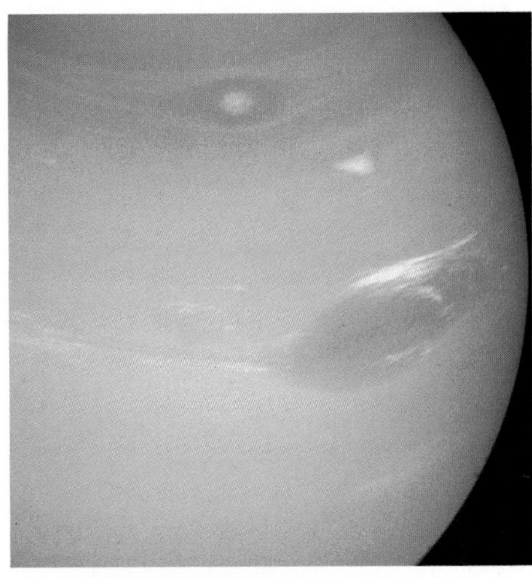

Figure 29–10. The Great Dark Spot on Neptune is a giant storm, similar to the Great Red Spot on Jupiter.

◀ Engineers designed *Galileo's* probe to withstand the highest impact speed ever attained by a human-made object—about 170,000 km/h.

water and elements such as carbon and nitrogen. According to this model, Jupiter should have a higher ratio of water and these elements than does the sun. Although the probe detected levels of carbon and nitrogen consistent with the model, it found only one-fifth the amount of water contained in the sun.

After the probe mission, *Galileo* began its own two-year study of Jupiter, its moons, and its magnetosphere. It will complete 11 orbits in all, 10 of which include close flybys of several of Jupiter's many moons. *Galileo's* early findings include evidence of liquid water on the moon Europa and the discovery that the moon Io contains a huge iron core that accounts for half its diameter.

Galileo's mission ends with a final Europa flyby on November 6, 1997, although mission scientists will continue to operate the spacecraft for as long as its fuel and power supplies permit. Thereafter, *Galileo* will orbit Jupiter until it most likely hits one of Jupiter's moons or the planet itself.

▲ Signs of ice movement on the surface of Europa raises the possibility that liquid water may exist there.

How would you define a gravity-assist trajectory?

Voyager images indicate that Neptune has a very active weather system. The solar system's strongest winds, exceeding 1,000 km/hr, and an earth-sized storm, known as the Great Dark Spot, are characteristic of Neptune's dynamic atmosphere. The average temperature of Neptune is probably about –225°C.

Pluto

Pluto, the ninth planet from the sun, was discovered in 1930 after a long search. Clyde Tombaugh, an astronomer at Flagstaff Observatory in Arizona, had been looking for a Planet X. It was thought that Planet X could explain the wobbling in the orbits of Uranus and Neptune. What Tombaugh found was Pluto. But Pluto is far too small to cause the wobbling. In other words, the discovery of Pluto was accidental.

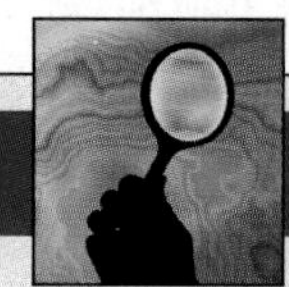

SMALL-SCALE INVESTIGATION

The Solar System

You can compare the relative sizes of the planets in the solar system and the distances among them with a simple model.

Materials

Planetary data table on pages 600–601; drawing compass; metric ruler; notebook paper; scissors; sheet of paper, about 1.5 m^2

Procedure

1. Draw a table like the one shown here. List the sun and nine planets in the first column.
2. Study Table 29–1 on pages 600–601. Using the scale 1 cm equals 10,000 km, calculate the scale diameter and distance from the sun for each planet. Complete your table.
3. Using the information in your table and a drawing compass, draw a to-scale circle on a sheet of paper to represent each planet. Cut out each circle, and label it with the name of the planet it represents.
4. Out of the large sheet of paper, cut a circle that approximates the scale diameter of the sun.
5. Compare the relative sizes of the sun and the planets you have constructed.

	Diameter (cm)	Distance from sun (cm)
Sun		
Mercury		
Venus		

6. Study the scale distance from the sun to each planet. Write a planetary-model plan that uses the scale distances you have calculated.

Analysis and Conclusions

1. Where is the greatest concentration of mass in the solar system?
2. Would it be practical to complete your model using the scale 1 cm equals 100,000 km for both distance from the sun and diameter?
3. Compare the distances between the inner planets with the distances between the outer planets.
4. Why are models of the solar system that are displayed in classrooms often inaccurate?

Figure 29–11. This illustration shows the relative position of the bodies in the solar system. Note, however, that the sizes of the sun and planets, as well as the relative distances between them, are not drawn to scale.

Pluto has a diameter of 2,445 km, making it the smallest planet in the solar system. Pluto rotates on its axis every six days, and its orbit period is almost 250 years. It orbits the sun in an unusually elongated ellipse, its distance from the sun varying from 4.4 billion km to 7.4 billion km and averaging 5.9 billion km. Pluto is sometimes inside the orbit of Neptune but is usually far beyond it.

Pluto appears to be made up mostly of frozen methane, rock, and ice, with an average temperature of between −208°C and −233°C. Infrared images of Pluto show that it has extensive methane ice caps and a very thin methane atmosphere. The Hubble Space Telescope clearly shows that Pluto has only one moon, Charon, which was discovered in 1978. Charon is half as large as Pluto. The two bodies orbit so close together that they appear to be a double-planet system.

The small size, unusual orbit, and comparatively large moon of Pluto once led some astronomers to believe that the planet was once a moon of Neptune and that the gravity of a passing object pulled Pluto away from Neptune and into its own orbit.

Section 29.3 Review

1. List the names of the five outer planets.
2. What makes Jupiter similar to the sun?
3. How dense is Saturn in comparison with other planets?
4. In what way is the axis of Uranus unusual?
5. Why is Pluto considered the ninth planet from the sun even though Neptune is sometimes farther from the sun?

Section Objectives

- **Describe the physical characteristics of asteroids and of comets.**
- **Compare and contrast meteoroids, meteorites, and meteors.**

29.4 Asteroids, Comets, and Meteoroids

In addition to the sun, the planets, and their moons, the solar system includes millions of smaller bodies of matter. Some of these bodies are just bits of dust or ice floating in space. Others are as big as small moons. Astronomers theorize that these small bodies were left over from the nebula from which the solar system formed.

Asteroids

The largest of the smaller bodies in the solar system are **asteroids,** or minor planets. Asteroids are fragments of rock that orbit the sun. Astronomers have observed more than 50,000 asteroids. Millions of asteroids may exist in the solar system. The orbits of most asteroids, like those of the planets, are ellipses. The largest known asteroid, Ceres, is about 1,000 km in diameter.

Most asteroids exist in a region between the orbits of Mars and Jupiter known as the **asteroid belt.** The asteroid belt begins about 100 million km beyond the orbit of Mars and stretches for about 150 million km toward the orbit of Jupiter.

Asteroids are usually classified into three types based on their composition. One type of asteroid is composed mostly of carbon materials. These materials give the asteroids a dark appearance. A second type is composed mostly of iron and nickel; these asteroids have a metallic appearance. The third and most common type of asteroid is made up mostly of silicate minerals. These asteroids look like ordinary earth rocks.

Figure 29–12. Asteroids are fragments of rock that orbit the sun. These photos of the asteroids Gaspra (left) and Ida (right) were taken by the spaceprobe *Galileo* on its way to a historic rendezvous with Jupiter.

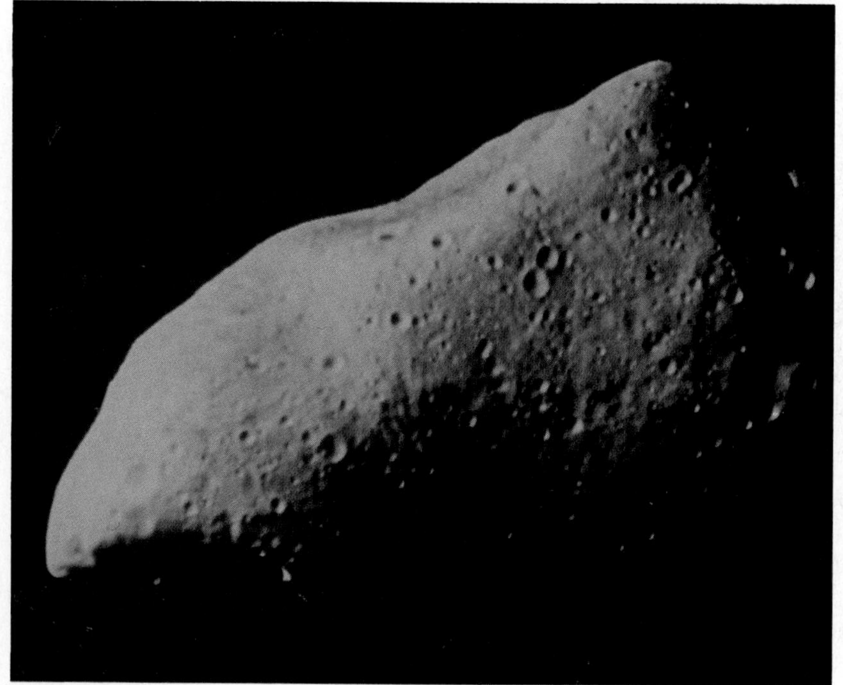

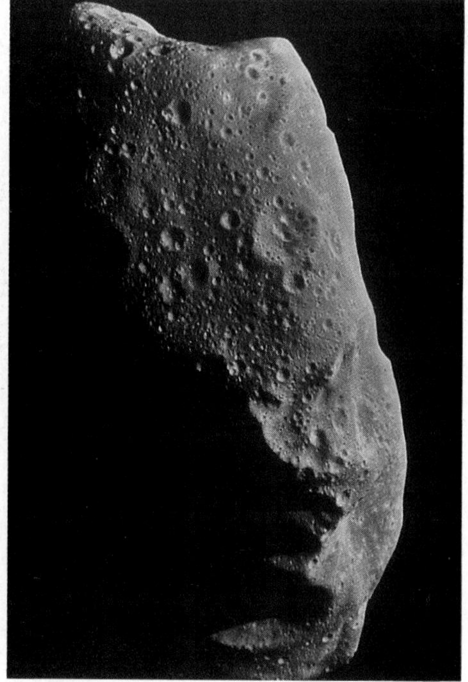

Many astronomers think that asteroids in the asteroid belt are the remains of planetesimals that were not able to form a planet because of the strong gravitational force of Jupiter. The composition of the asteroids suggests that they formed from the same materials as the planets. For example, iron is common in the cores of the inner planets, and silicate minerals are common in their crusts. Scientists estimate that the total mass of all asteroids, including Ceres, is not as great as that of the moon.

Some asteroids—the **Trojan asteroids** and the **earth-grazers**—orbit the sun but are not in the asteroid belt. The Trojan asteroids are concentrated in groups just ahead of and just behind the planet Jupiter. Earth-grazers have elongated elliptical orbits that sometimes bring them very close to the sun and the earth. Despite their name, earth-grazers do collide with the earth on occasion.

Comets

Comets orbit the sun in long ellipses. A comet is a body of rock, dust, methane, ammonia, and ice. The core, or *nucleus,* of a comet is made up of rock, metals, and ice and is usually between 1 km and 100 km in diameter. A spherical cloud of gas and dust—the **coma**—surrounds the nucleus. The coma can extend up to 1 million km from the nucleus. The bright appearance of a comet may be due to sunlight reflected by the coma. The nucleus and the coma together form the head of the comet.

The most spectacular part of a comet is its tail. The tail of a comet is the gas and dust that streams out from the head. The tail forms as sunlight and the solar wind—electrically charged particles

Figure 29–13. A comet, such as the Comet Hyakutake shown here in 1996, consists of a nucleus, a coma, and a tail.

released by the sun—pushes gas and dust away from the head of the comet. Thus, regardless of what direction the comet is traveling, its tail points away from the sun. Some of the larger comets have tails that are more than 80 million km long. Figure 29–13 shows the nucleus, coma, and tail of a typical comet.

Astronomers hypothesize that most comets originate in the **Oort cloud,** which was named after the Dutch astronomer Jan H. Oort. The Oort cloud is a spherical cloud of dust and ice that contains the nuclei of as many as a trillion comets. The bodies within the cloud circle the sun at a speed of up to only 140 m/sec, thus requiring a few million years to complete just one orbit. The cloud surrounds the solar system, beginning at a distance of one light-year from the sun and reaching halfway to the nearest star. The matter in the Oort cloud may have been left over from the formation of the solar system. The gravity of a star passing near the solar system may cause a comet within the cloud to fall into a long elliptical orbit. The orbit stretches from the Oort cloud to the sun and back to the cloud. Because the Oort cloud is spherical, comets are not oriented in any particular direction and, when released, can orbit the sun at any angle.

Some comets, called **long-period comets,** have periods of several thousand or even several million years. Other comets, called **short-period comets,** have periods of up to 100 years. The gravity of one of the outer planets, such as Jupiter, can affect these comets as they orbit the sun. Halley's comet, which appears about every 76 years, is a short-period comet. It last appeared in 1986. Approximately when should Halley's comet appear next?

Meteoroids

Figure 29–14. This meteorite, which is thought to have once been part of the planet Mars, was discovered in Antarctica in 1984. In 1996, scientists discovered fossil evidence of primitive microorganisms on the 4.5-billion-year-old rock, indicating that life may have existed on the "red planet" more than 3.6 billion years ago.

In addition to asteroids and comets, smaller bits of rock or metal called **meteoroids** move throughout the solar system. Most meteoroids are less than 1 mm in diameter. Such small meteoroids are pieces of matter that become detached from passing comets. Larger meteoroids—more than 1 cm in diameter—are produced by collisions between asteroids.

A meteoroid that enters the earth's atmosphere is called a **meteor.** As the meteor passes through the atmosphere, friction heats the meteor and slows it down. Most meteors burn up in the atmosphere before reaching the earth's surface, producing a bright streak of light. You may have seen a meteor, often called a *shooting star,* streaking through the sky. Meteoroids sometimes vaporize very quickly in a brilliant flash of light called a **fireball.** Observers on the earth can sometimes hear a loud noise as a fireball disintegrates.

Occasionally, a large number of small meteoroids enter the earth's atmosphere in a short period of time. Once in the earth's atmosphere, the meteors burn up, creating a **meteor shower.** During the most spectacular of these showers, several meteors may be visible every minute. Meteor showers occur at the same time each year when the earth intersects the orbits of comets that have left behind a

Figure 29–15. The straight vertical lines in this photograph are meteors. The lines appear thicker and brighter when meteors burn up in the earth's atmosphere. The curved lines are trails of stars created as the earth rotated during this long-exposure photograph.

trail of meteoroids. Astronomers estimate that about 1 million kg of matter from meteors falls to the earth each day.

Millions of meteors enter the earth's atmosphere each day. A few meteors do not burn up entirely in the atmosphere but fall to the earth. A meteor or any part of a meteor that is left after it hits the earth is called a **meteorite.** Most meteorites are small, with a mass of less than 1 kg. However, large meteorites occasionally strike the earth's surface with the force of an exploding bomb. These leave large craters. One of the best-known is Meteor Crater in Arizona, which was created by a meteor that struck the earth about 20,000 years ago. The meteor, as it entered the atmosphere, was 50 m in diameter and weighed 500,000 tons. It left a crater 1.3 km in diameter and 180 m deep, as well as about 25 tons of iron and nickel fragments spread out over the land.

Meteorites can be classified into three basic types: stony, iron, and stony-iron. **Stony meteorites** are similar in composition to the rocks found on the earth. Some stony meteorites contain carbon-bearing substances similar to the materials found in living organisms. Although most meteorites are stony, **iron meteorites** are easier to find. This is because stony meteorites look like ordinary earth rocks and often are not noticed. Iron meteorites have a distinctive metallic appearance. **Stony-iron meteorites** contain both iron and stone and are very rare.

Astronomers think that virtually all meteorites are asteroids and that the oldest meteorites are about 100 million years older than the earth or its moon. Therefore, meteorites can provide information about the composition of the solar nebula that existed before the earth and its moon formed. Some rare meteorites originated on the moon or Mars. How could they have reached the earth?

Section 29.4 Review

1. Between the orbits of what planets is the asteroid belt located?
2. In what direction does the tail of a comet point?
3. How is a meteor different from a meteorite?
4. In what way are the orbits of Trojan asteroids and earth-grazers unlike the orbits of other asteroids?

Chapter 29 Review

Key Terms

- asteroid (606)
- asteroid belt (606)
- astronomical unit (592)
- coma (607)
- comet (607)
- earth-grazer (607)
- ellipse (592)
- epicycle (591)
- fireball (608)
- focus (592)
- geocentric (591)
- heliocentric (591)
- inertia (593)
- iron meteorite (609)
- Jovian planet (598)
- long-period comet (608)
- meteor (608)
- meteor shower (608)
- meteorite (609)
- meteoroid (608)
- Oort cloud (608)
- orbit period (593)
- retrograde motion (591)
- short-period comet (608)
- stony meteorite (609)
- stony-iron meteorite (609)
- terrestrial planet (594)
- Trojan asteroid (607)

Key Concepts

Geocentric models of the solar system, such as those developed by Aristotle and Ptolemy, were replaced by the currently accepted heliocentric model proposed by Copernicus. **See page 591.**

Kepler's three laws describe the motion of the planets in their orbits around the sun. **See page 591.**

The four inner planets share similar characteristics and are called the terrestrial planets. **See page 594.**

Earth, Venus, and Mars have a history of geologic activity. **See page 596.**

Jupiter and Saturn rotate rapidly. **See page 598.**

Uranus is the third largest planet in the solar system. Neptune was named after the Roman god of the sea. Pluto is the smallest planet in the solar system. **See page 601.**

Most asteroids are found in the region between Mars and Jupiter known as the asteroid belt. Some comets orbit the sun in long periods, and others orbit in short periods. **See page 606.**

Meteoroids move throughout the solar system. When they enter the earth's atmosphere, they are meteors; when they strike the earth's surface, they are meteorites. **See page 608.**

Review

On your own paper, write the letter of the term that best completes each of the following statements.

1. Ptolemy modified Aristotle's model of the universe to include
   a. Oort clouds. b. retrograde motion.
   c. comets. d. shooting stars.
2. Copernicus's model of the solar system differed from Ptolemy's because it was
   a. geocentric. b. lunocentric. c. ethnocentric. d. heliocentric.
3. Kepler's first law states that each planet orbits the sun in a path called
   a. an ellipse. b. a circle. c. an epicycle. d. a period.
4. Kepler's law that describes how fast planets travel at different points in their orbits is called the law of
   a. ellipses. b. equal speed.
   c. equal areas. d. periods.
5. The weak magnetic field around Mercury suggests
   a. volcanic activity.
   b. a dense atmosphere.
   c. a core of molten iron.
   d. that it is located close to the sun.

6. The planet that rotates in a direction that is opposite the direction of the other planets is
 a. Mercury. b. Venus. c. the earth. d. Mars.
7. The tilt of the axis of Mars is nearly the same as that of
 a. Mercury. b. Venus. c. the earth. d. Jupiter.
8. The planet that rotates faster than any other planet in the solar system is
 a. the earth. b. Jupiter. c. Uranus. d. Pluto.
9. The most distinctive feature of Jupiter is its
 a. Great Red Spot. b. Great Dark Spot.
 c. rings. d. elongated orbit.
10. All of the outer planets in the solar system are large except
 a. Saturn. b. Uranus. c. Neptune. d. Pluto.
11. The asteroid belt exists in a region between the orbits of
 a. Mercury and Venus. b. Venus and the earth.
 c. the earth and Mars. d. Mars and Jupiter.
12. The composition of asteroids suggests that they are
 a. small moons.
 b. fragments of planetesimals.
 c. the nuclei of comets.
 d. environments that possibly can support life.
13. Meteorites can provide information about
 a. the composition of the solar nebula before the earth and its moon formed.
 b. the size of the earth.
 c. the destiny of the solar system.
 d. the size of the universe.

Critical Thinking

On your own paper, write answers to the following questions.

1. Assume that an intelligent life-form exists on Pluto—the planet with the longest orbit period in the solar system. Would astronomers on Pluto be likely to propose a heliocentric model of the solar system? Explain your answer.
2. If you know the distance from the sun to a planet, what other information can you determine about the orbit of the planet? Explain your answer.
3. The surfaces of some asteroids reflect only small amounts of light. Other asteroids reflect up to 40 percent of the light falling on them. Of what materials would each type of asteroid probably be composed?
4. Concept map ? ? ? By constructing a **concept map,** you make connections that illustrate relationships among certain terms. How would doing so assist your understanding of this chapter?

Application

1. Suppose that a new planet has just been discovered. It has no rings or moons and has a surface pitted with impact craters. In what group of planets do you think this planet is located? Explain how you know.
2. What type of core do you predict that the new planet mentioned in Question 1 will have?
3. Suppose you live in an unglaciated area and have found a chunk of rock that you suspect might be a stony meteorite. What data would help you verify your hypothesis?

Extension

1. Do library or Internet research on the discovery and characteristics of Pluto. Collect evidence both supporting and opposing its classification as a planet. Present your findings to the class in the form of a debate, first taking one side, then the other. Ask the class to vote on the issue.
2. Talk with the director of a local planetarium or natural history museum about meteorites that have been found in your area. If possible, go to look at any specimens that may be on display. Find out the age and composition of the meteorites, and classify them according to the system discussed in the chapter. Report your findings to the class.

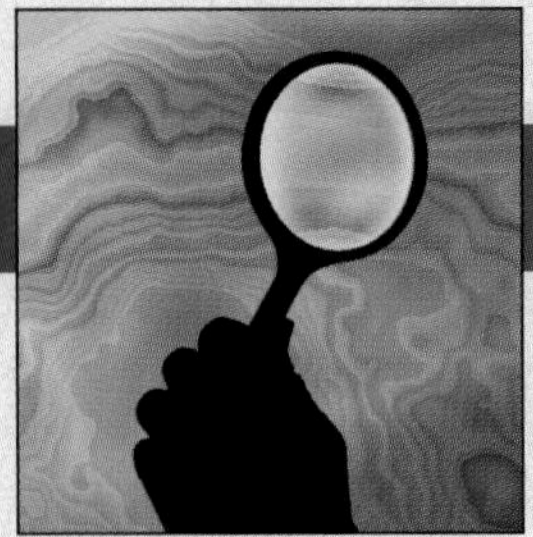

IN-DEPTH INVESTIGATION

Crater Analysis

Materials

- meter stick
- water
- scissors
- plaster of Paris
- shoe box
- 6 toothpicks
- marbles, 1 of them large
- protractor
- masking tape
- marker
- safety goggles
- lab apron

Introduction

All the inner planets—Mercury, Venus, Earth, and Mars—have many features in common. They are all mostly solid rock with metallic cores; they have no rings and from zero to two moons each; and every inner planet has bowl-shaped depressions called *impact craters.* Impact craters are formed as a result of collisions between the planets and rocky objects traveling through space. Most of these collisions took place during the formation of the solar system. Mercury's entire surface is covered with these craters, while very few are still evident on the surface of the earth. Many of the moons of the inner and outer planets also are heavily cratered.

In this investigation, you will experiment with making craters to discover the effect of speed and projectile angle on the craters formed.

Prelab Preparation

1. Review Section 29.2, page 594.
2. Review the guidelines for eye safety.
3. Place the top of a toothpick in the center of a 6-cm-long piece of masking tape. Fold the tape in half around the toothpick so that they form a small flag and flagpole. On the flag, write the letter *A.* Repeat this procedure with each of the other toothpicks, labeling them *B* through *F.*

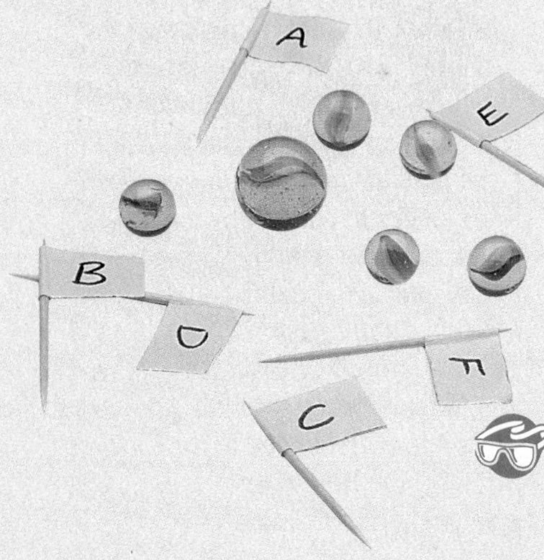

Procedure

1. Mix the plaster of Paris with water, according to instructions. Spread your mixture in the bottom of the shoe box. Make your surface about 4 cm thick. The surface should be as smooth as possible.
2. Allow your surface to dry until it is no longer soupy, but not yet rigid.
3. **CAUTION**: *Put on your safety goggles.* Drop a marble onto the plaster from a height of 50 cm above the surface. Quickly remove the marble, without damaging the crater if possible. Place flag *A* next to the crater to label it crater *A.*

4. Repeat Step 3 with a marble dropped from a height of 1 m and another dropped from a height of 25 cm. Use the flags to mark craters *B* (1-m drop) and *C* (25-cm drop).
5. Repeat Step 3 using the large marble dropped from a height of 1 m. Label the crater formed *D*.
6. Using your protractor as a guide, have your partner tilt the box at a 30-degree angle to the table. Be sure to hold the box steady. Then drop a marble vertically from a height of 50 cm. Label the crater formed *E*.
7. Repeat Step 5 using an angle of 45 degrees. Label the crater *F*.
8. Allow the plaster of Paris to harden. Write a description of each crater and the surrounding area.

Analysis and Conclusions

1. Which crater was formed by the marble with the highest velocity? What is the effect of velocity on the characteristics of the crater formed?
2. Study the shapes of craters *A, E,* and *F*. How does the angle of incidence affect the shape of the craters formed?
3. Compare craters *A, B,* and *D*. How do they differ? What caused this difference? Is the difference in the masses of the objects a factor? Explain.

Extension

Which type of crater formation have we not studied? Devise a method of studying this cratering process using the model you made.

Chapter 30

Moons and Rings

On July 20, 1969, astronaut Neil Armstrong stepped onto the surface of the moon. This first visit to another body in the solar system marked a new era in history and provided an excellent opportunity for firsthand study. This chapter discusses the earth's moon and the moons and rings of the other planets.

Chapter Outline

Astronauts reached the moon for the first time on July 20, 1969.

30.1 The Earth's Moon

Section Objectives

- **List the five kinds of lunar surface features.**
- **Describe the interior of the moon.**
- **Summarize the four stages in the development of the moon.**

A body that orbits a larger body is called a *satellite*. In 1957 the Soviet Union launched *Sputnik 1,* the earth's first artificial satellite. In 1958 the United States launched its first satellite, *Explorer 1*. A natural satellite of any planet is called a *moon*. The earth's natural satellite and closest neighbor is referred to as *the moon*.

If you were to visit the moon, you would immediately notice that the force of gravity of the moon differs from that of the earth. The mass of an object determines its gravitational force. Since the moon has less mass than the earth does, its gravity is weaker. Its surface gravity is about one-sixth the surface gravity of the earth. As a result, a person who weighs 600 newtons (N) on the earth would weigh about 100 N on the moon. As you may recall from Chapter 2, on the earth's surface, one kilogram of mass weighs about 10 N.

Between 1969 and 1972, the United States sent six *Apollo* spacecraft to the moon. The astronauts that landed on the moon found that the weak gravity affected the way they moved on the moon. Instead of walking, they bounced along in hops and leaps.

Gravity on the moon has never been strong enough to hold gases. Therefore, the moon has no atmosphere and cannot support life. Because the moon has no atmosphere to act as insulation, the temperature variation is far greater than it is on the earth. The temperature ranges from 134°C during the day to −170°C at night. The dramatic temperature shifts are due in part to slow rotation. The moon completes a rotation on its axis once every 29.5 days. Thus each lunar day is equal to about four weeks on earth.

The Lunar Surface

The moon is close enough to the earth to be seen easily without a telescope. The next closest object to the earth is Venus. The distance between Venus and the earth is more than 100 times the distance between the moon and the earth.

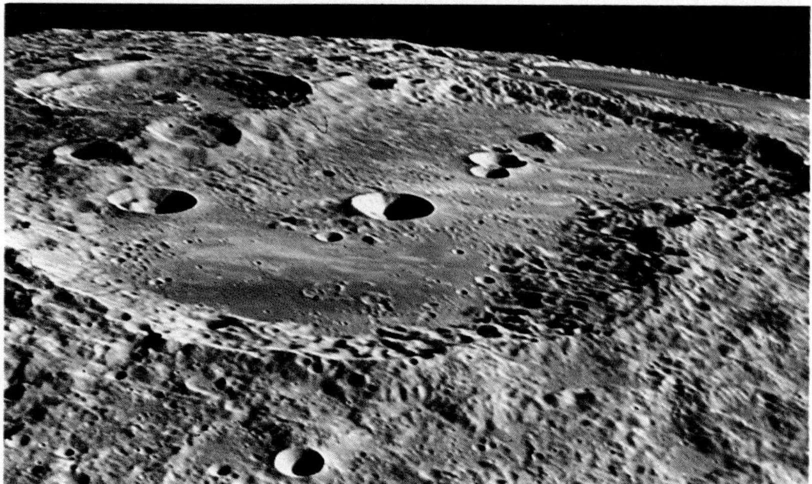

Figure 30–1. A clear but distant view of the moon is visible with unaided eyes from the earth. Space exploration provides detailed views, such as this photograph taken from the *Apollo 15* spacecraft in 1971.

Figure 30–2. The moon's rilles are long, deep channels that run through the maria (left). From a distance, the maria appear to be dark areas (right). The light areas are rough, cratered lunar highlands.

An observer on the earth can see light and dark patches on the moon. The light areas are rough highlands composed of light-colored rock. The dark areas, called **maria** [(MAHR-ee-uh); sing., *mare* (mahr-AY)] are smooth and reflect little light. *Mare* is the Latin word for "sea." Early astronomers thought that the dark areas they saw on the moon's surface were bodies of water. Today astronomers know that the maria are plains of dark solidified lava. These lava plains are the remains of ancient volcanic eruptions on the surface of the moon.

Long, deep channels called **rilles** run through the maria. A rille looks somewhat like a dry riverbed. However, there is no evidence that running water has ever existed on the moon. Some rilles resemble narrow trenches and are as much as 240 km long.

Craters

The surface of the moon is covered with bowl-shaped depressions called **craters.** Most of these were formed 4 billion years ago. At that time, the moon, like other bodies in the solar system, was bombarded with chunks of debris left over from the formation of the solar system. As this debris struck the moon, it pushed surface material up and out, forming craters. Debris that struck the moon in the more recent past, had an effect similar to that of an object striking the surface of a pool or lake. The displaced materials settled in streaks, called **rays,** that extended radially outward from the impact site.

Some craters have diameters as large as 250 km. The entire state of Connecticut could easily fit into some of the moon's craters. The largest craters are named for famous scholars and scientists, like

Tycho and *Copernicus*. The lunar surface also has millions of smaller craters. Many of these craters overlap one another. Some smaller craters are located entirely within larger craters.

Geologists suggest that the earth once had many craters similar to moon craters. The earth's craters have been almost completely eroded by the forces of water, wind, and ice. Because the moon lacks these erosional agents, its surface has changed little since it was formed.

Lunar Rocks

Meteorites of all sizes struck the moon. They crushed much of the rock on the surface into dust and small fragments. Today almost all of the lunar surface is covered by a layer of this dust and rock. The material in this layer is called *regolith*. The depth of the regolith layer varies from 1 m to 6 m. The number of meteorites that reached the surface of the moon was greater than the number that reached the earth's surface. Why was this?

Lunar rocks contain many of the same elements found in the earth's rocks, but in very different amounts. Near the lunar surface, igneous rocks, like those in the earth's crust, are composed mainly of oxygen and silicon. Rocks from the lunar highlands are light-colored, coarse-grained rocks called *anorthosites*. These rocks are rich in calcium and aluminum. Rocks from the maria are dark-colored, fine-grained basalts, which contain large amounts of titanium, magnesium, and iron.

One type of rock found in both maria and the highlands is *breccia*. Breccia contains fragments of other rocks that have melted together and were formed when meteorites struck the moon. The force of the impact broke up rocks, and the heat from the impact melted the fragments together.

Lunar rocks contain only small amounts of some elements commonly found in the rocks on earth. For example, lunar surface rocks are deficient in elements with low melting points, such as sodium. These elements may have boiled off when the moon was still molten. Also, lunar rocks do not contain water.

The surface rocks on the moon are about as dense as those on the surface of the earth. Yet the overall density of the moon is only

Figure 30–3. Analysis indicates that the rock shown at left is 4.3 billion to 4.5 billion years old—the oldest rock found on the moon. The varied texture of the rock indicates that it has a complicated history. The rock shown at right is the largest rock sample collected on the *Apollo 15* mission. The rock, a basalt, measures 30 cm × 15 cm × 15 cm.

three fifths the density of the earth. The difference in overall density indicates that the interior of the moon is less dense than the interior of the earth.

The Interior of the Moon

Most of the information that astronomers have gathered about the interior of the moon comes from seismographs. These instruments were placed on the moon by the Apollo astronauts. Seismographs have recorded numerous weak quakes on the moon, which are similar to earthquakes. They have also revealed that, on the side facing the earth, the crust of the moon is 60 km thick. On the side facing away from the earth, it is up to 100 km thick. The surface on the far side is mountainous, with only a few small maria. The crust appears to consist of materials similar to that of the rocks in the highlands.

Beneath the crust of the moon is the mantle, which is made up of dense rock that is probably rich in silica, magnesium, and iron. The mantle reaches to a depth of about 1,000 km.

The lower portion of the mantle may be slightly molten. The moon may even have a small iron core. Scientists estimate that the

SCIENCE & TECHNOLOGY

Mining on the Moon

Ever since astronauts first landed on the moon, scientists have been working on ways to build and operate permanent lunar bases.

The biggest obstacle to building lunar bases is the cost. Shipping costs for construction materials and life-support systems would be "out of this world." Therefore, scientists are developing ways of producing the necessary materials by mining the moon's own natural resources.

One of the most promising lunar mining methods uses *electrolysis,* a process that changes the chemical composition of a substance by passing an electrical current through it. Scientists hope to use electrolysis to extract iron and oxygen from rocks similar to the boulder the astronaut is examining at left.

To test the procedure, scientists created a silicate rock like the kind commonly found on the moon. After melting the rock, they placed a positive electrode and a negative electrode into the molten rock. An electrical current between the two electrodes caused iron to separate out at one electrode and oxygen to separate out at the other electrode. On the moon, the extracted iron could be used to manufacture steel for building structures, while the oxygen could be used to support human life.

Researchers envision using a solar-powered laser satellite to provide energy for everything

radius of the core is less than 700 km. The moon has almost no overall magnetic field, although there are small areas of local magnetism, suggesting that the core is probably solid.

Development of the Moon

The study of moon rocks and data gathered by astronauts has provided evidence to help astronomers understand the history of the moon. Astronomers generally support the *giant-impact hypothesis* about the first stage of its origin. This hypothesis suggests that the moon formed when a Mars-sized body struck the earth early in the history of the solar system. The collision caused an ejection of fragments into orbit around the earth. These fragments eventually joined to form the moon. Most of the ejected materials were from the silica-rich mantles of the earth and the colliding body, while dense, metallic core materials remained in the earth. The hypothesis is supported by the silica-rich samples of moon rocks brought back by Apollo astronauts.

Scientists generally agree about the subsequent stages of the development of the moon. In the second stage of the development, the

from lunar mining operations to human habitats. The satellite would use a large solar concentrator and lasers to beam energy to the moon's surface. There, the laser light would be converted into electricity. Three laser satellites orbiting the moon would make power reception possible anywhere on the surface.

Mining the moon may not occur for many years, if ever, although scientists continue to develop the technology to make it possible. In addition to developing the relevant technology, many more geological studies of the moon are necessary before undertaking such a project. If mining the moon were profitable, however, it would help secure funding for future space exploration by opening the door to private commercial ventures.

How might a lunar mining operation use solar energy?

▲ This artist's conception of a lunar mining operation depicts the production of liquid oxygen. Ilmenite, an oxygen-rich mineral, is being mined from the lunar soil.

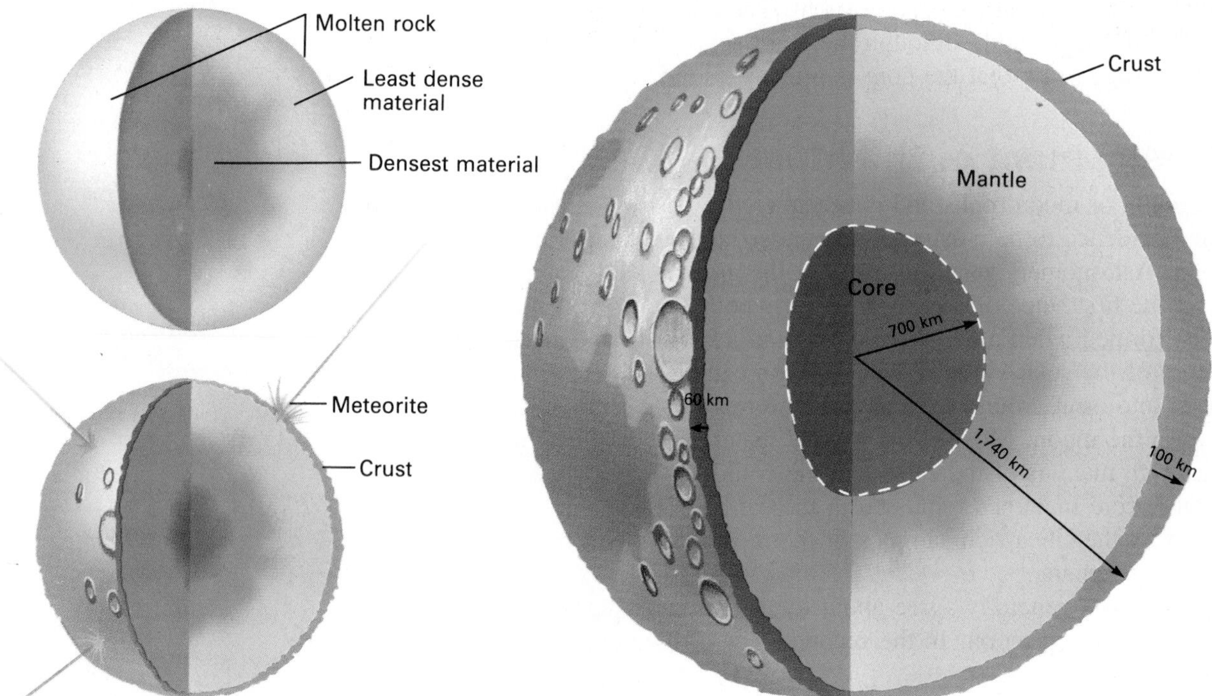

Figure 30–4. After the moon formed, it was covered by a layer of white-hot molten rock (top, left). As the moon cooled, a crust formed over the molten rock (bottom, left). By 3 billion years ago, the moon looked much as it does today (right).

surface of the moon was covered by an ocean of hot molten rock. In time, the minerals in the molten rock separated. The densest materials moved to the center and formed a small core. The least dense materials formed an outer layer. Materials of intermediate density settled between the core and the outer layer, forming the mantle.

The third stage in the development of the moon began as the outer surface of the moon cooled, forming a thick, solid crust over the molten rock. At this time, many pieces of the debris leftover from the formation of the solar system struck the moon, producing its cratered surface. Some of the meteorites that hit the moon broke through the crust. Molten rock from beneath the crust flowed to the surface and partially filled the craters, forming the smooth maria.

Sometime before 3 billion years ago, the number of loose particles in the solar system decreased. Therefore, less material hit the lunar surface, and cratering decreased. During this most recent stage of lunar development, virtually all geologic activity stopped because of cooling. Rayed craters were formed by the most recent meteor impacts. The moon today looks almost exactly as it did 3 billion years ago. Therefore, the moon is a valuable source of information about the conditions that existed in the solar system long ago.

Section 30.1 Review

1. Describe the maria that cover most of the surface of the moon.
2. How thick is the crust of the moon on the side facing the earth?
3. During which stage of development did the maria form?
4. How would the surface of the moon differ today if meteorites had continued to hit it at the same rate as they did 3 billion years ago?

30.2 Movements of the Moon

Section Objectives

- **Describe the orbit of the moon around the earth.**
- **Explain why eclipses occur.**

To observers on the earth, the moon appears to orbit the earth. To an observer in outer space, however, the earth and its moon appear to orbit each other. The earth and its moon together would appear to form a single system that orbits the sun. The center of this earth-moon system, not the center of the earth, follows a smooth orbit around the sun.

The mass of the moon is only 1/80 that of the earth. Therefore, the balance point of the earth-moon system is not halfway between the centers of the two bodies. Rather, the balance point of the system is located within the earth's interior due to the earth's greater mass, as shown in Figure 30–5.

Lunar Orbit and Rotation

The orbit of the moon around the earth forms an ellipse, not a circle. Therefore the distance between the earth and its moon varies. When the moon is farthest from the earth, it is at **apogee.** When the moon is closest to the earth, it is at **perigee.** The average distance of the moon from the earth is 384,000 km.

The rotation of the earth on its axis causes the moon to appear to rise and set. If you watch the moon rise on successive nights, you will note that it rises and sets approximately 50 minutes later each night. While the earth completes one rotation, the moon has moved

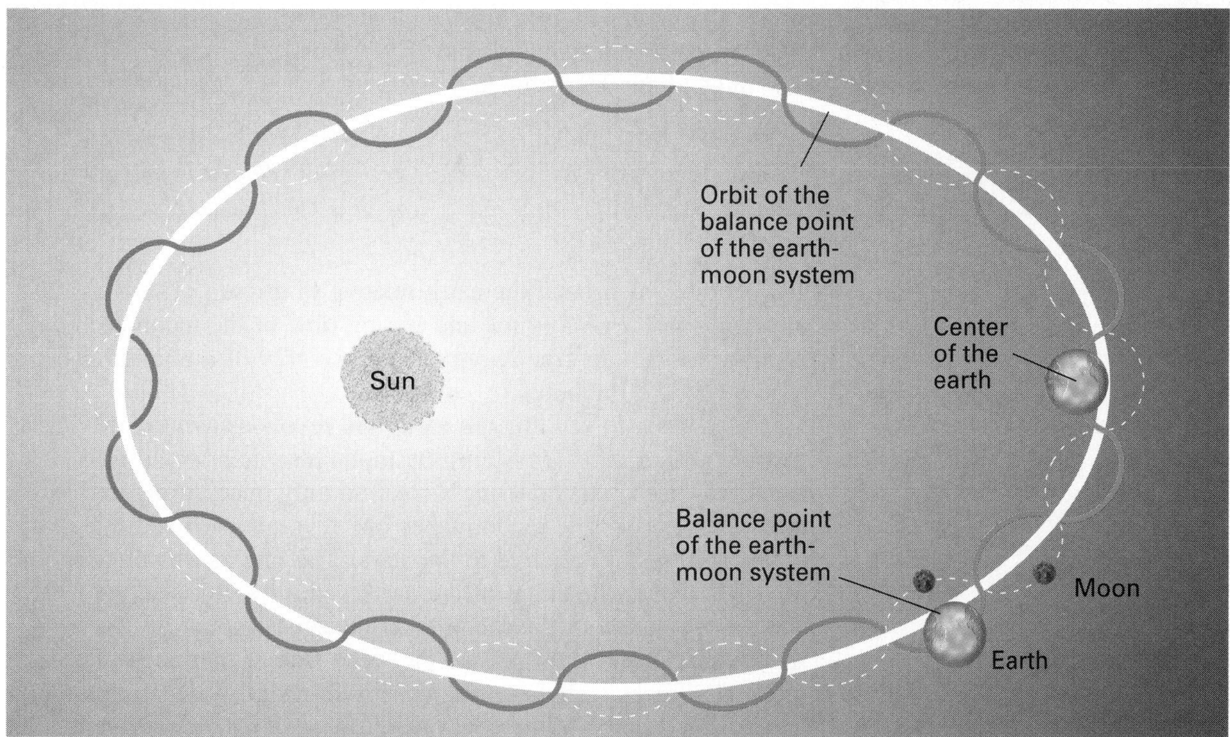

Figure 30–5. The balance point of the earth-moon system is in the interior of the earth. Only the balance point of the earth-moon system orbits the sun in a smooth ellipse.

EarthBeat

Effects of Tides

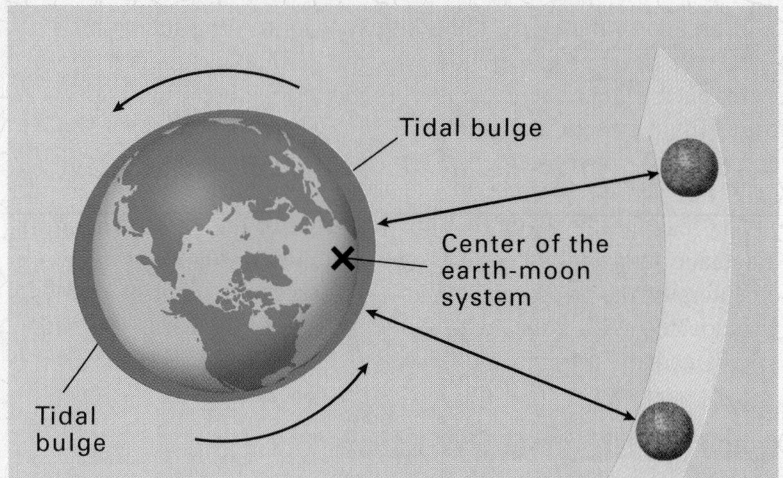

In Chapter 22, you read about how the moon and the sun cause tides on the earth. The gravitational attraction between the earth and the moon causes water levels to rise and fall along the shores of the ocean.

The tides also affect the moon. The earth's tides influence the speed at which the moon orbits the earth and the distance between the earth and the moon. The tides create a bulge of land and water on one side of the earth. As the earth rotates, this bulge moves eastward, slightly ahead of its expected position directly under the moon. As the bulge moves forward, the gravitational attraction between it and the moon pulls the moon along. The speed of the moon in its orbit around the earth increases. As the speed of the moon increases, the moon moves slightly farther away from the earth.

The tides caused by the moon also affect the speed of the earth's rotation. As the water moves to form the tides, it creates friction against the ocean floor. This friction slows the earth's rotation. Consequently, each rotation of the earth takes slightly longer than the previous one. Astronomers predict that 1,000 years from now the length of each day will be 0.2 seconds longer than it is now.

At this rate, in about how many years will the day become one minute longer than it is now?

1/29 of the way around the earth relative to the sun. Therefore, the difference in the rising time and setting time of the moon each day is equivalent to the time required to make 1/29 of a rotation. That time is about 50 minutes.

In addition to orbiting the earth and revolving around the sun as part of the earth-moon system, the moon also spins on its axis. It spins very slowly, completing a rotation only once during each orbit around the earth. The moon makes one revolution around the earth in about 27.3 days relative to the stars. The rate at which the moon rotates is due to the effect of the earth's gravity. Because the rotation and the revolution of the moon take the same amount of time, observers on the earth always see the same side of the moon. The moon also rocks slightly on its axis. As a result, observers on the earth can see about 59 percent of the surface of the moon.

Eclipses

All the bodies moving around the sun, including the earth and its moon, cast long shadows into space. An **eclipse** occurs when one planetary body passes through the shadow of another. Shadows cast by the earth and the moon have two parts. In the inner, cone-shaped part of the shadow, the **umbra,** sunlight is completely blocked. In the outer part of the shadow, the **penumbra,** sunlight is only partially blocked.

Solar Eclipses

When the moon is between the earth and the sun, the shadow of the moon may fall upon the earth, causing a **solar eclipse.** People within the umbra see a *total solar eclipse*. That is, the sun's light is completely blocked by the moon. The umbra falls on the area of the earth that lies directly in line with the moon and the sun. Outside the umbra, but within the penumbra, people see a *partial solar eclipse*. The penumbra falls on the area immediately surrounding the umbra.

The umbra of the moon is too small to make a large shadow on the earth's surface. Consequently, a total eclipse of the sun covers only a small part of the earth and is only seen by people in that particular part of the earth. The earth's rotation causes the shadow to move rapidly across the earth's surface. Therefore, a total solar eclipse never lasts more than seven minutes at any location.

At times the umbra of the moon is too short to reach the earth. If the moon is at or near apogee when it comes directly between the earth and the sun, its shadow does not reach the earth. If the umbra fails to reach the earth, a ring-shaped *annular eclipse* occurs. In an annular eclipse, the sun is not completely blocked out. Instead, a thin ring of sunlight is visible around the outer edge of the moon.

Figure 30–6. A total solar eclipse is a dramatic event (top). During a solar eclipse, the shadow of the moon falls on the earth (bottom).

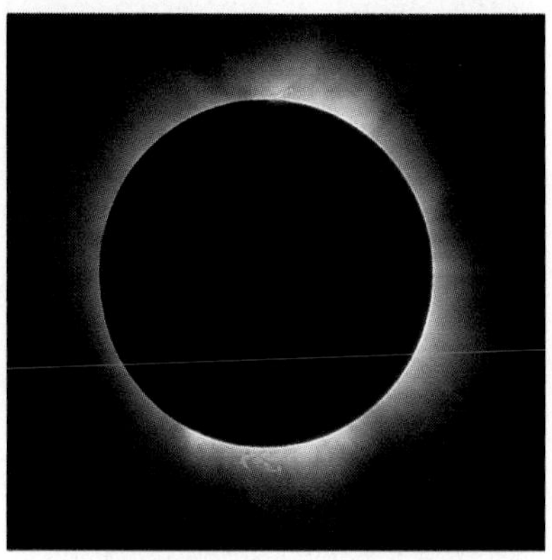

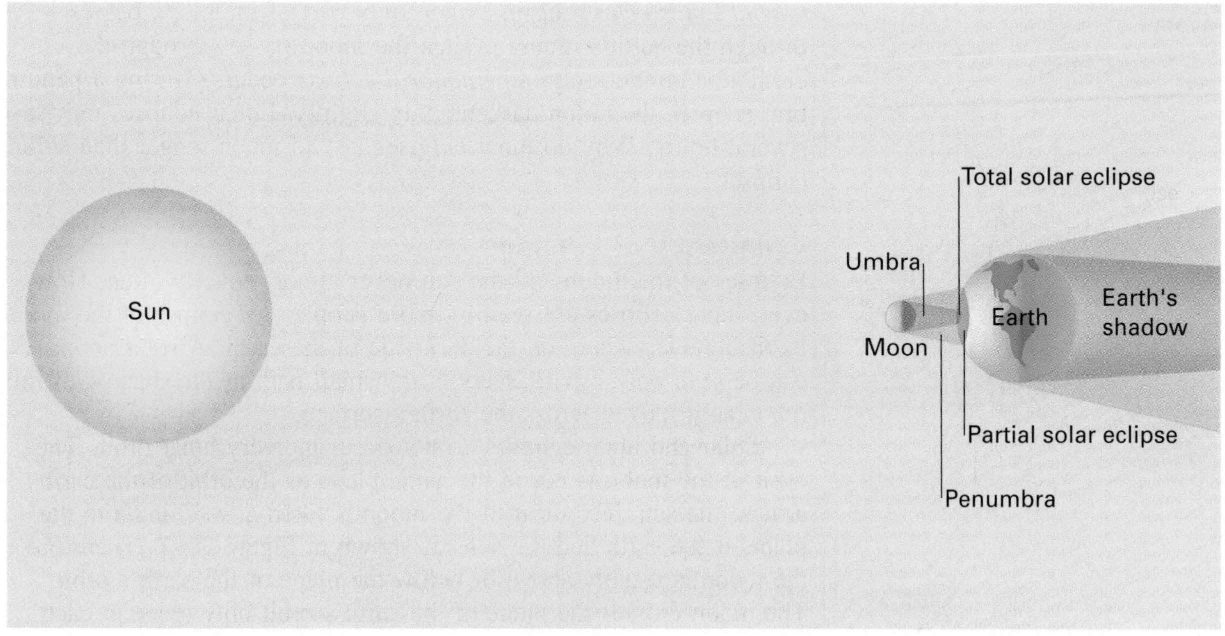

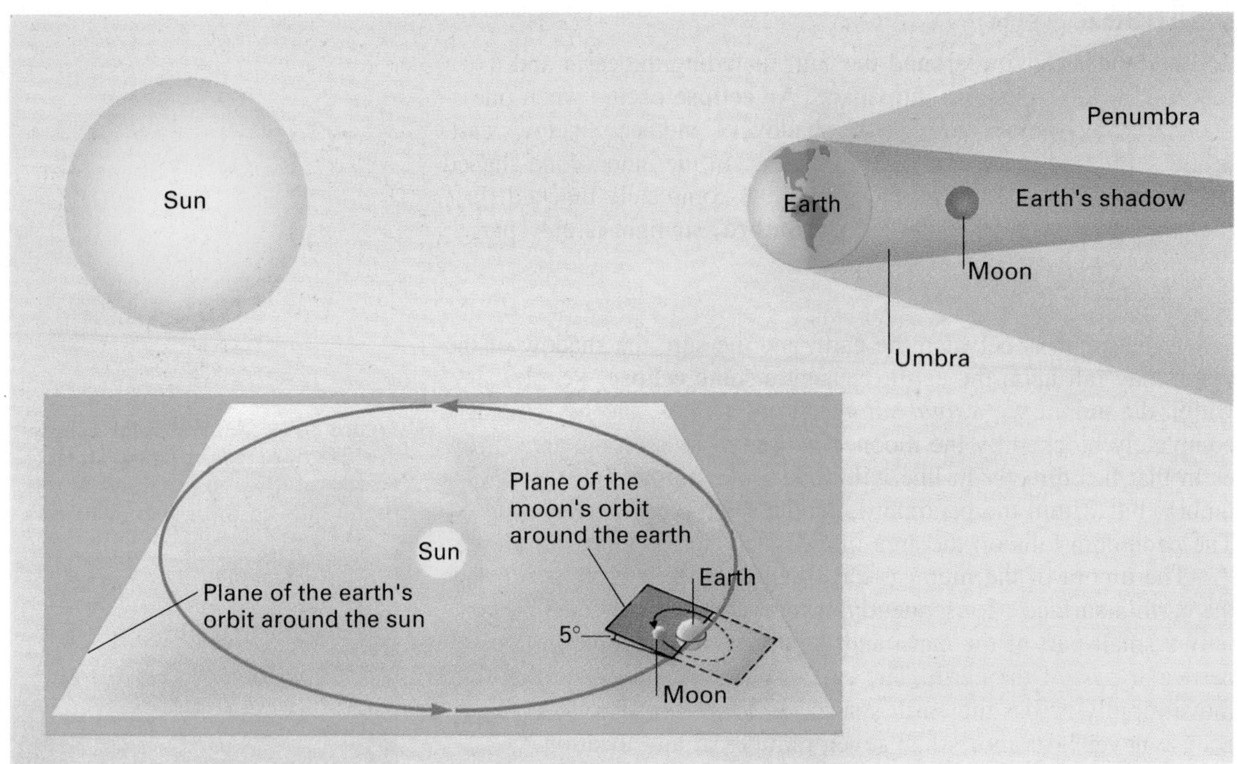

Figure 30–7. During a lunar eclipse, the earth's shadow falls on the moon (top). The plane of the moon's orbit around the earth is tilted at a 5° angle to the plane of the earth's orbit around the sun (bottom). Eclipses occur only when the moon crosses the plane of the earth's orbit at full moon.

Lunar Eclipses

Solar eclipses occur when the moon is between the earth and the sun. As the moon orbits the earth, the earth's shadow can cause eclipses on the moon. A **lunar eclipse** occurs when the earth is positioned between the moon and the sun, and the earth's shadow crosses the lighted half of the moon. To produce a *total lunar eclipse,* the moon must pass completely into the earth's umbra. A *partial lunar eclipse* occurs when only part of the moon passes through the earth's umbra. When the moon passes through the earth's penumbra only, a *penumbral eclipse* occurs. During a penumbral eclipse, the moon darkens only slightly. Lunar eclipses may last several hours. Why do lunar eclipses last so much longer than solar eclipses?

Frequency of Eclipses

Eclipses of the moon and the sun occur almost equally often. However, lunar eclipses are seen by more people. An eclipse of the moon is visible everywhere on the dark side of the earth. A solar eclipse can be seen only by observers in the small path of the shadow of the moon as it moves across the earth's surface.

Solar and lunar eclipses do not occur in every lunar orbit. The orbit of the moon is not in the same plane as the orbit of the earth around the sun. The orbit of the moon is tilted at a 5° angle to the plane of the earth and the sun, as shown in Figure 30–7. Therefore the moon is usually above or below the plane of the earth's orbit. The moon crosses the plane of the earth's orbit only twice in each

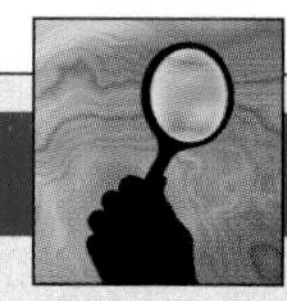

SMALL-SCALE INVESTIGATION

Eclipses

As the earth and moon revolve around the sun, each casts a shadow into space. An eclipse occurs when one planetary body passes through the shadow of another body. You can demonstrate how an eclipse occurs by using clay models of planetary bodies.

Materials

modeling clay, sheet of notebook paper, penlight or small flashlight, metric ruler

Procedure

1. Make two clay balls, one about 4 cm in diameter and one about 1 cm in diameter.
2. Position the balls about 15 cm apart on the sheet of paper, as shown in the photo.
3. Turn off any nearby lights. Place the penlight approximately 15 cm in front of and almost on level with the large ball. Shine the light on the large ball. Sketch your model, noting the effect of the beam of light.
4. Reverse the positions of the two balls and repeat Step 3. Sketch your model, again noting the effect of the beam of light. You may need to prop up the smaller ball to center the shadow on the larger ball.

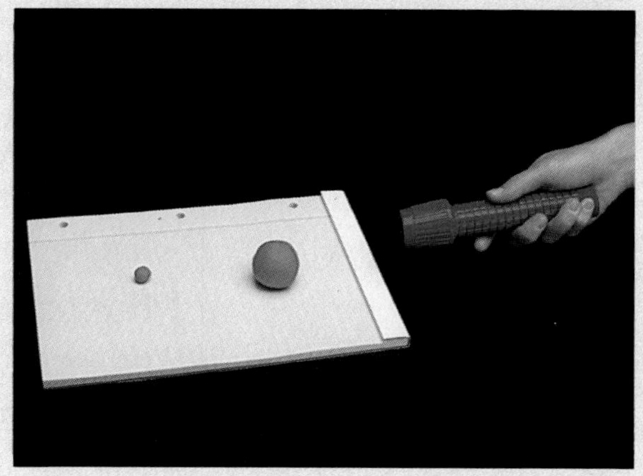

Analysis and Conclusions

1. What planetary body does the large clay ball represent? the small clay ball? the penlight?
2. As viewed from the earth, what event did you represent with your model in Step 3? as viewed from the moon?
3. As viewed from the earth, what event did you represent with your model in Step 4? as viewed from the moon?
4. What are the frequencies of the solar eclipse and the lunar eclipse in your model? How are the frequencies of these eclipses inaccurate? Explain your answer.

revolution around the earth. If this crossing occurs when the moon is between the earth and the sun, a solar eclipse will occur. If this crossing occurs when the earth is between the moon and the sun, a lunar eclipse may occur. Usually, however, no eclipse occurs when the moon crosses the plane of the earth's orbit.

Section 30.2 Review

1. Why do observers on the earth always see the same side of the moon?
2. Why does the moon rise and set about 50 minutes later each successive night?
3. What conditions must be present for a solar eclipse to occur?
4. Why doesn't a lunar eclipse occur every time the moon revolves around the earth in its monthly orbit?

Section Objectives

- **Describe the phases of the moon.**
- **Explain how calendars are based on the movements of the earth and the moon.**

30.3 The Lunar Cycle

On some nights the moon shines brightly enough for you to read a book by its light. But moonlight is not produced by the moon. The moon reflects light from the sun. As the moon revolves around the earth, different parts of the lighted side of the moon face the earth. Thus, as shown in Figure 30–8, the shape of the visible portion of the moon varies. These varying shapes, lighted by reflected sunlight, are called **phases** of the moon.

Phases of the Moon

When the moon is between the sun and the earth, the side of the moon facing the earth is unlighted. At these times the moon is in the **new moon** phase. During the new moon phase there is no lighted area of the moon visible from the earth.

As the moon continues to move in its orbit around the earth, part of its lighted half becomes visible. When the size of the visible portion of the moon is increasing, the moon is **waxing.** When a sliver of the moon becomes visible from the earth, the moon enters the *waxing-crescent* phase.

When the moon has moved through one-quarter of its orbit after the new moon phase, the moon looks like a semicircle. Half of the lighted side of the moon is facing the earth. When a waxing moon becomes a semicircle, it enters the *first-quarter* phase. When the visible portion of the moon is larger than a semicircle and still increasing, the moon is in the *waxing-gibbous* phase.

The moon continues to wax until it appears as a full circle. At **full moon** the earth is between the sun and the moon, as shown in Figure 30–8. Consequently, the entire half of the moon reflecting the light of the sun is visible from the earth.

After the full moon phase, the portion of the moon visible from the earth decreases. The moon is then **waning.** After the full-moon phase, the moon enters the *waning-gibbous* phase. In this phase the visible portion of the moon is still larger than a semicircle, but this portion is decreasing in size. When the visible portion of the moon becomes a semicircle, the moon enters the *last-quarter* phase. When only a sliver of the moon is visible, the moon enters the *waning-crescent* phase.

After the waning-crescent phase, the moon again moves between the earth and the sun, as shown in Figure 30–8. The moon becomes a new moon, and the cycle of phases begins again.

In the crescent phases before and after a new moon, only a small part of the moon shines brightly. However, the rest of the moon is not completely dark. It shines dimly from sunlight that reflects first off the earth's clouds and oceans and then reflects off the moon. Sunlight that is reflected off the earth is called **earthshine.**

Although the moon revolves around the earth in 27.3 days, a longer period of time is needed for it to go through a complete cycle

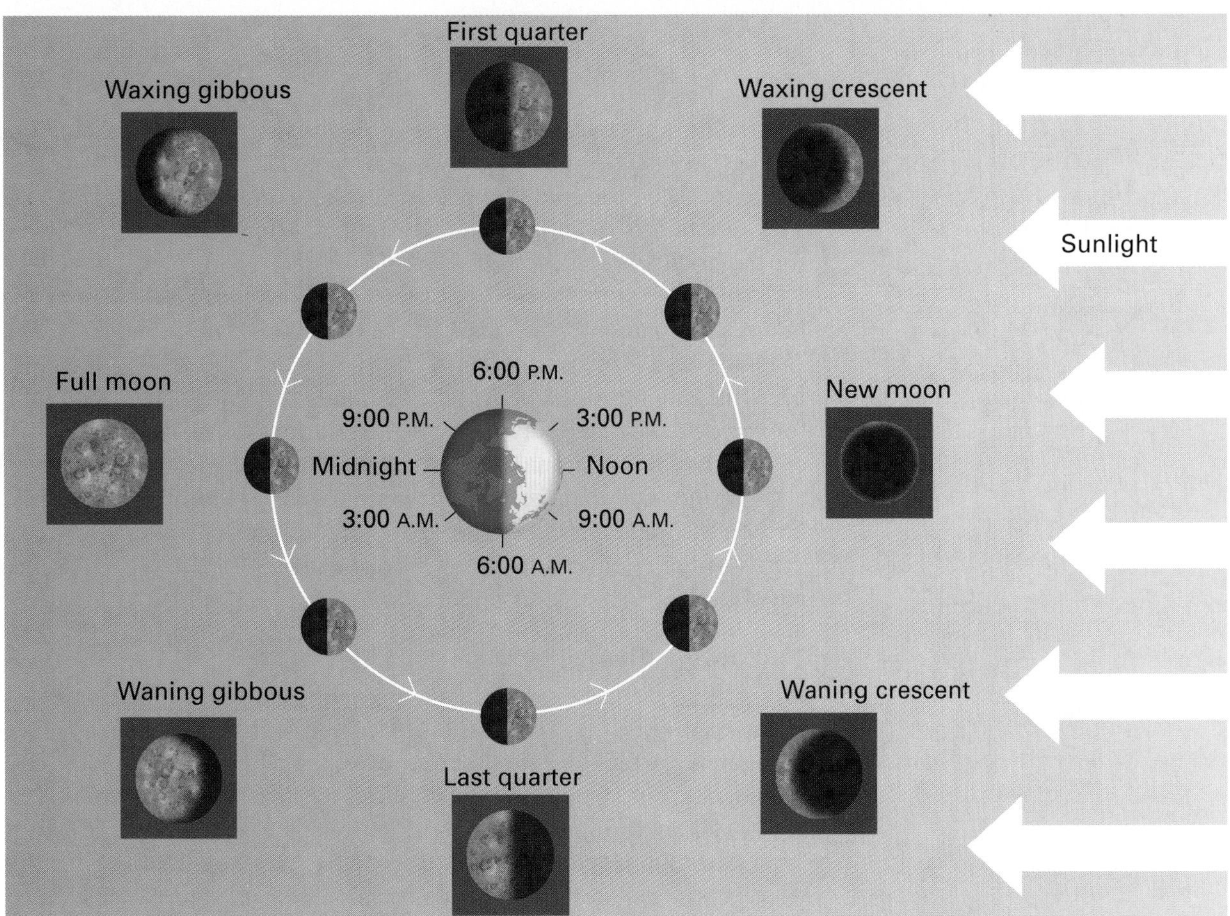

Figure 30–8. The moon goes through phases as the lighted portion of the moon visible from the earth changes during the lunar orbit.

of phases. The period from one new moon to the next is 29.5 days. This 2.2-day difference is due to the orbiting of the earth-moon system around the sun. In the 27.3 days in which the moon orbits the earth, the two bodies move slightly farther along their orbit around the sun. Therefore, the moon must go a little farther to get directly between the earth and the sun. This is the position of the moon in the new moon phase. About 2.2 days are needed for the moon to travel this extra distance.

The Calendar

For a long time, people were able to measure the passage of time by keeping track of the changing phases of the moon. Eventually many societies developed systems of measuring the passage of time called **calendars.**

The three basic units of most calendars—day, month, and year—are determined by the movements of the earth and the moon. A **day** is the time required for the earth to make one rotation on its axis, about 24 hours. A **month** is the time required for the moon to go through one cycle of phases as it orbits the earth, or 29.5 days. A **solar year** is the time required for the earth to make one orbit around the sun, about 365.24 days.

The different lengths of time that are needed for the movements of the earth and the moon cause problems in creating any calendar. First, the earth does not make a whole number of rotations on its axis in the time that it makes one orbit around the sun. A calendar year of 365 days is too short, while a calendar year of 366 days is too long. Therefore the number of days in a calendar year must vary. The variation is to compensate for the extra one-quarter day required for the earth to orbit the sun.

The second problem is that the earth does not make a whole number of rotations on its axis in the time the moon orbits the earth. A month of 29 days is too short, but a month of 30 days is too long. Thus the number of days in a month must vary in order to account for the extra one-half day in a cycle of the phases of the moon.

Third, the moon makes between 12 and 13 orbits around the earth in the time that the earth orbits the sun. Therefore the number of months in every year cannot exactly correspond to actual movements of the earth and its moon. To remain accurate, a calendar must account for these differences.

The Julian Calendar

In 738 B.C. the leader Romulus supposedly introduced the first calendar used in Rome. It consisted of a year of 304 days divided into 10 months. Although later rulers made various modifications in the calendar, it was about 3 months behind the seasons at the time of Julius Caesar's reign. For example, the winter solstice occurred in September. In 46 B.C. Caesar's astronomers revised the calendar into what is now called the **Julian calendar.** They divided the year into 12 months. Eleven of those months had 30 or 31 days. The other month, February, had 29 days. Later, one day was moved from February to August. This left February with only 28 days. Every fourth year was a **leap year,** a year with an extra day in it. During leap years, February had 29 days. Having a leap year was necessary to make the average length of a year 365.25 days.

According to the Julian calendar, which was used for over 1,500 years, the year was 365.25 days long. This, however, was about 11 minutes longer than the actual solar year. Each year the difference between the calendar and the solar year increased.

Figure 30–9. The Aztecs, who lived in what is now Mexico, used a stone calendar, such as the one shown here, to measure the passage of time.

The Gregorian Calendar

By 1580, the calendar was about ten days ahead of the seasons. To correct this discrepancy, Pope Gregory XIII ordered that ten days be dropped from the month of October in 1582. This made up for the error that had built up over previous centuries.

To correct the calendar for the future, the pope ordered a revision of the Julian calendar. The revised system, called the **Gregorian calendar,** is the calendar currently used in most of the world. Compared with the Julian calendar, the Gregorian calendar has three fewer days every 400 years. Leap years normally come in years divisible by 4. However, years ending in *00* that are not divisible by 400 are not leap years. For example, 1700, 1800, and 1900 are not divisible by 400, so none of these years were leap years. The year 1600 ends in *00* and is divisible by 400. Therefore, 1600 was a leap year. The year 2000 is also a leap year.

January

S	M	T	W	T	F	S
1	2	3	4	5	6	7
8	9	10	11	12	13	14
15	16	17	18	19	20	21
22	23	24	25	26	27	28
29	30	31				

February

S	M	T	W	T	F	S
			1	2	3	4
5	6	7	8	9	10	11
12	13	14	15	16	17	18
19	20	21	22	23	24	25
26	27	28	29	30		

March

S	M	T	W	T	F	S
					1	2
3	4	5	6	7	8	9
10	11	12	13	14	15	16
17	18	19	20	21	22	23
24	25	26	27	28	29	30

April

S	M	T	W	T	F	S
1	2	3	4	5	6	7
8	9	10	11	12	13	14
15	16	17	18	19	20	21
22	23	24	25	26	27	28
29	30	31				

May

S	M	T	W	T	F	S
			1	2	3	4
5	6	7	8	9	10	11
12	13	14	15	16	17	18
19	20	21	22	23	24	25
26	27	28	29	30		

June

S	M	T	W	T	F	S	
					1	2	
3	4	5	6	7	8	9	
10	11	12	13	14	15	16	
17	18	19	20	21	22	23	
24	25	26	27	28	29	30	W

July

S	M	T	W	T	F	S
1	2	3	4	5	6	7
8	9	10	11	12	13	14
15	16	17	18	19	20	21
22	23	24	25	26	27	28
29	30	31				

August

S	M	T	W	T	F	S
			1	2	3	4
5	6	7	8	9	10	11
12	13	14	15	16	17	18
19	20	21	22	23	24	25
26	27	28	29	30		

September

S	M	T	W	T	F	S
					1	2
3	4	5	6	7	8	9
10	11	12	13	14	15	16
17	18	19	20	21	22	23
24	25	26	27	28	29	30

October

S	M	T	W	T	F	S
1	2	3	4	5	6	7
8	9	10	11	12	13	14
15	16	17	18	19	20	21
22	23	24	25	26	27	28
29	30	31				

November

S	M	T	W	T	F	S
			1	2	3	4
5	6	7	8	9	10	11
12	13	14	15	16	17	18
19	20	21	22	23	24	25
26	27	28	29	30		

December

S	M	T	W	T	F	S	
					1	2	
3	4	5	6	7	8	9	
10	11	12	13	14	15	16	
17	18	19	20	21	22	23	
24	25	26	27	28	29	30	W

Figure 30–10. In the World calendar, a world day follows December 30. Every four years an additional world day follows June 30.

Compared with previous calendars, the Gregorian calendar is so accurate that the calendar year is only 26 seconds longer than the solar year. As a result, the current Gregorian calendar accumulates an error of less than one day over a period of 3,000 years.

Current Calendar Reform

Many reforms have been proposed to simplify the current calendar. Some of these recommend that all months and years begin on the same day of the week. They also try to give all months nearly the same number of days.

One proposed calendar is the called the *Thirteen-Month calendar.* Each of the thirteen months would contain four weeks. A new month, called *Sol,* would come between June and July. One day, called a *year day,* belonging to no week or month, would fall at the end of each year. Every four years, a leap-year day would be added at the end of Sol.

Another proposal is the World calendar. This calendar would have 12 months of either 30 or 31 days. A world day would be placed at the end of each year. A leap-year day would be added at the end of June every four years.

Section 30.3 Review

1. Describe the relative locations of the sun, the earth, and the moon during a new moon.
2. To an observer on the earth, how does the appearance of the moon change when it is waxing?
3. List three problems in devising any calendar based on the movements of the earth and the moon.
4. How many fewer days are in the Gregorian calendar than in the Julian calendar?
5. If the earth were to orbit the sun in 365.20 days, how often would a leap year be necessary?

Section Objectives

- **Compare the characteristics of the two moons of Mars.**
- **Compare the Galilean moons and the rings of Jupiter with the moons and rings of the other outer planets.**

30.4 Satellites of Other Planets

Until the 1600's, astronomers thought that the earth was the only planet with a moon. In 1610, however, Galileo discovered moons orbiting Jupiter. He also noticed what later were discovered to be the rings of Saturn. Since the time of Galileo, astronomers have found that all planets except Mercury and Venus have moons. In addition, Saturn, Jupiter, Uranus, and Neptune have rings.

Moons of Mars

Mars has two moons, Phobos and Deimos. They revolve around Mars quite rapidly. Phobos completes its orbit in 7 hours and 40 minutes. Deimos completes its orbit in about 30 hours. The two moons of Mars may be the remains of a single moon that collided with a large meteoroid.

The moons of Mars differ in shape and in size from the earth's moon. Phobos and Deimos are irregularly shaped chunks of rock, whereas the earth's moon is spherical. Phobos is only 27 km across at its longest place and 19 km at its shortest. Deimos is about 15 km across at its longest place and 11 km at its shortest. Either moon could fit within a medium-size crater on the earth's moon.

The surfaces of Phobos and Deimos are dark like the maria on the earth's moon. Both of these moons have many craters. Phobos has one crater that is 8 km wide. This crater covers a large portion of the entire surface of Phobos. The large number of craters suggests that these moons have been hit by many meteorites and asteroids and are therefore fairly old.

Figure 30–11. Phobos, one of the two moons of Mars, is small and irregularly shaped.

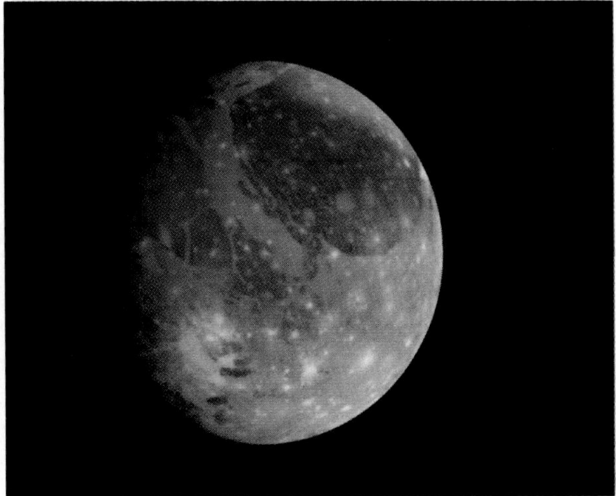

Ring and Moons of Jupiter

Jupiter has both a ring and many moons. The ring is about 6,000 km wide and less than 30 km thick. It is made up of very small particles of dark rock. Jupiter also has at least 16 moons. Most of the moons of Jupiter are very small. All except 4 moons are less than 200 km in diameter. The 4 largest satellites of Jupiter are called the **Galilean moons,** because they were first seen by Galileo in 1610.

Io, the Galilean moon closest to Jupiter, is about the size of the earth's moon. It is the only satellite in the solar system known to have active volcanoes. In addition, Io has lava flows, volcanic sulfur deposits, and large volcanic craters. The volcanoes on Io erupt with rocks and gases containing sulfur compounds. Thus, astronomers infer that the atmosphere of Io is composed of sulfur dioxide. The presence of these compounds and its volcanic activity make Io one of the most colorful moons in the solar system. Its colors include brilliant yellow, orange, and red. The surface of Io is covered with layers of sulfur and frozen sulfur dioxide. The interior of Io may still be molten rock. Data from the *Galileo* space probe indicate that Io has a giant iron core and may possess its own magnetic field.

Europa is the second closest Galilean moon to Jupiter. It is also about the size of the earth's moon, but it is less dense. Astronomers think that Europa has a rock core covered with a crust of ice about 100 km thick. An ocean of water may exist under this blanket of ice. If so, simple forms of life, similar to those in Antarctica, could exist there. Astronomers, however, have no evidence that life actually does exist on Europa.

Ganymede is the third Galilean moon from Jupiter. It is the largest moon in the solar system, larger even than the planet Mercury. However, the density of Ganymede is very low. This moon is probably composed of ice and rock. Ganymede has dark, crater-filled areas, but it also has lighter areas. These light areas show marks that seem to be long ridges and valleys. In 1996, *Galileo* provided evidence supporting the existence of a magnetic field around Ganymede, a magnetic field within Jupiter's own powerful magnetic field.

Figure 30–12. The four Galilean moons of Jupiter are Io, Europa, Ganymede, and Callisto. They are the largest moons of Jupiter. Io, shown left, is the only satellite in the solar system known to have active volcanoes. Ganymede, shown right, is the largest moon in the solar system.

INVESTIGATE!

To learn more about Jupiter's moons, try the In-Depth Investigation on pages 636–637.

Callisto is the farthest of the four Galilean moons from Jupiter. Callisto is similar to Ganymede in size, density, and composition. However, it has a much rougher surface than Ganymede does. Callisto may be one of the most densely cratered moons in the solar system.

Rings and Moons of Saturn

The rings of Saturn are much larger and brighter than those of Jupiter, Uranus, or Neptune. Each of the several rings circling Saturn is divided into hundreds of smaller ringlets. The ringlets are probably composed of billions of pieces of rock and ice. The pieces range in size from particles the size of dust to chunks the size of a house. Each piece follows its own individual orbit around Saturn. Altogether, the system of rings is about 67,000 km wide and less than a hundred meters thick.

Astronomers think that the rings of Saturn formed in one of two ways. One hypothesis is that the rings formed when a body orbiting Saturn was torn apart by the gravity of the planet. A second hypothesis is that the material in the ring was unable to condense into a moon while the other moons around Saturn were forming. In each case, astronomers are unsure whether the bits of matter in the rings are the fragments of a destroyed body or simply loose material.

In addition to the billions of pieces of material in the rings, Saturn has at least 20 moons. Most of them are small, icy bodies with many craters. However, 5 of the moons of Saturn are fairly large. One, called Titan, has a diameter of more than 5,000 km. Unlike most of the moons of Saturn, Titan has a thick atmosphere that is composed mainly of nitrogen. The atmosphere is so thick that it conceals the surface of the moon.

Figure 30–13. In addition to its rings, Saturn has at least 20 moons. Titan (top right) is Saturn's largest moon and the only one with a substantial atmosphere. The moons Mimas, Tethys, Dione, Enceladus, and Rhea (clockwise from bottom right) are also shown in this mosaic taken from several photographs.

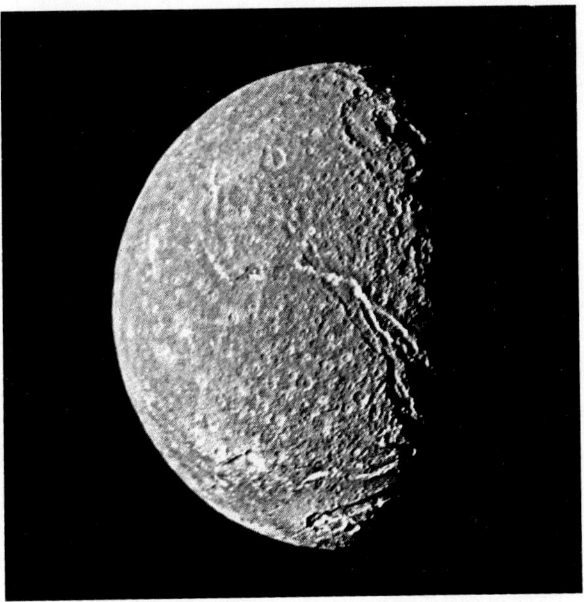

Figure 30–14. Ariel (left) and Titania (right) are two of the five larger moons that orbit Uranus along with 10 smaller moons.

Satellites of Uranus, Neptune, and Pluto

Uranus has at least 15 small moons. The 4 largest, Oberon, Titania, Umbriel, and Ariel, were known by the mid-1800's. A fifth, Miranda, was found in 1948. The other 10 moons have been detected since 1985 by the U.S. space probe *Voyager 2*. Uranus also has at least 11 small rings. These rings are composed of small pieces of black rock. The inner rings are very narrow, varying from 0.5 km to 12 km wide. The outermost ring is the largest. It is between 10 km and 100 km wide.

Neptune has eight known moons, among which are Triton and Nereid. Triton, an icy moon, is unusual because it travels from east to west as it revolves around Neptune. Most planets and moons orbit from west to east. Some astronomers think that Triton, because of its unusual orbit, is not an original moon of Neptune. Triton was probably captured by the gravity of Neptune after forming elsewhere. The diameter of Triton, 4,800 km, is about 20 times larger than that of Nereid. Nereid orbits Neptune in the same direction that Neptune rotates. Neptune also has a thin set of rings.

Pluto, which is the smallest planet, has one moon. The moon, named Charon, was discovered in 1978. Charon is about half as large as Pluto and completes its orbit around Pluto in 6.4 days, the same length as a day on Pluto. Consequently, Charon is always above the same spot on Pluto.

Section 30.4 Review

1. Describe the size and shape of Phobos and Deimos.
2. List the four moons of Jupiter noticed by Galileo.
3. Which four planets have rings?
4. How does the number of craters on Phobos and Deimos suggest that these moons are very old?

Chapter 30 Review

Key Terms

- apogee (621)
- calendar (627)
- crater (616)
- day (627)
- earthshine (626)
- eclipse (623)
- full moon (626)
- Galilean moon (631)
- Gregorian calendar (628)
- Julian calendar (628)
- leap year (628)
- lunar eclipse (624)
- maria (616)
- month (627)
- new moon (626)
- penumbra (623)
- perigee (621)
- phase (626)
- ray (616)
- rille (616)
- solar eclipse (623)
- solar year (627)
- umbra (623)
- waning (626)
- waxing (626)

Key Concepts

The lunar surface features have changed little since they formed. **See page 617.**

Astronomers have explored the structure of the interior of the moon using instruments that measure seismic waves. **See page 618.**

After the moon formed, it separated into layers and cooled. Eventually it became geologically inactive. **See page 620.**

The moon spins on its axis once during each orbit of the earth. **See page 621.**

Eclipses occur when one planetary body passes through the shadow of another. **See page 623.**

As the moon orbits the earth, the size of the lighted portion facing the earth increases and decreases. **See page 626.**

Calendars are based on the rotation of the earth and on the revolutions of the earth around the sun, and the moon around the earth. **See page 627.**

Phobos and Deimos, the moons of Mars, differ in shape and size from the earth's moon. **See page 630.**

The Galilean moons are the four largest satellites of Jupiter. Jupiter, Saturn, Uranus, and Neptune all have rings. **See page 631.**

Review

On your own paper, write the letter of the term that best completes each of the following statements.

1. Dark areas on the moon that are smooth and reflect little light are called
   a. rilles. b. maria. c. rays. d. breccia.
2. Most of the information astronomers have gathered about the interior of the moon has come from
   a. telescopes. b. satellites.
   c. spectrographs. d. seismographs.
3. Soon after the moon formed, it was covered with
   a. water. b. anorthosites.
   c. frozen hydrogen. d. molten rock.
4. In the most recent stage in the development of the moon,
   a. the densest material sank to the core.
   b. the crust began to break.
   c. the earth's gravity captured the moon.
   d. the number of meteorites hitting the moon decreased.

5. The moon is closest to the earth at
 a. new moon. b. full moon. c. perigee. d. apogee.

6. During each orbit around the earth, the moon spins on its axis
 a. 1 time. b. about 27 times.
 c. about 29 times. d. 365 times.

7. In a lunar eclipse, the moon
 a. casts a shadow on the earth.
 b. is in the earth's shadow.
 c. is between the earth and the sun.
 d. blocks part of the sun from view.

8. When the size of the visible portion of the moon is decreasing, the moon is
 a. full. b. annular. c. waxing. d. waning.

9. In the crescent phases, the entire moon shines dimly because of
 a. light produced by the earth.
 b. sunlight reflected off the earth.
 c. hydrogen fusion in the core of the moon.
 d. energy produced by the rotation of the moon.

10. The two moons of Mars are
 a. Io and Europa. b. Titan and Charon.
 c. Phobos and Deimos. d. Triton and Nereid.

11. Compared with the other moons of Jupiter, the four Galilean moons are
 a. larger. b. farther from Jupiter.
 c. denser. d. younger.

12. The rings of Saturn are probably composed of
 a. regolith.
 b. small pieces of black rock.
 c. several hundred small moons.
 d. billions of pieces of ice and rock.

Critical Thinking

On your own paper, write answers to the following questions.

1. How would the craters on the moon be different today if the moon had developed an atmosphere that had wind and contained water?
2. If meteorites had stopped hitting the moon before the outer surface of the moon cooled, why would the maria not have developed?
3. Suppose that the moon spun twice on its axis during each orbit around the earth. How would study of the moon from the earth be easier?
4. Venus does not cause a solar eclipse even though it passes between the earth and the sun. Instead, it appears as a black dot moving across the face of sun. Explain why this happens.
5. Would a satellite orbiting the earth go through phases like those of the moon. Explain your answer.

Application

1. Your friend who just got a new telescope wants to look at Venus during its fullest phase. Explain why that is impossible.
2. Imagine a planet that orbits the sun in 100 days and has one moon. The moon goes through a complete set of phases in 20 days. On this planet, how many months would one year have?
3. Concept map ? ? ? Using any of the new terms listed on the previous page, construct a **concept map** starting with the term "Time." See how many terms you can include.

Extension

1. Do research on how one of the following ancient or traditional cultures explained solar eclipses: Chinese, Persian, Greek, Roman, Mayan, Sioux, Aboriginal, or Tahitian. Present your findings to the class.
2. Imagine that astronomers have discovered a planet that spins on its axis in 20 hours and orbits the sun in 245 days and 4 hours. The imaginary planet has one moon that goes through a complete set of phases in 24 days and 10 hours. Design a calendar for this planet. Indicate how many months are in a year, how many days are in each month, and how often leap years occur.
3. Develop a display, based on the latest research, about one of the moons of Jupiter, Saturn, Uranus, or Neptune. Include both text and images.

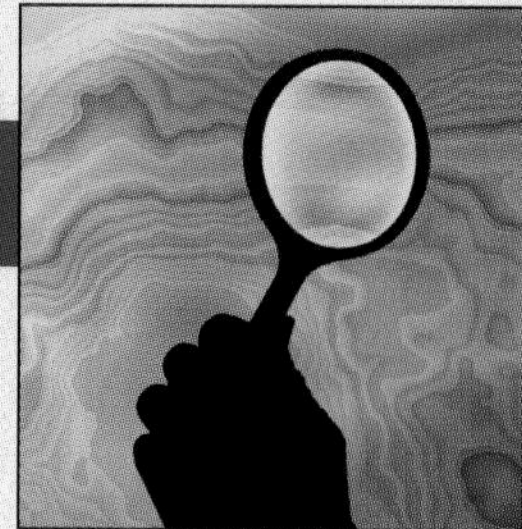

IN-DEPTH INVESTIGATION

Galilean Moons of Jupiter

Materials

- metric ruler
- calculator

Introduction

A German astronomer named Johannes Kepler developed three laws that explained most aspects of planetary motion. The third law—the law of periods—explained the relation between a planet's distance from the sun and the planet's period of orbit. The orbit period is the time required for the planet to make a complete revolution around the sun. According to the law of periods, the cube of the average distance of the planet from the sun is proportional to the square of the planet's period. It can be expressed mathematically as $K \times r^3 = p^2$, where r is the distance from the sun, p is the period, and K is a constant. Kepler's third law also may be applied to moons orbiting a planet, where r is the distance of a moon to the planet and p is the moon's period, or the time required to make one revolution around the planet.

In this investigation, you will verify that the orbital motions of Jupiter's moons obey Kepler's third law.

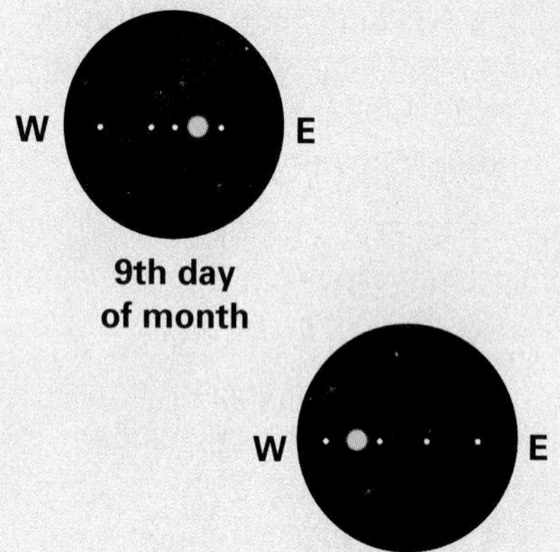

Prelab Preparation

1. Review Section 30.4, pages 631–632. Also review Section 29.1, page 593.
2. Find out what is meant by a constant and a variable.

Procedure

1. Two telescope eyepiece views at the left show how Jupiter and its four largest, or Galilean, moons appear through a telescope on the earth at midnight on the 9th and 19th of a month. Compare these illustrations with the chart below, which shows the path of each moon as it orbits Jupiter during the same month. The central horizontal band on the

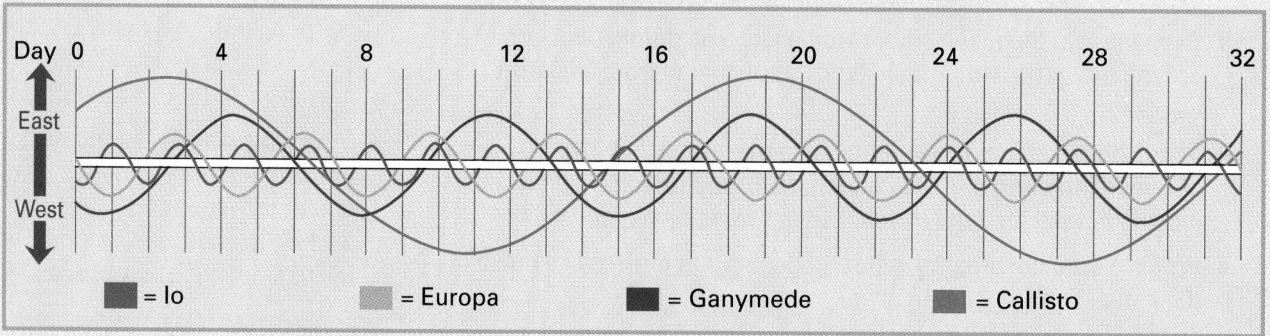

chart represents Jupiter. When a moon's path crosses in front of this band, the moon is in front of the planet. When a moon's path crosses behind this band, the moon is behind Jupiter.

a. List the days when each of Jupiter's moons crosses in front of the planet.

b. List the days when each of the moons is behind Jupiter.

2. Use the data in the table below to test Kepler's third law. Calculate p^2 and r^3 for each of the planets. Record your results in a table of your own. Then calculate K for each planet using Kepler's third law, $K = p^2/r^3$. Record your results in your table. Is K a constant?

Planet	r (in billions of kms)	p (in earth years)	r^3	p^2	K
Mercury	0.058	0.24			
Venus	0.108	0.62			
Earth	0.150	1			
Mars	0.228	1.88			
Jupiter	0.778	11.86			
Saturn	1.427	29.46			
Uranus	2.869	84.01			
Neptune	4.486	164.8			
Pluto	5.890	247.7			

3. Draw Jupiter and its moons as they would appear from the earth at midnight on the 2nd and 26th of the month.

4. Draw Jupiter's moons on the first day of the month that all four moons are on the same side of the planet. Give the date.

5. Give a date when only two moons will be visible. Name the two visible moons.

6. Follow each moon's motion on the chart. Find the length of time, in earth days, required for each moon to orbit Jupiter. To do this, measure the time between two points when the moon is in exactly the same position on the same side of Jupiter. Record your answers in a table listing the moons, r (in earth days), p (in cm), r^3, p^2, and K.

7. Measure the scale distance between the maximum outward swing of each moon and the center of Jupiter in centimeters. Record your answers in your table.

8. Square each period measurement and record the answer in your table. Cube each distance measurement and record your answer.

9. Use your results to test Kepler's third law. Because $K = p^2/r^3$, divide p^2 by r^3 for each moon to find K. Record your results in your table.

Analysis and Conclusions

1. Will you see all four of Jupiter's largest moons each time you look at Jupiter through a telescope or binoculars? Explain.

2. Jupiter's moons look like dots in a telescope. You cannot tell them apart by their appearance. If you had no charts, how could you identify each moon?

3. After you solve for K for each moon, study your results. Is K a constant?

Stars In Your Eyes

Humans have been observing stars and other celestial bodies for thousands of years. The star charts that appear on page 561 are the result of those observations. Refer to those charts, as well as the ones that appear on page 670, as you answer the following questions.

1. Explain why star charts have the east and west directions reversed. (Hint: Imagine lying flat on your back so that your feet are pointing south. Would the sun set over your right hand or over your left hand?)

2. If you wanted to hold the book upright in front of you and compare the star charts to the stars above your head, which direction should you face?

3. Why are the star charts different for the different seasons?

4. For many years, the best way to navigate at night was by using the stars as a guide. Many people used Polaris, the North Star, to orient themselves. This would not have worked for people all over the world, however. Why not?

Look at the star charts on the next page. Chart 1 is similar to the bottom half of the summer constellation chart on page 561. Chart 2 is the same as Chart 1, except that the constellation markings have been left out.

5. Name three constellations that appear in Chart 1 but not on page 561.

6. What constellation is just to the west of Lyra?

7. The stars could have been grouped into constellations differently, and in other cultures they have been. Using Chart 2, pick a group of stars, and sketch them onto a separate piece of paper. Connect them into a new constellation, name your constellation, and explain how you chose the name.

Going Further

To keep informed about current astronomical events, try visiting the Henry Buhl, Jr. Planetarium and Observatory on-line at www.csc.clpgh.org/csc_pages/homepage/planet/planet.htm.

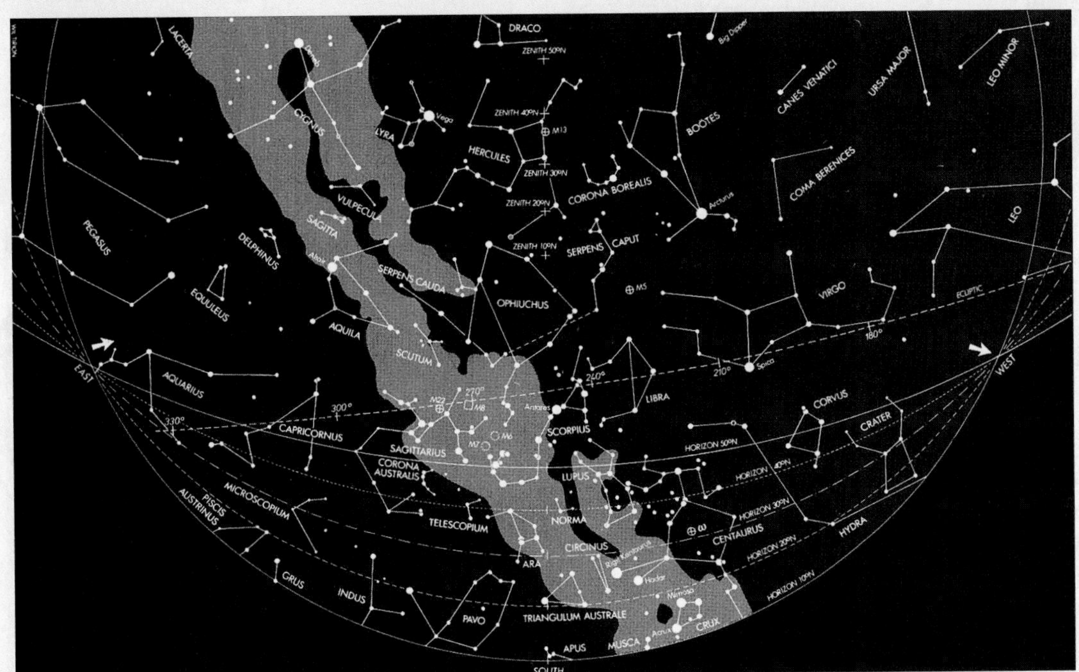

Chart 1

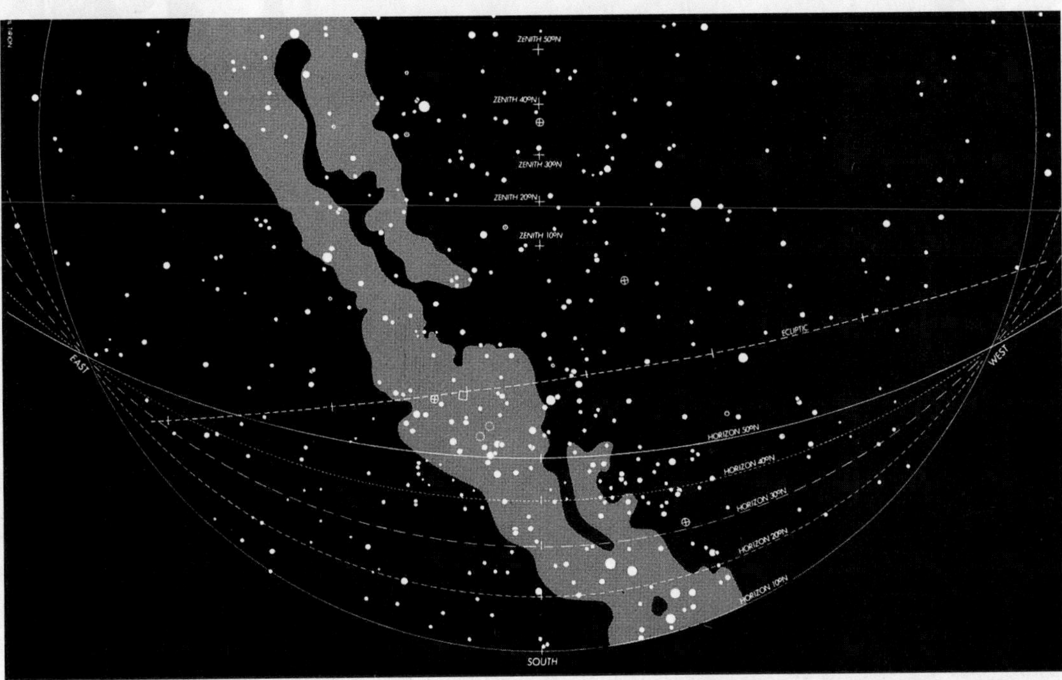

Chart 2

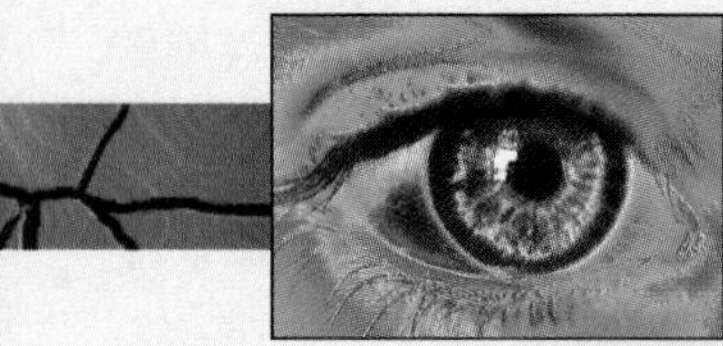

EYE ON THE ENVIRONMENT

The Great Junkyard in the Sky

Take a moment to think about the following question: If you were riding in a spaceship, quietly floating in orbit around the earth, what would you see? Does your answer include the moon, the earth's surface, and millions of stars? Perhaps you thought of a comet or another planet in our solar system. Well, how about chunks of orbiting space trash?

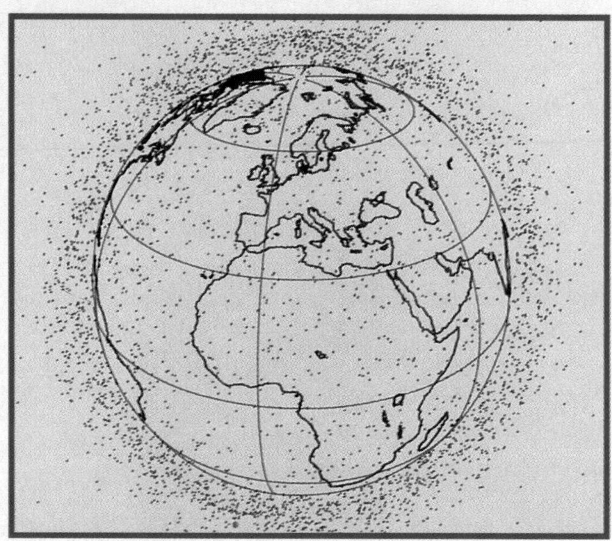

▲ **This diagram shows all of the objects known to be orbiting the earth a decade ago. Today, many more are known.**

Space Junk

For more than 40 years, people have been launching various types of satellites into orbit. Many of those satellites worked perfectly for as long as they were needed, while some malfunctioned almost immediately. Although they have varied drastically in size, shape, and function, almost all of those satellites had one thing in common—once they began orbiting the earth, they stayed in orbit. In fact, many of them are still out there. That has created a problem that continues to grow over time.

Studies of orbiting space trash have revealed some staggering facts. As early as the 1970's, the United States Space Command had records of about 7,000 orbiting objects that were larger than a softball. At least twice that many were discovered when astronomers looked for them using ground-based telescopes. As these studies continued in the years that followed, astronomers were able to pinpoint hundreds of thousands of objects measuring 1–2 cm in diameter. While that may seem small, these objects move at average speeds of more than 12,000 km/hr! (Consider that most bullets travel at less than 3,000 km/hr.)

A Planetary Problem

Today, astronomers are aware that millions of objects orbit the earth. Most of them are tiny, such as flecks of paint, but others are large enough to be seen from earth with a normal telescope. When these objects collide, the pieces may break apart and continue in orbit. In some cases, abandoned rockets explode when their leftover fuel ignites. Over time, more and more bits of space junk are created.

This orbiting space junk presents a big problem when we want to launch a new satellite. The satellite's path must be carefully planned so that the satellite does not collide with any orbiting debris. Such a collision could easily destroy the satellite, resulting in a huge loss of time and money. The risks are even higher for space travel because human lives are at stake. Thankfully, the planning has been successful. Many astronomers are worried, however, that it is only a matter of time before a major disaster occurs.

Unless obsolete equipment is reclaimed or burns up in the atmosphere, it remains in orbit as space junk. ▼

▲ **While most space junk is found in orbit around the earth, some is also found on the moon, where equipment, such as this lunar rover, was abandoned after the Apollo missions were completed.**

Reducing the Risks

Scientists are working on ways to reduce the risks posed by existing space trash. For one thing, scientists are finding better ways to keep an eye on the orbiting debris. For example, astronomers are using orbiting telescopes to pinpoint the trash and to provide other important data. Better data about the orbiting debris makes planning the paths of functional satellites much easier.

Scientists are also making sure that new satellites are "disposed of" when their jobs are done. This is usually accomplished by using one of two methods. In the first method, the satellites are moved into higher orbits when they are no longer functional. Since few satellites operate in these higher orbits, the risk of future collisions is reduced. In the second method, the satellites are moved closer in. When the satellites begin to enter the upper atmosphere, they are incinerated as a result of friction. As long as they burn up before they reach the earth's surface, they pose little risk to people or other satellites.

Think About It

Why would orbiting telescopes work better to track orbiting debris than ground-based telescopes?

Extension: Orbiting telescopes are used for a variety of purposes. Find out what they are used for, and report your findings to the class.

Long-Range Investigations

Scientific investigations that lead to important discoveries are almost never short-term. Usually these investigations last months, years, and even decades before results are considered complete and dependable. Investigations in earth science are no exception. The long-range investigations included in this section will give you practical experience investigating earth science the way earth scientists do—over extended periods of time. You will observe changes over time, keep detailed records of your observations, and then draw conclusions from your data. By following these steps, you will learn firsthand what it is like to be an earth scientist.

Safety First!

Many of the long-range investigations require you to make field trips to an observation site or to conduct your activities outdoors. Advance planning is essential. You should plan carefully for these investigations and be certain that you are aware of the safety guidelines that must be followed. The following are general guidelines for fieldwork.

While conducting fieldwork,

find out about on-site hazards before setting out. Determine whether there are likely to be poisonous plants or dangerous animals where you are going, and know how to identify them. Also, find out about other hazards, such as steep or slippery terrain.

wear protective clothing. Dress in a manner that will keep you warm, comfortable, and dry. Wear sunglasses, a hat, gloves, rain gear, etc., to suit local weather conditions. Wear waterproof shoes if you will be near water.

do not approach or touch wild animals unless you have permission from your teacher. Avoid animals that may sting, bite, scratch, or otherwise cause injury.

do not touch wild plants or pick wildflowers without permission from your teacher. Many wild plants can be irritating or toxic. Never taste any wild plant.

do not wander away from others. Do not go beyond where you can be seen or heard. Travel with a partner at all times.

report any hazards or accidents to your teacher immediately. Even if an incident seems unimportant, let your teacher know.

consider the safety of the ecosystem you will be visiting as well as your own safety. Do not remove anything from a field site without your teacher's permission. Stay on trails when possible to avoid trampling delicate vegetation. Never leave garbage behind at a field site. Strive to leave natural areas just as you found them.

For additional information about safety in the lab and in the field, refer to page xviii in the front of this book. Don't take any chances with safety!

LONG-RANGE INVESTIGATION 1

Positions of Sunrise and Sunset

Introduction

You are probably aware that the sun rises in the east and sets in the west each day. What may not be obvious to you is that the sun rises and sets at a different position along the horizon each day, following a specific pattern. As the position of sunrise and sunset changes, the amount of sunlight that an area receives also changes. In this investigation, you will observe the changes in the sun's position along the horizon at sunrise and sunset.

Materials

- bearing chart
- glue
- pen
- hand drill
- magnetic compass
- scissors
- graph paper
- poster board
- wooden tongue depressor
- paper fastener
- paper

Suggested Schedule

You will be making two observations on or near the 21st of each month for approximately 8 or 9 months (depending on school year).

Advance Preparation

Construct the bearing chart before taking any measurements. Copy the bearing chart from page 676 onto your own piece of paper and glue the copy to a piece of poster board. When the glue is dry, trim the excess poster board away. Take the tongue depressor and drill a 1/4-in. hole in its center. **CAUTION:** *If you are using hand or power tools, it is best to allow your parents or some other adult skilled in the tool's operation to help you.* Place the tongue depressor in the center of the bearing chart and fasten the two together with a paper fastener, as shown in the figure at right.

Procedure

1. On a cloudless morning just before sunrise, set the bearing chart on a level place where no buildings or trees can block your view of sunrise and sunset. **CAUTION:** *Although sunlight is less intense at sunrise and sunset, you should never stare at the sun for extended periods of time.*
2. Using the magnetic compass, determine the direction of north. Set the bearing chart so that 0° is pointing north and 180° is pointing south. *Note: Have the chart face the same direction for every observation, even months from now.* Try to align an

edge of the chart with some permanent object near your observation point.

3. Once the bearing chart is in place and properly aligned, measure the direction of sunrise. *Note: Sunrise occurs at the instant the top of the sun appears on the horizon.* Without looking at the sun, point the tongue depressor toward the sun. This can be done by using the shadow of the edge of the stick. Record the number of degrees in your lab notebook in a table like the one shown below. That evening, measure the direction of sunset, recording the number of degrees on your data table. *Note: Sunset occurs when the top edge of the sun drops below the horizon.*

4. Repeat Step 3 for each month's observation.

Date	Position of Sunrise (in degrees)	Position of Sunset (in degrees)

5. On a sheet of graph paper, prepare a graph similar to the one shown above. On the y-axis, plot the position of the sun at sunrise in degrees from 230° to 310° with 270° in the center of the graph. Use the x-axis for each date of observation. Prepare a second graph to plot the position of the sunset for each date in degrees, this time from 60° to 120° with 90° in the middle.

6. On each graph, connect the points with a smooth line. You might have to estimate the points within each month.

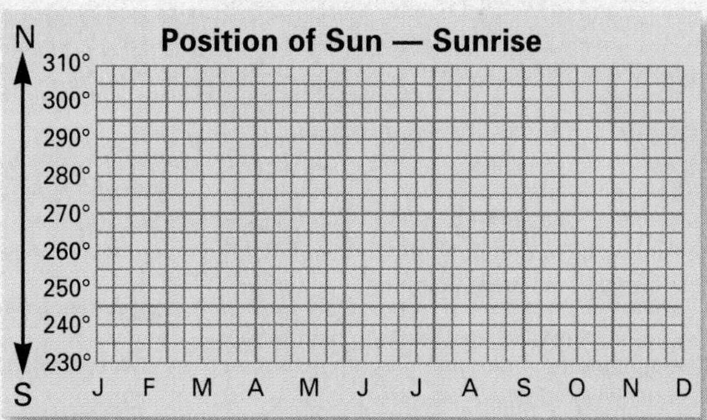

Note: On the equinoxes, the sun will rise due east and set due west. On the solstices, the sunrise and sunset will be shifted from these directions. The amount of the maximum shift will depend on the observer's latitude.

Analysis and Conclusions

Using the two graphs you have completed, answer the following questions.

1. On which date does the sun (a) rise the greatest number of degrees north of east? (b) set the greatest number of degrees north of west?

2. On which date does the sun (a) rise the greatest number of degrees south of east? (b) set the greatest number of degrees south of west?

3. On which date(s) does the sun (a) rise approximately 90° due east? (b) set approximately 270° due west?

4. In which season is (a) the sunrise north of east? (b) the sunset north of west?

5. What general statement can be made about the pattern of sunrise and sunset as shown by the graphs?

6. What causes the change in position of the sun as viewed from Earth?

LONG-RANGE INVESTIGATION 2

Water Clarity

Introduction

Submerged objects can be seen at a greater depth in clear, clean water than in water with a high concentration of suspended particles. The amount of suspended particles in the water, such as sand or silt, is referred to as* turbidity*. Several factors can cause a change in the turbidity of a body of water, such as a lake, bay, or river. These factors may include pollutants, the amount of runoff, turbulence in the water, and recent storms. If an object is lowered into water until it cannot be seen, the depth at which the object is not visible can be used in an indirect way to measure the turbidity of the water. A Secchi disc is a black-and-white disc that can be used to test water clarity. In this investigation, you will discover how changes in water clarity relate to changes in the environment.

Materials

- nylon cord
  1 10-m length
  3 1-m lengths
- 2 rolls of colored plastic tape, different colors
- 3 lead sinkers (100–200 g) or other weights
- meter stick
- sandpaper
- Secchi disc or 1/4-in. plywood or tempered hardwood disk
- paint
- graph paper

Suggested Schedule

Plan to make two or three measurements per week for three to four weeks. You will spend about 15 minutes at the test site each time. You and your partner should prepare a schedule listing every date on which you plan to make observations. Making your observations within a day or two of the approved schedule will be satisfactory.

Advance Preparation 

1. Locate a sturdy pier where the disc can be lowered into relatively deep water. **CAUTION:** *You should not go to the pier alone. Always work with your partner or an adult. Wear a life jacket while working near water.*
2. A Secchi disc can be made from 1/4-in. plywood or 1/4-in. tempered hardwood. Refer to the diagram as you construct your Secchi disc. Cut a circle of wood about 20 cm in diameter. **CAUTION:** *If you are using hand or power tools, it is best to allow your parents or some other adult skilled in the tool's operation to*

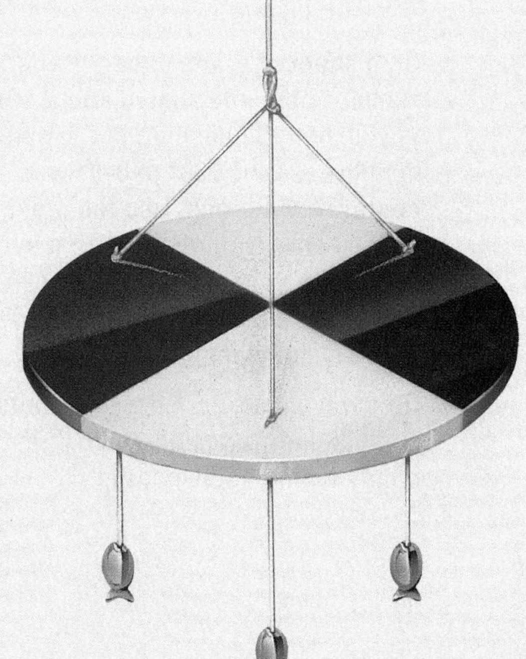

help you. Sand the edges and give the disc two or three coats of white enamel paint. Divide the disc into equal quarters, as shown in the illustration. Paint two opposite quarters black. Drill three 3/16-in. holes in the disc 120° apart and 2 cm from the edge. Tie a fixed loop at the end of the 10-m length of nylon cord. Tie each of the three 1-m cords to the loop. Tie knots on each 1-m cord at equal distances from the loop. From the painted side of the disc, thread the 1-m cords up to the knot. Tie the loose end of each 1-m cord to a lead sinker. In each cord, tie a knot very close to the bottom of the disc so that the disc is held tightly between both knots. Adjust the knots so that the disc hangs perpendicular to the 10-m cord when suspended.

3. To calibrate the 10-m cord, measure 1 m from the disc and wrap a 5-cm piece of colored tape around the cord at that point. Place similar pieces of tape 2, 3, 4, 5, and 6 m from the disc. Using the second color of tape, mark the cord at 20-cm intervals between the meter marks.

Procedure

1. Go to the end of the pier, and select a spot where the disc will not be blocked by the glare of the sun's reflection on the water's surface. *Note: The water must be relatively calm for the disc to remain in a horizontal position.*
2. Slowly lower the disc into the water. As you lower the disc, keep count of the 1-m and 20-cm marks. When the disc can no longer be seen, record the depth. Slowly raise the disc until it just becomes visible again. Record this depth, and then calculate the average depth for this trial. Record your results in a table of your own with columns for the date, trial number, and average depth.
3. Raise the disc about 1 m and repeat Step 2 twice.
4. Calculate the average of the three trials and record that in your table as well.
5. Record observable factors that may change the turbidity of the water, such as wind speed and direction, wave height, tide position (whether it is low or high, flooding or ebbing), a rainstorm the previous day, and so on.
6. Repeat Steps 1–5 on each of your scheduled days.
7. When all your observations are complete, prepare a line graph of your results. Let the horizontal axis of the graph represent the date of each observation and let the vertical axis represent the average depth of visibility in meters. Plot the average depth for each date of your observations.

Analysis and Conclusions

1. Does the graph rise and fall regularly?
2. Does there seem to be a pattern to your graph?
3. If your graph shows a pattern, what is the amount of time from one high or low point to the next high or low point?
4. Do factors such as rain, winds, or tides affect water clarity? Hypothesize why these factors may change the water turbidity. Explain your answer.
5. As the depth at which the disc disappears increases, what can be stated about the amount of suspended particles in the water?
6. On which day was the water turbidity the highest? Hypothesize why the water turbidity was so high. Explain your answer.
7. What conclusion can you make about water clarity at your site?

LONG-RANGE INVESTIGATION 3

Tides at the Shoreline

Introduction

Most people are aware that there are usually two high and two low tides each day. However, most people are unaware of several other aspects associated with tides. The time between consecutive high tides or low tides, called the* tidal period, *is about $12\frac{1}{2}$ hours. The difference in the height of the water between high and low tide is the* tidal range. *You may have heard of the very large tidal range associated with several places on earth, such as the Bay of Fundy in Nova Scotia.

In this investigation, you will measure the tidal range for your location. You also will observe whether the tidal range changes in some predictable way.

Materials

- string or fishing line (7 m long)
- lead sinker (100–200 g) or some other weight
- paper clip
- meter stick
- graph paper

Suggested Schedule

You will be making measurements on or near the dates of the new, first-quarter, full, and third-quarter moons of each month for two months. Plan to make your measurements within a day of each of these dates. The best time to make your measurements is on a Saturday or Sunday, when you are less likely to be rushed and can take careful measurements. You will need to visit your site two times each scheduled day. Once you learn the measurement technique, the amount of time at the site will be about 15 minutes.

Advance Preparation

1. Read Section 22.3, pages 439–443.
2. Consult a calendar, your newspaper, or an almanac to determine the dates of the new, first-quarter, full, and third-quarter moons over a two-month period.
3. Find the times of the high and low tides on the dates of each lunar phase in Step 2 by consulting a tide table or your local newspaper. Depending on the time, make measurements at high tide and at the following low tide, or measure at low tide and at the following high tide.
4. Locate a pier where there is usually little wave action. **CAUTION:** *You should not go to the pier alone. Always work with a partner or an adult. Wear a life jacket while working near water.*
5. Calculate an average. To find the average distance from the pier to the water's surface at high tide or at low tide, add the trials together, and divide this sum by the number of trials performed. Calculate the average distance to the water's surface for the following sample trials.

   a. High tide—Trial 1, 2.5 m; Trial 2, 2.0 m; Trial 3, 2.2 m
   b. Low tide—Trial 1, 4.1 m; Trial 2, 4.2 m; Trial 3, 4.3 m

Date	Time	Lunar Phase	Distance to Water (m)								Tidal Range
			High Tide				Low Tide				
			Trial 1	Trial 2	Trial 3	Average	Trial 1	Trial 2	Trial 3	Average	

6. Calculate tidal range. To find the tidal range, add the two average-distance figures together and divide by 2. Calculate the tidal range of the average high- and low-tide measurements in Step 5.

Procedure

1. Make the first measurement at about the time of high or low tide on the date of one of the lunar phases described earlier. **CAUTION:** *Do not make observations in any type of stormy weather.*
2. Go out on the pier. **CAUTION:** *You should not go to the water alone. Always work with a partner or an adult. Wear a life jacket while working near the water.* Tie the 200-g weight to the string. Lower the weight until it just touches the water. If there are waves, try to position the weight so that it is out of the water when the trough of a wave passes and under the water at the crest of the wave. When the weight is positioned correctly, attach the paper clip to the string at a point that is level with the top of the railing or the deck of the pier, as shown in the figure below.

3. Pull up the string and measure the distance from the paper clip to the bottom of the weight. Record this data in a table similar to the one shown on the previous page.
4. Repeat Steps 2 and 3 two more times.
5. Calculate the average distance of the three heights, and record the average in your table.
6. Return to the site at the following low or high tide, and repeat Steps 2–5.
7. Calculate the tidal range for the half tidal cycle you measured. Record this measurement in your table.
8. Repeat Steps 2–7 for each lunar phase for a two-month period.
9. After you have collected all your data over the two-month period, prepare a line graph of the tidal ranges for each date of observation. Use the horizontal axis of the graph to record the phase of the moon. Below each phase, record the date on which you took your measurements. Use the vertical axis of the graph to represent the tidal range (in meters) for that date.

Analysis and Conclusions

1. What causes the tides?
2. During which phase(s) of the moon is the tidal range the greatest? What was the tidal range?
3. During which phase(s) of the moon is the tidal range the smallest? What was the tidal range?
4. What local factors, if any, might affect the tidal range?
5. Describe the pattern of your graph. Did the pattern change in some regular, predictable way? If so, explain why this pattern exists.

LONG-RANGE INVESTIGATION 4

Precipitation and the Water Table

Introduction

Fresh water is essential for agriculture, industry, and household usage. Fresh water is also needed as drinking water. The primary source of fresh water is found underground, in spaces within soil and rock. The upper level of the water that fills these spaces is called the **water table.** *The rock or soil through which water can pass easily is called an* **aquifer.** *The water table moves upward or downward in the aquifer, depending on the amount of precipitation as well as on consumer use. In some areas of the country, wells tap into underground water reserves and pump the water above ground level for a variety of uses. When not heavily used, the water in a well is at the same level as the local water table. Spring-fed ponds work like natural wells, exposing ground water at the surface. The upper level of water in spring-fed lakes and ponds is the water table for that location. In this investigation, you will discover how the level of the water table varies with the amount of precipitation.*

Materials

- string or cord (11 m)
- masking tape
- rain gauge or coffee can
- graph paper
- meter stick
- small wooden block

Suggested Schedule

During this investigation, you will keep a daily log of the amount of precipitation. You will also visit a lake or pond near your home or school once a week for about two months. Check with your teacher for the best month to begin. In many areas, March is the best starting month because of spring rains and winter snow melt.

Advance Preparation

1. Find a location on a lake or pond where you can perform your investigation. **CAUTION:** *You should not go to the lake or pond alone. Always work with your partner or an adult. Wear a life jacket while working near water.* Make sure that the lake or pond you choose does not spill over the edge of its bank. If the lake or pond is on private property, be sure to get permission before you begin work.
2. Tie one end of the 11-m cord around the wooden block. Make knots in the string at 1-m intervals beginning at the bottom of the block. Make four knots to start. You can add more knots later, if necessary.
3. Obtain an empty coffee can or rain gauge.
4. Learn about the type of soil and rock found in your area. (a) What type of soil and rock is found in your area? (b) What type of soil is best to hold water? (c) What kinds of rock form good aquifers?

Procedure

1. Begin collecting precipitation data at least one week before your first visit to the body of water you selected. Set an empty coffee can or rain gauge in an open area. After a rainfall, measure the depth of the water in the can or gauge in millimeters with the meter stick. If it snows, bring the can or gauge indoors and let the snow melt before measuring the depth. Record the date and the amount of precipitation. Empty the can or gauge and put it back for the next day. You should record the amount of precipitation each day. For days without any precipitation, indicate no precipitation on your record.
2. On the first day you visit the lake or pond, select a point on the bank as the reference point for your measurements of the water table. Always measure from the same location.
3. Lower the block until it just touches the water.
4. With a piece of tape, mark the point on the string that is level with your reference point.
5. Raise the string and count the number of knots (meters) between the tape and the bottom of the block. Use the meter stick to measure the part of the 1 m remaining.
6. Record the distance to the water's surface and the date in a table with columns for date, distance to water level, date of last rainfall, and amount of rainfall.
7. Repeat Steps 3–6 every week, according to your schedule.
8. On a sheet of graph paper, make a bar graph of precipitation for each day from one week before your first water-table measurement until the final day of observation. Some days will have no rainfall, but include them also. Use the vertical axis of the graph to represent daily precipitation (in millimeters) and the horizontal axis to represent the dates of your observations.
9. On another sheet of graph paper, make a line graph of the water level covering the same dates as the precipitation graph. Use the vertical axis to represent the water levels (in meters). Use the horizontal axis for the dates. Draw a series of straight lines connecting the plotted points from the first date to the last date.

Analysis and Conclusions

1. On which date was the water table the highest? the lowest?
2. Compare the date of the highest water table with the amount of rainfall during the preceding days or weeks. What does your comparison indicate?
3. Compare the date of the lowest water table with the amount of rainfall during the preceding days or weeks. What does your comparison indicate?
4. Why might it take some time for rainfall to affect the water level in a lake or pond?
5. On hot days, some rainfall may evaporate before it soaks into the ground. Has the local temperature affected the water table? Explain.
6. People generally use more water in summer than in winter. How might this affect the water table in an area where wells are the source of water?
7. If nearby hills were snow-covered when you began your observations, how would the melting snow have affected your results?

Long-Range Investigation 5

Air Pollution Watch

Introduction

Any material that is foreign to the environment is called a **pollutant.** *If certain types of pollutants in the air are found in high concentrations, they may be dangerous to the general health and well-being of those in the area. The types of substances that are considered to be air pollutants vary widely. Some examples are dust and smoke particles, pollen, mold spores, and waste gases. If the tiny particles remain suspended in the air for long periods of time, they are called* **particulates.**

Wind direction has an effect on the number of particulates in the air. For example, if a source of particulates is found in only one area, it is likely that there will be a larger number of particulates in the air when the wind is blowing from that direction. If the wind is coming from the opposite direction, there will be fewer particulates in the air.

In this investigation, you will collect and view a few types of particulates using basic laboratory equipment. These particulates will stick to a greasy surface. A microscope slide spread with a thin coating of petroleum jelly can be used to collect particulates. A microscope can then be used to view and count them.

Materials

- magnetic compass
- microscope
- slide box
- grease pencil
- 8 or more microscope slides
- wooden post (1.5-m high) with 4 flat surfaces
- hammer
- graph paper
- masking tape or rubber bands
- petroleum jelly

Suggested Schedule

You and your partner should prepare a schedule that lists dates on which you plan to make observations for this investigation. Give this schedule to your teacher for approval. Keep a copy of the approved schedule for yourself. You will examine the slides at an outdoor site every day for about two weeks. Allow about 30 minutes to one hour each day to make the observations and view the slides.

Advance Preparation

1. Select a collection site that is in an open area, where the wind can blow past the site from every direction. A large field or pasture is ideal. You will need a four-sided post, such as a 4 × 4, that is at least 1 m above ground level. At the location, drive the post firmly into the ground. **CAUTION:** *Injuries can result from improper techniques for driving a fence post into the ground. If you are inexperienced at this, request assistance from an adult, or find a post or platform that has already been set up.*
2. Practice making thin smears of petroleum jelly on glass slides.
3. Make a table similar to the one below for your observations.

Day	Date	Wind direction	Slide direction	Number of particulates 1	2	3	4	5	Total
			N						
1			S						
			E						
			W						

Procedure

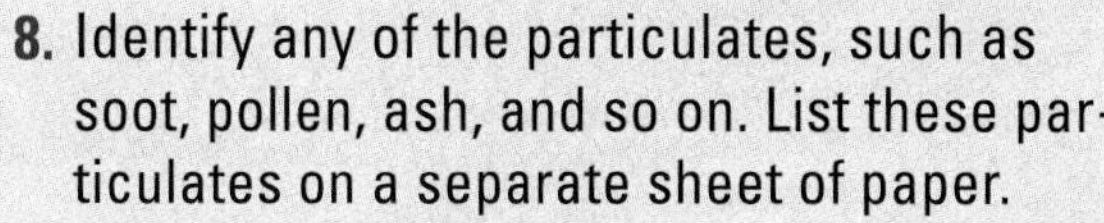

1. Place one slide on each of the four sides of your post. Attach the slides to the post by using tape or rubber bands. Refer to the figure at right.
2. Using your finger, spread a *thin* film of petroleum jelly on one side of the slide. Be sure to spread the petroleum jelly so that it is very thin and even. Try to avoid touching the slide after you spread the jelly on it.
3. Using a compass, establish north, south, east, and west directions from the slide's location. By watching objects move in the wind, such as a flag, you can determine the direction the wind is blowing. Record the wind direction and date in your table.
4. Look for any unusual phenomena that might affect the air quality, such as smoke-stacks with smoke, barbecues burning, or heavy traffic. Record these observations as well as the day's weather conditions.
5. Return to the site the following day. Remove the slides and, with a grease pencil, mark the direction that each slide faced on the back of the slide. Place each slide carefully in the slide box. *Note: Be sure not to touch the greased surface.*
6. Place freshly prepared greased slides on the post and record the wind direction in your table.
7. Examine each slide under the microscope at 100X. Focus on one section of the slide that you have *chosen at random.* Count the number of particulates you observe in the section. Record this number under Column 1 of your table.
8. Identify any of the particulates, such as soot, pollen, ash, and so on. List these particulates on a separate sheet of paper.
9. Move the slide to examine another section that you have chosen at random. Count the number of particulates in this section. Record the number in the next column of your table.
10. Repeat Step 9 three more times, for a total of five observations. Record the total number of particulates counted in the five sections.
11. Repeat Steps 6 through 10 each weekday for two weeks.
12. When you have finished examining the slides and recording your results, total the number of particulates counted for each of the wind directions.
13. Construct a bar graph from the observation data in your table. On the vertical axis, plot the total number of particulates obtained for the past five days, and on the horizontal axis, plot the day and wind direction.

Analysis and Conclusions

1. For your location, is there any one wind direction or group of wind directions that resulted in more particulates than any others?
2. Could you identify any of the particulates, such as soot, pollen, pieces of leaves, ash, and so on? If so, list some of the identifiable particulates.
3. What are possible sources of these particulates?
4. What would your results likely be if you set up this investigation near a populated urban area?
5. Did weather conditions have any effect on your data results or patterns? Explain.

LONG-RANGE INVESTIGATION 6

Correlating Weather Variables

Introduction

Can you predict what the weather will be tomorrow? You can describe the present weather conditions with some degree of accuracy, but predicting future weather conditions presents a greater challenge.

To make a weather prediction, you need to identify weather patterns that exist in your area. Weather records are used to identify relationships between variables. Although weather cannot be predicted with complete accuracy, the probability of certain weather conditions occurring can be established.

In this investigation, you will examine how the relationships between certain weather variables aid in the prediction of the weather. Then you will gather and organize weather information in a data table and present the information as a graph. Finally, you will analyze the relationships among the data collected and make predictions about the weather.

Materials

- aneroid barometer or barograph
- Celsius thermometer
- magnetic compass
- graph paper

Suggested Schedule

You will measure and record weather variables every day for a month. You will collect data in the morning, late afternoon, or early evening at approximately the same time each day.

Advance Preparation

Using graph paper, make two graph charts as shown at right. Make sure each chart extends for 15 days for a total of 30 days. Enter the dates for the time frame during which you plan to do the investigation.

	Day	A.M.	P.M.	A.M.	P.M.
Temperature (°C)	30				
	20				
	10				
	0				
	–10				
Pressure (mb)	1035				
	1013				
	991				
Wind and sky cover		○	○	○	○
Present weather					

Procedure

1. The following observations should be recorded in a data table with a column for the date and split columns for morning and afternoon observations of temperature (°C), barometric pressure (mb), wind direction, cloud cover, and present weather conditions.

- **Temperature:** Using a thermometer, measure and record the air temperature to the nearest degree Celsius.

Your thermometer should be placed in the shade and not exposed to precipitation. Wait at least three minutes before you read the temperature. Then plot a point for the temperature recorded on your graph. Connect the points with a smooth line.

- **Barometric Pressure:** Using an available barometer, record the barometric pressure to the nearest tenth of a millibar. If your barometer is calibrated in "inches of mercury," change inches to millibars using the conversion scale on page 672. *Note: Barometric pressure is also called "atmospheric pressure" or "air pressure."* Plot a point for the barometric pressure and connect the points.
- **Wind Direction:** Use a weather vane or observe objects moved by the wind to determine the wind direction. *Note: wind is named according to the direction from which it blows.* If you are observing an object moved by the wind, use the compass to help determine direction. Wind direction is shown by drawing a straight line off the circle that points in the direction the wind is coming from. Choose from the following symbols:

  north: ⦵ south: ⚲
  east: ○— west: —○
  northeast: southeast:
  northwest: southwest:

- **Cloud Cover:** Estimate the amount of sky covered by clouds. Using the following choices, shade in the circles at the bottom of your chart as follows:

  Clear: ○
  (no cover)
  Scattered clouds: ⦶
  (10 percent or less of the sky is covered)
  Partly cloudy: ◑
  (between 10 and 70 percent covered)
  Overcast with Openings: ⬤
  (almost complete sky cover)
  Overcast: ●
  (100 percent sky cover)

- **Present Weather:** Determine the present weather conditions using the following choices. Draw the correct symbol in the appropriate space in your chart.

  fog: ≡ rain shower: ▽
  drizzle: ❟ thunderstorm:
  rain: • snow shower: ▽
  snow: ✳ hail: △
  sleet: △ haze: ∞

Analysis and Conclusions

1. Referring to your graph, on how many days was the temperature falling?
2. How many of the days with falling temperature also had rising barometric pressure?
3. In general, what is the relationship between temperature and pressure?
4. How many days had falling barometric pressure?
5. What sky cover is usually associated with falling pressure?
6. What wind direction is usually associated with falling barometric pressure?
7. What weather conditions are usually associated with high barometric pressure?
8. What weather conditions are usually associated with low barometric pressure?
9. How do the relationships between certain weather variables help to predict the weather?

Long-Range Investigation 7

Weather Forecasting

Introduction

Every three hours, the National Weather Service produces a weather map. These maps are compiled from data received from about 800 weather stations located in the United States, in foreign countries, and on ships and buoys at sea. Daily newspapers often summarize the most recent weather data in the form of a national weather map. Weather information from all sources is plotted on blank national maps. The data usually include temperature, precipitation, cloud cover, and barometric pressure. Points of equal barometric pressure are connected by lines called* isobars. *The pattern of isobars provides meteorologists with a picture of where high-pressure and low-pressure systems are located. Weather fronts can then be identified, providing even more information about weather conditions in regions of the country.

In this investigation, you will use a series of daily weather maps to track the movements of weather systems and then make a prediction of future weather conditions for a given location.

Materials

- 2 sequences of 5 consecutive daily weather maps
- colored pencils

Suggested Schedule

This investigation can be conducted over two one-week periods: one week in the winter and another week in late spring.

Advance Preparation

1. Read Section 23.3, pages 469–473; Section 25.1, pages 499–501; and Section 25.4, pages 512–517.

2. Check your local or national newspapers to see which one of them includes a daily weather map. You will notice a variety of symbols on the weather maps.

a. What do the symbols represent? Compare the symbols used on the weather map with the symbols shown on page 513. Are the symbols used on the different maps similar? Why?

b. How is a low-pressure system indicated on the weather map in the newspaper? If isobars are plotted on the map, note that barometric pressure decreases toward the center of the low.

c. How is a high-pressure system indicated on the weather map in the newspaper? If isobars are plotted, note that barometric pressure increases toward the center of the high.

3. Make at least one copy of the data map on the bottom of page 675.

4. Use the following formula to calculate the average velocity of the high- or low-pressure center in kilometers per day. The average velocity equals the total distance traveled divided by the number of days traveled, or

$$\text{Average velocity} = \frac{\text{Total distance traveled}}{\text{Number of days}}$$

Procedure

1. Obtain a weather map from a newspaper or from the National Weather Service. Cut out the map and write the date on the map. Make a data table similar to the sample shown below. Fill in your table with the information on the weather map.

	Week One (winter)					Week Two (spring)		
	1	2	3	4	5	1	2	3
Temperature								
Barometric pressure								
Barometric trend (R = rising; F= falling; S = steady)								
Wind direction								
Cloud cover (C = clear; Cl = cloudy; PC = partly cloudy; O = overcast)								
Present weather (rain, sleet, snow, etc.)								
Prediction								

2. On your copy of the blank weather map, write an "L" at the location of the low-pressure center that you observed on your daily weather map in Step 1. Circle it with a colored pencil and label the circle with the date.
3. Write an "H" on your map at the location of the high-pressure center. Circle it with a second colored pencil and label the circle with the date.
4. Repeat Steps 1–3 using four consecutive daily weather maps that follow the first day's map. Use the same colors for each symbol as was used for the first day.
5. Draw arrows from each day's position to the next for each high-pressure center and for each low-pressure center.
6. Answer Questions 1–6 in Analysis and Conclusions.
7. In the spring, repeat the entire investigation.
8. After completing both your winter and spring observations, answer Questions 7–9 in Analysis and Conclusions.

Analysis and Conclusions

1. In which general direction do the pressure centers over the United States move?
2. From your calculations, what is the average rate of movement, in kilometers per day, of low- and high-pressure centers in winter?
3. Predict where the low- and high-pressure centers will be located on the day following the date of the last map in your series.
4. Describe the general weather conditions associated with regions of low and high atmospheric pressure.
5. Based on your series of daily weather maps, predict the weather for your hometown on the fifth day of the series. Write a forecast and fill in the data table with your estimates of weather conditions.
6. Compare your prediction with the daily weather map for the fifth day. Check the accuracy of your prediction. What factors could have resulted in errors in your prediction?
7. From your calculations, what is the average rate of movement, in kilometers per day, of low- and high-pressure centers in the spring?
8. After completing both winter and spring observations, compare the rate of movement of pressure systems during the different seasons.
9. How can you predict the weather using a series of daily weather maps?

LONG-RANGE INVESTIGATION 8

Comparing Climate Features

Introduction

A graph of the monthly temperatures and amounts of precipitation for a region is called a **climatograph.** *When the monthly temperatures and amounts of precipitation are graphed for an entire year, the graph will be similar to one of the seven world climates. In this investigation, you will use climate data to compare your local climate with that of other regions of the United States. You will then make comparisons of your graphed data with the different world climates and develop a conclusion about the type of climate in your location.*

Materials

- daily weather reports
- almanac
- atlas
- graph paper
- thermometer (optional)
- rain gauge (optional)
- metric ruler

Suggested Schedule

You will be keeping a daily temperature and precipitation log. You should begin on the first day of October and continue until the first day of April. You should record data every day until you prepare your conclusions.

Advance Preparation

1. Read the descriptions of the major climates of the United States in Section 26.2, pages 529–535.
2. On your graph paper, make eight copies of the blank climatograph shown on the next page. You will need these copies to analyze several climates, including the climate of the area in which you live.
3. Using an almanac, look up climate data for cities in the United States. Select one city from each of the following seven regions: New England, the Gulf Coast, the Midwest, the Southwest, the Pacific Northwest, the interior of Alaska, and the Hawaiian Islands. An atlas will show the location of each city. You may select cities in additional locations for more detailed comparisons. Look up the average monthly temperatures and precipitation for each city you choose. Most likely, the data will be given in units of degrees Fahrenheit and inches, rather than in metric units. Record the data just as it is presented in your reference source. Record the units as well.

Procedure

A. Obtain climate information for your city or town by following these steps:

1. Listen to or watch a daily weather report for your locality, or find this information in a daily newspaper. If you are adventurous, you can try to keep your own records using a thermometer and a rain gauge.

2. Beginning on the first day of October, keep a daily record of the high and low temperatures and any precipitation that occurs.

3. Calculate the average temperature for each day by dividing the sum of the high and low temperatures by two. Record.

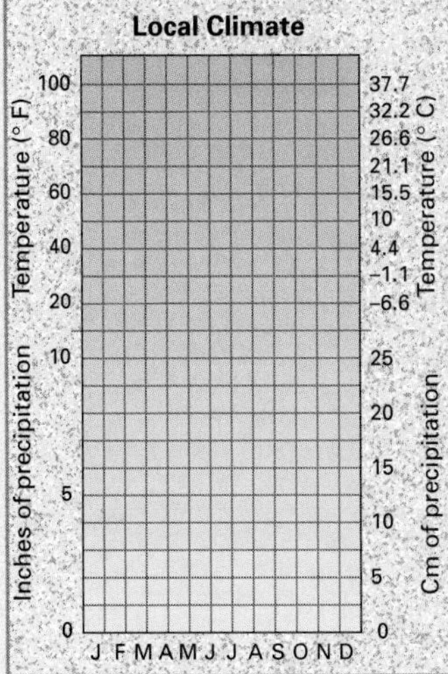

4. At the end of each month, record the average monthly temperature and the total monthly precipitation as given by the weather report, or calculate the average by dividing the sum of the daily averages by the number of days in the month.

B. When you have finished your observations at the end of March, complete the climatographs as follows:

1. Label each blank climatograph with the name of a different location you chose. Label the eighth climatograph with the name of your locality.

2. If your recorded temperatures and precipitation are in metric units, refer to the scales along the right-hand side of the climatograph; for English (American) units, refer to the scales along the left-hand side.

3. Starting with the climatograph for one of your chosen locations, plot the temperature for the month of January by placing a small dot at the intersection of the line for January and the corresponding temperature. Plot the average monthly precipitation using the same method on the precipitation scale.

4. Repeat Step 3 for each month's data for that location. When you are finished, connect the temperature points in order of the consecutive months. Then connect the precipitation points in order of consecutive months.

5. Repeat Steps 3 and 4 for the remaining chosen cities and for your locality using the data you have gathered.

Analysis and Conclusions

1. Compare each of the seven climatographs that you prepared with the sample climatographs for the major world climates on page 673. Identify the climate type or combination of climates for each location you selected. Explain those features of your climatograph that helped you classify each region.

2. Using the climatograph for your locality, classify your regional climate. Explain the features of your climatograph that helped you choose the climate type. From your experience in previous years or from information found in reference sources, explain whether you think this year's data is typical for your region.

3. Refer to Chapter 26 for other systems to classify climates. How would each of the climatographs, including the one for your region, fit into such systems? Using the new systems, reclassify each region for which you have data, and give your reasons.

4. In this investigation, you compared the climates by looking at average precipitation and temperature. What other factors might affect the climate of an area? Give examples to illustrate each factor.

5. Bermuda, a small island in the Atlantic Ocean, is at about the same latitude as St. Louis, Missouri, which lies in the middle of a continent. In which of the two locations does the temperature vary least from month to month? Explain the cause of the more moderate temperature pattern.

Long-Range Investigation 9

Planetary Motions

Introduction

While observing the evening sky over a period of time, you might have noticed points of light that look like stars but have patterns of movement that are quite different. These objects do not maintain a fixed position with respect to the celestial sphere as do the stars. These objects are the planets. As viewed from the earth, the planets show a relatively strange pattern in their motion at various times. In this investigation you will observe the planet Mars over a period of several months and use your observations to draw conclusions about planetary motion.

Materials

- constellation charts
- flashlight
- metric ruler
- celestial sphere model (optional)
- magnetic compass

Suggested Schedule

You will observe Mars in the night sky on the 1st and 15th of each month for eight months. If the nights are not clear, make your observations as close to these dates as possible.

Advance Preparation

1. Check newspapers to find the time that Mars will be visible. This information is found in the weather section of most newspapers.
2. Obtain constellation charts, or copy the one given on the next page. Practice locating various constellations visible in the Northern Hemisphere.
3. Obtain data of the position of Mars throughout last year from an astronomical yearbook. This book can be found in most libraries.
4. Find an open area that gives you a clear view of the eastern, southern, and western sky.

Procedure

1. For each night at the same scheduled time, locate Mars in the night sky. Mars will have a dull red appearance.

Estimating Angular Distance in Degrees
Hold your hand at arm's length.

1°

5°

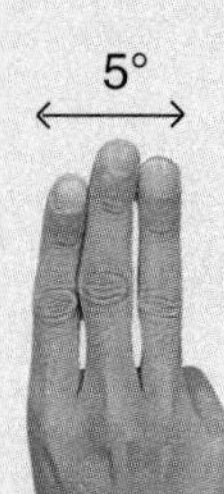

10°

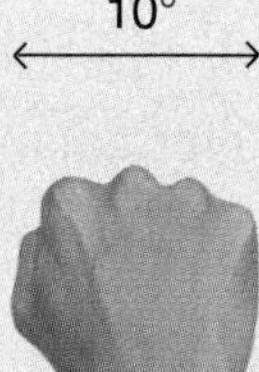

15°

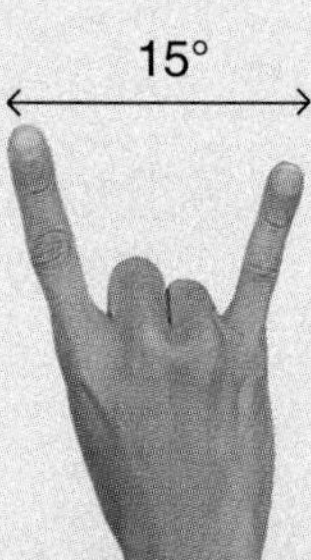

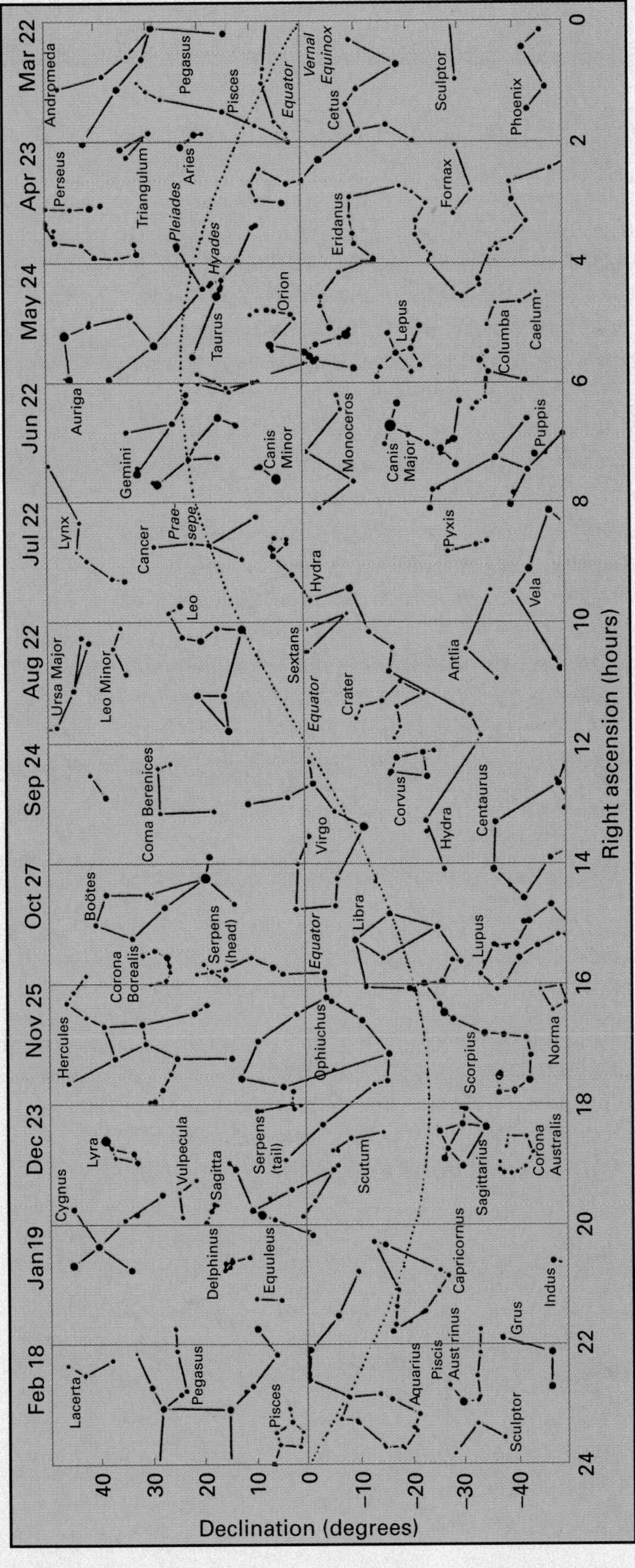

2. Use your compass to position yourself facing south, and observe Mars in its position against the background stars.
3. Estimate the planet's angular distance in degrees from a star or constellation by using the method illustrated on the bottom of the previous page.
4. Locate this star or constellation on your star chart. Draw Mars in the appropriate position in relationship to the star or constellation you located and label the planet's position with the date.
5. Compare the apparent brightness of Mars with that of the background stars. Record your observation on a separate sheet of paper.
6. Repeat Steps 1–5 on the 1st and 15th of each month for the next seven months. Plot the position of Mars for each scheduled observation on your constellation chart.
7. After each observation, draw an arrow from each position of Mars to the next. This will show Mars's apparent path against the background stars.

Analysis and Conclusions

1. In which months does Mars appear highest in the night sky? lowest?
2. In which direction does Mars appear to move across the sky?
3. Describe the apparent path of motion of Mars plotted on your star chart.
4. Will Mars be in the same position one year from today? Explain.
5. Referring to your observations of the apparent brightness of Mars throughout the investigation, what can you infer about the distance between Mars and the earth?

LONG-RANGE INVESTIGATION 10

Apparent Motions of the Moon

Introduction

The moon is the earth's nearest neighbor in space. While it is the sun that is the dominant object in the daytime sky, it is the moon that brightens the sky many nights each month. Because it reflects light from the sun, the moon can be seen from the earth. The amount of the lighted portion of the moon visible from the earth causes the moon to appear to change shape throughout the month. Each month, the moon progresses from new moon to full moon and back to new moon again, appearing to increase in size (wax) and then decrease in size (wane). Not only does the moon change appearance from night to night, but the time and location of moonrise and moonset vary from one night to the next.

In this investigation, you will observe the moon's appearance and altitude throughout its phases. Using two simple measurements, you will plot the location of the moon in the sky from one night to the next. You will identify each phase, determine the times of moonrise and moonset, and determine the apparent path of the moon for one day.

Materials

- flashlight
- paper clip
- protractor
- magnetic compass
- thread (15 cm)
- graph paper

Suggested Schedule

You will observe the moon at the same time every clear evening for approximately one month. Begin your observations three days after a new moon.

Advance Preparation

1. Construct a sextant.
   a. Cut the thread to a length of approximately 15 cm.
   b. Tie the paper clip to one end of the thread.
   c. Fasten the other end of the thread to the center point of the protractor. (If your protractor has a hole at this point, tie the thread. If not, use a small piece of tape to hold it in place.)
2. To read a sextant, refer to the figure at right. In the diagram, angle *X* represents the altitude of the moon above the horizon. This is equal to angle *Y* on the protractor. Practice finding angle *Y* by making observations of the moon before you begin this investigation. Refer to Step 2 of the Procedure section of this investigation.

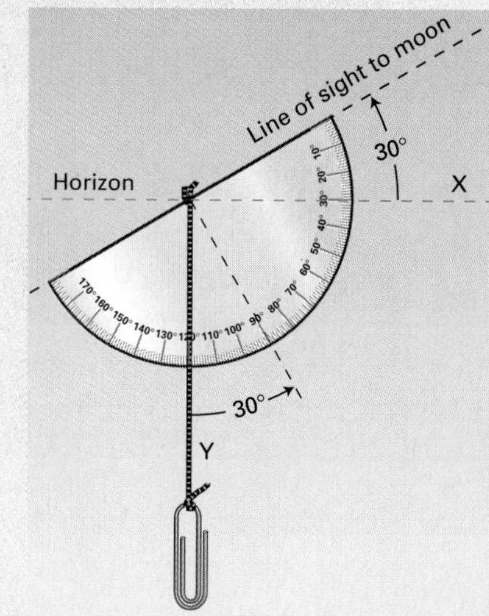

Procedure

1. Each evening at your scheduled time, observe the moon, then enter the date and phase name into a table with columns for date, phase, altitude (in degrees), and azimuth (in degrees). You can use the illustrations on page 627 as a reference for identifying the phases.

2. Each night at the same scheduled time, determine the moon's altitude. The altitude of any celestial object is its angular distance from the horizon, measured in degrees.
   a. Choose a location that gives you a clear view of the eastern, southern, and western sky. Using the compass, position yourself facing south.
   b. Sight the center of the moon along the straight edge of the protractor, as shown in the illustration.
   c. Let the thread with the paper clip hang freely.
   d. When the thread becomes motionless, press it against the protractor with your finger. Read the angle corresponding to angle *Y* in the illustration. Subtract 90° from the reading you obtain. This is the moon's altitude for this time on the assigned night. Record your reading in your data table.

3. After you have recorded the moon's altitude, you will measure its *azimuth.* The azimuth is the distance measured in degrees from due north along the observer's horizon. This measurement is used by astronomers to locate objects in the sky.
   a. Face the moon in the same position you used to determine altitude.
   b. Hold your compass level and line up the needle with 0°.
   c. Once the needle stops moving, read the number of degrees from north (0°) that you are facing. Record this measurement on your data table.

4. Make a graph with altitude on the vertical axis and azimuth on the horizontal axis, similar to the one shown below. Plot the position of the moon in the sky using the altitude and azimuth measured in Steps 2 and 3.

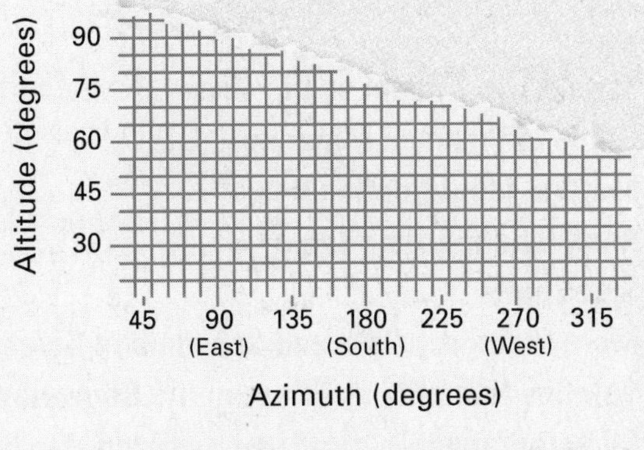

Position of the Moon in the Sky

5. On the graph where you plotted the position of the moon in Step 4, label the date next to the plotted position.

6. Repeat Steps 1–5 for each of your scheduled nightly observations.

Analysis and Conclusions

1. What sequence of the moon's phases did you observe?

2. a. You observed the moon each night at about the same time. Referring to your graph, what did you notice about the position of the moon during each observation?
   b. What might you conclude about the time of moonrise from night to night?

3. Describe the changes in the moon's appearance from the new moon to the full moon.

4. How does the moon's appearance and altitude at a particular time vary from one new moon to the next?

Contents

The Metric System and SI Units

The Système Internationale, or SI, has been the international standard for measurement since 1960. SI is a decimal system. The ease of converting among units in the SI system is a great advantage over the English system. For example, to convert kilometers to meters, you can multiply by 1,000 or simply move the decimal point three places to the right: 2.8 km equals 2,800 m. In contrast, to convert miles to yards in the English system, you must multiply the number of miles by 1760: 2.8 miles equals 4,928 yards.

SI Units

Length		SI-English equivalents	
kilometer (km)	= 1,000 m	1 km = 0.621 mile (mi.)	1 mi. = 1.61 km
meter (m)	= 100 cm	1 m = 3.28 feet (ft.)	1 yd. = 0.915 m
centimeter (cm)	= 0.01 m	1 cm = 0.394 inch (in.)	1 ft. = 0.305 m
millimeter (mm)	= 0.001 m	1 mm = 0.039 in.	1 in. = 2.54 cm
micrometer (µm)	= 0.000001 m		
nanometer (nm)	= 0.000000001 m		
Area			
square kilometer (km^2)	= 100 hectares	1 km^2 = 0.3861 mi^2	1 mi^2 = 2.590 km^2
hectare (ha)	= 10,000 m^2	1 ha = 2.471 acre (a)	1 a = 0.4047 ha
square meter (m^2)	= 10,000 cm^2	1 m^2 = 1.1960 yd^2	1 yd^2 = 0.8361 m^2
square centimeter (cm^2)	= 100 mm^2	1 cm^2 = 0.155 in^2	1 in^2 = 6.452 cm^2
Mass			
kilogram (kg)	1,000 g	1 kg = 2.205 pounds (lb.)	1 lb. = 0.4536 kg
gram (g)	1,000 mg	1 g = 0.0353 ounce (oz.)	1 oz. = 28.35 g
milligram (mg)	0.001 g		
microgram (µg)	0.000001 g		
Volume of solids			
cubic meter (m^3)	1,000,000 cm^3	1 m^3 = 1.3080 yd^3	1 yd^3 = 0.7646 m^3
cubic centimeter (cm^3)	1,000 mm^3	1 m^3 = 35.315 ft^3	1 ft^3 = 0.0283 m^3
		1 cm^3 = 0.0610 in^3	1 in^3 = 16.387 cm^3
Volume of liquids			
kiloliter (kL)	1,000 liters	1 kL = 264.17 gallons (gal.)	1 gal. = 3.785 L
liter (L)	1,000 mL	1 L = 1.06 quarts (qt.)	1 qt. = 0.94 L
milliliter (mL)	0.001 L	1 mL = 0.034 fluid ounces (fl. oz.)	1 pint (pt.) = 0.47 L
microliter (µL)	0.000001 L		1 fl. oz. = 29.57 mL
Temperature			
Celsius (°C) = $\frac{5}{9}$ (°F – 32)		0°C = 32°F = Freezing point of water at sea level	
Fahrenheit (°F) = $\frac{9}{5}$ (°C + 32)		100°C = 212°F = Boiling point of water at sea level	

Guide to Common Minerals

Mineral	Chemical formula	Color	Luster
Apatite	$Ca_5(OH,F,Cl)(PO_4)_3$	Green, blue, violet, brown, colorless	Glassy
Bauxite	$Al(OH)_3$	White or tan	Dull to earthy
Biotite	$K(Mg,Fe)_3AlSi_3O_{10}(OH)_2$	Black, brown, dark green	Pearly, glassy
Calcite	$CaCO_3$	Colorless or white to tinted	Glassy
Chalcopyrite	$CuFeS_2$	Brass yellow	Metallic
Corundum	Al_2O_3	Gray, red (ruby), blue (sapphire)	Brilliant to glassy
Dolomite	$CaMg(CO_3)_2$	Pink, white, gray, brown	Glassy or pearly
Fluorite	CaF_2	Light green, yellow, bluish green, other colors	Glassy
Galena	PbS	Lead gray	Metallic
Garnet	$Fe_3Al_2(SiO_4)_3$	Dark red	Glassy to resinous
Graphite	C	Black to gray	Metallic or earthy
Gypsum	$CaSO_4{\cdot}2H_2O$	White, pink, gray, colorless	Glassy, pearly, or silky
Halite	$NaCl$	Colorless to gray	Glassy
Hematite	Fe_2O_3	Reddish brown to black	Metallic to earthy
Hornblende	$(Ca,Na)_{2\text{-}3}(Mg,Fe,Al)_5Si_6(Si,Al)_2O_{22}(OH)_2$	Dark green, brown, black	Glassy, silky
Magnetite	Fe_3O_4	Iron black	Metallic
Malachite	$CuCO_3{\cdot}Cu(OH)_2$	Green	Silky
Muscovite	$KAl_2Si_3O_{10}(OH)_2$	Colorless to light gray or brown	Glassy or pearly
Olivine	$(Mg,Fe)_2SiO_4$	Olive green	Glassy
Orthoclase	$KAlSi_3O_8$	Colorless, white, pink, various colors	Glassy to pearly
Plagioclase	$(Na,Cl)(Al,Si)_4O_8$	Blue-gray to white	Glassy
Pyrite	FeS_2	Brass yellow	Metallic
Quartz	SiO_2	Colorless, white, any color when not pure	Glassy, waxy
Sphalerite	ZnS	Red-brown to black	Resinous
Talc	$Mg_3Si_4O_{10}(OH)_2$	Gray, white, green	Pearly to waxy

Streak	Hardness	Specific gravity	Cleavage, fracture, and other properties
White or pale red-brown	4.5–5	3.1	No cleavage; conchoidal to uneven fracture
Colorless	1.5–3.5	2.5	No cleavage, aluminum ore
White to gray	2.2–3	2.7–3.2	One cleavage plane (basal); thin flexible sheets
White	3	2.7	Three cleavage planes not at right angles (rhombohedral), double refraction
Greenish black	3.5–4.5	4.2	Indistinct cleavage, uneven fracture, copper ore
None	9	3.9–4.1	No cleavage; hexagonal crystal form
White to pale gray	3.5–4	3.9–4.2	Six cleavage planes (rhombohedral); subconchoidal fracture
White	4	3.2	Eight cleavage planes (octahedral)
Lead gray to black	2.5	7.4–7.6	Three cleavage planes at right angles (cubic), lead ore
White	6.5–7.5	4.2	No cleavage, 12- or 24-sided crystals
Black	1–2	2.3	One cleavage plane (basal), scales, soft, flaky
White	1–2.5	2.2–2.4	One cleavage plane (perfect) yielding thin layers, conchoidal and fibrous fracture
White	2.5–3	2.2	Three cleavage planes at right angles (cubic), saline taste, cubic crystals
Red to red-brown	5.5–6.5	5.25	No cleavage, uneven fracture, iron ore
Pale green or white	5–6	3.2	Two cleavage planes at 56° and 124° (prismatic), six-sided crystals
Black	5–6	5.2	In two planes at or near 90°, uneven fracture, some magnetic varieties
Emerald green	3.5–4	4	No cleavage, uneven splintery fracture
White	2–2.5	2.7–3	One cleavage plane, waxy feel, thin flexible sheets
White to pale green	6.5–7	3.2–3.3	Conchoidal to uneven fracture
White	6	2.6	Two cleavage planes at nearly right angles; octahedral parting
White	6	2.6–2.7	Two cleavage planes at 86° and 94°, striations
Greenish, brownish, black	6–6.5	5	No cleavage; conchoidal to uneven fracture, cubic crystals
Colorless, white	7	2.6	No cleavage; conchoidal fracture, six-sided crystals
Reddish-brown	3.5	4	Six cleavage planes at 120° (dodecahedral), zinc ore
White	1	2.7–2.8	One cleavage plane (basal), soapy or greasy feel,

Mineral Identification Key

Non-metallic light color	**Scratches glass**	**Cleavage**	White or pink, two cleavage planes at nearly right angles, H—6, S—white	**Orthoclase**
		No cleavage	Glassy luster, transparent to opaque, six-sided crystals, H—7,S—white, conchoidal fracture	**Quartz**
	Does not scratch glass	**Cleavage**	Colorless to gray, glassy luster, three cleavage planes at right angles, H—2.5–3, S—white	**Halite**
			Colorless to tinted, three cleavage planes not at right angles, double image when you look through it, H—3, S—white	**Calcite**
			White to pink to colorless, one good cleavage plane, small pieces, flexible, H—1–2.5, S—white	**Gypsum**
			White to green, soapy feel, one cleavage plane, thin scales, H—1,S—white	**Talc**
			Colorless to light gray or brown, one cleavage plane, thin sheets, H—2–2.5, S—white	**Muscovite**
Non-metallic dark color	**Scratches glass**	**Cleavage**	Dark green, brown, or black, two cleavage planes at 56° and 124°, H—5–6, S—white	**Hornblende**
		No cleavage	Tinted red, dull luster, fracture could be taken for cleavage, H—6.5, S—white	**Garnet**
	Does not scratch glass	**Cleavage**	Black to dark brown, one cleavage plane, thin sheets, H—2.2–3, S—white to gray	**Biotite**
		No cleavage	Reddish brown to black, metallic to earthy luster, H—5.5–6.5, S—red to red-brown	**Hematite**
Metallic luster	**Black to dark green streak**		Iron black, some magnetic varieties, H—5–6, S—black	**Magnetite**
			Black to gray, greasy feel, one cleavage plane, soft, flaky, H—1–2, S—black	**Graphite**
			Brass yellow, uneven fracture, cubic crystals, H—6–6.5, S—greenish black	**Pyrite**
			Lead gray, very heavy, three cleavage planes at right angles, H—2.5, S—lead gray to black	**Galena**

for Chapter 9 In-Depth Investigation

Rock Identification Table

Description	Rock class	Rock name
Coarse-grained; mostly light in color—shades of pink, gray, and white are common.	Igneous	Granite
Coarse-grained; mostly dark in color; much heavier that granite or diorite	Igneous	Gabbro
Fine-grained; dark in color; often rings like a bell when struck with a hammer	Igneous	Basalt
Light to dark in color; many holes—spongy appearance; light in weight, may float in water	Igneous	Pumice
Light to dark in color; glassy luster—sometimes translucent; conchoidal features	Igneous	Obsidian
Coarse-grained; foliated; layers of different minerals often give a banded appearance	Metamorphic	Gneiss
Coarse-grained; foliated; quartz abundant, commonly contains garnet; flaky minerals	Metamorphic	Schist
Fine-grained; foliated; cleaves into thin flat plates	Metamorphic	Slate
Coarse-grained; nonfoliated; reacts with acid, effervesces	Metamorphic	Marble
Fine-grained; soft and porous; normally white or buff color	Sedimentary	Chalk
Coarse-grained, over 2 mm; rounded pebbles; some sorting—clay and sand can be seen	Sedimentary	Conglomerate
Medium-grained, 1/16 to 2 mm; mostly quartz fragments—surface feels sandy	Sedimentary	Sandstone
Microscopic grains; clay composition; smooth surface—hardened mud appearance	Sedimentary	Shale
Coarse to medium grained; well-preserved fossils are common; soft—can be scratched with a knife; occurs in many colors but usually white-gray; reacts with acid	Sedimentary	Crystalline limestone
Coarse to fine-grained; cube-shaped crystals; normally colorless; does not react with acid	Sedimentary	Halite

for Chapter 10 In-Depth Investigation

Star Charts

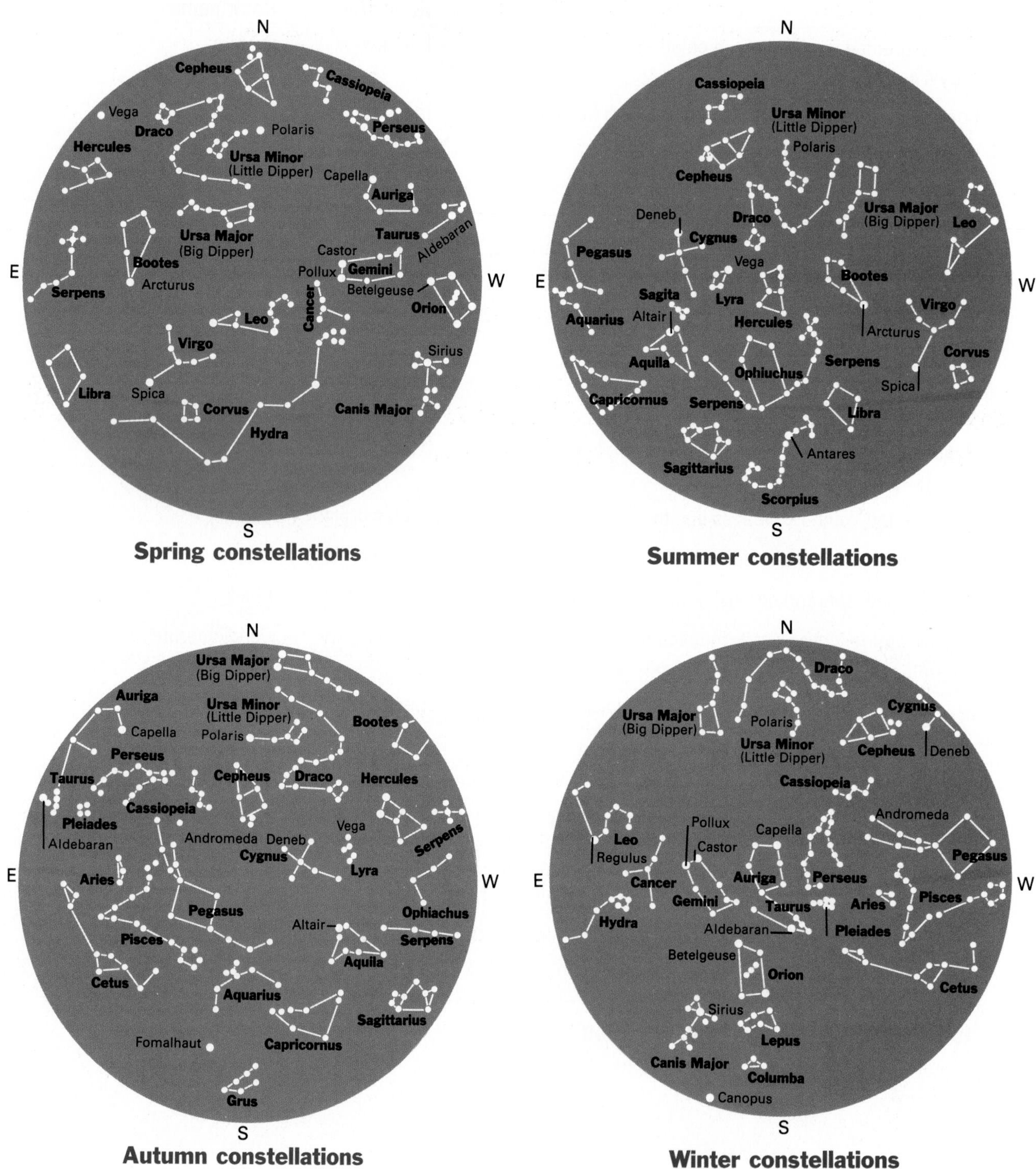

Spring constellations

Summer constellations

Autumn constellations

Winter constellations

More Planetary Data

Planet	Distance from sun (AU)	Orbit period (earth years)	Mean orbital speed (km/s)	Inclination of orbit to ecliptic (°)	Diameter (in relation to the earth)	Mass (in relation to the earth)	Surface gravity (in relation to the earth)
Mercury	0.39	0.24	47.9	7.0	0.38	0.06	0.38
Venus	0.72	0.62	35.0	3.4	0.95	0.82	0.91
Earth	1.00	1.00	29.8	0.0	1.00	1.00	1.00
Mars	1.52	1.88	24.1	1.8	0.53	0.11	0.38
Jupiter	5.20	11.86	13.1	1.3	11.2	317.8	2.53
Saturn	9.54	29.46	9.6	2.5	9.41	94.3	1.07
Uranus	19.19	84.07	6.8	0.8	4.11	14.6	0.92
Neptune	30.06	164.82	5.4	1.8	3.81	17.2	1.18
Pluto	39.53	248.6	4.7	17.2	0.17	0.003	0.09

Solar Eclipses from Present Through the Year 2030

Date	Duration of totality (min)	Location
Feb. 26, 1998	4.4	Central America
Aug. 11, 1999	2.6	Central Europe, Central Asia
June 21, 2001	4.9	Southern Africa
Dec. 4, 2002	2.1	South Africa, Australia
Nov. 23, 2003	2.0	Antarctica
April 8, 2005	0.7	South Pacific Ocean
March 29, 2006	4.1	Africa, Asia Minor, Russia
Aug. 1, 2008	2.4	Arctic Ocean, Siberia, China
July 22, 2009	6.6	India, China, South Pacific
July 11, 2010	5.3	South Pacific Ocean, Southern South America
Nov. 13, 2012	4.0	Northern Australia, South Pacific
Nov. 3, 2013	1.7	Atlantic Ocean, Central Africa
March 20, 2015	4.1	North Atlantic, Arctic Ocean
March 9, 2016	4.5	Indonesia, Pacific Ocean
Aug. 21, 2017	**2.7**	**Pacific Ocean, U.S.A., Atlantic Ocean**
July 2, 2019	4.5	South Pacific, South America
Dec. 14, 2020	2.2	South Pacific, South America, South Atlantic Ocean
Dec. 4, 2021	1.9	Antarctica
April 20, 2023	1.3	Indian Ocean, Indonesia
April 8, 2024	**4.5**	**South Pacific, Mexico, Eastern U.S.A.**
Aug. 12, 2026	2.3	Arctic, Greenland, North Atlantic, Spain
Aug. 2, 2027	6.4	North Africa, Arabia, Indian Ocean
July 22, 2028	5.1	Indian Ocean, Australia, New Zealand
Nov. 25, 2030	3.7	South Africa, Indian Ocean, Australia

Relative Humidity (in percentage)

Dry-bulb temperature (°C)	Difference in temperature (°C)									
	1.0	2.0	3.0	4.0	5.0	6.0	7.0	8.0	9.0	10.0
10	88	77	66	55	44	34	24	15	6	—
11	89	78	67	56	46	36	27	18	9	—
12	89	78	68	58	48	39	29	21	12	—
13	89	79	69	59	50	41	32	23	15	7
14	90	79	70	60	51	42	34	26	18	10
15	90	80	71	61	53	44	36	27	20	13
16	90	81	71	63	54	46	38	30	23	15
17	90	81	72	64	55	47	40	32	25	18
18	91	82	73	65	57	49	41	34	27	20
19	91	82	74	65	58	50	43	36	29	22
20	91	83	74	66	59	51	44	37	31	24
21	91	83	75	67	60	53	46	39	32	26
22	92	83	76	68	61	54	47	40	34	28
23	92	84	76	69	62	55	48	42	36	30
24	92	84	77	69	62	56	49	43	37	31
25	92	84	77	70	63	57	50	44	39	33
26	92	85	78	71	64	58	51	46	40	34
27	92	85	78	71	65	58	52	47	41	36
28	93	85	78	72	65	59	53	48	42	37
29	93	86	79	72	66	60	54	49	43	38
30	93	86	79	73	67	61	55	50	44	39
31	93	86	80	73	67	61	56	51	45	40
32	93	86	80	74	68	62	57	51	46	41
33	93	87	80	74	68	63	57	52	47	42
34	93	87	81	75	69	63	58	53	48	43
35	94	87	81	75	69	64	59	54	49	44
36	94	87	81	75	70	64	59	54	50	45
37	94	87	82	76	70	65	60	55	51	46
38	94	88	82	76	71	66	60	56	51	47
39	94	88	82	77	71	66	61	57	52	48
40	94	88	82	77	72	67	62	57	53	48

◄ **for Chapter 24 In-Depth Investigation**

Barometric Conversion Scale

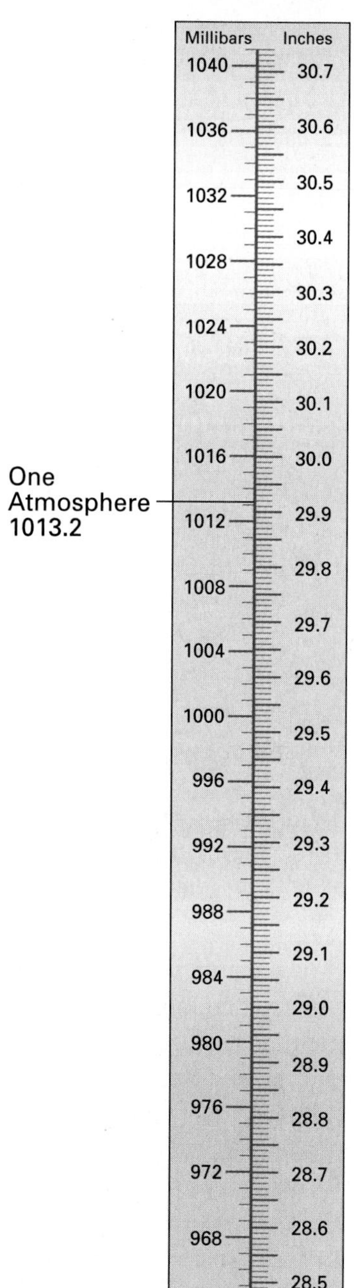

▲ **for Long-Range Investigation 6**

Climatographs

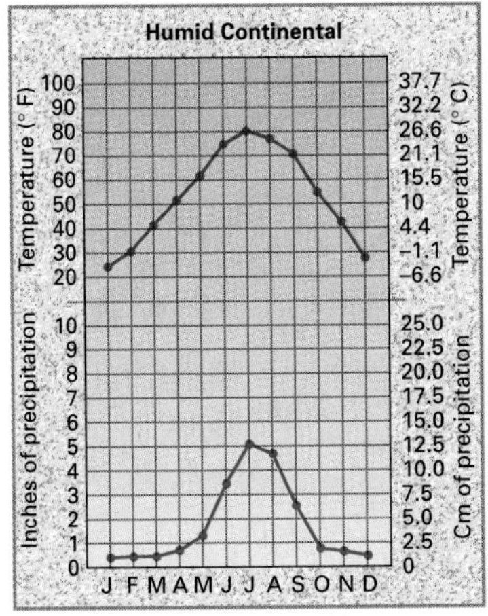

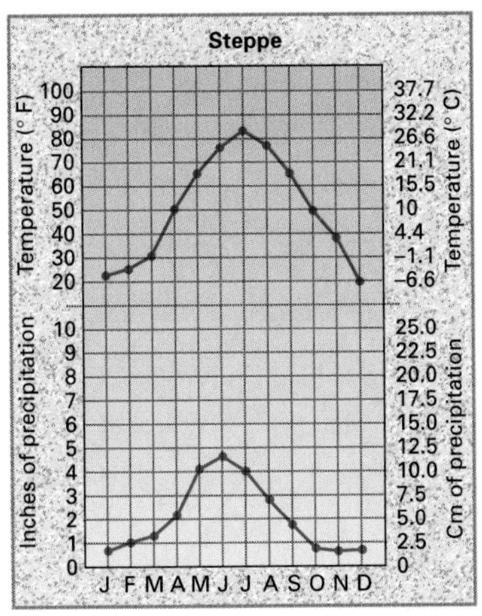

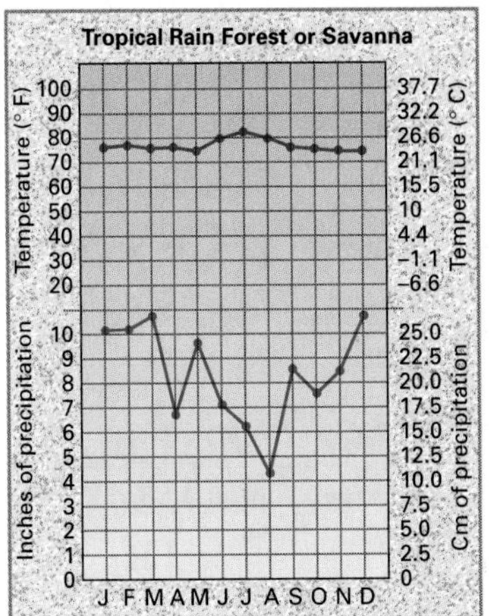

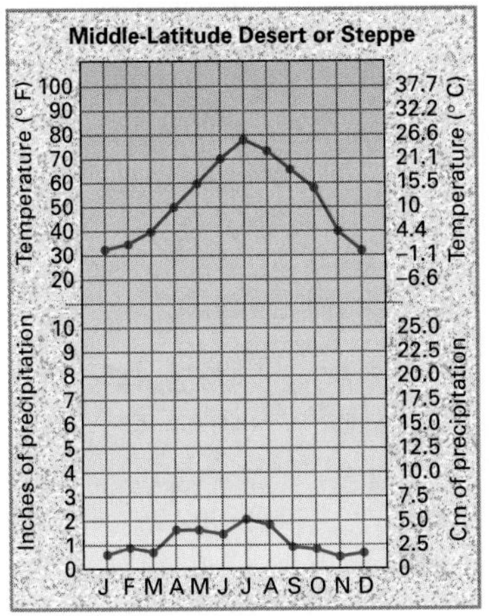

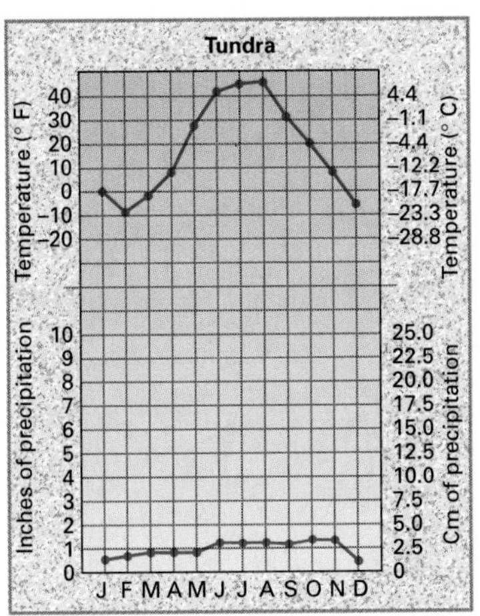

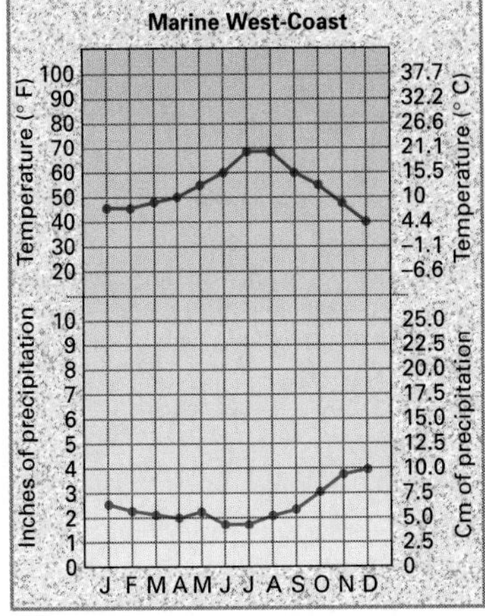

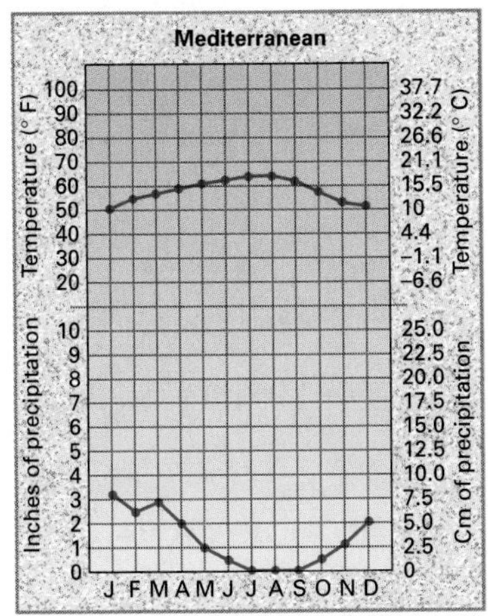

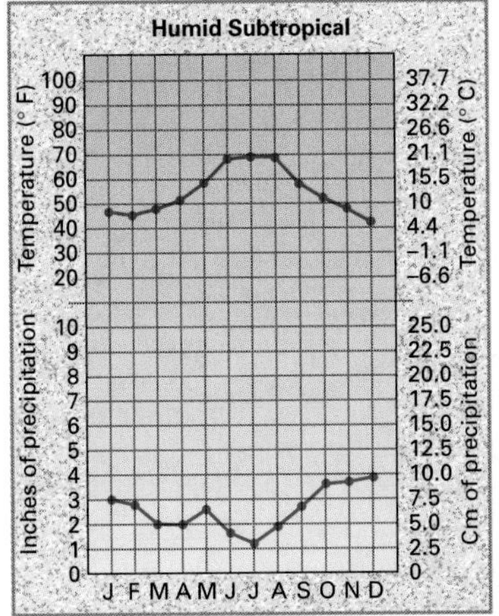

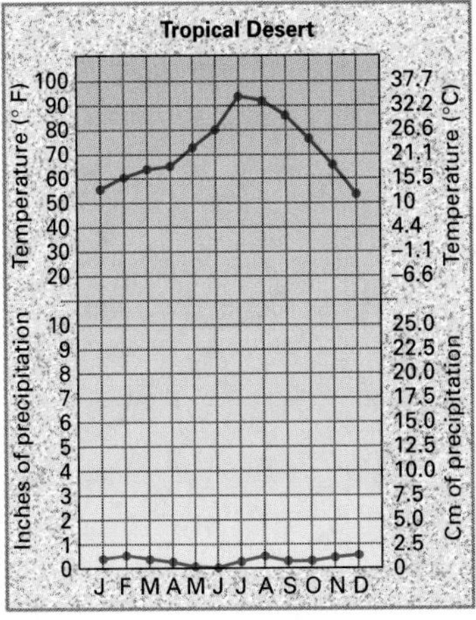

for Long-Range Investigation 8

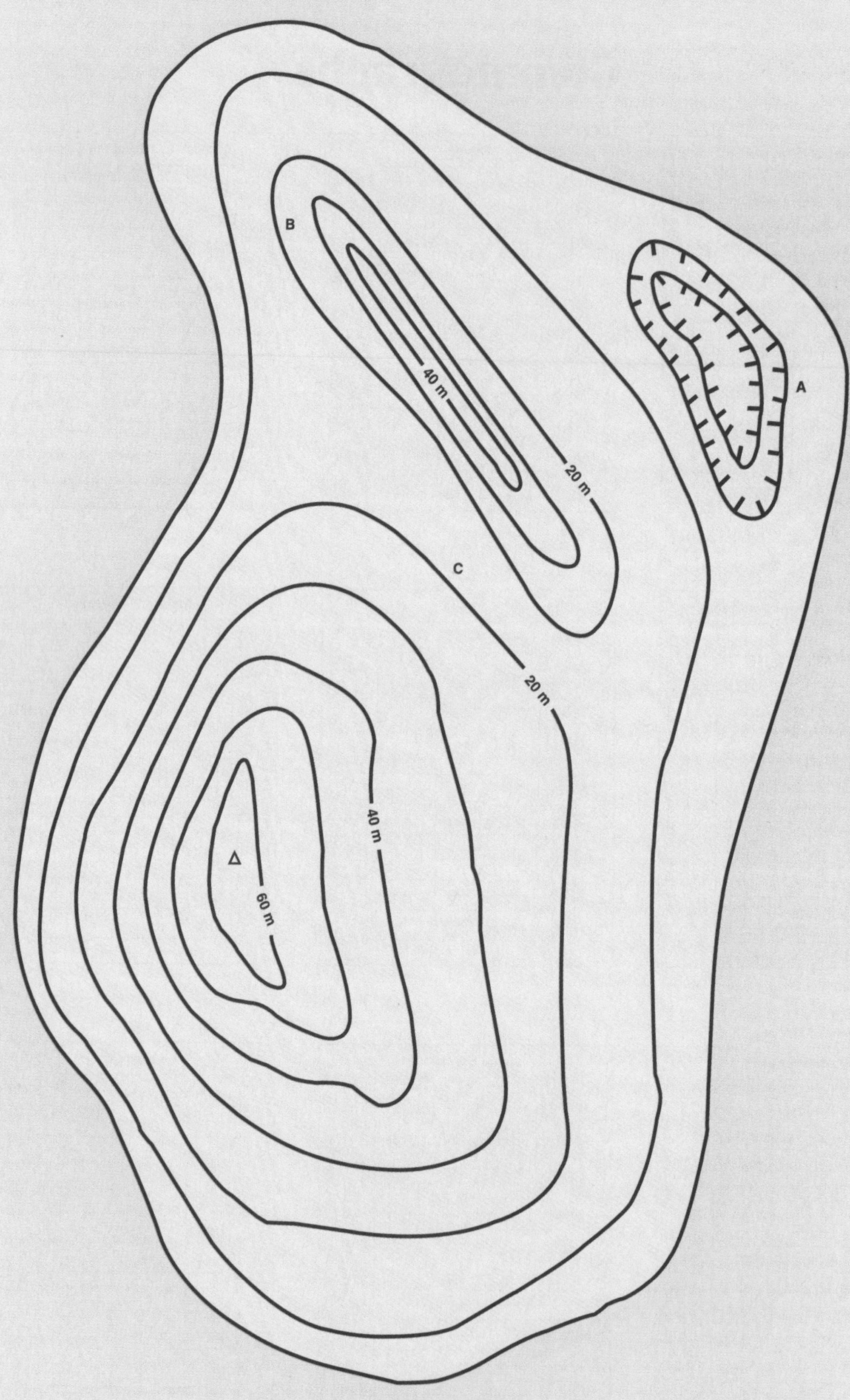

for Chapter 3 In-Depth Investigation

Weather Maps of the United States

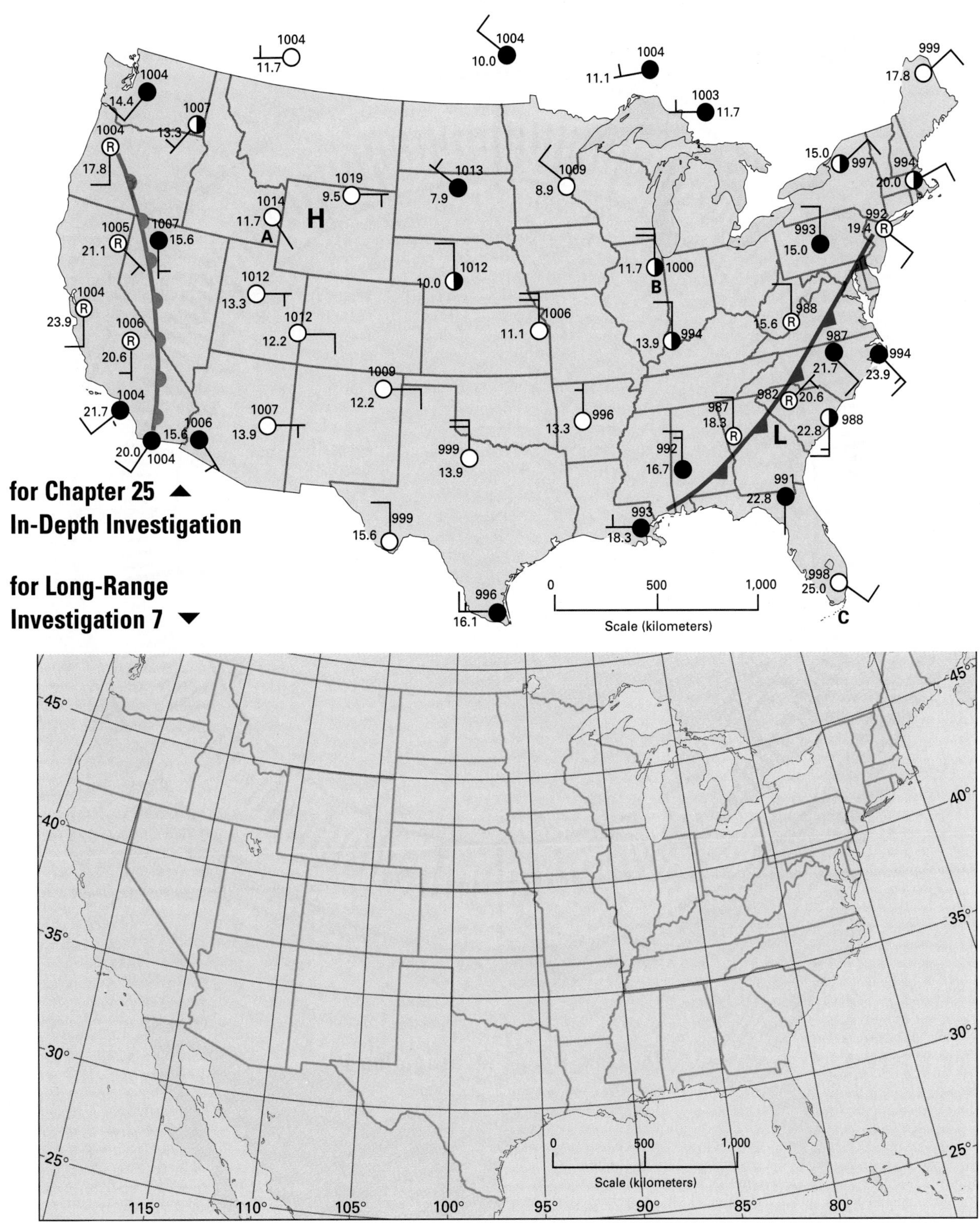

Bearing Chart

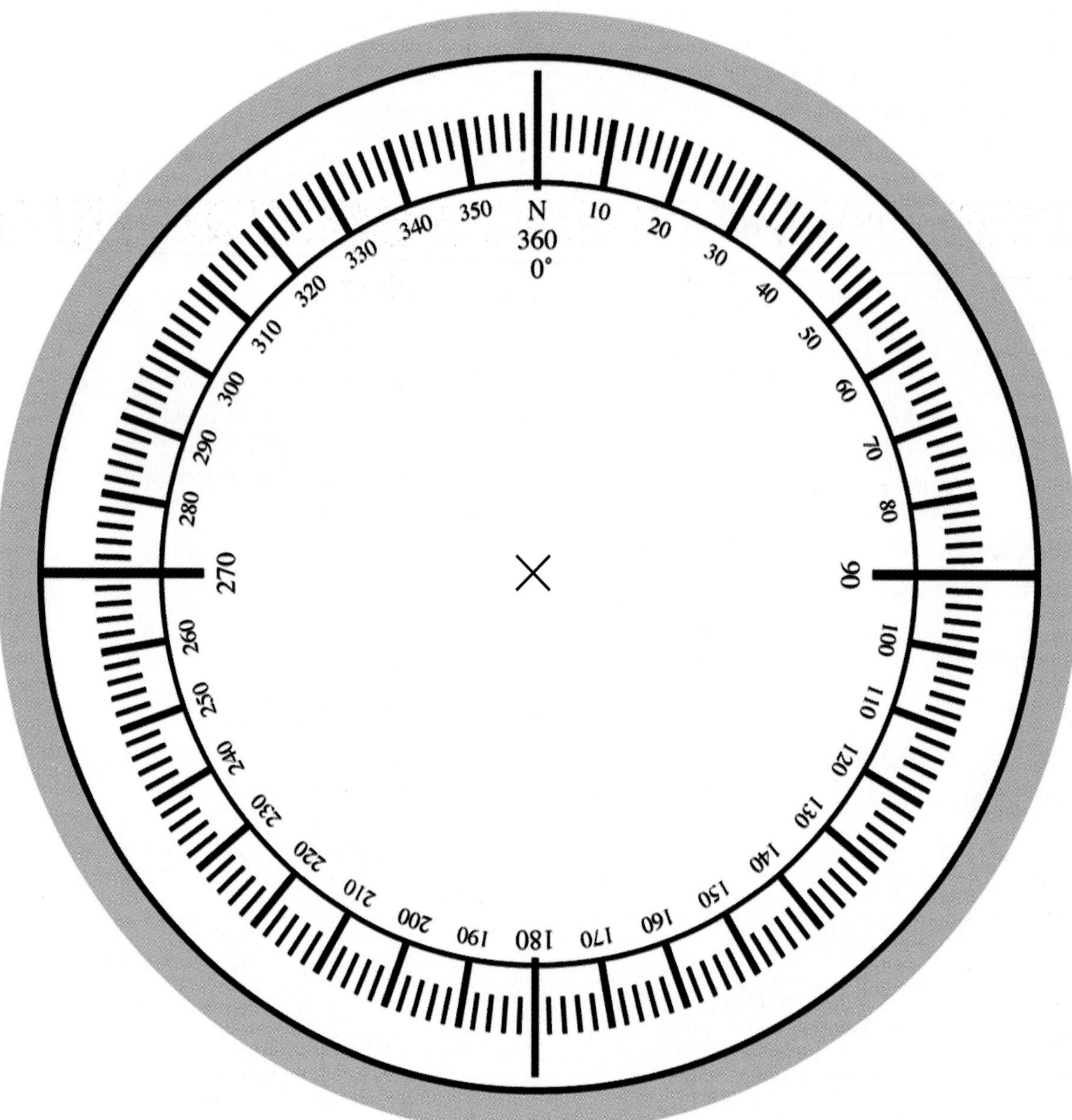

for Long-Range Investigation 1

Analyzing Science Terms

You can often unlock the meaning of an unfamiliar science term by analyzing its word parts. Prefixes and suffixes, for example, each carry a meaning that derives from a word root—usually Latin or Greek. The prefixes and suffixes listed below provide clues to the meanings of many science terms.

Science Terms

Word part or root	Meaning	Application
astr-, aster-	star	astronomy
bar-, baro-	weight, pressure	barometer
batho-, bathy-	depth	batholith, bathysphere
circum-	around	circum-Pacific, circumpolar
-cline	lean, slope	anticline, syncline
eco-	environment	ecology, ecosystem
epi-	on	epicenter
ex-, exo-	out, outside of	exosphere, exfoliation, extrusion
geo-	earth	geode, geology, geomagnetic
-graph	write, writing	seismograph
hydro-	water	hydrosphere
hypo-	under	hypothesis
iso-	equal	isoscope, isostasy, isotope
-lith, -lithic	stone	Neolithic, regolith
-logy	study	ecology, geology, meteorology
magn-	great, large	magnitude
mar-	sea	marine
meta-	among, change	metamorphic, metamorphism
micro-	small	microquake
-morph, -morphic	form, shape	metamorphic
nebula-	mist, cloud	nebula
neo-	new	Neolithic
paleo-	old	paleontology, Paleozoic
ped-, pedo-	ground, soil	pediment
peri-	around	perigee, perihelion
seism-, seismo-	shake, earthquake	seismic, seismograph
sol-	sun	solar, solstice
spectro-	look at, examine	spectroscope, spectrum
-sphere	ball, globe	geosphere, lithosphere
strati-, strato-	spread, layer	stratification, stratovolcano
terra-	earth, land	terracing, terrane
thermo-	heat	thermosphere, thermometer
top-, topo-	place	topographic
trop-, tropo-	turn, respond to	tropopause, troposphere

GLOSSARY

The glossary contains all of the key terms in the text and their definitions. The number of the page where each term is introduced in the text is enclosed in parentheses after the definition.

A

aa jagged chunks of lava formed by rapid cooling on the surface of a lava flow (120)
abrasion mechanical weathering in which rocks collide and scrape against each other, wearing away the exposed surfaces (221)
absolute age actual age of an object (327)
absolute magnitude brightness of a star as it would appear if located 32.6 light-years from the earth (553)
abyssal plain extremely level area of the deep ocean basin (396)
abyssal zone benthic environment that extends from beneath the bathyal zone to a depth of 6,000 m (416)
adiabatic describing a change in temperature resulting from the expansion or compression of air (484)
advection fog condensation of water vapor that results from the cooling of warm, moist air as it moves across a cold surface (488)
advective cooling decrease in the temperature of a mass of air that results as it moves over a cold surface (486)
aftershock tremor that follows and is smaller than a major earthquake (99)
air mass large body of air with uniform temperature and moisture content (499)
albedo percent of solar radiation reflected by a surface (466)
alloy solid solution of two or more metals (151)
alluvial fan fan-shaped deposit of sediments at the base of a slope on land (252)
amber hardened tree sap in which fossils may be preserved (335)
anemometer instrument used to measure wind speed (509)
angular unconformity boundary between horizontal and tilted layers of rock (325)
anthracite hardest form of coal (199)
anticline upcurved fold in horizontal rock layers (85)
anticyclone storm that spirals outward from a high-pressure center (505)
aphelion point in the orbit of a planet at which the planet is farthest from the sun (29)
apogee point in the orbit of a satellite at which the satellite is farthest from the earth (621)
apparent magnitude brightness of a star as it appears from the earth (552)
aquaculture farming of the ocean (420)
aquifer body of rock that can store much water and from which water flows freely (261)
arête sharp, jagged ridge formed between cirques (282)
artesian formation sloping layer of permeable rock sandwiched between two layers of impermeable rock and exposed at the surface (267)
artesian spring natural flow of water to the earth's surface from an artesian formation (267)
artesian well hole dug through the cap rock of an artesian formation through which water flows freely, with no pumping necessary (267)
asteroid fragment of rock that orbits the sun (606)
asteroid belt the region between the orbits of Mars and Jupiter in which most asteroids are found (606)
asthenosphere zone of mantle beneath the lithosphere that consists of slowly flowing solid rock (72)
astronomical unit average distance between the earth and the sun, approximately 149.5 million km (592)
astronomy study of the universe beyond the earth (6)
atmosphere thick blanket of gases surrounding the earth (7)
atmospheric pressure ratio of the weight of the air to the area of the surface on which it presses (457)
atoll nearly circular coral reef surrounding a shallow lagoon (311)
atom smallest unit of an element (140)
atomic number number of protons in an atom (141)
aurora sheets of colored light produced by a magnetic storm in the earth's upper atmosphere (579)
autumnal equinox beginning of the fall season (31)
axis imaginary straight line running through the earth from pole to pole (23)

B

background radiation low levels of energy evenly distributed throughout the universe (16)
barometer instrument that measures atmospheric pressure (458)
barred spiral galaxy type of spiral galaxy with a bar of stars that runs through its center (563)
barrier island long, narrow ridge of sand that lies parallel to the shore (310)

barrier reef type of coral reef that surrounds the remnant of a partially submerged volcanic island (311)

basal slip movement of a glacier caused by the melting of ice in contact with the ground (279)

batholith largest type of igneous intrusion, covering over 100 square kilometers and reaching a depth of thousands of meters (180)

bathyal zone benthic environment that begins at the end of the continental shelf and extends to a depth of 4,000 m (415)

bathyscaph self-propelled, free-moving submersible used for deep-ocean research (391)

bathysphere spherical submersible that remains attached to a research ship for communications and support (390)

beach deposit of rock fragments along an ocean shore or a lakefront (304)

bedding plane boundary between two sedimentary rock layers (324)

benthic environment major division of ocean environment that includes the five bottom zones: intertidal, sublittoral, bathyal, abyssal, and hadal (415)

benthos organisms that live on the ocean floor (415)

berm raised midsection of a beach, the part above which is usually used for recreation (305)

big bang theory theory that all matter and energy in the universe was compressed into an extremely small volume that suddenly, billions of years ago, began expanding in all directions (17)

bimetal thermometer instrument used to measure temperature, consisting of a bar made of two strips of different metals that curves when heated and straightens when cooled (509)

binary star pair of stars that revolve around each other (564)

biodegradable able to be broken down into component parts by microorganisms (8)

biosphere ecosystem encompassing all the life on earth and the physical environment that supports it (7)

bituminous coal soft coal (199)

black hole hole in space with a gravity so great that not even light can escape, formed by the collapse of a very large supernova (560)

bora cold northern wind that blows down the mountains of Greece and the Balkan nations toward the Adriatic Sea (528)

breaker foamy mass of water that washes onto the shore (436)

breccia elastic sedimentary rock composed of angular fragments cemented together by minerals (182)

Bright Angel shale rock layer of the Grand Canyon deposited during the Cambrian Period (374)

butte elevated, narrow, flat-topped area (237)

C

calcareous ooze type of ooze that is mostly calcium carbonate (401)

caldera large basin-shaped depression formed when an explosion destroys the upper part of a volcanic cone (124)

calendar system used to measure the passage of time (627)

Canadian Shield exposed portion of the craton around which North America has been built up (370)

cap rock top layer of impermeable rock in an artesian formation (267)

capillary fringe region of soil just above the water table that receives moisture from the zone of saturation by capillary action (263)

carbonation chemical weathering process in which minerals react with carbonic acid (222)

carbonization process in which plant materials are changed into carbon (198)

cartography science of map making (46)

cementation process in which dissolved minerals left by water passing through sediments bind the sediments together (182)

Cenozoic Era the most recent geologic era, beginning 65 million years ago; the Age of Mammals (346)

channel path that a stream follows (247)

chemical bond force that holds together the atoms that make up a compound (148)

chemical formula symbols indicating the elements a compound contains and the relative number of each element (150)

chemical property characteristic that describes how a substance interacts with other substances to produce different kinds of matter (139)

chemical sedimentary rock rock formed from minerals that have been dissolved in water (182)

chemical weathering process in which rock is broken down as a result of chemical reactions (219)

chinook warm, dry wind that flows down the eastern slope of the Rocky Mountains (528)

chromosphere thin layer of the sun's atmosphere that lies above the photosphere and glows with a reddish light; color sphere (575)

cinder cone steep-sloped deposit of solid fragments ejected from a volcano (122)

circumpolar describing any star that is always visible in the night sky and, from the Northern Hemisphere, can be seen circling Polaris (549)

cirque bowl-shaped depression produced by a valley glacier (282)

cirrus cloud feathery cloud composed of ice crystals that has the highest altitude of any cloud in the sky (486)

clastic sedimentary rock rock made up of fragments from pre-existing rocks (182)

cleavage splitting of a mineral along smooth, flat surfaces (164)

climate general weather conditions over many years (455)

cloud seeding addition of freezing nuclei to supercooled clouds in an attempt to induce or increase precipitation (492)

coalescence combination of different-sized cloud droplets to form larger droplets (491)

Coconino sandstone rock layer of the Grand Canyon deposited during the Permian Period (375)

cold front boundary formed where a cold air mass overtakes and lifts a warm air mass (502)

coma spherical cloud of gas and dust that surrounds the nucleus of a comet (607)

comet body of rock, dust, methane, ammonia, and ice that revolves around the sun in a long, elliptical orbit (607)

compaction process in which air and water are squeezed out of sediments, resulting in the formation of sedimentary rock (182)

composite cone also called *stratovolcano,* steep-sloped volcanic deposit with alternating layers of hardened lava flows and tephra (123)

compound two or more atoms that have been chemically combined (147)

compression stress that squeezes crustal rocks together (84)

concretion nodule of rock with a different composition from that of the main rock body (186)

condensation process by which water vapor changes to liquid water (244)

condensation nuclei solid particles in the atmosphere, such as ice and dust, that provide the surfaces on which water vapor condenses (484)

conduction type of energy transfer in which vibrating molecules pass heat along to other vibrating molecules by direct contact (468)

cone of depression lowered area of a water table produced by pumping water from a well (266)

conglomerate sedimentary rock composed of rounded gravel or pebbles cemented together by minerals (182)

conic projection map projection in which the meridians converge at the poles; the parallels appear as equally spaced, concentric curves (47)

constellation pattern of stars (561)

contact metamorphism change in the structure and mineral composition of rock surrounding an igneous intrusion (187)

continental crust material that makes up landmasses (72)

continental drift hypothesis stating that the continents once formed a single landmass, broke up, and drifted to their present locations (67)

continental ice sheet mass of ice that covers large land areas (278)

continental margin part of the ocean floor that is made of continental crust (393)

continental polar describing a cold, dry air mass that forms over land in polar regions (500)

continental rise accumulation of sediments at the base of a continental slope (395)

continental shelf edge of a continent covered by shallow ocean water (393)

continental slope steep incline at the edge of a continental shelf (394)

continental tropical describing a warm, dry air mass that forms over land in tropical regions (500)

contour interval difference in elevation between one contour line and the next (51)

contour line line on a map connecting points with the same elevation (50)

convection transfer of heat through the movement of a fluid material (74)

convection cell looping pattern of flowing air (470)

convection current movement in a fluid caused by uneven heating (74)

convective cooling decrease in the temperature of a mass of air that results as the air rises and expands (484)

convective zone region around the sun's radiative zone in which moving gases transfer energy (574)

convergent boundary border formed by the direct collision of two lithospheric plates (73)

coprolite fossilized waste material from an animal (337)

coral reef ridgelike coastal feature made of millions of coral skeletons (311)

core center of a planetary body, such as the earth (24)

core sample cylindrical sample of sediments from the deep-ocean floor (399)

Coriolis effect deflection of wind and ocean currents caused by the earth's rotation (428)

corona outermost layer of the sun's atmosphere; crown (575)

covalent bond bond based on the attraction between atoms that share electrons (150)

covalent compound compound formed from atoms that share electrons (150)

crater funnel-shaped pit at the top of a volcanic cone (124); bowl-shaped depression on the surface of a planetary body (616)

craton large area of Precambrian rocks found on all continents (366)

creep slow downhill movement of weathered rock material (236)

crest highest point of a wave (433)

crevasse large crack that forms on the surface of a glacier (280)

crude oil unrefined petroleum (200)

crust outermost zone of the solid earth (24)

crystal natural solid substance that has a definite geometric shape (159)

cumulus cloud thick, billowy cloud that forms above stratus clouds and below cirrus clouds (486)

current steady movement in one direction, such as that of water in the ocean (427)

D

day time required for the earth to make one rotation on its axis, about 24 hours (627)

daylight saving time system in which clocks are set one hour ahead of standard time from April to October (34)

deep current streamlike movement of water beneath the surface of the ocean (427)

deep ocean basin part of the ocean floor made of oceanic crust (393)

deflation most common form of wind erosion in which fine, dry soil particles are blown away (298)

deflation hollow shallow depression left after the wind has eroded a layer of exposed soil (298)

deformation bending, tilting, and breaking of the earth's crust (83)

delta fan-shaped deposit of sediments at the mouth of a stream (252)

density ratio of the mass of a substance to its volume, expressed as g/cm^3 (167)

depression contour contour line with short, straight lines drawn along the inside of the loop pointing toward its center that indicates a depression (55)

desalination process of removing salt from ocean water (246)

desert pavement surface of closely packed small rocks left after the top layer of soil has been removed by deflation (298)

dew type of condensation formed when air that is in contact with a cool surface loses heat until it reaches saturation (483)

dew point temperature to which air must be cooled to become saturated (483)

diatomic consisting of two atoms (147)

dike igneous intrusion that cuts across rock layers (181)

discharge volume of water moved by a stream within a given time (249)

disconformity boundary between layers of rock that have not been deposited continuously (325)

distillation process in which ocean water is heated until it evaporates in order to separate fresh water from dissolved salts (417)

divergent boundary boundary formed by two lithospheric plates that are moving apart (73)

divide elevated region that separates two watersheds (247)

doldrums narrow zone of low air pressure at the equator characterized by weak and undependable winds (471)

dome mountain landform created when molten rock pushes up rock layers on the earth's surface and the layers then are worn away in places, leaving separate high peaks (93)

Doppler effect apparent shift in the wavelength of energy, such as a sound wave or a light wave, emitted by a source moving away from or toward an observer (15)

double refraction property exhibited by transparent minerals that produce a double image of any object viewed through them (168)

drift weak, slow-moving ocean current (429)

drumlin long, low, tear-shaped mound of till (285)

dune mound of windblown sand (299)

E

earth science study of the earth and the universe around it (3)

earth-grazer asteroid that orbits the sun in an elongated ellipse that may pass close to the earth and sun (607)

earthquake vibration of the earth's crust (99)

earthshine sunlight reflected off the surface of the earth (626)

eclipse passing of one planetary body through the shadow of another (623)

ecology study of the complex relationships between living things and their environment (7)

ecosystem community of organisms and the environment they inhabit (7)

elastic rebound theory theory that rocks that are strained past a certain point will fracture and spring back to their original shape (99)

electrical thermometer instrument used to measure temperature based on the increased flow of electricity through certain materials when the materials are heated (509)

electromagnetic spectrum complete range of wavelengths of radiation (463)

electron subatomic particle with a negative electrical charge (141)

electron cloud region of space around the nucleus of an atom in which electrons may be found (141)

element substance that cannot be broken down into a simpler form by ordinary chemical means (15)

elevation height above sea level (50)

ellipse oval whose shape is determined by two points within the figure (592)

elliptical galaxy type of galaxy with a very bright center that contains little dust and gas and is spherical to disklike in shape (563)

emergent coastline coast along which sea level falls or the land rises (310)

energy level arrangement of electrons within the electron cloud of an atom (147)

epicenter point on the earth's surface directly above the focus of an earthquake (99)

epicycle small circular motion of the planets within their orbits proposed as an explanation for retrograde motion (591)

epoch subdivision of a geologic period (346)

era largest unit of geologic time (346)

erosion process by which the products of weathering are transported (231)

erratic large boulder transported and deposited by a glacier (284)

esker long, winding ridge of gravel and coarse sand deposited by a glacier (286)

estuary wide, shallow bay formed where ocean water submerges the mouth of a river and where salt water and fresh water mix (309)

evaporites sedimentary rocks formed from minerals left after water evaporates (183)

evapotranspiration process by which water enters the atmosphere; evaporation and transpiration combined (243)

evolution change of living things over time (334)

exfoliation process in which sheets of rock peel or flake as a result of weathering (219)

exosphere layer of the atmosphere above the ionosphere that merges with interplanetary space (461)

experimentation process by which a scientific procedure is carried out according to certain guidelines (12)

extrusive igneous rocks rocks formed from molten lava that hardens on the earth's surface (178)

F

fault break in rock along which rocks on either side of the break move (86)

fault plane surface of a fault along which movement of rocks occurs (86)

fault zone group of interconnected faults (101)

fault-block mountain mountain formed where faulting breaks the earth's crust into large blocks and the blocks are uplifted and tilted (92)

felsic lava silica-rich lava (120)

fetch distance that wind can blow across open water (435)

fiord narrow, deep, steep-walled bay formed by flooding of a glacial valley due to a rise in sea level (310)

fireball brilliant flash of light produced by a meteor that vaporizes quickly (608)

firn grainy ice in a glacier that has been partially melted and refrozen (277)

fissure crack in a rock surface through which lava flows (119)

floodplain part of the valley floor that may be covered with water during a flood (253)

fluorescence ability to glow under ultraviolet light (168)

focus area along a fault at which slippage first occurs, initiating an earthquake (99); one of two points within an ellipse that determines the shape of the figure (592)

foehn warm, dry wind that flows down the slopes of the Alps (528)

folded mountain landform created when tectonic movements bend and uplift rock layers (90)

folding permanent deformation or bending of a rock under stress (85)

foliated describing a metamorphic rock with visible parallel bands (187)

footwall rock below a fault plane (86)

fossil trace or remains of a plant or an animal in sedimentary rock (186)

fossil fuel fuel formed from the remains of living organisms, such as coal, petroleum, and natural gas (198)

fracture break in rock along which there is no movement (86)

fracture zone faults running perpendicular to a mid-ocean ridge (398)

freezing nuclei condensation nuclei with a crystalline structure like that of ice (491)

fringing reef type of coral reef that forms around a volcanic island (311)

front boundary between air masses of different densities (502)

frost ice crystals formed when the dew point is below 0°C and water vapor directly enters the solid state (483)

full moon phase of the moon during which the entire half of the moon facing the earth is visible (626)

G

galaxy large-scale group of stars (562)

Galilean moon any one of the four largest satellites of Jupiter, which were first seen by Galileo (631)

gas physical form of matter that does not have a definite volume or shape (145)

gastrolith fossilized stone found within the digestive system of a dinosaur or other reptile (337)

gemstone nonmetallic mineral that is brilliant and colorful when cut (196)

geocentric earth-centered (591)

geologic column arrangement of rock layers based on the ages of the rocks (345)

geology study of the origin, history, and structure of the solid earth and the processes that shape it (4)

geomagnetic pole point on the earth's surface above a pole of the earth's imaginary internal magnet (45)

geosphere the solid earth (7)

geosynchronous orbit orbit directly above the earth's equator and moving in the direction of the earth's rotation (35)

geothermal energy energy contained in and available from water heated by magma or gases within the earth (205)

geyser hot spring that erupts periodically (268)

giant very large, cool, bright star (554)

glacial drift sediments deposited by a glacier (285)

glacier mass of moving ice (277)

glaze ice thick layer of sheet ice formed when rain freezes as it contacts a surface (489)

globular cluster spherically shaped group of hundreds of stars located around the core of the Milky Way Galaxy (564)

gnomonic projection map projection in which the parallels appear as unevenly spaced, concentric circles, the meridians appear as straight lines radiating from a central point, and all other great circles appear as straight lines (47)

Gondwanaland southern landmass that broke away from Pangaea and later formed South America, Africa, India, Australia, and Antarctica (367)

graben long, narrow valley formed by faulting and downward slippage of a crustal block (92)

gradient change in elevation over a distance (249)

gravity force of attraction between all matter in the universe (27)

great circle any circle that divides the globe in half (44)

greenhouse effect process by which the atmosphere traps infrared rays over the earth's surface (467)

Gregorian calendar revision of the Julian calendar by Pope Gregory XIII; currently used in most of the world (628)

ground moraine unsorted material left beneath a glacier when the ice melts (285)

groundwater water that soaks deep into soil and rock (243)

Gulf Stream swift, warm Atlantic current that flows from the Gulf of Mexico, around Florida, and up the east coast of North America (428)

guyot flat-topped, submerged seamount (398)

gyre huge circle of moving ocean water formed as a result of the wind belts and the Coriolis effect (428)

H

H-R diagram Hertzsprung-Russell diagram; graph showing the relationship of the surface temperature and absolute magnitude of a star (553)

hadal zone benthic environment of the ocean deeper than 6,000 m (416)

hail type of precipitation in the form of lumps of ice (490)

hair hygrometer instrument used to measure relative humidity, based on the fact that human hair stretches as humidity increases (482)

half-life time required for half the mass of a radioactive element to decay into its daughter elements (330)

hanging valley small abandoned glacial valley suspended on a mountain above the main glacial valley (283)

hanging wall rock above a normal fault plane (86)

hard water water that contains relatively large amounts of dissolved minerals (269)

hardness measure of the ability of a mineral to resist scratching (165)

headland change-resistant projection of rock out from shore into the water (303)

headward erosion lengthening and branching of a stream (247)

headwaters beginning of a stream (249)

heliocentric sun-centered (591)

Hermit shale rock layer of the Grand Canyon deposited during the Permian Period (375)

horizon layer of a soil profile (228)

horn sharp, pyramid-like peak formed where several arêtes join (282)

horse latitudes subtropical high-pressure belt of air, around 30° latitude (471)

hot spot area of volcanism within a lithospheric plate (119)

hot spring hot groundwater that rises to the surface before cooling (267)

humid continental climate middle-latitude climate occurring between 35° and 50° north latitude, with warm, humid summers and cold winters (534)

humid subtropical climate very wet middle-latitude climate that occurs in southeastern coastal areas between 30° and 40° north and south latitude, with warm, humid summers and generally mild winters (534)

humidity amount of water vapor in the atmosphere (480)

humus dark, organic material formed from the decayed remains of plants and animals (227)

hurricane severe storm that develops over tropical oceans, with strong winds that spiral in toward the intensely low-pressure storm center (505)

hydrocarbon compound made up of atoms of carbon and hydrogen (198)

hydroelectric energy energy produced by running water (206)

hydrolysis chemical reaction between water and another substance (221)

hydrosphere all the earth's water (7)

hypothesis possible explanation of a problem that is based on facts (11)

ice age long period of climatic cooling during which ice sheets cover large areas of the earth's surface (289)

ice wedging mechanical weathering caused by the freezing and thawing of water that seeps into cracks in rocks (219)

igneous rock rock formed from cooled and hardened magma (175)

impermeable rock or sediment through which water cannot flow (263)

index contour every fifth contour line on a topographic map that is printed bolder for reference (53)

index fossil guide fossil; fossil found in the rock layers of only one geologic age and is used to establish the relative age of the rock layers (338)

inertia tendency of a moving body to remain in motion or a stationary body to remain at rest until an outside force acts on it (593)

inorganic not made up of living organisms or the remains of living organisms (157)

intensity amount of damage caused by an earthquake (104)

internal plastic flow slow movement of a glacier in which ice crystals slip over each other (279)

International Date Line line running from north to south through the Pacific Ocean where the date changes from one day to the next (34)

intertidal zone benthic environment that lies between the low-tide and high-tide lines (415)

intrusive igneous rocks rocks formed from the cooling of magma beneath the earth's surface (178)

invertebrate animal without a backbone (349)

ion atom or group of atoms that carries an electrical charge (149)

ionic bond bond in which electrons are transferred from one atom to another (148)

ionic compound compound formed through the transfer of electrons (148)

ionosphere lower region of the thermosphere, at an altitude of 80 to 550 km (461)

iron meteorite type of meteorite made of iron, with a characteristic metallic appearance (609)

irregular galaxy type of galaxy with no identifiable shape and an uneven distribution of stars within it (563)

island arc chain of volcanic islands formed along an ocean trench (74)

isobar line drawn on a weather map connecting points of equal atmospheric pressure (513)

isostasy balancing of the forces pressing up and down on the earth's crust (83)

isostatic adjustment up-and-down movements of the earth's crust to reach isostasy (83)

isotope atom of an element that has the same atomic number but different atomic mass than another atom of that element (145)

jet streams bands of high-speed, high-altitude westerly winds (472)

Jovian planet any one of the first four outer planets—Jupiter, Saturn, Uranus, or Neptune—with properties similar to those of Jupiter (598)

Julian calendar calendar devised by Julius Caesar's astronomers consisting of 12 months; 11 with 30 or 31 days, one with 28 days, and an extra day added every four years (628)

Kaibab limestone rock layer of the Grand Canyon deposited during the Permian Period (375)

karst topography region where the effects of chemical weathering due to groundwater, such as sinkholes and caverns, are clearly visible (271)
kettle depression in a glacial outwash plain (286)

L

L wave surface or long wave; the slowest wave generated by an earthquake and the last to be recorded by a seismograph (103)
laccolith flat-bottomed intrusion that pushes overlying rock layers into an arc (181)
lagoon narrow region of shallow water between a barrier island and the shore (310)
landform physical feature of the earth's surface (236)
landslide sudden movement of loose rock and soil down a slope (233)
lapilli tephra particles between 2 mm and 64 mm in diameter (121)
latent heat energy stored in molecules (480)
lateral moraine unsorted material deposited along the sides of a valley glacier (285)
laterite thick infertile soils produced in tropical climates (229)
latitude angular distance north or south of the equator (43)
Laurasia northern landmass that broke away from Pangaea and later formed North America and Eurasia (367)
lava magma that reaches the earth's surface (117)
lava plateau raised, flat-topped area made of layers of hardened lava (181)
law of crosscutting relationships principle that a fault or intrusion is always younger than the rock layers it cuts through (326)
law of gravitation principle that the force of attraction between two objects depends on the masses and the distance between the objects (28)
law of superposition principle that a sedimentary rock layer is older than the layers above it and younger than the layers below it (324)
leaching process in which water carries dissolved minerals to lower layers of rock (221)
leap year year with an extra day in it, occurring every four years (628)
legend list of map symbols and their meanings (48)
light-year distance that light travels in one year, about 9.5 trillion km (550)
lignite brown coal (199)
liquid physical form of matter with a definite volume but no definite shape (145)
lithosphere thin outer shell of the earth consisting of the crust and the rigid upper mantle (72)
lode deposit formed by thick mineral veins (195)
loess thick, yellowish deposit of windblown dust (301)
longitude angular distance east or west of the prime meridian (44)
long-period comet comet with a period of several thousand or million years (608)
lunar eclipse passing of the earth between the moon and the sun during which the earth's shadow crosses the lighted half of the moon (624)
luster light reflected from the surface of a mineral (164)

M

mafic lava dark-colored lava rich in magnesium and iron (120)
magma liquid rock produced deep inside the earth (117)
magnetic declination angle between the direction of the earth's geographic pole and the direction in which a compass needle points (45)
magnetosphere a region of space that is affected by the earth's magnetic field (26)
main-sequence star star with characteristics that place it within a band running through the middle of the H-R diagram (554)
mantle zone of rock below the earth's crust (24)
map projection flat map that represents a three-dimensional curved surface (46)
maria (sing: mare) dark areas of smooth, dry, solidified lava on the moon that reflect little light (616)
marine west-coast climate wet middle-latitude climate that occurs in western coastal areas located between 40° and 60° latitude, with relatively cool summers and mild winters (533)
maritime polar describing a cold, moist air mass that is formed over the ocean in polar areas (500)
maritime tropical describing a warm, moist air mass that is formed over the ocean in tropical areas (500)
mass movement movement of rock fragments down a slope (233)
mass number sum of the numbers of protons and neutrons in an atom (141)
matter substance that takes up space and has mass (139)
mean sea level point midway between the highest and lowest tide levels of the ocean (50)
meander wide curve in a stream channel (251)
measurement comparison of a property of an object or phenomenon with a standard unit (9)
mechanical weathering process that changes the physical form of rocks (219)

medial moraine ridge of unsorted glacial material along the center of a valley glacier (285)

Mediterranean climate middle-latitude climate that occurs in coastal areas located between 30° and 40° latitude, with dry summers and wet winters (533)

meltwater melted ice flowing from a glacier (285)

Mercalli scale scale that expresses the intensity of an earthquake with a Roman numeral and a description (104)

Mercator projection map projection in which the meridians appear as straight, parallel, evenly spaced lines and form a grid with the parallels, which appear as straight, parallel, and unevenly spaced lines (47)

meridian semicircle on the earth that runs from pole to pole (43)

mesa elevated, flat-topped area smaller than a plateau (237)

mesopause upper boundary of the mesosphere, marked by an increase in temperature (461)

mesosphere coldest layer of the atmosphere that extends upward from the stratosphere to an altitude of about 80 km (461)

Mesozoic Era geologic era that lasted from 245 million to 65 million years ago; the Age of Reptiles (346)

metamorphic rock rock formed from other rocks as a result of intense heat, pressure, or chemical processes (176)

metamorphism changing of one type of rock to another by heat, pressure, and chemical processes (187)

meteor meteoroid that enters the earth's atmosphere (608)

meteor shower phenomenon caused by the burning up of large numbers of meteors as they enter the earth's atmosphere (608)

meteorite meteor or part of a meteor left after it hits the earth's surface (609)

meteoroid small bit of rock or metal moving through the solar system left by a comet or produced by a collision between asteroids (608)

meteorology study of the earth's atmosphere (5)

microquake earthquake with a magnitude less than 2.5 on the Richter scale (104)

Mid-Atlantic Ridge undersea mountain range with a steep, narrow valley along its center (68)

middle-latitude climates climates with a maximum average temperature of 18°C in the coldest month and a minimum average temperature of 10°C in the warmest month (529)

middle-latitude desert climate middle-latitude climate that is very dry, with both a cold winter and a warm to very hot summer (533)

middle-latitude steppe climate middle-latitude climate with slightly more precipitation than a middle-latitude desert climate and a high yearly temperature range (533)

mid-ocean ridges system of undersea mountain ranges that wind around the earth (68)

Milankovitch theory theory that small, regular changes in the earth's orbit and in the tilt of the earth's axis caused the ice ages (290)

mineral natural inorganic, crystalline solid found in the earth's crust (157)

mineralogist scientist who specializes in the study of minerals (163)

mistral strong, cold, northern wind that blows down the Alps toward the Mediterranean Sea (528)

mixture material that contains two or more substances that are not chemically combined (151)

Moho the Mohorovičić discontinuity, boundary between the earth's crust and mantle (24)

Mohs hardness scale standard against which the hardness of a mineral is tested (165)

molecule smallest complete unit of a compound (147)

monadnock knob of rock that protrudes above a peneplain (236)

monocline gently dipping bend in horizontal rock layers (85)

monsoon seasonal wind that blows toward the land in summer, bringing heavy rains, and away from the land in the winter, bringing dry weather (527)

month time required for the moon to go through one set of phases as it orbits the earth, about 29.5 days (627)

moon body that is smaller than a planet and orbits the planet (580)

mountain belt group of large mountain systems (88)

mountain range group of adjacent mountains with the same general shape and structure (88)

mountain system group of adjacent mountain ranges (88)

Muav limestone rock layer of the Grand Canyon deposited during the Cambrian Period (374)

mud fine particles of rock combined with water (401)

mud pot weathered rock around a hot spring that mixes with the hot water to form liquid clay that bubbles at the surface (268)

mudflow rapidly moving large mass of mud (233)

mummification preservation of a dead organism by drying (335)

N

natural bridge arch of rock formed by groundwater erosion (271)

natural levee raised riverbank that results when a river deposits its load at the river's edges (254)

neap tide tide with minimum daily tidal range that occurs during the first and third quarters of the moon (440)

nebula dark cloud of gas and dust in space; first stage in the development of a star (555)

nebular theory theory that the sun and the planets condensed out of a spinning cloud of gas and dust (580)

nekton forms of ocean life that swim, such as fish, dolphins, and squid (415)

neritic zone pelagic environment above the sublittoral zone filled with marine life (415)

neutron subatomic particle with no electrical charge (141)

neutron star collapsed core of a supernova consisting of a small, extremely dense ball of neutrons (559)

new moon phase of the moon during which the side of the moon facing the earth is unlighted (626)

nitrogen cycle process in which nitrogen moves from the air to the soil to animals and back to the air (457)

nodule lump of minerals on the ocean floor (401)

nonconformity unconformity in which stratified rock rests on unstratified rock (325)

nonrenewable resource substance of limited supply that cannot be replaced (195)

nonsilicate mineral mineral that does not contain silicon (158)

normal fault fault in which the hanging wall moves down relative to the footwall (86)

nova white dwarf star that explodes as it cools, temporarily becoming thousands of times brighter (559)

nuclear fission splitting of the nucleus of a large atom into smaller nuclei (202)

nuclear fusion combination of the nuclei of small atoms to form a larger nucleus (203)

nucleus region in the center of an atom that contains the protons and neutrons (141)

O

observation act of using the senses to gather information (9)

occluded front boundary formed where a fast-moving cold air mass overtakes and lifts a warm air mass, completely cutting it off from the ground (503)

ocean floor continental crust and oceanic crust that lie beneath the ocean (390)

ocean trench deep valley in the ocean floor that forms along a subduction zone (73)

oceanic crust material that makes up the ocean floor (72)

oceanic zone pelagic environment that extends seaward beyond the continental shelf (416)

oceanography study of the earth's oceans (5)

Oort cloud spherical cloud of dust and ice surrounding the solar system that may contain as many as a trillion comets (608)

ooze soft organic sediment on the ocean floor (401)

open cluster loosely shaped group of hundreds of stars (564)

orbit period time required for a planet to make one revolution around the sun (593)

ordinary spring natural flow of groundwater to the earth's surface (266)

ordinary well hole dug below the water table that fills with groundwater (266)

ore deposit of minerals from which metals and nonmetals can be profitably removed (195)

organic sedimentary rock rock formed from the remains of organisms (182)

outcrop area of exposed rock (374)

outwash plain deposit of stratified drift in front of a glacier (286)

oxbow lake water remaining in an isolated meander in a floodplain (251)

oxidation chemical combination of metallic elements with oxygen (222)

ozone form of atmospheric oxygen that has three atoms per molecule (455)

P

P wave primary wave; the fastest wave generated by an earthquake and the first to be recorded by a seismograph (103)

Pacific Ring of Fire major earthquake zone that forms a ring around the Pacific Ocean (100)

pack ice floating layer of ice that completely covers an area of the ocean surface (410)

pahoehoe solidified mafic lava with a wrinkled surface (120)

paleontologist scientist who studies fossils (334)

paleontology study of fossils (334)

Paleozoic Era geologic era that followed Precambrian time, lasting from 570 million to 245 million years ago (346)

Pangaea single landmass thought to have been the origin of all continents (67)

Panthalassa giant ocean surrounding Pangaea (67)

parallax method of determining the distance from the earth to a star based on the shift in the apparent position of the star when viewed from different angles (550)

parallel any circle that runs east and west around the earth parallel to the equator (43)

peat brownish-black material produced by partial decomposition of plant remains (198)

pelagic environment major division of ocean environment that includes the two water zones: neritic and oceanic (415)

peneplain low, almost level surface of a mountain in its old stage (236)

penumbra outer part of the shadow cast by the earth or the moon in which sunlight is only partially blocked (623)

perched water table secondary water table formed by a layer of impermeable rock above the main water table (264)

perigee point in the orbit of a satellite at which it is closest to the earth (621)

perihelion point in the orbit of a planet at which it is closest to the sun (29)

period subdivision of a geologic era (346)

periodic table system for classifying the elements (141)

permeability the ease with which water flows through the open spaces in a rock or sediment (262)

petrification process in which organic materials are replaced by minerals (335)

petrochemical chemical derived from petroleum (200)

phase varying shape of the visible portion of the moon (626)

phosphorescence ability to glow during and after exposure to ultraviolet light (168)

photosphere innermost layer of the solar atmosphere; light sphere (574)

physical property characteristic that is observable in a substance without changing the chemical composition of the substance (139)

phytoplankton microscopic ocean plants (414)

pillow lava lava that flows out of fissures on the ocean floor and cools rapidly in rounded shapes (120)

placer deposit fragments of native metals that are concentrated in layers at the bottom of a stream bed (196)

planet any one of the nine major bodies that orbit the sun (580)

planetary nebula expanding shell of gases shed by a dying star (558)

planetesimal small body of matter that formed in the outer regions of the solar nebula while the sun was forming in its center (580)

plankton free-floating, microscopic ocean plants and animals (414)

plate tectonics theory that the lithosphere is made up of plates that float on the asthenosphere and that the plates possibly are moved by convection currents (72)

plateau large area of flat-topped rocks high above sea level (91)

polar climates climates with a maximum average monthly temperature of 10°C (529)

polar easterlies weak global winds located north of 65° north latitude and south of 65° south latitude that flow away from the poles (472)

polar front boundary at which cold polar air meets the warmer air of the middle latitudes (503)

polar orbit orbit that passes over the earth's North and South poles (35)

pollution contamination of the environment with waste products or impurities (8)

polyconic projection map made by fitting together a series of conic projections of adjoining areas (48)

porosity percentage of open spaces in a rock or sediment (261)

porphyry igneous rock composed of large and small crystals (178)

Precambrian time earliest and longest geologic era, lasting from 4.6 billion to 570 million years ago (346)

precession slow circular motion of the earth's axis as it turns in space that traces a circle in space every 26 thousand years (32)

precipitation process by which water falls from clouds to the earth as rain, snow, sleet, and hail (244)

prime meridian the meridian that passes through Greenwich, England, designated as 0° (44)

principle of uniformitarianism theory that geologic processes at work in the present were also at work in the past (323)

prominence cloud of glowing gases that arches high above the sun's surface (577)

proton subatomic particle with a positive electrical charge (141)

protoplanet large body of matter that formed from the coalescence of planetesimals in the solar nebula (580)

protostar center of a shrinking, spinning nebula; the second stage in the development of a star (555)

psychrometer instrument used to measure relative humidity (480)

pulsar neutron star that emits two beams of radiation that sweep across space (560)

pyroclastic material also called tephra, all of the rock fragments ejected from a volcano (121)

Q

quasar starlike object that gives off radio waves and X rays (565)

R

radar device that can detect objects and weather conditions in the upper atmosphere by sending and receiving radio waves (511)

radiation fog condensation of water vapor that results from the cooling of air that is in contact with the ground (488)

radiative zone region surrounding the core of the sun in which energy is transferred in the form of electromagnetic waves (574)

rain gauge instrument used to measure the amount of rainfall (493)

ray streak of displaced rock material radiating from a crater (616)

red shift apparent lengthening of the light waves emitted by a star moving away from the earth (550)

Redwall limestone rock layer of the Grand Canyon deposited during the Mississippian Period (374)

refraction bending of a light ray as it passes from one substance to another (168); bending of a wave as it reaches shallow water (437)

regional metamorphism metamorphism that affects rocks over large areas during periods of tectonic activity (187)

regolith layer of weathered rock fragments covering much of the earth's surface (227)

rejuvenated describing a river with a gradient that has been made steeper by a movement of the earth's crust (251)

relative age age of an object compared with the ages of other objects (324)

relative humidity ratio of the amount of water vapor in the air to the amount of water vapor the air can hold when saturated (480)

relief difference in elevation between the highest and lowest points of an area (51)

renewable resource substance that can be replaced (195)

retrograde motion apparent periodic reversal in the motion of some planets as viewed from the earth (591)

reverse fault fault in which the hanging wall moves up relative to the footwall (86)

revolution movement of a planet around the sun (29)

Richter scale scale that expresses the magnitude of an earthquake (104)

rift valley steep, narrow valley formed as lithospheric plates separate (73)

rille long, deep channel that runs through the maria on the moon (616)

rip current swift movement of water caused by the return of water to the ocean through channels in underwater sand bars (437)

roche moutonnée rounded knob of rock produced by glacial erosion (283)

rock cycle series of processes in which rock changes from one type to another and back again (176)

rockfall fall of rock from a steep cliff (233)

rock-forming mineral any common mineral that forms the rocks of the earth's crust (157)

rotation spinning of a planet on its axis (29)

runoff water that flows over the land into streams and rivers (243)

S

S wave secondary wave; a wave generated by an earthquake and the second to be recorded by a seismograph (103)

salinity number of grams of dissolved salt in 1 kg of ocean water (408)

saltation movement of sand by short jumps, caused by wind or water (249)

sand bar long ridge of sand deposited offshore (305)

satellite object in orbit around a body with a larger mass (35)

saturated describing air that contains all the water vapor it can hold at a specific temperature (480)

scale relationship between distance shown on a map and actual distance (49)

scientific law rule that correctly describes a natural phenomenon (14)

scientific methods organized, logical approaches to scientific research (9)

seafloor spreading movement of the ocean floor away from either side of a mid-ocean ridge (70)

seamount isolated volcanic mountain on the ocean floor (398)

sediment fragments that result from the breaking of rocks, minerals, and organic matter (175)

sedimentary rock rock formed from hardened deposits of sediment (175)

seismic gap zone of rock in which a fault is locked and unable to move and in which no major earthquake has occurred for at least 30 years (110)

seismic wave vibration that travels through the earth (23)

seismograph instrument used to detect and record seismic waves (103)

shadow zone location on the earth's surface where no seismic waves or only P waves can be detected (25)

shearing stress that pushes rocks in opposite horizontal directions (84)

sheet erosion process in which parallel layers of topsoil are stripped away, exposing the surface of the underlying subsoil or partially weathered bedrock (232)

shield cone volcanic deposit of hardened lava with a broad base and gentle slopes (122)

shoreline place where the ocean and the land meet (302)

short-period comet comet with a period of up to 100 years (608)

silicate mineral mineral that contains atoms of silicon and oxygen (158)

siliceous ooze type of ooze that is mostly silicon dioxide (401)

silicon-oxygen tetrahedron four oxygen atoms arranged in a pyramid with one silicon atom in the center (161)

sill sheet of hardened magma that forms between and parallel to layers of rock (181)

sinkhole circular depression caused when the roof of a cavern collapses (270)

sleet ice pellets that form when rain falls through a layer of freezing air (489)

slump downhill movement of a large block of soil under the influence of gravity (234)

smog air pollution formed from a mixture of dust and chemicals (151)

snowfield almost motionless mass of permanent snow and ice (277)

snowline elevation above which ice and snow remain throughout the year (277)

soft water water that contains few dissolved minerals (269)

soil profile cross section of soil layers and bedrock (228)

solar collector device for capturing solar energy (204)

solar eclipse passing of the moon between the earth and the sun during which the shadow of the moon falls on the earth (623)

solar flare sudden, violent eruption of electrically charged atomic particles from the sun's surface (578)

solar nebula cloud of gas and dust that developed into the solar system (580)

solar system the sun and the bodies that revolve around it (580)

solar wind electrically charged atomic particles that stream out into space through holes in the sun's corona (575)

solar year time required for the earth to make one orbit around the sun, about 365.24 days (627)

solid physical form of matter with a definite shape and volume (145)

solifluction slow downslope flow of wet, muddy topsoil over frozen or clay-rich subsoil (235)

solution mixture in which one substance is uniformly dispersed in another substance (151)

sonar acronym for sound navigation and ranging, method of mapping the ocean floor using reflected sound waves (391)

sorting uniformity in the size of the particles of a rock or sediment (261)

specific heat amount of heat needed to raise the temperature of 1g of a substance 1°C (526)

specific humidity actual amount of moisture in the air (483)

spectroscope instrument that splits white light into a band of colors (15)

spectrum band of the various colors of light (14)

spiral galaxy type of galaxy with a nucleus of bright stars and flattened arms that swirl around the nucleus (563)

spit long, narrow deposit of sand connected at one end to the shore (306)

spring tide tide with maximum daily tidal range that occurs during the new and full moons (440)

squall line long line of heavy thunderstorms that may occur just ahead of a fast-moving cold front (502)

stalactite cone-shaped calcite deposit suspended from the ceiling of a cavern (270)

stalagmite cone-shaped calcite deposit built up from the floor of a cavern (271)

standard atmospheric pressure the atmospheric pressure measured at sea level; 760 mm of mercury (459)

standard time zone one of 24 regions of the earth in which noon is set as the time when the sun is highest over the center of the region (33)

star body of gases that gives off a tremendous amount of radiant energy in the form of light and heat (547)

station model cluster of weather symbols plotted on a map indicating the weather conditions at a particular reporting station (512)

stationary front boundary formed where two air masses meet and neither is displaced (503)

steam fog condensation of water vapor that results when cool air moves over warm water (488)

stock igneous intrusion with an area less than 100 square kilometers (181)

stony meteorite most common type of meteorite, similar in composition to rocks found on the surface of the earth (609)

stony-iron meteorite rare type of meteorite that contains both iron and stone (609)

strain change in shape and volume of rocks that occurs due to stress (84)

stratification layering of sedimentary rock (185)

stratified drift glacial deposit that has been sorted and layered by the action of streams or meltwater (285)

stratopause high-temperature zone that marks the upper boundary of the stratosphere (460)

stratosphere layer of the atmosphere that extends upward from the troposphere to an altitude of 50 km; contains most atmospheric ozone (460)

stratovolcano also called *composite cone,* steep-sloped volcanic deposit with alternating layers of hardened lava flows and tephra (123)

stratus cloud cloud with a sheetlike or layered form that is the lowest cloud in the sky (486)

streak color of a mineral in powder form (164)

stream load sediments carried by a stream (249)

stream piracy capture of a stream in one watershed by a stream in another watershed (247)

stress force that causes pressure in rocks of the earth's crust (84)

strike-slip fault fault in which the rock on either side of a fault plane slides horizontally (87)

subarctic climate type of polar climate that occurs in areas between 55° and 65° north latitude, with little precipitation and a large yearly temperature range (532)

subduction zone region where one lithospheric plate moves under another (73)

sublimation process in which a solid changes directly into a vapor (479)

sublittoral zone shallow benthic environment that is continuously submerged and that contains the largest number of benthos (415)

submarine canyon deep valley in the continental slope and shelf (395)

submergent coastline coast along which sea level rises or the land sinks (308)

submersible underwater research vessel (390)

subpolar low belt of low air pressure at about 60° north and 60° south latitude (472)

summer solstice the beginning of summer (31)

sunspot cool, dark area of gas within the photosphere caused by powerful magnetic fields (576)

sunspot cycle periodic variation in the number of sunspots that occurs approximately every 11 years (577)

Supai formation rock layers of the Grand Canyon deposited during the Pennsylvanian Period (374)

supercooling process in which water droplets are induced to remain liquid at temperatures below 0°C (491)

supergiant extremely large, giant star (554)

supernova star that blows apart with a tremendous explosion (559)

surface current streamlike movement of water on or near the surface of the ocean (427)

swell one of a group of long, rolling waves that are all the same size (434)

syncline downcurved fold in horizontal rock layers (85)

T

talus pile of rock fragments that accumulates at the base of a slope (233)

Tapeats sandstone rock layer of the Grand Canyon deposited during the Cambrian Period (374)

temperature inversion atmospheric condition in which warm air traps cooler air near the earth's surface (462)

temperature range difference between the highest and lowest temperatures of a particular time period (523)

tension stress that pulls rocks apart (84)

tephra also called *pyroclastic material,* all the rock fragments ejected from a volcano (121)

terminal moraine till deposited at the leading edge of a melting glacier (286)

terrane piece of land with a geologic history distinct from that of the surrounding land (76)

terrestrial planet any one of the four planets closest to the sun—Mercury, Venus, Earth, and Mars—with properties similar to those of the earth (594)

theory hypothesis or set of hypotheses supported by the results of experimentation and observation (14)

theory of evolution theory that organisms change over time and that new organisms are derived from ancestral types (347)

theory of suspect terranes theory that continents are a patchwork of pieces of land that have individual geologic histories (76)

thermocline zone of rapid temperature change that begins just below the surface of the ocean (410)

thermograph instrument that measures temperature changes by recording the movement of the bar of a bimetal thermometer (509)

thermosphere the atmospheric layer above the mesosphere (461)

thrust fault type of reverse fault in which the fault plane is nearly horizontal rather than vertical (86)

thunderstorm storm accompanied by thunder, lightning, and strong winds (506)

tidal bore surge of water that rushes upstream in a river as the tide rises (443)

tidal current movement of water toward and away from the coast due to the rise and fall of the tides (443)

tidal flat muddy or sandy part of a lagoon that is visible at low tide (310)

tidal oscillation slow, rocking motion of ocean water that occurs as tidal bulges move around the earth (442)

tidal range difference between the levels of the high and low tides at a specific location (440)

tide daily change in the level of the ocean surface (439)

till unsorted rock material deposited by a glacier (285)

tombolo ridge of sand that connects an island to the mainland (306)

topographic map map that shows the surface features of the earth (50)

topography surface features of the earth (50)

tornado whirling, funnel-shaped cyclone (507)

Toroweap formation rock layers of the Grand Canyon deposited during the Permian Period (375)

trace fossil fossil trace left by an ancient organism, such as a track, footprint, boring, or burrow (336)

trade winds global winds flowing toward the equator between 30° and 0° latitude (470)

transform fault boundary boundary formed where two lithospheric plates slide past each other (74)

travertine form of calcite that is deposited in terraces around the mouths of hot springs (268)

trench deep valley in the ocean floor (395)

tributary feeder stream that flows into a main stream (247)

Trojan asteroid asteroid that orbits the sun just ahead of or behind the planet Jupiter (607)

tropical climates climates with a minimum average monthly temperature of 18°C (529)

tropical desert climate dry, warm climate that occurs in regions about 20° to 30° north and south of the equator (529)

tropical rain forest climate warm, humid climate that occurs within 5° to 10° on either side of the equator (529)

tropical savanna climate tropical climate located between tropical rain forest and tropical desert climates, producing very wet summers and very dry winters (529)

tropopause upper boundary of the troposphere in which the temperature remains almost constant (460)

troposphere atmospheric layer closest to the earth's surface where nearly all weather occurs (460)

trough lowest point between two wave crests (433)

true north direction of the geographic North Pole (45)

tsunami giant ocean wave that often occurs after a major earthquake with an epicenter on the ocean floor (106)

tundra climate polar climate that occurs in areas near the ocean at the latitude of the Arctic Circle, with a small yearly temperature range and very little precipitation (532)

turbidity current dense current that carries large amounts of sediment down the continental slopes (395)

typhoon hurricane that forms over the Pacific Ocean (505)

U

umbra inner, cone-shaped part of the shadow cast by the earth or the moon in which sunlight is completely blocked (623)

unconformity break in the geologic record created when rock layers are removed by erosion (325)

undertow irregular current that pulls water from a beach back to the ocean (437)

unfoliated describing a metamorphic rock without visible bands (187)

upslope fog condensation of water vapor that results from the lifting and adiabatic cooling of air rising up a slope of land (488)

upwelling process in which surface water moves farther out into the ocean and deep water moves upward to replace the surface water (414)

valley glacier long, narrow, wedge-shaped mass of ice that usually moves through a mountain valley (278)

variable factor in an experiment that can be changed (12)

varve annual layer of sedimentary deposit on a lake bed (328)

vein narrow band of mineral deposits in rock (195)

vent opening through which molten rock flows onto the earth's surface (117)

ventifact any stone smoothed by wind abrasion (298)

vernal equinox the beginning of spring (32)

vertebrate animal with a backbone (350)

Vishnu schist the bottom-most and oldest rock layer of the Grand Canyon (374)

volcanic ash tephra particles between 0.25 mm and 2 mm in diameter (121)

volcanic block the largest tephra, formed from solid rock blasted from a fissure (122)

volcanic bomb large, spindle-shaped clot of lava thrown out of a volcano (122)

volcanic dust tephra particles less than 0.25 mm in diameter (121)

volcanic mountain mountain formed when molten rock erupts onto the earth's surface (92)

volcanic neck solidified central vent of a volcano (181)

volcanism any activity that includes the movement of magma toward or onto the earth's surface (117)

volcano lava and tephra built up on the earth's surface around a vent (117)

W

waning describing the phase of the moon during which the size of its visible portion is decreasing (626)

warm front boundary formed where a warm air mass overtakes and rises over a cold air mass (502)

water budget gains and losses of water from a region (244)

water cycle continuous movement of water from the air to the earth and back again (243)

water gap deep notch left where a stream erodes through a mountain as it is uplifted (249)

water table upper surface of the zone of saturation (263)

watershed land from which water runs off into a stream (247)

waterspout tornado that occurs over the ocean (508)

wave periodic up-and-down movement of water (433)

wave cyclone large storm that develops along cold or stationary fronts, with winds that spiral in toward a central region of low air pressure (503)

wave height vertical distance between the crest and the trough of a wave (433)

wave period time required for a complete wavelength to pass a given point (433)

wave-built terrace extension of a wave-cut terrace that results from deposition of eroded material offshore (303)

wave-cut terrace nearly level platform of rock left beneath the water after the erosion of a sea cliff (303)

wavelength distance between wave crests (14)

waxing describing the phase of the moon during which the size of its visible portion is increasing (626)

weather general condition of the atmosphere at a particular time and place (455)

weathering change in the physical form or chemical composition of rock materials exposed at the earth's surface (219)

westerlies global winds located between 40° and 60° latitude that flow from the southwest in the Northern Hemisphere and from the northwest in the Southern Hemisphere (471)

white dwarf small, hot, dim star (554)

whitecap crest of a wave that is blown off by high winds (435)

wind gap water-eroded notch in a mountain through which water no longer flows (249)

wind vane instrument used to determine the direction of the wind (509)

winter solstice the beginning of winter (31)

Z

zone of aeration upper region of groundwater between the water table and the earth's surface (263)

zone of saturation lower region of groundwater where all the pore spaces in a rock or sediment are filled with water (263)

zooplankton microscopic ocean animals (414)

INDEX

Italicized page numbers denote definitions; boldface page numbers denote illustrations.

A

B

C

D

E

F

G

H

M

N

O

P

Q

R

S

T

U

V

W

X

Y

Z

Art Credits

Abbreviations used: (t) top, (c) center, (b) bottom, (l) left, (r) right, (bckgd) background, (bdr) border.

Table of Contents: Page v, Uhl Studio; vii, Joseph LeMonnier/Melissa Turk & The Artist Network; viii, Uhl Studio; xii, Precision Graphics.

Introducing Earth Science: Page xvi, Mountain High Maps, Copyright 1993 Digital Wisdom, Inc.

Unit 1: Page 4, Uhl Studio; 12, Precision Graphics; 14, Uhl Studio; 16, Precision Graphics; 23, Ligature, Inc.; 24, Ligature, Inc.; 25(t), Precision Graphics; 25(b), Uhl Studio; 28, Precision Graphics; 29, Precision Graphics; 30, George Kelvin; 31, Uhl Studio; 32, George Kelvin; 34, George Kelvin; 35, Ligature, Inc.; 40, Uhl Studio; 43, George Kelvin; 44(t)(b), George Kelvin; 45(l), George Kelvin; 45(r), Ligature, Inc.; 46(l), Ligature, Inc.; 46(r), George Kelvin; 47(l)(r), Ligature, Inc.; 48(all), Ligature, Inc.; 50(t), Precision Graphics; 50(c)(b)(r), Ligature, Inc.; 52(t), Ligature, Inc.; 53, Uhl Studio; 55(t), Precision Graphics; 55(b), Ligature, Inc.; 62, Uhl Studio.

Unit 2: Page 71, Uhl Studio; 73(t)(b), Uhl Studio; 74(t)(b), Uhl Studio; 77 (t)(b), Uhl Studio; 83(l)(r), Uhl Studio; 84(all), Precision Graphics; 85(all), Precision Graphics; 86(all), Precision Graphics; 91, Uhl Studio; 97, Uhl Studio; 99, Uhl Studio; 100(r), Uhl Studio; 103, Ligature, Inc.; 104 (t)(b), Doug Walston; 114, Doug Walston; 115(r), Doug Walston; 117, Ligature, Inc.; 118(l)(r), Uhl Studio; 119, Uhl Studio; 122(all), Uhl Studio; 130, Doug Walston.

Unit 3: Page 139(r), Uhl Studio; 141, Precision Graphics; 144, Academy Arts, Inc.; 145, Precision Graphics; 146, Precision Graphics; 149, Precision Graphics; 150, Precision Graphics; 154, Uhl Studio; 155, Uhl Studio; 161, Precision Graphics; 162(all), Precision Graphics; 164, Sarah Woodward; 166, Precision Graphics; 167, Doug Walston; 177, Precision Graphics; 181, Precision Graphics; 193, Doug Walston; 196(l)(r), Precision Graphics; 200(t)(b), Joseph LeMonnier/Melissa Turk & The Artist Network; 201(l)(r), Ligature, Inc.; 203, Precision Graphics; 207, Uhl Studio; 213, Geosystems.

Unit 4: Page 225, Precision Graphics; 227, Uhl Studio; 229, Uhl Studio; 230, Uhl Studio; 243, Sarah Woodward; 245(all), Precision Graphics; 251(all), Ligature, Inc.; 259, Doug Walston; 261(all), Uhl Studio; 263(tr)(tc), Uhl Studio; 263(bl), Joseph LeMonnier/Melissa Turk & The Artist Network; 264, Joseph LeMonnier/Melissa Turk & The Artist Network; 266, Joseph LeMonnier/Melissa Turk & The Artist Network; 267, Joseph LeMonnier/Melissa Turk & The Artist Network; 268, Joseph LeMonnier/ Melissa Turk & The Artist Network; 275, Doug Walston; 279, Uhl Studio; 282(all), Uhl Studio; 285, Uhl Studio; 287(all), Geosystems; 289, Geosystems; 291, Ligature, Inc.; 294, Doug Walston; 295, Uhl Studio; 297, Ligature, Inc.; 299(all), Sarah Woodward; 305, Sarah Woodward; 306, Sarah Woodward; 307, Ligature, Inc.; 308, Sarah Woodward; 311(all), Sarah Woodward; 317, Doug Walston; 318(r), Uhl Studio.

Unit 5: Page 324, Joseph LeMonnier/Melissa Turk & The Artist Network; 325(all), Joseph LeMonnier/Melissa Turk & The Artist Network; 326, Joseph LeMonnier/Melissa Turk & The Artist Network; 329, Ligature, Inc.; 333(r), Ligature, Inc.; 345, Joseph LeMonnier/Melissa Turk & The Artist Network; 346, Joseph LeMonnier/Melissa Turk & The Artist Network; 352, Precision Graphics; 353, Precision Graphics; 354, Joseph LeMonnier/ Melissa Turk & The Artist Network; 355, Steve Roberts/Morgan Cain & Associates; 359, Doug Walston; 365, Geosystems; 367(t)(b), Geosystems; 368(t)(b) Geosystems; 369, Geosystems; 370, Uhl Studio; 371, Ligature, Inc.; 372, Uhl Studio; 375, Joseph LeMonnier/Melissa Turk & The Artist Network; 380, Joseph LeMonnier/Melissa Turk & The Artist Network; 381, Joseph LeMonnier/Melissa Turk & The Artist Network; 383 (t)(b), Doug Walston; 384, Joseph LeMonnier/Melissa Turk.

Unit 6: Page 389, Geosystems; 391, Ligature, Inc.; 393, Uhl Studio; 395, Uhl Studio; 405, Doug Walston; 409, Geosystems; 410, Ligature, Inc.; 414, Uhl Studio; 416, Joseph LeMonnier/Melissa Turk & The Artist Network; 418, Uhl Studio; 425(t)(b), Doug Walston; 428, Joseph LeMonnier/Melissa Turk & The Artist Network; 429, Geosystems; 431, Joseph LeMonnier/ Melissa Turk & The Artist Network; 432, Joseph LeMonnier/ Melissa Turk & The Artist Network; 433, Precision Graphics; 434, Precision Graphics; 436, Precision Graphics; 439, Joseph LeMonnier/Melissa Turk & The Artist Network; 446, Doug Walston; 447, Uhl Studio; 448, Uhl Studio; 450, Joseph LeMonnier/Melissa Turk & The Artist Network.

Unit 7: Page 456, Precision Graphics; 457, Precision Graphics; 460, Ligature, Inc.; 461, Ligature, Inc.; 462(t)(b), Precision Graphics; 463, Digital Art; 465, Joseph LeMonnier/Melissa Turk & The Artist Network; 466, Precision Graphics; 467, Joseph LeMonnier/Melissa Turk & The Artist Network; 469, Joseph LeMonnier/Melissa Turk & The Artist Network; 472, Joseph LeMonnier/Melissa Turk & The Artist Network; 473(t)(b), Uhl Studio; 477(t)(b), Doug Walston; 479, Precision Graphics; 484, Uhl Studio; 487, Uhl Studio; 491(t), Precision Graphics; 496, Uhl Studio; 500, Geosystems; 502(all), Joseph LeMonnier/Melissa Turk & The Artist Network; 504, Joseph LeMonnier/Melissa Turk & The Artist Network; 505, Precision Graphics; 509, Ligature, Inc.; 513, Ligature, Inc.; 514, Geosystems; 524(t)(b), Precision Graphics; 526, Joseph LeMonnier/Melissa Turk & The Artist Network; 527, Doug Walston; 528, Uhl Studio; 529, Precision Graphics; 534, Geosystems; 538(t)(b), Uhl Studio; 539, Uhl Studio; 541, Geosystems; 543, Uhl Studio.

Unit 8: Page 550, Precision Graphics; 553, Ligature, Inc.; 554, Stephen Durke/Washington-Artists' Representative, Inc.; 560, Uhl Studio; 561(l)(r), Ligature, Inc.; 568, Uhl Studio; 569, Uhl Studio; 572, Precision Graphics; 575, Tony Randazzo/American Artists Rep., Inc.; 581(all), Precision Graphics; 584(l)(r), Precision Graphics; 588, Uhl Studio; 604, Doug Walston; 620, Precision Graphics; 621, Joseph LeMonnier/Melissa Turk & The Artist Network; 622, Joseph LeMonnier/Melissa Turk & The Artist Network; 623, Precision Graphics; 624, Precision Graphics; 627, Joseph LeMonnier/Melissa Turk & The Artist Network; 629, Ligature, Inc.

Long-Range Investigations: Page 646, Uhl Studio; 649, Uhl Studio; 653, Uhl Studio; 659, Joseph LeMonnier/Melissa Turk & The Artist Network; 662, Joseph LeMonnier/Melissa Turk & The Artist Network.

Reference Section: Page 664, Joseph LeMonnier/Melissa Turk & The Artist Network; 672, Joseph LeMonnier/Melissa Turk & The Artist Network; 673, Joseph LeMonnier/Melissa Turk & The Artist Network; 675, Geosystems.

Photo Credits

Abbreviations used: (t) top, (c) center, (b) bottom, (l) left, (r) right, (bckgd) background, (bdr) border.

Cover: (tl), John Lund/Tony Stone Images; (tr), Anglo-Australian Observatory, photograph by David Malin; (c), NASA; (bl), SuperStock; (br), Telegraph Colour Library/FPG.

Title Page: NASA.

Front Matter: Page iii, HRW photo by Sam Dudgeon.

Table of Contents: Page iv, SOHO (ESA & NASA) and LASCO Consortium; v, Ben Simmons/The Stock Market; vi, Breck Kent; vii, C. C. Lockwood/Animals Animals/Earth Scenes; viii, Gary Ladd; ix(t), Field Museum of Natural History, Chicago; ix(b), Ron & Valerie Taylor/Bruce Coleman, Inc.; x(t), Marilyn Gartman/Photri/MGA; x(b), John Mead/ Science Photo Library/Photo Researchers, Inc.; xi(t), Alan Moller/Weather-stock; xi(b), E. I. Robinson/Photo Researchers, Inc.; xii, John Sanford/ Astrostock; xv, Alan Carey/The Image Works.

Introducing Earth Science: Page xvi(bl), Laurence Parent; xvii(tr), Jean Miele/The Stock Market; xvii(br), U.S. Geological Survey; xvii(tl), F. S. Westmorland/Photo Researchers, Inc.; xvii(bl), Courtesy Robert Sager.

Unit 1: Page xxii(inset), Jeffrey Rotman; xxii(bckgd), Visuals Unlimited; 1(bl), U.S.G.S.; 1(right, top to bottom), Gerry Ellis/Ellis Nature Photography; NASA; Jim Brandenburg/Minden Pictures; 2, Jim

Brandenburg/Minden Pictures; 3, Ginzy Schaefer/Tony Stone Images; 5(l), Phil Degginger/Color-Pic; 5(r), Woods Hole Oceanographic Institute; 6(l), Alan Carey/The Image Works; 6(r), Milton Mann/Cameramann/MGA/Photri; 7, B. Miller/Terraphotographics/BPS; 8(tl), Doug Wechsler/Animals Animals/Earth Scenes; 8(cl), Adele Hodde/Gamma Liaison; 9, David Falconer/West Stock; 10(tl), Bruce Bohor/U.S.G.S.; 10(bl), HRW photo by Sam Dudgeon; 11, National Geophysical Data Center; 13, Francois Gohier/Photo Researchers, Inc.; 15, WARD'S Natural Science Establishment, Inc./Wabash Instruments; 17, HRW photo by Ralph Brunke; 18, Jim Brandenburg/Minden Pictures; 20, HRW photos by Peter van Steen; 22, NASA; 26(l), Ken Pierce, David Swift; 26(br), Energy Technology Visuals; 27, Grant Heilman; 33, William Means/Tony Stone Images; 36, NASA/Phototake; 37, HRW photo by Ralph Brunke; 38, NASA; 41, HRW photo by Peter van Steen; 42, Gerry Ellis/Ellis Nature Photography; 49, Simon & Schuster, Inc.; 52, Courtesy NovAtel Communications, Ltd.; 54, HRW photo by Ralph Brunke; 56, Gerry Ellis/Ellis Nature Photography; 58, HRW photo by Peter van Steen; 59, Trails Illustrated/© National Geographic Society; 60 (inset), HRW photo by Peter van Steen; 60(bckgd), Laurence Parent; 61, U.S.G.S.; 62, Robert Goldstrom/The Newborn Group; 63, (c) The Nature Conservancy.

Unit 2: Page 64(both), Laurence Parent; 65(bl), Courtesy Waverly J. Person; 65(right, top to bottom), Will & Deni McIntyre/Photo Researchers, Inc.; Tom Carroll/Phototake; David Weintraub/Photo Researchers, Inc.; Barbara von Hoffmann/Tom Stack & Associates; 66, Tom Carroll/Phototake; 68, Charles Palek/Animals Animals/Earth Scenes; 69(l), Kenneth Garrett/Woodfin Camp & Associates; 69(r), W. R. Normack/U.S.G.S.; 70, Pete Turner/The Image Bank; 75, HRW photo by Ralph Brunke; 76, Bob Roback; 78, Tom Carroll/Phototake; 82, Barbara von Hoffmann/Tom Stack & Associates; 85, Breck P. Kent/Animals Animals/Earth Scenes; 87, HRW photo by Ralph Brunke; 89, Galen Rowell/Mountain Light Photography; 90(l), NASA; 90(r), University of Texas at Austin; 91, Dr. Randy Marrett/The University of Texas at Austin; 92(tl), Jerome Wyckoff/Animals Animals/Earth Scenes; 92(r), Reagan Bradshaw; 92(bl), Byron Augustin/Tom Stack & Associates; 94, Barbara von Hoffmann/Tom Stack & Associates; 96, HRW photo by Peter van Steen; 98, Willi & Deni McIntyre/Photo Researchers, Inc.; 103, John Yates/Tony Stone Images; 105, HRW photo by Ralph Brunke; 107, Asahi Shimbun; 108(t), Dr. Wayne Thatcher/Gamma/Liaison; 108(b), B. F. Bohor/U.S.G.S; 109, G. R. Roberts; 110, George Hall/Woodfin Camp & Associates; 112, Will & Deni McIntyre/Photo Researchers, Inc.; 116, David Weintraub/Photo Researchers, Inc.; 118, David Falconer/West Stock; 120, Laurence Parent; 121(l), Phil Deggenger/Tony Stone Images; 121(r), Kal Muller/Woodfin Camp & Associates; 122(t), Jeff Greenberb/MRp/Visuals Unlimited; 122(c), Manfred Gottschalk/Tom Stack & Associates; 122(b), Brian Lovell/Nawrocki Stock Photos; 123, HRW photo by Ralph Brunke; 124, Greg Vaughn/Tom Stack & Associates; 125, Wolfgang Kaehler; 126, 127, NASA; 128, David Weintraub/Photo Researchers, Inc.; 132, Ben Simmons/The Stock Market; 134(l), Philippe Bourseiller/Hoa-Qui Agence Photographique; 134(r), Shawn Henry/SABA.

Unit 3: Page 136(inset), Farrel Grehan/Photo Researchers, Inc.; 136 (bckgd), Bill Bachman/Photo Researchers, Inc.; 137(bl), U.S.G.S.; 137(right, top to bottom), David Muench; E. R. Degginger/Animals Animals/Earth Scenes; Gary Ladd; Spencer Swanger/Tom Stack & Associates; 138, Spencer Swanger/Tom Stack & Associates; 140(t), George Whiteley/Photo Researchers, Inc.; 140(b), HRW photo by Russell Dian/Ralph Brunke; 144, Fermilab Visual Media Services; 145, Fermilab Visual Media Services; 151(l), T. J. Florian/Nawrocki Stock Photo; 151(r), Grant Heilman Photography; 152, Spencer Swanger/Tom Stack & Associates; 156, E. R. Degginger/Animals Animals/Earth Scenes; 157(l), Breck P. Kent/Animals Animals/Earth Scenes; 157(r), Dr. E. R. Deggenger/Color-Pic; 157(c), Andrew J. Martinez/Photo Researchers, Inc.; 158(tl), E. R. Degginger/Animals Animals/Earth Scenes; 158(tr), Joyce Photographics/Photo Researchers, Inc.; 158(cl), Breck Kent; 158(cr, bl), E. R. Degginger/Color-Pic; 158(br), Breck Kent/Animals Animals/Earth Scenes; 159(tl, tr), E. R. Degginger/Color-Pic; 159(cl), Joyce Photographics/Photo Researchers, Inc.; 159(cr), Renée Purse/Photo Researchers, Inc.; 159(bl), Breck Kent; 159(br), Breck P. Kent/Animals Animals/Earth Scenes; 160(t), Colin Mulvany; 160 (b), J. Steere/Nawrocki Stock Photo; 161, Ted Coringley/Nawrocki Stock Photo; 163(l), Mary A. Root/Root Resources; 163(r), Runk/Schoenberger/Grant Heilman Photography; 164(both), 165, 168(all), E. R. Degginger/Color-Pic; 169, Roger Beck/Nawrocki Stock Photo; 170, E. R. Degginger/Animals Animals/Earth Scenes; 172, HRW photo by Peter van Steen; 174, David Muench; 175, 176(l, r), E. R. Degginger/Color-Pic; 176(c), 178, Joyce Photographics/Photo Researchers, Inc.; 179(all), Breck Kent; 180, HRW photo by Ralph Brunke; 181, Francois Gohier/Photo Researchers, Inc.; 182 (l), Breck Kent/Animals Animals/Earth Scenes; 182(r), Ruth Dixon; 183(t), A. J. Copley/Visuals Unlimited; 183(b), J.N.A. Lott/Terraphotographics/BPS; 184, NOAA/NESDIS; 185(tr), Joe Viesti/FPG; 185(bl), Photo Researchers, Inc.; Earth Satellite Corporation/Science Photo Library; 186(tl), Steve Traudt/Tom Stack & Associates; 186(tr), Jim Brandenburg/Tony Stone Images; 186(cl), Breck Kent; 188(tl), A. J. Copley/Visuals Unlimited; 188 (tc), E. R. Degginger/Animals Animals/Earth Scenes; 188(tr), Breck Kent; 188(bl), Breck P. Kent/Animals Animals/Earth Scenes; 188(br), E. R. Degginger/Animals Animals/Earth Scenes; 189(tr), Martin Rogers/Tony Stone Images; 189(bl), NASA; 190, David Muench; 193, Charles D. Winters/Photo Researchers, Inc.; 194, Gary Ladd; 195, William E. Ferguson; 198, 199, G. R. Roberts; 202, Pierre Kopp/WestLight; 204, William E. Ferguson; 205, HRW photo by Ralph Brunke; 206, National Renewable Energy Laboratory/U.S. Department of Energy; 208, Gary Ladd; 210, HRW photo by Peter van Steen; 211, 214, E. R. Degginger/Color-Pic; 215, Eugene Richards/Magnum.

Unit 4: Page 216(inset), Porterfield-Chickering/Photo Researchers, Inc.; 216(bckgd), Jeff Gnass; 217(bl), National Park Service; 217(right, top to bottom), Laurence Parent; Laurence Parent; Royce L. Bair/The Stock Solution; Nancy Simmerman/Tony Stone Images; William E. Ferguson; 218, Laurence Parent; 219, Nancy Simmerman/Tony Stone Images; 220, HRW photo by Ligature; 221, Jeff Gnass; 222(tr), Willard Clay/ Tony Stone Images; 222(bl), Mireille Vautier/Woodfin Camp & Associates; 223, J. N. A. Lott/Terraphotographics/BPS; 224, Doris DeWitt/Tony Stone Images; 226(l), New York Historical Society; 226(r, c), Bernard Boutrit/Woodfin Camp & Associates; 228, William E. Ferguson; 231, John Gerlach/Animals Animals/Earth Scenes; 232(bl), B. Crader/Hillstrom Stock Photography; 232(br), Larry LeFever/Grant Heilman Photography; 233, Breck P. Kent/Animals Animals/Earth Scenes; 234(l), Soil Conservation Service; 234(r), Cameron Davidson/Tony Stone Images; 235, John Running/Tony Stone Images; 236(bl), Gerry Ellis/Ellis Nature Photography; 236(br), Ruth Dixon; 237(tl), Jim Brandenburg/Tony Stone Images; 237(tr), Gary Ladd; 238, Laurence Parent; 241, HRW photo by Peter van Steen; 242, Laurence Parent; 244(bl), William Boehm/West Stock; 244(br), Breck Kent; 246, Gary Milburn/Tom Stack & Associates; 247, C. C. Lockwood/Animals Animals/Earth Scenes; 248, HRW photo by Ralph Brunke; 249, G. R. Roberts; 250 (bl), Thomas Wanstall/The Image Works; 250(br), 252(both), G. R. Roberts; 253, D. W. Hamilton/The Image Bank; 254, Randall Hyman; 255(tl), E. R. Degginger/Color-Pic; 255(tr), Suzi Barnes/Tom Stack& Associates; 256, Laurence Parent; 258, HRW photo by Peter van Steen; 260, Royce L. Bair/ The Stock Solution; 262, HRW photo by Ralph Brunke; 265, William E. Ferguson; 269, John Lemker/Animals Animals/Earth Scenes; 270, Leif Skoogfors/Woodfin Camp & Associates; 271, Gary Ladd; 272, Royce L.Bair/The Stock Solution; 274, HRW photo by Peter van Steen; 276, Nancy Simmerman/Tony Stone Images; 277, Wilson Goodrich/MGA/Photri; 278 (tl), Nancy Simmerman/Tony Stone Images; 278(tr), Fred Hirschmann/Ken Graham Agency; 280, Galen Rowell/Mountain Light; 281(both), Reprinted with permission from Science Magazine Vol 262, December 1993 © 1998 American Association for the Advancement of Science; 283, Galen Rowell/ Mountain Light; 284, HRW photo by Ralph Brunke; 285, Kevin Schafer/Peter Arnold, Inc.; 286, George Gersler/Photo Researchers, Inc.; 290, Bob Waterman/WestLight; 292, Nancy Simmerman/Tony Stone Images; 296, William E. Ferguson; 298, E. R. Degginger/Color-Pic; 300, HRW photo by Ralph Brunke; 301, Grant Heilman/Grant Heilman Photography; 302, Thomas Wanstall/The Image Works; 303, Laurence Parent; 304, Phil Degginger/Tony Stone Images; 309, Jonathan Blair/Woodfin Camp & Associates; 310, Reagan Bradshaw; 312, William E. Ferguson; 314, 315, HRW photos by Sam Dudgeon; 319(tr), Laurence Parent; 319(bc), Donald Specker/Animals Animals/Earth Scenes.

Unit 5: Page 320(inset), Doug Wechsler/Animals Animals/Earth Scenes; 320(bckgd), Gary Ladd; 321(bl), Courtesy of Richard Kerr; 321(right, top to bottom), Gary Ladd; Breck Kent; Chuck Place; 322, 323, Gary Ladd; 327, Mary Root/Root Resources; 328, Runk/Schoenberger/Grant Heilman Photography; 330, 331, I. S. Williams/Research School of Earth Sciences/The Australian National University; 332, HRW photo by Ralph Brunke; 333, Edward Ross; 334, Richard Kolar/Animals Animals/Earth Scenes; 335, Field Museum of Natural History, Chicago; 336(both), Gary Ladd; 337, Grant Heilman/Grant Heilman Photography; 338, Louise Brown/Root Resources; 339, Eastcott/Momatiuk/The Image Works; 340, Gary Ladd; 342(tr), Jane Burton/Bruce Coleman, Inc.; 342(bl), Dr. E. R. Degginger/Color-Pic; 343

(tc), John Cancalosi/Peter Arnold, Inc.; 343(br), Phil Degginger/Tony Stone Images; 344, Breck Kent; 347, Kevin Schafer/Tom Stack & Associates; 348, Breck Kent; 349, Grant Heilman/Grant Heilman Photography; 350, The University of Texas at Austin; 351(cr), Bill Steele/A+C Anthology; 351(br), Andrew Leitch/(c) 1992 The Walt Disney Co. Reprinted with permission of *Discover* Magazine; 353, E. R. Degginger/Animals Animals/Earth Scenes; 356, E. R. Degginger/Color-Pic; 357, Alan G. Nelson/Hilstrom Stock Photo; 358, Tom McHugh/Photo Researchers, Inc./Natural History Museum of Los Angeles County; 360, Breck Kent; 362, 363, HRW photos by Peter van Steen; 364, Chuck Place; 366, HRW photo by Ralph Brunke; 373, Tim Thompson; 374, James Tallon/Outdoor Exposures; 376(l), Field Museum of Natural History, Chicago; 376(r), S. McNutt/Phototake; 377, Frank Straub/ The Picture Cube; 378, Chuck Place; 382, Betty Press/Animals Animals/ Earth Scenes; 383, Gary Retherford/Photo Researchers, Inc.; 384, 385, Prof. Andrew Hill/Yale University.

Unit 6: Page 386(inset), Andrew J. Martinez/Photo Researchers, Inc.; 386 (bckgd), John Telford; 387(right, top to bottom), Fred Bavendam/Peter Arnold, Inc.; Fred Bavendam/Peter Arnold, Inc.; Chris Arend; 387(bl), Courtesy of Bonnie McGregor; 388, Fred Bavendam/Peter Arnold, Inc.; 390, Jason Ion/Woods Hole Oceanographic Institute; 391, Rod Catavah/Woods Hole Oceanographic Institute; 392, HRW Photo by Ralph Brunke; 394, The Image Works Archives; 396(l), Chris Newbert/Marine Life Photography; 396(r), Werner Braun/FPG; 397, Stuart Westmorland; 398, Marie Tharp; 399, Al Driscoll © Woods Hole Oceanographic Institute; 400(bl), M. Abbey/Photo Researchers; 400(br), Manfred Kage/Peter Arnold; 401, Steinhart/Photo Researchers; 402, Fred Bavendam/Peter Arnold, Inc.; 404, HRW photo by Peter van Steen; 406, Fred Bavendam/Peter Arnold, Inc.; 411, HRW photo by Ralph Brunke; 412, Al Grotell; 413, Ron & Valerie Taylor/Bruce Coleman, Inc.; 415(both), E. R. Degginger/Color-Pic; 419, Doug Hicks; 420, University of Hawaii; 421, Tom Pantages; 422, Fred Bavendam/Peter Arnold, Inc.; 424, HRW photo by Peter van Steen; 426, Chris Arend; 430, E. R. Degginger/Animals Animals/Earth Scenes; 435, HRW Photo by Ralph Brunke; 436, Marilyn Gartman; 437, John Shelton; 440, Kerr Illustration; 441, Greg Vaughn/Pacific Stock; 442(both), Breck Kent/Animals Animals/Earth Scenes; 443, NHPA; 444, Chris Arend; 449, © 1995, David T. Sandwell and Walter H. F. Smith, NOAA Geosciences; 450, Peter Southwick/Stock Boston; 451, Michael Dwyer/Stock Boston.

Unit 7: Page 452(inset), Gary Ladd; 452(bckgd), HRW photo by Stock Editions; 453(bl), National Weather Service; 453(right, top to bottom), Daryl Balfour/Tony Stone Images; Gary Milburn/Tom Stack & Associates; Darryl Torckler/Tony Stone Images; Pierre Kopp/Westlight; 454, Pierre Kopp/ WestLight; 458, David Fritts/Animals Animals/Earth Scenes; 458, HRW Photo by Ralph Brunke; 459, Breck Kent; 464, John Livzey/Dot, Inc.; 470, Robert Freerk/Tony Stone Images; 471, Ruth Dixon; 474, Pierre Kopp/ WestLight; 476, HRW photo by Peter van Steen; 478, Darryl Torckler/Tony Stone Images; 481, Breck Kent; 482, HRW Photo by Ralph Brunke; 483(tr), Craig Blouin/Manhattan Views; 483(cr), Randy Duchane/Manhattan Views; 485, Phillip Amdal/West Stock; 486(tl), Darryl Torckler/Tony Stone Images; 486(cl), John Mead/Science Photo Library/Photo Researchers, Inc.; 486(bl), Kent Wood/Weatherstock; 488, Thomas Wanstal/The Image Works; 489, Robert Essey/Manhattan Views; 490, Nuridsany & Perennou/Photo Researchers, Inc.; 491, Royce Bair/The Stock Solution; 492, Gary Ladd; 492(inset), National Optical Astronomy Observatory; 493, HRW photo by Victoria Smith; 494, Darryl Torckler/Tony Stone Images; 498, Gary Milburn/Tom Stack & Associates; 499, The Stock Market; 506, J. Christopher/Weatherstock; 508, E. R. Degginger/Color-Pic; 510, HRW photo by Ralph Brunke; 511, Lawrence Migdale; 515, Alan Moller/Weatherstock; 516(l), Lawrence Migdale; 516(r), Waverly Pearson/National Earthquake Information Center; 517, James Kay/The Stock Solution; 518, Gary Milburn/Tom Stack & Associates; 522, Daryl Balfour/Tony Stone Images; 525, Brian Munoz/Scripps Institution of Oceanography; 530, Chip Isinhart/Tom Stack& Associates; 531, Michael Fogden/Animals Animals/ Earth Scenes; 533, Chuck Place; 536, Daryl Balfour/Tony Stone Images; 540, Nawrocki Stock Photo; 542, Dee Breger/Lamont-Doherty Earth Observatory.

Unit 8: Page 544(inset), NASA; 544(bckgd), Palomar Observatory/ California Institute of Technology; 545 (bl), NASA; 545(right, top to bottom), NASA; Mark Martin/NASA/Photo Researchers; NASA; Robert Williams and the Hubble Deep Field Team (STScI) and NASA; 546, Robert Williams and the Hubble Deep Field Team (STScl) and NASA; 548, Anglo-Australian Telescope Board; 549, National Optical Astronomy Observatories; 549(tl), Nicholas Foster/The Image Bank; 551, HRW photo by Ralph Brunke; 552(both), Yerkes Observatory; 555, NASA; 556, NASA image, courtesy of Lunar and Planetary Institute; 557, NASA; 558, Hansen Planetarium; 559, NOAO/Science Photo Library/Photo Researchers, Inc.; 562, National Optical Astronomy Observatories; 563(l), Jerry Lodriguss/ Photo Researchers, Inc.; 563(r), Carnegie Institution Observatories/ Astrostock; 563(c), Kitt Peak National Observatory/ AstroStock; 565, National Optical Astronomy Observatories; 566, Robert Williams and the Hubble Deep Field Team (STScI) and NASA; 570, NASA; 573, HRW Photo by Ralph Brunke; 574, Dan McCoy/Rainbow; 576, Dr. W. C. Livingston/ National Optical Astronomy Observatories/ National Solar Observatory and the National Science Foundation; 577, NASA; 578, SOHO (ESA & NASA) and LASCO Consortium; 579, Jeff Apoian/Nawrocki Stock; 582, E.I. Robinson/Photo Researchers, Inc.; 582(l), Dan Coyro; 582(r), Two-Twenty-Two Productions; 585, SOHO (ESA & NASA) and LASCO Consortium; 586, NASA; 590, Mark Martin/NASA/Photo Researchers, Inc.; 592, Ligature; 594, JPL/NASA; 595, NASA; 596, JPL/NASA; 597(l), Philip James/University of Toledo, Steven Lee/ University of Colorado, Boulder, and NASA; 597(r), JPL/NASA/Astrostock; 598, 599, JPL/NASA; 600, NASA/Peter Arnold, Inc.; 601, JPL/NASA; 602, NASA; 603, JPL/ AstroStock; 603(l), Julian Baum/Science Photo Library/ Photo Researchers, Inc.; 605, David A. Hardy/Science Photo Library/Photo Researchers, Inc.; 606(l), NASA/JPL/Astrostock; 606(r), NASA/ JPL/ Astrostock; 607, John Chumack/Photo Researchers, Inc.; 608, NASA; 609, Pearson/Milon/Science Photo Library/Photo Researchers, Inc.; 610, Mark Martin/NASA/Photo Researchers, Inc.; 612, HRW photo by Peter van Steen; 613, HRW photo by Peter van Steen; 614, 615 616(l), NASA; 616(r), John Sanford/AstroStock; 617, 618, NASA; 623, Jerry Lodriguss/Photo Researchers, Inc.; 625, HRW photo by Ralph Brunke; 628, Aldona Sabalis/Photo Researchers, Inc.; 630, NASA image, courtesy of Lunar and Planetary Institute; 631(l), JPL/NASA/ Astrostock; 631(r), 632, 633, JPL/NASA; 634, NASA; 638, Earth and Sky; 639(both), Sky maps by Wil Tirion from A Field Guide to the Stars and Planets 2/e. ©1983 by The Estate of Donald H. Menzel and by Jay M. Pasachoff. Reprinted by permission of Houghton Mifflin Company. All rights reserved; 640, NASA; 641(l), NASA/ Peter Arnold, Inc.; 641(r), NASA; 692(r), JPL/NASA/AstroStock.